Stochastic Methods in Asset Pricing

Stochastic Methods in Asset Pricing

Andrew Lyasoff

The MIT Press
Cambridge, Massachusetts
London, England

This book is typeset by the author with TEX and Emacs Lisp.
Printed and bound in the United States of America.

Library of Congress Cataloging-in-Publication Data

Names: Lyasoff, Andrew, author.
Title: Stochastic methods in asset pricing / Andrew Lyasoff.
Description: Cambridge, MA : MIT Press, [2017] | Includes bibliographical references and index.
Identifiers: LCCN 2017000433 | ISBN 9780262036559 (hardcover : alk. paper)
Subjects: LCSH: Securities–Prices–Mathematical models. | Stochastic processes.
Classification: LCC HG4636 .L93 2017 | DDC 332.63/2–dc23
LC record available at https://lccn.loc.gov/2017000433

10 9 8 7 6 5 4 3 2 1

"The special sphere of finance within economics is the study of allocation and deployment of economic resources, both spatially and across time, in an uncertain environment. To capture the influence and interaction of time and uncertainty effectively requires sophisticated mathematical and computational tools. Indeed, mathematical models of modern finance contain some truly elegant applications of probability and optimization theory. These applications challenge the most powerful computational technologies. But, of course, all that is elegant and challenging in science need not also be practical; and surely, not all that is practical in science is elegant and challenging. Here we have both. In the time since publication of our early work on the option-pricing model, the mathematically complex models of finance theory have had a direct and wide-ranging influence on finance practice. This conjoining of intrinsic intellectual interest with extrinsic application is central to research in modern finance."

— Robert C. Merton
(from his Nobel Prize lecture, December 9, 1997)

"There's No Such Thing as a Free Lunch"

— Milton Friedman
(title of his 1975 book)

CONTENTS

Preface

There is no doubt that asset pricing is one manifestly important subject, and packing in a single course of study, or a book, most concepts from probability and stochastic processes that asset pricing – especially asset pricing in continuous time – builds upon is one nearly impossible project. Nevertheless, the present book is the outcome from my attempt at one such project. It grew out of lecture notes that I developed over the course of many years and is faithful to its title: in most part it is a book about stochastic methods used in the domain of asset pricing, not a book on asset pricing, despite the fact that some important aspects of this discipline are featured prominently, and not just as an aside.

What follows in this space is the material that I typically cover in a two-semester first-year graduate course (approximately 90 lecture hours). The book is therefore meant for readers who are not familiar with measure-theoretic probability, stochastic calculus, and asset pricing, or perhaps readers who are well versed in stochastic calculus, but are not familiar with its connections to continuous-time finance. Compared to other classical texts that pursue similar objectives – and the book (Karatzas and Shreve 1991) is the first to come to mind – the present one starts from a much earlier stage, namely from measure-theoretic probability and integration (and includes also a brief synopsis of analysis and topology). At the same time it is less rigorous, in the sense that many fundamental results are stated without proofs, which made it possible to keep the scope of the material quite broad and to include some advanced topics such as Lévy processes and stochastic calculus with jumps. In doing this, I attempted the impossible task of introducing in a more or less rigorous fashion some advanced features of local martingales and semimartingales, and connecting those features with the principles of asset pricing, while leaving out most of the proofs. My insistence that models involving processes with jumps must be covered made this task all the more difficult.

By and large, in the second decade of the twenty-first century probability theory and stochastic processes can be considered as well developed and mature mathematical disciplines. However, while indispensable for modeling and understanding of random phenomena, including those encountered in finance and economics, the language and the tools that these disciplines provide are yet to become commonplace – as, say, calculus, statistics, or linear algebra already are commonplace. In some sense, this book is an attempt to make the theory of stochastic processes a bit less exotic and a bit more accessible. The reason why achieving such an objective is not easy is that the terminology, theory, methods, and ideas, from the general theory of stochastic processes in continuous time – expounded in (Doob 1953), (Meyer 1966), (Itô and McKean 1974), (Dellacherie and Meyer 1980), (Jacod and Shiryaev 1987), and (Revuz and Yor 1999) – is intrinsically complex and, yes, technical and at times overwhelming. Thus it is impossible to develop an introduction to this subject for first-year graduate students without "glossing over" some of the technical concepts involved. As a result, most introductory texts on the subject, including the present one, differ mainly in what is being "glossed over" and how. My particular choice is based on the recognition that there is a point at which playing down too many connections and asking for too much suspension of disbelief could make a learner of the field feel lost.

Most of the material covered in the present book can be found – and in a much more rigorous and detailed form – in a number of classical texts (see below). My goal was to compile a coherent single overview, with at least some attempt at pedagogy. By way of an example, discrete-time martingales are first introduced as a consequence of market equilibrium, and are then connected with the stochastic discount factors, before the general definition is given. In addition, there are more than 450 exercises spread throughout the text (many accompanied with hints), which are meant to help the reader develop better understanding and intuition. A very brief synopsis of analysis and topology is included in appendix A, while other preparatory material is included at the very beginning – I hope these additions will make following the main part of the book a bit easier. The recent results on general market equilibrium for incomplete financial markets from (Dumas and Lyasoff 2012), and on the no-arbitrage condition from (Lyasoff 2014) were incorporated in the book as well. Most of the concrete models discussed in the book (e.g., models of stochastic volatility) were borrowed from research papers, which are quoted in the text. While concrete practical applications are not in the focus of the book, a small number of such applications are detailed, and some of the related (working) computer code is included in appendix B. To the best of my knowledge, the numerical method and the program described in section B.3 are new.

This book can be used in a number of ways. The obvious one is to cover the entire material in the course of two semesters. For a one-semester course, one could consider covering chapters 1 through 14 – perhaps omitting 5, 7, and possibly 12. But on the other hand, for students who have already studied measure-theoretic probability and integration, one would start from chapter 5 and spend more time on chapters 13 and 14 (perhaps with greater emphasis on practical applications). It is also possible to design a short course in probability theory (with or without applications to finance) based on chapters 1 through 5, or a short course on stochastic calculus for continuous processes based on chapters 8 through 12, or a short course on stochastic calculus with jumps and Lévy processes based on chapters 15 and 16, or a short course on optimal control based on chapters 17 and 18. Finally, one could design a short introductory course in asset pricing, meant for an audience familiar with all mathematical aspects of probability, stochastic processes and optimal control, by following chapters 6, 13, 14, and 18.

For readers who intend to enter the field of finance, it would be beneficial to read this book in parallel with, for example, (Cochrane 2009), (Duffie 2010), and (Föllmer and Schied 2011), and then follow up with at least some of the following books: (Aït-Sahalia and Jacod 2014), (Crépey 2012), (Duffie 2011), (Duffie and Singleton 2012), (Glasserman 2004), (Jeanblanc et al. 2009), (Karatzas and Shreve

1998), (Musiela and Rutkowski 2005), and (Carmona and Tehranchi 2007) – the choices are many and very much depend on personal preferences and professional objectives.

The readers of this book are assumed to be firmly in the know about undergraduate analysis, linear algebra, and differential equations. Familiarity with real analysis at greater depth – say, at the level of (Rudin 1976) – would be very beneficial. Some basic knowledge of finance and economics – say, by studying (Fabozzi et al. 2010), (Kreps 2013), and (Mishkin 2015) – would be beneficial as well.

In my work toward this book I have benefited enormously from my professional interactions with (in a somewhat chronological order): Jordan Stoyanov, Svetoslav Gaidow, Mark Davis, Yuri Kabanov, Alexander Novikov, Albert Shiryaev, Paul-André Meyer, Hans Föllmer, Ioannis Karatzas, Martin Goldstein, Daniel W. Stroock, Persi Diaconis, Ofer Zeitouni, Moshe Zakai, James Norris, Marc Yor, Robert C. Merton, Steven Ross, Peter Carr, and Bernard Dumas. I am greatly indebted to all of them, and am also obliged to the following indispensable sources: (Meyer 1966), (Itô and McKean 1974), (Liptser and Shiryaev 1974), (Dellacherie and Meyer 1980), (Stroock 1987), (Jacod and Shiryaev 1987), (Merton 1992), (Karatzas 1996), (Karatzas and Shreve 1998), (Revuz and Yor 1999), (Dudley 2002), (Duffie 2010), (Föllmer and Schied 2011), and not the least to (Knuth 1984). There are many other sources that are just as relevant, and I acknowledge that the ones on which this book builds in terms of content and style – and these are also the sources most often cited throughout the text – reflect my personal bias, which in turn reflects the circumstances under which I was introduced to the subject.

In the last several years I was blessed with lots of extraordinarily motivated students. I am grateful to many of them who spotted numerous typos, errors, and omissions in earlier versions of this book, and extend my special thanks to: Xue Bai, Ge (Dovie) Chu, Yi (Frances) Ding, Juntao (Eric) Fang, Xinwei (Richard) Huang, Tong Jin, Quehao (Tony) Li, Cheng Liang, Knut Lindaas, Dhruv Madeka, Wayne Nielsen, Sunjoon Park, Hao (Sophia) Peng, Maneesha Premaratne, Ao (Nicholas) Shen, Yu Shi, Mingyan Wang, Lyan Wong, Weixuan Xia, Qing (Agatha) Xu, and Wenqi Zhang.

Special thanks are due to the MIT Press and my editors Virginia Crossman and Emily Taber for their interest in this project, seemingly unlimited patience, and relentless pursuit of perfection.

The debt of gratitude that I owe to my wife Hannelore is yet another example of a nonmeasurable set. In the course of many years she made it possible for me to work on my projects with only a minimal distraction from other matters in life. Without her never failing support and encouragement this book would not have been written.

Andrew Lyasoff
July 31, 2017

Notation

¤	generic currency
\$	dollar (AUD/CAD/USD)
€	euro (EUR)
£	pound sterling (GBP)
¥	yen/yuan (YEN/CNY)
$\wedge$	the infimum
$\vee$	the supremum
$\{a\}$	singleton; set that has only one element, a
$\emptyset$	the empty set
2^S	all subsets of the set S
$S^{\mathbb{R}_+}$	all functions from $\mathbb{R}_+$ into the set S
I	identity matrix
$I_{n\otimes n}$	n-by-n identity matrix
$A^\intercal$	transpose of matrix A
$x \in \mathbb{R}^n$	vector-column (list) $x = (x_1, \dots, x_n)$
$x^\intercal$	vector-row $x^\intercal = (x_1, \dots, x_n)^\intercal$
$\deg(P)$	degree of polynomial P
$\mathrm{Dgn}(a)$	diagonal matrix with diagonal = vector a
$\mathrm{len}(a)$	length of vector a
$f \circ g$	composition of f and g $(f \circ g)(x) = f\big(g(x)\big)$
$f \restriction A$	restriction of function f to set A
$\partial_i f$	partial derivative in the ith arg. of f
∂_x	first derivative in x (same us $\frac{\partial}{\partial x}$)
$\partial_{x,y}$	same as $\partial_x \partial_y = \frac{\partial^2}{\partial x \partial y}$
∂_x^2	same as $\partial_{x,x}$
$\partial^k f$	kth derivative in the last argument of f
∂f	same as $\partial^1 f$
$\mathbb{C}$	complex numbers
$\imath$	the imaginary unit $\imath \in \mathbb{C}$, $\imath^2 = -1$
$\mathrm{Re}(z)$	real part of $z \in \mathbb{C}$
$\mathrm{Im}(z)$	imaginary part of $z \in \mathbb{C}$
$\bar{z}$	complex conjugate of $z \in \mathbb{C}$
$\mathbb{I}, \mathbb{J}$	generic index sets
$\mathbb{N}$	the set of natural numbers $\{0, 1, 2, \dots\}$
$\bar{\mathbb{N}}$	the set $\mathbb{N} \cup \{+\infty\}$
$\mathbb{N}_{++}$	the set of strictly positive elements of $\mathbb{N}$
$\bar{\mathbb{N}}_{++}$	the set $\mathbb{N}_{++} \cup \{+\infty\}$
$\mathbb{N}_{\mid n}$	the set $\{0, 1, \dots, n\}$
$\mathbb{Q}$	rational real numbers
$\mathbb{Q}_+$	nonnegative elements of $\mathbb{Q}$
$\mathbb{Q}_{++}$	strictly positive elements of $\mathbb{Q}$
$\mathbb{R}$	real numbers $]-\infty, +\infty[$
$\bar{\mathbb{R}}$	extended real line $\mathbb{R} \cup \{-\infty\} \cup \{+\infty\}$
$]a, b[$	open interval
$]a, b]$	left-open right-closed interval
$[a, b[$	left-closed right-open interval
$\mathbb{R}_+$	the interval $[0, +\infty[$
$\mathbb{R}_-$	the interval $]-\infty, 0]$
$\bar{\mathbb{R}}_+$	the interval $[0, +\infty]$
$\mathbb{R}_{++}$	the interval $]0, +\infty[$
$\bar{\mathbb{R}}_{++}$	the interval $]0, +\infty]$
$\mathbb{Z}$	real integers (+, −, or 0)
x^+	$x \vee 0,\ x \in \mathbb{R}$
x^-	$-(x \wedge 0),\ x \in \mathbb{R}$
$\lfloor x \rfloor$	$\vee\{n \in \mathbb{Z} : n \le x\}$ (integer part of x)
$\lceil x \rceil$	$\wedge\{n \in \mathbb{Z} : n \ge x\}$ (integer ceiling of x)
$\log(x)$	natural logarithm of x
$\mathrm{sign}(x)$	$= 1$ for $x \in [0, +\infty]$; $= -1$ for $x \in [-\infty, 0[$
$\mathbb{R}^{m\otimes n}$	real m-by-n matrices

$\mathbb{R}^n$	Euclidean space with dimension n	$\vec{1}$	the vector $(1, \dots, 1) \in \mathbb{R}^n$
$\lvert\cdot\rvert$	Euclidean norm in $\mathbb{R}^n$ (absolute value in $\mathbb{R}$)	$\vec{0}$	the zero in a vector space (in $\mathbb{R}^n$: $\vec{0} = (0, \dots, 0)$)
$\lVert\cdot\rVert$	norm or seminorm in a vector space	$\imath$	identity function on $\mathbb{R}_+$ ($\imath(t) = t$, $t \in \mathbb{R}_+$)

CONVENTIONS ABOUT MULTIVARIATE OBJECTS: A vector $x \in \mathbb{R}^n$ is understood to be a column with entries x_i, $1 \leq i \leq n$. A list of the form $(x_1, \dots, x_n)$ is understood to be a column – not a row! – and may be expressed more succinctly by the token x, as in $x \in \mathbb{R}^n$. If A is a matrix, then its entry positioned in the ith row and the jth column is $A^{i,j}$.

The symbol $|a|$ denotes: the absolute value of a, if $a \in \mathbb{R}$ or $a \in \mathbb{C}$ (that is, $|a| = \sqrt{\mathrm{Re}(a)^2 + \mathrm{Im}(a)^2}$); the Euclidean norm of a, if a is a vector; the Frobenius norm (alias: Hilbert-Schmidt norm) of a (that is, $|a| = \sqrt{\mathrm{Trace}(a^\intercal a)}$) if a is a matrix with dimensions (m, n).[1]

If a and b are vectors of the same dimension and orientation (both are columns or both are rows), then all binary operations $a + b$, $a - b$, ab, and a/b are understood as element-by-element operations, that is, if $a, b \in \mathbb{R}^n$, then $a/b = (a_1/b_1, \dots, a_n/b_n)$, $ab = (a_1 b_1, \dots, a_n b_n)$ – and so on. The element-by-element product ab is called the *arithmetic product* of $a, b \in \mathbb{R}^n$. If a and b are matrices of compatible dimensions, then ab is understood to be the usual matrix product of a and b. If ab can be interpreted as a matrix product, then ab is always understood to be the matrix product of a and b. If a and b are both columns or both rows of equal dimensions, then ab has no meaning as a matrix product, and is understood to be the arithmetic product of a and b. An example: if $a, b \in \mathbb{R}^n$, then $a^\intercal b$ is understood as a matrix product (between a row and a column), resulting in the scalar $a \cdot b$ (the dot product between a and b), but ab and $a^\intercal b^\intercal$ are understood as arithmetic products. If one of the symbols a and b is a scalar and the other one is a vector in either orientation, then the product ab is understood as the result of the usual multiplication of a vector by a scalar (in this case ab has no meaning as a matrix or an arithmetic product).

The integration domain in a multivariate integral of the form

$$\int_{a_n}^{b_n} \cdots \int_{a_2}^{b_2} \int_{a_1}^{b_1} f(x_1, x_2, \dots, x_n) \mathrm{d}x_1 \mathrm{d}x_2 \dots \mathrm{d}x_n$$

is understood as $\{x \in \mathbb{R}^n : x_1 \in]a_1, b_1], x_2 \in]a_2, b_2], \dots x_n \in]a_n, b_n]\}$.

[1] The Frobenius norm is the same as the Euclidean norm of $a \in \mathbb{R}^{m \otimes n}$ treated as a vector with dimension $m \times n$. It is named after the German mathematician Ferdinand Georg Frobenius (1849–1917).

Preliminaries

The following notions and related notation are assumed familiar: set, elements of a set ($a \in A$), subset and (strict or non-strict) inclusion of a set ($\subset$, $\subseteq$), the empty set $\emptyset$ (subset of any set), in addition to all basic operations with sets: union $\cup$, intersection $\cap$, and complement $(\cdot)^{\complement}$.

If A and B are any sets, the Cartesian product $A\times B$ is the collection (set) of all ordered pairs (a, b), for all possible choices of $a \in A$ and $b \in B$. Given any set A, the subsets of the Cartesian product $A\times A$ are also called *relations* in A. Relations are usually expressed by symbols such as $\succeq$, or by other similar (generally, non–symmetric) tokens. As an example, the symbolic expression $a \succeq b$ is understood to have the same meaning as the symbolic expression $(a, b) \in \mathscr{R}$, for an appropriate subset $\mathscr{R} \subset A\times A$. If the connection between the symbol $\succeq$, understood as a relation in A, and the associated subset $\mathscr{R} \subset A\times A$ must be emphasized, we can express the set $\mathscr{R}$ as $(\succeq)$. If $\succeq$ is a relation in A, then $\preceq$ is understood as the mirror image of $\succeq$, in that the set $(\preceq) \subset A\times A$ is the mirror image of the set $(\succeq) \subset A\times A$ across the diagonal of $A\times A$. To put it another way, $a \preceq b$ if and only if $b \succeq a$. If $\succeq$ is a relation in A, then $\not\succeq$ is understood to be the relation $(\succeq)^{\complement} \subset A\times A$. The relation $\succeq$ is said to be: *reflexive* if $a \succeq a$ for any $a \in A$, *transitive* if $a, b, c \in A$ and $a \succeq b$ and $b \succeq c$ imply $a \succeq c$, and *symmetric* if $a, b \in A$ and $a \succeq b$ imply $b \succeq a$. For example, $(\simeq) \overset{\text{def}}{=} (\succeq) \cap (\preceq) \subset A\times A$ is a symmetric relation. Symmetric relations are usually expressed by symbols that look symmetric, such as $=$, $\approxeq$, $\sim$, $\simeq$. A relation that is simultaneously reflexive, transitive, and symmetric is called *equivalence relation*.

The following properties of the operations $\cup$ and $\cap$, involving a set B and an arbitrary collection of sets $(A_i)_{i\in\mathbb{I}}$ ($\mathbb{I}$ is an arbitrary index set) are well known and easy to verify:

$$B \cap \left(\cup_{i\in\mathbb{I}} A_i\right) = \cup_{i\in\mathbb{I}}(B \cap A_i), \quad (\cup_{i\in\mathbb{I}} A_i)^{\complement} = \cap_{i\in\mathbb{I}}(A_i^{\complement}), \quad (\cap_{i\in\mathbb{I}} A_i)^{\complement} = \cup_{i\in\mathbb{I}}(A_i^{\complement}).$$

The set difference $A\backslash B$ is defined as $A \cap B^{\complement}$ – this is the collection of all elements of A that are not elements of B. Two sets A and B are said to be *disjoint* if they do not have a common element, that is, $A \cap B = \emptyset$. A nonempty set that contains only one element, a, is called a *singleton* and is denoted by $\{a\}$.

The most important sets in this space are the set of natural numbers $\mathbb{N} = \{0, 1, \ldots\}$, the set of all integer numbers $\mathbb{Z} = \{\ldots, -1, 0, 1, \ldots\}$ (alias: the integer lattice), the set of real numbers $\mathbb{R}$, the set of rational numbers $\mathbb{Q}$, and the set of complex numbers $\mathbb{C}$. The complex number $\sqrt{-1}$ will be denoted by $\imath$, and by $\mathbb{Z}_+$, $\mathbb{R}_+$, and $\mathbb{Q}_+$ ($\mathbb{Z}_{++}$, $\mathbb{R}_{++}$, and $\mathbb{Q}_{++}$) we denote the subsets of all nonnegative (strictly positive) elements of $\mathbb{Z}$, $\mathbb{R}$, and $\mathbb{Q}$ (so that $\mathbb{N} = \mathbb{Z}_+$). With a slight abuse of the language "positive" is understood to mean "≥ 0," and a positive number which is not 0 is said to be *strictly positive*. Similarly, "negative" is understood to mean "≤ 0," and a negative number which is not 0 is said to be *strictly negative*.

The reader is also expected to be familiar with the notion of "function from a set A into a set B." Such objects are commonly expressed as $f : A \mapsto B$. The set A is the domain (for f) and the set B is the range set (aliases: range space, destination set, target space, target set) for f. We say "f takes values in B," or "f is a B-valued function on A," or "f is a function with values in B, defined on A." The range of the function $f : A \mapsto B$ is the set

$$\mathrm{Ran}(f) \stackrel{\text{def}}{=} \big\{ f(a) : a \in A \big\} \subseteq B ,$$

and if $\mathrm{Ran}(f) = B$ we say that f is a function *onto* B. The function $f : A \mapsto B$ is said to be one-to-one if it is onto and is such that $f(a)$ and $f(a')$ are two identical elements of B if and only if a and a' are two identical elements of A.

Depending on the context, a set may be referred to as *space*, which, generally, understates some additional attributes, other than those associated with the basic structure of a set. The unwritten – and never strictly followed – rule is to denote sets with symbols such as A, B, C, etc., and denote spaces with symbols such as $\mathbb{R}$, $\mathbb{C}$, Ω, $\mathbb{X}$, $\mathbb{Y}$, or other more complex strings of symbols. Sometimes a function may be referred to as *mapping*.[2] Other common aliases for "function" are *functional* and *operator*. A function (mapping) from the space $\mathbb{X}$ into the space $\mathbb{Y}$ may be expressed also as

$$\mathbb{X} \ni x \rightsquigarrow f(x) \in \mathbb{Y} .$$

One of the most rudimentary $\mathbb{R}$-valued functions defined on a set $\mathbb{X}$ are the *indicator functions*. Often called simply *indicators*, these functions have the form

$$\mathbb{X} \ni x \rightsquigarrow 1_A(x) = \begin{cases} 0 & \text{if } x \notin A , \\ 1 & \text{if } x \in A , \end{cases}$$

for some fixed subset $A \subseteq \mathbb{X}$.

Sequences of elements chosen from some space $\mathbb{X}$ can be treated as functions (mappings) from $\mathbb{N}$ into $\mathbb{X}$:

$$\mathbb{N} \ni i \rightsquigarrow f(i) \in \mathbb{X} .$$

Usually we write x_i instead of $f(i)$ and express the sequence as x_i, $i = 0, 1, \ldots$, or as $(x_i)_{i \in \mathbb{N}}$, or as $(x_i \in \mathbb{X})_{i \in \mathbb{N}}$, or simply as $(x_i \in \mathbb{X})$, or even as (x_i), if the missing symbols are understood from the context. If the set of natural numbers, $\mathbb{N}$, is replaced with some general index set $\mathbb{I}$, then a function from $\mathbb{I}$ into $\mathbb{X}$ is called *generalized sequence* (indexed, or labeled, by the set $\mathbb{I}$) and is expressed, for example, as $(x_i \in \mathbb{X})_{i \in \mathbb{I}}$.

[2] The word "mapping" has the advantage of being a verbal noun, formed from the verb "map." This allows one to use "map" and "mapping" as in the phrase "we draw to get a drawing." Unfortunately, in most languages there is no verb that can be linked to the noun "function" in the same way.

If $f\colon \mathbb{X} \mapsto \mathbb{Y}$ is any function and $B \subseteq \mathbb{Y}$ is any subset of $\mathbb{Y}$, the preimage of B under f, denoted $f^{-1}(B)$, is the subset of X defined as

$$f^{-1}(B) \stackrel{\text{def}}{=} \{x \in \mathbb{X} : f(x) \in B\}.$$

If $A \subseteq \mathbb{X}$ is any subset of $\mathbb{X}$, the image of A under f, denoted $f(A)$, is the subset of $\mathbb{Y}$ defined as

$$f(A) \stackrel{\text{def}}{=} \{f(x) \in \mathbb{Y} : x \in A\}.$$

All results from analysis, probability, and measure that are used in this space in one form or another depend in a crucial way on the axiom of choice, which we reproduce next for completeness of the exposition, and also as a "statement of full disclosure."

(0.1) THE AXIOM OF CHOICE: Let $\mathbb{X}$ be any set and let $\mathcal{S}$ be the collection of all nonempty subsets of $\mathbb{X}$. Then there is a function, $f\colon \mathcal{S} \mapsto \mathbb{X}$, such that $f(A) \in A$ for every $A \in \mathcal{S}$. ○

(0.2) CARDINALITY OF A SET: If A and B are two sets such that there is a mapping $f\colon A \mapsto B$ that is one-to-one, then A is understood to have just as many elements as B, and B is understood to have just as many elements as A. If a one-to-one mapping $f\colon A \mapsto B$ exists, we say that A and B have the same cardinality and write $|A| = |B|$. The cardinality of the set $\mathbb{N}_{|n} \stackrel{\text{def}}{=} \{0, 1, \ldots, n\}$, $n \in \mathbb{N}$, is $|\mathbb{N}_{|n}| = n + 1$,[3] and the cardinality of the set of all natural numbers $\mathbb{N}$ is called *aleph-null* and is usually expressed $\aleph_0$, understood to be another notation for $|\mathbb{N}|$. The cardinality of the set $\mathbb{R}$ is strictly larger than $\aleph_0$, as will be detailed below, and is usually expressed as $\mathfrak{c}$, understood to be another notation for $|\mathbb{R}|$.[4] The cardinality of the empty set $\emptyset$ is 0 and we have $n < \aleph_0 < \mathfrak{c}$ for any $n \in \mathbb{N}$. If we write $|A| = n$ for some set A and some $n \in \mathbb{N}$, then this means that A has n elements, which is to say, as many elements as $\mathbb{N}_{|n-1}$ if $n \in \mathbb{N}_{++}$, or 0 elements if $A = \emptyset \Leftrightarrow |A| = 0$. ○

(0.3) COUNTABLE SETS: A nonempty set $\mathbb{X}$ (it could be any set) is said to be *countable* if there is a function from $\mathbb{N}$ onto $\mathbb{X}$. To put it another way, $\mathbb{X}$ is countable if it is possible to attach to every natural number $i \in \mathbb{N}$ an element $x_i \in \mathbb{X}$ so that every element of $\mathbb{X}$ is attached to at least one natural number from $\mathbb{N}$. If the nonempty set $\mathbb{X}$ is not finite (does not have finitely many elements) and is countable, it is said to be *countably infinite*. A set is countably infinite if and only if it has cardinality equal to $\aleph_0$. A set that is not countable is said to be uncountable. ○

Clearly, a finite nonempty set is always countable, and if $\mathbb{X} \neq \emptyset$ is countable, then its elements can be arranged in a finite or infinite sequence of distinct elements. The converse is also true: the collection of all elements in any finite or infinite sequence represents a countable set. Why is it important for us to know

[3] The natural numbers $n \in \mathbb{N}$ can be viewed as "expressions of cardinality."

[4] If the continuum hypothesis is imposed as an axiom (there is no cardinality that stands between $\aleph_0$ and $\mathfrak{c}$), then the notation $\aleph_1$ is used instead of $\mathfrak{c}$.

whether it is possible to arrange the elements of a given set $\mathbb{X}$ in a sequence or not? The reason is that in many practical situations it is necessary to attach "a mass," say $\rho(x) \in \mathbb{R}_{++}$, to all elements $x \in \mathbb{X}$. Furthermore, we would like, at least heuristically, phrases like "the total mass of $\mathbb{X}$," or "the total mass of the subset $A \subset \mathbb{X}$," to be meaningful. In particular, for the total mass of $\mathbb{X}$ we would like to write $\sum_{x\in\mathbb{X}} \rho(x)$. If the elements of $\mathbb{X}$ can be arranged in a sequence, say $(x_i)_{i\in\mathbb{N}}$, that is, if the set (space) $\mathbb{X}$ is countable, then $\sum_{x\in\mathbb{X}} \rho(x)$ can be understood as the infinite series $\sum_{i\in\mathbb{N}} \rho(x_i)$. However, if $\mathbb{X}$ is uncountable, in general, the expression $\sum_{x\in\mathbb{X}} \rho(x)$ can no longer be understood as an infinite series. It is not very difficult to show that – this is simple but important – every nonempty subset of a countable set is also countable. Thus, if $\mathbb{X}$ is countable, then the expression $\sum_{x\in A} \rho(x)$ would be perfectly meaningful for every subset $A \subset \mathbb{X}$, $A \neq \emptyset$.

(0.4) EXERCISE: An experiment consists of tossing a coin countably infinitely many times. An *elementary outcome* from this experiment is an infinite sequence of the symbols H and T, which we can write as $\big(\varepsilon_i \in \{H,T\}\big)_{i\in\mathbb{N}}$. Prove that the collection, Ω_∞, of all such sequences is uncountable.

HINT: Consider proving the statement by contradiction. Specifically, prove that for any sequence $x = (x_k \in \Omega_\infty)_{k\in\mathbb{N}}$, however chosen, there must be at least one element of Ω_∞ that is not contained in the sequence x. ○

(0.5) EXERCISE: Prove that the open interval $]0,1[$ has the same cardinality as the entire real line $\mathbb{R}$.

HINT: Construct a one-to-one mapping from $]0,1[$ onto $]0,\pi[$ and a one-to-one mapping from $]0,\pi[$ onto $\mathbb{R}$. ○

(0.6) EXERCISE: Give an example of a one-to-one mapping from the interval $[0,1]$ onto the interval $[0,1[$. Then construct a one-to-one mapping from the closed interval $[0,1]$ onto the open interval $]0,1[$.

HINT: For $x = \frac{1}{n}$, for some $n \in \mathbb{N}_{++}$, define $f(x) = \frac{1}{n+1}$. For any $x \in [0,1]$ that cannot be expressed as $\frac{1}{n}$ for some $n \in \mathbb{N}_{++}$, define $f(x) = x$. ○

In exercise (1.69) we will see that, treated as a set (of real numbers), the interval $[0,1]$ contains a subset that has the same cardinality as the set of sequences Ω_∞ that was introduced in exercise (0.4), leading to the conclusion that $[0,1]$, and therefore also $\mathbb{R}$, are uncountable sets of identical cardinality.

(0.7) EXERCISE: Obviously, the set $\mathbb{N}$ is countable by definition, but is the set $\mathbb{N}\times\mathbb{N}$ countable? What about the set $\mathbb{N}\times\cdots\times\mathbb{N}$ (n-times)? Prove that the set of all rational numbers is countable. Prove that the union of any countable collection of countable sets is a countable set. Prove that the set of all points in $\mathbb{R}^2$ with rational coordinates is countable. ○

Throughout the sequel we will invariably use the notions *limit*, *convergence*, and *continuity* more or less in the way these notions are known to us from calculus, except that we will use them in the context of spaces that are more general than the spaces $\mathbb{R}$ or $\mathbb{R}^n$. However, for the most part, this "general" context will differ from the one encountered in calculus only in the notation: one merely needs to replace the real line $\mathbb{R}$ (or the space $\mathbb{R}^n$) with some other space, $\mathbb{X}$, and replace the usual Euclidean distance $d(x, y) = |x - y|$ between two real numbers $x, y \in \mathbb{R}$, or vectors $x, y \in \mathbb{R}^n$, with some more general distance, $(x, y) \rightsquigarrow d(x, y)$, between two elements $x, y \in \mathbb{X}$. Appendix A contains a brief synopsis of the essential results from analysis and topology that are used throughout this book.

(0.8) SPACES OF CONTINUOUS FUNCTIONS: If $\mathbb{X}$ and $\mathbb{Y}$ are any two topological spaces, then the space of continuous function from $\mathbb{X}$ into $\mathbb{Y}$ will be denoted by $\mathscr{C}(\mathbb{X};\mathbb{Y})$. If the range space $\mathbb{Y}$ happens to be the real line $\mathbb{R}$, or, more generally, the Euclidean space $\mathbb{R}^n$, then it becomes possible to prepend "function" with the qualifier "bounded." A *bounded function* is a function the absolute value (or Euclidean norm) of which is bounded, in that there is a constant $C \in \mathbb{R}_{++}$ (possibly depending on the function) such that $|f(x)| \leq C$ for all $x \in \mathbb{X}$. The space of all bounded and continuous functions from $\mathbb{X}$ into $\mathbb{R}^n$ will be expressed as $\mathscr{C}_b(\mathbb{X};\mathbb{R}^n)$, and we note that $\mathscr{C}_b(\mathbb{X};\mathbb{R}^n)$ and $\mathscr{C}(\mathbb{X};\mathbb{R}^n)$ are the same if $\mathbb{X}$ happens to be compact. The space $\mathscr{C}_b(\mathbb{X};\mathbb{R}^n)$ is⊸ (A.46) complete for the uniform distance

$$d_\infty(f,g) \stackrel{\text{def}}{=} \|f - g\|_\infty = \sup\{|f(x) - g(x)| : x \in \mathbb{X}\}, \quad f, g \in \mathscr{C}_b(\mathbb{X};\mathbb{R}^n),$$

and is therefore a Banach space (with norm $\|\cdot\|_\infty$). This Banach space is separable (admits a countable dense subset) if $\mathbb{X}$ happens to be separable metric space (e.g., if $\mathbb{X}$ is a subset of $\mathbb{R}^n$). By way of an example, given any finite interval $[a, b] \subseteq \mathbb{R}$, the (countable) set of all polynomials with rational coefficients can be shown to be dense in $\mathscr{C}([a, b];\mathbb{R}^n)$ for the uniform distance.

In many concrete applications we need to work with the space $\mathscr{C}(\mathbb{R}_+;\mathbb{R}^n)$, which is different from $\mathscr{C}_b(\mathbb{R}_+;\mathbb{R}^n)$. As the uniform distance is no longer a well-defined distance function on $\mathscr{C}(\mathbb{R}_+;\mathbb{R}^n)$, we need to work with the locally uniform distance, defined as

$$d_{\text{loc},\infty}(f,g) \stackrel{\text{def}}{=} \sum\nolimits_{k \in \mathbb{N}_{++}} 2^{-k}\Big(1 \wedge \sup\{|f(x) - g(x)| : x \in [0, k]\}\Big)$$

for any $f, g \in \mathscr{C}(\mathbb{R}_+;\mathbb{R}^n)$. A sequence of function converges in the locally uniform distance if and only if it converges in the uniform distance on every compact interval $[a, b] \subset \mathbb{R}_+$, and it is not difficult to see that $\mathscr{C}(\mathbb{R}_+;\mathbb{R}^n)$ is a separable metric space for the locally uniform distance.

If $\mathscr{D} \subseteq \mathbb{R}^m$ is some connected open domain, then $\mathscr{C}^k(\mathscr{D};\mathbb{R}^n)$ will denote the space of all $\mathbb{R}^n$-valued functions on $\mathscr{D}$ the coordinates of which have *continuous* mixed derivatives (in all combinations of the arguments) of order at least $k \in \mathbb{N}_{++}$. If these functions and derivatives are required also to be bounded, then the associated space will be expressed as $\mathscr{C}_b^k(\mathscr{D};\mathbb{R})$. The space $\mathscr{C}_b^k(\mathscr{D};\mathbb{R}^n)$ can be endowed with a norm that makes it into a Banach space. With appropriate localization (see above), this norm gives rise to a distance on $\mathscr{C}^k(\mathscr{D};\mathbb{R}^n)$ that makes it into a Polish space. The space $\mathscr{C}^{2,1}(\mathscr{D}\times\mathbb{R}_+;\mathbb{R}^n)$, consists of all $\mathbb{R}^n$-valued functions of the arguments $(x,t)\in\mathscr{D}\times\mathbb{R}_+$ the coordinates of which admit a continuous first derivative in the variable $t\in\mathbb{R}_+$ and admit continuous mixed derivatives of order at least 2 in all combinations of the variables $x=(x_1,\dots,x_n)\in\mathscr{D}$. If these functions and derivatives are also required to be bounded, then the associated space will be expressed as $\mathscr{C}_b^{2,1}(\mathscr{D}\times\mathbb{R}_+;\mathbb{R}^m)$. ○

In some financial models it is essential not to insist on continuous realizations, despite the continuous-time paradigm. Thus, we must be able to work also with spaces of functions that are allowed to be discontinuous, and, clearly, to develop useful models, we must impose some requirements on the types of discontinuity that we are going to allow. For reasons that will become clear later in the exposition, the natural classes of functions to be used in stochastic modeling are classes of right-continuous functions that admit left limits.

(0.9) Continue à droite, limite à gauche (càdlàg): This French phrase translates to "right-continuous with left limits," and, the efforts of some authors to introduce the acronym RCLL notwithstanding, the most commonly used acronym for "right-continuous with left limits" is cadlag, or càdlàg.[5] The space of càdlàg functions from an interval $I\subseteq\mathbb{R}$ into $\mathbb{R}^n$ will be expressed as $\mathbb{D}(I;\mathbb{R}^n)$. If I is a compact interval, then $\mathbb{D}(I;\mathbb{R}^n)$ is complete for the uniform distance, and is therefore a Banach space for the norm $\|\cdot\|_\infty$. However, it is well known that this Banach space fails to be separable, and this makes the construction of probabilistic structures on that space both difficult and impractical. In order to turn $\mathbb{D}(I;\mathbb{R}^n)$ into a separable metric space, we need to find a metric that is, on the one hand, sufficiently weaker than the uniform distance to make separability possible, and, on the other hand, sufficiently strong to be still useful. Such a metric was proposed by Skorokhod (1956) and is known as the *Skorokhod distance*.[6] In what follows we will adopt the description of the Skorokhod distance (and the related tools and nomenclature) from (Jacod and Shiryaev 1987). Let $\mathfrak{A}$ denote the collection of all functions $\alpha\colon\mathbb{R}_+\mapsto\mathbb{R}_+$ that are: (a) strictly increasing without bound ($\lim_{t\to\infty}\alpha(t)=\infty$), (b) continuous, and (c) start from 0 ($\alpha(0)=0$). Given any $\alpha\in\mathfrak{A}$, let

$$|||\alpha||| \overset{\text{def}}{=} \sup\left\{ \left|\log\left(\frac{\alpha(t)-\alpha(s)}{t-s}\right)\right| : 0\le s<t<\infty\right\},$$

and given any $k\in\mathbb{N}_{++}$, let

$$h_k(t) \overset{\text{def}}{=} \begin{cases} 1 & \text{if } t\le k, \\ k+1-t & \text{if } k<t<k+1, \\ 0 & \text{if } t\ge k+1. \end{cases}$$

[5] In English texts it is often acceptable – even among French authors – to write cadlag instead of càdlàg. In this space we prefer to use càdlàg.

[6] Named after the Ukrainian mathematician Anatoliy Volodymyrovych Skorokhod (1930–2011).

The Skorokhod distance between two functions, $f, g \in \mathbb{D}(\mathbb{R}_+; \mathbb{R}^n)$, is given by

$$d_{\mathfrak{s}}(f,g) \stackrel{\text{def}}{=} \sum_{k=1}^{\infty} 2^{-k}\Big(1 \wedge \inf_{\alpha \in \mathfrak{A}} \Big\{ |||\alpha||| + \sup_{t \in \mathbb{R}_+} \big| h_k(\alpha(t)) f(\alpha(t)) - h_k(t) g(t) \big| \Big\}\Big).$$

Endowed with this distance, the space $\mathbb{D}(\mathbb{R}_+; \mathbb{R}^n)$ is called the *Skorokhod space*. The topology associated with the metric $d_{\mathfrak{s}}(\cdot,\cdot)$ is strictly weaker than the topology associated with the metric $d_{\mathrm{loc},\infty}(\cdot,\cdot)$ (since $\imath \in \mathfrak{A}$), and the Skorokhod space $\mathbb{D}(\mathbb{R}_+; \mathbb{R}^n)$ is a complete separable metric space for the distance $d_{\mathfrak{s}}(f,g)$ – see (Jacod and Shiryaev 1987, VI 1.33, 1.43). ○

(0.10) TEST FUNCTIONS: There are situations where we need to work with functions that have all the smoothness and continuity properties that a function can have. Specifically, we will need to apply certain concepts and operations that, in one form or another, involve "derivatives" of functions that may not be even continuous (see below), or will need to approximate non-smooth functions with smooth ones. The so-called *test functions* are defined as mappings $\varphi\colon \mathbb{R}^n \mapsto \mathbb{R}$ such that: (a) φ admits partial derivatives of all orders and in all combinations of its arguments *everywhere* in $\mathbb{R}^n$, and (b) φ vanishes outside some compact set, namely, there is a compact set $K \subset \mathbb{R}^n$ such that $\varphi(x) = 0$ for all $x \in \mathbb{R}^n \backslash K$. Property (b) is often stated as "φ has compact support." Test functions do exist, as is evidenced by the following classical example:[7]

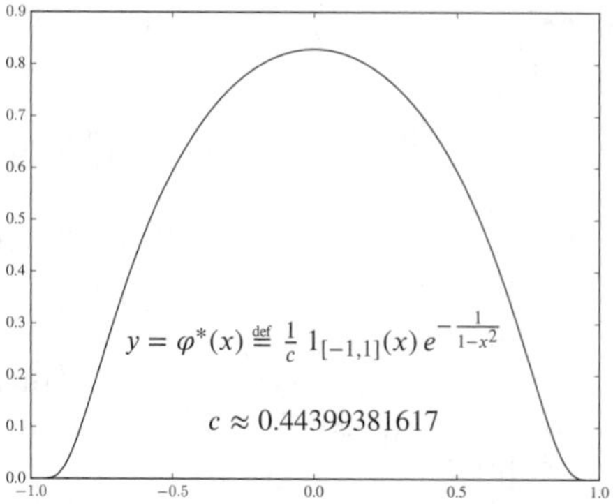

FIGURE 0.1
Positive test function that integrates to 1.

The only points at which the differentiability of this function may seem questionable are $x = 1$ and $x = -1$. However, a straightforward computation gives

$$\lim_{x \nearrow 1} \partial^k \varphi^*(x) = 0 \quad \text{and} \quad \lim_{x \searrow -1} \partial^k \varphi^*(x) = 0 \quad \text{for every } k \in \mathbb{N}.$$

The space of all test functions defined on $\mathbb{R}^n$ – that is, the collection of all infinitely differentiable functions from $\mathbb{R}^n$ into $\mathbb{R}$ that have compact support – we de-

[7] See (Rudin 1986). This example can be found in most textbooks covering functional analysis.

note by $\mathscr{C}_K^\infty(\mathbb{R}^n;\mathbb{R})$. If $\mathscr{D}$ is some connected domain inside $\mathbb{R}^n$, then $\mathscr{C}_K^\infty(\mathscr{D};\mathbb{R})$ will denote the space of all test functions with compact support contained in $\mathscr{D}$. ○

(0.11) MOLLIFICATION AND APPROXIMATION WITH TEST FUNCTIONS: The fact that a positive test function that integrates to 1 exists allows one to *mollify* non-smooth functions: given any piecewise continuous function $f\colon]a,b[\ \mapsto \mathbb{R}$ (the interval $]a,b[$ may be the entire real line $\mathbb{R}$) such that $\int_a^b |f(x)|\,\mathrm{d}x < \infty$, and given any sufficiently small $\varepsilon > 0$, we set (with φ^* taken from figure 0.1)

$$f_\varepsilon(x) \stackrel{\text{def}}{=} \int_{-1}^{+1} \varphi^*(u) f(x-\varepsilon u) 1_{[a+\varepsilon,b-\varepsilon]}(x-\varepsilon u)\,\mathrm{d}u\,, \quad x\in\mathbb{R}\,,$$

which function is easily seen to be infinitely smooth (admits derivatives of all orders). Moreover, if the function f happens to be uniformly continuous on some open interval $I \subset\ ]a,b[$, then $f_\varepsilon \to f$ as $\varepsilon\to 0$ uniformly on any closed interval contained in I, as we now illustrate in the case $f(x) \stackrel{\text{def}}{=} 1_{[-.5,.5]}(x)$, $x\in\mathbb{R}$:

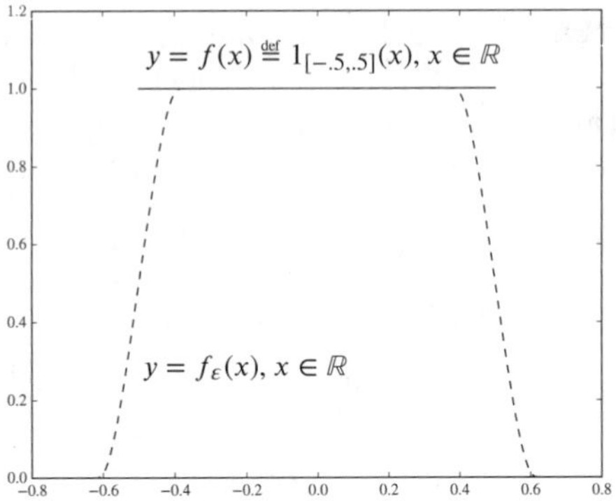

FIGURE 0.2

Mollification of the indicator $1_{[-.5,.5]}$.

○

(0.12) GENERALIZED DERIVATIVES AND δ-FUNCTIONS: All functions from the space $\mathscr{C}_K^\infty(\mathbb{R};\mathbb{R})$ are automatically continuous and bounded, and the same is true for their derivatives. If the function $f\colon\mathbb{R}\mapsto\mathbb{R}$ has a continuous first derivative, then

$$\int_{-\infty}^{+\infty} \partial f(x)\varphi(x)\,\mathrm{d}x = -\int_{-\infty}^{+\infty} f(x)\partial\varphi(x)\,\mathrm{d}x \quad \text{for every } \varphi\in\mathscr{C}_K^\infty(\mathbb{R};\mathbb{R})\,.$$

Since the integral on the right side would be perfectly meaningful for a very large class of functions that may not be differentiable in the usual sense, one can define the so-called *generalized derivative* of f as the linear functional on the space $\mathscr{C}_K^\infty(\mathbb{R};\mathbb{R})$ given by $\varphi \rightsquigarrow -\int f\partial\varphi$, as long as this last integral exists, which is a substantially less stringent requirement for f than differentiability. For example, the function $f(x) = 1_{[a,+\infty[}(x)$, $x\in\mathbb{R}$, is not differentiable at $x=a$, yet the quantity

$$-\int_{-\infty}^{+\infty} f(x)\partial\varphi(x)\,\mathrm{d}x = -\int_{a}^{+\infty} \partial\varphi(x)\,\mathrm{d}x = \varphi(a)$$

is perfectly meaningful for every $\varphi \in \mathscr{C}_K^{\infty}(\mathbb{R};\mathbb{R})$. Consequently, the generalized derivative of $f = 1_{[a,+\infty[}$ can be identified as the evaluation mapping

$$\mathscr{C}_K^{\infty}(\mathbb{R};\mathbb{R}) \ni \varphi \rightsquigarrow \epsilon_a(\varphi) \stackrel{\text{def}}{=} \varphi(a) \in \mathbb{R}\,.$$

It is quite common to express the (generalized) derivative $\partial_x 1_{[a,+\infty[}(x)$ as a hypothetical function $x \rightsquigarrow \delta_a(x)$ – often referred to as Dirac's delta[8] – for which we can write, formally,

$$\int_{-\infty}^{+\infty} f(x)\delta_a(x)\,\mathrm{d}x = \epsilon_a(f) \equiv f(a)$$

for any finite function $f\colon \mathbb{R} \mapsto \mathbb{R}$. ○

(0.13) GAMMA FUNCTION: The *gamma function* is given by

$$\mathbb{R}_{++} \ni p \rightsquigarrow \Gamma(p) = \int_0^{\infty} u^{p-1}e^{-u}\,\mathrm{d}u\,.$$

It has the property $\Gamma(n+1) = n!$ for any $n \in \mathbb{N}$, and also the property $\Gamma(p) = \int_0^{\infty} c^p u^{p-1}e^{-cu}\,\mathrm{d}u$ for any $c \in \mathbb{R}_{++}$ (the result of a straightforward change of variables). The *upper incomplete gamma function* is given by

$$\mathbb{R}_{++}\times\mathbb{R}_{+} \ni (p,x) \rightsquigarrow \Gamma(p,x) = \int_x^{\infty} u^{p-1}e^{-u}\,\mathrm{d}u\,.$$

It has the property (the result of a straightforward integration by parts)

† $$\Gamma(p+1,x) = e^{-x}p!\sum\nolimits_{i=0}^{p}\frac{x^i}{i!}\,, \quad p \in \mathbb{N}\,, \quad x \in \mathbb{R}_{+}\,.$$

The *lower incomplete gamma function* is given by

$$\mathbb{R}_{++}\times\mathbb{R}_{+} \ni (p,x) \rightsquigarrow \int_0^{x} u^{p-1}e^{-u}\,\mathrm{d}u = \Gamma(p) - \Gamma(p,x)\,.$$

Both the gamma function and the upper incomplete gamma function are implemented as standard functions on most computing systems. On some systems the lower incomplete gamma function is implemented as well under the name γ_{greek}. Unfortunately, the upper incomplete gamma function is sometimes referred to as "the incomplete gamma function," which is ambiguous. ○

We conclude our list of preliminaries with the following classical result:

(0.14) ARITHMETIC-GEOMETRIC MEAN INEQUALITY: For any $x_i \in \mathbb{R}_{+}$, $1 \le i \le n < \infty$, one has

$$(x_1 x_2 \ldots x_n)^{1/n} \le \frac{x_1 + x_2 + \ldots + x_n}{n}\,. \quad ○$$

[8] Named after the English physicist Paul Dirac (1902–1984).

1

Probability Spaces and Related Structures

Quantifying the chances for various random events to occur is somewhat similar to measuring the area, or the volume, or the mass, of some portion of the physical space. Matters become considerably more involved in random experiments that have infinitely many possible outcomes. The reason is that not all pieces from the Euclidean space can be measured, or weighted. The present chapter develops the classical measure-theoretic paradigm of probability theory. In addition, it introduces the coin toss space and the random walk.

1.1 Randomness in the Financial Markets

The term *random* (alias: *stochastic*) is usually associated with experiments such as rolling a die, flipping a coin, or playing roulette in a casino. The random phenomena studied in this book do have something in common with such experiments, but it is important to acknowledge that there are fundamental differences between the way in which randomness manifests itself in a casino and the way in which it manifests itself in the practice of finance. While some may share the view that all financial markets are essentially large-scale casinos – somewhat similar to those found in Monte Carlo or Las Vegas – even a superficial first look would reveal that financial markets are more like casinos in which the uncertainty in the payoffs has a lot to do with the way the players interact, and, most important, markets do not "favor the house." Unlike the interactions between the patrons of a casino and its owners, in aggregate, the interactions between the market participants do not amount to a zero-sum game. Among other things, financial markets offer a risk premium that no casino can do. This has a lot to do with the important social function that financial interactions have: they are a medium for economic activity and technological innovation, and are thus instrumental for the creation of new wealth in the economy as a whole. In short, casinos do not create new wealth, but financial markets do, and this is what allows for the generation of a risk premium.

Even a cursory glance through history would reveal that financial innovation and technological innovation have been moving mostly in a lockstep. In the realm of finance, uncertainty and risk are not merely faced and played against; rather, uncertainty and risk are shared and transformed. This makes the interaction between asset pricing and probability theory especially interesting. We begin our exploration of this interaction with a simple illustration: a snapshot of the daily closing prices of the financial security with ticker symbol TSLA from its very first day of trading (June 29, 2010) until June 29, 2016.[9]

FIGURE 1.1
Daily closing prices for TSLA

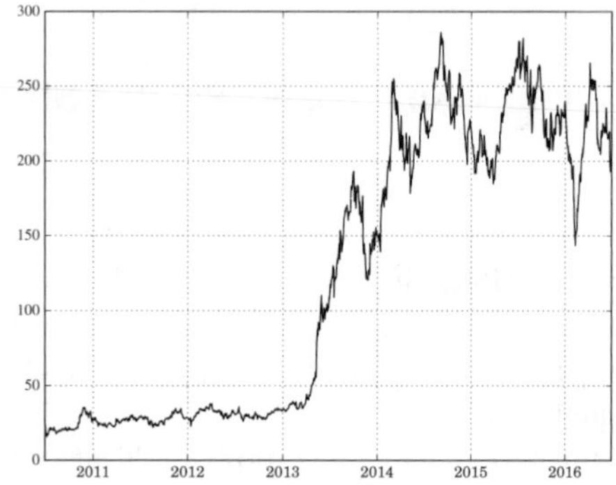

We see that on a micro scale the prices of TSLA move in a way that appears to be unpredictable and chaotic; yet on a macro scale one can still recognize certain trends and patterns. To better understand the random nature of figure 1.1, consider the daily net returns from TSLA for the same period:

FIGURE 1.2
Daily net returns for TSLA

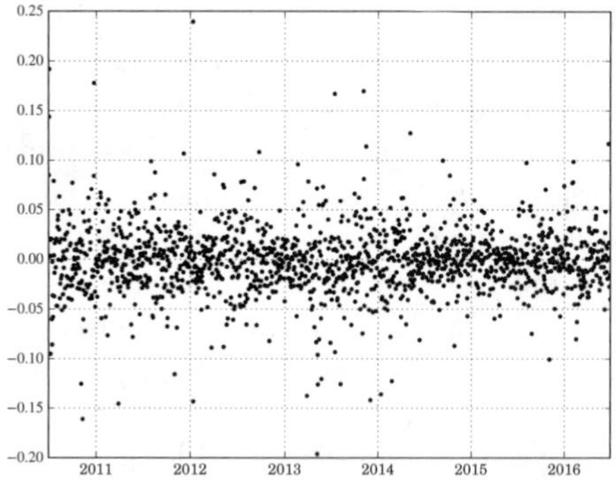

In some sense, the actual spot price at which a particular asset is traded can be viewed as a mere accounting convention, and, typically, what matters the most in financial interactions are the actual returns. There are 1,512 data points in the last plot, and the pattern that we see seems to resemble – somewhat – the outcome from flipping a coin 1,512 times (except that the "coin" does not produce just two values: head and tail). But are the daily returns really unrelated to one another, in

[9] The computer code used to access the relevant market data and create the plots included in the present section can be found in section B.1 in the appendix.

the same way in which the results from flipping a coin are typically assumed to be? What if there is a small interdependence among those daily returns that is visually imperceptible? How does one describe such a dependence? Clearly, these are questions that cannot be answered by simply observing the plots just shown. One of our objectives in this space is to develop a framework within which seemingly innocent questions of this nature can be understood – and eventually answered.

Still, a quick look at figure 1.2 can already tell us quite a bit. In particular, we see that the realized returns tend to appear much more frequently near 0, and daily returns outside the $\pm 5\%$ range are not very common. Yet on May 8, 2013, the TSLA stock dropped by more 19% (the largest daily net-loss in the sample), while on January 12, 2012, the TSLA stock jumped by more than 24% (the largest daily net-gain in the sample). One way to quantify the intuitive notions of "tends to" and "occurs more frequently" is to build a histogram from the observed returns. This comes down to dividing the range of the observed returns into equally spaced bins (intervals) and counting how many of the realized returns fall into every individual bin. However, to do so, one must first decide on the size of the bins, and this decision is not as straightforward as it may appear at first. We have yet to develop the tools that would allow us to address such matters, but at this point we could simply state the Freedman–Diaconis rule for choosing the width of the bins:[10] if the size of the sample is n and the inter-quartile range of the sample, defined as the difference between the 75th percentile in the sample and the 25th percentile, is IQR, then the width of the bins is to be set to $w = 2(IQR)n^{-(1/3)}$.

At this junction we must note that the Freedman–Diaconis rule is based on the assumption that the observations in the sample are not influencing one another, and this is not an assumption that we can commit to at this point. In any case, ignoring this matter and carrying on with building the histogram would result in this plot:

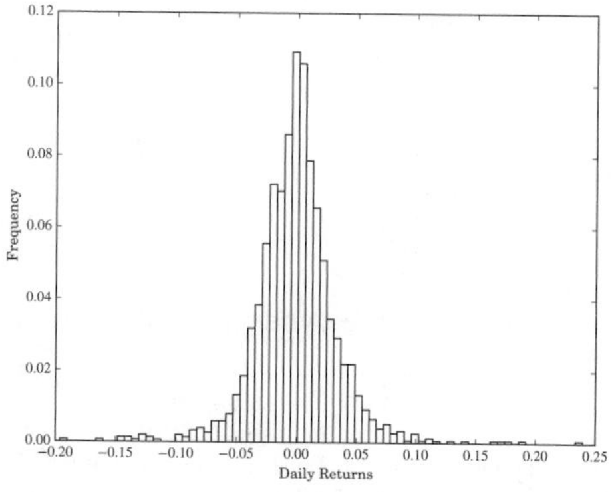

FIGURE 1.3
Freedman–Diaconis histogram for the returns in figure 1.2

[10] Several other rules are also commonly used. The Freedman–Diaconis rule is meant to minimize the mean-square distance to the true density of the distribution – this concept will be developed later.

But we must still ask the question: if it is unclear whether the daily returns influence one another or not, is the histogram above still meaningful? Perhaps it could be made meaningful if we were to make certain corrections in the sample, but what kind of corrections? Before we turn to the general methodology that would allow us to address such matters – known as probability theory – we are going to meet yet another source of randomness, which is exclusive to the realm of asset pricing.

1.2 A Bird's-Eye View of the One-Period Binomial Model

It is hard to think of a better starting point for our exploration of probability theory and its connections with asset pricing than the model developed by Cox et al. (1979). What follows next is a brief outline of a particularly simple implementation of this model. One important point that we are about to illustrate is that the notion of probability is deeply rooted in the very foundations of asset pricing.

The one-period binomial model postulates a financial market with only two securities: a risky stock and a riskless bond. The time variable has only two values $t = 0$ (the beginning of the period) and $t = 1$ (the end of the period). The state of the economy at $t = 0$ (the present) is known and fixed, but there are two possible states of the economy at $t = 1$: H ("Head") and T ("Tail"). To put it another way, the state of the economy at time $t = 1$ is determined from "the flipping of a coin," the outcome of which is only known at time $t = 1$. We stress that we are not imposing the assumption that the coin must be symmetric, in that we do not insist that the states H and T must be equally likely to occur; in fact, whether H is overwhelmingly more likely to show up than T, or the other way around, would be irrelevant for the analysis that we are about to carry out. In any case, the prices of the stock and the bond depend on the state of the economy as follows:[11]

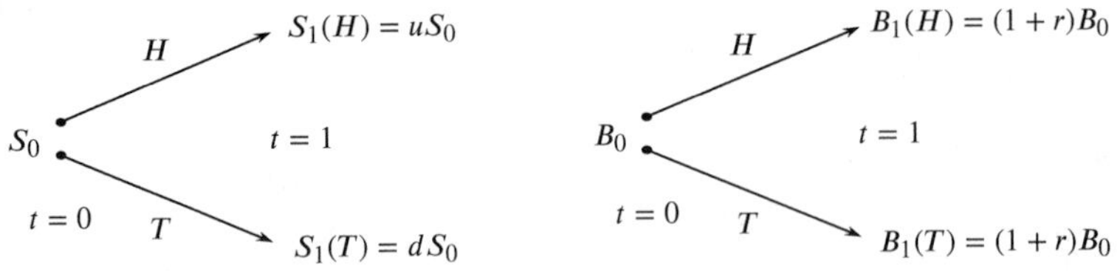

FIGURE 1.4 Price dynamics of the stock S and the bond B.

[11] Unless noted otherwise, prices of securities are assumed to be expressed in units of some generic currency. In an attempt to make the present text more international, we use the generic currency symbol ¤, rather than €, £, \$, ¥, etc. Generally, we are going to avoid using the currency symbol, and write S or V instead of ¤S or ¤V. The convention in some countries is to write the currency symbol after the numerical value, while in other countries it is the other way around – this is true even among countries using the same currency (most notably the €). In this space we prefer to place the currency symbol before the numerical value – in those rare occasions where using this symbol is helpful.

The givens in the model are S_0, B_0, d, u, and the interest rate r. We suppose that

$$0 < d \leq (1+r) \leq u \,, \qquad e_1$$

and the financial meaning of these relations should be clear: if $(1+r) > u$, investing in the riskless bond at $t = 0$ would be preferable regardless of how the uncertainty may get resolved at $t = 1$, and, similarly, if $(1+r) < d$, the risky stock would still be preferable at $t = 0$ no matter what the future may bring at $t = 1$.

Consider now a market position that consists of x shares of the stock and y shares of the bond ($-\infty < x < +\infty$ and $-\infty < y < +\infty$).[12] The value of this position is:

$$\begin{aligned} \text{at } t = 0 \text{ (now):} \quad & V_0 = xS_0 + yB_0 \\ \text{at } t = 1 \text{ (in the future):} \quad & V_1(\omega) = xS_1(\omega) + yB_1(\omega)\,, \end{aligned}$$

where $\omega \in \{H, T\}$.

(1.1) EXERCISE: Find the values for x and y that give:
(a) $V_1(H) = 1$ and $V_1(T) = 0$;
(b) $V_1(H) = 0$ and $V_1(T) = 1$. ○

The solution to part (a) in the last exercise must be of the form

$$x = x_H^* \equiv x_H^*(S_0, B_0, u, d, r) \quad \text{and} \quad y = y_H^* \equiv y_H^*(S_0, B_0, u, d, r)\,,$$

and similarly, the solution to part (b) must be of the form

$$x = x_T^* \equiv x_T^*(S_0, B_0, u, d, r) \quad \text{and} \quad y = y_T^* \equiv y_T^*(S_0, B_0, u, d, r)\,.$$

As a result, by the law of one price,[13]

$$\Phi_0^H \stackrel{\text{def}}{=} x_H^* S_0 + y_H^* B_0$$

gives the price at time $t = 0$ (now) of a security that promises to pay at time $t = 1$ (in the future) ¤1 in state H and ¤0 in state T. Similarly,

$$\Phi_0^T \stackrel{\text{def}}{=} x_T^* S_0 + y_T^* B_0$$

gives the price at time $t = 0$ (now) of a security that promises to pay at time $t = 1$ (in the future) ¤0 in state H and ¤1 in state T. We see how the information encrypted in the spot prices S_0 and B_0 can be interpreted as information about the way in which the market participants are pricing and trading the uncertain future states of the economy.

(1.2) EXERCISE: Calculate Φ_0^H and Φ_0^T as functions of S_0, B_0, d, u, and r. ○

[12] It is assumed that the reader is familiar with the practice of long or short sales of securities.

[13] The law of one price demands that securities that have identical payoffs in every possible future state of the economy must trade at the same price. Later in this space we are going to see that this proposition is the simplest form of the no arbitrage condition.

(1.3) EXERCISE: Given an arbitrary $a \in \mathbb{R}$, find x and y so that $V_1(H) = a$ and $V_1(T) = 0$. Show that the solution is $x = ax^*_H$ and $y = ay^*_H$. Similarly, find x and y so that $V_1(H) = 0$ and $V_1(T) = a$, and show that the solution is given by $x = ax^*_T$ and $y = ay^*_T$. ○

As an immediate corollary of the last exercise (and the law of one price), a security that promises to pay ¤a in state H and ¤b in state T, for every possible choice of $a, b \in \mathbb{R}$, must be priced at time $t = 0$ at:

$$a\Phi_0^H + b\Phi_0^T . \qquad \mathrm{e}_2$$

Indeed, the payoffs from such a security at time $t = 1$ are identical (in either state H or T) to the payoffs from the portfolio (created at time $t = 0$) that holds $ax^*_H + bx^*_T$ shares of the stock and $ay^*_H + by^*_T$ shares of the bond. As there are only two uncertain states in the future, H and T, the expression in e_2 gives the price of any security with any uncertain payoff whatsoever.

(1.4) ARROW–DEBREU SECURITIES:[14] The solution to exercise (1.2) reveals that Φ_0^H and Φ_0^T depend on d, u, and r, but not on S_0 and B_0. The securities that Φ_0^H and Φ_0^T represent are known as *Arrow–Debreu securities* – a tool that will be introduced in a more general setting later in the sequel. But even at this early stage of our exploration of the mechanics of asset pricing, it is abundantly clear that the Arrow–Debreu securities are one crucially important tool. ○

(1.5) EXAMPLE: A call option with strike price K, chosen so that $S_1(T) < K < S_1(H)$, has a payoff at $t = 1$ of $S_1(H) - K \equiv u\,S_0 - K$ in state H, and payoff of 0 in state T. The price of this option at $t = 0$ must be

$$(u\,S_0 - K)\,\Phi_0^H . \quad ○$$

(1.6) COMMENTS: (a) All Arrow–Debreu securities in this model are redundant, in that these securities can be replicated in terms of financial positions that consist of holdings (long or short) only in the bond and the stock. To put it another way, securities with payoffs that are exactly identical to the payoffs from the Arrow–Debreu securities can be recreated from securities that are traded in the market. In fact, any security other than the bond and the stock, with any payoff whatsoever, is going to be redundant in this sense.

(b) The prices of the Arrow–Debreu securities can be derived from the prices of the stock and the bond, and this derivation can be done without any knowledge of how likely it is for the economy to be in state H or T. Indeed, using e_2, after completing exercises (1.1) and (1.2), one can produce the price of any uncertain payoff without any familiarity with the notion of "probability." ○

Still, are there actual probabilities to be found somewhere in this construction?

[14] Named after the American economist Kenneth Joseph Arrow and the French-American economist Gérard Debreu (1921–2004) – both recipients of the Sveriges Riksbank Prize in Economic Sciences in Memory of Alfred Nobel, respectively, in 1972 and 1983.

(1.7) EXERCISE: Using simple algebra, prove the relation

$$\Phi_0^H + \Phi_0^T = \frac{1}{1+r}. \quad \bigcirc$$

(1.8) RISK-NEUTRAL PROBABILITIES: With the last relation in mind, we can set $\tilde{p} = (1+r)\Phi_0^H$, $\tilde{q} = (1+r)\Phi_0^T$ and observe that $\tilde{p} + \tilde{q} = 1$. Thus, $\tilde{p}$ and $\tilde{q}$ can be thought of as "probabilities" attached, respectively, to the uncertain outcomes H and T. We call $\tilde{p}$ and $\tilde{q}$ *risk-neutral probabilities*, and, after completing exercise (1.2), arrive at the following formulas

† $$\tilde{p} = \frac{(1+r)-d}{u-d} \quad \text{and} \quad \tilde{q} = \frac{u-(1+r)}{u-d},$$

which give us yet another reason for imposing the conditions in e_1. The reason for calling $\tilde{p}$ and $\tilde{q}$ "risk-neutral" will become at least partly clear in the next exercise – we will return to this concept later in the sequel. The Arrow–Debreu securities can now be linked to the risk-neutral probabilities and the interest rate r through the relations

$$\Phi_0^H = \frac{\tilde{p}}{1+r} \quad \text{and} \quad \Phi_0^T = \frac{\tilde{q}}{1+r}. \quad \bigcirc$$

(1.9) EXERCISE: Using elementary algebra, prove that

$$S_0 = \frac{1}{1+r}\big(S_1(H)\,\tilde{p} + S_1(T)\,\tilde{q}\big) \quad \text{and} \quad B_0 = \frac{1}{1+r}\big(B_1(H)\,\tilde{p} + B_1(T)\,\tilde{q}\big). \quad \bigcirc$$

(1.10) STATE-PRICE DENSITY (SPD): If the probability for state H to occur in the real world is p and the probability for state T to occur is $q = 1 - p$ (for now, as a first step, we adopt the folklore interpretation of "probability"), then the ratios Φ_0^H/p and Φ_0^T/q can be cast, respectively, as

$$\frac{\psi_1(H)}{\psi_0} \quad \text{and} \quad \frac{\psi_1(T)}{\psi_0},$$

where the list of values $\{\psi_0, \psi_1(H), \psi_1(T)\}$ (there is one value attached to every node on the tree) gives the so-called state-price density process – an important tool in asset pricing, which will be introduced later in the sequel. This list (it will be called "process" soon) is determined up to a multiplicative factor, usually by postulating $\psi_0 = 1$. In terms of the SPD, the relations in (1.9) can be written as

$$S_0 = S_1(H)\frac{\psi_1(H)}{\psi_0}\,p + S_1(T)\frac{\psi_1(T)}{\psi_0}\,q$$

and

$$B_0 = B_1(H)\frac{\psi_1(H)}{\psi_0}\,p + B_1(T)\frac{\psi_1(T)}{\psi_0}\,q.$$

Here we remark that, unlike in (1.9), these expressions for the spot prices of the two assets at time $t = 0$ do involve the real-world probabilities, not the risk-neutral probabilities $\tilde{p}$ and $\tilde{q}$ that can be extracted from the asset prices and the payoffs alone by way of a straightforward algebra. $\bigcirc$

* * *

Our main goal in this space is to develop a paradigm in which one can lift the relations and the connections discovered in the present section to models with a very large number of assets that exist for many (possibly very short) periods of time, and in a world in which the uncertainty is much more complex than flipping a single coin just once. One may be tempted to think that the passage from small models to large – perhaps very, very large – is merely a matter of increased computing power, but this would be more or less like trying to model the universe by tracking the dynamics of every elementary particle and pretending that such a task may be accomplished by way of improving the available computing technology. The need for a qualitatively different paradigm is clear.

1.3 Probability Spaces

The main building block in any probabilistic model is the choice of the *sample space*, which is commonly denoted by Ω. As a mathematical object Ω is simply a nonempty set, and it could be *any* nonempty set. It can be thought of as a complete list of all possible uncertain outcomes from a particular experiment. These outcomes are often called *elementary outcomes* and are usually expressed as $\omega \in \Omega$. Here are some examples:

$\Omega = \{H, T\}$.

$\Omega = \{1, 2, 3, 4, 5, 6\}$.

$\Omega = \{HHH, HHT, HTH, THH, TTT, TTH, THT, HTT\}$.

Ω = all possible arrangements of a deck of 52 playing cards; the cardinality of this set is $|\Omega| = 52!$ (how large is this number?).

Ω = all possible *unordered* sets of 13 cards, sampled without replacement from a standard deck of 52 playing cards; the cardinality of this set is $|\Omega| = \binom{52}{13} \equiv \frac{52!}{13!39!} = \frac{52\times\cdots\times 40}{1\times 2\times\cdots\times 13}$.

$\Omega = \mathbb{N} \equiv \{0, 1, 2, \ldots\}$ (the space of all nonnegative integers) – this sample space is *countably infinite.*

Ω = the space of all *infinite* sequences of the symbols H and T – this sample space is *uncountably infinite.*

$\Omega = [0, 1]$ – this sample space is *uncountably infinite.*

$\Omega = \mathbb{R}$, $\Omega = \mathbb{R}_+$, $\Omega = \mathbb{R}_{++}$, $\Omega = [0, 1]\times[0, 1]$, $\Omega = \mathbb{R}\times\mathbb{R}$, $\Omega = \mathbb{R}_+\times\mathbb{R}_+$, $\Omega = \mathbb{R}_{++}\times\mathbb{R}_{++}$ – all examples of uncountably infinite sample spaces.

$\Omega = \mathscr{C}([0, T]; \mathbb{R})$ for some $T > 0$, or $\Omega = \mathscr{C}(\mathbb{R}_+; \mathbb{R})$ – these are spaces of continuous sample paths.

$\Omega = \mathbb{D}([0, T]; \mathbb{R})$ for some $T > 0$, or $\Omega = \mathbb{D}(\mathbb{R}_+; \mathbb{R})$ – these are spaces of càdlàg sample paths.

A random event is any subset of the sample space Ω, which is to say any collection of possible outcomes $\omega \in \Omega$. A random event $A \subseteq \Omega$ is said to occur if the actual (physically realized) outcome from the experiment, $\omega^* \in \Omega$, is a member of A, a feature expressed as $\omega^* \in A$. If the event $A \subseteq \Omega$ does not occur, then its complement $A^{\complement} \equiv \Omega \backslash A$ must occur, for the relation $\omega^* \notin A$ is the same as $\omega^* \in A^{\complement}$. Extreme cases of events are the event that never occurs, represented by the empty set $\emptyset \subseteq \Omega$, and the event that always occurs, represented by the set $\Omega \subseteq \Omega$ (the complement of $\emptyset$). By the very definition of "complement" – the operation expressed as $(\cdot)^{\complement}$ – we have $(A^{\complement})^{\complement} = A$ for any $A \subseteq \Omega$.

If $A_i \subseteq \Omega$, $i \in \mathbb{I}$, is any – and not necessarily countable – family of events, then the event "at least one A_i occurs" can be expressed as $\cup_{i \in \mathbb{I}} A_i$, and the event "every A_i occurs" can be expressed as $\cap_{i \in \mathbb{I}} A_i$. The event "at least one $A_i{}^{\complement}$ occurs" is the same as "$\cap_{i \in \mathbb{I}} A_i$ does not occur," and expressed more succinctly this connection comes down to

$$\bigcup_{i \in \mathbb{I}} \left(A_i{}^{\complement}\right) = \left(\bigcap_{i \in \mathbb{I}} A_i\right)^{\complement}.$$

It is easy to see that this relation is equivalent to

$$\left(\bigcup_{i \in \mathbb{I}} A_i\right)^{\complement} = \bigcap_{i \in \mathbb{I}} \left(A_i{}^{\complement}\right).$$

Now we turn to the important notion of *probability*. Heuristically, the sample space Ω can be viewed as a weightless medium that can absorb "very tiny particles." One can then imagine an experiment in which a very large number of such "very tiny particles," of finite total mass, are dropped in some random fashion into the medium Ω. After the experiment is complete and there are no particles left to drop, the sample space Ω will no longer be a weightless medium, and some parts of it may be heavier than others, in that, throughout the experiment, some parts of Ω may have absorbed more particles than others. The probability, $\mathsf{P}(A)$, of a random event $A \subseteq \Omega$ can now be imagined as the total mass of particles absorbed in A, expressed as a proportion of the total mass of particles dropped into all of Ω. This proportion can be thought of as a measure of the propensity of the experiment to choose A each time a particle is being dropped. It can also be thought of as a measure of the likelihood of any one particle to end up being dropped in A. For example, one may interpret the histogram in figure 1.3 as a distribution of a unit mass over the real line $\mathbb{R}$, in which the base of each bin has a mass equal to the area of the rectangle right above it. One could then imagine the limit histogram obtained by letting the sample size n and the number of bins increase to infinity. Since the total mass distributed over the sample space Ω is strictly positive and finite, and since we are only interested in what proportion of that mass is spread over a particular event $A \subseteq \Omega$, we are free to choose the unit for measuring mass in any way that we like. The most convenient choice is to set that unit to be the total mass of Ω, which would then make "proportion of mass" the same as "mass." As a result of this convention, we have $\mathsf{P}(\Omega) = 1$. It makes perfect sense to impose that the empty set $\emptyset \subseteq \Omega$ does not have any mass, and that therefore $P(\emptyset) = 0$, but there may be other pieces of Ω without a mass, which is to say there may be a set $A \subseteq \Omega$ that is nonempty and still such that $P(A) = 0$.

It is clear from the foregoing that the notion of probability measure – or simply probability – over the sample space Ω must be defined as a real-valued function that takes values in the interval $[0, 1]$. The domain of this function – in other words, the collection of objects that the probability function can take as arguments – is some family, $\mathscr{A}$, of subsets of Ω. For this reason, a probability measure is sometimes called *probability set function.* Such a function may be expressed as $P\colon \mathscr{A} \mapsto [0, 1]$, or as $P(A) \in [0, 1]$, $A \in \mathscr{A}$. If the sample space Ω and the family of subsets $\mathscr{A}$ are understood from the context, we can simply refer to "the probability P." Now we must address two important questions:

(a) What properties should we postulate for the probability set function P and its domain $\mathscr{A}$?

(b) If we postulate properties that are consistent with our practical needs, can we always take the domain $\mathscr{A}$ to include all possible subsets of the sample space Ω? In other words, can we assign – and in a meaningful way – probabilities to all random events?

In order to develop some basic intuition, we now turn to concrete examples. The simplest cases are the ones where Ω is finite and nonempty. If the sample space is finite, one can attach a probability mass, $P(\{\omega\}) \in [0, 1]$, to every elementary outcome $\omega \in \Omega$, treated as a singleton $\{\omega\} \subseteq \Omega$, and in such a way that

$$\sum_{\omega \in \Omega} P(\{\omega\}) = 1\,.$$

The probability mass of *any* set $A \subseteq \Omega$ is then $P(A) = \sum_{\omega \in A} P(\{\omega\})$. In particular, one can set $P(\{\omega\}) = 1/|\Omega|$ (recall that $|\Omega|$ denotes the cardinality – in this case, the number of elements – of Ω) for any $\omega \in \Omega$. Thus, if the unit probability mass is distributed uniformly (or evenly, or equally) throughout the sample space Ω – which is another way of saying that all elementary outcomes $\omega \in \Omega$ are equally likely to occur – then, given any event $A \subseteq \Omega$, one can write (recall that $|\emptyset| = 0$)

$$P(A) = \frac{|A|}{|\Omega|}\,.$$

To put it another way, if the sample space Ω is finite and the probability mass is distributed evenly over Ω, then the problem of calculating $P(A)$ comes down to *counting* the elements of A.

(1.11) EXAMPLE: If Ω is the collection of all possible unordered sets of 13 cards, sampled without replacement from a standard deck of 52 playing cards, then Ω would be a finite set with a total of $\binom{52}{13}$ elementary outcomes. Probability on Ω can be defined simply by placing $\binom{52}{13}^{-1}$ units of mass at every elementary outcome.

Given any event $A \subseteq \Omega$, the total mass of A is then

$$P(A) = \sum_{\omega \in A} \binom{52}{13}^{-1} = \frac{|A|}{\binom{52}{13}} .$$

As the summation is always finite for every choice of the set A, the probability measure of any subset of Ω – any subset whatsoever – is well defined. In particular, if A is the event that there are four aces in the sample, then $|A| = \binom{48}{9}$. The probability to get four aces is therefore $\binom{48}{9}/\binom{52}{13} = 11/4165 \approx 0.00264105642256903$. ○

(1.12) EXERCISE: In the setting of the last example, calculate the probability for at least one ace in a hand of 13 cards. ○

(1.13) EXAMPLE: Suppose that $\Omega \stackrel{\text{def}}{=} \mathbb{N} \equiv \{0, 1, 2, \ldots\}$ so that every elementary outcome $\omega \in \Omega$ can be identified as a finite nonnegative integer. Fix some $c \in \mathbb{R}_{++}$ and suppose that $e^{-c}\frac{c^{\omega}}{\omega!}$ units of mass are placed at the integer ω for every $\omega \in \Omega$. As this sample space is countable, one has

$$\sum_{\omega \in \Omega} e^{-c}\frac{c^{\omega}}{\omega!} = e^{-c} \sum_{i \in \mathbb{N}} \frac{c^{i}}{i!} = 1 ,$$

and the quantity

$$P(A) = \sum_{\omega \in A} e^{-c}\frac{c^{\omega}}{\omega!}$$

is well defined for every $A \subseteq \Omega$ (with the usual understanding that $\sum_{\omega \in \emptyset}$ always yields 0). The key here is that all random events $A \subseteq \Omega$ are countable sets, since Ω is countable. ○

(1.14) EXERCISE: In the context of the last example, calculate $P(A)$, where $A = \{n \in \mathbb{N} : n \text{ is even}\}$. ○

(1.15) EXERCISE: Let $\Omega = \{1, 2, \ldots\}$. Suppose that the mass placed at the integer ω is $\frac{C}{\omega^2}$ for every $\omega \in \Omega$. Choose the constant C so that the total mass of Ω equals 1 – just as in (1.13). Would it be possible to make the same choice if the mass placed at the integer ω is $\frac{C}{\omega}$? ○

Just as one would expect, if the sample space Ω is uncountably infinite, then measuring the volume (or the area) of a set is not as straightforward. This phenomenon is manifested in the following remarkable result.

(1.16) THE BANACH–TARSKI PARADOX:[15] (Banach and Tarski 1924) A solid ball of radius 1 inside $\mathbb{R}^3$ can be broken into finitely many pieces, which can then be reassembled only by using rotation and translation (in other words, without squeezing or expanding) into two different balls, both of radius 1. ○

One conclusion from this result is that the notion of "volume" (or "mass") cannot be defined, as the intuition may suggest, in some sort of a uniform (meaning,

[15] Named after the Polish mathematicians Stefan Banach (1892–1945) and Alfred Tarski (1901–1983).

invariant under translation and rotation) fashion if this notion is to be attached to all possible subsets $A \subseteq \mathbb{R}^3$. There would be no paradox if the pieces into which the ball is broken cannot be measured (have no volume), since being able to assemble two balls from volumless pieces cannot be interpreted as the identity $1 = 1 + 1$. It turns out that the matter of existence of sets that cannot be measured is deeply rooted in the very foundation of mathematics, and is intimately related to the axiom of choice. This was clarified by Solovay (1970), who showed that if the axiom of choice is dispensed with, while all other axioms of set theory are kept in place, then it becomes possible to measure all sets of real numbers. Unfortunately, dispensing with the axiom of choice entails dispensing with many important tools from mathematics, which explains why it is generally preferable to work in a world where not all sets can be measured, rather than a world where the axiom of choice does not hold. In any case, nonmeasurable sets were first mentioned by Vitali (1905),[16] and other examples can be found in most texts on measure theory – see (Dudley 2002, 4.2.1) and (Halmos 1950). A more recent discovery of nonmeasurable sets, which is more closely related to the discussion in the present chapter, is due to Blackwell and Diaconis (1996).

What the existence of nonmeasurable sets (that is, the adoption of the axiom of choice) means for us is that in models where the sample space Ω is uncountably infinite, in general, we will not be able to define the probability set function on the entire family, 2^{Ω}, of all possible random events – unless, of course, we define probability in a trivial and not very useful fashion. Thus, some exceptional situations aside, in general, the domain, $\mathscr{A}$, of the probability P is going to be a particular collection of random events, that is, the inclusion $\mathscr{A} \subset 2^{\Omega}$ is going to be strict. There are also other practical reasons as to why one may want to restrict the domain of P that have to do with the modeling of the information arrival dynamics – we will come to this later.

Those intrinsic restrictions on the domain of P notwithstanding, to be able to build useful models, we still must endow this domain with certain attributes. Specifically, if A is an event from the domain $\mathscr{A}$, we would like its complement, $A^{\complement}$, to be in the domain $\mathscr{A}$, too, so that if we can measure the probability for A to occur, we should be able to measure also the probability for A to not occur. In addition, we would like the union and the intersection of any sequence of events from $\mathscr{A}$ to belong to $\mathscr{A}$, so that if we can measure the probabilities of the events in a particular sequence of events, we should then be able to measure also the probability for at least one of the events in the sequence to occur, as well as the probability for all events in the sequence to occur. Plainly, to come up with a useful definition of "probability," we must insist that its domain is closed under complement and countable union (in which case it is automatically closed under countable intersection), or closed under complement and countable intersection (in which case it is

[16] G. Vitali's argument is detailed in (Stroock 1999, 2.1.17).

automatically closed under countable union). We could, of course, demand that the domain $\mathscr{A}$ be closed under any (not necessarily countable) union and intersection, but this would be too much to ask for, in the sense that the probabilities that we can define on such domains would be essentially trivial and uninteresting – except when $|\Omega| < \infty$. To be precise, we postulate that the domain $\mathscr{A}$ has the structure of a σ-field – an object that we introduce next – and, following the well-established tradition, from now on we will write $\mathcal{F}$ instead of $\mathscr{A}$ if the structure of a σ-field is assumed.

(1.17) DEFINITION: Let Ω be any nonempty set. A nonempty collection of subsets, $\mathcal{F} \subseteq 2^{\Omega}$, is said to be a σ-field (alias: σ-algebra) if the following three axioms are satisfied:

1. $\emptyset \in \mathcal{F}$,
2. $A \in \mathcal{F}$ implies $A^{\complement} \in \mathcal{F}$,
3. $\bigcup_{i \in \mathbb{N}} A_i \in \mathcal{F}$ for any sequence $(A_i \in \mathcal{F})_{i \in \mathbb{N}}$.

If the last axiom is replaced with

3′. $\bigcup_{i \in \mathbb{N}_{|n}} A_i \in \mathcal{F}$ for any $n \in \mathbb{N}$ and any finite sequence $(A_i \in \mathcal{F})_{i \in \mathbb{N}_{|n}}$,

then we say that $\mathcal{F}$ is an algebra, in which case we may write $\mathscr{A}$ instead of $\mathcal{F}$. ○

(1.18) EXERCISE: Let Ω be any nonempty set and let $\mathcal{F} \subseteq 2^{\Omega}$ be any σ-field. Prove that for every nonempty subset $\emptyset \neq A \subseteq \Omega$, which may or may not belong to $\mathcal{F}$, the family

$$A \cap \mathcal{F} \stackrel{\text{def}}{=} \{A \cap B \colon B \in \mathcal{F}\} \subseteq 2^A$$

constitutes a σ-field over the set A (treated as a sample space). It is called the *induced σ-field* on A by $\mathcal{F}$. ○

(1.19) EXERCISE AND DEFINITION: A collection, $\mathscr{R} \subseteq 2^{\Omega}$, of subsets of some nonempty set Ω is said to be a *ring* if $\emptyset \in \mathscr{R}$ and $A, B \in \mathscr{R}$ implies both $A \cup B \in \mathscr{R}$ and $A \backslash B \in \mathscr{R}$. Prove that if $\mathscr{R}$ is a ring, then the family of sets

$$\mathscr{A} \stackrel{\text{def}}{=} \mathscr{R} \cup \{A^{\complement} \colon A \in \mathscr{R}\}$$

is an algebra. Conclude that if $\Omega \in \mathscr{R}$, then the ring $\mathscr{R}$ is an algebra. Verify that every algebra is a ring. ○

(1.20) EXERCISE: Prove that a nonempty collection of subsets, $\mathscr{A} \subseteq 2^{\Omega}$, is an algebra if and only if it is closed under finite union and complement, that is, $A, B \in \mathscr{A}$ implies $A \cup B \in \mathscr{A}$, and $A \in \mathscr{A}$ implies $A^{\complement} \in \mathscr{A}$. ○

(1.21) EXERCISE: Prove that if $\Omega \neq \emptyset$ and $\mathcal{F}$ is any σ-field of subsets of Ω, then the following properties are in force:

(a) $\Omega \in \mathcal{F}$;

(b) $\bigcap_{i \in \mathbb{N}} A_i \in \mathcal{F}$ and $\bigcup_{i \in \mathbb{N}} A_i \in \mathcal{F}$ for any sequence $(A_i \in \mathcal{F})_{i \in \mathbb{N}}$;

(c) $\bigcup_{i \in \mathbb{N}_{|n}} A_i \in \mathcal{F}$ and $\bigcap_{i \in \mathbb{N}_{|n}} A_i \in \mathcal{F}$ for any $n \in \mathbb{N}$ and $A_i \in \mathcal{F}$, $i \in \mathbb{N}_{|n}$. ○

Due to reasons that will become clear later, we are going to need to work with *families of σ fields* – this has to do with the modeling of information arrival. While the need to work with such objects is yet to come, now is the time to introduce the related concepts and notation.

(1.22) EXERCISE: Prove that the intersection of *any* (countable or uncountable) family of σ-fields (over one and the same sample space Ω) is always a σ-field, and explain why the same claim cannot be made about a union of σ-fields – even about the union of just two σ-fields. ○

(1.23) THE SMALLEST σ-FIELD: Let Ω be any nonempty set and let $\mathscr{K} \subseteq 2^{\Omega}$ be any collection of subsets of Ω. Since the collection 2^{Ω} of all subsets of Ω is a σ-field, we see that the family of all σ-fields over Ω that include $\mathscr{K}$ cannot be empty, regardless of how $\mathscr{K}$ is chosen. According to (1.22) the intersection of all σ-fields that include $\mathscr{K}$ is a σ-field. It is the smallest σ-field over Ω that includes $\mathscr{K}$ and is commonly denoted by $\sigma(\mathscr{K})$. In what follows we are going to use (usually without further notice) the following general convention: by

$$\sigma\big(\langle\text{list of objects or conditions}\rangle\big)$$

we will denote the smallest σ-field that includes all of the listed objects, or satisfies all of the listed conditions. We stress that the smallest σ-field always exists and can be identified as an intersection of some collection of σ-fields over Ω. In particular, if $(\mathscr{F}_i)_{i \in \mathbb{I}}$ is any family of σ-fields over Ω, indexed by any set $\mathbb{I}$, then the smallest σ-field that includes every $\mathscr{F}_i$ will be expressed as

$$\vee_{i \in \mathbb{I}} \mathscr{F}_i \stackrel{\text{def}}{=} \sigma\big(\mathscr{F}_i : i \in \mathbb{I}\big). \quad ○$$

We are now ready to give the rigorous definition of "probability measure," often called "probability." As in much of the literature related to probability theory, we follow the axiomatic approach of Kolmogorov (1933).

(1.24) PROBABILITY AND MEASURE: Let Ω be any nonempty set and let $\mathcal{F}$ be any σ-field of subsets of Ω. A function $P: \mathcal{F} \mapsto [0, \infty]$ is said to be a *probability measure* if the following two axioms are satisfied:

1. $P(\Omega) = 1$;
2. $P(\emptyset) = 0$ and for any sequence $(A_i \in \mathcal{F})_{i \in \mathbb{N}}$ of disjoint sets, that is, sets chosen so that $A_i \cap A_j = \emptyset$ for $i \neq j$, the following condition is satisfied:

$$P\big(\cup_{i \in \mathbb{N}} A_i\big) = \sum_{i \in \mathbb{N}} P(A_i).$$

The feature encrypted into the second axiom is often referred to as *countable additivity*, or *σ-additivity*, and if the objects Ω, $\mathcal{F}$, and P satisfy the two axioms above, then the triplet

$$(\Omega, \mathcal{F}, P)$$

is called *probability space*. If only the second axiom is satisfied, in which case $P(\Omega) = \infty$ becomes possible, the function $P: \mathcal{F} \mapsto [0, \infty]$ is said to be a *measure* on $\mathcal{F}$, and if the first axiom is replaced by $P(\Omega) < \infty$, then P is said to be a *finite measure.* If P is a measure and there is a sequence $(A_i \in \mathcal{F})_{i \in \mathbb{N}}$, such that $P(A_i) < \infty$ for every i and $\cup_{i \in \mathbb{N}} A_i = \Omega$, then the measure P is said to be *σ-finite.* Unless specified otherwise, all measure that we are going to encounter in this space will be assumed σ-finite by default. ○

(1.25) EXERCISE AND REMARK: Prove that the condition $P(\emptyset) = 0$ in axiom 2 from (1.24) is redundant if P is a measure and there is a set $\emptyset \neq A \in \mathcal{F}$ such that $P(A) \neq \infty$. In particular, $P(\emptyset) = 0$ would be redundant if axiom 1 is in force, or if the measure P is σ-finite. This was the main reason for the condition $P(\emptyset) = 0$ not to be given the status of a stand-alone axiom.

HINT: Apply the remaining part of axiom 2 (without the condition $P(\emptyset) = 0$) to the sequence $\{A, \emptyset, \emptyset, \dots\}$, for any $\emptyset \neq A \in \mathcal{F}$ with $P(A) \neq \infty$. ○

(1.26) SOME COMMENTS AND CONVENTIONS: A probability measure is a special case of a finite measure. The letter P is usually reserved for probability measures, however, probability measures may be denoted also by Q, $\tilde{P}$, $\tilde{Q}$ and other such symbols. General measures, whether finite or not, are usually denoted by m, λ, μ, and other such symbols. Another common – but not universally followed – convention is to use a different symbol for the sample space Ω if the underlying measure on that space is not a probability measure. In particular, if $\mathscr{X}$ is any nonempty set, $\mathscr{S}$ is any σ-field of subsets of $\mathscr{X}$, and μ is any measure on $\mathscr{S}$, then the triplet $(\mathscr{X}, \mathscr{S}, \mu)$ is called *measure space* and the pair $(\mathscr{X}, \mathscr{S})$ is called *measurable space.* Strictly speaking, measures are defined on σ-fields, but it is acceptable to say that "P is a probability measure on Ω," or that "μ is a measure on $\mathscr{X}$," as in most situations the underlying σ-field is understood from the context. ○

(1.27) ANALYTIC SETS AND CAPACITIES: Throughout most of the literature involving anything related to probability and measure, it is generally assumed that a particular measure is well defined as long as one can specify a σ-field and a mapping from it into $\bar{\mathbb{R}}_+$ that satisfies the relevant axioms from (1.24). Unfortunately, this setup alone is inadequate for a number of important applications, including some very basic ones, as will be evidenced in (2.12). It turns out that a truly useful measure-theoretic setup has to be based on the so-called *analytic sets* and the notion of *capacity* associated with a particular σ-field. In this space we are going to follow the well-established tradition of glossing over such matters, and point to the standard reference (Meyer 1966, III) for further details. The canonical workaround is to complete the σ-field as described in (1.48), and this operation can be introduced with a very minimal amount of technicalities. ○

(1.28) EXERCISE: Let $(\mathscr{X}, \mathscr{S}, \mu)$ be any measure space and let $\emptyset \neq A \in \mathscr{S}$ be any nonempty element of $\mathscr{S}$. Prove that the set function$^{\circ\!\!\rightarrow(1.18)}$

$$A \cap \mathscr{S} \ni B \rightsquigarrow \mu(B)$$

is a measure on the σ-field $A \cap \mathscr{S}$. It is called *the measure induced on A by μ* and is nothing but the restriction $\mu \restriction (A \cap \mathscr{S})$. Since there is no risk of ambiguity, this restriction is still denoted by μ, and the measure space $(A, A \cap \mathcal{S}, \mu\lceil(A \cap \mathcal{S}))$ is often written more succinctly in the form $(A, A \cap \mathcal{S}, \mu)$.

Conversely, show that if μ is a measure on $A \cap \mathcal{S}$, for some set $A \in \mathcal{S}$, then μ gives rise to a measure on $\mathcal{S}$ given by $\mu(B) = \mu(B \cap A)$ for any $B \in \mathcal{S}$. ○

(1.29) EXERCISE: Let $(\Omega, \mathcal{F}, P)$ be any probability space, and let $(A_i \in \mathcal{F})_{i \in \mathbb{N}}$ be any sequence of random events in that space. Prove that:

(a) If $n \in \mathbb{N}$ is arbitrarily chosen, then

$$P\left(\cup_{i \in N_{|n}} A_i\right) \leq \sum\nolimits_{i \in N_{|n}} P(A_i)\,.$$

(b) If $A_i \subseteq A_{i+1}$ for every $i \in \mathbb{N}$ and $A = \cup_{i \in \mathbb{N}} A_i$, then

$$P(A) = \lim\nolimits_i P(A_i)\,.$$

(c) If $A_i \supseteq A_{i+1}$ for every $i \in \mathbb{N}$ and $A = \cap_{i \in \mathbb{N}} A_i$, then

$$P(A) = \lim\nolimits_i P(A_i)\,.$$

Prove that these properties still hold if P is a finite measure, but give a counterexample showing that (c) may fail if P is σ-finite but not finite. ○

(1.30) CONTINUITY OF PROBABILITY MEASURES: Condition (b) in (1.29) may be interpreted as *inner continuity*, while condition (c) may be interpreted as *outer continuity* of the probability measure P. In other words, the axioms that we postulated in (1.24) imply the (two-sided) continuity of the probability measure. One can actually show that the second axiom is equivalent to the requirement that the probability measure is continuous in the sense of (b). ○

Certain measures can be defined on the full σ-field 2^Ω, as the next exercise details.

(1.31) EXERCISE (UNIT POINT-MASS MEASURES): Given a nonempty set Ω, a *unit point-mass measure* on Ω is any set function of the form

$$2^\Omega \ni A \rightsquigarrow \epsilon_{\omega_*}(A) \stackrel{\text{def}}{=} 1_A(\omega_*) \quad \text{for some fixed } \omega_* \in \Omega\,.$$

Equivalently, $\epsilon_{\omega_*}(A) = 1$ if $\omega_* \in A$ and $\epsilon_{\omega_*}(A) = 0$ if $\omega_* \notin A$. Prove that, as defined above, ϵ_{ω_*} is indeed a measure – in fact, a probability measure – on the full σ-field 2^Ω for any $\omega_* \in \Omega$. Conclude that for any finite subset $\{\omega_1, \dots, \omega_n\} \subseteq \Omega$ the set function $A \rightsquigarrow \mu(A) \stackrel{\text{def}}{=} \epsilon_{\omega_1}(A) + \cdots + \epsilon_{\omega_n}(A)$ is a (finite) measure on the full σ-field 2^Ω. A *point-mass measure* is any positive multiple of some unit point-mass measure. ○

1.4 Coin Toss Space and Random Walk

(1.32) FROM PARTITIONS TO σ-FIELDS: There are different ways in which one can arrive at a σ-field, but one particularly interesting and instructive construction is by partitioning the sample space Ω. A *partition* of Ω is any finite collection, $\mathscr{P}$, of nonempty subsets of Ω, characterized by the property that every $\omega \in \Omega$ belongs to exactly one element of $\mathscr{P}$. In other words, a partition of Ω is any finite collection of sets, $\mathscr{P} = (A_i \subseteq \Omega)_{i \in \mathbb{N}_{|n}}$, such that $\cup_{i \in \mathbb{N}_{|n}} A_i = \Omega$, $A_i \neq \emptyset$, and $A_i \cap A_j = \emptyset$ for all $i, j \in \mathbb{N}_{|n}$ with $i \neq j$. Let $\mathcal{F}$ denote the collection of all sets inside Ω that can be formed by taking unions of elements of the partition $\mathscr{P}$. There are 2^{n+1} such unions, or, to put it another way, $\mathcal{F}$ has cardinality $|\mathcal{F}| = 2^{n+1}$ (recall that $\mathbb{N}_{|n}$ has cardinality $|\mathbb{N}_{|n}| = n + 1$). Indeed, while forming a union of elements of $\mathscr{P}$, one has 2 choices for every set A_i whether to include A_i in the union or not. If no set is included, the resulting union is understood to be the empty set $\emptyset$, and if every set A_i is included, the resulting union is Ω. Thus $\emptyset$ and Ω both belong to $\mathcal{F}$, and it is not difficult to check that $\mathcal{F}$ satisfies the axioms for a σ-field and that all elements of the partition $\mathscr{P}$ are also elements of $\mathcal{F}$ (so that $\mathscr{P} \subseteq \mathcal{F}$). We will say that the σ-field $\mathcal{F}$ is generated by the partition $\mathscr{P}$, or that $\mathscr{P}$ generates $\mathcal{F}$, and will express this connection as $\mathcal{F} = \sigma(\mathscr{P})$. ○

(1.33) EXAMPLE: Let Ω be some nonempty set with partition $\mathscr{P} = \{A, B, C\}$. Thus, the subsets $A, B, C \subseteq \Omega$ are nonempty, $A \cap B = A \cap C = B \cap C = \emptyset$, and $A \cup B \cup C = \Omega$. Then the σ-field generated by $\mathscr{P}$ is

$$\sigma(\mathscr{P}) = \{\emptyset, \Omega, A, B, C, A \cup B, A \cup C, B \cup C\} .$$

Clearly, $\mathscr{P}_0 = \{A, A^{\complement}\}$ is another partition of Ω and (notice that $A^{\complement} = B \cup C$)

$$\sigma(\mathscr{P}_0) = \{\emptyset, \Omega, A, A^{\complement}\} .$$

In this example the partition $\mathscr{P}$ is a *refinement* of the partition $\mathscr{P}_0$, in that every element of $\mathscr{P}_0$ can be expressed as a union of (generally finer) elements of $\mathscr{P}$. We also say that $\mathscr{P}_0$ is *coarser* than $\mathscr{P}$, or that $\mathscr{P}$ is *finer* than $\mathscr{P}_0$. If $\mathscr{P}$ is a refinement of $\mathscr{P}_0$, then $\mathscr{P}_0 \subseteq \sigma(\mathscr{P})$, and therefore $\sigma(\mathscr{P}_0) \subseteq \sigma(\mathscr{P})$. ○

(1.34) FINITELY GENERATED σ-FIELDS: It is not difficult to see that every σ-field $\mathcal{F}$ that has cardinality $|\mathcal{F}| < \infty$ (as a set of sets) is generated by a unique partition (in particular, the cardinality of any finite σ-field can only be of the form 2^n for some $n \in \mathbb{N}_{++}$). The elements of this partition are called *atoms* of $\mathcal{F}$. These atoms are the smallest nonempty sets that belong to $\mathcal{F}$, with "smallest" understood in the sense that there is no element of $\mathcal{F}$ that is included strictly in the atom. To put it another way, $A \in \mathcal{F}$ is an atom if and only if $A \cap B$ is either the empty set $\emptyset$ or is A itself for every $B \in \mathcal{F}$.

In conclusion, finite σ-fields and partitions come down to the same thing. One may ask, then, why introduce partitions at all? The reason is – and we are about to illustrate this – that although infinite σ-fields cannot be related to partitions directly, in many applications an infinite σ-field arises as the limit of an ever increasing sequence of finite σ-fields, which is to say, arises as the limit of an ever refining sequence of partitions. Such constructions can be viewed as models of *flow of information*. The atoms in the partition are the basic elements (or tokens) from which a particular model can be built. The events in the σ-field are all the "phrases" that one can form – based on the "grammar rules" for σ-fields – from the collection of atoms (tokens). The finer the atoms become (or, the more refined the model), the more configurations one can observe and the more statements one can make about what is observed. To put it another way, the finer the atoms in the partition $\mathscr{P}$, the more informed and more precise the model based on $\mathscr{P}$. The most informed partitions are the ones in which every atom is a singleton $\{\omega\}$. The configurations that can be observed with a given partition can also be observed with any finer partition, but not the other way around. ○

(1.35) EVENT TREES AND REFINING CHAINS OF PARTITIONS: It is very common in finance to model certain stochastic variables of interest (stock prices, say) as event trees. In turn, event trees can be interpreted as refining chains of partitions of the sample space, and this interpretation is closer to the tools that we are going to need and are about to develop. A refining chain of partitions is any finite or infinite sequence of partitions $(\mathscr{P}_i)$ such that $\mathscr{P}_0 = \{\Omega\}$ and, for every i, $\mathscr{P}_{i+1}$ is a refinement of $\mathscr{P}_i$, in that every element (atom) of $\mathscr{P}_i$ is the union of elements (atoms) of $\mathscr{P}_{i+1}$ (equivalently, every atom from $\mathscr{P}_{i+1}$ is included in precisely one atom from $\mathscr{P}_i$). This connection is the same as $\mathscr{P}_i \subseteq \sigma(\mathscr{P}_{i+1})$. Any such chain of partitions can be viewed as an event tree in the obvious way: the elements of $\mathscr{P}_i$ are the level-i nodes on the tree, and the spikes (or branches) that emanate from every level-i node lead to those level $(i+1)$-nodes (that is, elements of $\mathscr{P}_{i+1}$) that are contained inside the level-i node, treated as an atom from $\mathscr{P}_i$. ○

(1.36) EXERCISE: Consider the following chain of partitions of Ω:

$$\mathscr{P}_0 = \{\Omega\},$$
$$\mathscr{P}_1 = \{A_1, A_2\},$$
$$\mathscr{P}_2 = \{A_{1,1}, A_{1,2}, A_{2,1}, A_{2,2}, A_{2,3}\},$$
$$\mathscr{P}_3 = \{A_{1,1,1}, A_{1,1,2}, A_{1,2,1}, A_{1,2,2}, A_{1,2,3}, A_{2,1}, A_{2,2,1}, A_{2,2,2}, A_{2,2,3}, A_{2,3}\},$$

where the addition of a subscript means partitioning of the respective subset; for example, $A_{1,1} \cup A_{1,2} = A_1$, $A_{1,2,1} \cup A_{1,2,2} \cup A_{1,2,3} = A_{1,2}$, etc. Plot the event tree associated with the chain of partitions $\{\mathscr{P}_0, \mathscr{P}_1, \mathscr{P}_2, \mathscr{P}_3\}$. ○

(1.37) REMARK: Partitions are an especially convenient way to define probability measures. Specifically, any prescription for a probability measure on $\sigma(\mathscr{P})$ can be given as an assignment $P\colon \mathscr{P} \mapsto [0,1]$ such that $\sum_{A\in\mathscr{P}} P(A) = 1$. Indeed, the probabilities of the atoms $A \in \mathscr{P}$ automatically determine the probabilities of the events that belong to the σ-field $\sigma(\mathscr{P})$, as every member of $\sigma(\mathscr{P})$ is a finite union of non-overlapping atoms $A \in \mathscr{P}$. This is the most straightforward (and common) way to introduce probability measures on finite σ-fields. Eventually, such a construction can be lifted from finite to infinite σ-fields, though, typically, such a "lift" is highly nontrivial, as will be evidenced below. ○

(1.38) EXERCISE: (follow-up to (1.36)) There are 10 level-3 nodes in the event-tree introduced in (1.36), or, to put it another way, the partition $\mathscr{P}_3$ consists 10 non-overlapping subsets of Ω. Suppose that all elements of $\mathscr{P}_3$ have equal probability, so that $P(A) = \frac{1}{10}$ for every $A \in \mathscr{P}_3$. Calculate the probabilities of the following events from the σ-field $\sigma(\mathscr{P}_3)$:

$$A_1, \quad A_2, \quad A_{1,2} \cup A_2, \quad A_1 \cup A_{2,1} \cup A_{2,2,2} \cup A_{2,2,3}\,. \quad ○$$

* * *

We now turn to the coin toss space and the construction of the random walk. The coin toss space Ω_∞ consists of all possible infinite sequences that can be created from the tokens $+1$ and -1, whereby a generic element $\omega \in \Omega_\infty$ has the form $\omega = \big(\omega_i \in \{-1, 1\}\big)_{i \in \mathbb{N}_{++}}$. The elements of the sequence ω are thought of as the outcomes from tossing coins, and the space Ω_∞ is the collection of all possible configurations that can occur as the result of the tossing of countably infinitely many distinguished and ordered coins. Although the coins tossed are countably many, the associated space of configurations Ω_∞ is uncountably infinite, as we established in (0.4). Given any $\omega \in \Omega_\infty$, by $\mathcal{X}_i(\omega) \stackrel{\text{def}}{=} \omega_i$ we will denote the symbol (identified as either $+1$ or -1) in position $i \in \mathbb{N}_{++}$ of the sequence ω. The mappings $\mathcal{X}_i \colon \Omega_\infty \mapsto \{-1, 1\}$, $i \in \mathbb{N}_{++}$, are the so-called canonical coordinate mappings on Ω_∞, and we set, formally, $\mathcal{X}_0(\omega) = 0$. Every sequence $\omega \in \Omega_\infty$ can be viewed as a prescription for a trajectory of the random walk, as we detail next.

Consider the movement on the lattice $\mathbb{Z}$ of a particle (better yet, think of the log of a stock price) that starts its trip inside $\mathbb{Z}$ from position 0 in period $t = 0$, and then changes its position in all subsequent periods $t \in \mathbb{N}_{++}$ in such a way that in period t the particle moves to the right (from where it happens to be in period $t - 1$) by one unit of distance if $\mathcal{X}_t(\omega) = +1$, and moves to the left by one unit of distance if $\mathcal{X}_t(\omega) = -1$. The positions of the particle in periods $t \in \mathbb{N}$ are given by the sequence $\mathcal{Z} = \big(\mathcal{Z}_t(\omega) \in \mathbb{Z}\big)_{t \in \mathbb{N}}$ such that

$$\mathcal{Z}_t(\omega) = \sum\nolimits_{i \in \mathbb{N}_{|t}} \mathcal{X}_i(\omega)\,, \quad t \in \mathbb{N}\,.$$

We call this sequence *random walk* associated with the outcome $\omega \in \Omega_\infty$.

If we are only observing the movement of the particle from period 0 until period $t \in \mathbb{N}_{++}$, then we are only able to see the first t tokens in the sequence ω. Thus, if we stop observing in period t (or, for some reason, cannot look inside the sequence ω deeper than position t), then there would be 2^t possible configurations that we can see, despite the fact that the effects that we are observing are caused by infinitely many sequences (configurations) $\omega \in \Omega_\infty$. For, if we can only see the first t symbols in a given sequence ω, we are unable to distinguish between two sequences $\omega, \omega' \in \Omega$ that differ only after position t, that is, $\mathcal{X}_i(\omega) = \mathcal{X}_i(\omega')$ for all $i \in \mathbb{N}_{|t}$ and $\mathcal{X}_j(\omega) \neq \mathcal{X}_j(\omega')$ for at least one $j > t$. We can now partition Ω_∞ into 2^t disjoint sets, and thus create a partition, $\mathscr{P}_t$, determined by the property that for every atom $A \in \mathscr{P}_t$ and every $\omega \in A$, we have $\omega' \in A$ if and only if $\mathcal{X}_i(\omega') = \mathcal{X}_i(\omega)$ for all $i \in \mathbb{N}_{|t}$. To put it another way, an atom in $\mathscr{P}_t$ contains all sequences in Ω_∞ the first t tokens in which coincide with a particular string of t tokens selected from the set $\{+1, -1\}$. As there are 2^t such strings, there are 2^t atoms in the partition $\mathscr{P}_t$. For $t = 0$, we declare $\mathscr{P}_t$ to be the trivial partition $\mathscr{P}_0 = \{\Omega_\infty\}$. We can now define the (finite) σ-fields $\mathscr{F}_t = \sigma(\mathscr{P}_t)$, $t \in \mathbb{N}$. It is straightforward to check that $\mathscr{F}_0$ is the trivial σ-field $\{\emptyset, \Omega_\infty\}$, and also that $\mathscr{P}_{t+1}$ is a (strict) refinement of $\mathscr{P}_t$, hence $\mathscr{F}_t \subset \mathscr{F}_{t+1}$, for every $t \in \mathbb{N}$. As a result, we are able to produce a (strictly) refining sequence of partitions $(\mathscr{P}_t)_{t \in \mathbb{N}}$, corresponding to the (strictly) increasing sequence of σ-fields $(\mathscr{F}_t)_{t \in \mathbb{N}}$ (such a sequence is called a filtration, but we postpone the introduction of this terminology).

Our main task now is to endow the coin toss space with a probability measure. Intuitively, if the coins are symmetric, and the tosses are random and without influence from one another, all configurations $\omega \in \Omega_\infty$ should be equally likely to occur. But if we were to assign the same probability, $P(\omega) = p$, to every configuration ω, as the configurations are infinitely many, the only practical choice would be $p = 0$ and this is not very practical. Clearly, the construction of a probability measure on Ω_∞ that we have in mind must be approached in some other way. First, we must postulate that the events in every σ-field $\mathscr{F}_t$, for every choice of $t \in \mathbb{N}$, must have probabilities assigned to them. As we have just realized, the obvious assignment of a probability measure to the σ-field $\mathscr{F}_t$, that would be consistent with our heuristic understanding of the nature of the coins, is to declare that all atoms $A \in \mathscr{P}_t$ have the same probability. This leaves us with the only choice $P_t(A) = 2^{-t}$, $A \in \mathscr{P}_t$, and this choice uniquely determines a probability measure on $\mathscr{F}_t$, still denoted by P_t.

(1.39) EXERCISE: Outline the intuition behind the definition $P_t(A) = 2^{-t}$ for all $A \in \mathscr{P}_t$. Why is it reasonable to impose that all atoms in the partition $\mathscr{P}_t$ must have the same probability? ○

A careful reader may have noticed a compatibility issue that we must address right away: since $\mathscr{F}_t \subset \mathscr{F}_{t+1}$, $P_{t+1}(A)$ is well defined for every $A \in \mathscr{F}_t$, but is it the same as $P_t(A)$? In other words, is the probability measure $P_{t+1} \restriction \mathscr{F}_t$ the same as P_t?

(1.40) EXERCISE: Prove that all probability measures P_t, $t \in \mathbb{N}$, are compatible, in the sense that for every fixed $u > t$, $P_u(A) = P_t(A)$ for every $A \in \mathscr{F}_t$.

HINT: It is enough to show that $P_{t+1}(A) = P_t(A)$ for any $A \in \mathscr{P}_t$. ○

Since the measures P_t, $t \in \mathbb{N}$, are compatible, we can assign a probability, $P_*(A)$, to every event $A \in \mathscr{A} \overset{\text{def}}{=} \cup_{t\in\mathbb{N}}\mathscr{F}_t$ by declaring that $P_*(A) = P_t(A)$ for any $t \in \mathbb{N}$ such that $A \in \mathscr{F}_t$. There are two fundamental reasons why the assignment $P_*\colon \mathscr{A} \mapsto [0,1]$ that we just introduced is inadequate for what we need. First, as it stands, P_* is only defined for the events in the family $\mathscr{A}$, which includes only those random events that can be described in terms of finitely many positions in the (infinite) sequence $\omega \in \Omega$; in other words, P_* can be attached only to events produced by observing the random walk during some finite, however long, period of time. To see why this is inadequate, consider the following events:

$$\{\omega \in \Omega\colon \text{the random walk } \big(\mathcal{Z}_t(\omega) \in \mathbb{Z}\big)_{t\in\mathbb{N}} \text{ crosses 0 infinitely often}\};$$
$$\{\omega \in \Omega\colon \text{the random walk } \big(\mathcal{Z}_t(\omega) \in \mathbb{Z}\big)_{t\in\mathbb{N}} \text{ never reaches 100}\};$$
$$\{\omega \in \Omega\colon \text{the random walk } \big(\mathcal{Z}_t(\omega) \in \mathbb{Z}\big)_{t\in\mathbb{N}} \text{ crosses 100 infinitely often}\}.$$

Clearly, such events are important from a practical point of view and, for this reason, we would like to be able to attach probabilities to them – in some way. However, at this point we can't, for to say the least, none of them belongs to the family $\mathscr{A}$. Indeed, none of them can be claimed to belong to $\mathscr{F}_t$ for some, however large, finite $t \in \mathbb{N}$, which is to say that none of them belongs to $\cup_{t\in\mathbb{N}}\mathscr{F}_t$. Another problem is that the family of sets $\mathscr{A}$ is an algebra (see the next exercise), but is not a σ-field, which means that we do not have at our disposal all the operations with events that a basic probabilistic model would typically involve.

(1.41) EXERCISE: With $\mathscr{F}_t \overset{\text{def}}{=} \sigma(\mathscr{P}_t)$, $t \in \mathbb{N}$, prove that the family $\mathscr{A} \overset{\text{def}}{=} \cup_{t\in\mathbb{N}}\mathscr{F}_t$ is$^{\multimap(1.17)}$ an algebra (in particular,$^{\multimap(1.19)}$ a ring), but is not a σ-field. ○

Clearly, we must find an interesting σ-field that includes $\mathscr{A}$. The keyword in the last sentence is "interesting," for a trivial choice is readily available: we can always take the full σ-field 2^{Ω_∞} – the σ-field that includes all subsets of Ω_∞. Unfortunately, 2^{Ω_∞} is way too large to be useful: we cannot define interesting probability measures on 2^{Ω_∞}, let alone extend P_* from $\mathscr{A}$ to 2^{Ω_∞}. A much more realistic approach would be to work with the smallest σ-field that includes $\mathscr{A}$, that is,$^{\multimap(1.23)}$ the σ-field $\sigma(\mathscr{A}) \equiv \vee_{t\in\mathbb{N}}\mathscr{F}_t$. Thus, we now need to develop the tools that will allow us to extend the measure P_* from $\mathscr{A}$ to $\sigma(\mathscr{A})$, that is, will allow us to construct a probability measure on $\vee_{t\in\mathbb{N}}\mathscr{F}_t$ that coincides with P_* on $\cup_{t\in\mathbb{N}}\mathscr{F}_t$.

(1.42) CARATHEODORY'S EXTENSION THEOREM:[17] (Caratheodory 1918)[18] Let $\mathbb{X}$ be any set, let $\mathscr{R}$ be any collection of subsets of $\mathbb{X}$ that constitutes a ring, and let $\mu\colon \mathscr{R} \mapsto \bar{\mathbb{R}}_+$ be any function such that $\mu(\emptyset) = 0$ and

† $$\mu\big(\cup_{i\in\mathbb{N}} A_i\big) = \sum_{i\in\mathbb{N}} \mu(A_i),$$

[17] Named after the German mathematician Constantin Caratheodory (1873–1950).

[18] The reader may consult (Dudley 2002, 3.1.4, 3.1.10).

for every choice of the sequence $(A_i \in \mathscr{R})_{i \in N}$ such that $A_i \cap A_j = \emptyset$ for $i \neq j$ and $\cup_{i \in N} A_i \in \mathscr{R}$. Then there is a measure defined on $\mathscr{S} \stackrel{\text{def}}{=} \sigma(\mathscr{R})$, the smallest σ-field that includes $\mathscr{R}$, such that its restriction to $\mathscr{R} \subseteq \mathscr{S}$ is precisely μ. Furthermore, if $\mathscr{R}$ is an algebra (that is, $\mathscr{X} \in \mathscr{R}$) and μ is σ-finite (in particular, if μ is finite), then there is precisely one measure on $\mathscr{S}$ that coincides with μ on $\mathscr{R} \subseteq \mathscr{S}$, that is, the extension of μ from $\mathscr{R}$ to $\mathscr{S}$ becomes unique. ○

(1.43) EXERCISE: In the setting of (1.42), prove that if μ is finite everywhere in $\mathscr{R}$, that is, $\mu \colon \mathscr{R} \mapsto \mathbb{R}_+$ – in particular, if μ is a probability set function, that is, $\mathscr{X} \in \mathscr{R}$ and $\mu \colon \mathscr{R} \mapsto [0,1]$ – then the conditions in Caratheodory's theorem are equivalent to: $\mu(\emptyset) = 0$ and

$$\lim_i \mu(A_i) = 0\,,$$

for every sequence $(A_i \in \mathscr{R})_{i \in N}$ such that $A_i \supseteq A_{i+1}$, and $\cap_{i \in N} A_i = \emptyset$. ○

(1.44) REMARKS: Since measures are countably additive set functions and their domains are σ-fields, if it is at all possible to extend a set function, μ, from some family of sets, $\mathscr{R}$, to some σ-field that contains $\mathscr{R}$, then μ must be already countably additive on $\mathscr{R}$. Thus, condition $(1.42^\dagger)$ is minimal, in the sense that if this condition is violated, then the extension to a measure would not be possible. In addition, the family on which μ is originally defined must be sufficiently rich for the extension to a σ-field to be possible. In this sense, the requirement in Caratheodory's theorem for the given collection of sets $\mathscr{R}$ to be a ring is not accidental. ○

Back to the coin toss space, now we would like to extend the probability measure P_* from the algebra $\mathscr{A}$, which contains all finite-time events, to the σ-field $\mathscr{F}_\infty \stackrel{\text{def}}{=} \sigma(\mathscr{A})$ by using the extension theorem from (1.42). However, verifying that the probability set function P_* satisfies the conditions in Caratheodory's theorem is far from obvious. This verification is essentially a variation of a result known as Ionescu Tulcea's theorem – see (Ionescu Tulcea 1949-1950) – and is offered as an exercise next (a detailed outline of the main steps in the solution is included at the end of this section).

(1.45) EXERCISE: Let $(A_i \in \mathscr{A})_{i \in N}$ be any sequence of events in the algebra $\mathscr{A}$ that has the following two properties: (a) the sequence is monotone and decreasing, in that $A_i \supseteq A_{i+1}$ for all $i \in \mathbb{N}$, and (b) it is such that $\cap_{i \in N} A_i = \emptyset$. Prove that $\lim_i P_*(A_i) = 0$.

HINT: Try to solve this problem without consulting (1.52) below. ○

In conjunction with Caratheodory's extension theorem, the result from the last exercise guarantees that there is a unique probability measure, P_∞, defined on the σ-field $\mathscr{F}_\infty \stackrel{\text{def}}{=} \sigma(\mathscr{A})$ that coincides with P_* on $\mathscr{A}$: $P_\infty(A) = P_*(A)$ for all $A \in \mathscr{A}$. Thus, the triplet

$$\left(\Omega_\infty, \mathscr{F}_\infty, P_\infty\right)$$

is a well-defined probability space – a particularly interesting and important probability space in the domain of asset pricing.

(1.46) EXERCISE: Prove that for any $\omega \in \Omega_\infty$, the singleton $\{\omega\}$ is a measurable set (that is $\{\omega\} \in \mathscr{F}_\infty$) and that $P_\infty(\{\omega\}) = 0$.

HINT: Find a decreasing sequence of events from $\mathscr{F}_\infty$, the probabilities of which converge to 0 and the intersection of which is precisely $\{\omega\}$. ○

As we have just established, the probability measure P_∞ is *nonatomic* in the sense of the following definition:

(1.47) NON-ATOMIC AND PURELY ATOMIC MEASURES: Let $(\mathbb{X}, \mathscr{S}, \mu)$ be any measure space such that $\{x\} \in \mathscr{S}$ for any $x \in \mathbb{X}$. Then the measure μ is said to be *nonatomic* if $\mu(\{x\}) = 0$ for every $x \in \mathbb{X}$. It is said to be *purely atomic* if there is a countable set $\mathscr{X} \subset \mathbb{X}$ such that

$$\mu(A) = \sum\nolimits_{x \in \mathscr{X} \cap A} \mu(\{x\}) \quad \text{for every } A \in \mathscr{S},$$

with the understanding that the summation yields 0 if $\mathscr{X} \cap A = \emptyset$. ○

(1.48) SETS OF MEASURE 0 AND COMPLETION OF THE σ-FIELD: Contrary to what one might think, sets of measure 0 are very important. In fact, many important objects (the stochastic integral, for example), built on a particular measure space $(\mathbb{X}, \mathscr{S}, \mu)$, actually depend only on the collection of μ-null sets inside $\mathbb{X}$. Such objects depend on the measure μ, but only to the extent to which μ determines which sets are negligible, so that if one is to replace μ with any other measure that designates the same exact family of sets as negligible, then the constructed object will not change. The formal definition of μ-null set (alias: μ-negligible set) is the following: a set $\mathcal{E} \subset \mathbb{X}$ (which may not be in $\mathscr{S}$) is said to be a *μ-null set* if there is a set $E \in \mathscr{S}$ such that $\mu(E) = 0$ and $E \supseteq \mathcal{E}$. The collection of all μ-null sets we denote by $\mathscr{N}_\mu(\mathscr{S})$ and stress that, generally, $\mathscr{N}_\mu(\mathscr{S})$ would not be included in $\mathscr{S}$. Intuitively, it should be possible to extend μ for sets $B \subset \mathbb{X}$ that are not in $\mathscr{S}$ but are still μ-almost in $\mathscr{S}$, in the sense that one can find some set $A \in \mathscr{S}$ such that $A \,\Delta\, B \in \mathscr{N}_\mu(\mathscr{S})$, where

$$A \,\Delta\, B \overset{\text{def}}{=} (A \backslash B) \cup (B \backslash A)$$

is the symmetric set-difference between A and B. Specifically, one should be able to write (without violating the countable additivity axiom) $\mu(B) = \mu(A)$, even if $B \notin \mathscr{S}$, as long as $A \in \mathscr{S}$ and $A \,\Delta\, B \in \mathscr{N}_\mu(\mathscr{S})$. This intuition is correct: it is not difficult to check that the collection of all subsets of $\mathbb{X}$ that are μ-almost in $\mathscr{S}$ (obviously, this collection includes $\mathscr{S}$) is a σ-field, which we are going to denote by $\tilde{\mathscr{S}}^\mu$, or by $\tilde{\mathscr{S}}$ if the measure μ is understood from the context. It is also not difficult to check that $\tilde{\mathscr{S}}^\mu$ can be identified as⊶(1.23) the σ-field $\mathscr{S} \vee \mathscr{N}_\mu(\mathscr{S})$ (the smallest σ-field that contains both $\mathscr{S}$ and $\mathscr{N}_\mu(\mathscr{S})$). The measure μ, which was originally defined only on $\mathscr{S}$, has a unique extension to a (countably-additive) measure on the larger σ-field $\tilde{\mathscr{S}}^\mu$ – see (Dudley 2002, 3.3.2, 3.3.3). The transition from $(\mathbb{X}, \mathscr{S}, \mu)$ to the larger space $(\mathbb{X}, \tilde{\mathscr{S}}^\mu, \mu)$ is known as *completion of the measure* and $(\mathbb{X}, \tilde{\mathscr{S}}^\mu, \mu)$ is called *complete measure space*. The phrase "$(\mathbb{X}, \mathscr{S}, \mu)$ is a complete measure space" is understood to mean that all μ-null sets are already included in $\mathcal{S}$, even though we do not write $\tilde{\mathcal{S}}$ or $\tilde{\mathcal{S}}^{\mu}$.

Of course, the completion can be applied to probability measures as well: given any probability space $(\Omega, \mathcal{F}, P)$, we can complete $\mathcal{F}$ with all P-null sets and thus obtain $(\Omega, \tilde{\mathcal{F}}^P, P)$. We may say "$(\Omega, \mathcal{F}, P)$ is a complete probability space," implying that $\mathcal{F} = \tilde{\mathcal{F}}^P$ ($\mathcal{F}$ includes all P-null sets). If $A \in \tilde{\mathcal{F}}^P$ and $P(A) = 1$, then we say "A occurs P-almost surely" (abbreviated as "P-a.s."), or "A is an almost sure (a.s.) event" (relative to P).

If $(\mathcal{X}, \mathcal{S}, \mu)$ is any measure space, and a particular property holds for all $x \in \mathcal{X} \backslash A$ for some μ-negligible set $A \subset \mathcal{X}$ (that is, the property holds everywhere in $\mathcal{X}$ except on some μ-negligible set), we say that this property holds μ-almost everywhere (abbreviation: μ-a.e.) in X, or simply almost everywhere (a.e.) if μ and $\mathcal{X}$ are understood. In the case of a probability measure P, instead of "P-a.e." it is very common to say "P-a.s.," or simply "a.s." if the probability measure is understood.

Last, we note that, in practically all spaces that are of interest to us, the completion of the σ-field is intimately connected with the notion of *analytic sets*, which lies outside of our main concerns. We will revisit this connection briefly in (2.12) and realize that at least one crucially important result related to projections of measurable sets simply cannot hold without completing the σ-field. ○

Now we have a framework in which – in principle, at least – we can assign probabilities to events associated with observing the random walk during infinitely many time periods. The first step toward the study of such events is to give precise meaning to the attribute "occurs infinitely often." Since $\mathcal{F}_\infty$ is a σ-field, given any sequence of events $(A_i \in \mathcal{F}_\infty)_{i \in \mathbb{N}}$, the event

$$\limsup_i A_i \stackrel{\text{def}}{=} \cap_{i \in \mathbb{N}} \cup_{j \geq i} A_j \qquad e_1$$

belongs to $\mathcal{F}_\infty$ and therefore has a well-defined probability.

(1.49) EXERCISE: Explain why the random event in e_1 can be described as "infinitely many events A_i occur" and can be expressed as $\{A_i \text{ i.o.}\}$. ○

A powerful tool for analyzing the probability of "infinitely often" is provided by the following classical result.

(1.50) FIRST[19] BOREL–CANTELLI LEMMA:[20] (Dudley 2002, 8.3.4) Let $(\mathcal{X}, \mathcal{S}, \mu)$ be any measure space and let $(A_i \in \mathcal{S})_{i \in \mathbb{N}}$ be any sequence of sets from $\mathcal{S}$. Then

$$\sum_{i \in \mathbb{N}} \mu(A_i) < \infty \quad \text{implies} \quad \mu\big(\limsup_i A_i\big) = 0 . \qquad ○$$

[19] The qualifier "first" suggests that there is also a "second" Borel–Cantelli lemma. We will discover⊸ (4.29) later on that the second Borel–Cantelli lemma is especially useful, but in order to formulate this result we need to introduce the notion of independent random events first.

[20] Named after the French mathematician Émile Borel (1871–1956) and the Italian mathematician Francesco Paolo Cantelli (1875–1966).

The more interesting for us special case of the first Borel–Cantelli lemma is the one where the measure space $(\mathscr{X}, \mathscr{S}, \mu)$ is some probability space $(\Omega, \mathcal{F}, P)$.

(1.51) EXERCISE: Prove – yet again, but this time using the first Borel–Cantelli lemma – that the probability to toss a coin infinitely many times and to never observe "heads," after some, however large, finite number of tosses, is 0. ○

There are other – and much more interesting – applications of the First Borel–Cantelli lemma, but before we can get to those applications we need to delve a bit deeper into the theory of probability.

We conclude this section with the promised outline of Ionescu Tulcea's result.

(1.52) EXTENSIVE HINTS TO EXERCISE (1.45): The proof can be carried out by contradiction as follows:

Step 1: If the sequence $(A_i \in \mathscr{A})_{i \in \mathbb{N}}$ satisfies the assumptions in (1.45), then $P_*(A_i) \geq P_*(A_{i+1})$ for all $i \in \mathbb{N}$ and $\lim_i P_*(A_i)$ always exists. Suppose that

$$\lim_i P_*(A_i) = \varepsilon > 0\,.$$

Step 2: By the very definition of $\mathscr{A}$, if $A \in \mathscr{A}$, then $A \in \sigma(\mathscr{P}_t)$ for some $t \in \mathbb{N}$. If there is an integer $t \in \mathbb{N}$ such that $A_i \in \sigma(\mathscr{P}_t)$ for all $i \in \mathbb{N}$, then the relation above would not be possible since $P_*(A_i) \geq 2^{-t}$. Prove that if no such $t \in \mathbb{N}$ exists, then one can find a sequence of integers $0 \leq t_0 \leq t_1 \leq \ldots \to \infty$ such that $A_i \in \sigma(\mathscr{P}_{t_i})$ for all $i \in \mathbb{N}$.

Step 3: Prove that $\lim_i P_*(A_i \cap E)$ exists for any fixed atom $E \in \mathscr{P}_{t_0}$ and

$$\lim_i P_*(A_i) = \sum_{E \in \mathscr{P}_{t_0}} \Big(\lim_i P_*(A_i \cap E)\Big)\,.$$

In particular, it must be that $\lim_i P_*(A_i \cap E) > 0$ for at least one $E \in \mathscr{P}_{t_0}$, and for any such E, it must be that $E \subseteq A_0$ (for, otherwise, $P_*(A_i \cap E) = 0$ for all $i \in \mathbb{N}$).

Step 4: Let $E_0 \in \mathscr{P}_{t_0}$ be such that $E_0 \subseteq A_0$ and $\lim_i P_*(A_i \cap E_0) > 0$. Now apply the argument from step 3 to the sequence $(A_{i+1} \cap E_0)_{i \in \mathbb{N}}$, and conclude that there is an atom $E_1 \in \mathscr{P}_{t_1}$ such that $E_1 \subseteq E_0 \cap A_1$ and

$$\lim_i P_*(A_i \cap E_0 \cap E_1) > 0\,.$$

Similarly, there is an atom $E_2 \in \mathscr{P}_{t_2}$ with $E_2 \subseteq E_1 \cap E_0 \cap A_2$ and

$$\lim_i P_*(A_i \cap E_0 \cap E_1 \cap E_2) > 0\,.$$

Continuing by induction, construct a sequence of atoms $(E_i \in \mathscr{P}_{t_i})_{i \in \mathbb{N}}$ with $E_0 \subseteq A_0$ and $E_i \subseteq E_{i-1} \cap \ldots \cap E_0 \cap A_i$ for all $i \in \mathbb{N}_{++}$. In particular, it must be that $E_i \subseteq A_i$ for every $i \in \mathbb{N}$.

Step 5: Argue that for any sequence of atoms $(E_i \in \mathscr{P}_{t_i})_{i \in \mathbb{N}}$, chosen so that $E_0 \supset E_1 \supset \ldots$, there is a unique $\omega^* \in \Omega_\infty$ such that $\cap_{i \in \mathbb{N}} E_i = \{\omega^*\}$. Conclude that if the construction of the sequence $(E_i \in \mathscr{P}_{t_i})_{i \in \mathbb{N}}$ in the previous step would be possible, then the relation $\cap_{i \in \mathbb{N}} A_i = \emptyset$ would not be. ○

1.5 Borel σ-Fields and Lebesgue Measure

We now turn to the important task of introducing measurability structure and measure on the real line $\mathbb{R}$. We will follow the same general strategy that allowed us to introduce measurability structure on the coin toss space in the previous section. First, we must identify an easy to describe collection of "elementary" sets, the measurability of which cannot be dispensed with, keeping in mind$^{\multimap (1.44)}$ that this collection must be sufficiently broad. We must then decide what the measure of those sets should be, keeping in mind that countable additivity must hold. In a first step toward this program, we postulate that all left-open and right-closed intervals of the form $]a, b]$, for arbitrary $a, b \in \mathbb{R}$, must be measurable (if $a \geq b$, the set $]a, b]$ is understood to be the empty set $\emptyset$). As it stands, the collection of such intervals is not sufficiently rich for our purpose. For this reason, we now expand our initial choice of measurable sets to include all finite unions of disjoint finite intervals of the form $]a, b]$, and denote the collection of all such unions by $\mathscr{R}$. Thus, a generic element $A \in \mathscr{R}$ is a set of the form

$$A =]a_1, b_1] \cup]a_2, b_2] \cup \ldots \cup]a_n, b_n] \qquad \mathrm{e}_1$$

for some choice of the integer $n \in \mathbb{N}_{++}$ and the reals $-\infty < a_1 \leq b_1 \leq a_2 \leq b_2 \leq \ldots \leq a_n \leq b_n < +\infty$. Just as the notation suggests, the family $\mathscr{R}$ is a ring (the reader is invited to verify this claim) and is therefore sufficiently rich for our purpose. Specifically, by Caratheodory's extension theorem any countably additive set function on $\mathscr{R}$ would give rise to a unique measure on the σ-field generated by $\mathscr{R}$, which σ-field we denote by $\sigma(\mathscr{R})$, and remind that $\sigma(\mathscr{R})$ is the intersection of all σ-fields over $\mathbb{R}$ that include $\mathscr{R}$ and this intersection is nothing but the smallest σ-field that includes the ring $\mathscr{R}$.

(1.53) THE BOREL σ-FIELD[21] ON $\mathbb{R}$ AND $\bar{\mathbb{R}}$: The σ-field generated by all sets of the form e_1 is known as the *Borel σ-field* (alias: the *Borel σ-algebra*) on the real line $\mathbb{R}$, and is denoted by $\mathcal{B}(\mathbb{R})$. The elements of $\mathcal{B}(\mathbb{R})$ are said to be Borel sets, and any measure defined on $\mathcal{B}(\mathbb{R})$ is said to be a Borel measure on $\mathbb{R}$.

The Borel σ-field on the extended real line $\bar{\mathbb{R}}$, denoted by $\mathcal{B}(\bar{\mathbb{R}})$, can be introduced in the obvious way by amending $\mathcal{B}(\mathbb{R})$ with all sets of the form: $B \cup \{+\infty\}$, or $B \cup \{-\infty\}$, or $B \cup \{-\infty\} \cup \{+\infty\}$, for some choice of $B \in \mathcal{B}(\mathbb{R})$ (the reader is invited to check that this extended collection indeed constitutes a σ-field). ○

(1.54) EXERCISE: Prove that $\mathcal{B}(\mathbb{R})$ can be described equivalently as the σ-field over $\mathbb{R}$ generated by all open intervals $]a, b[$, for all $-\infty < a < b < +\infty$. ○

(1.55) BOREL σ-FIELDS ON METRIC SPACES: If a set $\mathbb{X}$ is endowed with the structure of a metric space – or, more generally, endowed with topology – the collection of all open subsets of $\mathbb{X}$ (the topology) is a well-defined family of subsets. The smallest σ-field on $\mathbb{X}$ that contains that family – to put it another way, the

[21] Named after the French mathematician Émile Borel (1871–1956).

intersection of all σ-fields that include the topology – is called the *Borel σ-field* on $\mathbb{X}$ and is denoted by $\mathcal{B}(\mathbb{X})$. The elements of any Borel σ-field are called *Borel sets*, and a measure defined on some Borel σ-field is called *Borel measure*. It is clear from the last exercise that this definition is compatible with the definition of the Borel σ-field on the real line $\mathbb{R}$ that we introduced earlier.

The distance in any metric space $\mathbb{X}$ gives rise to a topology – and therefore also to a Borel structure – on any subset $A \subseteq \mathbb{X}$, and it is not difficult to see that $\mathcal{B}(A)$ (the Borel σ-field on A, treated as a stand-alone metric space) coincides with$^{\multimap (1.18)}$ the induced σ-field $A \cap \mathcal{B}(\mathbb{R})$. Thus, given any $I \subseteq \bar{\mathbb{R}}$ it would be consistent for us to set $\mathcal{B}(I) \overset{\text{def}}{=} I \cap \mathcal{B}(\bar{\mathbb{R}})$.

We emphasize that the Borel σ-field depends on the choice of the metric only to the extent to which the metric determines the topology (so that metrics that are equivalent in topological sense generate identical Borel structures), and when we write $\mathcal{B}(\mathbb{X})$ we assume that the topology is understood from the context. For example, $\mathscr{C}(\mathbb{R}_+;\mathbb{R}^n)$ is understood$^{\multimap (0.8)}$ to have the structure of a metric space provided by the locally uniform distance $d_{\mathrm{loc},\infty}(\cdot,\cdot)$. Similarly, the space $\mathbb{D}(\mathbb{R}_+;\mathbb{R}^n)$ is understood$^{\multimap (0.9)}$ to have the structure of a metric space provided by the Skorokhod distance $d_{\mathfrak{s}}(\cdot,\cdot)$. Thus, the metrics $d_{\mathrm{loc},\infty}(\cdot,\cdot)$ and $d_{\mathfrak{s}}(\cdot,\cdot)$ will be automatically understood in any reference to "Borel sets" or "measures" associated with the spaces $\mathscr{C}(\mathbb{R}_+;\mathbb{R}^n)$ and $\mathbb{D}(\mathbb{R}_+;\mathbb{R}^n)$. ○

(1.56) EXERCISE: Prove that if $\mathbb{X}$ is a metric space, then every singleton $\{x\}$, for any $x \in \mathbb{X}$, is a Borel set. ○

Now we turn to the construction of Borel measures, that is, measures defined on Borel σ-fields. Naturally, Borel measures on the real line are going to be of prime interest for us. As we have already seen, a Borel measure on $\mathbb{R}$ can be introduced as a countably additive set function on the ring $\mathscr{R}$. Because of the special structure of $\mathscr{R}$ (every set in $\mathscr{R}$ is a finite union of disjoint intervals $]a,b]$), any prescription for an additive set function on $\mathscr{R}$ comes down to the assignment of a positive real value to every interval $]a,b]$ in a way that is consistent, in that the value assigned to $]a,b]$ must be the sum of the values assigned to $]a,c]$ and $]c,b]$ for every choice of $a \leq c \leq b$.[22] The assignment of a value (measure) to the interval $]a,b]$ that comes to mind first is

$$\Lambda_*(]a,b]) = b - a \equiv \imath(b) - \imath(a)\,.$$

[22] This step involves a technical detail that is generally innocuous, in that it is easy to resolve, but must be kept in mind, still: one and the same set in $\mathscr{R}$ can be expressed as the union of disjoint intervals $]a,b]$ in many – in fact, infinitely many – different ways. Thus, we must ensure that lifting the notion of "measure" from the intervals $]a,b]$ to all the elements of $\mathscr{R}$ is invariant under the various representations of the sets in $\mathscr{R}$ as finite unions of disjoint intervals $]a,b]$.

However, for reasons that will become clear very soon, we need a slightly more general assignment, namely

$$\mu_*(]a,b]) = F(b) - F(a), \qquad \text{e}_2$$

for some increasing and right-continuous function $F\colon \mathbb{R} \mapsto \mathbb{R}$ – an object that we introduce next.

(1.57) INCREASING FUNCTIONS: Let $I \subseteq \bar{\mathbb{R}}$ be any nonempty interval. The function $F\colon I \mapsto \mathbb{R}$ is said to be *increasing* if $F(a) \leq F(b)$ for every choice of $a < b$, $a, b \in I$, and this is the same as saying that F is *nondecreasing*. If $F(a) < F(b)$ for every $a, b \in I$, $a < b$, we say that $F(\cdot)$ is *strictly increasing*. The domain of an increasing function can be extended to the left and right end-points of the interval I, I_l and I_r (if those are not contained in I) in the obvious way by setting $F(I_l) \overset{\text{def}}{=} \lim_{x \searrow I_l} F(x)$ and $F(I_r) \overset{\text{def}}{=} \lim_{x \nearrow I_r} F(x)$. Furthermore, the limits $F(x-) \overset{\text{def}}{=} \lim_{a \nearrow x} F(a)$ and $F(x+) \overset{\text{def}}{=} \lim_{a \searrow x} F(a)$ are also well-defined and finite for any $I_l < x < I_r$. Thus, if the increasing function F happens to be right-continuous then it is automatically càdlàg, and if it happens to be left-continuous then it is automatically càglàd (continue à gauche, limite à droite, that is, left-continuous with right limits). ○

(1.58) EXERCISE: Prove that if the set function $\mu_*\colon \mathscr{R} \mapsto \mathbb{R}_+$, defined through the relation e₂ for some increasing function $F\colon \mathbb{R} \mapsto \mathbb{R}$, can be claimed to be countably additive, then the function $F(\cdot)$ must also be right-continuous, that is to say, must be an increasing càdlàg function. ○

In fact, the statement in the last exercise holds also in the opposite direction, as we now detail.

(1.59) BOREL AND LEBESGUE MEASURES: A crucially important and highly nontrivial result from measure theory – see (Dudley 2002, 3.1.3), for example – states the following: if $F\colon \mathbb{R} \mapsto \mathbb{R}$ is *any* increasing càdlàg function, then the set function $\mu_*\colon \mathscr{R} \mapsto \mathbb{R}_+$ given by e₂ is countably additive on the ring $\mathscr{R}$. Caratheodory's extension theorem then guarantees that there is a unique measure on the Borel σ-field $\mathcal{B}(\mathbb{R})$ that coincides with μ_* on $\mathscr{R}$. We will denote this measure by μ, or, depending on the context, by μ_F, or by $\mathrm{d}F$, if the dependence on the function F must be emphasized. The measure μ can be treated as a measure on $\mathcal{B}(\bar{\mathbb{R}})$ in the obvious way by postulating $\mu(\{-\infty\}) = \mu(\{+\infty\}) = 0$ – or by attaching other values to the atoms $\{-\infty\}$ and $\{+\infty\}$. If no values are implied by the context, the default assumption is that these two atoms have measure 0.

The celebrated *Lebesgue measure*[23] is simply the measure on $\mathcal{B}(\mathbb{R})$ associated with the increasing function $F = \imath$. Throughout this space we denote this measure by Λ, and use the same token for its$^{\multimap (1.28)}$ restriction to the various intervals $I \subset \mathbb{R}$, or, more generally, to the various Borel sets $B \in \mathcal{B}(\mathbb{R})$.

[23] Named after the French mathematician Henri Lebesgue (1875–1941).

To summarize: every increasing càdlàg function gives rise to a unique Borel measure on $\mathbb{R}$, and every Borel measure on $\mathbb{R}$ can be associated with an increasing càdlàg function through the relation e_2. Since the right side of e_2 would not change if the function $F(\cdot)$ is shifted by a constant, the association between Borel measures and increasing càdlàg functions is not one-to-one. It could be made one-to-one if, for example, the measure μ is finite ($\mu(\mathbb{R}) < \infty$) by imposing the requirement $F(-\infty) = 0$, which then uniquely determines F from the relation $F(x) = \mu(]-\infty, x])$, $x \in \mathbb{R}$. ○

(1.60) PRODUCTS OF MEASURES: Let $(\mathbb{X}, \mathcal{S}, \mu)$ and $(\mathbb{Y}, \mathcal{T}, \nu)$ be any two measure spaces, endowed with σ-finite measures μ and ν. Consider the collection of all rectangular subsets (alias: rectangles) of the Cartesian product $\mathbb{X}\times\mathbb{Y}$, namely,

$$A\times B \overset{\text{def}}{=} \left\{(x, y) \in \mathbb{X}\times\mathbb{Y} : x \in A,\ y \in B\right\}$$

for all choices of $A \in \mathcal{S}$ and $B \in \mathcal{T}$. Given a rectangle $A\times B$, define its product measure by

$$(\mu\times\nu)_*(A\times B) \overset{\text{def}}{=} \mu(A)\nu(B)\,.$$

Just as before, the definition of the set function $(\mu\times\nu)_*$ can be extended in a consistent fashion to a set function defined on the collection of all finite unions of disjoint rectangles. This collection again constitutes a ring, and this ring generates a σ-field on the Cartesian product $\mathbb{X}\times\mathbb{Y}$, which σ-field we call *the product of $\mathcal{S}$ and $\mathcal{T}$* and denote by $\mathcal{S}\otimes\mathcal{T}$. It again turns out – see (Dudley 2002, 4.4), for example – that $(\mu\times\nu)_*$ is σ-additive and, by Caratheodory's theorem, extends to a unique measure on the σ-field $\mathcal{S}\otimes\mathcal{T}$. We denote this measure by $\mu\times\nu$ and call it *the product of μ and ν*. The extension of $(\mu\times\nu)_*$ to the product σ-field $\mathcal{S}\otimes\mathcal{T}$ exists even without the requirement for μ and ν to be σ-finite, but without this requirement the extension is no longer unique – we refer the reader to (ibid., 4.4) for further details and for complete proofs.

Of course, a measure defined on the product σ-field $\mathcal{S}\otimes\mathcal{T}$ may not be a product measure. It is easy to check that if m is any measure on $\mathcal{S}\otimes\mathcal{T}$ then for any fixed $A \in \mathcal{S}$ the mapping $\mathcal{T} \ni B \rightsquigarrow m(A\times B)$ is a measure on $\mathcal{T}$, which measure we denote by $m(A, \cdot)$. Similarly, for any fixed $B \in \mathcal{T}$ the mapping $\mathcal{S} \ni A \rightsquigarrow m(A\times B)$ is a measure on $\mathcal{S}$, which measure we denote by $m(\cdot, B)$. Thus, given any $A \in \mathcal{S}$ and any $B \in \mathcal{T}$, the token $m(A, B)$ has the same meaning as $m(A\times B)$.

The product of any finite number of σ-fields and measures will be understood in the obvious way by induction: if $(\mathbb{X}_i, \mathcal{S}_i, \mu_i)$, $1 \le i \le n$, is any finite collection of measure spaces, then

$$\mathcal{S}_1 \otimes \ldots \otimes \mathcal{S}_i \overset{\text{def}}{=} \left(\mathcal{S}_1 \otimes \ldots \otimes \mathcal{S}_{i-1}\right) \otimes \mathcal{S}_i$$

$$\text{and} \quad \mu_1\times \ldots \times\mu_i \overset{\text{def}}{=} \left(\mu_1\times \ldots \times\mu_{i-1}\right)\times\mu_i \quad \text{for } i = 2, \ldots, n\,.$$

It is easy to see that both products $\otimes$ and $\times$ are associative. ○

(1.61) EXERCISE AND DEFINITION: Prove that the product σ-field

$$\mathcal{B}(\mathbb{R})\otimes \ldots \otimes \mathcal{B}(\mathbb{R}) \quad (n \text{ factors}, n \in \mathbb{N}_{++})$$

can be identified with the Borel σ-field on $\mathbb{R}^n$, understood relative to the usual Euclidean distance in $\mathbb{R}^n$. The measure $\Lambda\times \ldots \times\Lambda$ (n factors) is still called the *Lebesgue measure on* $\mathbb{R}^n$ and will be denoted by Λ^n, or, with a slight abuse of the notation, by Λ again, if the dimension is implied by the context. ○

(1.62) INVARIANCE: It would not be a huge exaggeration to say that the Lebesgue measure is the most important measure in the universe. The reason is that – see (Stroock 1999, 2.2) – up to a multiplicative constant, it is the only uniform measure on $\mathbb{R}^n$ that has a finite value on every finite rectangle, with "uniform" understood to mean that the measure is invariant under isometry, that is, invariant under any transformation of $\mathbb{R}^n$ that preserves the Euclidean distance.[24] To be precise, the Lebesgue measure of the set $A \in \mathcal{B}(\mathbb{R}^n)$ is the same as the Lebesgue measure of the set

$$x + UA \stackrel{\text{def}}{=} \{x + Ua : a \in A\},$$

for every choice of the vector $x \in \mathbb{R}^n$ and the orthogonal matrix $U \in \mathbb{R}^{n\otimes n}$.

It is interesting that the measure P_∞ that was constructed on the coin toss space Ω_∞, is also uniform in the following sense: for every fixed $A \in \mathscr{F}_\infty$ and $\omega^* \in \Omega_\infty$, $P_\infty(\omega^*\times A) = P_\infty(A)$, where $\omega^*\times A \stackrel{\text{def}}{=} \{\omega^*\times\omega : \omega \in A\}$, with the understanding that for $\omega = \{\varepsilon_0, \varepsilon_1, \ldots\}$ and $\omega^* = \{\varepsilon_0^*, \varepsilon_1^*, \ldots\}$ the product $\omega^*\times\omega$ is defined as (recall that $\varepsilon_i, \varepsilon_i^* \in \{-1, +1\}$) $\omega^*\times\omega = \{\varepsilon_0^*, \varepsilon_0, \varepsilon_1^*, \varepsilon_1, \ldots\}$. ○

(1.63) EXERCISE: Prove that the Lebesgue measure Λ (or, more generally, Λ^n) is nonatomic: for any $x \in \mathbb{R}$, one has $\{x\} \in \mathcal{B}(\mathbb{R})$ and $\Lambda(\{x\}) = 0$. Conclude that for every $-\infty < a < b < +\infty$,

$$\Lambda\big([a,b]\big) = \Lambda\big(]\,a,b]\big) = \Lambda\big([a,b[\big) = \Lambda\big(]\,a,b[\big).$$

Prove that if $C \subset \mathbb{R}$ is any countable set inside $\mathbb{R}$, then $C \in \mathcal{B}(\mathbb{R})$ and $\Lambda(C) = 0$. Conclude that the set $\mathbb{Q}$ of all rational numbers is a Borel set and $\Lambda(\mathbb{Q}) = 0$. Prove that the set of all points in $\mathbb{R}^2$ that have rational coordinates is a Borel set and Lebesgue measure of that set is 0. ○

We saw in (1.28) that the restriction of a measure to a particular measurable subset turns that subset into a measure space. In many situations it is useful to carry out this procedure in the opposite direction: a measure defined on a particular subset can be treated as a measure on the whole space. For example, the probability measure P that was constructed in (1.13) on the set $\mathbb{N}$ can be treated as a purely atomic Borel measure on $\mathbb{R}_+$ given by

$$\mathsf{P}(A) = \sum\nolimits_{x\in A\cap\mathbb{N}} e^{-\lambda}\frac{\lambda^x}{x!}, \qquad A \in \mathcal{B}(\mathbb{R}_+),$$

[24] Any isometry in $\mathbb{R}^n$ is a composition of a rotation, that is, multiplication by an orthogonal matrix $U \in \mathbb{R}^{n\otimes n}$, and translation by a vector $x \in \mathbb{R}^n$.

with the understanding that the summation yields 0 if $A \cap \mathbb{N} = \emptyset$. This measure can be expressed equivalently as an infinite sum of point-mass measures, namely,

$$P = \sum_{x \in \mathbb{N}} e^{-\lambda} \frac{\lambda^i}{i!} \epsilon_x .$$

Now we have two different measures on $\mathbb{R}_+$: the probability measure P and the Lebesgue measure Λ. The probability P is amassed (or, supported) on the set of integers $\mathbb{N} \subset \mathbb{R}$, in the sense that $P(\mathbb{R}_+ \backslash \mathbb{N}) = 0$ (and therefore $P(\mathbb{N}) = 1$). To put it another way, the measure P will not suffer if all noninteger numbers are removed from the real line $\mathbb{R}$. On the other hand, $\Lambda(\mathbb{N}) = 0$, so that the Lebesgue measure will not suffer if all integer numbers are removed from the real line $\mathbb{R}$. The measures Λ and P are then *singular* in the sense of the following definition:

(1.64) SINGULAR MEASURES: Let $(\mathcal{X}, \mathcal{S})$ be any measurable space and let μ and ν be any two measures on $\mathcal{S}$, such that $\mu(\mathcal{X}) > 0$ and $\nu(\mathcal{X}) > 0$. We say that the measures μ and ν are singular (notation: $\mu \perp \nu$) if there is a set $A \in \mathcal{S}$ such that $\mu(A) = \nu(A^{\complement}) = 0$. ○

(1.65) EXERCISE: Prove that if μ is any⊸(1.47) purely atomic measure on $\mathbb{R}$, then $\mu \perp \Lambda$. Explain how one can construct a purely atomic measure Q on $\mathbb{R}$ that is supported on the (countable) set of rational numbers $\mathbb{Q}$, in that $Q(\mathbb{R} \backslash \mathbb{Q}) = 0$. ○

The last exercise leads us to the question: can a nonatomic measure be singular to Λ? The answer is positive, and there is a particularly instructive example, which, in addition to answering the question, reveals the intrinsic nature of measures defined on uncountably infinite spaces. Our interest in this example and the associated construction is motivated also by its connection with the coin toss space, which is an important source of intuition and insight for practically everything that follows in this space.

Recall that the binary expansion of a number $x \in [0, 1]$ has the form

$$x = 0.\varepsilon_0\varepsilon_1 \ldots \quad \Longleftrightarrow \quad x = \sum_{i \in \mathbb{N}} \frac{\varepsilon_i}{2^{i+1}} ,$$

for some sequence $(\varepsilon_i \in \{0, 1\})_{i \in \mathbb{N}}$. At the same time, the ternary expansion of $x \in [0, 1]$ has the form

$$x = 0.\delta_0\delta_1 \ldots \quad \Longleftrightarrow \quad x = \sum_{i \in \mathbb{N}} \frac{\delta_i}{3^{i+1}} ,$$

for some sequence $(\delta_i \in \{0, 1, 2\})_{i \in \mathbb{N}}$. Since the elements of the coin toss space Ω_∞ are nothing but sequences of the form $\omega = (\omega_i \in \{-1, +1\})_{i \in \mathbb{N}}$, we can identify Ω_∞ with the space of binary sequences (that is, sequences of 0's and 1's) simply by interpreting the symbol -1 as 0.[25] Because of this connection, one may be tempted to think that the coin toss space Ω_∞ differs from the interval $[0, 1]$ only

[25] One can always transform the sequence ω_i into $\varepsilon_i = (1 + \omega_i)/2$, or transform the sequence ε_i into $\omega_i = 2\,\varepsilon_i - 1$.

in the notation, and that therefore the measure P_∞ can be treated also as a measure on $[0, 1]$. Nevertheless, the next exercise will make it clear that such a line of reasoning is fundamentally flawed.

(1.66) EXERCISE: Prove that any number of the form $m/2^n$, for any choice of $1 \leq m \leq 2^n - 1$ and $n \in \mathbb{N}_{++}$, has two different binary expansions and, similarly, any number of the form $m/3^n$, for any $1 \leq m \leq 3^n - 1$ and $n \in \mathbb{N}_{++}$, has two different ternary expansions. Show that all binary expansions $0.0\varepsilon_1 \ldots$ correspond to points in the interval $[0, \frac{1}{2}]$, all binary expansions $0.1\varepsilon_1 \ldots$ correspond to points in the interval $[\frac{1}{2}, 1]$, all binary expansions $0.00\varepsilon_2 \ldots$ correspond to points in the interval $[0, \frac{1}{4}]$, all binary expansions $0.01\varepsilon_2 \ldots$ correspond to points in the interval $[\frac{1}{4}, \frac{1}{2}]$ – and so on. Next, show that all ternary expansions $0.0\delta_1 \ldots$ correspond to points in the interval $[0, \frac{1}{3}]$, all ternary expansions $0.1\delta_1 \ldots$ correspond to points in the interval $[\frac{1}{3}, \frac{2}{3}]$, all ternary expansions $0.2\delta_1 \ldots$ correspond to points in the interval $[\frac{2}{3}, 1]$, all ternary expansions $0.00\delta_2 \ldots$ correspond to points in the interval $[0, \frac{1}{9}]$, all ternary expansions $0.01\delta_2 \ldots$ correspond to points in the interval $[\frac{1}{9}, \frac{2}{9}]$ – and so on.

HINT: The main point in this exercise is to realize that the end points of the mentioned intervals have multiple expansions. For example, the ternary expansion of $\frac{1}{3}$ could be $0.02222\ldots$, but it could be also $0.10000\ldots$; the ternary expansion of $\frac{2}{3}$ could be $0.111111\ldots$, but it could be also $0.20000\ldots$ – and so on. ○

(1.67) COIN TOSS SPACE AND THE CANTOR SET: The Cantor set[26] C is the intersection, $C = \cap_{n=0}^{\infty} C_n$, of the sets in the decreasing sequence of Borel subsets of $[0, 1]$, namely, $C_0 \supset C_1 \supset C_2 \supset C_3 \supset \ldots$, which sets are defined recursively as follows: for $i = 0$, $C_i = [0, 1]$, and for $i \geq 1$, C_i is obtained by removing the middle third of every one of the intervals that C_{i-1} consists of, with the understanding that all removed intervals are open. As an illustration,

$$C_0 = [0, 1],$$

$$C_1 = \left[0, \frac{1}{3}\right] \cup \left[\frac{2}{3}, 1\right],$$

$$C_2 = \left[0, \frac{1}{9}\right] \cup \left[\frac{2}{9}, \frac{1}{3}\right] \cup \left[\frac{2}{3}, \frac{7}{9}\right] \cup \left[\frac{8}{9}, 1\right],$$

$$C_3 = \left[0, \frac{1}{27}\right] \cup \left[\frac{2}{27}, \frac{1}{9}\right] \cup \left[\frac{2}{9}, \frac{7}{27}\right] \cup \left[\frac{8}{27}, \frac{1}{3}\right] \cup \left[\frac{2}{3}, \frac{19}{27}\right]$$
$$\cup \left[\frac{20}{27}, \frac{7}{9}\right] \cup \left[\frac{8}{9}, \frac{25}{27}\right] \cup \left[\frac{26}{27}, 1\right],$$

$$\ldots$$

[26] Named after the German mathematician Georg Cantor (1845–1918).

It becomes clear from exercise (1.66) why the Cantor set $C = \cap_{n=0}^{\infty} C_n$ can be identified as the collection of those $x \in [0,1]$ that have special ternary expansion of the form $x = 0.\delta_0\delta_1\ldots$ with $\delta_i \in \{0,2\}$ for any $i \in \mathbb{N}$, that is to say, can be written only in terms of the tokens 0 and 2, even though the expansion is ternary. What this means is that every 0-and-2 ternary expansion yields an element of the Cantor set C, and for every element of the Cantor set there is precisely one ternary expansion written only in terms of the tokens 0 and 2 (though the same element may have other ternary expansions that involve also the token 1). As an example, if $x \in C_1$, the first token in the expansion of x can be taken to be 0 if $x \in [0, 1/3]$ and 2 if $x \in [2/3, 1]$; the second token in the expansion of any $x \in C_2$ can be taken to be 0 if $x \in [0, 1/9] \cup [2/3, 7/9]$ and 2 if $x \in [2/9, 1/3] \cup [8/9, 1]$ – and so on.

Since we can always interpret the token 0 as -1 and the token 2 as $+1$, we can identify the Cantor set C with the coin toss space Ω_∞. One immediate consequence from this observation is that the Cantor set C is uncountably infinite. Certain technical points notwithstanding, it is not difficult to check that the σ-field $\mathscr{F}_\infty$, which we constructed in Ω_∞ and now can treat as a σ-field over C, is nothing but the induced σ-field $C \cap \mathcal{B}([0,1])$, and this is nothing but the Borel σ-field on C treated as a metric space. Therefore P_∞ can be treated as a measure on $C \cap \mathcal{B}([0,1]) = \mathcal{B}(C)$. ○

(1.68) EXERCISE: Prove that the Cantor set C is a Borel set ($C \in \mathcal{B}([0,1])$) and that $\Lambda(C) = 0$. Construct a nonatomic probability measure P on $\mathcal{B}([0,1])$ which is supported by the Cantor set C, in that $P(C^{\complement}) = 0$, and is therefore singular to Λ ($P \perp \Lambda$).

HINT: The measure P that you need is essentially P_∞, except that P_∞ is defined on $C \cap \mathcal{B}([0,1])$, while P must be defined on $\mathcal{B}([0,1])$. ○

(1.69) EXERCISE: Prove that the set of real numbers $[0,1]$ is an uncountably infinite. Conclude that the same property must hold also for the set of all reals $\mathbb{R}$. ○

(1.70) EXERCISE: Consider the graph of the function $y = x^2$ as a subset of $\mathbb{R}^2$:

$$\Gamma = \{(x, x^2) : -\infty < x < \infty\} \subset \mathbb{R}^2 .$$

Prove that $\Gamma \in \mathcal{B}(\mathbb{R}^2)$ and compute $\Lambda^2(\Gamma)$. ○

(1.71) EXERCISE: Prove that

$$B = \{(x, y) : 0 \le x \le 1, 0 \le y < x^2\} \subset \mathbb{R}^2 .$$

is a Borel set, that is, $B \in \mathcal{B}(\mathbb{R}^2)$, and compute $\Lambda^2(B)$. ○

2

Integration

In one form or another, the computation of the "average value" of a random variable comes down to the computation of an integral. However, as will be detailed shortly, the familiar Riemann integral is inadequate for this type of averaging. This chapter is a brief résumé of the theory of integration developed by H. Lebesgue, with focus on its applications to the theory of probability.

2.1 Measurable Functions and Random Variables

In many practical situations the sample space Ω is simply too big and too complex to allow for direct observations of the elementary outcomes $\omega \in \Omega$. What one might be able to observe is some measurement X that depends on the (unobservable) outcome ω and yields a scalar value that can be expressed as $X = X(\omega)$ (think of ω as being the state of the universe, including the positions of all stars and constellations in the sky, the mood of the public, etc., and think of $X(\omega)$ as the most recent stock quote for AAPL). Instead of a scalar value, the result of the measurement, $X(\omega)$, could be a vector object, or perhaps some more complex object, say, an entire trajectory $X(\omega) \in \mathscr{C}(\mathbb{R}_+; \mathbb{R})$ (e.g., the outcome from a continuously monitored stock price). Thus, the measurement X maps the sample space Ω into some concrete – and much more manageable – space, such as $\mathbb{X} = \mathbb{R}$, or $\mathbb{X} = \mathbb{R}^n$, or $\mathbb{X}$ = some other metric space. To put it another way, the measurement X can be modeled as a function from the domain Ω into the range space $\mathbb{X}$, and can be expressed as $X\colon \Omega \mapsto \mathbb{X}$, or as

$$\Omega \ni \omega \rightsquigarrow X(\omega) \in \mathbb{X}$$

if we want to emphasize in the notation the domain and the range of X.

(2.1) FIRST EXAMPLES: We have already encountered functions of the form above in our study of the coin toss space Ω_∞. Indeed, for any fixed $t \in \mathbb{N}_{++}$, the coordinate mapping

† $$\Omega_\infty \ni \omega \rightsquigarrow \mathcal{X}_t(\omega) \in \{-1, +1\}$$

returns the symbol (either -1 or $+1$) located in position t of the sequence ω. Similarly, the function

$$\Omega_\infty \ni \omega \rightsquigarrow \mathcal{Z}_t(\omega) \in \big\{ -t, -t+2, \ldots, t-2, t \big\} \subset \mathbb{Z}$$

returns the position of the random walk after t periods. In both cases we can write $\mathcal{X}_t(\omega) \in \mathbb{R}$ and $\mathcal{Z}_t(\omega) \in \mathbb{R}$, if there is no need to be too specific about the range. We can also map each ω into the entire trajectory of the random walk, which is an infinite sequence from the space[27] $\mathbb{Z}^{\mathbb{N}}$ (so, now $\mathbb{X} = \mathbb{Z}^{\mathbb{N}} \subset \mathbb{R}^{\mathbb{N}}$):

$$\Omega_\infty \ni \omega \rightsquigarrow \bar{\mathcal{Z}}(\omega) \overset{\text{def}}{=} \{\mathcal{Z}_0(\omega), \mathcal{Z}_1(\omega), \mathcal{Z}_2(\omega), \dots\} \in \mathbb{Z}^{\mathbb{N}}. \quad \circ$$

Now we turn to the special case where the range space is the real line $\mathbb{R}$ – this is the situation that we will encounter most often. If ω is sampled randomly from the set Ω, then the result (or realization) $X(\omega) \in \mathbb{R}$ becomes random as well. In particular, if $]a, b]$ is some interval of interest, then $X(\omega)$ may or may not belong to that interval. The phrase "the measurement falls into $]a, b]$" describes a random event, namely,

$$X^{-1}(]a, b]) \overset{\text{def}}{=} \{\omega \in \Omega : X(\omega) \in]a, b]\} \subseteq \Omega,$$

which we will often abbreviate as $\{X \in]a, b]\}$. If Ω is endowed with a σ-field $\mathcal{F}$ and a probability measure P on $\mathcal{F}$, then we can measure the probabilities of the events $\{X \in]a, b]\}$, provided that $X^{-1}(]a, b]) \in \mathcal{F}$ for all choices of $a, b \in \mathbb{R}$, or, which is the same, $X^{-1}(]-\infty, a]) \in \mathcal{F}$ for all choices of $a \in \mathbb{R}$. It turns out that this requirement is equivalent to the seemingly stronger requirement[28] that $X^{-1}(B) \in \mathcal{F}$ for every choice of $B \in \mathcal{B}(\mathbb{R})$, where

$$X^{-1}(B) \overset{\text{def}}{=} \{\omega \in \Omega : X(\omega) \in B\} \equiv \{X \in B\}.$$

(2.2) EXERCISE: Let Ω be any nonempty set, let $(\mathbb{X}, \mathcal{S})$ be any measurable space, and let $X : \Omega \mapsto \mathbb{X}$ be any function from Ω into $\mathbb{X}$. Prove that

$$\dagger \qquad X^{-1}(\mathcal{S}) \overset{\text{def}}{=} \{X^{-1}(B) : B \in \mathcal{S}\} \subseteq 2^{\Omega}$$

is a σ-field in Ω.

HINT: Prove that $f^{-1}(\cup_{i \in \mathbb{N}} B_i) = \cup_{i \in \mathbb{N}} f^{-1}(B_i)$ for every choice of the sequence $(B_i \in \mathcal{S})_{i \in \mathbb{N}}$. $\circ$

Now we come to an important definition.

(2.3) MEASURABLE FUNCTIONS AND RANDOM VARIABLES: Given any two measurable spaces, $(\Omega, \mathcal{F})$ and $(\mathbb{X}, \mathcal{S})$, and a function $X : \Omega \mapsto \mathbb{X}$, we say that X is $\mathcal{S}/\mathcal{F}$-measurable if the σ-field $X^{-1}(\mathcal{S})$ from $(2.2^\dagger)$ is included in the σ-field $\mathcal{F}$, or $X^{-1}(\mathcal{S}) \subseteq \mathcal{F}$ for short. The $\mathcal{S}/\mathcal{F}$-measurability of X is equivalent to the condition $X^{-1}(B) \in \mathcal{F}$ for every $B \in \mathcal{S}$. If the σ-field in the range space $\mathbb{X}$ is understood from the context, we may say "$\mathcal{F}$-measurable" instead of "$\mathcal{S}/\mathcal{F}$-measurable," and

[27] Recall that if A and B are sets, then B^A is the collection of all functions from A to B. In particular, $\mathbb{R}^{\mathbb{N}}$ is the space of all sequences of real numbers and $\mathbb{Z}^{\mathbb{N}} \subset \mathbb{R}^{\mathbb{N}}$ is the collection of all integer-valued sequences.

[28] This should not be surprising given that, as we have already seen, the family of intervals $]a, b] \subseteq \mathbb{R}$ generates $\mathcal{B}(\mathbb{R})$.

if both $\mathcal{S}$ and $\mathcal{F}$ are understood we may simply say "measurable." The phrase "X is $\mathcal{S}/\mathcal{F}$-measurable" will be abbreviated as $X \prec \mathcal{S}/\mathcal{F}$, or simply as $X \prec \mathcal{F}$ if the σ-field in the range space cannot be mistaken.

In the most common situations the range $\mathbb{X}$ is endowed with topology, and, usually, this topology is automatically understood (as in any of the cases $\mathbb{X} = \mathbb{R}$, $\mathbb{X} = \mathbb{R}^n$, $\mathbb{X} = \mathbb{R}_+$, or $\mathbb{X} = \mathbb{R}^n_+$, which we will encounter most often). In such situations, unless specified otherwise, the σ-field $\mathcal{S}$ is automatically understood to be the Borel σ-field $\mathcal{B}(\mathbb{X})$, and we may say "X is Borel-measurable," understood to mean "X is $\mathcal{B}(\mathbb{X})/\mathcal{F}$-measurable."

If the measurable space $(\Omega, \mathcal{F})$ is endowed with a probability measure P, so that $(\Omega, \mathcal{F}, P)$ is a probability space, then it is common to refer to all $\mathcal{B}(\bar{\mathbb{R}})/\mathcal{F}$-measurable functions $X: \Omega \mapsto \bar{\mathbb{R}}$ as *random variables*, but the term *random variable*, often abbreviated as r.v., may be given more liberally to measurable functions defined on $(\Omega, \mathcal{F}, P)$ and taking values in some more exotic metric space, such as $\mathscr{C}(\mathbb{R}_+; \mathbb{R}^n)$, or $\mathbb{D}(\mathbb{R}_+; \mathbb{R}^n)$. Random variables that take values in $\mathbb{R}$ are said to be finite, and random variables with values in $\mathbb{R}^n$ may be called *random vectors*. A random variable that takes values in some compact subset of the real line $\mathbb{R}$, or, more generally, of $\mathbb{R}^n$, is said to be *bounded*. We stress that without any qualifiers, the default meaning of "random variable" is a function that can take the values $+\infty$ and/or $-\infty$, possibly on a set that has strictly positive probability. ○

In most situations the attribute "measurable" is too complex to establish directly from the definition – or even to picture. For this reason, we need a tool that would allow us to conclude that every set $X^{-1}(B)$, for every choice of $B \in \mathcal{S}$, belongs to $\mathcal{F}$, but without the need to identify directly the sets B and $X^{-1}(B)$. Our next result, a somewhat simplified version of (Dudley 2002, 4.1.6), shows that it is enough to establish that $X^{-1}(B) \in \mathcal{F}$ only for certain special sets $B \in \mathcal{S}$, and still conclude that $X^{-1}(B) \in \mathcal{F}$ for all sets $B \in \mathcal{S}$.

(2.4) THEOREM: Let $(\Omega, \mathcal{F})$ and $(\mathbb{X}, \mathcal{S})$ be any two measurable spaces and suppose that the σ-field $\mathcal{S}$ is generated by some family of sets $\mathcal{C} \subseteq 2^{\mathbb{X}}$, in that $\mathcal{S}$ is the smallest σ-field over $\mathbb{X}$ that includes $\mathcal{C}$. Then the function $X: \Omega \mapsto \mathbb{X}$ is $\mathcal{S}/\mathcal{F}$-measurable if and only if $X^{-1}(A) \in \mathcal{F}$ for every $A \in \mathcal{C}$.

PROOF: The collection of sets $\mathscr{T} \overset{\text{def}}{=} \{A \subseteq \mathbb{X} : X^{-1}(A) \in \mathcal{F}\}$ is a σ-field in $\mathbb{X}$. Indeed, $X^{-1}(\mathbb{X}) = \Omega \in \mathcal{F}$, so that $\mathbb{X} \in \mathscr{T}$. Furthermore, if $A \in \mathscr{T}$, then $X^{-1}(A^{\complement}) = (X^{-1}(A))^{\complement} \in \mathcal{F}$, so that $A^{\complement} \in \mathscr{T}$, and if $A_i \in \mathscr{T}$ for every $i \in \mathbb{N}$, then $X^{-1}\left(\cup_{i \in \mathbb{N}} A_i\right) = \cup_{i \in \mathbb{N}} X^{-1}(A_i) \in \mathcal{F}$, so that $\cup_{i \in \mathbb{N}} A_i \in \mathscr{T}$. By assumption, $\mathcal{C} \subseteq \mathscr{T}$ and $\mathcal{S}$ is generated by $\mathcal{C}$. This implies $\mathcal{S} \subseteq \mathscr{T}$, which means $X^{-1}(\mathcal{S}) \subseteq \mathcal{F}$. ○

The power of the last result will be illustrated in the next two exercises and also in exercise (2.19).

(2.5) EXERCISE: Let $(\Omega, \mathcal{F})$, $(\mathbb{X}, \mathcal{S})$, and $(\mathbb{Y}, \mathcal{T})$ be any three measurable spaces and let $f : \Omega \mapsto \mathbb{X}\times\mathbb{Y}$ be any function. Prove that f is $(\mathcal{S}\otimes\mathcal{T})/\mathcal{F}$-measurable if and only if $f^{-1}(A\times B) \in \mathcal{F}$ for every $A \in \mathcal{S}$ and every $B \in \mathcal{T}$. ○

(2.6) EXERCISE: Given a measurable space $(\Omega, \mathcal{F})$, prove that $X : \Omega \mapsto \mathbb{R}$ is Borel-measurable if and only if $X^{-1}(]-\infty, a]) \in \mathcal{F}$ for every $a \in \mathbb{R}$. The same claim can be made with $X^{-1}(]-\infty, a]) \in \mathcal{F}$ replaced with $X^{-1}(]-\infty, a[) \in \mathcal{F}$, or, indeed, replaced with $X^{-1}(I) \in \mathcal{F}$ for all intervals $I \subset \mathbb{R}$ of a particular type (open, closed, or only half-open). ○

(2.7) EXERCISE: Let $(\Omega, \mathcal{F})$, $(\mathbb{X}, \mathcal{S})$, and $(\mathbb{Y}, \mathcal{T})$ be any three measurable spaces, and suppose that the mappings $X : \Omega \mapsto \mathbb{X}$ and $Y : \mathbb{X} \mapsto \mathbb{Y}$ are, respectively, $\mathcal{S}/\mathcal{F}$-measurable and $\mathcal{T}/\mathcal{S}$-measurable. Prove that the mapping $Z \stackrel{\text{def}}{=} Y \circ X$ (the composition of Y and X), given by

$$\Omega \ni \omega \rightsquigarrow Z(\omega) = (Y \circ X)(\omega) \equiv Y(X(\omega)) \in \mathbb{Y},$$

is $\mathcal{T}/\mathcal{F}$-measurable. Plainly, the composition of measurable functions is a measurable function. ○

(2.8) EXERCISE: Let $(\Omega, \mathcal{F})$, $(\mathbb{X}, \mathcal{S})$, and $(\mathbb{Y}, \mathcal{T})$ be any three measurable spaces and let $f : \Omega\times\mathbb{X} \mapsto \mathbb{Y}$ be any $\mathcal{T}/(\mathcal{F}\otimes\mathcal{S})$-measurable function. Prove that for any fixed $x^* \in \mathbb{X}$, the mapping $\omega \rightsquigarrow f(\omega, x^*)$ is a $\mathcal{T}/\mathcal{F}$-measurable function from Ω into $\mathbb{Y}$.

HINT: Using theorem (2.4), show that $\omega \rightsquigarrow (\omega, x^*)$ is an $(\mathcal{F}\otimes\mathcal{S})/\mathcal{F}$-measurable function from Ω into $\Omega\times\mathbb{X}$. ○

(2.9) σ-FIELDS GENERATED BY FUNCTIONS: If Ω is any set, $(\mathbb{X}, \mathcal{S})$ is any measurable space, and $X : \Omega \mapsto \mathbb{X}$ is any function, then the σ-field $X^{-1}(\mathcal{S})$ from $(2.2^\dagger)$ is the smallest σ-field in Ω for which X can be declared measurable. Any other σ-field $\mathcal{F}$ in Ω must include $X^{-1}(\mathcal{S})$ if the property "X is $\mathcal{S}/\mathcal{F}$-measurable" is to hold. If the measurable space $(\mathbb{X}, \mathcal{S})$ is understood, we may refer to $X^{-1}(\mathcal{S})$ as "the σ-field generated by X" and may write $\sigma(X)$ instead of $X^{-1}(\mathcal{S})$. If $X_i : \Omega \mapsto \mathbb{X}$, $i \in \mathbb{I}$, is any family of functions, then the σ-field $\sigma(X_i : i \in \mathbb{I})$ is understood to be the smallest σ-field in Ω that contains $\sigma(X_i)$ for every $i \in \mathbb{I}$, or, equivalently, understood to be the smallest σ-field in Ω for which all X_i, $i \in \mathbb{I}$, are measurable. We say that $\sigma(X_i : i \in \mathbb{I})$ is the σ-field generated by the family $\{X_i : i \in \mathbb{I}\}$. ○

(2.10) EXAMPLE: Let $(\mathbb{X}, \mathcal{S})$ and $(\mathbb{Y}, \mathcal{T})$ be any two measurable spaces and let π_X and π_Y denote the usual projections (alias: canonical coordinate mappings) of $\mathbb{X}\times\mathbb{Y}$ onto, respectively, $\mathbb{X}$ and $\mathbb{Y}$, that is, $\pi_{\mathbb{X}}(x, y) = x$ and $\pi_{\mathbb{Y}}(x, y) = y$ for any $(x, y) \in \mathbb{X}\times\mathbb{Y}$. It is straightforward to check that $\pi_{\mathbb{X}}$ is $\mathcal{S}/\mathcal{S}\otimes\mathcal{T}$-measurable and $\pi_{\mathbb{Y}}$ is $\mathcal{T}/\mathcal{S}\otimes\mathcal{T}$-measurable. It is just as easy to check that $\mathcal{S}\otimes\mathcal{T} = \sigma(\pi_{\mathbb{X}}, \pi_{\mathbb{Y}})$, which is to say that the product σ-field $\mathcal{S}\otimes\mathcal{T}$ can be defined equivalently as the σ-field generated by the projections $\pi_{\mathbb{X}}$ and $\pi_{\mathbb{Y}}$. This feature extends for products of any finite number of measurable spaces in the obvious way. ○

(2.11) EXERCISE (MEASURABILITY OF THE DIAGONAL): Prove that the diagonal in the product $\mathbb{R}\times\mathbb{R}$ is a Borel set in $\mathbb{R}^2$. Explain why, more generally, given any$^{\circ\!\!-\!\bullet\,(A.11),\,(A.22)}$ separable Hausdorff space $\mathbb{X}$, one can claim that the diagonal in the product $\mathbb{X}\times\mathbb{X}$ is a Borel set (for the product topology). Give an example of a set $\mathbb{X}$ and two σ-fields on $\mathbb{X}$, $\mathscr{S}$ and $\mathscr{T}$, such that the diagonal in $\mathbb{X}\times\mathbb{X}$ is not an element of $\mathscr{S}\otimes\mathscr{T}$.

HINT: The diagonal in $\mathbb{R}\times\mathbb{R}$ is the complement of the union of all rectangles $I_1\times I_2$ where I_1 and I_2 are both finite non-overlapping open intervals with rational endpoints. As for the counterexample, take $\mathbb{X}$ to be any finite set of at least 2 elements, let $\mathscr{S}$ be the trivial σ-field $\mathscr{S} = \{\emptyset, \mathbb{X}\}$, and let $\mathscr{T} = \mathscr{S}$. Argue that $\mathscr{T}\otimes\mathscr{T}$ is the trivial σ-field on $\mathbb{X}\times\mathbb{X}$ and explain why it cannot contain the diagonal. In fact, if $\mathscr{S}$ is chosen to be the full σ-field $2^{\mathbb{X}}$ and $\mathscr{T}$ is chosen to be the trivial σ-field $\mathscr{T} = \{\emptyset, \mathbb{X}\}$, it will still be the case that the diagonal in $\mathbb{X}\times\mathbb{X}$ cannot belong to the product σ-field $\mathscr{S}\otimes\mathscr{T}$. ○

(2.12) PROJECTIONS OF MEASURABLE SETS: Contrary to what one might be tempted to think, the projection into $\mathbb{R}$ of a Borel subset of $\mathbb{R}^2$, that is, of a set chosen from $\mathcal{B}(\mathbb{R})\otimes\mathcal{B}(\mathbb{R})$, need not be a Borel set in $\mathbb{R}$, that is, need not be an element of $\mathcal{B}(\mathbb{R})$, and this matter is somewhat delicate.[29] In the early stages of development of the theory of integration it was believed that Lebesgue (1905) had established that the projection of any $A \in \mathcal{B}(\mathbb{R})\otimes\mathcal{B}(\mathbb{R})$ into $\mathbb{R}$ is an element of $\mathcal{B}(\mathbb{R})$, but eventually Souslin (1917) found an error in Lebesgue's argument, and showed that, as it stands, this claim is false. The work of Souslin motivated the development of what is now known as the theory of analytic sets, which lies outside of our main concerns and we refer the curious reader to (Dellacherie and Meyer 1975, III) and (Dudley 2002, ch. 13) for details and complete proofs. Summarized in only a few words, the story is this: The projection into $\mathbb{R}$ of any Borel set in $\mathbb{R}^2$ is an analytic set in $\mathbb{R}$, but it could be an analytic set that is not a Borel set. Fortunately, under certain conditions, all analytic sets for a particular σ-field belong to$^{\circ\!\!-\!\bullet\,(1.48)}$ the completion of that σ-field, and we have the following deep result from measure theory – see (Revuz and Yor 1999, I (4.14)); (Dellacherie and Meyer 1975, III 13,33):

THEOREM: Let $(\Omega, \mathcal{F}, P)$ be any *complete* probability space ($\mathcal{F}$ contains all P-null sets) and let $\mathbb{S}$ be any$^{\circ\!\!-\!\bullet\,(A.42)}$ Polish space. Then the canonical projection from $\Omega\times\mathbb{S}$ into Ω of any set from $\mathcal{F}\otimes\mathcal{B}(\mathbb{S})$ is a set from $\mathcal{F}$.

It may appear that the scope of this theorem does not transcend the domain of measure theory, but we will discover later on that it is actually instrumental in the study of stopping times, which, in turn, are instrumental in continuous-time finance and optimal control. ○

[29] Of course, the projections are $\mathcal{B}(\mathbb{R})/(\mathcal{B}(\mathbb{R})\otimes\mathcal{B}(\mathbb{R}))$-measurable mappings (the pre-image under any of the projection mappings of a Borel set in $\mathbb{R}$ is always a Borel set in $\mathbb{R}^2$), as we saw in (2.10).

Intuitively, the σ-field $\sigma(X)$ can be thought of as "information about the function $X\colon \Omega \mapsto \mathbb{X}$." If this intuition is correct, then it should be possible to describe every $\sigma(X)$-measurable function $Y\colon \Omega \mapsto \mathbb{R}$ only in terms of X. The mathematical metaphor for "describe" in this case boils down to being able to point to an $\mathcal{S}$-measurable function $f\colon \mathbb{X} \mapsto \mathbb{R}$ such that $Y = f \circ X$, that is, such that $Y(\omega) = f(X(\omega))$ for every $\omega \in \Omega$. As the next result demonstrates, this intuition is entirely correct.

(2.13) THEOREM: (Dudley 2002, 4.2.8) Let Ω be any nonempty set, let $(\mathbb{X}, \mathcal{S})$ be any measurable space, and let $X\colon \Omega \mapsto \mathbb{X}$ be any function. Then the function $Y\colon \Omega \mapsto \mathbb{R}$ is $\sigma(X)$-measurable, that is, $X^{-1}(\mathcal{S})$-measurable, if and only if there is an $\mathcal{S}$-measurable function $f\colon \mathbb{X} \mapsto \mathbb{R}$ such that $Y = f \circ X$. ○

In order to be useful, the attribute "measurable" must be stable under the common operations with functions, and we now turn to some general results that provide such features.

(2.14) BASIC OPERATIONS: If X and Y are any two measurable functions on $(\Omega, \mathcal{F})$ with values in some Euclidean space (or separable Banach space), then every linear combination $aX + bY$, $a, b \in \mathbb{R}$, is also a measurable function (with values in the same Euclidean or Banach space). This property holds also for generic ($\bar{\mathbb{R}}$-valued) random variables as long as the linear combination does not involve the indeterminate operation $\infty - \infty$ (e.g., if $a, b \in \mathbb{R}_+$ and the sets $\{X = -Y = \infty\}$ and $\{Y = -X = \infty\}$ are both empty). In addition, if X and Y are any two (not necessarily finite) random variables, then $X \wedge Y \stackrel{\text{def}}{=} \min\{X, Y\}$ and $X \vee Y \stackrel{\text{def}}{=} \max\{X, Y\}$ are also random variables (not necessarily finite). All these properties readily follow from the definitions in (2.3). ○

(2.15) SIMPLE FUNCTIONS: Let $(\Omega, \mathcal{F})$ and $(\mathbb{X}, \mathcal{S})$ be any two measurable spaces and let X be any mapping from Ω into $\mathbb{X}$. We say that X is *simple* if its effective range is finite, that is, there is a *finite* nonempty set of distinct elements of $\mathbb{X}$, $\mathcal{X} = \{a_0, a_1, \dots, a_n\} \subseteq \mathbb{X}$, such that $X(\omega) \in \mathcal{X}$ for every $\omega \in \Omega$ – or, equivalently, $X^{-1}(\mathcal{X}) = \Omega$. A simple mapping of this form is measurable if and only if $X^{-1}(\{x\}) \in \mathcal{F}$ for every $x \in \mathbb{X}$. If the range space $\mathbb{X}$ is the extended real line $\bar{\mathbb{R}}$, then X is simple and measurable if and only if it can be expressed in the form

$$\Omega \ni \omega \rightsquigarrow X(\omega) = \sum\nolimits_{i \in \mathbb{N}_{|n}} a_i 1_{A_i}(\omega)$$

for some choice of the scalars $a_i \in \bar{\mathbb{R}}$ and the *disjoint* sets $A_i \in \mathcal{F}$, $i \in \mathbb{N}_{|n}$, chosen so that $\cup_{i \in \mathbb{N}_{|n}} A_i = \Omega$. ○

Since limits of random variables are instrumental for both the practice and theory of probability, it is important to establish that the attribute "measurable" survives the operation lim.

(2.16) LIMITS OF MEASURABLE FUNCTIONS: (Dudley 2002, 4.2.2) Let $(\Omega, \mathcal{F})$ be any measurable space and let $(X_i)_{i \in \mathbb{N}}$ be some sequence of $\mathcal{F}$-measurable functions on Ω with values in some$^{\circ\!\!-\!\bullet\,(A.42)}$ Polish space $\mathbb{S}$, such that for any fixed $\omega \in \Omega$ the sequence $(X_i(\omega))_{i \in \mathbb{N}}$ converges in the metric of $\mathbb{S}$. Then the function

$$\Omega \ni \omega \rightsquigarrow X(\omega) \stackrel{\text{def}}{=} \lim_i X_i(\omega) \in \mathbb{S}$$

is $\mathcal{F}$-measurable. If the range space is $\bar{\mathbb{R}}$ instead of $\mathbb{S}$, in conjunction with (2.14), this result specializes to the following: given any sequence, $(X_i)_{i \in \mathbb{N}}$, of $\mathcal{F}$-measurable $\bar{\mathbb{R}}$-valued functions on Ω, the functions

$$\Omega \ni \omega \rightsquigarrow \limsup_i X_i(\omega) \equiv \lim_i \Big(\sup_{j \geq i} X_j(\omega)\Big) \in \bar{\mathbb{R}}$$

$$\text{and} \quad \Omega \ni \omega \rightsquigarrow \liminf_i X_i(\omega) \equiv \lim_i \Big(\inf_{j \geq i} X_j(\omega)\Big) \in \bar{\mathbb{R}}$$

are both $\mathcal{F}$-measurable. ○

As many modeling tools and ideas are easier to understand and develop for simple functions first, it becomes imperative to establish that certain measurable functions – random variables, say – can be expressed as limits of simple functions. It would not be very wrong to think of "measurable" as tantamount to "limit of simple functions," but such an interpretation has certain limitations and must be made precise. The result that we state next is a somewhat simplified version of (Dudley 2002, 4.2.7).

(2.17) APPROXIMATION WITH SIMPLE FUNCTIONS: Given any measurable space $(\Omega, \mathcal{F})$, any Polish space $\mathbb{S}$, and any $\mathcal{F}$-measurable function $X : \Omega \mapsto \mathbb{S}$, one can find a sequence of simple $\mathcal{F}$-measurable functions $X_i : \Omega \mapsto \mathbb{S}$, $i \in \mathbb{N}$, such that the sequence $(X_i(\omega))_{i \in \mathbb{N}}$ converges to $X(\omega)$ at every $\omega \in \Omega$. ○

Another intuitive and crucially important feature is the Borel-measurability of continuous functions, which is automatic, as we now detail.

(2.18) EXERCISE: Prove that every continuous function $f : \mathbb{R} \mapsto \mathbb{R}$ is Borel-measurable. Prove that if X is a random variable on $(\Omega, \mathcal{F}, P)$, then e^X and $|X|$ are also random variables. If X is a strictly positive random variable, then $1/X$ and $\log(X)$ are random variables as well.

HINT: What would be the natural approximation of a given continuous function $f : \mathbb{R} \mapsto \mathbb{R}$ with finite linear combinations of indicators of the form $1_{]a,b]}$? ○

(2.19) EXERCISE: Let $(\mathbb{X}, \mathcal{T})$ and $(\mathbb{Y}, \mathcal{U})$ be any two topological spaces. Prove that any function from $\mathbb{X}$ to $\mathbb{Y}$ that is continuous for the topologies $\mathcal{U}$ and $\mathcal{T}$ must be measurable for the Borel σ-fields that these two topologies generate.

HINT: Instead of relying on the explicit structure of the range space, as the hint to the previous exercise suggests, consider using theorem (2.4). ○

2.2 Distribution Laws

(2.20) DISTRIBUTION OF RANDOM VARIABLES: Let $(\Omega, \mathcal{F}, P)$ be any probability space and let X be any $\mathbb{R}$-valued random variable on that space. It is a simple matter to check that the assignment

$$\mathcal{B}(\mathbb{R}) \ni B \rightsquigarrow \mathcal{L}_X(B) \overset{\text{def}}{=} P\big(X^{-1}(B)\big)$$

defines a probability measure on $\mathcal{B}(\mathbb{R})$. This probability measure is called *the distribution law of* X, or simply *the law of* X. The law $\mathcal{L}_X : \mathcal{B}(\mathbb{R}) \mapsto [0,1]$ can be identified as the composition of $X^{-1} : \mathcal{B}(\mathbb{R}) \mapsto \mathcal{F}$ and $P : \mathcal{F} \mapsto [0,1]$, that is, $\mathcal{L}_X = P \circ X^{-1}$.

There is nothing special about $\mathcal{B}(\mathbb{R})$ in this construction: if $(\mathcal{X}, \mathcal{S})$ is any measurable space and if $X : \Omega \mapsto \mathcal{X}$ is any $\mathcal{S}/\mathcal{F}$-measurable function, then $\mathcal{L}_X(B) = P(X^{-1}(B))$, $B \in \mathcal{S}$, is a well-defined probability measure on $\mathcal{S}$. To put it another way, a random variable always induces a probability measure on the space where it takes its values. Figuratively speaking, every random variable "pushes" the probability measure from the probability space on which it is defined onto the space where it takes its values, and the result of that "push" is its distribution law. ○

(2.21) EXERCISE: Consider the coin toss space $(\Omega_\infty, \mathcal{F}_\infty, P_\infty)$ and let $\mathcal{Z}_{10}(\omega)$ denote the position of the random walk after 10 periods, associated with the outcome $\omega \in \Omega_\infty$. Prove that $\mathcal{Z}_{10}$ is a well-defined r.v. on Ω_∞ and give its distribution law. Calculate $P_\infty(\{\mathcal{Z}_{10} \geq 0\})$ and $P_\infty(\{2 \leq \mathcal{Z}_{10} < 5\})$. ○

Now we turn to some common methods for specifying distribution laws of real-valued random variables. Any such law is a probability measure on $\mathcal{B}(\mathbb{R})$, and,⊸ (1.59) as we already know, there is a one-to-one correspondence between probability measures on $\mathcal{B}(\mathbb{R})$ and increasing càdlàg functions $F : \mathbb{R} \mapsto \mathbb{R}$ such that $F(-\infty) = 0$ and $F(+\infty) = 1$.

(2.22) DISTRIBUTION FUNCTIONS: Let X be any random variable and let $\mathcal{L}_X$ denote its distribution law. The distribution function of X (alias: the cumulative distribution function of X) is given by

$$\mathbb{R} \ni x \rightsquigarrow F_X(x) = \mathcal{L}_X(]-\infty, x]) \equiv P(X \in]-\infty, x]) \in [0,1] .$$

Thus, the law of X can be expressed as $\mathcal{L}_X = \mathrm{d}F_X$, and if the random variable X is understood from the context, we may write $\mathcal{L}$ and F instead of $\mathcal{L}_X$ and F_X. ○

(2.23) EXERCISE: Give the distribution function of the random variable $\mathcal{Z}_{10}$ from exercise (2.21). ○

(2.24) EXERCISE: Prove that the distribution function, F_X, associated with any random variable X, is càdlàg, increasing, and such that $\lim_{x \searrow -\infty} F_X(x) = 0$ and $\lim_{x \nearrow +\infty} F_X(x) = 1$. In addition, prove that F_X is continuous (that is, it is left-continuous) if and only if the distribution law $\mathcal{L}_X$ is nonatomic. Is it possible for the distribution law $\mathcal{L}_X$ and the Lebesgue measure Λ to be singular to one another ($\mathcal{L}_X \perp \Lambda$) if the distribution function F_X is continuous? ○

(2.25) DISCRETE DISTRIBUTION LAWS: As we are already aware, a probability measure over $\mathbb{R}$ could be purely atomic. A random variable with purely atomic distribution law is said to be *discrete*. Thus, a discrete random variable is any random variable X that is defined on some probability space $(\Omega, \mathcal{F}, P)$ and has the following property: there is a countable set $\mathcal{X} \subset \mathbb{R}$ such that $P(\{X \in \mathcal{X}\}) = 1$ and $P(\{X = x\}) > 0$ for every $x \in \mathcal{X}$. Because summation over a countable set is always meaningful, for any such random variable we have

$$F_X(x) = \sum_{y \in \mathcal{X} \cap\,]-\infty, x]} P(\{X = y\}), \quad x \in \mathbb{R},$$

and, more generally,

$$\mathcal{L}_X(B) = \sum_{y \in \mathcal{X} \cap B} P(\{X = y\}) \quad \text{for any } B \in \mathcal{B}(\mathbb{R}).$$

The distribution function F_X above is a pure jump càdlàg function with jump of size $P(\{X = x\})$ at any $x \in \mathcal{X}$. ○

(2.26) SOME COMMON DISCRETE DISTRIBUTIONS: (a) uniform distribution with parameter $N \in \mathbb{N}$:

$$P\{X = i\} = \frac{1}{N+1}, \quad i \in \{0, 1, \dots, N\};$$

(b) binomial distribution with parameters (N, p), $N \in \mathbb{N}_{++}$, $p \in\,]0, 1[$:

$$P\{X = i\} = \binom{N}{i} p^i (1-p)^{N-i}, \quad i \in \{0, \dots, N\};$$

(c) geometric distribution with parameter $p \in\,]0, 1[$:

$$P\{X = i\} = p(1-p)^i, \quad i \in \mathbb{N};$$

(d) hyper-geometric distribution with parameters $M, N, n \in \mathbb{N}_{++}$, $n \le M$:

$$P\{X = i\} = \frac{\binom{N}{i}\binom{M}{n-i}}{\binom{M+N}{n}}, \quad i \in \{0, \dots, \min(N, n)\};$$

(e) Poisson distribution with parameter $\lambda \in \mathbb{R}_{++}$:

$$P\{X = i\} = e^{-\lambda}\frac{\lambda^i}{i!}, \quad i \in \mathbb{N}. \quad ○$$

Certain continuous distribution laws on the real line $\mathbb{R}$ can be identified by their probability density functions.

(2.27) PROBABILITY DENSITIES: A probability density for the random variable X is any integrable function $\phi_X : \mathbb{R} \mapsto \mathbb{R}_+$, if one exists, such that

$$F_X(x) = \int_{-\infty}^{x} \phi_X(u)\,\mathrm{d}u \quad \text{for every } x \in \mathbb{R}\,.$$

Clearly, such a connection would be possible only if the distribution of X is nonatomic, but, as we know from exercise (2.24), the fact that X has a nonatomic law does not guarantee the existence of a density. As usual, we will write ϕ instead of ϕ_X if X cannot be mistaken.

Without the power of the Lebesgue integral we are confined to densities ϕ that are piecewise continuous (see the examples below), in which case the integral can be understood as a Riemann integral. Another limitation of the Riemann integral is that it allows one to identify the probability $P(X \in]a, b])$ as the integral $\int_a^b \phi(u)\,\mathrm{d}u$ (if a density exists), but, in general, the probability $P(X \in B)$ cannot be expressed as a Riemann integral for a generic Borel set $B \in \mathcal{B}(\mathbb{R})$, even if the law of X has a smooth density. ○

Even without the tools of Lebesgue's theory of integration, we can still write down the most important probability densities encountered in practice.

(2.28) SOME COMMON PROBABILITY DENSITIES: (a) uniform distribution on $]a, b[$, $-\infty < a < b < \infty$ (notation: $\mathcal{U}(]a, b[)$):

$$\phi(x) = \begin{cases} 0 & \text{for } x \leq a\,, \\ \dfrac{1}{b-a} & \text{for } a < x < b\,, \\ 0 & \text{for } x \geq b\,; \end{cases}$$

(b) exponential distribution with parameter $c \in \mathbb{R}_{++}$:

$$\phi(x) = \begin{cases} 0 & \text{for } x < 0\,, \\ c\,e^{-c\,x} & \text{for } x \geq 0\,; \end{cases}$$

(c) standard normal (Gaussian) distribution[30] (notation: $\mathcal{N}(0, 1)$):

$$\phi(x) = \frac{1}{\sqrt{2\,\pi}}\, e^{-\frac{x^2}{2}}\,, \quad x \in \mathbb{R}\,;$$

(d) normal (Gaussian) distribution with parameters $a \in \mathbb{R}$ and and $\sigma \in \mathbb{R}_{++}$ (notation: $\mathcal{N}(a, \sigma^2)$):

$$\phi(x) = f_{a,\sigma^2}(x) \stackrel{\text{def}}{=} \frac{1}{\sigma\sqrt{2\,\pi}}\, e^{-\frac{(x-a)^2}{2\,\sigma^2}}\,, \quad x \in \mathbb{R}\,;$$

[30] Named after the German mathematician Johann Carl Friedrich Gauss (1777–1855).

(e) Cauchy distribution[31] (the cumulative distribution function for this law is $F(x) = \frac{1}{2} + \frac{1}{\pi}\arctan(x)$):

$$\phi(x) = \frac{1}{\pi(1+x^2)}, \qquad x \in \mathbb{R};$$

(f) gamma distribution with parameters $c, p \in \mathbb{R}_{++}$:

$$\phi(x) = \begin{cases} 0 & \text{for } x \le 0, \\ \frac{c^p}{\Gamma(p)} x^{p-1} e^{-cx} & \text{for } x > 0; \end{cases}$$

(g) inverse gamma distribution with parameters $c, p \in \mathbb{R}_{++}$ (the distribution law of $1/X$ for X distributed with gamma law of parameters (c, p)):

$$\phi(x) = \begin{cases} 0 & \text{for } x \le 0, \\ \frac{c^p}{\Gamma(p)} x^{-p-1} e^{-c/x} & \text{for } x > 0; \end{cases}$$

(h) beta distribution with parameters $\alpha, \beta \in \mathbb{R}_{++}$:

$$\phi(x) = \begin{cases} 0 & \text{for } x < 0, \\ \frac{\Gamma(\alpha+\beta)}{\Gamma(\alpha)\Gamma(\beta)} x^{\alpha-1}(1-x)^{\beta-1} & \text{for } 0 \le x \le 1, \\ 0 & \text{for } x > 1; \end{cases}$$

(i) non-central chi-squared density with parameters $d \in \mathbb{R}_{++}$ (degrees of freedom) and $\lambda \in \mathbb{R}$ (non-centrality parameter): $\phi(x) = 0$ for $x \le 0$ and

$$\begin{aligned} \phi(x) &= e^{-\frac{1}{2}(x+\lambda)} \frac{x^{\frac{d}{2}-1}}{2^{d/2}} \sum_{n=0}^{\infty} \left(\frac{\lambda}{4}\right)^n \frac{x^n}{n!\,\Gamma(n+d/2)} \\ &= e^{-\frac{1}{2}(x+\lambda)} \frac{x^{\frac{d}{4}-\frac{1}{2}}}{2\lambda^{\frac{d}{4}-\frac{1}{2}}} I_{\frac{d}{2}-1}(\sqrt{\lambda x}) \quad \text{for } x \in \mathbb{R}_{++}, \end{aligned}$$

where

$$I_\nu(z) = \left(\frac{z}{2}\right)^\nu \sum_{n=0}^{\infty} \frac{z^{2n}}{2^{2n} n!\,\Gamma(\nu+n+1)}$$

is the modified Bessel function[32] with parameter ν. The class of probability distribution laws of this type is denoted by $\chi^2(d, \lambda)$. If $d \in \mathbb{N}_{++}$, then $\chi^2(d, \lambda)$ is the distribution law of $\sum_{i=1}^d X_i^2$ for $X_i \in \mathcal{N}(\mu_i, 1)$, with $\lambda = \mu_1^2 + \ldots + \mu_d^2$. ○

(2.29) EXERCISE: Consider the probability space $([0,1], \mathcal{B}([0,1]), \Lambda)$ and the random variables $X(\omega) = \omega$ and $Y(\omega) = 1 - \omega$, $\omega \in [0,1]$. Prove that X and Y are

[31] Named after the French mathematician Augustin-Lois Cauchy (1789–1857).

[32] Named after the German astronomer and mathematician Friedrich Wilhelm Bessel (1784–1846).

both uniformly distributed in $[0, 1]$. Prove that $Z(\omega) = a + \omega(b - a)$, $\omega \in [0, 1]$, is uniformly distributed in $[a, b]$ for every choice of $a, b \in \mathbb{R}$, $a < b$. ○

(2.30) EXERCISE: Consider the coin toss space $(\Omega_\infty, \mathcal{F}_\infty, P_\infty)$ and let

$$Y(\omega) = \sum_{t \in \mathbb{N}_{++}} \frac{(1 + \mathcal{X}_t(\omega))/2}{2^t}, \qquad \omega \in \Omega_\infty,$$

where⊸• (2.1†) $\mathcal{X}_t$, $t \in \mathbb{N}$, are the coordinate mappings on Ω_∞. Prove that the random variable Y is uniformly distributed in $[0, 1]$. ○

2.3 Lebesgue Integral[33]

As we noted earlier, one of the disadvantages of the Riemann integral[34] is that even if the law of the r.v. X admits density, the probability of the event $\{X \in B\}$, with a generic choice of $B \in \mathcal{B}(\mathbb{R})$, cannot be obtained, as the intuition suggests, by integrating (in the sense of Riemann) the density over the set B. Such an integral would be meaningful only of B is an interval (or a union of finitely many intervals), in which case we can write

$$P(X \in]a, b]) = \int_{]a,b]} \phi_X(x)\,\mathrm{d}x = \int_{-\infty}^{+\infty} \phi_X(x) 1_{]a,b]}(x)\,\mathrm{d}x\,. \qquad e_1$$

The problem is that $\int_{-\infty}^{+\infty} \phi_X(x) 1_B(x)\,\mathrm{d}x$ can be understood in the sense of Riemann only if the function

$$\mathbb{R} \ni x \rightsquigarrow \phi_X(x) 1_B(x) \in \mathbb{R}$$

is piecewise continuous, and this property can hold only for some rather special choices of the Borel set B. To see why this situation leaves a lot do be desired, suppose that the density ϕ_X is constant on the set B; say, just for simplicity, $\phi_x = 1$ for all $x \in B$. If $B =]a, b]$, the integral in e_1 is just the area of the rectangle

$$A =]a, b] \times [0, 1] \equiv \{(x, y) : a < x \le b,\ 0 \le y \le 1\},$$

and this area is nothing but the Lebesgue measure of A, that is, $\Lambda^2(A)$. However, the rectangle $A = B \times [0, 1]$ belongs to $\mathcal{B}(\mathbb{R}^2)$, and therefore the quantity $\Lambda^2(B \times [0, 1])$ is meaningful for every choice of $B \in \mathcal{B}(\mathbb{R})$. It makes perfect sense, then, to postulate that

$$\int_{-\infty}^{\infty} 1_B(x)\,\mathrm{d}x = \Lambda^2(B \times [0, 1]) = \Lambda(B) \quad \text{for every } B \in \mathcal{B}(\mathbb{R}),$$

regardless of whether the function $x \rightsquigarrow 1_B(x)$ is piecewise continuous or not. Thus, the integral of the indicator of any Borel set has a well understood meaning, which is on the one hand intuitive and obvious, and on the other hand intractable in

[33] Named after the French mathematician Henri Lebesgue (1875–1941).

[34] Named after the German mathematician Georg Friedrich Bernhard Riemann (1826–1866).

the sense of Riemann. The basic intuition and motivation behind Lebesgue's construction of the integral is in the simple observation that we just made. Its extension for general measurable functions obtains by way of approximating such functions with linear combinations of indicators, and, as we are about to realize, such an approximation is much more straightforward – and, in fact, considerably less technical – than the approximation with step-functions, on which Riemann's construction of the integral is based. What follows next is a brief summary of Lebesgue's construction of the integral.

(2.31) LEBESGUE INTEGRAL OF INDICATOR FUNCTIONS: For any $B \in \mathcal{B}(\mathbb{R})$, the Lebesgue integral of the indicator function

$$\mathbb{R} \ni x \rightsquigarrow 1_B(x) = \begin{cases} 0 & \text{if } x \notin B\,, \\ 1 & \text{if } x \in B\,, \end{cases}$$

is given by

$$\int_{\mathbb{R}} 1_B(x)\Lambda(\mathrm{d}x) \overset{\text{def}}{=} \Lambda(B)\,.$$

This integral may be expressed more succinctly as $\int_{\mathbb{R}} 1_B \,\mathrm{d}\Lambda$ and we stress that, depending on the choice of B, it may equal $+\infty$. The same integral may be written also as $\int_{\mathbb{R}} 1_B(x)\mathrm{d}x$ if it is interpreted as a Lebesgue–Stieltjes integral – this nomenclature will be clarified later. ○

(2.32) LEBESGUE INTEGRAL OF SIMPLE FUNCTIONS: Let⊸ (2.15) $f\colon \mathbb{R} \mapsto \mathbb{R}$ be any simple function with $\mathrm{Ran}(f) = \{a_0, a_1, \dots, a_m\}$ for some distinct reals $a_i \in \mathbb{R}$, $1 \leq i \leq m$. Suppose further that f is Borel measurable, which is the same as the requirement that all sets

$$A_i \overset{\text{def}}{=} f^{-1}(\{a_i\}) = \{x \in \mathbb{R}\colon f(x) = a_i\}\,, \quad i \in \mathcal{N}_{|m}\,,$$

belong to $\mathcal{B}(\mathbb{R})$. Any such function f can be expressed in the form

† $$f(x) = \sum_{i \in \mathcal{N}_{|m}} a_i\, 1_{A_i}(x)\,, \quad x \in \mathbb{R}\,.$$

If the function f is positive ($\mathrm{Ran}(f) \subset \mathbb{R}_+$), the integral of f (in the sense of Lebesgue) against the measure Λ is defined as

$$\int_{\mathbb{R}} f(x)\Lambda(\mathrm{d}x) = \sum_{i \in \mathcal{N}_{|m}} a_i \int_{\mathbb{R}} 1_{A_i}(x)\Lambda(\mathrm{d}x) = \sum_{i \in \mathcal{N}_{|m}} a_i\, \Lambda(A_i)\,,$$

with the understanding that $a_i\,\Lambda(A_i) = 0$ if $a_i = 0$ and $\Lambda(A_i) = \infty$ (we want the integral of the function 0 to be 0 over any set). In other words,

‡ $$\int_{\mathbb{R}} f(x)\Lambda(\mathrm{d}x) = \sum_{i \in \mathcal{N}_{|m}} a_i\, \Lambda\big(f^{-1}(\{a_i\})\big)\,.$$

The reason for insisting that f must be positive is clear: we would like the last expression to be meaningful even if some its terms equal ∞, and for this to be the case we cannot allow indeterminate expressions of the form $\infty - \infty$ to show up inside the summation.

It is not difficult to check that any finite linear combination of indicator functions, that is, any function of the form

$$f(x) = \sum_{j \in N_{|n}} c_j \, 1_{B_j}(x), \quad x \in \mathbb{R},$$

for some (any) choice of the integer $n \in \mathbb{N}_{++}$, some (any) choice of the (not necessarily disjoint) Borel sets $B_j \in \mathcal{B}(\mathbb{R})$, and some (any) choice of the (not necessarily distinct) scalars $c_j \in \mathbb{R}$, is a simple function in the sense of (2.15), in that it can be rewritten in the above form but with disjoint sets B_j and distinct scalars c_j, that is, can be cast as a "true" simple function of the form †. Furthermore, it is easy to check that if all c_j's are positive, then the integral of f, as defined in ‡ for "true" simple functions of the form †, can be expressed (without an explicit reference to the special form †) as

$$\int_{\mathbb{R}} f(x)\Lambda(\mathrm{d}x) = \sum_{j \in N_{|n}} c_j \int_{\mathbb{R}} 1_{B_j}(x)\mathrm{d}\Lambda(x) = \sum_{j \in \mathbb{N}_{|n}} c_j \Lambda(B_j).$$

Such a representation is consistent with the intuition: the integral should operate on functions linearly, and its value should depend on the function itself, and not on the way the function is written in terms of indicators. ○

(2.33) COMMENTS: As we can see from (2.32‡), Lebesgue's construction of the integral is based on quantization of the range of the function, followed by measuring the associated preimages in the domain. In contrast, Riemann's construction is based on quantization of the domain, and this requires the domain to be some interval inside $\mathbb{R}$, or perhaps some rectangle in $\mathbb{R}^n$. The advantages of Lebesgue's approach are many. In the first place, all that this construction requires for the domain of the function is to have the structure of a measure space, so that the sets $f^{-1}(\{a_i\})$ could be measured – one does not need to know anything about these sets, save for their measure. In fact, nothing will change in (2.32) if we replace the domain of the function f, namely $(\mathbb{R}, \mathcal{B}(\mathbb{R}), \Lambda)$, with any other measure space $(\mathcal{X}, \mathcal{S}, \mu)$, or, in particular, with any probability space $(\Omega, \mathcal{F}, P)$. Furthermore, the approximation of measurable functions with simple ones is completely elementary, and, as we are about to realize, extending the definition in (2.32‡) to all positive measurable functions, defined on any measure space, is entirely straightforward. As a result, compared to the Riemann integral, operating with Lebesgue integrals is considerably easier and less technical. ○

(2.34) INTEGRATION OF POSITIVE BOREL FUNCTIONS: Armed with the intuition developed from the integration of simple functions, now we are ready to construct the integral of any Borel-measurable function $f : \mathbb{R} \mapsto \bar{\mathbb{R}}_+$ – yes, any positive (and not necessarily finite) Borel function has a well-defined integral in the sense of Lebesgue! Given any integer $n \in \mathbb{N}_{++}$, we can partition the range $\bar{\mathbb{R}}_+$ into the intervals, $(n\,2^n) + 1$ in number,

$$\dagger \quad I_k^{(n)} = [k\,2^{-n}, (k+1)\,2^{-n}[\,, \quad 0 \le k \le (n\,2^n) - 1 \quad \text{and} \quad I_{n\,2^n}^{(n)} = [n, \infty]\,,$$

and define the simple functions

$$f_n(x) = \sum\nolimits_{k=0}^{n\,2^n} (k\,2^{-n})\, 1_{f^{-1}(I_k^{(n)})}(x)\,, \qquad n \in \mathbb{N}\,,$$

with (Lebesgue) integrals

$$J_n \stackrel{\text{def}}{=} \int_{\mathbb{R}} f_n(x) \Lambda(\mathrm{d}x) = \sum\nolimits_{k=1}^{n\,2^n} (k\,2^{-n}) \Lambda(f^{-1}(I_k^{(n)}))\,.$$

The following properties are now straightforward:

(a) $f_n(x) \le f(x)$ for any $x \in \mathbb{R}$ for any $n \in \mathbb{N}_{++}$;
(b) $0 \le f(x) - f_n(x) \le 2^{-n}$ for all $x \in f^{-1}([0, n[)$ for any $n \in \mathbb{N}_{++}$;
(c) if $x \in \mathbb{R}$ and $f(x) < \infty$, then $0 \le f(x) - f_n(x) \le 2^{-n}$ for all sufficiently large n, so that $\lim_n f_n(x) = f(x)$;
(d) if $x \in \mathbb{R}$ and $f(x) = \infty$, then $f_n(x) = n \to \infty = f(x)$, that is, $\lim_n f_n(x) = f(x)$ again;
(e) the sequence $(J_n \in \mathbb{R}_+)_{n \in \mathbb{N}_{++}}$ is increasing.

As long as we are willing to accept ∞ as a legitimate limit, then any increasing sequence of real numbers would have a limit. In particular, $\lim_n J_n$ always exists, and since the sequence of simple functions $(f_n)_{n \in \mathbb{N}}$ approximates f (in the sense of (a), (b), (c) and (d) above), we can define the Lebesgue integral of f as

$$\int_{\mathbb{R}} f(x) \Lambda(\mathrm{d}x) \stackrel{\text{def}}{=} \lim_{n \to \infty} \int_{\mathbb{R}} f_n(x) \Lambda(\mathrm{d}x)\,.$$

Of course, the limit on the right-side – and therefore the integral on the left side – may take the value ∞. We again emphasize that this construction can be applied to any Borel-measurable function $f : \mathbb{R} \mapsto \bar{\mathbb{R}}_+$. It is not difficult to see that if the function $f : \mathbb{R} \mapsto \bar{\mathbb{R}}_+$ happens to be simple, then the construction that we just outlined would yield the same result as in (2.32‡). ○

(2.35) EXERCISE: Let $f : \mathbb{R} \mapsto \bar{\mathbb{R}}_+$ be any Borel-measurable function. Prove that:

(a) if $\Lambda(\{x \in \mathbb{R} : f(x) > 0\}) = 0$, then $\int_{\mathbb{R}} f(x) \Lambda(\mathrm{d}x) = 0$;
(b) if $\Lambda(\{x \in \mathbb{R} : f(x) = +\infty\}) > 0$, then $\int_{\mathbb{R}} f(x) \Lambda(\mathrm{d}x) = \infty$;
(c) if $\Lambda(\{x \in \mathbb{R} : f(x) > 0\}) > 0$, then $\int_{\mathbb{R}} f(x) \Lambda(\mathrm{d}x) > 0$. ○

(2.36) INTEGRATION OF BOREL FUNCTIONS: This step is completely straightforward. Any Borel function $f : \mathbb{R} \mapsto \bar{\mathbb{R}} \equiv [-\infty, +\infty]$ can be written as the difference of two positive functions:

$$f(x) = f^+(x) - f^-(x), \quad x \in \mathbb{R},$$

where

$$f^+(x) = \max(f(x), 0) \quad \text{and} \quad f^-(x) = \max(-f(x), 0) \quad \text{for all } x \in \mathbb{R}\,.$$

Just by definition we have $f^+(x) \geq 0$, $f^-(x) \geq 0$ and, clearly, $f^+(x)$ and $f^-(x)$ cannot be simultaneously strictly positive. Thus, since $f(x) = f^+(x) - f^-(x)$, most basic intuition suggests that we should be able to write

$$\int_{\mathbb{R}} f(x)\Lambda(dx) = \int_{\mathbb{R}} f^+(x)\Lambda(dx) - \int_{\mathbb{R}} f^-(x)\Lambda(dx).$$

The only problem that we may encounter is that there is nothing to prevent both integrals on the right to be infinite, in which case the difference becomes indeterminate. If this happens to be the case, then we say that the Lebesgue integral of the function f is indeterminate, or that it does not exist, or that it is not defined – otherwise, we say that the integral exists, or that it is defined. To put it another way, $\int_{\mathbb{R}} f(x)\Lambda(dx)$ is defined (or exists), if and only if either $\int_{\mathbb{R}} f^+(x)\Lambda(dx) < \infty$ or $\int_{\mathbb{R}} f^-(x)\Lambda(dx) < \infty$. If the integrals $\int_{\mathbb{R}} f^+(x)\Lambda(dx)$ and $\int_{\mathbb{R}} f^-(x)\Lambda(dx)$ are both finite, or, equivalently, if $\int_{\mathbb{R}} |f(x)|\Lambda(dx) < \infty$, then the function f is said to be *integrable*, and we note that "integrable" is a stronger requirement than "the integral is defined". ○

(2.37) INTEGRATION ON MEASURE SPACES: If we replace in the foregoing the triplet $(\mathbb{R}, \mathcal{B}(\mathbb{R}), \Lambda)$ with some other (arbitrary) measure space $(\mathcal{X}, \mathcal{S}, \mu)$, then the introduction of the Lebesgue integral on $\mathcal{X}$ with respect to μ would come down to repeating the construction of the Lebesgue integral on $\mathbb{R}$ with respect to Λ, after merely renaming some of the symbols involved. Indeed, any $\mathcal{S}$-measurable function $f : \mathcal{X} \mapsto \bar{\mathbb{R}}_+$ can be approximated with simple functions in the same exact fashion, and we can again define (the intervals $I_k^{(n)}$ are as in (2.34†))

$$\int_{\mathcal{X}} f(x)\mu(dx) \stackrel{\text{def}}{=} \lim_n \sum_{k=1}^{n2^n} (k\,2^{-n})\,\mu(f^{-1}(I_k^{(n)})).$$

The integral of any $\mathcal{S}$-measurable function $f : \mathcal{X} \mapsto [-\infty, +\infty]$ is once again understood as

$$\int_{\mathcal{X}} f(x)\mu(dx) = \int_{\mathcal{X}} f^+(x)\mu(dx) - \int_{\mathcal{X}} f^-(x)\mu(dx),$$

as long as the right side is not indeterminate, in which case we say that "the integral exists." Just as before, we say that "f is integrable," or that "f is μ-integrable," if $\int_{\mathcal{X}} |f(x)|\mu(dx) < \infty$.

If $A \in \mathcal{S}$ is any set, then $\int_A f(x)\mu(dx)$ can be understood as an integral on the measure space $(A, A \cap \mathcal{S}, \mu)$. It is straightforward to check that this integral is defined if and only if the integral of $1_A f$ is defined (on $\mathcal{X}$), and if this condition holds, we have

$$\int_A f(x)\mu(dx) = \int_{\mathcal{X}} 1_A(x) f(x)\mu(dx). \quad ○$$

(2.38) NOTATION FOR THE INTEGRAL: Throughout most of the literature one can find two slight variations of the notation for the Lebesgue integral. The common notation is $\int_{\mathbb{X}} f(x)\mathrm{d}\mu(x)$, but in the literature on Poisson random measures and Lévy processes the same object is usually expressed as $\int_{\mathbb{X}} f(x)\mu(\mathrm{d}x)$. This discrepancy may have to do with the need to integrate on product spaces, as in $\int_{\mathbb{X}\times\mathbb{Y}} f(x,y)\mu(\mathrm{d}x,\mathrm{d}y)$, or in $\int_{\mathbb{Y}} f(y)\mu(A,\mathrm{d}y)$ for some fixed measurable set A. In any case, since Poisson random measures and Lévy processes will eventually enter the present exposition, we prefer to use $\mu(\mathrm{d}x)$ throughout. This will also allow us to distinguish between the Lebesgue integral and the Lebesgue–Stieltjes integrals (see below), although making this distinction is neither particularly useful nor is it our objective. In general, it is preferable to refrain from including the argument of the function in the notation for the integral. For this reason, wherever possible, we will write $\int_{\mathbb{X}} f\,\mathrm{d}\mu$ instead of $\int_{\mathbb{X}} f(x)\mathrm{d}\mu(x)$ or $\int_{\mathbb{X}} f(x)\mu(\mathrm{d}x)$. Unfortunately, the traditional notation for some standard functions (e.g., e^x, x^2, etc.) involves the argument explicitly.

Just for the record, some authors write simply $\mu(f)$ instead of $\int_{\mathbb{X}} f(x)\mu(\mathrm{d}x)$. Such a convention is both intuitive and elegant, but unfortunately it is not widely adopted. One notable exception is the unit point-mass measure, $\mu = \epsilon_a$, $a \in \mathbb{X}$, which we identify as the evaluation mapping $f \rightsquigarrow \epsilon_a(f) = f(a)$ (formally, $\epsilon_a(f) = \int_{\mathbb{X}} f(x)\,\epsilon_a(\mathrm{d}x) = f(a)$). If $\mathbb{X} = \mathbb{R}$, the unit point-mass measure ϵ_a can be connected with Dirac's delta⊶(0.12) through the relation $\epsilon_a(\mathrm{d}x) = \delta_a(x)\mathrm{d}x$.[35] ○

(2.39) LEBESGUE–STIELTJES INTEGRALS: We have seen earlier⊶(1.59) that every increasing càdlàg function $F\colon \mathbb{R} \mapsto \mathbb{R}$ gives rise to a Borel measure on the real line $\mathbb{R}$. Since the Lebesgue integral against any Borel measure on $\mathbb{R}$ is meaningful, the integral against the measure associated with any increasing càdlàg function F would be meaningful, too – as a Lebesgue integral. This integral is called the *Lebesgue–Stieltjes integral* and is denoted by $\int_{\mathbb{X}} f(x)\mathrm{d}F(x)$. The measure associated with the increasing càdlàg function F is often written as $\mathrm{d}F$, and the Lebesgue–Stieltjes integral with respect to F, that is, the Lebesgue integral with respect to $\mathrm{d}F$, we will often abbreviate as $\int_{\mathbb{X}} f\,\mathrm{d}F$.

Since the Lebesgue measure on the real line $\mathbb{R}$, which we denote by Λ, is precisely the Borel measure associated with the increasing function $\imath(x) = x$, the Lebesgue integral $\int_{\mathbb{R}} f(x)\Lambda(\mathrm{d}x)$ can be understood also as a Lebesgue–Stieltjes integral against the function $\imath$, and can be written equivalently as $\int_{\mathbb{R}} f\,\mathrm{d}\imath$, or as $\int_{\mathbb{R}} f(x)\mathrm{d}x$. This last notation is the one commonly used for the Riemann integral, but there should be no confusion, since it is not too difficult to see that any function that is integrable in the sense of Riemann is also integrable in the sense of Lebesgue, and the two types of integrals coincide – as long as both are meaningful. ○

[35] Some authors use δ_a to denote both the unit point-mass measure ϵ_a and the associated Dirac delta function. However, just in general, the distinction between probability distributions and their densities is necessary, even though on the real line both objects represent one and the same generalized function. In any case, it is very common to visualize $\delta_a(x)$ as a Gaussian density with vanishing variance, or, formally, $\delta_a(x) = \lim_{\sigma\searrow 0} f_{a,\sigma^2}(x)$.

(2.40) EXERCISE: Prove that the function

$$\mathbb{R} \ni x \rightsquigarrow f(x) \stackrel{\text{def}}{=} \begin{cases} 1 & \text{if } x \in \mathbb{Q}, \\ 0 & \text{if } x \in \mathbb{R}\backslash\mathbb{Q}, \end{cases}$$

is Borel-measurable. Show that $\int_0^1 f(x)\mathrm{d}x$ is well-defined as a Lebesgue integral (or as Lebesgue-Stieltjes integral), but not as a Riemann integral. Calculate the values of the integrals $\int_{[0,1]} f(x)\Lambda(\mathrm{d}x)$ and $\int_{\mathbb{R}} f(x)\Lambda(\mathrm{d}x)$. ○

(2.41) THE SPACES $\mathscr{L}^0$ AND $\mathscr{L}^1$: Let $(\mathbb{X}, \mathcal{S}, \mu)$ be any measure space. The space of all $\mathcal{S}$-measurable functions from $\mathbb{X}$ into $\bar{\mathbb{R}} \equiv [-\infty, +\infty]$ we denote by $\mathscr{L}^0(\mathbb{X}, \mathcal{S}, \mu)$. The space of all $\mathcal{S}$-measurable functions from $\mathbb{X}$ into $\bar{\mathbb{R}} \equiv [-\infty, +\infty]$ that are⊶(2.36) μ-integrable we denote by $\mathscr{L}^1(\mathbb{X}, \mathcal{S}, \mu)$. These spaces can be expressed more succinctly as $\mathscr{L}^0(\mu)$ and $\mathscr{L}^1(\mu)$, or simply as $\mathscr{L}^0$ and $\mathscr{L}^1$, if the space, the σ-field, and the measure are all understood.

The definitions of the spaces $\mathscr{L}^0$ and $\mathscr{L}^1$ can be extend in the obvious way for functions $f : \mathbb{X} \mapsto \bar{\mathbb{R}}^n$. To be precise, if f_i, $1 \le i \le n$, denote the coordinates of f (treated as functions into $\bar{\mathbb{R}}$), the relations $f \in \mathscr{L}^0$ and $f \in \mathscr{L}^1$ are understood, respectively, as $f_i \in \mathscr{L}^0$ and $f_i \in \mathscr{L}^1$ for all $1 \le i \le n$. If the range space has to be emphasized we will write $\mathscr{L}^0(\mathbb{X}, \mathcal{S}, \mu; \bar{\mathbb{R}}^n)$ and $\mathscr{L}^1(\mathbb{X}, \mathcal{S}, \mu; \bar{\mathbb{R}}^n)$, with the understanding that if it is not explicitly specified, the range space is understood to be $\bar{\mathbb{R}}$. Given any $f \in \mathscr{L}^1(\mathbb{X}, \mathcal{S}, \mu; \bar{\mathbb{R}}^n)$, the integral $\int_{\mathbb{X}} f \,\mathrm{d}\mu$ will be understood as the vector in $\mathbb{R}^n$ with components $\int_{\mathbb{X}} f_i \,\mathrm{d}\mu$, $1 \le i \le n$.

More generally, given any metric space $\mathbb{Y}$, we denote by $\mathscr{L}^0(\mathbb{X}, \mathcal{S}, \mu; \mathbb{Y})$ the space of all $\mathcal{B}(\mathbb{Y})/\mathcal{S}$-measurable functions $f : \mathbb{X} \mapsto \mathbb{Y}$, and, given a Banach space $\mathbb{B}$ with norm $\|\cdot\|$, we denote by $\mathscr{L}^1(\mathbb{X}, \mathcal{S}, \mu; \mathbb{B})$ the space of all $f \in \mathscr{L}^0(\mathbb{X}, \mathcal{S}, \mu; \mathbb{B})$ such that

$$\int_{\mathbb{X}} \|f(x)\| \,\mu(\mathrm{d}x) < \infty . \quad ○$$

(2.42) EXERCISE: Prove that if $f \in \mathscr{L}^1(\mathbb{X}, \mathcal{S}, \mu)$, then

$$\Big|\int_{\mathbb{X}} f(x)\mu(\mathrm{d}x)\Big| \le \int_{\mathbb{X}} |f(x)|\mu(\mathrm{d}x) . \quad ○$$

(2.43) PROPERTIES OF THE LEBESGUE INTEGRAL:[36] Let $(\mathbb{X}, \mathcal{S}, \mu)$ be any (arbitrary) measure space.

1. If f is any $\mathcal{S}$-measurable function from $\mathbb{X}$ into $\bar{\mathbb{R}} = [-\infty, +\infty]$, then:
   (a) for any given $a \in \mathbb{R}\backslash\{0\}$, $\int_{\mathbb{X}} f \,\mathrm{d}\mu$ is defined if and only if $\int_{\mathbb{X}} (a\, f)\mathrm{d}\mu$ is defined, and if either integral is defined, then

$$\int_{\mathbb{X}} (a\, f)\mathrm{d}\mu = a \int_{\mathbb{X}} f \,\mathrm{d}\mu ;$$

[36] Some of the properties listed here involve the results from exercise (2.35), which are not specific to the Borel σ-field and the Lebesgue measure.

(b) $f \geq 0$ μ-a.e. implies that $\int_{\mathbb{X}} f \, d\mu$ is defined (that is, it is enough for the function to be positive μ-a.e., as opposed to everywhere);
(c) $f = 0$ μ-a.e. implies $\int_{\mathbb{X}} f \, d\mu = 0$;
(d) $f \geq 0$ μ-a.e. and $\mu(f = +\infty) > 0$ implies $\int_{\mathbb{X}} f \, d\mu = +\infty$;
(e) $f \geq 0$ μ-a.e. and $\mu(f > 0) > 0$ implies $\int_{\mathbb{X}} f \, d\mu > 0$ (this is both straightforward and important).

2. If f and g are any two $\mathcal{S}$-measurable functions from $\mathbb{X}$ into $\bar{\mathbb{R}}$, such that either $\int_{\mathbb{X}} f^+ \, d\mu$ and $\int_{\mathbb{X}} g^+ \, d\mu$ are both finite, or $\int_{\mathbb{X}} f^- \, d\mu$ and $\int_{\mathbb{X}} g^- \, d\mu$ are both finite, then $\int_{\mathbb{X}} (f + g) \, d\mu$ is defined[37] and

$$\int_{\mathbb{X}} f \, d\mu + \int_{\mathbb{X}} g \, d\mu = \int_{\mathbb{X}} (f + g) \, d\mu .$$

3. If f and g are any two $\mathcal{S}$-measurable functions from $\mathbb{X}$ into $\bar{\mathbb{R}}$, such that $f \leq g$ μ-a.e. and $\int_{\mathbb{X}} g \, d\mu$ and $\int_{\mathbb{X}} f \, d\mu$ are defined, then $\int_{\mathbb{X}} f \, d\mu \leq \int_{\mathbb{X}} g \, d\mu$.
4. $\mathscr{L}^0(\mathbb{X}, \mathcal{S}, \mu)$ and $\mathscr{L}^1(\mathbb{X}, \mathcal{S}, \mu)$ are both linear spaces: if f and g are members of the either space, then so is also $a\,f + b\,g$ for any $a, b \in \mathbb{R}$.
5. The integration operator $f \rightsquigarrow \int_{\mathbb{X}} f \, d\mu$ acts on $\mathscr{L}^1(\mathbb{X}, \mathcal{S}, \mu)$ linearly: if $f, g \in \mathscr{L}^1(\mathbb{X}, \mathcal{S}, \mu)$ and $a, b \in \mathbb{R}$, then $a\,f + b\,g \in \mathscr{L}^1(\mathbb{X}, \mathcal{S}, \mu)$ and

$$a \int_{\mathbb{X}} f \, d\mu + b \int_{\mathbb{X}} g \, d\mu = \int_{\mathbb{X}} (a\,f + b\,g) \, d\mu . \quad \circ$$

(2.44) EXERCISE: Let $(\mathbb{X}, \mathcal{S}, \mu)$ be any measure space and let f and g be any two $\mathcal{S}$-measurable functions from $\mathbb{X}$ into $\bar{\mathbb{R}}$. Prove that:

(a) if $f = g$ μ-a.e., then $\int_{\mathbb{X}} f \, d\mu$ is defined if and only if $\int_{\mathbb{X}} g \, d\mu$ is defined, and, when defined, the two integrals must be equal;
(b) if $f \leq g$ μ-a.e. and $\int_{\mathbb{X}} f \, d\mu$ and $\int_{\mathbb{X}} g \, d\mu$ are both defined, then $\int_{\mathbb{X}} f \, d\mu \leq \int_{\mathbb{X}} g \, d\mu$;
(c) if $f \leq g$ μ-a.e., $\int_{\mathbb{X}} f \, d\mu$ is defined, and $\int_{\mathbb{X}} f \, d\mu > -\infty$, then $\int_{\mathbb{X}} g \, d\mu$ must be defined, too, and $\int_{\mathbb{X}} f \, d\mu \leq \int_{\mathbb{X}} g \, d\mu$. $\circ$

(2.45) EQUIVALENCE CLASSES: One interesting and important feature of the Lebesgue integral is that its value is invariant under any modification of the integrand on a set of measure 0. One consequence from this feature is that $\int_{\mathbb{X}} f \, d\mu$ is meaningful even if f is defined only outside some μ-null set in $\mathbb{X}$. Thus, the functions that the Lebesgue integral accepts as integrands only need to be defined a.e. (with respect to the integrator). To put it another way, the integration operation $f \rightsquigarrow \int_{\mathbb{X}} f \, d\mu$ does not act on functions f, but rather, it acts on equivalence classes of functions, the equivalence between two functions (that are defined μ-a.e.) being understood as the property that there is an μ-negligible set outside of which the two functions are both defined and coincide. $\circ$

[37] The reader is invited to check that $(f + g)^+ \leq f^+ + g^+$ and $(f + g)^- \leq f^- + g^-$.

(2.46) EXPECTATION OF RANDOM VARIABLES: Let $(\Omega, \mathcal{F}, P)$ be any probability space and let X be any r.v. on Ω (that is, X is a measurable function from Ω into $\bar{\mathbb{R}}$). The *expectation* of X, also called the *mean value of* X and denoted by $\mathsf{E}[X]$, is defined as the Lebesgue integral of the function $X\colon \Omega \mapsto \mathbb{R}$ with respect to to the probability measure P:

$$\mathsf{E}[X] \stackrel{\text{def}}{\equiv} \int_\Omega X(\omega) P(\mathrm{d}\omega) \equiv \int_\Omega X \,\mathrm{d}P .$$

In general, this integral may or may not be defined, which is to say that the expectation may or may not exist. In many situations we will need to work with more than one probability measure defined on the same σ-field – this is particularly important in finance. If we need to remove any possible ambiguity about the probability measure with respect to which the expectation is taken, we may write $\mathsf{E}^P[X]$ instead of $\mathsf{E}[X]$.

If X is a discrete random variable with distribution law amassed on the countable set $\mathcal{X} \subset \mathbb{R}$, just as in (2.25), then

$$\mathsf{E}[X] = \sum\nolimits_{x\in\mathcal{X}} x\, P(\{X = x\}) .$$

The last relation shows that the expectation of X has the meaning of the *average value of* X: it is the weighted sum of all possible values that the random variable can take, with weights equal to the respective probabilities. Of course, the summation symbol would be meaningless if the effective range of X is uncountably infinite, but, nevertheless, as long as it exists, the interpretation of $\mathsf{E}[X]$ as "the average value of X" remains. In some sense, the Lebesgue integral can be seen as a tool that allows one to "average" the values of X even if the range of those values is uncountably infinite. ○

(2.47) EXERCISE: Consider the coin toss space $(\Omega_\infty, \mathscr{F}_\infty, P_\infty)$, and let $\mathcal{Z}_t(\omega)$ denote the position of the random walk in period t, associated with the outcome $\omega \in \Omega_\infty$. Calculate the expected value $\mathsf{E}[\mathcal{Z}_t]$ (that is, $\mathsf{E}^{P_\infty}[\mathcal{Z}_t]$). ○

(2.48) EXERCISE: Calculate $\mathsf{E}[X]$, assuming that X has geometric law with parameter $p \in\,]0,1[$. A symmetric coin is tossed until it falls "heads." How many times is the coin tossed on average? ○

(2.49) EXERCISE: Calculate $\mathsf{E}[X]$, assuming that X has Poisson distribution law with parameter $\lambda > 0$. ○

(2.50) EXERCISE: Give an example of a random variable $X = X(\omega)$, such that $0 < X(\omega) < \infty$ for every $\omega \in \Omega$, and yet $\mathsf{E}[X] = +\infty$.

HINT: Consider the distribution on $\mathbb{N}_{++}$ given by $P(X = i) = C/i^2$, $i \in \mathbb{N}_{++}$, for some appropriate choice of the constant $C > 0$. What can you say about $\mathsf{E}[X]$ if X has Cauchy distribution, or inverse gamma distribution with parameters $c > 0$ and $p = 2$? ○

2.4 Convergence of Integrals

Throughout this section $(\mathcal{X}, \mathcal{S}, \mu)$ will be some fixed measure space, and measurability will be understood relative to $\mathcal{S}$. Our main goal is to understand whether it is possible to exchange the operations lim and $\int$, that is, to understand whether it is possible, under certain (to be determined) conditions, to write

$$\lim_i \int_{\mathcal{X}} f_i \,\mathrm{d}\mu = \int_{\mathcal{X}} (\lim_i f_i) \mathrm{d}\mu\,. \qquad \mathrm{e}_1$$

Developing the tools that allow one to justify such identities is certainly important, especially in the context of stochastic calculus and mathematical finance, as many objects of interest in these fields arise as limits of functions, and many quantities of interest are essentially averages (integrals) of observed values. To gain some insight into the nature of e_1, we first look at some examples.

(2.51) EXAMPLES: Consider the special case where $\mathcal{X} =]0,1]$, $\mathcal{S} = \mathcal{B}(]0,1])$, and $\mu = \Lambda$ (the Lebesgue measure). Define $f_i(x) = i\, 1_{]0,1/i]}(x)$, $x \in]0,1]$ for any $i \in \mathbb{N}_{++}$. Clearly, $\lim_i f_i(x) = 0$ for any $x \in]0,1]$, and yet $\int_{]0,1]} f_i(x)\mathrm{d}x{=}1$ for any $i \in \mathbb{N}_{++}$. We see that in this case e_1 fails, as it reads $1 = 0$.

Nothing would change if we were to replace the interval $]0,1]$ with $[0,1]$ and set $f_i(x) = i\, 1_{[0,1/i]}(x)$, $x \in [0,1]$. In this case $\lim_i f_i(x)$ is either 0, if $x \in]0,1]$, or is $+\infty$, if $x = 0$. The limiting function is again 0, except on the set $\{0\}$, where it equals $+\infty$. But the Lebesgue integral of such a function is still 0, as the set $\{0\}$ has Lebesgue measure 0. Therefore e_1 would still read $1 = 0$.

There is hope, however. If we were to define $f_i(x) = 1_{[0,1/2+1/i]}(x)$, $x \in [0,1]$, then $\lim_i f_i(x) = 1_{[0,1/2]}(x)$. So, now e_1 becomes $\lim_i(\frac{1}{2} + \frac{1}{i}) = \frac{1}{2}$, which is acceptable. ○

(2.52) CONVERGENCE A.E. AND A.S.: Since an integrand in any Lebesgue integral only needs to be defined a.e. (almost everywhere) in order for the relation e_1 to be meaningful, it would be enough for us to require that the functions f_i converge a.e., that is, there is a set $E \in \mathcal{S}$ with $\mu(E) = 0$, such that the sequence $(f_i(x))_{i\in\mathbb{N}}$ is defined and converges for any $x \in \mathcal{X}\setminus E$. We call such a convergence "convergence a.e." If the measure μ happens to be a probability measure, we use a.s. (almost surely) instead of a.e. ○

(2.53) MONOTONE CONVERGENCE THEOREM: (Dudley 2002, 4.3.2) Let $(f_i)_{i\in\mathbb{N}}$ be any sequence of measurable functions from $\mathcal{X}$ into $\bar{\mathbb{R}}$. Suppose that this sequence is monotone a.e., in that $f_i \leq f_{i+1}$ a.e. for every $i \in \mathbb{N}$, and let $f = \lim_i f_i$ (defined a.e.). If $\int_{\mathcal{X}} f_0 \,\mathrm{d}\mu$ is defined and $\int_{\mathcal{X}} f_0 \,\mathrm{d}\mu > -\infty$,[38] then $\lim_i \int_{\mathcal{X}} f_i \,\mathrm{d}\mu = \int_{\mathcal{X}} f \,\mathrm{d}\mu$. ○

[38] This condition implies that the integrals $\int_{\mathcal{X}} f_i \,\mathrm{d}\mu$ and $\int_{\mathcal{X}} f \,\mathrm{d}\mu$ are all defined and $\int_{\mathcal{X}} f_i \,\mathrm{d}\mu \leq \int_{\mathcal{X}} f \,\mathrm{d}\mu$ – see exercise (2.44).

(2.54) REMARK: The most common application of the monotone convergence theorem is with $f_0 = 0$, in which case all functions in the sequence are positive (the sequence must still be monotone) and the condition "$\int_{\mathbb{X}} f_0 \, d\mu$ is defined and $\int_{\mathbb{X}} f_0 \, d\mu > -\infty$" is satisfied automatically. ○

(2.55) COUNTEREXAMPLE: Take $\mathbb{X} = \mathbb{R}_+$, $\mu = \Lambda$, and let $f_i = -1_{[i,\infty[}$. Obviously, with this choice $\int_{\mathbb{R}_+} f_i \, d\Lambda = -\infty$ for any $i \in \mathbb{N}$, and yet $f_i \to 0$ everywhere (monotonically). The monotone convergence theorem fails because the condition $\int_{\mathbb{R}_+} f_0 \, d\Lambda > -\infty$ fails. ○

(2.56) FATOU'S LEMMA:[39] (Dudley 2002, 4.3.3) This result is a corollary from the monotone convergence theorem. Let $(f_i)_{i \in \mathbb{N}}$ be any sequence of measurable functions from $\mathbb{X}$ into $\bar{\mathbb{R}}_+$ (note that all functions are positive). Then

$$\int_{\mathbb{X}} (\liminf_i f_i) \, d\mu \leq \liminf_i \int_{\mathbb{X}} f_i \, d\mu . \quad ○$$

(2.57) AN APPLICATION: Let $(f_i)_{i \in \mathbb{N}}$ be any sequence of measurable functions from $\mathbb{X}$ into $\bar{\mathbb{R}}_+$, and suppose that $f_i(x) \to f(x)$ for μ-a.e. $x \in \mathbb{X}$. Now we have $\liminf_i f_i = f$ and Fatou's lemma gives

$$\int_{\mathbb{X}} f \, d\mu \leq \sup_{i \in \mathbb{N}} \inf_{j \geq i} \int_{\mathbb{X}} f_j \, d\mu \leq \sup_{i \in \mathbb{N}} \int_{\mathbb{X}} f_i \, d\mu .$$

Plainly, the integral of the limit of the functions (if a limit exists) can be no larger than the largest of the integrals in the sequence. The first example in (2.51) illustrates the last relation and shows that the inequalities can be strict. In that example we have $f = 0$ and $\int_{\mathbb{X}} f_i \, d\mu = 1$ for any i. ○

While the monotone convergence theorem is a powerful and useful tool, it may be inadequate for many practical applications, in which the monotonicity condition is simply not available. One widely used condition that guarantees that the interchange between the operations lim and $\int$ can take place is provided by the following powerful result:

(2.58) THE DOMINATED CONVERGENCE THEOREM: (Dudley 2002, 4.3.5) Let $(f_i)_{i \in \mathbb{N}}$ and g be functions from the space $\mathcal{L}^1(\mathbb{X}, \mathcal{S}, \mu)$. Suppose that the following property is satisfied for any $i \in \mathbb{N}$: $|f_i(x)| \leq g(x)$ for μ-a.e. $x \in \mathbb{X}$. If $\lim_i f_i(x)$ exists for μ-a.e. $x \in \mathbb{X}$, and if f_* is any function on $\mathbb{X}$, chosen so that $f_* = \lim_i f_i$ μ-a.e., then it must be that $f_* \in \mathcal{L}^1(\mathbb{X}, \mathcal{S}, \mu)$ and $\lim_i \int_{\mathbb{X}} f_i \, d\mu = \int_{\mathbb{X}} f_* \, d\mu$. ○

If the measure μ happens to be finite ($\mu(\mathbb{X}) < \infty$), the most common (though less interesting) use of the dominated convergence theorem is with a function g chosen to be some constant $c \in \mathbb{R}_{++}$. In (2.51) we have $1_{[0,1/2+1/i]} \leq 1$ (the constant 1 is in $\mathcal{L}^1$ if $\mathbb{X} = [0, 1]$ and $\mu = \Lambda$) and $\lim_i 1_{[0,1/2+1/i]} = 1_{[0,1/2]}$, which explains why $\lim_i \int_{[0,1]} 1_{[0,1/2+1/i]} \, d\Lambda = \int_{[0,1]} 1_{[0,1/2]} \, d\Lambda = ½$.

[39] Named after the French mathematician and astronomer Pierre Joseph Louis Fatou (1878–1929).

2.5 Integration Tools

All results presented in this section are standard and can be found in most textbooks on real analysis and integration. For brevity, we are not going to provide separate references, but note that complete proofs and details on the material covered in this section can be found, for example, in (Dudley 2002) and (Stroock 1999). We begin with the classical result that tells us under what condition the order of integration in multiple integrals can be interchanged.

(2.59) TONELLI–FUBINI'S THEOREM:[40] Let $(\mathscr{X}, \mathscr{S}, \mu)$ and $(\mathscr{Y}, \mathscr{T}, \nu)$ be any (arbitrarily chosen) measure spaces with σ-finite measures, and let f be any $\mathscr{S}\otimes\mathscr{T}$-measurable function from $\mathscr{X}\times\mathscr{Y}$ into $\bar{\mathbb{R}}$, which is either positive, or is integrable with respect to the product measure $\mu\times\nu$. Then the integral $\int_{\mathscr{X}} f(x, y)\mu(\mathrm{d}x)$ exists for ν-a.e. $y \in \mathscr{Y}$, the integral $\int_{\mathscr{Y}} f(x, y)\nu(\mathrm{d}y)$ exists for μ-a.e. $x \in \mathscr{X}$, the integral of the function $x \rightsquigarrow \int_{\mathscr{Y}} f(x, y)\nu(\mathrm{d}y)$ for the measure μ exists, the integral of the function $y \rightsquigarrow \int_{\mathscr{X}} f(x, y)\mu(\mathrm{d}x)$ for the measure ν exists (in particular, these two functions are, respectively μ-a.e. measurable and ν-a.e. measurable), and the following two identities hold:

$$\int_{\mathscr{X}\times\mathscr{Y}} f(x, y)(\mu\times\nu)(\mathrm{d}x, \mathrm{d}y) = \int_{\mathscr{X}}\Big(\int_{\mathscr{Y}} f(x, y)\nu(\mathrm{d}y)\Big)\mu(\mathrm{d}x)$$
$$= \int_{\mathscr{Y}}\Big(\int_{\mathscr{X}} f(x, y)\mu(\mathrm{d}x)\Big)\nu(\mathrm{d}y). \quad \circ$$

(2.60) EXERCISE: Compute the integral $\int_{\mathbb{R}_+^2} e^{-2x-3y}\Lambda^2(\mathrm{d}x, \mathrm{d}y)$ in two ways:

$$\text{as} \quad \int_{\mathbb{R}_+}\Big(\int_{\mathbb{R}_+} e^{-2x-3y}\mathrm{d}y\Big)\mathrm{d}x \quad \text{and as} \quad \int_{\mathbb{R}_+}\Big(\int_{\mathbb{R}_+} e^{-2x-3y}\mathrm{d}x\Big)\mathrm{d}y.$$

Explain why Tonelli–Fubini's theorem can be applied in this case. $\circ$

The following classical result from analysis plays an instrumental role in probability theory.

(2.61) THEOREM (CHANGE OF VARIABLES): Let $(\mathscr{X}, \mathscr{S}, \mu)$ be any measure space, let $(\mathscr{Y}, \mathscr{T})$ be any measurable space, let X be any $\mathscr{T}/\mathscr{S}$-measurable function from $\mathscr{X}$ into $\mathscr{Y}$, and let f be any $\mathscr{T}$-measurable function from $\mathscr{Y}$ into $\bar{\mathbb{R}}$. Then[41]

$$\int_{\mathscr{X}} (f \circ X)\mathrm{d}\mu = \int_{\mathscr{Y}} f\,\mathrm{d}(\mu \circ X^{-1}),$$

in the sense that if the integral on the either side exists, then the integral on the other side exists and the identity holds. $\circ$

[40] Named after the Italian mathematicians Leonida Tonelli (1885–1946) and Guido Fubini (1879–1943).

[41] The first integral can be written as $\int_{\mathscr{X}} f(X(x))\mu(\mathrm{d}x)$ and recall that $\mu \circ X^{-1}$ is the measure on $\mathscr{T}$ given by $(\mu \circ X^{-1})(A) = \mu(X^{-1}(A))$ for all $A \in \mathscr{T}$.

(2.62) COMPUTATION OF EXPECTED VALUES: Suppose that $(\Omega, \mathcal{F}, P)$ is a probability space and let X be any finite real-valued random variable on Ω. Recall that the distribution law of X is the probability measure on $\mathbb{R}$ given by $\mathcal{L}_X = P \circ X^{-1}$. If f is any Borel function from $\mathbb{R}$ into $\mathbb{R}$, then $f \circ X$ is just another random variable on Ω, which can be expressed as $\Omega \ni \omega \rightsquigarrow f \circ X(\omega) \equiv f(X(\omega)) \in \mathbb{R}$. It is very common to denote this r.v. by $f(X)$. By the change of variables theorem, the expectation of $f(X)$ can be computed as

† $$\begin{aligned} \mathsf{E}[f(X)] &= \int_\Omega f(X(\omega))P(\mathrm{d}\omega) \\ &\equiv \int_\Omega f \circ X \,\mathrm{d}P = \int_{\mathbb{R}} f \,\mathrm{d}\mathcal{L}_X \equiv \int_{\mathbb{R}} f(x)\mathcal{L}_X(\mathrm{d}x). \end{aligned}$$ ○

(2.63) ABSOLUTE CONTINUITY AND EQUIVALENCE: Let μ and ν be any two measures on the measurable space $(\mathcal{X}, \mathcal{S})$. We say that ν is absolutely continuous with respect to μ (notation: $\nu \lesssim \mu$) if every μ-null set from $\mathcal{S}$ is also a ν-null set, that is, if $\mu(A) = 0$, for some $A \in \mathcal{S}$, implies $\nu(A) = 0$. If $\nu \lesssim \mu$ and $\mu \lesssim \nu$ — in other words, if μ and ν share the same family of null-sets — then we say that the measures μ and ν are equivalent and express this feature as $\mu \approx \nu$. ○

As it will become clear later in this space, absolute continuity and equivalence are instrumental for many important aspects of asset pricing, and so is also the object that we introduce next.

(2.64) THEOREM (RADON–NIKODYM DERIVATIVES[42]): Let $(\mathcal{X}, \mathcal{S})$ be any measurable space and let μ and ν be any two measures on $(\mathcal{X}, \mathcal{S})$ such that $\nu \lesssim \mu$. Then there is a function $R \in \mathscr{L}^1(\mathcal{X}, \mathcal{S}, \mu)$ such that

† $$\nu(A) = \int_A R \,\mathrm{d}\mu \quad \text{for every } A \in \mathcal{S}.$$

We will often express the last relation more succinctly as $\nu = R \odot \mu$. Any function $R \in \mathscr{L}^1(\mathcal{X}, \mathcal{S}, \mu)$ such that $\nu = R \odot \mu$ is said to be *Radon–Nikodym derivative* (or *density*) of ν with respect to (or relative to) μ, and is expressed as $R = \mathrm{d}\nu/\mathrm{d}\mu$.[43] ○

(2.65) EXERCISE: With the notation as in (2.64), suppose that f is some measurable function from $\mathcal{X}$ into $\mathbb{R}$ such that $\int_{\mathcal{X}} f \,\mathrm{d}\nu$ is defined. Prove that $\int_{\mathcal{X}} f\, R \mathrm{d}\mu$ is also defined and

$$\int_{\mathcal{X}} f \,\mathrm{d}\nu = \int_{\mathcal{X}} f\, R \mathrm{d}\mu .$$

[42] Named after the Austrian mathematician Johann Radon (1887–1956) and the Polish mathematician Otto Marcin Nikodym (1887–1974).

[43] This notation is ambiguous, in that, strictly speaking, $\mathrm{d}\nu/\mathrm{d}\mu$ stands for an equivalence class from the space $L^1(\mathcal{X}, \mathcal{S}, \mu)$, and by $R = \mathrm{d}\nu/\mathrm{d}\mu$ we mean to say that R is one (arbitrarily chosen) representative of that class.

This justifies the (synonymous) relations $\nu = R \odot \mu$, $\mathrm{d}\nu = R\mathrm{d}\mu$, and $R = \mathrm{d}\nu/\mathrm{d}\mu$.

HINT: It is enough to consider the case $f \geq 0$. Establish the result for simple functions first, and then use the monotone convergence theorem. ○

(2.66) EXERCISE: Suppose that $\nu \precsim \mu$ and let $R = \mathrm{d}\nu/\mathrm{d}\mu$ be the Radon–Nikodym derivative introduced in (2.64). Prove that:

(a) $R \geq 0$ μ-a.e.;

(b) R is μ-a.e. unique (if ($2.64^{\dagger}$) holds also with R' in place of R, then $R = R'$ μ-a.e.);

(c) $\nu \approx \mu$ if and only if $\mathrm{d}\nu/\mathrm{d}\mu > 0$ μ-a.e.;

(d) if $\nu \approx \mu$, $R = \mathrm{d}\nu/\mathrm{d}\mu$, and $R' = \mathrm{d}\mu/\mathrm{d}\nu$, then $R' = 1/R$ ν-a.e. (or, which is the same, μ a.e.). ○

(2.67) COMPUTATION OF EXPECTED VALUES CONTINUED: With the notation as in (2.62), we say "the law of X is absolutely continuous," or "X has absolutely continuous distribution," if $\mathcal{L}_X \precsim \Lambda$ (the distribution law of X is absolutely continuous with respect to the Lebesgue measure). By the Radon–Nikodym theorem from (2.64), if this property holds, then there is a function $\phi_X \in \mathscr{L}^1(\Lambda)$, called *density of* X, that can be identified as the Radon–Nikodym derivative $\frac{\mathrm{d}\mathcal{L}_X}{\mathrm{d}\Lambda}$, that is, $\mathcal{L}_X = \phi_X \odot \Lambda$. Using the result from (2.65) and the relation ($2.62^{\dagger}$), we can compute the expectation of the r.v. $f(X)$ as

$$\mathsf{E}[f(X)] = \int_{\mathbb{R}} f\,\mathrm{d}\mathcal{L}_X = \int_{\mathbb{R}} f\,\phi_X\,\mathrm{d}\Lambda \equiv \int_{\mathbb{R}} f(x)\phi_X(x)\mathrm{d}x\,.$$

If the random variable cannot be mistaken, we may write $\mathcal{L}$ and ϕ instead of $\mathcal{L}_X$ and ϕ_X. ○

(2.68) EXERCISE: Calculate $\mathsf{E}[X]$ and $\mathsf{E}[X^2]$ for each of the absolutely continuous distributions listed in (2.28). ○

(2.69) EXERCISE: Calculate $\mathsf{E}[e^{-X}]$ if the law of X is: (a) exponential with parameter $\lambda > 0$; (b) Gaussian with parameters $a \in \mathbb{R}$ and $\sigma \in \mathbb{R}_{++}$. ○

(2.70) STANDARD DEVIATION AND VARIANCE: Given a random variable X, if the expectation $\mathsf{E}[X]$ is defined and finite, we can define the *variance of* X as

$$\mathrm{Var}(X) \stackrel{\text{def}}{=} \mathsf{E}\big[\big(X - \mathsf{E}[X]\big)^2\big] = \mathsf{E}[X^2] - \mathsf{E}[X]^2\,,$$

and define the *standard deviation of* X as $\mathrm{Var}(X)^{1/2} \equiv \sqrt{\mathrm{Var}(X)}$, understood to be the positive root of $\mathrm{Var}(X)$. The variance and the standard deviation measure the degree of dispersion in the range of values that X takes, as we detail below. The main reason for introducing the standard deviation is that it depends linearly on X: if $a \in \mathbb{R}$, then $\mathrm{Var}(aX)^{1/2} = |a|\mathrm{Var}(X)^{1/2}$, whereas $\mathrm{Var}(aX) = a^2\mathrm{Var}(X)$. ○

(2.71) EXERCISE: Calculate the standard deviation and the variance for each of the absolutely continuous distributions listed in (2.28). Explain how the values of the parameters affect the standard deviation and the shape of the density. ○

(2.72) THE SPACES $\mathscr{L}^p$ AND L^p: Given any measure space $(\mathbb{X}, \mathcal{S}, \mu)$ and any $p \in \mathbb{R}_{++}$, the space $\mathscr{L}^p(\mathbb{X}, \mathcal{S}, \mu)$ is defined as the collection of all measurable functions $f : \mathbb{X} \mapsto \bar{\mathbb{R}}$ such that

$$\int_{\mathbb{X}} |f|^p \,\mathrm{d}\mu < \infty .$$

This is consistent with our previous definition⊸ (2.41) of the space $\mathscr{L}^1(\mathbb{X}, \mathcal{S}, \mu)$. We may write for simplicity $\mathscr{L}^p(\mu)$, $\mathscr{L}^p(\mathbb{X})$, or even just $\mathscr{L}^p$, if the missing symbols are understood from the context and cannot be mistaken. If $\mathbb{X}$ is a subset of $\mathbb{R}$ or of $\mathbb{R}^n$, a missing $\mathcal{S}$ will be understood to be the Borel σ-field on $\mathbb{X}$, and a missing μ will be understood to be the Lebesgue measure.

Notice that although we allow functions from $\mathscr{L}^p$ to take the values $\pm\infty$, the definition of $\mathscr{L}^p$ implies that all functions from that space are a.e. finite. Furthermore, it is often convenient to extend the definition of $\mathscr{L}^p$ to include functions that are defined only a.e.

Similarly, if f takes values in $\bar{\mathbb{R}}^n$ instead of $\bar{\mathbb{R}}$ and $|f|$ is understood to be the Euclidean norm of f, or, more generally, if f is a measurable function with values in some Banach space $\mathbb{B}$ with norm $\|\cdot\|$, we denote by $\mathscr{L}^p(\mathbb{X}, \mathcal{S}, \mu; \mathbb{B})$ the space of all $\mathbb{B}$-valued and Borel-measurable functions on $\mathbb{X}$ such that the $\mathbb{R}_+$-valued function $\|f\|^p$ is μ-integrable. If the range space $\mathbb{B}$ is missing from the notation – say, if we write $\mathscr{L}^p(\mu)$, or just $\mathscr{L}^p$ – then the range space will be understood to be the extended real line $\bar{\mathbb{R}}$.

Given any $p \in \mathbb{R}_{++}$, the L^p-seminorm in the space $\mathscr{L}^p(\mu)$ is defined as

$$\|f\|_p \stackrel{\text{def}}{=} \left(\int_{\mathbb{X}} |f|^p \,\mathrm{d}\mu \right)^{1/p}, \qquad f \in \mathscr{L}^p(\mu) .$$

Of course, one must verify that $\|\cdot\|_p$ so defined has the properties of a seminorm. The key step in this verification is⊸ (2.77) the Minkowski–Riesz inequality. The reason why $\|\cdot\|_p$ is not a norm is that $\|f - g\|_p = 0$ for every choice of f and g such that $f = g$ μ-a.e., even if f and g are different functions from $\mathscr{L}^p(\mu)$. The remedy is to consider functions that are identical μ-a.e. in $\mathbb{X}$ as indistinguishable, representing, so to speak, one and the same "almost function" (several alternative terms, most notably "functionoid," have been proposed, but are hardly ever used in the literature). To be precise, we must form the space $L^p(\mathbb{X}, \mathcal{S}, \mu)$, the elements of which are μ-equivalence classes of functions (that is, equivalence classes for the relation "identity μ-a.e. in $\mathbb{X}$") inside the space $\mathscr{L}^p(\mathbb{X}, \mathcal{S}, \mu)$. Just as in the case of the spaces $\mathscr{L}^p(\mathbb{X}, \mathcal{S}, \mu)$, we may write simply $L^p(\mu)$, $L^p(\mathbb{X})$, or L^p, if the missing symbols are understood from the context. Usually, when we operate with the elements of $L^p(\mu)$ we pretend that these objects are just functions. Such an illusion is innocuous if, for example, one is referring to $\|f\|_p$ or to $\int_{\mathbb{X}} |f|^p \,\mathrm{d}\mu$ for some $f \in L^p(\mu)$, since such quantities are the same across all members of the equivalence class $f \in L^p(\mu)$. Generally, however, if f is identified as an element of $L^p(\mu)$ and the measure μ is nonatomic, we must avoid expressions of the form $f(x)$ for $x \in \mathbb{X}$, since one can find functions in the class f that take any value from the extended real line $\bar{\mathbb{R}}$ at the point x. To put it another way, typically, if f is an equivalence class, then $f(x)$ would not be a well-defined quantity for a fixed $x \in \mathbb{X}$. Nevertheless, we may still write $f(x)$ – say, under a particular integral – with the understanding that in any such instance f is treated as a generic (arbitrary) function from the equivalence class f.

Given a Banach space $\mathbb{B}$, the symbol $L^p(\mathbb{X}, \mathcal{S}, \mu; \mathbb{B})$ will be understood in the obvious way as the collection of all μ-equivalence classes of functions in the space $\mathcal{L}^p(\mathbb{X}, \mathcal{S}, \mu; \mathbb{B})$.

In (2.41) we introduced the spaces $\mathcal{L}^0$, which are simply spaces of measurable functions. These spaces cannot be endowed with a seminorm, but the respective spaces of equivalence classes, which we again denote by L^0, are perfectly meaningful. We are going to discover in (4.8) that the spaces $\mathcal{L}^0(\Omega, \mathcal{F}, P)$ can be endowed with a pseudometric, which again becomes a metric if it is restricted to $L^0(\Omega, \mathcal{F}, P)$. ○

(2.73) THE SPACES $\mathcal{L}^\infty$ AND L^∞: The space $\mathcal{L}^\infty(\mathbb{X}, \mathcal{S}, \mu)$ – which may again be abbreviated as $\mathcal{L}^\infty(\mu)$, $\mathcal{L}^\infty(\mathbb{X})$, or simply as $\mathcal{L}^\infty$ – consists of all measurable functions from $\mathbb{X}$ into $\mathbb{R}$ that are essentially bounded. A measurable function $f : \mathbb{X} \mapsto \mathbb{R}$ is said to be *essentially bounded* if $|f| \le M$ μ-a.e. for some finite constant $M \in \mathbb{R}_{++}$. The smallest constant that bounds f μ-a.e. is

$$\|f\|_\infty = \inf \left\{ M \in \mathbb{R}_{++} : |f| \le M \ \ \mu\text{-a.e.} \right\}$$

and is called the *L^∞-norm of f*. The notion "essentially bounded" and the L^∞-norm are perfectly meaningful for functions with values in any Banach space $\mathbb{B}$, in particular, with values in $\mathbb{R}^n$. If we want to emphasize the target space in the notation we will write $\mathcal{L}^\infty(\mathbb{X}, \mathcal{S}, \mu; \mathbb{B})$.

It is straightforward to check that if the measure μ is finite ($\mu(\mathbb{X}) < \infty$), then $\mathcal{L}^\infty(\mu) \subseteq \mathcal{L}^p(\mu)$ for any $p \in [0, \infty[$, but it is easy to see that if $\mu(\mathbb{X}) = \infty$ then this property cannot hold (e.g., none of the spaces $\mathcal{L}^p(\mathbb{R}, \mathcal{B}(\mathbb{R}), \Lambda)$, $p \in [1, \infty[$, includes the constant functions on the real line).

Similarly to the spaces L^p, $p \in \mathbb{R}_+$, the space $L^\infty(\mathbb{X}, \mathcal{S}, \mu)$ (or $L^\infty(\mu)$, or L^∞) is defined as the collection of all μ-equivalence classes inside the space (of actual functions) $\mathcal{L}^\infty(\mathbb{X}, \mathcal{S}, \mu)$. ○

The remarkable result that we state next reveals a particularly interesting and useful features of the L^∞-norm.

(2.74) THEOREM: (Stroock 1999, 6.2.22) Let $(\mathbb{X}, \mathcal{S}, \mu)$ be any measure space and suppose that $f \in \cap_{1 \le p \le \infty} \mathcal{L}^p(\mu)$. Then

$$\|f\|_\infty = \lim_{p \to \infty} \left(\int_{\mathbb{X}} |f|^p \, d\mu \right)^{1/p} .$$

In particular, if $\mu(\mathbb{X}) < \infty$, then this relation holds for every $f \in \mathcal{L}^\infty(\mu)$. ○

Remarkably, according to the last theorem, the calculation of the maximum of a function can be accomplished simply by way of integrating (in other words, by way of averaging) sufficiently high powers of the function – and nothing else.

(2.75) EXERCISE: Using numerical integration, show that the integrals

$$\left(\int_{-1}^{1}(1-x^2)^{1000}\,\mathrm{d}x\right)^{1/1000} \quad \text{and} \quad \left(\int_{-\infty}^{\infty}(e^{-1000\,x^2})\mathrm{d}x\right)^{1/1000}$$

are both close to 1, which is the maximum of $(1-x^2)$ for $-1 \le x \le 1$, and is also the maximum of e^{-x^2} on the domain $-\infty < x < \infty$. ○

(2.76) THE SPACES $\mathscr{L}^p_{\mathrm{loc}}(\mathbb{R}_+)$ AND $L^p_{\mathrm{loc}}(\mathbb{R}_+)$: Recall that $\mathscr{L}^p([0,T])$ is a shorthand for $\mathscr{L}^p([0,T],\mathcal{B}([0,T]),\Lambda)$ and define $\mathscr{L}^p_{\mathrm{loc}}(\mathbb{R}_+)$ to be the space of all Borel functions $f\colon \mathbb{R}_+ \mapsto \mathbb{R}$ whose restrictions $f\restriction[0,T]$ belong to $\mathscr{L}^p([0,T])$ for every $T \in \mathbb{R}_{++}$. Accordingly, $L^p_{\mathrm{loc}}(\mathbb{R}_+)$ is defined as the space of all Λ-equivalence classes inside $\mathscr{L}^p_{\mathrm{loc}}(\mathbb{R}_+)$. The spaces $L^p_{\mathrm{loc}}(\mathbb{R}_+)$ can be metrized by the distance

$$d_{\mathrm{loc},p}(f,g) \overset{\mathrm{def}}{=} \sum\nolimits_{n\in\mathbb{N}_{++}} 2^{-n}\left(1\wedge \left\|f\restriction[0,n]-g\restriction[0,n]\right\|_p\right),$$

which makes⊸ (2.82) $L^p_{\mathrm{loc}}(\mathbb{R}_+)$ into a Polish space for every $p \in [1,+\infty]$. The convergence $f_i \to g$ in $d_{\mathrm{loc},p}(\cdot,\cdot)$ is equivalent to the convergence $f_i\restriction[0,T] \to g\restriction[0,T]$ in $L^p([0,T])$ for every fixed $T \in \mathbb{R}_{++}$.

The analogous spaces of Borel functions $f\colon \mathbb{R}_+ \mapsto \mathbb{R}^n$ will be denoted by $\mathscr{L}^p_{\mathrm{loc}}(\mathbb{R}_+;\mathbb{R}^n)$ and $L^p_{\mathrm{loc}}(\mathbb{R}_+;\mathbb{R}^n)$. These spaces can be introduced as above, or, equivalently, can be understood as spaces of $\mathbb{R}^n$-valued functions, the coordinates of which belong to $\mathscr{L}^p_{\mathrm{loc}}(\mathbb{R}_+)$ and $L^p_{\mathrm{loc}}(\mathbb{R}_+)$, respectively. ○

(2.77) BASIC INEQUALITIES: (a) *Hölder's inequality:*[44] Suppose that $p,q \in [1,\infty]$ are chosen so that $\frac{1}{p}+\frac{1}{q}=1$ (that is, $q=\frac{p}{p-1}$ and $p=\frac{q}{q-1}$, with the understanding that $1/\infty = 0$), let $f \in \mathscr{L}^p(\mathbb{X},\mathcal{S},\mu)$, and let $g \in \mathscr{L}^q(\mathbb{X},\mathcal{S},\mu)$. Then $fg \in \mathscr{L}^1(\mathbb{X},\mathcal{S},\mu)$ and

$$\left|\int_{\mathbb{X}} f\,g\,\mathrm{d}\mu\right| \le \|f\,g\|_1 \le \|f\|_p\|g\|_q\,.$$

(b) *Cauchy–Bunyakovski–Schwarz inequality* (a special case of Hölder's inequality with $p=q=2$):[45] If $f,g \in \mathscr{L}^2(\mathbb{X},\mathcal{S},\mu)$, then $f\,g \in \mathscr{L}^1(\mathbb{X},\mathcal{S},\mu)$ and

$$\left|\int_{\mathbb{X}} f\,g\,\mathrm{d}\mu\right| \le \|f\,g\|_1 \le \|f\|_2\|g\|_2\,.$$

[44] Named after the German mathematician Otto Hölder (1859–1937).

[45] Named after the French mathematician Augustin-Louis Cauchy (1789–1857), the Ukrainian mathematician Viktor Yakovlevich Bunyakovsky (1804–1889), and the German mathematician Karl Hermann Amandus Schwarz (1843–1921).

(c) *Minkowski–Riesz inequality:*[46] If $p \in [1, \infty]$ and $f, g \in \mathscr{L}^p(\mathcal{X}, \mathcal{S}, \mu)$, then $f + g \in \mathscr{L}^p(\mathcal{X}, \mathcal{S}, \mu)$ and

$$\|f + g\|_p \leq \|f\|_p + \|g\|_p .$$

In particular, $\mathscr{L}^p(\mathcal{X}, \mathcal{S}, \mu)$ is a linear space and $\|\cdot\|_p$ is a seminorm for any $p \in [1, \infty]$.

(d) *Jensen's inequality:*[47] Let $]a, b[\subseteq \mathbb{R}$ be any open interval, let $f :]a, b[\mapsto \mathbb{R}$ be any convex function, and let X be any random variable on $(\Omega, \mathcal{F}, P)$ (P is now a probability measure) such that $P(X \in]a, b[) = 1$ and $\mathsf{E}[|X|] < \infty$. Then $\mathsf{E}[X] \in]a, b[$, $\mathsf{E}[f(X)]$ is defined with $\mathsf{E}[f(X)] \in]-\infty, \infty]$, and $f(\mathsf{E}[X]) \leq \mathsf{E}[f(X)]$.

(e) *The Bienaymé–Chebyshev inequality:*[48] Let X be any r.v. on $(\Omega, \mathcal{F}, P)$ such that $X \geq 0$ P-a.s. Then

$$P(X \geq \varepsilon) \leq \mathsf{E}[X]/\varepsilon \quad \text{for any } \varepsilon \in \mathbb{R}_{++} .$$

PROOF: $\mathsf{E}[X] \geq \mathsf{E}[X\, 1_{\{X \geq \varepsilon\}}] \geq \varepsilon\, \mathsf{E}[1_{\{X \geq \varepsilon\}}] = \varepsilon\, P(X \geq \varepsilon)$. ○

(2.78) COMMENTS: (a) Let X be any random variable defined on $(\Omega, \mathcal{F}, P)$. With $f(x) = |x|$ and $f(x) = x^2$, $x \in \mathbb{R}$, Jensen's inequality gives[49]

$$\mathsf{E}[X]^2 \leq \mathsf{E}[|X|]^2 \leq \mathsf{E}[X^2] ,$$

or, equivalently, $|\mathsf{E}[X]| \leq \mathsf{E}[|X|] \leq \mathsf{E}[X^2]^{1/2}$ (which is trivial if $\mathsf{E}[|X|] < \infty$ and $\mathsf{E}[X^2] = \infty$ – see the next exercise). If $\mathsf{E}[X^2] < \infty$, then $|\mathsf{E}[X]| \leq \mathsf{E}[X^2]^{1/2}$ follows from the Cauchy–Bunyakovski–Schwarz inequality, which says that

$$\big|\mathsf{E}[X\, 1]\big| \leq \mathsf{E}[|X\, 1|] \leq \mathsf{E}[X^2]^{1/2}\, \mathsf{E}[1^2]^{1/2} .$$

In particular, we see that for probability measures, or, more generally, for finite measures μ, the space $\mathscr{L}^2(\mu)$ is included into $\mathscr{L}^1(\mu)$, and, as we are about to realize, this inclusion is typically strict.

(b) Chebyshev's inequality is often stated in the following form: for *any* r.v. X and any $\varepsilon \in \mathbb{R}_{++}$,

$$P(|X| \geq \varepsilon) \leq \mathsf{E}[X^2]/\varepsilon^2 .$$

---

[46] Named after the German mathematician Hermann Minkowski (1864–1909) and the Hungarian mathematician Frigyes Riesz (1880–1956).

[47] Named after the Danish mathematician Johan Jensen (1859–1925).

[48] First formulated by the French mathematician Irénée-Jules Bienaymé (1796–1878) and later reformulated and proved in a more precise form by the Russian mathematician Pafnuty Chebyshev (1821–1894). Throughout the literature this inequality is mostly known as "Chebyshev's inequality."

[49] The first inequality is actually elementary, since $\mathsf{E}[X]^2 \equiv \big|\mathsf{E}[X]\big|^2 \leq \mathsf{E}[|X|]^2$ – see exercise (2.42).

The proof is essentially the same: $\mathsf{E}[X^2] \geq \mathsf{E}[X^2\, 1_{\{|X| \geq \varepsilon\}}] \geq \varepsilon^2\, \mathsf{E}[1_{\{|X| \geq \varepsilon\}}]$. In particular, if $\mathsf{E}[|X|] < \infty$, then

$$\dagger \qquad P\Big(\mathsf{E}[X] - \varepsilon < X < \mathsf{E}[X] + \varepsilon\Big) \geq 1 - \mathrm{Var}(X)/\varepsilon^2\,.$$

Thus, Chebyshev's inequality gives an estimate of the probability for the random variable not to deviate beyond certain threshold from its mean. It illustrates the effect that the variance has on this probability: when the variance decreases to 0 the probability converges to 1 for every, however small, choice of $\varepsilon \in \mathbb{R}_{++}$. ○

(2.79) EXERCISE: Give an example of a random variable $X > 0$ such that $\mathsf{E}[X]$ exists and is finite, but $\mathsf{E}[X^2] = +\infty$. ○

(2.80) EXERCISE: Suppose that X is a standard normal r.v. ($X \in \mathcal{N}(0,1)$) and let $Y = e^X$. Given any fixed $p \in \mathbb{R}_+$, consider the function $f_p(x) \stackrel{\text{def}}{=} x^p$, $x \in \mathbb{R}_{++}$, and compute $\mathsf{E}[f_p(Y)]$ as an explicit function of the parameter p. Using the explicit expression for $\mathsf{E}[f_p(Y)]$, determine the range of values for the parameter p for which Jensen's inequality $f_p(\mathsf{E}[Y]) \leq \mathsf{E}[f_p(Y)]$ holds and the range of values for which this inequality fails. Explain your result. ○

(2.81) EXERCISE: Given a r.v. X with distribution law $\mathcal{N}(0, \sigma^2)$, calculate the probability $P(-3\,\sigma \leq X \leq 3\,\sigma)$ with 10 digits of precision and compare this "exact" value with the lower bound provided by Chebyshev's inequality in the form (2.78†). ○

(2.82) THEOREM (COMPLETENESS OF L^p): (Dudley 2002, 5.2.1) Given any measure space $(\mathcal{X}, \mathcal{S}, \mu)$ and any $p \in [1, \infty]$ the space (of μ-equivalence classes) $L^p(\mathcal{X}, \mathcal{S}, \mu)$ is complete, and is therefore a Banach space, for the norm $\|\cdot\|_p$. If the space $\mathcal{X}$ is a complete separable metric space (that is, a Polish space) and $\mathcal{S}$ is the Borel σ-field on $\mathcal{X}$, then the space $L^p(\mathcal{X}, \mathcal{S}, \mu)$ is also separable (and is therefore a Polish space). In addition, if $\mathcal{X}$ is chosen to be $\mathbb{R}^n$ (or some sub-domain inside $\mathbb{R}^n$), $\mathcal{S} = \mathcal{B}(\mathbb{R}^n)$, and $\mu = \Lambda$ (the Lebesgue measure), then the space of test functions $\mathscr{C}_K^\infty(\mathbb{R}^n; \mathbb{R})$ – or, to be precise, the space of equivalence classes associated with test functions – is a dense subset of $L^p(\mathbb{R}^n, \mathcal{B}(\mathbb{R}^n), \Lambda)$. ○

(2.83) L^2 AS A HILBERT SPACE: With $p = 2$ the norm in the Banach space $L^p(\mathcal{X}, \mathcal{S}, \mu)$ can be associated with the inner product

$$f \cdot g = \int_{\mathcal{X}} f g \,\mathrm{d}\mu\,, \quad f, g \in L^2(\mathcal{X}, \mathcal{S}, \mu)\,,$$

the existence and finiteness of the integral being guaranteed by Hölder's inequality, and○→ (A.56) the polarization identity

$$f \cdot g = \frac{1}{4}\Big(\|f + g\|_2^2 - \|f - g\|_2^2\Big) \quad \text{for all } f, g \in L^2(\mathcal{X}, \mathcal{S}, \mu)$$

is in force. Thus, the space $L^2(\mathcal{X}, \mathcal{S}, \mu)$ has the structure of a Hilbert space. It is a separable Hilbert space if, for example, $\mathcal{X} \subseteq \mathbb{R}^n$, $\mathcal{S} = \mathcal{B}(\mathcal{X})$ and $\mu = \Lambda$. ○

(2.84) L^p-NORMS AND MOMENTS OF RANDOM VARIABLES: L^p-spaces associated with probability measures will play an important role in what follows. What makes probability measures special is that the constant 1 (that is, any constant function on Ω) is automatically in $\mathscr{L}^p(\Omega, \mathcal{F}, P)$ for any $p \in [1, \infty]$ and $\|1\|_p = 1$. By Hölder's inequality, given any $X \in \mathscr{L}^0(\Omega, \mathcal{F}, P)$, any $p \in [1, \infty[$, and any $\varepsilon \in \mathbb{R}_{++}$, we have

$$\mathsf{E}\left[|X|^p 1\right] \leq \mathsf{E}\left[\left(|X|^p\right)^{1+\varepsilon}\right]^{\frac{1}{1+\varepsilon}} \mathsf{E}\left[1^{(1+\varepsilon)/\varepsilon}\right]^{\frac{\varepsilon}{1+\varepsilon}} = \mathsf{E}\left[|X|^{p(1+\varepsilon)}\right]^{\frac{1}{1+\varepsilon}},$$

which can be rearranged as

$$\mathsf{E}\left[|X|^p\right]^{\frac{1}{p}} \leq \mathsf{E}\left[|X|^{p(1+\varepsilon)}\right]^{\frac{1}{p(1+\varepsilon)}} \quad \Longleftrightarrow \quad \|X\|_p \leq \|X\|_{p(1+\varepsilon)}.$$

There are two important conclusions to be made from this relation. The first one is that $\|X\|_p$ is an increasing function of p, and the second one is that

$$\mathscr{L}^p(\Omega, \mathcal{F}, P) \subseteq \mathscr{L}^q(\Omega, \mathcal{F}, P) \quad \text{for } p \geq q \geq 1.$$

Thus, given any $p \in [1, \infty[$ and any $X \in \mathscr{L}^p(\Omega, \mathcal{F}, P)$, the expectation $\mathsf{E}[|X|^p]$ is finite. It is called the pth *moment* (or, *moment of order* p) of X. The phrase "X admits finite moments up to order p" is just another way of saying that $X \in \mathscr{L}^p(\Omega, \mathcal{F}, P)$, and therefore $X \in \mathscr{L}^q(\Omega, \mathcal{F}, P)$ for all $q \in [1, p]$). It is straightforward to check that the relations among the $\mathscr{L}^p$-spaces that we just established also hold for the respective L^p-spaces of equivalence classes. ○

(2.85) REMARK (SQUARE INTEGRABLE R.V.): From a practical point of view, spaces of r.v. that admit finite moments of order 2 (alias: square integrable r.v., or r.v. with finite second moment), and also the convergence in the L^2-norm (alias: convergence in mean-square), are very important. This has a lot to do with the fact that $L^2(\Omega, \mathcal{F}, P)$ has the structure of a Hilbert space. For example, if $(X_i)_{i \in \mathbb{N}}$ is any sequence of square integrable r.v., then⊸ (A.62) the convergence of that sequence in L^2-norm is equivalent to the convergence of the (scalar) double sequence $\left(\mathsf{E}[X_i X_j] \in \mathbb{R}\right)_{i \in \mathbb{N}, j \in \mathbb{N}}$. ○

The next exercise is an illustration of the fact that $L^1(\Omega, \mathcal{F}, P)$ is bigger than $L^2(\Omega, \mathcal{F}, P)$.

(2.86) EXERCISE: Give an example of a function $f\colon]0, 1] \mapsto \mathbb{R}_{++}$ such that

$$\int_{]0,1]} f \, \mathrm{d}\Lambda < \infty, \quad \text{while} \quad \int_{]0,1]} f^2 \, \mathrm{d}\Lambda = \infty.$$

Explain the relation $L^q(\Omega, \mathcal{F}, P) \supseteq L^p(\Omega, \mathcal{F}, P)$ for $p \geq q \geq 1$ only in terms of the basic properties of the Lebesgue integral, without relying on Hölder's inequality.

HINT: Whether or not $\mathsf{E}[|X|]$ is finite depends only on the r.v. $X 1_{|X|>1}$. ○

2.6 The Inverse of an Increasing Function

Given any random variable X, its$^{\circ\!\to\bullet\,(2.22)}$ distribution function

$$\mathbb{R} \ni x \rightsquigarrow F_X(x) \in [0,1]$$

is$^{\circ\!\to\bullet\,(1.57)}$ increasing, and if it happens to be strictly increasing and continuous, then its inverse (understood in the usual way) $]0,1[\, \ni y \rightsquigarrow F_X^{-1}(y) \in \mathbb{R}$ would be well defined, since in this case F_X is a one-to-one mapping from $\mathbb{R}$ onto $]0,1[$. However, as we are well aware, F_X may not be strictly increasing and may not be continuous – say, if the distribution law of X contains atoms. Nevertheless, a distribution function can always be inverted and this operation is quite useful in a number of applications as we now detail.

(2.87) DEFINITION: Let $I \subseteq \bar{\mathbb{R}}$ be any (finite or infinite) left-closed interval (possibly $[-\infty,\infty[$) and let $F\colon I \mapsto \mathbb{R}$ be any$^{\circ\!\to\bullet\,(1.57)}$ *increasing* function. The *span* of F is defined as the set

$$\mathrm{Span}(F) \stackrel{\text{def}}{=} \left\{ y \in \mathbb{R} \colon y \leq F(x) \ \text{ for some } \ x \in I \right\}$$

and the inverse of F is defined as the function

$$\mathrm{Span}(F) \ni y \rightsquigarrow F^{-1}(y) \stackrel{\text{def}}{=} \inf \left\{ x \in I \colon F(x) \geq y \right\}.$$

Notice that $F^{-1}(y)$ is guaranteed to be an element of I since I is left-closed.[50] ○

(2.88) EXERCISE: Calculate the inverse of the distribution function $F_X^{-1}(y)$ in the following cases: (a) X is a standard normal ($\mathcal{N}(0,1)$) random variable; (b) X is a binomial random variable with parameters $N = 3$ and $p = \frac{1}{2}$. ○

Now we turn to some important features of the inverse that we defined in (2.87). In what follows $I \subseteq \bar{\mathbb{R}}$ stands for a generic left-closed interval.

(2.89) EXERCISE: Let $F\colon I \mapsto \mathbb{R}$ be any increasing and *càdlàg* function. Prove that for any $x \in I$ and any $y \in \mathrm{Span}(F)$ the relations $x \geq F^{-1}(y)$ and $F(x) \geq y$ are equivalent.

HINT: Argue that for any $y \in \mathrm{Span}(F)$ the set $A(y) = \left\{ x \in I \colon F(x) \geq y \right\}$ is a nonempty left-closed interval. What is the left end-point of this interval? Explain why the requirement for $F(\cdot)$ to be càdlàg is needed. ○

(2.90) EXERCISE: Let $F\colon I \mapsto \mathbb{R}$ be any increasing and càdlàg function. Prove that the function

$$\mathrm{Span}(F) \ni y \rightsquigarrow F^{-1}(y) \in I,$$

as defined in (2.87), is left-continuous and increasing (thus càglàd).

HINT: The sets $A(y)$ from the hint to the previous exercise decrease as y increases. Argue that if $y_n \nearrow y$ then $A(y) = \cap_n A(y_n)$. ○

---

[50] As was noted in (1.57), an increasing function defined on a left-open interval can be always extended to an increasing function on the closure of the interval. In particular, any distribution function $F_X(\cdot)$ can be treated as a function on $[-\infty,\infty[$ by postulating $F_X(-\infty) = 0$.

(2.91) EXERCISE: Let $F\colon I \mapsto \mathbb{R}$ be any increasing and *continuous* function. Prove that for any $x \in \mathbb{R}$ one has

$$F(x) \stackrel{\text{def}}{=} \inf\left\{y \in \mathrm{Span}(F)\colon F^{-1}(y) \geq x\right\},$$

where F^{-1} is defined as in (2.87). To put it another way, the inverse of the inverse of the function is the function itself, but we stress that the continuity of F (in addition to the attribute "increasing") is essential for this claim – explain why.

HINT: What can you say about

$$\sup\left\{y \in \mathrm{Span}(F)\colon F^{-1}(y) < x\right\}? \quad \circ$$

(2.92) MONTE CARLO SIMULATION OF DISTRIBUTION LAWS: Suppose that Y is any random variable that is uniformly distributed in the interval $]\,0,1[$, or $Y \in \mathcal{U}(]\,0,1[)$ for short. The computation of the distribution function of the random variable $\tilde{X} \stackrel{\text{def}}{=} F_X^{-1}(Y)$ is straightforward and yields:

$$F_{\tilde{X}}(x) = P[F_X^{-1}(Y) \leq x] = P[F_X(x) \geq Y] = F_X(x).$$

Consequently, the law of $\tilde{X}$ is the same as the law of X, or $\mathcal{L}_{\tilde{X}} = \mathcal{L}_X$ for short. In particular, any probability law whatsoever can be generated as a functional transformation of the uniform law $\mathcal{U}(]\,0,1[)$. In other words, if we have a computing technology that would allow us to sample from the law $\mathcal{U}(]\,0,1[)$, then the same technology would allows us to sample from any other distribution law. The importance of this observation cannot be overstated, but one must be aware of certain pitfalls in the practical implementation of this approach to simulation. Indeed, no matter how well designed, a random number generator is only a pseudo-random number generator, which is to say it produces samples that may resemble random samples really well, but are actually not random at all. The fact that virtually all existing computing systems can work only with fixed – and thus limited – precision arithmetic adds an extra layer of inaccuracy to the picture. Even if one has at hand a pseudo-random number generator designed in such a way that these issues are still tolerable when sampling from the uniform distribution, such a generator cannot be aware of what functional transformations the output may be passed through. This is clearly a problem, because many distributions of common use are such that the inverse $F_X^{-1}(\cdot)$ has a vertical asymptote at at least one of the end-points of the interval $]0,1[$ – as is the case with the Gaussian distribution, for example. This means that in the evaluation of $F_X^{-1}(Y)$ any errors in the numerical expression for Y do get multiplied by an arbitrarily large factor whenever Y takes values that are sufficiently close to the end-points of $]0,1[$. For this and several other reasons, most computing systems provide separate random number generators for the distribution laws considered to be of common use – including, of course, the Gaussian distribution. Nevertheless, the relation

$$\mathcal{L}_X = \mathcal{L}_{F_X^{-1}(Y)} \quad \text{for any } Y \in \mathcal{U}(]\,0,1[)$$

is very useful in both theory and practice. Its origins can be traced back to (von Neumann 1951) – see (Glasserman 2004) for more details. ○

(2.93) EXERCISE: By calling a random number generator for the uniform distribution $\mathcal{U}(]0,1[)$ 10 times, produce the values $u_i \in]0,1[$, $1 \le i \le 10$. Now consider a *discrete* random variable Z that takes each of the values $F_X^{-1}(u_i)$, $1 \le i \le 10$, with one and the same probability $\frac{1}{10}$. Plot the distribution function F_Z together with the distribution function F_X in each of the following two cases: (a) $X \in \mathcal{N}(0,1)$; (b) X has a binomial law with $N = 3$ and $p = \frac{1}{2}$. ○

3

Absolute Continuity, Conditioning, and Independence

What distinguishes probability theory from measure theory and integration are the notions of conditional probability and independence. Our goal in this chapter is to develop these important concepts and to review some classical tools and results that are related to them.

3.1 Quasi-invariance of the Gaussian Distribution under Translation

(3.1) COMPUTATION OF DENSITIES: We saw earlier$^{\multimap (2.67)}$ that if the distribution law of the r.v. X admits probability density ϕ and $f\colon \mathbb{R} \mapsto \mathbb{R}$ is a Borel-measurable function such that $\mathsf{E}[f(X)]$ is defined, then

$$\dagger \qquad \mathsf{E}[f(X)] = \int_{\mathbb{R}} f(x)\phi(x)\mathrm{d}x\,.$$

Although the justification of this feature lies outside of our main concerns, it is important to note that † actually *characterizes* the density ϕ in the following sense: if one can establish that there is a measurable function $\phi\colon \mathbb{R} \mapsto \mathbb{R}_+$ such that † is satisfied for every test function $f \in \mathscr{C}_K^\infty(\mathbb{R};\mathbb{R})$, then one can conclude that the distribution law of X is absolutely continuous and its density is precisely ϕ. As an application of this result, suppose that it is somehow known that the r.v. X admits density ϕ, and the objective is to determine whether the r.v. $Y = h(X)$ also admits density, for some choice of the Borel function $h\colon \mathbb{R} \mapsto I$, with values in the (finite or infinite) interval $I = \,]a, b[\, \subseteq \mathbb{R}$. If $\varphi \in \mathscr{C}_K^\infty(\mathbb{R};\mathbb{R})$ is any test function, then

$$\mathsf{E}[\varphi(Y)] = \mathsf{E}[\varphi(h(X))] = \int_{\mathbb{R}} \varphi(h(x))\,\phi(x)\mathrm{d}x\,.$$

If the function $h(\cdot)$ is invertible, in that there is a Borel function $g\colon I \mapsto \mathbb{R}$ such that $g(h(x)) = x$ for Λ-a.e. $x \in \mathbb{R}$, then,$^{\multimap (2.61)}$ by the change of variables rule,

$$\mathsf{E}[\varphi(Y)] = \int_{\mathbb{R}} \varphi(h(x))\,\phi(g(h(x))\Lambda(\mathrm{d}x) = \int_I \varphi(y)\,\phi(g(y))\,(\Lambda \circ h^{-1})(\mathrm{d}y)\,.$$

If the function h is strictly monotone and continuous, if the identity $g(h(x)) = x$ holds for all $x \in \mathbb{R}$, and, finally, if the derivative ∂g exists and is continuous everywhere in I, then for any interval $]\,y, y+\varepsilon]$ we can write

$$(\Lambda \circ h^{-1})(]\,y, y+\varepsilon]) = |g(y+\varepsilon) - g(y)| = |\partial g(y)|\,\varepsilon + o(\varepsilon)\,;$$

in other words – at least formally – $(\Lambda \circ h^{-1})(\mathrm{d}y) = |\partial g(y)|\,\mathrm{d}y$. One can show that these relations are not only formal, and, furthermore

$$\mathsf{E}[\varphi(Y)] = \int_I \varphi(y)\,\phi(g(y))\,|\partial g(y)|\,\mathrm{d}y\,.$$

Since this relation holds for every test function φ, it leads to the conclusion that the distribution law of Y, $\mathcal{L}_Y$, admits density ($\mathscr{L}_Y \precsim \Lambda$) given by

$$I \ni y \rightsquigarrow \frac{\mathrm{d}\mathcal{L}_Y}{\mathrm{d}\Lambda}(y) = \phi(g(y))\,|\partial g(y)|\,. \quad \circ$$

(3.2) EXAMPLE: If $X \in \mathcal{N}(0,\sigma^2)$ and $a \in \mathbb{R}$, then the argument of (3.1) with $h(x) = x + a$, $x \in \mathbb{R}$ and $g(y) = y - a$, $y \in \mathbb{R}$ would tell us that $Y = X + a$ has absolutely continuous law with density (in this case $\partial g = 1$)

$$\mathbb{R} \ni y \rightsquigarrow \ell_{a,\sigma^2}(y) = \frac{1}{\sqrt{2\pi\sigma^2}}\,e^{-\frac{(y-a)^2}{2\sigma^2}} \equiv \ell_{0,\sigma^2}(y-a) \in \mathbb{R}_{++}\,,$$

which is consistent with what we already know about the Gaussian denisty. $\circ$

(3.3) EXERCISE: Using the argument of (3.1), compute the density of $Y = e^X$, where $X \in \mathcal{N}(a,\sigma^2)$. N.B.: The law of Y is called *log-normal law* and is widely used in many applications – especially in the domain of finance. $\circ$

(3.4) PARALLEL SHIFTS OF GAUSSIAN LAWS AND EQUIVALENT TRANSFORMATIONS OF THE MEASURE: Suppose that X is some r.v. on $(\Omega, \mathcal{F}, P)$ with distribution law $\mathcal{N}(0,\sigma^2)$ and let $\theta \in \mathbb{R}$ be arbitrarily chosen. As we just realized, the r.v. $Y = X - \theta$ has density

$$\ell_{-\theta,\sigma^2}(x) = \frac{1}{\sqrt{2\pi\sigma^2}}\,e^{-\frac{(x+\theta)^2}{2\sigma^2}} = \frac{1}{\sqrt{2\pi\sigma^2}}\,e^{-\frac{x^2}{2\sigma^2}}\,e^{-\frac{\theta x}{\sigma^2}-\frac{\theta^2}{2\sigma^2}}\,,$$

which can be written more succinctly in the form $\ell_{-\theta,\sigma^2}(x) = \ell_{0,\sigma^2}(x)\,\rho_{\theta,\sigma^2}(x)$, with $\rho_{\theta,\sigma^2}(x) \overset{\text{def}}{=} e^{-\frac{\theta x}{\sigma^2}-\frac{\theta^2}{2\sigma^2}}$. Plainly, the density of $X - \theta$ obtains by multiplying the density X by the factor $\rho_{\theta,\sigma^2}(x)$. To put it another way, in terms of distribution laws, shifting the distribution of X by $(-\theta)$ comes down to multiplying the density of X by $\rho_{\theta,\sigma^2}(x)$. In any case, if $f\colon \mathbb{R} \mapsto \mathbb{R}$ is any bounded Borel function, then

$$\begin{aligned}\int_{\mathbb{R}} f(x)\ell_{0,\sigma^2}(x)\mathrm{d}x &= \mathsf{E}\big[f(X)\big] = \mathsf{E}\big[f(Y+\theta)\big]\\ &= \int_{\mathbb{R}} f(x+\theta)\ell_{0,\sigma^2}(x)\,\rho_{\theta,\sigma^2}(x)\mathrm{d}x\\ &= \mathsf{E}\big[f(X+\theta)\rho_{\theta,\sigma^2}(X)\big]\,.\end{aligned}$$

†

With $f \equiv 1$ the last relation gives $\mathsf{E}[\rho_{\theta,\sigma^2}(X)] = 1$. As the random variable $Z \overset{\text{def}}{=} \rho_{\theta,\sigma^2}(X)$ is strictly positive and integrates to 1, we can define a new probability measure on $\mathcal{F}$, namely, $Q \stackrel{\text{def}}{=} Z \odot P$, which is$^{\circ\!\!-\!\bullet\,(2.66\,\text{c})}$ equivalent to P ($Q \approx P$). In particular, equation † (in which $\mathsf{E} \equiv \mathsf{E}^P$) is telling us$^{\circ\!\!-\!\bullet\,(2.65)}$ that

$$\mathsf{E}^Q\big[f(X+\theta)\big] = \mathsf{E}^P\big[f(X+\theta)Z\big] = \mathsf{E}^P\big[f(X)\big] = \int_{\mathbb{R}} f(x) f_{0,\sigma^2}(x)\,\mathrm{d}x\,.$$

We see that $X + \theta$ has the same density under Q, namely f_{0,σ^2}, as the density of X under P. In other words, if we simultaneously shift the random variable X and replace the probability measure with an equivalent one accordingly, then nothing in our statistical observations over X would change. ○

(3.5) REMARK: The above line of reasoning can be seen from yet another angle. Since under Q the law of $X + \theta$ is $\mathcal{N}(0, \sigma^2)$, the law of X (under Q) must be $\mathcal{N}(-\theta, \sigma^2)$ – and, of course, the law of X under P is $\mathcal{N}(0, \sigma^2)$, which is how X was initially chosen. Thus, switching from Q to P preserves the normality of X (in fact, it even preserves the variance), but annihilates the mean value of X. To be precise, the equivalent transformation of the probability measure Q that annihilates the mean value of X (equal to $(-\theta)$ under Q) is achieved with the Radon–Nikodym derivative

$$\frac{\mathrm{d}P}{\mathrm{d}Q} = \frac{1}{Z} = e^{\theta\frac{X}{\sigma^2}+\frac{1}{2}\frac{\theta^2}{\sigma^2}} = e^{\theta\frac{X+\theta}{\sigma^2}-\frac{1}{2}\frac{\theta^2}{\sigma^2}}\,.$$

To put it in yet another way, the Radon–Nikodym derivative

$$Z = \frac{\mathrm{d}Q}{\mathrm{d}P} = e^{-\theta\frac{X}{\sigma^2}-\frac{1}{2}\frac{\theta^2}{\sigma^2}}$$

can be seen as a prescription for an equivalent transformation of P that annihilates the mean of $X+\theta$ (equal to θ under P). More generally, if X is a Gaussian r.v. with law $\mathcal{N}(a, \sigma^2)$, an equivalent change of the underlying measure that preserves the normality of X, and also preserves the variance σ^2, but shifts the mean a by the amount $\theta \in \mathbb{R}$ (that is, under the new measure the expectation of X is $a + \theta$), is given by the following Radon–Nikodym derivative:

$$e^{\theta\frac{(X-a)}{\sigma^2}-\frac{1}{2}\frac{|\theta|^2}{\sigma^2}}\,. \quad ○$$

(3.6) EXERCISE: Let $X \in \mathcal{N}(a, \sigma^2)$ and let $\theta \in \mathbb{R}$. Prove that

$$\mathsf{E}\big[e^{-\theta\frac{X-a}{\sigma^2}-\frac{1}{2}\frac{|\theta|^2}{\sigma^2}}\big] = \mathsf{E}\big[e^{\theta\frac{X-a}{\sigma^2}-\frac{1}{2}\frac{|\theta|^2}{\sigma^2}}\big] = 1$$

for every choice of the parameters $a, \theta \in \mathbb{R}$ and $\sigma \in \mathbb{R}_{++}$.

HINT: The solution comes down to calculating the integrals involved. ○

3.2 Moment-Generating Functions, Laplace, and Fourier Transforms

Given any $\bar{\mathbb{R}}$-valued r.v. X, we can define the function

$$\mathbb{R} \ni t \rightsquigarrow m(t) \stackrel{\text{def}}{=} \mathsf{E}\big[e^{t\,X}\big] \in \bar{\mathbb{R}}_+ .$$

This function is called the *moment-generating function of* X. Since $e^{tX} > 0$, the moment-generating function is well defined for all $t \in \mathbb{R}$. Of course, it may happen that $m(t) = +\infty$ for some values of t. Suppose that for some $\varepsilon \in \mathbb{R}_{++}$ one can claim that $m(t) < \infty$ for all $t \in [-\varepsilon, \varepsilon]$, that is, $m(\cdot)$ is finite in some neighborhood of 0. Formal differentiation at $t \in \,]-\varepsilon, \varepsilon[$ gives

$$\partial^n m(t) = \mathsf{E}\big[X^n\, e^{t\,X}\big] \quad \Rightarrow \quad \partial^n m(0) = \mathsf{E}\big[X^n\big] .$$

While this relation may justify the term "moment-generating function," it remains unclear if it is anything more than a formal manipulation. For, to say the least, there is no reason for an arbitrarily chosen X to admit moments of all orders.

To gain some understanding of the situation just described, given an arbitrary $t \in [-\varepsilon, \varepsilon]$ and arbitrary $n \in \mathbb{N}_{++}$, let

$$S_{n,t} \stackrel{\text{def}}{=} \sum\nolimits_{k \in \mathbb{N}_{|n}} \frac{t^k\, X^k}{k!} ,$$

and observe that

$$|S_{n,t}| \le \sum\nolimits_{k \in \mathbb{N}_{|n}} \frac{|t|^k\, |X|^k}{k!} \le \sum\nolimits_{k \in \mathbb{N}} \frac{|t|^k\, |X|^k}{k!} = e^{|tX|} \le e^{|\varepsilon X|} \le e^{-\varepsilon X} + e^{\varepsilon X} ,$$

and that $\lim_n S_{n,t} = \sum_{k \in \mathbb{N}} \frac{t^k X^k}{k!} = e^{tX}$. Since the expectation $\mathsf{E}\big[e^{-\varepsilon X} + e^{\varepsilon X}\big] = m(-\varepsilon) + m(\varepsilon)$ is finite by assumption, the preceding relations imply that

$$\sum\nolimits_{k \in \mathbb{N}_{|n}} \frac{|t|^k\, \mathsf{E}[|X|^k]}{k!} \le m(-\varepsilon) + m(\varepsilon) < \infty \quad \text{for all } t \in [-\varepsilon, \varepsilon] \text{ and } n \in \mathbb{N}_{++} .$$

In particular, $\mathsf{E}[|X|^n] < \infty$ for any $n \in \mathbb{N}_{++}$. Thus, our assumption that there is an $\varepsilon \in \mathbb{R}_{++}$ such that $m(t) < \infty$ for all $t \in [-\varepsilon, \varepsilon]$ leads to the conclusion that X has finite moments of all orders. Furthermore, under the same assumption, a straightforward application of the dominated convergence theorem yields

$$m(t) = \mathsf{E}[e^{tX}] = \mathsf{E}[\lim_n S_{n,t}] = \lim_n \sum\nolimits_{k \in \mathbb{N}_{|n}} \frac{t^k\, \mathsf{E}[X^k]}{k!} = \sum\nolimits_{k \in \mathbb{N}} t^k\, \frac{\mathsf{E}[X^k]}{k!} ,$$

and we already know that this last power series converges absolutely for every $t \in [-\varepsilon, \varepsilon]$. Recall that a power series can always be differentiated term-by-term in the interior of its interval of convergence, and, in fact, the coefficients in the expansion can be expressed in terms of the derivatives of the sum at 0 by the formula $a_k = \partial^k m(0)/k!$. As a result, we have $\partial^k m(0) = \mathsf{E}[X^k]$ for any $k \in \mathbb{N}$, so that the calculation of the moments of X reduces to the differentiation of the moment-generating function at 0. We stress that this feature comes as a result of our assumption that the moment-generating function is finite in some (however small) neighborhood of 0.

(3.7) EXERCISE: Calculate the moment-generating functions associated with the Gaussian and Poisson distributions. ○

Closely related to the moment-generating function is the Laplace transform[51] of the random variable X, which is simply the moment-generating function of $(-X)$ and is given by

$$\mathbb{R} \ni t \rightsquigarrow \ell(t) = \mathsf{E}\big[e^{-tX}\big] = \int_{\mathbb{R}} e^{-tx} \mathcal{L}_X(\mathrm{d}x) \in \bar{\mathbb{R}}_+ .$$

The Laplace transform of a positive r.v. is always finite in the domain $\bar{\mathbb{R}}_+$, and this is the case that we are about to examine more closely. More generally, given any (not necessarily finite) Borel measure μ on $\mathbb{R}_+$, one can define its Laplace transform as

$$\mathbb{R}_{++} \ni t \rightsquigarrow \ell(t) = \int_{\mathbb{R}_+} e^{-tx} \mu(\mathrm{d}x) \in \bar{\mathbb{R}}_+ . \qquad \mathrm{e}_1$$

If this function happens to be finite, it is not only infinitely differentiable on $\mathbb{R}_{++}$, but, in fact, has the property

$$(-1)^k \partial^k \ell(t) \geq 0 \quad \text{for all } t \in \mathbb{R}_+ \quad \text{for all } k \in \mathbb{N} . \qquad \mathrm{e}_2$$

Curiously – see (Feller 1971, XIII.4)[52] – a (finite) function $\ell \colon \mathbb{R}_{++} \mapsto \mathbb{R}_+$ can be expressed in the form e_1, for some choice of the measure μ, if and only if it has the property e_2, and can be expressed in the form e_1 for some choice of a probability measure μ if and only if it has the property e_2 and the property $\lim_{t\searrow 0} \ell(t) = 1$. A positive r.v. X has a finite kth moment if and only if the kth derivative of its Laplace transform $\partial^k \ell(t)$ is such that $\lim_{t\searrow 0} \partial^k \ell(t)$ exists and is finite, and if this last property holds, then one has the following connection (ibid., XIII.2):

$$\mathsf{E}\big[X^k\big] = (-1)^k \lim_{t\searrow 0} \partial^k \ell(t) .$$

The Laplace transform is a particularly useful tool in all fields related to probability theory due to the following remarkable result (ibid., XIII.4 thm 2).

(3.8) AN INVERSION FORMULA: Let $\ell \colon \mathbb{R}_{++} \mapsto \mathbb{R}_+$ be any function of the form e_1 for some measure μ. Then, for any $x \in \mathbb{R}_+$ with $\mu(\{x\}) = 0$,

$$\mu([0,x]) = \lim_{u\to\infty} \sum_{k=0}^{\lfloor ux \rfloor} \frac{(-u)^k}{k!} \partial^k \ell(u) . \quad ○$$

In particular, we see that the distribution law of a positive r.v. is completely determined by its Laplace transform. The power of this result cannot be overstated: it allows one to establish identities in law by way of Laplace-transform calculus.

[51] Named after the French mathematician, physicist, and astronomer Pierre-Simon Laplace (1749–1827).

[52] This result is attributed to the Russian mathematician Sergei Natanovich Bernstein (1880–1968).

(3.9) EXERCISE: Calculate the Laplace transform of the random variable X^2 for $X \in \mathcal{N}(a, \sigma^2)$. ○

(3.10) FOURIER TRANSFORM:[53] Closely related to the Laplace transform, and also to the notion of characteristic function introduced in (3.12), is the Fourier transform of a function $f \in \mathscr{L}^1(\mathbb{R}^n; \mathbb{C})$, defined as the complex-valued function

† $$\mathbb{R}^n \ni y \rightsquigarrow (\mathbb{F}f)(y) \stackrel{\text{def}}{=} \frac{1}{(2\pi)^{n/2}} \int_{\mathbb{R}^n} e^{-\imath x \cdot y} f(x) \mathrm{d}x \in \mathbb{C}\,.$$

We stress that although the definition of $\mathbb{F}f$ can be extended to include functions $f \in \mathscr{L}^2(\mathbb{R}^n; \mathbb{C})$ (see below), this extension cannot be attained directly from † because a function $f \in \mathscr{L}^2(\mathbb{R}^n; \mathbb{C})$ may not be in $\mathscr{L}^1(\mathbb{R}^n; \mathbb{C})$, in which case the integral in † may not be meaningful (as a Lebesgue integral). It is also clear from its very definition that $\mathbb{F}f$ depends only on the equivalence class of f (for the Lebesgue measure). ○

(3.11) EXTENSION AND REMARK: The Fourier transform $\mathbb{F}$, which maps functions on $\mathbb{R}^n$ into functions on $\mathbb{R}^n$, is an instrumental tool in analysis, and is covered in practically any textbook on this subject – see (Reed and Simon 1980), or (Rudin 1991), for example. In what follows the Fourier transform will be used almost exclusively in situations where the function f is going to be some probability density on the real line $\mathbb{R}$. More generally, the Fourier transform $\mathbb{F}$ can be defined as a transformation that maps probability distribution laws $\mathcal{L}$ over $\mathbb{R}^n$ into complex functions on $\mathbb{R}^n$ through the relation

$$\mathbb{R}^n \ni y \rightsquigarrow (\mathbb{F}\mathcal{L})(y) \stackrel{\text{def}}{=} \frac{1}{(2\pi)^{n/2}} \int_{\mathbb{R}^n} e^{-\imath x \cdot y} \mathcal{L}(\mathrm{d}x) \in \mathbb{C}\,. \quad ○$$

For reasons that may have to do with tradition, instead of working Fourier transforms of probability distribution laws in $\mathbb{R}^n$, it is most common in probability theory to work with the so-called characteristic functions, which are nothing but $(2\pi)^{n/2}(\mathbb{F}\mathcal{L})(-y)$ for a given law $\mathcal{L}$. For the purpose of completeness of the exposition and better clarity, we now give an independent definition, despite the obvious tautology.

(3.12) CHARACTERISTIC FUNCTIONS: If $\mathcal{L}$ is any probability distribution law over $\mathbb{R}^n$, then the following function is well defined:

$$\mathbb{R}^n \ni u \rightsquigarrow H(u) \stackrel{\text{def}}{=} \int_{\mathbb{R}} e^{\imath\, u \cdot x} \mathcal{L}(\mathrm{d}x) \equiv \int_{\mathbb{R}^n} \big(\cos(u \cdot x) + \imath \sin(u \cdot x)\big) \mathcal{L}(\mathrm{d}x) \in \mathbb{C}\,.$$

It is called the *characteristic function of* $\mathcal{L}$, or the *characteristic function of the* $\mathbb{R}^n$*-valued random variable* ξ, if $\mathrm{Law}(\xi) = \mathcal{L}$. In this later case we can write $H(u) = \mathsf{E}[e^{\imath\, u \cdot \xi}]$. If the dependence on the probability law or random variable must be emphasized in the notation, we will write $H_{\mathcal{L}}$ or H_ξ.

[53] Named after the French mathematician Jean-Baptiste Joseph Fourier (1768–1830).

The term "characteristic function" is justified by the fact that every probability distribution is uniquely determined by its characteristic function, in the sense that any two probability distributions over $\mathbb{R}^n$ that share the same characteristic function must be identical (Dudley 2002, 9.5.1). This result can be seen also as a consequence of the features of the Fourier transform listed in (3.14) below. In particular, we see that any *finite* Borel measure μ on $\mathbb{R}^n$ is uniquely determined by the values of the integrals $\int_{\mathbb{R}^n} f(x)\mu(dx)$ for all choices of $f \in \mathscr{C}_b(\mathbb{R}^n;\mathbb{R})$.[54] In fact, a finite measure, μ, on $\mathbb{R}^n$ would be fully determined by the values $\int_{\mathbb{R}^n} f(x)\mu(dx)$, for all choices of the test function $f \in \mathscr{C}_K^\infty(\mathbb{R}^n;\mathbb{R})$. We will take up our study of characteristic functions in section 16.2 in connection with the study of Lévy processes. ○

(3.13) CONNECTION WITH THE LAPLACE TRANSFORM: If the function f on the right side of (3.10†) happens to be the probability density of some positive r.v. X, then the Fourier transform $\mathbb{F}f$ can be linked to its Laplace transform $\ell(\cdot)$ through the relation $(\mathbb{F}f)(y) = \frac{1}{\sqrt{2\pi}}\ell(\imath y)$. Notice that if μ is a finite measure that is supported on $\mathbb{R}_+$, then its Laplace transform $\ell(t)$ would be meaningful for any complex $t \in \mathbb{C}$ such that $\mathrm{Re}(t) \geq 0$. ○

(3.14) FOURIER TRANSFORM (KEY FEATURES): Some of the most interesting and useful features of the Fourier transform can be traced back to the following result, known as Plancherel's theorem: if $f \in \mathscr{C}_K^\infty(\mathbb{R}^n;\mathbb{C})$ is any test function, then $\|f\|_2 = \|\mathbb{F}f\|_2$, where $\|\cdot\|_2$ is the Hilbert norm in the space $L^2(\mathbb{R}^n;\mathbb{C})$. In a more detailed form, Plancherel's theorem[55] can be stated as:

$$\|f\|_2^2 = \int_{\mathbb{R}^n} f(x)\overline{f(x)}\,dx = \int_{\mathbb{R}^n} (\mathbb{F}f)(y)\overline{(\mathbb{F}f)(y)}\,dy \quad \text{for any } f \in \mathscr{C}_K^\infty(\mathbb{R}^n;\mathbb{C}).$$

In particular, the mapping $\mathbb{F}: \mathscr{C}_K^\infty(\mathbb{R}^n;\mathbb{C}) \mapsto L^2(\mathbb{R}^n;\mathbb{C})$ is continuous relative to the norm $\|\cdot\|_2$ in both the domain and the range. As a result, since the space of test functions $\mathscr{C}_K^\infty(\mathbb{R}^n;\mathbb{C})$ is dense in $L^2(\mathbb{R}^n;\mathbb{C})$, the Fourier transform $\mathbb{F}$ can be extended to a unitary operator $\mathbb{F}: L^2(\mathbb{R}^n;\mathbb{C}) \mapsto L^2(\mathbb{R}^n;\mathbb{C})$, that is,

$$\int_{\mathbb{R}^n} \mathbb{F}(f)\overline{(\mathbb{F}g)}\,d\Lambda = \int_{\mathbb{R}^n} f\overline{g}\,d\Lambda \quad \text{for all } f,g \in L^2(\mathbb{R}^n;\mathbb{C}),$$

and we again stress that a function from the space $L^2(\mathbb{R}^n;\mathbb{C})$ may not be in the space $L^1(\mathbb{R}^n;\mathbb{C})$. Moreover, Plancherel's theorem states that the adjoint transformation is given by

$$\mathbb{R}^n \ni y \rightsquigarrow (\mathbb{F}^*f)(x) \stackrel{\text{def}}{=} \frac{1}{(2\pi)^{n/2}} \int_{\mathbb{R}^n} e^{\imath x\cdot y} f(y)\,dy \in \mathbb{C}.$$

[54] More generally, one can make this claim with $\mathbb{R}^n$ replaced by a generic topological space. This follows from the Stone–Daniell theorem – see (Dudley 2002, 4.5.2).

[55] Named after the Swiss mathematician Michel Plancherel (1885–1967).

As it stands, $\mathbb{F}^*f$ is well defined only for functions $f \in \mathscr{L}^1(\mathbb{R}^n;\mathbb{C})$, but it is again the case that this definition can be extended for all $f \in \mathscr{L}^2(\mathbb{R}^n;\mathbb{C})$ – and by the same mechanism. The transformation $\mathbb{F}^*$ is *adjoint* to $\mathbb{F}$ (whence the token *) in the sense that if $f, g \in \mathscr{L}^2(\mathbb{R}^n;\mathbb{C})$, then

$$\int_{\mathbb{R}^n} f\,\overline{\mathbb{F}g}\,\mathrm{d}\Lambda = \int_{\mathbb{R}^n} (\mathbb{F}^*f)\,\overline{g}\,\mathrm{d}\Lambda\,.$$

The last relation has the following implication:

$$\int_{\mathbb{R}^n} f\,\overline{g}\,\mathrm{d}\Lambda = \int_{\mathbb{R}^n} (\mathbb{F}f)\overline{(\mathbb{F}g)}\,\mathrm{d}\Lambda = \int_{\mathbb{R}^n} (\mathbb{F}^*\mathbb{F}f)\,\overline{g}\,\mathrm{d}\Lambda \quad \text{for all } f, g \in L^2(\mathbb{R}^n;\mathbb{C}),$$

which, in turn, is telling us that the product $\mathbb{F}^*\mathbb{F}$ is the identity map on $L^2(\mathbb{R}^n;\mathbb{C})$. In other words, the adjoint $\mathbb{F}^*$ is nothing but the inverse of $\mathbb{F}$, treated as an action on the domain $L^2(\mathbb{R}^n;\mathbb{C})$. In fact, $\mathbb{F}^*\mathbb{F}f$ can be identified as f (that is, as the equivalence class of f) as long as $\int_{\mathbb{R}^n} |(\mathbb{F}f)(y)|\,\mathrm{d}y < \infty$. Moreover, if $\mathcal{L}$ is a distribution law over $\mathbb{R}^n$ such that $\int_{\mathbb{R}^n} |(\mathbb{F}\mathcal{L})(y)|\,\mathrm{d}y < \infty$, then $\mathcal{L}$ admits density, which can be expressed as

$$\varphi(x) = \frac{1}{(2\pi)^{n/2}} \int_{\mathbb{R}^n} e^{\imath x\cdot y}(\mathbb{F}\mathcal{L})(y)\,\mathrm{d}y\,, \quad x \in \mathbb{R}^n\,. \quad \circ$$

3.3 Conditioning and Independence

(3.15) BASIC RULES AND DEFINITIONS: Given a probability space $(\Omega, \mathcal{F}, P)$ and any two random events $A, B \in \mathcal{F}$ with $P(B) > 0$, the *conditional probability of A given B* is defined as

$$\dagger \qquad P(A|B) = \frac{P(A \cap B)}{P(B)}\,.$$

The intuition behind this definition is that if the event B is known to have occurred (which is what "given B" translates to) – that is, if the random outcome ω is known to be in B – then B becomes the sample space and the σ-field of observable random events becomes $B \cap \mathcal{F}$ (if we know that "B occurs," then "A occurs" becomes "$A \cap B$ occurs"). One can then define a probability measure on $B \cap \mathcal{F}$ by simply normalizing the probabilities of the events in the induced σ-field $B \cap \mathcal{F}$ by the probability of B. Clearly, comparing the conditional probability $P(A|B)$ with the unconditional probability $P(A)$ reveals how much information about the random event A is to be found in the random event B. For example, if $P(A|B) = 1$ or $P(A|B) = 0$, then B is fully informed about A, but if $P(A|B) = P(A)$, then B contains no information about A.

It is quite common to write AB instead of $A \cap B$, and, with this convention in mind, $\dagger$ can be cast as

$$P(AB) = P(A|B)\,P(B)\,,$$

which is easy to extend by induction to what is known as the *product probability formula*:

$$P(A_1A_2\cdots A_n) = P(A_1)P(A_2|A_1)P(A_3|A_1A_2)\cdots P(A_n|A_1A_2\cdots A_{n-1})$$

for every choice of $A_1,\dots,A_n \in \mathcal{F}$ with $P(A_1\cdots A_{n-1}) > 0$.

If $\mathscr{P} = \{H_1,\dots,H_n\}$ is some measurable partition of Ω, then

$$P(A) = \sum_{i=1}^{n} P(AH_i) = \sum_{i=1}^{n} P(A|H_i)P(H_i) \quad \text{for all } A \in \mathcal{F}.$$

The last relation is known as the *total probability formula.* It gives rise to the important *Bayes' formula:*[56] if $A \in \mathcal{F}$ and $P(A) > 0$, then

$$P(H_i|A) = \frac{P(A|H_i)}{\sum_{k=1}^{n} P(A|H_k)P(H_k)}\,P(H_i) \quad \text{for all } i \in \{1,\dots,n\}. \quad \circ$$

The next three classical exercises (found in many introductory texts) illustrate the nature of these formulas.

(3.16) EXERCISE: Box 1 contains two red and one white balls and box 2 contains two white and one red balls. A ball is selected at random from box 1 and is moved to box 2, after which a ball is selected at random from box 2. (a) What is the probability that the ball selected from box 2 is red? (b) If the ball selected from box 2 is found to be red, how likely is it that the ball moved from box 1 to box 2 was also red? ○

(3.17) EXERCISE: Box 1 contains two red and one white balls, box 2 contains two white and one red balls and box 3 contains two red and two white balls. A ball is selected at random from box 1 and is moved to box 2, after which a ball is selected at random from box 2 and is moved to box 3. Finally, one ball is selected at random from box 3. What is the probability that the balls selected from boxes 1, 2, and 3 are all red? ○

(3.18) EXERCISE: A student is taking a test. One of the questions on the test is a multiple choice question with n possible answers ($n \geq 2$), only one of which is correct. If the student knows the correct answer, they would always answer correctly. If, however, the student does not know the correct answer, then they would choose one of the answers at random and in such a way that all answers have equal probability to get selected. Before coming to the test the student was able to learn only 70% of the material (the probability that they know how to answer any one question is 0.7). (a) What is the probability that the student would answer the multiple choice question correctly? (b) If it is known that the student has answered the multiple choice question correctly, what is the probability that the student knew the answer? How do the answers to (a) and (b) change if n increases to ∞? ○

[56] Named after the English statistician Rev. Thomas Bayes (1701–1761).

(3.19) JOINT DISTRIBUTION LAWS: Let X and Y be any two $\mathbb{R}$-valued random variables on $(\Omega, \mathcal{F}, P)$. The pair (X, Y) can be treated as a random vector, that is, as a single $\mathbb{R}^2$-valued random variable (a $\mathcal{B}(\mathbb{R}^2)/\mathcal{F}$-measurable function from Ω into $\mathbb{R}^2$). This random variable pushes the probability measure P from Ω into a probability measure measure on the σ-field $\mathcal{B}(\mathbb{R}^2)$, which probability measure we denote by $\mathcal{L}_{X,Y}$. The measure $\mathcal{L}_{X,Y}$ is called the *joint distribution law*, or simply *the joint law*, of X and Y. Of course, X and Y have their own distribution laws, $\mathcal{L}_X$ and $\mathcal{L}_Y$, defined on $\mathcal{B}(\mathbb{R})$. The laws $\mathcal{L}_X$ and $\mathcal{L}_Y$ can be recovered from the joint law $\mathcal{L}_{X,Y}$ via the relations

$$\mathcal{L}_X(A) = \mathcal{L}_{X,Y}(A \times \mathbb{R}), \quad A \in \mathcal{B}(\mathbb{R}), \quad \text{and} \quad \mathcal{L}_Y(B) = \mathcal{L}_{X,Y}(\mathbb{R} \times B), \quad B \in \mathcal{B}(\mathbb{R}).$$

More generally, any probability law μ defined on $\mathcal{B}(\mathbb{R}^2)$ gives rise to the so-called *marginal distribution laws* $\mathcal{L}_1(A) \stackrel{\text{def}}{=} \mu(A \times \mathbb{R})$, $A \in \mathcal{B}(\mathbb{R})$, and $\mathcal{L}_2(B) \stackrel{\text{def}}{=} \mu(\mathbb{R} \times B)$, $B \in \mathcal{B}(\mathbb{R})$. In particular, the distribution laws of X and Y can be identified as the marginal laws of their joint distribution law $\mathcal{L}_{X,Y}$.

Similar to the distribution laws of $\mathbb{R}$-valued random variables, the joint law $\mathcal{L}_{X,Y}$ can be characterized in terms of the joint (cumulative) distribution function

$$F_{X,Y}(u, v) = \mathcal{L}_{X,Y}\big(]-\infty, u] \times]-\infty, v]\big), \quad (u, v) \in \mathbb{R}^2.$$

The properties of this distribution function are analogous to the properties of the univariate distribution functions (right-continuity, left limits, etc. – all of which follow directly from the continuity property of measures).

The notions of joint and marginal distribution laws extend in the obvious way for n-dimensional random vectors $(X_1, X_2, \ldots, X_n)$ and for probability distribution laws defined on $\mathcal{B}(\mathbb{R}^n)$. $\circ$

(3.20) JOINT AND MARGINAL DENSITIES: If the joint law of X and Y is absolutely continuous with respect to Λ^2 ($\mathcal{L}_{X,Y} \lesssim \Lambda^2$), we say that X and Y admit a joint density $\phi_{X,Y} = \mathrm{d}\mathcal{L}_{X,Y}/\mathrm{d}\Lambda^2$. By Tonelli–Fubini's theorem, the existence of joint density implies the existence of marginal densities, which can be recovered from the joint one through the relations

$$\phi_X(x) = \int_{\mathbb{R}} \phi_{X,Y}(x, y)\,\mathrm{d}y, \quad x \in \mathbb{R},$$

and

$$\phi_Y(y) = \int_{\mathbb{R}} \phi_{X,Y}(x, y)\,\mathrm{d}x, \quad y \in \mathbb{R}. \quad \circ$$

(3.21) INDEPENDENCE: The notion of independence is central in probability theory and its applications. Two events, A and B, are said to be *independent* (with respect to P) if $P(AB) = P(A)P(B)$, in which case $P(A|B) = P(A)$ if $P(B) > 0$, and $P(B|A) = P(B)$ if $P(A) > 0$. Heuristically speaking, the independence between A and B means that A contains no information about B and B contains no information about A.

Two real-valued random variables, X and Y, both defined on $(\Omega, \mathcal{F}, P)$, are said to be *independent* (with respect to P) if their joint law $\mathcal{L}_{(X,Y)}$ coincides with the product measure $\mathcal{L}_X \times \mathcal{L}_Y$ on $\mathbb{R}^2$ (the product of their individual distribution laws). Thus, the independence between X and Y amounts to

$$P\big((X,Y) \in A \times B\big) = P(X \in A)\,P(Y \in B) \quad \text{for all } A, B \in \mathcal{B}(\mathbb{R}).$$

Similarly, n real-valued random variables (defined on the same probability space), $X_1, \ldots, X_n$, are said to be *jointly independent* if their joint law in $\mathbb{R}^n$ coincides with the product of their respective individual distribution laws (in $\mathbb{R}$). The members of a given set of random variables, whether finite or infinite, are said to be *independent* if any finite subset of it consists of jointly independent random variables. The members of the same set are said to be *pairwise independent* if every two random variables in the set are independent. We stress that "joint independence" and "pairwise independence" are different notions, which coincide only in some very special cases – see the multivariate Gaussian distribution below. Without a qualifier, "independent" is understood to mean "jointly independent."

Given a probability space $(\Omega, \mathcal{F}, P)$, the sub-σ-fields $\mathscr{F}_1 \subset \mathcal{F}$ and $\mathscr{F}_2 \subset \mathcal{F}$ are said to be *independent* if $P(AB) = P(A)P(B)$ for every choice of $A \in \mathscr{F}_1$ and $B \in \mathscr{F}_2$. A random variable X on $(\Omega, \mathcal{F}, P)$ is said to be independent of the sub-σ-field $\mathscr{F}_0 \subset \mathcal{F}$ if the σ-fields $\sigma(X) \overset{\text{def}}{=} X^{-1}(\mathcal{B}(\mathbb{R}))$ and $\mathscr{F}_0$ are independent. Thus, the independence between the random variables X and Y is the same as the independence between the σ-fields $\sigma(X)$ and $\sigma(Y)$. In particular, the independence between the r.v. X and the random event $A \in \mathcal{F}$ is the same as the independence between the r.v. 1_A and the σ-field $\sigma(X)$. This property comes down to $P(A \cap \{X \in B\}) = P(A)P(X \in B)$ for every $B \in \mathcal{B}(\mathbb{R})$. ○

(3.22) EXERCISE: Given a probability space $(\Omega, \mathcal{F}, P)$ and an integer $n \in \mathbb{N}_{++}$, let $A_i \in \mathcal{F}$, $1 \leq i \leq n$. Prove that the independence of the events $A_1, \ldots, A_n$ is equivalent to the independence of the indicator random variables $1_{A_1}, \ldots, 1_{A_n}$. As a result, the notions "joint independence," "independence," and "pairwise independence" are all meaningful not just for collections of random variables, but also for collections of random events, as well as for collections of σ-fields. ○

The next two examples demonstrate that, generally, joint independence and pairwise independence are different attributes. These examples show up in many texts – one source is (Stoyanov 2014, pp. 13–15) – and are considered classical.[57]

(3.23) EXAMPLE (PAIRWISE, BUT NOT JOINT, INDEPENDENCE): Consider a tetrahedron the four sides of which are marked a, b, c, and abc. Consider an experiment that randomly selects one of the four sides, all sides having the same probability to get selected. Let S denote the selected side, and let A, B, and C, denote, respec-

[57] Several publications point to (Bernstein 1946) as the original source.

tively, the random events $\{a \in S\}$, $\{b \in S\}$, and $\{c \in S\}$. We have

$$P(A) = P(B) = P(C) = \frac{1}{2}, \quad P(AB) = P(AC) = P(BC) = \frac{1}{4},$$

and yet

$$P(ABC) = \frac{1}{4} \neq \frac{1}{8} = P(A)P(B)P(C). \quad \circ$$

(3.24) EXAMPLE (JOINT, BUT NOT PAIRWISE, INDEPENDENCE): Consider the sample space $\Omega = \{\omega_1, \omega_2, \omega_3, \omega_4, \omega_5\}$, and let P denote the probability measure on Ω given by

$$P(\omega_1) = \frac{2}{16}, \quad P(\omega_2) = P(\omega_3) = P(\omega_4) = \frac{3}{16}, \quad \text{and } P(\omega_5) = \frac{5}{16}.$$

For the events $A = \{\omega_1, \omega_2, \omega_3\}$, $B = \{\omega_1, \omega_2, \omega_4\}$, and $C = \{\omega_1, \omega_3, \omega_4\}$, we have

$$P(A) = P(B) = P(C) = \frac{1}{2}, \quad P(ABC) = P(\{\omega_1\}) = \frac{1}{8} = P(A)P(B)P(C),$$

and yet

$$P(AB) = P(AC) = P(BC) = \frac{5}{16} \neq \frac{1}{4}. \quad \circ$$

(3.25) EXERCISE: Let X and Y be any two random variables defined on the same probability space. (a) Assuming that X admits density ϕ_X and Y admits density ϕ_Y, prove that if X and Y are independent, then the joint law of X and Y admits density $\phi_{X,Y}(x, y) = \phi_X(x)\,\phi_Y(y)$, $(x, y) \in \mathbb{R}^2$. (b) Assuming that the joint law of X and Y admits density $\phi_{X,Y}$ and there are two Borel functions, $\phi_X : \mathbb{R} \mapsto \mathbb{R}_+$ and $\phi_Y : \mathbb{R} \mapsto \mathbb{R}_+$, such that $\phi_{X,Y}(x, y) = \phi_X(x)\,\phi_Y(y)$ for all $(x, y) \in \mathbb{R}^2$ and $\int_{\mathbb{R}} \phi_X(x)\,\mathrm{d}x = \int_{\mathbb{R}} \phi_Y(y)\,\mathrm{d}y = 1$, prove that X and Y are independent, and, furthermore, their respective distribution laws admit densities given by ϕ_X and ϕ_Y. $\circ$

(3.26) EXERCISE: Let X and Y be any two independent random variables from the space $\mathscr{L}^1(\Omega, \mathcal{F}, P)$. Prove that $XY \in \mathscr{L}^1(\Omega, \mathcal{F}, P)$ and $\mathsf{E}[XY] = \mathsf{E}[X]\mathsf{E}[Y]$.

HINT: What does Tonelli–Fubini's theorem say? $\circ$

(3.27) CONVOLUTION OF PROBABILITY DISTRIBUTIONS: Let X and Y be any two independent random variables with densities $\phi(x)$ and $\psi(x)$, $x \in \mathbb{R}$. For any bounded Borel function $f\colon \mathbb{R} \mapsto \mathbb{R}$,

$$\begin{aligned}\mathsf{E}\big[f(X+Y)\big] &= \int_{\mathbb{R}}\int_{\mathbb{R}} f(x+y)\phi(x)\psi(y)\,\mathrm{d}x\,\mathrm{d}y \\ &= \int_{\mathbb{R}}\Big(\int_{\mathbb{R}} f(x+y)\phi(x)\,\mathrm{d}x\Big)\psi(y)\,\mathrm{d}y\end{aligned}$$

(changing the variable in the inner integral $x + y \rightarrowtail z$)

$$\begin{aligned}&= \int_{\mathbb{R}}\Big(\int_{\mathbb{R}} f(z)\phi(z-y)\,\mathrm{d}z\Big)\psi(y)\,\mathrm{d}y \\ &= \int_{\mathbb{R}} f(z)\Big(\int_{\mathbb{R}} \phi(z-y)\psi(y)\,\mathrm{d}y\Big)\,\mathrm{d}z\,.\end{aligned}$$

This shows that the density of $X + Y$ must be given by the convolution of the two individual densities, which is defined as the function

$$\mathbb{R} \ni z \rightsquigarrow (\phi * \psi)(z) \overset{\text{def}}{=} \int_{\mathbb{R}} \phi(z-y)\psi(y)\,\mathrm{d}y\,. \quad \circ$$

(3.28) COVARIANCE: An important (and widely used in practice) characteristic is the so-called *covariance* between two random variables, X and Y, defined on the same probability space. The covariance is given by

$$\mathrm{Cov}(X, Y) \overset{\text{def}}{=} \mathsf{E}\big[\big(X - \mathsf{E}[X]\big)\big(Y - \mathsf{E}[Y]\big)\big]\,,$$

provided that all expectations on the right side exist.

If it exists, $\mathrm{Cov}(X, X)$ is nothing but the variance of X, which was introduced in (2.70) and was denoted by $\mathrm{Var}(X)$. Intuitively, $\mathrm{Var}(X)$ is a proxy for the degree of dispersion in a large independent sample from the distribution law of X. Just as an illustration, compare $\mathrm{Var}(X)$ with $\mathrm{Var}(Y)$, where X and Y are binary random variables chosen so that

$$\mathsf{P}(X = 1) = \mathsf{P}(X = -1) = \frac{1}{2} \quad \text{and} \quad \mathsf{P}(Y = 100) = \mathsf{P}(Y = -100) = \frac{1}{2}. \quad \circ$$

(3.29) EXERCISE: Let X and Y be any two r.v. with finite second moments (that is, $\mathsf{E}[X^2] < \infty$ and $\mathsf{E}[Y^2] < \infty$), defined on the same probability space.

(a) Prove that $\mathrm{Var}(X)$ and $\mathrm{Var}(Y)$ are well defined and finite.

(b) Prove that $\mathrm{Cov}(X, Y)$ is well defined and finite; in fact, prove that

$$\big|\mathrm{Cov}(X, Y)\big| \leq \mathrm{Var}(X)^{1/2}\mathrm{Var}(Y)^{1/2}.$$

(c) Prove that if, in addition, X and Y are independent, then $\mathrm{Cov}(X, Y) = 0$, which is the same property as $\mathsf{E}[X\,Y] = \mathsf{E}[X]\,\mathsf{E}[Y]$.

(d) Can one claim that $\mathrm{Cov}(X, Y) = 0$ implies that X and Y are independent?

REMARK: These features make it clear why the covariance is such a useful characteristic: if we know that $\mathrm{Cov}(X, Y) \neq 0$, then we know that X and Y are not independent. $\circ$

(3.30) EXERCISE: Prove that $\mathrm{Cov}(\cdot\,,\,\cdot)$ is a symmetric bilinear function on the product space $\mathscr{L}^2(\Omega,\mathcal{F},P)\times\mathscr{L}^2(\Omega,\mathcal{F},P)$. In addition, prove that if $\mathrm{Cov}(X,Y)$ exists, then $\mathrm{Cov}(X+c,Y)$ also must exist for any fixed $c\in\mathbb{R}$; in fact, prove that $\mathrm{Cov}(X+c,Y)=\mathrm{Cov}(X,Y)$. ○

If X is a random variable with a finite second moment and with $\mathrm{Var}(X)>0$, then the so-called standardized version of X is given by

$$\tilde{X} \stackrel{\text{def}}{=} \frac{X-\mathsf{E}[X]}{\sqrt{\mathrm{Var}(X)}}\,.$$

(3.31) EXERCISE: Prove that $\mathsf{E}[X^2]<\infty$ implies $\mathsf{E}[\tilde{X}^2]<\infty$, $\mathsf{E}[\tilde{X}]=0$, and $\mathrm{Var}(\tilde{X})=1$. ○

(3.32) CORRELATION: Given any two random variables, X and Y, defined on the same probability space and having finite second moments, the correlation coefficient $\mathrm{Cor}(X,Y)$ is given by

$$\mathrm{Cor}(X,Y) \stackrel{\text{def}}{=} \mathrm{Cov}(\tilde{X},\tilde{Y}) \equiv \mathsf{E}\left[\frac{X-\mathsf{E}[X]}{\sqrt{\mathrm{Var}(X)}}\,\frac{Y-\mathsf{E}[Y]}{\sqrt{\mathrm{Var}(Y)}}\right] \equiv \frac{\mathrm{Cov}(X,Y)}{\sqrt{\mathrm{Var}(X)\mathrm{Var}(Y)}}\,. \quad \circ$$

(3.33) EXERCISE: Let X and Y be any two random variables with finite second moments that are defined on the same probability space. Prove that $\mathrm{Cor}(X,Y)\in[-1,1]$ and that $|\mathrm{Cor}(X,Y)|=1$ implies that X and Y are linear transformations of one another; to be precise $\mathrm{Cor}(X,Y)=\pm1$ implies $\tilde{X}=\pm\tilde{Y}$.

HINT: $L^2(\Omega,\mathcal{F},P)$ is a Hilbert space and its subspace spanned by $X-\mathsf{E}[X]$ and $Y-\mathsf{E}[Y]$ is a Euclidean space (of dimension ≤ 2). ○

(3.34) EXERCISE: Suppose that the random variables $X_1,\ldots,X_n$ have finite second moments, and let $S_n \stackrel{\text{def}}{=} X_1+\cdots+X_n$. Prove that S_n also has a finite second moment and

$$\mathrm{Var}(S_n)=\sum_{i=1}^{n}\mathrm{Var}(X_n)+2\sum_{i=1}^{n-1}\sum_{j=i+1}^{n}\mathrm{Cov}(X_i,X_j)\,. \quad \circ$$

(3.35) CONDITIONING WITH RESPECT TO PARTITIONS: Let $(\Omega,\mathcal{F},P)$ be any probability space and let $\mathscr{P}=\{A_1,\ldots,A_n\}$ be any partition of Ω such that $\sigma(\mathscr{P})\subseteq\mathcal{F}$.[58] Let us denote the σ-field $\sigma(\mathscr{P})$ simply by $\mathcal{G}$ and observe that a random variable $Y:\Omega\mapsto\bar{\mathbb{R}}$ is $\mathcal{G}$-measurable if and only if Y is constant on every one of the atoms A_i, $1\leq i\leq n$. To put it another way, any $\mathcal{G}$-measurable random variable must be of the form

$$\dagger \qquad Y(\omega)=\sum_{i=1}^{n}c_i\,1_{A_i}(\omega)\,,\qquad \omega\in\Omega\,,$$

[58] Recall that, by the definition of "partition," all atoms A_i in a partition $\mathscr{P}$ must have strictly positive probabilities $P(A_i)>0$.

for some choice of the scalars $c_i \in \bar{\mathbb{R}}$. Thus, a $\mathcal{G}$-measurable r.v. is nothing but a function from $\mathscr{P}$ into $\bar{\mathbb{R}}$, and any function on $\mathscr{P}$ can also be treated as a function on Ω in the obvious way.

Next, suppose that X is some $\mathcal{F}$-measurable (notation $X \prec \mathcal{F}$) random variable that represents some quantity of interest. In addition, suppose that one cannot determine directly whether the events in $\mathcal{F}$ occur or not, while the events in $\mathcal{G}$ are observable. To be precise, once the outcome $\omega \in \Omega$ is realized, one can determine whether the event A has occurred or not (that is, whether $\omega \in A$ or $\omega \notin A$) as long as $A \in \mathcal{G}$, but not if $A \in \mathcal{F}$, $A \notin \mathcal{G}$. In such a situation one would like to replace the r.v. $X \prec \mathcal{F}$ with a r.v. $Y \prec \mathcal{G}$ – and without losing any of the properties of X as far as the interactions between X and $\mathcal{G}$ go. To be precise, assuming that $\mathsf{E}[X] \equiv \int_\Omega X \,\mathrm{d}P$ is defined, we would like to find some $Y \prec \mathcal{G}$ such that

$$\mathsf{E}[1_A X] \equiv \int_A X \,\mathrm{d}P = \int_A Y \,\mathrm{d}P \equiv \mathsf{E}[1_A Y] \quad \text{for every } A \in \mathcal{G}.$$

Actually, an r.v. $Y \prec \mathcal{G}$ that has this property is rather easy to find; in fact, there can be only one such r.v., namely the r.v. in † with

$$c_i = \frac{1}{P(A_i)} \int_{A_i} X \,\mathrm{d}P \equiv \frac{\mathsf{E}[1_{A_i} X]}{P(A_i)}, \quad 1 \le i \le n.$$

With this particular choice for the values c_i, the r.v. Y in † is called the *conditional expectation* of X relative to the σ-field $\mathcal{G}$ (or, given the σ-field $\mathcal{G}$), and is denoted by $\mathsf{E}[X \,|\, \mathcal{G}]$. We stress that the conditional expectation $\mathsf{E}[X \,|\, \mathcal{G}]$ does not represent a scalar value, and by its very construction it is a $\mathcal{G}$-measurable random variable. Heuristically speaking, $\mathsf{E}[X \,|\, \mathcal{G}]$ is the average of X, given the information encrypted in the partition $\mathscr{P}$. Indeed, $P(A_i)^{-1} P(\cdot)$ is a probability measure on the set A_i (the measure P conditioned to A_i) and $\mathsf{E}[1_{A_i} X]/P(A_i)$ is the expectation (average) of X (restricted to A_i) with respect to this conditional measure on A_i.

Since $\mathsf{E}[X \,|\, \mathcal{G}]$ can be treated as a function defined on $\mathscr{P}$ (a function that takes as arguments the atoms $A_i \in \mathscr{P}$), a more appropriate notation for the conditional expectation might be $\mathsf{E}[X \,|\, \mathscr{P}]$. In some cases we may indeed write $\mathsf{E}[X \,|\, \mathscr{P}]$ instead of $\mathsf{E}[X \,|\, \mathcal{G}]$ if we want to make the dependence on the partition explicit, but the main reason for using the symbol $\mathsf{E}[X \,|\, \mathcal{G}]$ is that the notion of conditional expectation with respect to a given σ-field is meaningful for *any* sub-σ-field $\mathcal{G} \subseteq \mathcal{F}$, not just for σ-fields generated by partitions in $\mathcal{F}$. The rather rudimentary construction that we just described is only a prelude to a much more general (and technical) definition of conditional expectations relative to generic sub-σ-fields of $\mathcal{F}$. Instead of getting sidetracked into the technical aspects of the most general construction, in what follows we will derive all the useful features of the conditional expectation operator only in the case of partitions, which is both elementary and intuitive, and will refer the reader to some standard texts – see (Dudley 2002) or (Parthasarathy 1977), for example – for the rigorous justification of the claim that the same exact features continue to be meaningful and hold for generic (not necessarily partition-generated) sub-σ-fields of $\mathcal{F}$. ○

(3.36) EXERCISE: It is easy to see that $\mathsf{E}[X \mid \mathcal{G}]$ is defined as long as $\mathsf{E}[X]$ is defined. Prove that in this later case $\mathsf{E}\big[\mathsf{E}[X \mid \mathcal{G}]\big]$ is defined as well.

HINT: If $\mathsf{E}[X]$ is defined, then $\mathsf{E}[X \mid \mathcal{G}]$ takes finitely many values either in the interval $]-\infty, \infty]$, or in the interval $[-\infty, \infty[$. ○

(3.37) EXERCISE: Prove that the conditional expectation of conditional expectation equals the conditional expectation. To be precise, let $\mathscr{P}'$ denote another partition from $\mathcal{F}$ that refines $\mathscr{P}$ and gives rise to the σ-field $\mathcal{G}' = \sigma(\mathscr{P}')$ (so that $\mathcal{G} \subseteq \mathcal{G}' \subseteq \mathcal{F}$), and let X be any r.v. such that $\mathsf{E}[X]$ is defined. Prove that (P-a.s.)

$$\mathsf{E}\big[\mathsf{E}[X \mid \mathcal{G}'] \mid \mathcal{G}\big] = \mathsf{E}[X \mid \mathcal{G}] . \quad ○$$

(3.38) EXERCISE: Prove that the conditional expectation relative to $\mathcal{G} = \sigma(\mathscr{P})$, as defined in (3.35), has the following properties (all relations between r.v. are understood P-a.s.):

(a) If $a, b \in \mathbb{R}$ and X and Y are any two integrable r.v., then

$$\mathsf{E}[aX + bY \mid \mathcal{G}] = a\,\mathsf{E}[X \mid \mathcal{G}] + b\,\mathsf{E}[Y \mid \mathcal{G}] .$$

(b) If the r.v. X and Y are such that $\mathsf{E}[X]$ and $\mathsf{E}[Y]$ are both defined and $X \le Y$ then $\mathsf{E}[X \mid \mathcal{G}] \le \mathsf{E}[Y \mid \mathcal{G}]$.

(c) If the r.v. X is such that $\mathsf{E}[X]$ is defined and Y is *any* $\mathcal{G}$-measurable r.v.,[59] then $\mathsf{E}[XY \mid \mathcal{G}] = Y\mathsf{E}[X \mid \mathcal{G}]$.

(d) If $]a, b[\subseteq \mathbb{R}$ is any open interval, the r.v. X is such that $P(X \in\,]a, b[) = 1$ and $\mathsf{E}[|X|] < \infty$, and the function $f :]a, b[\mapsto \mathbb{R}$ is convex and such that $\mathsf{E}[|f(X)|] < \infty$, then (Jensen's inequality for conditional expectations)

$$f\big(\mathsf{E}[X \mid \mathcal{G}]\big) \le \mathsf{E}[f(X) \mid \mathcal{G}] .$$

(e) If the r.v. X is independent of the σ-field $\mathcal{G}$ and such that $\mathsf{E}[X]$ is defined, then $\mathsf{E}[X \mid \mathcal{G}] = \mathsf{E}[X]$ (the conditional expectation relative to an independent σ-field is a constant r.v. equal to the unconditional expectation). ○

(3.39) CONDITIONING WITH RESPECT TO σ-FIELDS: Let $(\Omega, \mathcal{F}, P)$ be any probability space and let $\mathcal{G}$ be any sub-σ-field of $\mathcal{F}$. Just as we did in (3.35) in the special case where $\mathcal{G}$ is generated by a partition, given any $X \in \mathscr{L}^1(\Omega, \mathcal{F}, P)$, it is again possible to construct – see (Dudley 2002, 10.1), for example – a random variable $Y \prec \mathcal{G}$ (Y is measurable for $\mathcal{G}$), which is determined uniquely up to P-equivalence by the relation

$$\dagger \qquad \mathsf{E}[1_A X] \equiv \int_A X \,\mathrm{d}P = \int_A Y \,\mathrm{d}P \equiv \mathsf{E}[1_A Y] \quad \text{for every } A \in \mathcal{G} .$$

The fact that the r.v. Y can be determined from this relation only up to P-equivalence should be clear: nothing will change in † if we replace Y with another

[59] Note that there are no requirements for Y other than $Y \prec \mathcal{G}$.

$\mathcal{G}$-measurable r.v. that is identical to Y everywhere outside some P-negligible set from $\mathcal{G}$. It turns out (this is again a classical result from probability theory – see ibid) that all random variables $Y \prec \mathcal{G}$ that satisfy † for a particular (fixed) $X \in \mathscr{L}^1(\Omega, \mathcal{F}, P)$, represent an element of $L^1(\Omega, \mathcal{G}, P)$. This element is again called the *conditional expectation of X given $\mathcal{G}$*, and is again denoted by $\mathsf{E}[X\,|\,\mathcal{G}]$. Following the well-established tradition, we are going to treat $\mathsf{E}[X\,|\,\mathcal{G}]$ as a $\mathcal{G}$-measurable random variable – this convention is harmless as long as the operations that we subject the conditional expectation to are invariant under P-equivalence, but one must always keep in mind that $\mathsf{E}[X\,|\,\mathcal{G}](\omega)$ is not a well-defined quantity for a fixed $\omega \in \Omega$, even though we pretend that $\mathsf{E}[X\,|\,\mathcal{G}]$ as a "function" on Ω.

Of course, nothing will change in the relation † if we modify X on some P-negligible set from $\mathcal{F}$. This means that $\mathsf{E}[X\,|\,\mathcal{G}]$ (as an equivalence class) actually depends only on the equivalence class from $L^1(\Omega, \mathcal{F}, P)$ that contains X. As a result, $\mathsf{E}[\,\cdot\,|\,\mathcal{G}]$ can be treated as a linear operator of the form

$$L^1(\Omega, \mathcal{F}, P) \ni X \rightsquigarrow \mathsf{E}[X\,|\,\mathcal{G}] \in L^1(\Omega, \mathcal{G}, P)\,.$$

It is well known – see (Dudley 2002), for example – that the properties established in (3.37), and (3.38) for the case of σ-fields generated by partitions, actually hold for conditional expectations with respect to generic (arbitrary) sub-σ-fields $\mathcal{G} \subseteq \mathcal{F}$, as long as the random variables involved (X and Y) are integrable. The integrability requirement can be relaxed by following these standard steps. If $X \in L^0(\Omega, \mathcal{F}, P)$ is positive with probability 1, then by (3.38b) $\mathsf{E}[X \wedge n|\mathcal{G}]$, $n \in \mathbb{N}$, is an a.s. increasing sequence of r.v., so that $\lim_n \mathsf{E}[X \wedge n|\mathcal{G}]$ is a well-defined $\bar{\mathbb{R}}_+$-valued random variable (actually, an equivalence class for the measure P). Thus, if $X \in L^0(\Omega, \mathcal{F}, P)$ is such that either $\mathsf{E}[X^+|\mathcal{G}]$ or $\mathsf{E}[X^-|\mathcal{G}]$ is P-a.s. finite, then $\mathsf{E}[X\,|\,\mathcal{G}]$ can be defined P-a.e. in the obvious way as

$$\mathsf{E}[X\,|\,\mathcal{G}] = \mathsf{E}[X^+|\mathcal{G}] - \mathsf{E}[X^-|\mathcal{G}]\,.$$

If $Y \in L^0(\Omega, \mathcal{G}, P)$ – that is, is Y is *any* r.v. that is measurable for $\mathcal{G}$ – we will suppose that $\mathsf{E}[Y\,|\,\mathcal{G}]$ is defined and equals Y. In addition, if $X \in L^0(\Omega, \mathcal{F}, P)$ is such that that $\mathsf{E}[X\,|\,\mathcal{G}]$ is defined and if $Y, Z \in L^0(\Omega, \mathcal{G}, P)$, then we will suppose that $\mathsf{E}[XY + Z\,|\,\mathcal{G}]$ is defined and equals $X\mathsf{E}[X\,|\,\mathcal{G}] + Z$. ○

(3.40) EXTENDED CONDITIONAL EXPECTATION: It turns out that the conditional expectation $\mathsf{E}[X\,|\,\mathcal{G}]$ can be defined (again, as an equivalence class) in a way that is both meaningful and useful for *any* $X \in L^0(\Omega, \mathcal{F}, P)$ and for any sub-σ-field $\mathcal{G} \subseteq \mathcal{F}$. Specifically, Jacod and Shiryaev (1987) show that "it is very convenient" to work – in fact, develop the entire theory of stochastic processes – with the following extended version of the conditional expectation:

$$\mathsf{E}[X\,|\,\mathcal{G}] \stackrel{\text{def}}{=} \begin{cases} \mathsf{E}[X^+|\mathcal{G}] - \mathsf{E}[X^-|\mathcal{G}] & \text{on } \big\{\mathsf{E}[|X|\,|\mathcal{G}] < \infty\big\}, \\ +\infty & \text{on } \big\{\mathsf{E}[|X|\,|\mathcal{G}] = \infty\big\}, \end{cases}$$

which is indeed perfectly meaningful for *any* $X \in L^0(\Omega, \mathcal{F}, P)$. ○

(3.41) EXERCISE: (a) Prove that if $X \in \mathscr{L}^p(\Omega, \mathcal{F}, P)$ for some $p > 1$ and if $\mathcal{G} \subseteq \mathcal{F}$ is some sub-σ-field of $\mathcal{F}$, then $\mathsf{E}[X \,|\, \mathcal{G}] \in \mathscr{L}^p(\Omega, \mathcal{G}, P)$.

(b) Prove that the conditional expectation operator $\mathsf{E}[\cdot \,|\, \mathcal{G}]$ is continuous in L^p-norm for any $p > 1$: if the sequence $(X_i \in \mathscr{L}^p(\Omega, \mathcal{F}, P))_{i \in \mathbb{N}}$ converges in L^p-norm to $X_* \in \mathscr{L}^p(\Omega, \mathcal{F}, P)$, then $(\mathsf{E}[X_i \,|\, \mathcal{G}])_{i \in \mathbb{N}}$ converges in L^p-norm to $\mathsf{E}[X_* \,|\, \mathcal{G}]$.

(c) Prove that $\mathsf{E}[\cdot \,|\, \mathcal{G}]$ acts as an orthogonal projector from the Hilbert space $L^2(\Omega, \mathcal{F}, P)$ onto the Hilbert subspace $L^2(\Omega, \mathcal{G}, P)$: if $X \in \mathscr{L}^2(\Omega, \mathcal{F}, P)$, then $\mathsf{E}\big[\big(X - \mathsf{E}[X \,|\, \mathcal{G}]\big)Y\big] = 0$ for any $Y \in L^2(\Omega, \mathcal{G}, P)$, so that $X - \mathsf{E}[X \,|\, \mathcal{G}]$ is orthogonal to the subspace $L^2(\Omega, \mathcal{G}, P)$.

HINT: For (a) and (b) use Jensen's inequality from (3.38d). ○

(3.42) CONVERGENCE THEOREMS FOR CONDITIONAL EXPECTATIONS: All convergence theorems for integrals have analogues for conditional expectations. This should not be surprising, given that the construction of conditional expectations relative to partitions comes down to integration over the individual atoms in the partition. In what follows $\mathcal{G}$ denotes an arbitrary sub-σ-field of $\mathcal{F}$.

Monotone convergence: Let $(X_i)_{i \in \mathbb{N}}$ be any sequence of r.v. on $(\Omega, \mathcal{F}, P)$ such that $X_0 \geq 0$ P-a.s. and $X_i \leq X_{i+1}$ P-a.s. for any $i \in \mathbb{N}$ (the sequence is a.s. positive and monotone). Then $\lim_i \mathsf{E}[X_i \,|\, \mathcal{G}] = \mathsf{E}[\lim_i X_i \,|\, \mathcal{G}]$ P-a.s.

Dominated convergence: Let $(X_i)_{i \in \mathbb{N}}$ be any sequence of r.v. on $(\Omega, \mathcal{F}, P)$ and suppose that there is a r.v. $Y \in \mathscr{L}^1(\Omega, \mathcal{F}, P)$ such that $|X_i| \leq Y$ P-a.s. for any $i \in \mathbb{N}$. If $\lim_i X_i$ exists (a.s.) then $\lim_i \mathsf{E}[X_i \,|\, \mathcal{G}] = \mathsf{E}[\lim_i X_i \,|\, \mathcal{G}]$ P-a.s.

Fatou's lemma: If $(X_i)_{i \in \mathbb{N}}$ is any sequence of *positive* r.v. on $(\Omega, \mathcal{F}, P)$, then

$$\mathsf{E}\Big[\liminf_i X_i \;\Big|\; \mathcal{G}\Big] \leq \liminf_i \mathsf{E}\left[X_i \,|\, \mathcal{G}\right] . \quad ○$$

(3.43) CONDITIONING WITH RESPECT TO RANDOM VARIABLES: Given any random variable $X \in \mathscr{L}^1(\Omega, \mathcal{F}, P)$ and any random variable $Y \in \mathscr{L}^0(\Omega, \mathcal{F}, P)$, the conditional expectation $\mathsf{E}[X \,|\, Y]$ is understood as $\mathsf{E}[X \,|\, \sigma(Y)]$ – recall that $\sigma(Y) \stackrel{\text{def}}{=} Y^{-1}(\mathcal{B}(\mathbb{R}))$ is the σ-field generated by Y. Thus, $\mathsf{E}[X \,|\, Y]$ is an equivalence class of random variables that are measurable with respect to $\sigma(Y)$, and,⊸ (2.13) as we are well aware, every $\sigma(Y)$-measurable r.v. can be expressed in the form $\Phi \circ Y = \Phi(Y)$, for some Borel-function $\Phi \colon \mathbb{R} \mapsto \bar{\mathbb{R}}$. In particular, every element of the equivalence class $\mathsf{E}[X \,|\, Y]$ can be written in this form. ○

(3.44) EXERCISE: Let X and Y be two random variables with joint density $\phi_{X,Y}$, and suppose that $\mathsf{E}[|f(X,Y)|] < \infty$ for some Borel function $f \colon \mathbb{R}^2 \mapsto \mathbb{R}$. Prove that $\mathsf{E}[f(X,Y) \,|\, Y] = \Phi(Y)$, where the function $\Phi \colon \mathbb{R} \mapsto \mathbb{R}$ is given by

$$\Phi(y) = \frac{\int_{\mathbb{R}} f(x,y)\phi_{X,Y}(x,y)\,\mathrm{d}x}{\int_{\mathbb{R}} \phi_{X,Y}(x,y)\,\mathrm{d}x} .$$

HINT: Prove that if $g \colon \mathbb{R} \mapsto \mathbb{R}$ is any bounded Borel function, then

$$\mathsf{E}[\Phi(Y)\, g(Y)] = \mathsf{E}[f(X,Y)\, g(Y)] . \quad ○$$

(3.45) BAYES' FORMULA FOR CONDITIONAL EXPECTATIONS: Let $(\Omega, \mathcal{F}, P)$ be any probability space, let $\mathcal{G} \subseteq \mathcal{F}$ be some σ-field inside $\mathcal{F}$, let Q be another probability measure on $\mathcal{F}$ such that $Q \precsim P$, and let $R \overset{\text{def}}{=} \mathrm{d}Q/\mathrm{d}P$. Since⊸ (2.64) $R \in L^1(\Omega, \mathcal{F}, P)$, the conditional expectation $\mathsf{E}^P[R\,|\,\mathcal{G}]$ (understood relative to P) is defined. Let X be any random variable such that $\mathsf{E}^P[|X|R] < \infty$, and notice that this last relation is the same as $X \in L^1(\Omega, \mathcal{F}, Q)$. In particular, the conditional expectation $\mathsf{E}^Q[X\,|\,\mathcal{G}]$ (understood relative to Q) is defined and is given by

† $$\mathsf{E}^Q[X\,|\,\mathcal{G}] = \frac{1}{\mathsf{E}^P[R\,|\,\mathcal{G}]}\,\mathsf{E}^P[XR\,|\,\mathcal{G}]\,.$$

PROOF: Let Y be any bounded and $\mathcal{G}$-measurable r.v. Then

$$\begin{aligned}\mathsf{E}^Q\Big[\frac{Y}{\mathsf{E}^P[R\,|\,\mathcal{G}]}\,\mathsf{E}^P[XR\,|\,\mathcal{G}]\Big] &= \mathsf{E}^P\Big[\frac{YR}{\mathsf{E}^P[R\,|\,\mathcal{G}]}\,\mathsf{E}^P[XR\,|\,\mathcal{G}]\Big]\\ &= \mathsf{E}^P\Big[\frac{Y\,\mathsf{E}^P[R\,|\,\mathcal{G}]}{\mathsf{E}^P[R\,|\,\mathcal{G}]}\,\mathsf{E}^P[XR\,|\,\mathcal{G}]\Big] = \mathsf{E}^P\big[Y\,\mathsf{E}^P[XR\,|\,\mathcal{G}]\big]\\ &= \mathsf{E}^P[XYR] = \mathsf{E}^Q[XY]\,. \quad \circ\end{aligned}$$

(3.46) EXERCISE: Give a complete and detailed proof of (3.45†) without consulting with the proof above. Explain every step. ○

(3.47) EXERCISE: Explain why (3.45†) is often called "the Bayes' rule." ○

3.4 Multivariate Gaussian Distribution

(3.48) MULTIVARIATE GAUSSIAN DENSITY: Any symmetric and strictly positive definite matrix[60] $\Sigma \in \mathbb{R}^{k\otimes k}$ and any vector $m \in \mathbb{R}^k$ give rise to the function[61]

† $$\mathbb{R}^k \ni x \rightsquigarrow f_{m,\Sigma}(x) = \frac{1}{\sqrt{(2\pi)^n \det(\Sigma)}}\, e^{-\frac{1}{2}(x-m)^\intercal(\Sigma^{-1})(x-m)} \in \mathbb{R}_{++}\,,$$

which (see the next exercise) represents a probability density in $\mathbb{R}^k$, called *multivariate Gaussian density* with parameters m and Σ. Clearly, the density for the Gaussian law $\mathcal{N}(a, \sigma^2)$ that was introduced in (2.28d) is a special case of the above with $k = 1$, and with $m = a$ and $\Sigma = \sigma^2 > 0$ treated as scalar objects. The class of distribution laws in $\mathbb{R}^k$ that admit density of the form † we denote by $\mathcal{N}(m, \Sigma)$. ○

(3.49) EXERCISE: Prove that the function introduced in (3.48†) is indeed a probability density. In addition, prove that the attribute "multivariate Gaussian density" does not depend on the choice of the coordinate system, or on the choice of coordinate units. Explain how the parameters m and Σ change as the coordinate system and units change. ○

---

[60] Recall that a matrix $\Sigma \in \mathbb{R}^{k\otimes k}$ is said to be symmetric if $\Sigma = \Sigma^\intercal$, positive definite if $v^\intercal \Sigma v \geq 0$, and strictly positive definite if $v^\intercal \Sigma v > 0$ for any vector $\vec{0} \neq v \in \mathbb{R}^k$.

[61] Σ^{-1} stands for the matrix-inverse of Σ.

(3.50) GAUSSIAN FAMILIES, VECTORS, AND DISTRIBUTIONS: A finite family of $\mathbb{R}$-valued random variables, $(\xi_i)_{1 \le i \le n}$, all defined on the same probability space, is said to be a *Gaussian family* (alias: $(\xi_i)_{1 \le i \le n}$ are jointly Gaussian), or, equivalently, the random vector

$$X = (\xi_1, \dots, \xi_n)$$

is said to be a *Gaussian random vector*, if the following condition is met: X is distributed in some flat[62] of dimension $1 \le k \le n$ inside $\mathbb{R}^n$, with a multivariate Gaussian density (on that flat) that can be expressed in the form (3.48†) for some (arbitrary) choice of a Euclidean coordinate system and coordinate units. If all components of X, that is, all random variables $(\xi_i)_{1 \le i \le n}$, are independent and identically distributed (alias: i.i.d.) with law $\mathcal{N}(0, 1)$, then we say that X is a *standard Gaussian vector*.

A probability measure on $\mathbb{R}^n$, supported on some flat of dimension $1 \le k \le n$ (which flat could be the entire $\mathbb{R}^n$), and in such a way that on that flat it admits a multivariate Gaussian density of the form (3.48†), is called a *multivariate Gaussian measure*, or a *multivariate Gaussian distribution*. For brevity, in some cases we may drop "multivariate," if this attribute is understood from the context.

More generally, any (finite or infinite, including uncountably infinite) family of random variables, defined on the same probability space, is said to be a *Gaussian family* if any finite subfamily of it is Gaussian. ○

(3.51) EXERCISE: Show that if the random vector $X = (\xi_1, \dots, \xi_k)$ is distributed in $\mathbb{R}^k$ with Gaussian density of the form (3.48†), then

$$\mathsf{E}[\xi_i] = m_i \quad \text{and} \quad \mathrm{Cov}(\xi_i, \xi_j) = \Sigma^{i,j}, \quad 1 \le i, j \le k .$$

For this reason the matrix Σ and the vector m in (3.48†) are called, respectively, the *covariance matrix* and the *mean* (alias: *expectation*) *vector* of X. In particular, "X is a standard Gaussian vector" is a synonym for "X is distributed with law $\mathcal{N}(\vec{0}, I_{n \otimes n})$" (notation: $X \in \mathcal{N}(\vec{0}, I_{n \otimes n})$). ○

(3.52) SOME USEFUL FEATURES: The following properties are both very useful and straightforward to check:

(a) The random vector $X = (\xi_1, \dots, \xi_n)$ is Gaussian if and only if the $\mathbb{R}$-valued random variable $u \cdot X$ is Gaussian for every fixed $u \in \mathbb{R}^n$.

(b) If the random vector $X = (\xi_1, \dots, \xi_n)$ is Gaussian, then, for every choice of the integer $k \in \mathbb{N}_{++}$ and the vectors $u_1, \dots, u_k \in \mathbb{R}^n$, the random vector

$$(u_1 \cdot X, \dots, u_k \cdot X)$$

[62] A *flat* inside a Euclidean space is any subset of that space that is congruent to a Euclidean space with dimension at least 1. To put it another way, $V \subseteq \mathbb{R}^n$ is a flat if and only if $(V - a)$ is a linear subspace of $\mathbb{R}^n$ for any $a \in V$. For example, a flat in $\mathbb{R}^3$ could be a line, a plane (not necessarily containing the origin $\vec{0}$), or the entire $\mathbb{R}^3$.

is also Gaussian. In particular, if a family of random variables is Gaussian, then any sub-family of it is Gaussian as well.

(c) For a Gaussian family of random variables, pairwise independence implies joint independence.

(d) If the random variables ξ_i, $1 \leq i \leq 2n+1$, $n \in \mathbb{N}_{++}$, are jointly Gaussian with $\mathsf{E}[\xi_i] = 0$, then $\mathsf{E}\big[\xi_1 \cdots \xi_{2n+1}\big] = 0$ and

$$\mathsf{E}[\xi_1 \cdots \xi_{2n}] = \sum \mathsf{E}[\xi_{i_1}\xi_{j_1}] \cdots \mathsf{E}[\xi_{i_n}\xi_{j_n}],$$

where the summation is taken over all partitions of the set $\{1, 2, \dots, 2n\}$ into n disjoint and unordered pairs $\{i_1, j_1\}, \dots, \{i_n, j_n\}$ (there are exactly $(2n)!/(2^n n!)$ such partitions). ○

(3.53) EXERCISE: Assuming that the random variables ξ and η are jointly Gaussian, prove that ξ and η are independent if and only if $\mathrm{Cov}(\xi, \eta) = 0$. ○

(3.54) EXERCISE: Assuming that ξ is a standard normal r.v. ($\xi \in \mathcal{N}(0, 1)$), give an example of a nontrivial continuous function $f : \mathbb{R} \mapsto \mathbb{R}$ such that $\mathrm{Cov}(\xi, f(\xi)) = 0$. Conclude that, for a generic choice of the random variables ξ and η, the relation $\mathrm{Cov}(\xi, \eta) = 0$ does not imply independence between ξ and η. ○

(3.55) EXERCISE: Let $X \in \mathcal{N}(\vec{0}, I_{n\otimes n})$ be any standard Gaussian vector with dimension n and let $u, v \in \mathbb{R}^n$ be any two vectors of the same dimension as X. Prove that the (jointly Gaussian) random variables $\xi \stackrel{\text{def}}{=} u \cdot X$ and $\eta \stackrel{\text{def}}{=} v \cdot X$ have covariance $\mathsf{E}[\xi\eta] = u \cdot v$. ○

(3.56) EXERCISE: Let $X \in \mathcal{N}(\vec{0}, I_{d\otimes d})$ be any standard Gaussian vector with dimension d and let $U \in \mathbb{R}^{n\otimes d}$ be any orthogonal matrix, that is, any matrix with dimensions (n, d) such that $UU^\intercal = I_{n\otimes n}$. Prove that $Y \stackrel{\text{def}}{=} UX$ is a standard Gaussian vector with dimension n, that is, $Y \in \mathcal{N}(\vec{0}, I_{n\otimes n})$. ○

(3.57) USEFUL OBSERVATION: In the context of the last exercise, repeated independent realizations of the random vector Y, no matter how many, would not provide any information about the structure of the $n \times d$ matrix U, not even information about its second dimension d. A statistical study of Y, no matter how persistent, could only lead to the conclusion that Y is a standard Gaussian vector with dimension n. So to speak, information about the mechanism with which Y was produced, specifically, information about the matrix U, cannot be recovered from observations over Y. More generally, if $(U_i \in \mathbb{R}^{n\otimes d})_{i\in\mathbb{N}}$ is any (arbitrary) sequence of orthogonal matrices, and $(X_i)_{i\in\mathbb{N}}$ is any sequence of i.i.d. copies of a standard Gaussian vector $X \in \mathcal{N}(\vec{0}, I_{d\otimes d})$, then the sequence $(Y_i \stackrel{\text{def}}{=} U_i X_i)_{i\in\mathbb{N}}$ will have the statistical properties of an i.i.d. sequence of standard Gaussian vectors with dimension n. As a result, the (fully deterministic) sequence $(U_i)_{i\in\mathbb{N}}$ will be perfectly encrypted into the sequence $(Y_i)_{i\in\mathbb{N}}$ (and can be decrypted only by knowing the realization of $(X_i)_{i\in\mathbb{N}}$). ○

(3.58) EXERCISE: The statement in exercise (3.56) has the following generalization: Let $X \in \mathcal{N}(\vec{0}, I_{d\otimes d})$ and let $\Sigma \in \mathbb{R}^{n\otimes d}$ be any matrix with dimensions (n, d). Prove that $Y \stackrel{\text{def}}{=} \Sigma X$ is a Gaussian vector with distribution law $\mathcal{N}(\vec{0}, \Sigma\Sigma^{\intercal})$. ○

(3.59) EXERCISE: Suppose that the random variables ξ and η are jointly Gaussian with $\mathsf{E}[\xi] = 1$, $\mathsf{E}[\eta] = 2$, $\text{Var}(\xi) = 4$, $\text{Var}(\eta) = 9$, and $\text{Cov}(\xi, \eta) = 3$. Give the joint density of ξ and η. ○

(3.60) EXERCISE: For $\xi \in \mathcal{N}(0, \sigma^2)$, calculate $\mathsf{E}[\xi^4]$, $\mathsf{E}[\xi^6]$, and $\mathsf{E}[\xi^8]$. ○

(3.61) DEFINITION (SKEWNESS AND KURTOSIS): The kurtosis of a random variable X with a finite 4th moment is defined as the quantity

$$\mathsf{E}\left[\left(\frac{\xi - \mathsf{E}[\xi]}{\sqrt{\text{Var}(\xi)}}\right)^4\right] \equiv \frac{\mathsf{E}\left[\left(\xi - \mathsf{E}[\xi]\right)^4\right]}{\mathsf{E}\left[\left(\xi - \mathsf{E}[\xi]\right)^2\right]^2}.$$

This is simply the 4th standardized moment. The 3rd standardized moment (assuming that X has a finite 3rd moment), namely

$$\mathsf{E}\left[\left(\frac{\xi - \mathsf{E}[\xi]}{\sqrt{\text{Var}(\xi)}}\right)^3\right]$$

is called skewness of ξ. ○

(3.62) EXERCISE: Calculate the skewness and the kurtosis of a Gaussian r.v. with law $\mathcal{N}(a, \sigma^2)$. ○

(3.63) EXERCISE: Suppose that the r.v. X is exponentially distributed with parameter $c \in \mathbb{R}_{++}$. What is the skewness and the kurtosis of X, treated as functions of the parameter c? How does the shape of the density of X change as the kurtosis of X increases? ○

(3.64) EXERCISE: Suppose that X is a Gaussian r.v. with $\mathsf{E}[X] = 0$. Let $\sigma^2 = \mathsf{E}[X^2]$, and let $f\colon \mathbb{R} \mapsto \mathbb{R}$ be any function with continuous first derivative such that

$$\lim_{x\to\pm\infty} \partial f(x)\, e^{-x^2/\sigma^2} = 0\,.$$

As an example, the last condition would be satisfied if the derivative $x \rightsquigarrow \partial f(x)$ has at most polynomial growth at $\pm\infty$ (in particular, if the first derivative is bounded). Prove the following *integration by parts* relation

$$\mathsf{E}[f(X)\,X] = \mathsf{E}[\partial f(X)]\,\mathsf{E}[X^2]\,,$$

understood as the claim that the integrability of either of the random variables $f(X)\,X$ and $\partial f(X)$ implies the integrability of the other one and implies the identity. Explain why this identity deserves to be called "integration by parts." ○

(3.65) EXERCISE: Let $\{X, Y_1, \dots, Y_n\}$ be any collection of $n + 1$ jointly Gaussian random variables that are defined on the same probability space and are such that

$\mathsf{E}[X] = \mathsf{E}[Y_1] = \ldots = \mathsf{E}[Y_n] = 0$. Prove that if $f\colon \mathbb{R}^n \mapsto \mathbb{R}$ is any function with continuous partial derivatives that have at most polynomial growth at $\pm\infty$, then[63]

$$\mathsf{E}[f(Y_1,\ldots,Y_n)\,X] = \sum_{i=1}^{n} \mathsf{E}[\partial_i f(Y_1,\ldots,Y_n)]\,\mathsf{E}[Y_i X]\,.$$

Explain why all expectations in this identity are well defined.

HINT: The statement can be reduced to the one in the previous exercise. ○

(3.66) EXERCISE: Let $\{X, Y_1, \ldots, Y_n\}$ be any collection of $n+1$ jointly Gaussian random variables on $(\Omega, \mathcal{F}, P)$ such that $\mathsf{E}[X] = \mathsf{E}[Y_1] = \ldots = \mathsf{E}[Y_n] = 0$, and let $\mathcal{G} = \sigma(Y_1,\ldots,Y_n)$ be the σ-field generated by $(Y_i)_{1\le i\le n}$. Prove that $\mathsf{E}[X\,|\,\mathcal{G}]$ is nothing but the orthogonal projection in $L^2(\Omega,\mathcal{F},P)$ of the r.v. X onto the *linear* subspace

$$\big\{a_1Y_1 + \cdots a_nY_n \colon a = (a_1,\ldots,a_n) \in \mathbb{R}^n\big\}.$$

In particular, there is a vector $a^\circ = (a_1^\circ,\ldots,a_n^\circ) \in \mathbb{R}^n$ such that

$$\mathsf{E}[X\,|\,\mathcal{G}] = a_1^\circ Y_1 + \cdots + a_n^\circ Y_n\,.$$

HINT: Use the fact that for jointly Gaussian r.v. the attributes "uncorrelated" and "independent" are the same, and argue that the random variable $X - \mathsf{E}[X\,|\,\mathcal{G}]$ is independent of $\mathcal{G}$. ○

(3.67) CHOLESKY FACTORIZATION AND SIMULATION OF GAUSSIAN LAWS: Most computing systems have readily available standard functions that return pseudo-random i.i.d. sequences from the univariate Gaussian law $\mathcal{N}(a,\sigma^2)$, with user supplied values for the parameters $a,\sigma \in \mathbb{R}$. However, typically, samples from the multivariate Gaussian law $\mathcal{N}(m,\Sigma)$, for a given expectation vector m and covariance matrix Σ, will have to be made. According to exercise (3.58), sampling from such laws boils down to sampling from $\mathcal{N}(0,1)$, with which one can easily produce samples from $\mathcal{N}(\vec{0}, I)$, and extracting a "square root" from the matrix Σ, that is, finding a matrix S such that $SS^\intercal = \Sigma$. This is always possible to do as long as Σ is positive definite, but, in general, there are going to be infinitely many such roots. The matter is purely computational, and the efficiency of the numerical algorithm involved matters, since in many practical applications the covariance matrix Σ could be very large and not necessarily strictly positive definite. The most common approach is to use the so-called *Cholesky decomposition*:[64] for every positive definite matrix Σ there is a lower triangular matrix L that is unique if Σ is strictly positive definite, and is such that $\Sigma = LL^\intercal$. As most computing systems provide standard functions that take Σ as an argument and return the corresponding L from the Cholesky decomposition, at least one square root, namely $S = L$, is readily available. Thus, in order to generate a single sample from $\mathcal{N}(m,\Sigma)$, one must first

63 This relation is often referred to as "Stein's lemma."

64 Named after the French mathematician André-Louis Cholesky (1875–1918).

generate a sample $X \in \mathcal{N}(\vec{0}, I)$, by sampling len($X$)-times from $\mathcal{N}(0,1)$, and then transform X into $m + SX$, which is distributed with law $\mathcal{N}(m, SS^{\intercal} = \Sigma)$.

A (generally different) square root of Σ can be constructed just as easily from the eigenvalues and the eigenvectors of the matrix Σ, which are also readily available through standard functions that take Σ as an argument and return the eigenvectors arranged in a matrix U and the eigenvalues arranged in a vector λ. The matrix U is orthogonal ($UU^{\intercal} = U^{\intercal}U = I$) and the matrix Σ can be expressed as

$$\Sigma = UDU^{\intercal},$$

where $D = \mathrm{Dgn}(\lambda)$ is the diagonal matrix with diagonal given by the vector λ. This relation is nothing but the familiar spectral decomposition of the matrix Σ. Since $D = D^{\intercal}$, and since the eigenvalues of a positive definite matrix are positive, it is straightforward to check that $S = U\mathrm{Dgn}(\lambda^{1/2})$ is a square root of Σ. Numerical efficiency aside, one advantage of this method is that in situations where Σ is not strictly positive definite,[65] in which case some entries in λ may be 0, it allows one to identify the actual linear subspace on which the distribution is supported. In addition, the spectral decomposition of the covariance matrix is often needed for other reasons and may already be available.

Both methods are illustrated with concrete examples, including the relevant computer code, in appendix B.2. ○

3.5 Hermite–Gauss Quadratures

Strictly speaking, the topic reviewed in this section belongs to the realm of numerical analysis. Nevertheless, as many common financial technologies rely on some form of averaging against Gaussian distribution laws, we would be remiss if the most basic tools used in such calculations are left out.

A *quadrature rule* for a given measure μ on $\mathcal{B}(R)$ is a finite set of abscissas, $(x_i \in \mathbb{R})_{i \in \mathbb{N}_{|n}}$, and a set of associated weights, $(w_i \in \mathbb{R}_{++})_{i \in \mathbb{N}_{|n}}$, all chosen so that for any sufficiently "nice" function $f \colon \mathbb{R} \mapsto \mathbb{R}$ the linear combination

$$\sum\nolimits_{i \in \mathbb{N}_{|n}} w_i f(x_i)$$

provides a reasonably good approximation of the integral $\int_{\mathbb{R}} f(x)\mu(\mathrm{d}x)$.[66] At least intuitively, if the integrand f is continuous and integrable with respect to μ, the quadrature rule could be made arbitrarily precise if the number of abscissas, n, is to be increased arbitrarily. Clearly, quadrature rules must be designed so that for any fixed n the rule is as efficient as possible. In what follows we will be concerned

[65] In most practical applications the covariance matrix Σ obtains from statistical analysis of a particular data set, so that one cannot impose a priori what this matrix should look like.

[66] On most computing systems the numerical evaluation of integrals with continuous integrands essentially boils down to using one quadrature rule or another.

exclusively with the case where the measure μ is some Gaussian distribution law $\mathcal{N}(a,\sigma^2)$, in which case the integral $\int_{\mathbb{R}} f(x)\mu(\mathrm{d}x)$ is nothing but the expected value $\mathsf{E}[f(X)]$ with $X \in \mathcal{N}(a,\sigma^2)$. Since the law $\mathcal{N}(a,\sigma^2)$ can be identified as the law of $\sigma\tilde{X} + a$ for $\tilde{X} \in \mathcal{N}(0,1)$, it would be enough for us to develop quadrature rules only for the distribution law $\mathcal{N}(0,1)$. The first step in this program is to decide how the efficiency of a particular quadrature rule is to be understood. One common approach is to choose the abscissas $(x_i \in \mathbb{R})_{i\in\mathbb{N}_{|n}}$ and the weights $(w_i \in \mathbb{R}_{++})_{i\in\mathbb{N}_{|n}}$ so that the rule would give the exact value of the integral whenever f is a polynomial of degree $\deg(f)$ no greater than $k \in \mathbb{N}_{++}$ for the largest possible k (given the number of abscissas $n+1 \in \mathbb{N}_{++}$). Since any polynomial f with $\deg(f) \le n$ is completely determined by the values $(f(x_i))_{i\in\mathbb{N}_{|n}}$, one may suspect that the best (that is, the largest) k that can be achieved with $n+1$ abscissas is $k = n$. Somewhat surprisingly, it turns out that with $n+1$ abscissas one can design a quadrature rule which is exact for any polynomial of degree up to $k = 2n+1$. This involves a well-known trick – see (Abramowitz and Stegun 1962), for example – that utilizes some of the features of the Hermite polynomials,[67] which show up in many applications involving Gaussian random variables.

(3.68) HERMITE POLYNOMIALS: Throughout the literature one can find Hermite polynomials defined either as

$$h_n(x) \stackrel{\text{def}}{=} (-1)^n e^{\frac{1}{2}x^2}\,\partial_x^n\left(e^{-\frac{1}{2}x^2}\right) \quad\text{or as}\quad H_n(x) \stackrel{\text{def}}{=} (-1)^n e^{x^2}\,\partial_x^n\left(e^{-x^2}\right), \quad x \in \mathbb{R}.$$

The first form appears to be preferable to probabilists, while the second one appears to be preferable to physicists and engineers. The distinction between the two forms is inessential, however, since switching from one to the other is merely a matter of rescaling. To be precise, we have

$$h_n(x) = 2^{-n/2} H_n\left(x/\sqrt{2}\right) \quad\Leftrightarrow\quad H_n(x) = 2^{n/2} h_n\left(x\sqrt{2}\right), \quad x \in \mathbb{R}.$$

The polynomials $h_n(x)$, $n \in \mathbb{N}$, are often called *normalized Hermite polynomials* – they are "normalized" in that the leading coefficient for the highest power is 1. The list of the first six normalized Hermite polynomials is the following:

$$h_0(x) = 1,\ h_1(x) = x,\ h_2(x) = x^2 - 1,\ h_3(x) = x^3 - 3x,$$
$$h_4(x) = x^4 - 6x^2 + 3,\ h_5(x) = x^5 - 10x^3 + 15x.$$

The next exercise provides a tool that makes the expansion of this list completely straightforward. ○

(3.69) EXERCISE: Prove the relations: (a) $h_{n+1}(x) = xh_n(x) - \partial h_n(x)$, $n \in \mathbb{N}$; (b) $\partial h_{n+1}(x) = (n+1)h_n(x)$, $n \in \mathbb{N}$; (c) $\int_{\mathbb{R}} h_n(x)\ell_{0,1}(x)\mathrm{d}x = 0$, or, equivalently, $\mathsf{E}[h_n(X)] = 0$ for any $X \in \mathcal{N}(0,1)$ and any $n \in \mathbb{N}_{++}$. ○

[67] Named after the French mathematician Charles Hermite (1822–1901).

(3.70) EXERCISE: Prove that $h_n(x)$ has n distinct real roots $x_1 < x_2 < \cdots < x_n$ for any $n \in \mathbb{N}_{++}$.

HINT: Consider using induction, in conjunction with Rolle's theorem. ○

(3.71) EXERCISE: Prove that the polynomials $\left(\frac{1}{\sqrt{n!}} h_n\right)_{n \in \mathbb{N}}$ are orthonormal elements of the Hilbert space $L^2\big(\mathbb{R}, \mathcal{B}(\mathbb{R}), \mathcal{N}(0,1)\big)$, in that

$$\frac{1}{\sqrt{2\pi}} \int_{\mathbb{R}} h_n(x)^2 e^{-\frac{x^2}{2}} \, dx = n! \quad \text{and} \quad \frac{1}{\sqrt{2\pi}} \int_{\mathbb{R}} h_n(x) h_m(x) e^{-\frac{x^2}{2}} \, dx = 0 \,,$$

for any $m, n \in \mathbb{N}$, $m \neq n$. ○

(3.72) EXERCISE: Given any $n \in \mathbb{N}$, prove that any polynomial P of degree $\deg(P) \leq n$ can be expressed as a linear combination of the Hermite polynomials $\{h_0(x), h_1(x), \ldots, h_n(x)\}$. Conclude that if $\deg(P) \leq n$, then

$$\frac{1}{\sqrt{2\pi}} \int_{\mathbb{R}} h_{n+1}(x) P(x) e^{-\frac{x^2}{2}} \, dx = 0 \,.$$

In other words, for any $n \in \mathbb{N}$, the polynomial $h_{n+1}(x)$ is orthogonal to the linear subspace $\mathcal{P}_n \subset L^2\big(R, \mathcal{B}(R), \mathcal{N}(0,1)\big)$, spanned by all polynomials P with $\deg(P) \leq n$. ○

(3.73) THEOREM: The normalized Hermite polynomials $\left(\frac{1}{\sqrt{n!}} h_n\right)_{n \in \mathbb{N}}$ form an orthonormal basis in the Hilbert space $L^2\big(\mathbb{R}, \mathcal{B}(\mathbb{R}), \mathcal{N}(0,1)\big)$. This statement is equivalent to the claim that if the r.v. X, defined on $(\Omega, \mathcal{F}, P)$, has law $\mathcal{N}(0,1)$, then $\left(\frac{1}{\sqrt{n!}} h_n(X)\right)_{n \in \mathbb{N}}$ is an orthonormal basis in the Hilbert space $L^2(\Omega, \sigma(X), P)$. ○

Now we have available all of the ingredients needed to develop a quadrature rule that has the features mentioned earlier. To get started, given any $n \in \mathbb{N}$, choose an arbitrary polynomial Q with $\deg(Q) \leq 2n+1$, and observe that, generally, the polynomial division of Q by h_{n+1} would produce a remainder R, which is a polynomial with $\deg(R) \leq n$. Thus, we can write

$$Q(x) = P(x) h_{n+1}(x) + R(x) \,, \quad x \in \mathbb{R} \,,$$

for polynomials P and R with $\deg(P) \leq n$ and $\deg(R) \leq n$. Let $\{x_0, x_1, \ldots, x_n\}$ denote$^{\multimap (3.70)}$ the $n+1$ distinct real roots of $h_{n+1}(x)$, and observe that $Q(x_i) = R(x_i)$, $i \in \mathbb{N}_{|n}$, and that$^{\multimap (3.72)}$

$$\frac{1}{\sqrt{2\pi}} \int_{\mathbb{R}} Q(x) e^{-\frac{x^2}{2}} \, dx = \frac{1}{\sqrt{2\pi}} \int_{\mathbb{R}} R(x) e^{-\frac{x^2}{2}} \, dx \,.$$

Remarkably, now we see that any quadrature rule with abscissas $\{x_0, x_1, \ldots, x_n\}$ that is exact for polynomials of degree at most n would be exact also for polynomials of degree no greater than $2n+1$. The only remaining step is to attach weights to the abscissas $\{x_0, x_1, \ldots, x_n\}$ so that the quadrature rule would be exact for polynomials of degree no greater than n, and, according to (3.72), this is the same as developing a quadrature rule that is exact for each of the polynomials $\{h_0(x), h_1(x), \ldots, h_n(x)\}$. Thus, the weights $\{w_0, w_1, \ldots, w_n\}$ will have to solve the following system of $n+1$ equations:

$$\sum_{i \in \mathbb{N}_{|n}} w_i h_k(x_i) = \frac{1}{\sqrt{2\pi}} \int_{\mathbb{R}} h_k(x) e^{-\frac{x^2}{2}} \, \mathrm{d}x \,, \quad k = 0, 1, \ldots, n \,.$$

After some manipulation – perhaps with the aid of computer algebra – one can show that this system has a unique solution given by

$$w_i = \frac{n!}{(n+1)\, h_n(x_i)^2} \,, \quad i \in \mathbb{N}_{|n} \,.$$

To summarize: a quadrature rule with $n+1$ abscissas chosen to be the $n+1$ distinct real roots of $h_n(x)$, and with weights computed as above, would be exact for all polynomials of degree at most $2n+1$. As an aside, the above weights define a probability distribution on the set $\{x_0, x_1, \ldots, x_n\}$. The essence of this method, then, comes down to approximating the distribution $\mathcal{N}(0,1)$ with a special discrete probability distribution, amassed on the set of roots of the normalized Hermite polynomial $h_n(x)$.

4

Convergence of Random Variables

Intuitively, after flipping a coin 1,000 times one would expect the observed number of heads to be close to 500. Yet, it would be very surprising if this number is exactly equal to 500. The main objective in the present chapter is to develop the necessary tools that allow us to model, and eventually explain, such phenomena – and not just in the context of tossing a coin.

4.1 Types of Convergence for Sequences of Random Variables

Our goal in this section is to introduce the various types of convergence for sequences of random variables defined on a common probability space $(\Omega, \mathcal{F}, P)$. For simplicity, we will be concerned only with real-valued random variables, but emphasize that nothing will change if we replace everywhere the real line with some generic Polish space, and, accordingly, replace the Euclidean distance $|x-y|$ with the distance $d(x, y)$ in that space. The only exception is the convergence in distribution, for which the transition from the range space $\mathbb{R}$ to a generic Polish space $\mathbb{S}$ is nontrivial and somewhat technical, while the transition from $\mathbb{R}$ to $\mathbb{R}^n$ essentially comes down to adjusting the notation accordingly – see (4.18).

(4.1) CONVERGENCE ALMOST SURELY (A.S.): Let $X = (X_i)_{i \in \mathbb{N}}$ be *any* sequence of random variables on $(\Omega, \mathcal{F}, P)$. One can think of X as a family of scalar sequences labeled by the set Ω, namely, $X(\omega) = (X_i(\omega))_{i \in \mathbb{N}}$, $\omega \in \Omega$. It is not difficult to see that the collection of all $\omega \in \Omega$ at which $X(\omega)$ converges to a finite limit can be identified as the set on which the random variable $\lim_n \sup_{i,j \geq n} |X_i - X_j|$ takes the value 0. Similarly, the collection of all $\omega \in \Omega$ at which $X(\omega)$ converges to $+\infty$ (converges to $-\infty$) can be identified as the set on which the random variable $\lim_n \inf_{i \geq n} X_i$ (the random variable $\lim_n \sup_{i \geq n} X_i$) equals $+\infty$ (equals $-\infty$). Thus, whether empty or not, the set

$$\left\{\omega \in \Omega \colon \left(X_i(\omega)\right)_{i \in \mathbb{N}} \text{ has a limit in } \bar{\mathbb{R}}\right\}$$

is always measurable. If this set happens to be of full probability (its complement is P-negligible), then we say that the sequence X converges almost surely, or converges P-almost surely, which we abbreviate as "a.s.," or as "P-a.s." ○

(4.2) REMARKS: We stress that the a.s. convergence depends on the measure P only to the extent to which P determines which sets from the σ-field $\mathcal{F}$ are sets of measure zero. Thus, if we replace P with another probability measure, Q, which is equivalent to P ($Q \approx P$), then the a.s. convergence will not change: any sequence of random variables that converges a.s. with respect to P converges a.s. with respect to Q and vice versa. If, however, Q is absolutely continuous with respect to P ($Q \precsim P$), then Q may have a larger collection of null-sets than P. If this is the case, then it would be easier for the sequence X to converge a.s. with respect to Q rather than P. To put it another way, if $Q \precsim P$, then a.s. convergence with respect to P implies a.s. convergence with respect to Q, so that "convergence P-a.s." is a stronger property than "convergence Q-a.s." ○

Now we turn to another type of convergence that will be instrumental for much of what follows in this space. In some sense, the particular notion of convergence that we are about to introduce is what distinguishes probability theory from real analysis and other domains of mathematics.

(4.3) CONVERGENCE IN PROBABILITY: Let $X = (X_i)_{i \in \mathbb{N}}$ be some sequence of random variables on $(\Omega, \mathcal{F}, P)$, that is, some sequence in $\mathscr{L}^0(\Omega, \mathcal{F}, P)$. We say that X converges in probability if there is a random variable $X_* \in \mathscr{L}^0(\Omega, \mathcal{F}, P)$ such that the sequence (of real numbers)

† $$\big(P(|X_i - X_*| > \varepsilon)\big)_{i \in \mathbb{N}}$$

converges to 0 for every fixed $\varepsilon \in \mathbb{R}_{++}$. Any random variable X_* that has this property is said to be a limit in probability P, or a P-limit, of the sequence X. Clearly, if X_* is a P-limit and if $X'_* \in \mathscr{L}^0(\Omega, \mathcal{F}, P)$ is such that $X'_* = X_*$ P-a.s., then X'_* is a P-limit of X as well. The convergence in probability is therefore convergence in the space of P-equivalence classes of random variables, that is, convergence in the space $L^0(\Omega, \mathcal{F}, P)$. ○

The intuition behind the convergence in probability is clear: the probability for X_i to differ from its limit X_* by more than any fixed amount $\varepsilon \in \mathbb{R}_{++}$ could be made arbitrarily small if the index i is chosen to be sufficiently large. To a large extent, the importance of the convergence in probability can be attributed to the following remarkable result.

(4.4) THEOREM (CONVERGENCE IN PROBABILITY EXPLAINED IN TERMS OF CONVERGENCE A.S.): (Dudley 2002, 9.2.1) The sequence $X = (X_i)_{i \in \mathbb{N}}$ converges in probability if and only if every subsequence $(X_{i_j})_{j \in \mathbb{N}}$ contains a sub-subsequence, $(X_{i_{j_k}})_{k \in \mathbb{N}}$, which converges P-a.s. ○

(4.5) COMMENTS: There are two remarkable corollaries from the last theorem. The first one is that convergence in probability only depends on the collection of sets of probability 0. It does not depend on the actual probabilities of the events, except when those probabilities happen to be 0. Thus, if $Q \approx P$, then the convergence in probability Q is no different from the convergence in probability P. If, however, $Q \precsim P$, then convergence in probability P implies (but cannot be claimed to be equivalent to) convergence in probability Q. To put it another way, if $Q \lesssim P$, then "convergence in probability P" is a stronger property than "convergence in probability Q."

Another important and straightforward corollary from (4.4) is that convergence P-a.s. implies convergence in probability P. Generally, however, "convergence P-a.s." is a stronger property than "convergence in probability P," and the next exercise clarifies this claim. ○

(4.6) EXERCISE: Give an example of a sequence of random variables that converges in probability but not almost surely.

HINT: Take $\Omega = [0,1]$ endowed with the (uniform) Lebesgue measure and consider the sequence of the indicators of the intervals $[0,1]$, $[0,\frac{1}{2}]$, $[\frac{1}{2},1]$, $[0,\frac{1}{4}]$, $[\frac{1}{4},\frac{2}{4}]$, $[\frac{2}{4},\frac{3}{4}]$, $[\frac{3}{4},1]$, $[0,\frac{1}{8}]$, $[\frac{1}{8},\frac{2}{8}]$ .... ○

Nevertheless, there are situations in which convergence in probability implies convergence a.s., and the next exercise describes one such situation.

(4.7) EXERCISE: Suppose that the sequence of random variables $X = (X_i)_{i \in \mathbb{N}}$ is a.s. monotone, in that $X_i \leq X_{i+1}$ P-a.s. for all $i \in \mathbb{N}$ or $X_i \geq X_{i+1}$ P-a.s. for all $i \in \mathbb{N}$. In addition, suppose that for some r.v. X_*, $X_i \to X_*$ in probability P. Prove that the convergence $X_i \to X_*$ takes place also P-a.s.

HINT: The convergence in probability implies that there is a subsequence $(X_{i_j})_{j \in \mathbb{N}}$ that converges P-a.s. (to X_*). Argue that there is a set E with $P(E) = 0$ such that $(X_i(\omega))_{i \in \mathbb{N}}$ is a monotone sequence for every $\omega \in \Omega \setminus E$ and $(X_{i_j}(\omega))_{j \in \mathbb{N}}$ converges to $X_*(\omega)$ for every $\omega \in \Omega \setminus E$. Explain why this last property implies that $(X_i(\omega))_{i \in \mathbb{N}}$ converges to $X_*(\omega)$ for every $\omega \in \Omega \setminus E$. ○

(4.8) THE KY FAN METRIC: The importance of the Cauchy criterion in calculus is clear: it allows one to determine whether a sequence converges or not without the need to identify the limit explicitly. An analogous description of the convergence in probability is also possible. To formulate such a result, we must first introduce the right kind of "distance in probability P" between two random variables, X and Y. One such choice is the ingenious Ky Fan metric (or distance),[68] given by

$$\dagger \quad d_P(X,Y) = \inf\{\varepsilon \in \mathbb{R}_+ : P(|X-Y| > \varepsilon) \leq \varepsilon\}, \quad X,Y \in \mathcal{L}^0(\Omega,\mathcal{F},P).$$

It is well known – see (Dudley 2002, 9.2.2), for example, or (Fan 1944) – that the Ky Fan distance metrizes the convergence in probability: $X_i \to X_*$ in probability P if and only if $d_P(X_i, X_*) \to 0$. It is not difficult to check that the convergence in probability can also be metrized by any of the following two metrics:

$$d'(X,Y) = \mathsf{E}\left[\frac{|X-Y|}{1+|X-Y|}\right] \quad \text{or} \quad d''(X,Y) = \mathsf{E}\big[|X-Y| \wedge 1\big]. \quad ○$$

Metrics that characterize the convergence in probability are instrumental in risk management, specifically, in the study of risk measures. This topic lies outside

[68] Named after the American mathematician Ky Fan (1914–2010).

of our main concerns, and we refer the reader to, for example, (Rachev et al. 2011) for more details and references.

(4.9) CAUCHY CRITERION FOR CONVERGENCE IN PROBABILITY: (Dudley 2002, 9.2.3) The sequence of random variables $X = (X_i)_{i \in \mathbb{N}}$ converges in probability P if and only if

$$\lim_i \sup_{j \geq i} d_P(X_i, X_j) = 0\,,$$

that is, if and only if the sequence X is a Cauchy sequence for the distance $d_P(\cdot, \cdot)$. In particular, *$L^0(\Omega, \mathcal{F}, P)$ is a complete metric space for the Ky Fan distance.* ○

It is interesting that although the a.s. convergence is not metrizable (there is no metric on the space of random variables the convergence in which is equivalent to the a.s. convergence), it is still possible to establish the a.s. convergence within the sequence, that is, without identifying the limit. The precise result is the following:

(4.10) CAUCHY CRITERION FOR CONVERGENCE A.S.: (Dudley 2002, 9.2.4) The sequence of random variables $X = (X_i)_{i \in \mathbb{N}}$ converges P-a.s. if and only if

$$\lim_i P\Big(\sup_{j \geq i} |X_j - X_i| \geq \varepsilon\Big) = 0 \quad \text{for every } \varepsilon \in \mathbb{R}_{++}\,. \quad ○$$

(4.11) EXERCISE: Consider two sequences of random variables, $X = (X_i)_{i \in \mathbb{N}}$ and $Y = (Y_i)_{i \in \mathbb{N}}$, not necessarily defined on the same probability space, and suppose that these two sequences are statistically indistinguishable, in the sense that from observing either sequence it would not be possible to tell whether the observed sequence is X or Y.[69] Prove that if either of these two sequences converges in probability (converges a.s.), then the other one must also converge in probability (must converge a.s.).

HINT: Argue that $d_P(X_i, X_j) = d_P(Y_i, Y_j)$ for any $i, j \in \mathbb{N}$. ○

In many practical applications it is preferable to work with averages rather than probabilities, and, for this reason, it would be useful to describe the convergence of random variables in terms of moments. One major drawback from this approach is that, generally, such a convergence would not be invariant under equivalent changes of the probability measure – a type of invariance that is essential in asset pricing, and not only in asset pricing. Nevertheless, convergence described in terms of moments is still instrumental for both theory and practice. This is nothing but the convergence⊸ (2.82) in the Banach space $L^p(\Omega, \mathcal{F}, P)$, which is already familiar to us.

(4.12) L^p CONVERGENCE: Given any $p \in [1, \infty[$, the sequence of random variables $X = (X_i \in \mathscr{L}^p(\Omega, \mathcal{F}, P))_{i \in \mathbb{N}}$ is said to converge in L^p-sense, or in L^p-norm, if there is a random variable $X_* \in \mathscr{L}^p(\Omega, \mathcal{F}, P)$, such that

$$\lim_i \mathsf{E}[|X_i - X_*|^p] = 0\,. \quad ○$$

[69] This property is equivalent to the claim that for any finite set of integers $\{i_1, \ldots, i_k\}$, the random vectors $(X_{i_1}, \ldots, X_{i_k})$ and $(Y_{i_1}, \ldots, Y_{i_k})$ share the same distribution law in $\mathbb{R}^k$.

The L^p-convergence can be connected with the convergence in probability and the a.s. convergence, as we now detail.

(4.13) EXERCISE: Prove that for any $p \in [1, \infty[$ convergence in L^p-norm implies convergence in probability and, therefore, implies the existence of a subsequence that convergence a.s.

HINT: It is enough to consider the case $p = 1$ (why?). With $p = 1$, use Chebyshev's inequality. ○

There are situations where all that one can claim is convergence of the distribution laws, without any reference to probability measures and spaces – after all, all that one can deduce from practical experiments are the distribution laws of the observed variables. How should the convergence of the distribution laws $\mathcal{L}_{X_i}$ be understood? One may be tempted to define the convergence of $\mathcal{L}_{X_i}$ to mean that there is a distribution law $\mathcal{L}_*$ such that

$$\lim_i \mathcal{L}_{X_i}(B) = \mathcal{L}_*(B) \quad \text{for every} \quad B \in \mathcal{B}(\mathbb{R}), \qquad \mathrm{e}_1$$

but one quickly realizes that, however "intuitive," this approach is in fact naive. Indeed, most basic intuition would suggest that the unit point-mass distribution laws $\epsilon_{1/i}$ converge (as distributions) to ϵ_0, yet with $\mathcal{L}_{X_i} \stackrel{\text{def}}{=} \epsilon_{1/i}$ and $\mathcal{L}_* \stackrel{\text{def}}{=} \epsilon_0$ condition e_1 fails for $B =]-1, 0]$, or for $B =]0, 1[$. This simple observation shows that e_1 is simply too much to ask for, and, instead of insisting that this property must hold for every $B \in \mathcal{B}(\mathbb{R})$, it would be more practical to require that it holds only for Borel sets $B \subset \mathbb{R}$ such that $\mathcal{L}_*(\partial B) = 0$, where ∂B is the set of all border points of B, defined as⊸ (A.11) the set difference between the closure of B and the interior of B (so that ∂B is closed by definition), that is, $\partial B \stackrel{\text{def}}{=} \bar{B} \backslash B^\circ$. As an example, $\partial(]-1, 0]) = \{0\} \cup \{1\}$, $\partial(]0, 1[) = \{0\} \cup \{1\}$, and it is not difficult to see that e_1 indeed holds with $\mathcal{L}_{X_i} = \epsilon_{1/i}$ as long as 0 is not a border point for B. It turns out – see (Dudley 2002, 11.1.1) – that the requirement $\lim_i \mathcal{L}_{X_i}(B) = \mathcal{L}_*(B)$ for any $B \in \mathcal{B}(\mathbb{R})$ such that $\mathcal{L}_*(\partial B) = 0$ is equivalent to the weak convergence $\mathcal{L}_{X_i} \Rightarrow \mathcal{L}_*$ defined in (4.14) below. Furthermore, a result known as the *Helley–Bray theorem* – see (ibid., 11.1.2) – states that the weak convergence $\mathcal{L}_{X_i} \Rightarrow \mathcal{L}_*$ is equivalent to the claim that $F_{X_i}(x) \to F_*(x)$ for any $x \in \mathbb{R}$ at which the limiting distribution function $F_*(\cdot) \stackrel{\text{def}}{=} \mathcal{L}_*(]-\infty, \cdot\,])$ happens to be continuous.

(4.14) WEAK CONVERGENCE (DEFINITION): The sequence of probability measures $(P_i)_{i \in \mathbb{N}}$, all defined on $\mathcal{B}(\mathbb{R})$, is said to converge weakly to the probability measure P_*, also defined on $\mathcal{B}(\mathbb{R})$, or $P_i \Rightarrow P_*$ for short, if

$$\lim_i \int_{\mathbb{R}} f \, \mathrm{d}P_i = \int_{\mathbb{R}} f \, \mathrm{d}P_* \quad \text{for any} \quad f \in \mathscr{C}_b(\mathbb{R}; \mathbb{R}).$$

The random variables $(X_i)_{i \in \mathbb{N}}$ are said to converge in distribution (or converge in law) to the random variable X_* if $\mathcal{L}_{X_i} \Rightarrow \mathcal{L}_{X_*}$ (understood as week convergence of probability measures on $\mathcal{B}(\mathbb{R})$). ○

The next theorem shows that the convergence in distribution is the weakest form of convergence for random variables that we have encountered so far.

(4.15) THEOREM: (Dudley 2002, 9.3.5) Any sequence of random variables that converges in probability converges also in distribution. ○

(4.16) EXERCISE: Can a sequence of random variables converge in distribution but not in probability?

HINT: Let $X_1, X_2, \ldots$ be any sequence of independent and identically distributed (i.i.d.) random variables. Can this sequence converge in probability? Does it converge in distribution? ○

(4.17) CONVERGENCE OF CHARACTERISTIC FUNCTIONS: One immediate consequence of the weak convergence $\mathcal{L}_{X_i} \Rightarrow \mathcal{L}_{X_*}$ is that⊸(3.12) the characteristic functions of $\mathcal{L}_{X_i}$ converge to the characteristic function of $\mathcal{L}_{X_*}$ in pointwise sense everywhere in $\mathbb{R}$. It is important to identify conditions that would allow us to make this claim in the opposite direction (and to also clarify what "opposite direction" means). A standard tool in probability theory – often used in conjunction with the central limit theorem that will be introduced later in this chapter – is the following result, which we borrow from (Dudley 2002, 9.5.5):

THEOREM: Let $(\mathcal{L}_i)_{i \in \mathbb{N}}$ be any sequence of probability laws on $\mathbb{R}$, and let H_i denote the characteristic function of $\mathcal{L}_i$. Suppose that the collection of probability measures $(\mathcal{L}_i)_{i \in \mathbb{N}}$ is (see below) uniformly tight, and that there is a function $H_* : \mathbb{R} \mapsto \mathbb{R}$ such that $H_i(u) \to H_*(u)$ as $i \to \infty$ for every $u \in \mathbb{R}$. Then there is a probability law, $\mathcal{L}_*$, on $\mathbb{R}$ such that the characteristic function of $\mathcal{L}_*$ is exactly H_* and $\mathcal{L}_i \Rightarrow \mathcal{L}_*$.

Of course, we must define the meaning of "uniformly tight" and the definition is the following: a collection, $\mathbb{P}$, of probability measures, defined on the Borel σ-field in some topological space $\mathbb{X}$, is said to be *uniformly tight* if for any given (and fixed) $\varepsilon \in \,]0,1[$ one can find a compact set $K_\varepsilon \subseteq \mathbb{X}$ such that $P(K_\varepsilon) > 1-\varepsilon$ simultaneously for all $P \in \mathbb{P}$. Any sequence of distribution laws on $\mathbb{R}$ that converges weakly (to some distribution law of $\mathbb{R}$) is automatically uniformly tight – see (ibid., 9.3.4). A (single) probability measure, P, on $\mathbb{X}$ is said to be tight, if the singleton $\{P\}$, understood as a collection of measures, is uniformly tight. It should not come as a surprise that one of the simplest tools for establishing uniform tightness, used in many common situations, is Chebyshev's inequality.

A particularly interesting and useful variation of the theorem above is the following powerful result (ibid., 9.8.2):

LÉVY'S CONTINUITY THEOREM: Let $(\mathcal{L}_i)_{i \in \mathbb{N}}$ be any sequence of probability laws on $\mathbb{R}$ and let H_i denote⊸(3.12) the characteristic function of $\mathcal{L}_i$. Suppose that there is a function $H_* : \mathbb{R} \mapsto \mathbb{R}$ that is continuous at 0 and is such that $H_i(u) \to H_*(u)$ as $i \to \infty$ for every $u \in \mathbb{R}$. Then there is a probability law, $\mathcal{L}_*$, on $\mathbb{R}$ such that its characteristic function is exactly H_* and $\mathcal{L}_i \Rightarrow \mathcal{L}_*$. ○

(4.18) REMARK: All notions of convergence for sequences of random variables, together with all related results from this section, except for the Helley–Bray theorem, can be restated for $\mathbb{R}^n$-valued random variables in the obvious way. For example, in the Lévy theorem the continuity of the function H_* at 0 is to be understood as continuity at 0 in each coordinate, with the remaining coordinates being fixed. In fact, all results related to convergence a.s., or convergence in probability, including the definition of the Ky Fan distance, can be restated for random variables taking values in a generic Polish space $\mathbb{S}$ (complete separable metric space). In particular, the space $L^0(\Omega, \mathcal{F}, P; \mathbb{S})$ is a complete metric space for the Ky Fan distance. ○

(4.19) EXERCISE: Let $X = (X_i)_{i \in \mathbb{N}}$ and $Y = (Y_i)_{i \in \mathbb{N}}$ be any two sequences of real random variables that converge in distribution to the random variables X_* and Y_*, respectively. In addition, suppose that X_i and Y_i are independent for every $i \in \mathbb{N}$. Prove that the random variables X_* and Y_* are independent and the sequence of $\mathbb{R}^2$-valued r.v. $(X_i, Y_i)_{i \in \mathbb{N}}$ converges in distribution to (X_*, Y_*).

HINT: By the dominated convergence theorem, every characteristic function, in particular the one associated with (X_*, Y_*), must be continuous – this will be established again in exercise (16.19). Use the two-dimensional version of Lévy's continuity theorem to show that the characteristic function of (X_*, Y_*) is the product of the characteristic functions of X_* and Y_*. Conclude that the law of (X_*, Y_*) is the product of the laws of X_* and Y_*. ○

4.2 Uniform Integrability

Suppose that ξ is some $\mathbb{R}$-valued r.v. defined on $(\Omega, \mathcal{F}, P)$ and observe that the integrability of ξ – in other words, the property $\mathsf{E}[|\xi|] < \infty$ – is the same as $\mathsf{E}[|\xi| 1_{\{|\xi| \geq c\}}] < \infty$ for some, and therefore for any, constant $c \in \mathbb{R}_{++}$. Thus, if the r.v. ξ happens to be integrable, the dominated convergence theorem would imply that

$$\lim_{c \to \infty} \mathsf{E}\left[|\xi| 1_{\{|\xi| \geq c\}}\right] = 0 . \qquad \text{e}_1$$

In particular, the last condition implies that $\mathsf{E}\left[|\xi| 1_{\{|\xi| \geq c\}}\right] < \infty$ for any sufficiently large $c \in \mathbb{R}_{++}$, so that condition e_1 is just another way of saying that ξ is an integrable random variable.

Now suppose that Ξ is some (arbitrary) collection of random variables on $(\Omega, \mathcal{F}, P)$. If Ξ is an integrable family, in the sense that every member of Ξ is an integrable random variable, then condition e_1 must hold for every $\xi \in \Xi$. For reasons that will become clear soon, it is important to distinguish the case where the random variables in the family Ξ are not just all integrable, but are, so to speak, "integrable in an orchestra," in the sense that the convergence in e_1 is uniform for all $\xi \in \Xi$, which property can be expressed as

$$\lim_{c\to\infty} \sup_{\xi \in \Xi} \mathsf{E}\big[|\xi| 1_{\{|\xi| \geq c\}}\big] = 0 .$$

(4.20) DEFINITION: If the last condition is satisfied, then the family Ξ is said to be uniformly integrable (abbreviation: u.i.). ○

(4.21) EXERCISE: Prove that if the family Ξ is uniformly integrable, then there is a constant $C \in \mathbb{R}_{++}$ such that $\mathsf{E}[|\xi|] \leq C$ for any $\xi \in \Xi$. ○

The most common and straightforward way to test for uniform integrability is by using the following powerful result.

(4.22) THEOREM (SUFFICIENT CONDITION FOR U.I.): (Meyer 1966, II 22) Suppose that Ξ is some family of integrable random variables. Then the family Ξ is uniformly integrable if and only if there is a positive and increasing convex function $f : \mathbb{R}_+ \mapsto \mathbb{R}_+$ such that $\lim_{x\to\infty} \frac{f(x)}{x} = \infty$[70] and $\sup_{\xi \in \Xi} \mathsf{E}[f(|\xi|)] < \infty$. The "if" part in this statement holds without the requirement for f to be convex. ○

As an immediate and straightforward corollary of the last theorem, if the L^2-norms of all r.v. in the family Ξ are uniformly bounded, in that there is a constant $C > 0$ such that $\mathsf{E}[\xi^2] \leq C$ for every $\xi \in \Xi$, then the family Ξ can be declared uniformly integrable. The same claim can be made with the L^2-norm replaced by any L^p-norm with $p > 1$, but not with $p = 1$.

Another powerful result related to uniform integrability is the following.

(4.23) THEOREM: (Meyer 1966, II 21) Let $(\xi_i)_{i \in \mathbb{N}}$ be any sequence of integrable random variables that converges a.s. to a random variable ξ_*. Then $\mathsf{E}[|\xi_*|] < \infty$ and the convergence $\xi_i \to \xi_*$ takes place also in L^1-norm if and only if the family $(\xi_i)_{i \in \mathbb{N}}$ is uniformly integrable. Furthermore, if the random variables $(\xi_i)_{i \in \mathbb{N}}$ are all positive and $\mathsf{E}[\xi_*] < \infty$, then $(\xi_i)_{i \in \mathbb{N}}$ is a uniformly integrable family if and only if $\mathsf{E}[\xi_i] \to \mathsf{E}[\xi_*]$. If this last condition is satisfied, then $\xi_i \to \xi_*$ also in L^1-norm (due to the first assertion). ○

We see from the last result why uniform integrability is such a useful feature: to establish that the a.s. convergence $\xi_i \to \xi_*$ takes place also in L^1-norm it suffices to show that $(\xi_i)_{i \in \mathbb{N}}$ is u.i. – say, by using the result in (4.22). The next result reveals yet another important aspect of the notion "uniformly integrable," which is instrumental in the theory of stochastic processes and is often used without a warning or a reference.

(4.24) THEOREM: (Dellacherie 1972, V 9) Let ξ be any integrable random variable on $(\Omega, \mathcal{F}, P)$. Then the collection of all random variables on $(\Omega, \mathcal{F}, P)$ that can be expressed as conditional expectations of the form $\mathsf{E}[\xi \mid \mathcal{G}]$, for all possible choices of the sub-σ-field $\mathcal{G} \subseteq \mathcal{F}$, is a uniformly integrable family. ○

The next exercise describes a situation in which uniform integrability is rather easy to establish.

[70] That is, f converges to $+\infty$ faster than x, as with, say, $f(x) = x^2$, or $f(x) = e^x$.

(4.25) EXERCISE: Let $\mathscr{A}$ be any family of random variables and suppose that there is an integrable r.v., η, that dominates the family $\mathscr{A}$, in that $|\xi| \leq |\eta|$ P-a.s. for all $\xi \in \mathscr{A}$. Prove that the family $\mathscr{A}$ is uniformly integrable.

This statement admits the following generalization: if $\mathscr{A}$ and $\mathscr{B}$ are any two families of random variables, defined on the same probability space, if the family $\mathscr{B}$ is uniformly integrable, and if $\mathscr{B}$ dominates $\mathscr{A}$, in the sense that for every $\xi \in \mathscr{A}$ there is some $\eta \in \mathscr{B}$ such that $|\xi| \leq |\eta|$ (a.s.), then the family $\mathscr{A}$ must be uniformly integrable as well. ○

4.3 Sequences of Independent Random Variables and Events

In our study of the coin toss space $(\Omega_\infty, \mathscr{F}_\infty, P_\infty)$ we were able to construct the probability measure P_∞ without any reference to the notion of independence. The key step in the construction of P_∞ was to postulate that all sets in the partition $\mathscr{P}_t$, for any $t \in \mathbb{N}_{++}$, have the same probability equal to 2^{-t}. As we are about to see, this construction was only meant to ensure that the coordinate mappings $\mathcal{X}_t \colon \Omega_\infty \mapsto \{-1, +1\}$ are independent random variables – just as the outcomes from repeated coin tosses are assumed to be.

(4.26) EXERCISE: Consider the coin toss space $(\Omega_\infty, \mathcal{F}_\infty, P_\infty)$, which was introduced in section 1.4, and prove that for any $s, t \in \mathbb{N}$,

$$P_\infty\big(\mathcal{X}_{t+s} = +1 |\, A\big) = P_\infty\big(\mathcal{X}_{t+s} = -1 |\, A\big) = \frac{1}{2} \quad \text{for any } \; A \in \sigma(\mathscr{P}_t).$$

Conclude that the coordinate mappings $(\mathcal{X}_t)_{t \in \mathbb{N}}$ are independent – not just pairwise independent! – random variables on $(\Omega_\infty, \mathscr{F}_\infty, P_\infty)$. ○

One consequence of the last exercise is that for every choice of the sequence $\big(\varepsilon_t \in \{-1, +1\}\big)_{t \in \mathbb{N}}$, the events $A_t \overset{\text{def}}{=} \{\mathcal{X}_t = \varepsilon_t\}$, $t \in \mathbb{N}$, can be claimed to be independent – again, not just pairwise independent.

The purpose of this section is to investigate the behavior of sequences of independent random variables or events. To gain some insight into the nature of such sequences, observe that for some outcomes $\omega \in \Omega_\infty$ the series $\sum_{t=1}^{\infty} (\mathcal{X}_t(\omega)/t)$ may converge to a finite limit, while for other outcomes $\omega \in \Omega_\infty$ it may not converge, or it may converge to $+\infty$ or to $-\infty$. For example, $\sum_{n=1}^{\infty} \frac{1}{n} = \infty$, while $\sum_{n=1}^{\infty} \frac{(-1)^n}{n} = -\log(2)$. This observation raises the question: if the coin toss sequence $\omega \in \Omega_\infty$ is sampled at random from the probability law P_∞, what is the probability $P_\infty(A)$ of the event

$$A = \Big\{\omega \in \Omega_\infty \colon \sum\nolimits_{t \in \mathbb{N}_{++}} \frac{\mathcal{X}_t(\omega)}{t} \text{ converges to a finite limit} \Big\}? \qquad e_1$$

Somewhat surprisingly, we will establish in (4.36) that $P_\infty(A) = 1$. We have yet to develop the tools that will allow us to arrive at this conclusion, and are going to focus for now at a much more general (but less precise) result: for a random event such as the one in e_1, either $P_\infty(A) = 0$ or $P_\infty(A) = 1$. To be able to formulate such a result, consider an arbitrary sequence of independent random variables $(\xi_i)_{i \in \mathbb{N}}$ and define the *finite tail σ-fields* for this sequence as

$$\mathcal{T}_i \stackrel{\text{def}}{=} \sigma\{\xi_i, \xi_{i+1}, \dots\}, \quad i \in \mathbb{N}.$$

Then define the (infinite) *tail σ-field* as the intersection of all finite tail σ-fields:

$$\mathcal{T}_\infty \stackrel{\text{def}}{=} \bigcap\nolimits_{i \in \mathbb{N}} \mathcal{T}_i .$$

(4.27) EXERCISE: Prove that the random event e_1 belongs to the tail σ-field for the sequence of coordinate mappings $(\mathcal{X}_t)_{t \in \mathbb{N}}$ on Ω_∞. ○

Now we come to an important theorem:

(4.28) KOLMOGOROV'S 0–1 LAW: (Dudley 2002, 8.4.4) Let $(\xi_i)_{i \in \mathbb{N}}$ be any sequence of *independent* r.v. defined on $(\Omega, \mathcal{F}, P)$, let $\mathcal{T}_\infty \subseteq \mathcal{F}$ be the tail σ-field associated with this sequence, and let $A \in \mathcal{T}_\infty$ be any tail event (for the sequence $(\xi_i)_{i \in \mathbb{N}}$). Then either $P(A) = 0$ or $P(A) = 1$. ○

Closely related to the study of tail events is the following important result, which is the promised followup to (1.50):

(4.29) SECOND BOREL–CANTELLI LEMMA: (Dudley 2002, 8.3.4) Let $(\Omega, \mathcal{F}, P)$ be any probability space and let $(A_i \in \mathcal{F})_{i \in \mathbb{N}}$ be any sequence of *independent* events. Then⊶(1.49)

$$\sum\nolimits_{i \in \mathbb{N}} P(A_i) = \infty \quad \text{implies} \quad P(A_i \text{ i.o.}) = 1 . \quad ○$$

The next three exercises provide interesting – and, as we will soon realize, rather useful – applications of the first and the second Borel–Cantelli lemma.

(4.30) EXERCISE: Consider again the coin toss space $(\Omega_\infty, \mathcal{F}_\infty, P_\infty)$. Given a sequence $\omega = \big(\varepsilon_i \in \{-1, +1\}\big)_{i \in \mathbb{N}_{++}} \in \Omega_\infty$, a run of size $k \geq 2$ inside ω is any string of k consecutive tokens $\{\varepsilon_j, \varepsilon_{j+1}, \dots \varepsilon_{j+k-1}\}$, such that $\varepsilon_j = \varepsilon_{j+1} = \dots = \varepsilon_{j+k-1}$. Prove that for any fixed $k \geq 2$ the probability that a randomly chosen sequence $\omega \in \Omega_\infty$ has infinitely many runs of size k is equal to 1. Conclude that the probability for a randomly chosen sequence $\omega \in \Omega_\infty$ to contain infinitely many runs of any possible finite size is also equal to 1. ○

(4.31) EXERCISE: Let $(X_i)_{i \in \mathbb{N}}$ be any sequence of (not necessarily independent) r.v. on $(\Omega, \mathcal{F}, P)$, chosen so that $\sum_{i \in \mathbb{N}} \mathsf{E}[|X_i|] < \infty$. Prove that $\lim_i X_i = 0$ P-a.s.

HINT: The property $P(X_i \to 0) = 1$ is the same as the property

$$P\Big(\big\{|X_i| > \varepsilon\big\} \text{ i.o}\Big) = 0 \quad \text{for any fixed } \varepsilon > 0.$$

Use Chebyshev's inequality and⊶(1.50) the first Borel–Cantelli lemma. ○

(4.32) EXERCISE: Let $(X_i)_{i \in \mathbb{N}}$ be any sequence of (not necessarily independent) identically distributed r.v. on $(\Omega, \mathcal{F}, P)$, chosen so that $\mathsf{E}[|X_i|] < \infty$. Prove that $\lim_i\big(\frac{1}{i}X_i\big) \to 0$ P-a.s.

HINT: Use the hint to the previous exercise. Notice that $P(|X_i| = \infty) = 0$ and observe that, for any $\varepsilon > 0$ and any $i \in \mathbb{N}$,

$$\sum_{i \in \mathbb{N}} i\,\varepsilon\, P\big(i\,\varepsilon < |X_0| \leq (i+1)\,\varepsilon\big) \;\leq\; \mathsf{E}[|X_0|],$$

where

$$P\big(i\,\varepsilon < |X_0| \leq (i+1)\,\varepsilon\big) = P\big(|X_i| > i\,\varepsilon\big) - P\big(|X_{i+1}| > (i+1)\,\varepsilon\big).$$

Conclude that

$$\sum_{i \in \mathbb{N}_{++}} P\big(|X_i| > i\,\varepsilon\big) \;\leq\; \frac{\mathsf{E}[|X_0|]}{\varepsilon}. \quad \circ$$

The following two classical results are instrumental for the study of sums of independent random variables. Detailed proofs can be found in many texts in probability theory – see (Dudley 2002, sec. 9.7), for example.

(4.33) LÉVY'S EQUIVALENCE THEOREM: Let $(X_i)_{i \in \mathbb{N}}$ be any sequence of independent $\mathbb{R}$-valued random variables. Then the following statements are equivalent:

(a) The series $\sum_{i \in \mathbb{N}} X_i$ converges a.s.

(b) The series $\sum_{i \in \mathbb{N}} X_i$ converges in probability.

(c) The series $\sum_{i \in \mathbb{N}} X_i$ converges in distribution, in the sense that the sequence of partial sums $\big(\sum_{i \in \mathbb{N}_{|n}} X_i\big)_{n \in \mathbb{N}}$ converges in distribution.

If these statements are false, then $\sum_{i \in \mathbb{N}} X_i$ diverges a.s. $\circ$

(4.34) REMARK: The last result is interesting only if the random variables X_i can change signs. Indeed, if all X_i are positive, condition (a), and therefore also (b) and (c), would be satisfied in a trivial way, and without the requirement for the X_i's to be independent – see exercise (4.7). $\circ$

(4.35) THREE-SERIES THEOREM: Let $(X_i)_{i \in \mathbb{N}}$ be any sequence of independent $\mathbb{R}$-valued random variables. Then the equivalent conditions (a), (b), and (c) from (4.33) are also equivalent to the requirement that all of the following three series (of real numbers) converge:

$$\sum_{i \in \mathbb{N}} P(|X_i| \geq 1), \quad \sum_{i \in \mathbb{N}} \mathsf{E}\big[X_i 1_{\{|X_i| \leq 1\}}\big], \quad \sum_{i \in \mathbb{N}} \mathrm{Var}\big(X_i 1_{\{|X_i| \leq 1\}}\big). \quad \circ$$

(4.36) EXERCISE: Prove that the random event in e_1 has full probability, that is, $P_\infty(A) = 1$.

HINT: Use the three series theorem. What can you say about the variance of $\sum_{t \in \mathbb{N}_{++}} \frac{\mathcal{X}_t(\omega)}{t}$? $\circ$

4.4 Law of Large Numbers and the Central Limit Theorem

Let $X = (X_i)_{i \in \mathbb{N}_{++}}$ be any sequence of random variables on $(\Omega, \mathcal{F}, P)$ and let

$$S_n \overset{\text{def}}{=} X_1 + \cdots + X_n , \quad n \in \mathbb{N}_{++} .$$

(4.37) WEAK AND STRONG LAWS OF LARGE NUMBERS: The sequence X is said to satisfy the *weak law of large numbers* if there is a constant, $c \in \mathbb{R}$, such that $S_n/n \to c$ in probability P as $n \to \infty$. If the convergence $S_n/n \to c$ takes place P-a.s., then we say that X satisfies the *strong law of large numbers*. ○

(4.38) EXERCISE: Prove that if $\mathsf{E}[X_i^2] = 1$, $\mathsf{E}[X_i] = 0$, and $\mathsf{E}[X_i X_j] = 0$ for all $i, j \in \mathbb{N}_{++}$, $i \neq j$, then the sequence X satisfies the weak law of large numbers.

HINT: Show that $S_n/n \to 0$ in the L^2 sense. ○

(4.39) EXERCISE: Suppose that the random variables X_i, $i \in \mathbb{N}_{++}$, are independent and identically distributed (i.i.d.) with finite mean $a = \mathsf{E}[X_i]$ and finite variance $\sigma^2 = \mathrm{Var}(X_i)$. Prove that $S_n/n \to a$ in probability.

HINT: Apply the result from (4.38) to the sequence $\big((X_i - a)/\sigma\big)_{i \in \mathbb{N}_{++}}$. ○

In fact, the following much stronger result is in force (Dudley 2002, 8.3.5):

(4.40) STRONG LAW OF LARGE NUMBERS FOR I.I.D. SEQUENCES: Suppose that the random variables $(X_i)_{i \in \mathbb{N}_{++}}$ are independent and identically distributed with $\mathsf{E}[|X_i|] < \infty$ and let $a = \mathsf{E}[X_i]$. Then $S_n/n \to a$ with probability 1. Furthermore, if $\mathsf{E}[|X_i|] = \infty$, then with probability 1 the sequence $(S_n/n)_{n \in \mathbb{N}_{++}}$ does not converge to any finite limit. ○

The practical implications of the strong law of large numbers are very broad. As an example, suppose that X is any r.v. and let $f : \mathbb{R} \mapsto \mathbb{R}$ be any Borel-measurable function such that $\mathsf{E}[|f(X)|] < \infty$. In addition, suppose that one can generate any number of independent draws from the law $\mathcal{L}_X$; in other words, suppose that a perfect random numbers generator for the law of X is readily available. The strong law of large numbers then guarantees that if $x_1, \ldots, x_n$ are the results of n such (independent) draws from $\mathcal{L}_X$, then the arithmetic average

$$\frac{f(x_1) + \cdots + f(x_n)}{n}$$

is close to the expected value $\mathsf{E}[f(X)]$, provided that the size of the sample, n, is sufficiently large. This observation is the very basis for the Monte Carlo method – one of the most widely used computing technologies in finance and in many other branches of science and engineering.

However, as is often the case, solving one problem brings up another (usually, harder) problem. Indeed, while the theorem in (4.40) says that for large values of n the the quantity S_n/n is going to be close to the expected value a, the theorem is not telling us how is the random variable S_n/n distributed around the constant a for some fixed (and a priori chosen) sample size n. This is obviously a problem because if we were to consider the average in e₁ to be a reasonably good guess for the expected value $\mathsf{E}[f(X)]$, we would still like to know how big is the probability for this guess to be very wrong. It may be intuitive that by choosing the size of the sample, n, to be sufficiently large this probability could be made as small as desired, but at this point we have no way of knowing how large is large enough. To illustrate this observation, consider again the random walk $\mathcal{Z} = (\mathcal{Z}_t)_{t \in \mathbb{N}}$ defined on the coin toss space $(\Omega_\infty, \mathcal{F}_\infty, P_\infty)$. The law of large numbers tells us that $\mathcal{Z}_t/t \to 0$ with probability 1, but no matter how large is the number of periods t, the random variable $\mathcal{Z}_t/t$ can still take the values $+1$ or -1 with strictly positive probability (with probability 2^{-t} to be precise) and, clearly, $+1$ and -1 would be unacceptably crude estimates of the constant 0. At this point we have no tools at hand that would allow us to estimate the probability for the average $\mathcal{Z}_t/t$ to fall into other domains that are still away from 0. What is the probability $P_\infty\big(\mathcal{Z}_t/t \geq 1/2\big)$, for example? Thus, a general tool that would allow us to estimate such probabilities is instrumental for both theory and practice. The following fundamental result, which we replicate from (Dudley 2002, 9.5.6), provides one such tool:

(4.41) CENTRAL LIMIT THEOREM (CLT): Let $X = (X_i)_{i \in \mathbb{N}_{++}}$ be any sequence of independent and identically distributed (i.i.d.) random variables such that $\mathsf{E}[|X_i|^2] < \infty$, and let $a \overset{\text{def}}{=} \mathsf{E}[X_i]$ and $\sigma^2 \overset{\text{def}}{=} \operatorname{Var}(X_i)$. Then the sequence

$$\frac{S_n - n\,a}{\sigma\sqrt{n}} \equiv \frac{X_1 + \cdots + X_n - n\,a}{\sigma\sqrt{n}}, \qquad n \in \mathbb{N}_{++},$$

converges⊸• (4.14) in distribution to the standard normal law $\mathcal{N}(0,1)$. ○

Notice that if the law of $\frac{S_n - n\,a}{\sigma\sqrt{n}}$ is close to $\mathcal{N}(0,1)$, then the law of $\frac{S_n}{n} - a$ is close to $\mathcal{N}(0,\sigma^2/n)$; in other words, the average S_n/n approximates the mean a with Gaussian error of mean 0 and variance σ^2/n.

One truly remarkable aspect of the central limit theorem is the conclusion that the limiting distribution of the sequence $\frac{S_n - n\,a}{\sigma\sqrt{n}}$ does not depend on the shape of the distribution of the r.v. X_i. Somehow, any distribution whatsoever, however skewed, which has finite mean and variance, leads to a standard normal distribution after summation over independent samples and appropriate normalization! It is also remarkable that, while it converges in distribution, in general, the sequence $\frac{S_n - n\,a}{\sigma\sqrt{n}}$ cannot be claimed to converge in probability.[71]

(4.42) EXERCISE: Let $(\mathcal{X}_t)_{t \in \mathbb{N}}$ denote the usual coordinate mappings on the coin toss space Ω_∞ (recall that $\mathcal{X}_0 = 0$ by convention). Using Monte Carlo simulation, produce 100 independent observations of the r.v. $(\mathcal{X}_0 + \cdots + \mathcal{X}_{100})/10$. Plot the histogram from these 100 samples against the standard normal density. ○

[71] It is always possible to construct a special probability space $(\tilde{\Omega}, \tilde{\mathcal{F}}, \tilde{P})$, on which there is an i.i.d. sequence $(\tilde{X}_i)_{i \in \mathbb{N}_{++}}$ with $\mathcal{L}_{\tilde{X}_i} = \mathcal{L}_{X_i}$ such that $\tilde{S}_n/n \to a$ in probability $\tilde{P}$. In general, however, without reconstructing the probability space one can only claim that the convergence is in distribution.

(4.43) EXERCISE: Let $X \in \mathcal{N}(1,9)$. First calculate $\mathsf{E}[e^X]$ exactly. Then produce an estimate for $\mathsf{E}[e^X]$ from 50 independent random samples from the law $\mathcal{N}(1,9)$. Now produce 100 such estimates (each based on 50 independent samples from $\mathcal{N}(1,9)$). Plot the histogram of the respective errors (100 in number) against the density of the normal law $\mathcal{N}(0,9/50)$. ○

(4.44) SAMPLE MEAN AND VARIANCE: Let ξ be any r.v. with finite second moment ($\mathsf{E}[\xi^2] < \infty$). Set $a \stackrel{\text{def}}{=} \mathsf{E}[\xi]$, $\sigma^2 \stackrel{\text{def}}{=} \mathrm{Var}(\xi)$, and let $(X_i)_{i \in \mathbb{N}_{++}}$ be any i.i.d. sequence with $\mathcal{L}_{X_i} = \mathcal{L}_\xi$. The random variables

$$\overline{X}_n \stackrel{\text{def}}{=} \frac{1}{n}(X_1 + \cdots + X_n) \quad \text{and} \quad \frac{1}{n}\Big((X_1 - \overline{X}_n)^2 + \cdots + (X_n - \overline{X}_n)^2\Big), \qquad n \in \mathbb{N}_{++},$$

are called, respectively, *sample mean* and *sample variance*. The sample mean is an unbiased estimator of the mean value a, in that $\mathsf{E}[\overline{X}_n] = a$ for every $n \in \mathbb{N}_{++}$. However, the sample variance is not an unbiased estimator of σ^2, in that its expected value is not σ^2. The unbiased estimator of σ^2 is given by

$$\widehat{\sigma}_n^2 \stackrel{\text{def}}{=} \frac{1}{n-1}\Big((X_1 - \overline{X}_n)^2 + \cdots + (X_n - \overline{X}_n)^2\Big), \qquad n \geq 2. \quad ○$$

(4.45) EXERCISE: With the notation as in (4.44), prove that $\widehat{\sigma}_n^2$ is indeed an unbiased estimator of the variance σ^2, as claimed; that is, prove that $\mathsf{E}[\widehat{\sigma}_n^2] = \sigma^2$ for every $n \in \mathbb{N}_{++}$, $n \geq 2$.

HINT: First prove and then use the following identity

$$\frac{1}{n}\sum_{i=1}^n (X_i - a)^2 = (\overline{X}_n - a)^2 + \frac{1}{n}\sum_{i=1}^n (X_i - \overline{X}_n)^2 . \quad ○$$

(4.46) EMPIRICAL DISTRIBUTIONS: In many practical situations it is possible to observe a very large number of independent samples from a particular distribution law $\mathcal{L}_*$, for which no other information is available. The usual way to model such a scenario is to postulate that one observes sequentially the realizations of an i.i.d. sequence $X = (X_i)_{i \in \mathbb{N}_{++}}$ with $\mathcal{L}_{X_i} = \mathcal{L}_*$. If $(x_1, \ldots, x_n)$ is any concrete realization of the random vector $(X_1, \ldots, X_n)$, then one can construct the empirical measure

$$\nu_n \stackrel{\text{def}}{=} \sum_{i=1}^n \frac{1}{n}\epsilon_{x_i} ,$$

which is an equally weighted sum of n unit point-mass measures. The measure ν_n can be treated as a realization of the probability-law-valued random variable ($\nu_n(\omega)$ is a probability measure for every ω)

$$\nu_n(\omega)(B) = \sum_{i=1}^n \frac{1}{n}\epsilon_{X_i(\omega)}(B), \qquad B \in \mathcal{B}(\mathbb{R}) .$$

Intuitively, for large n the measure ν_n should be close to the law $\mathcal{L}_*$ – a phenomenon that we are already familiar with from exercise (2.93). This intuition is correct, but we need a precise statement. ○

(4.47) GLIVENKO–CANTELLI THEOREM:[72] (Dudley 2002, 11.4.2) With the notation as in (4.46), let F_* denote the distribution function associated with $\mathcal{L}_*$, and let $F_n(\omega)$ be the empirical distribution function associated with the (random) law $\nu_n(\omega)$, that is, $F_n(\omega)(x) = \nu_n(\omega)(]-\infty, x])$, $x \in \mathbb{R}$. Then P-a.s. the sequence of functions $(F_n(\omega))_{n \in \mathbb{N}_{++}}$ converges to the function F_* uniformly on $\mathbb{R}$, that is,

$$\lim_n \left(\sup_{x \in \mathbb{R}} \left|F_n(\omega)(x) - F_*(x)\right|\right) = 0 \quad \text{for } P\text{-a.e. } \omega \in \Omega . \quad \circ$$

(4.48) EXERCISE: Let $(\mathcal{X}_t)_{t \in \mathbb{N}}$ denote the usual coordinate mappings on the coin toss space (recall that $\mathcal{X}_0 = 0$ by convention). Using Monte Carlo simulation, produce 25 independent realizations of the random variable

$$\frac{\mathcal{Z}_{100}}{10} = \frac{\mathcal{X}_0 + \mathcal{X}_1 \cdots + \mathcal{X}_{100}}{10} .$$

Plot the empirical distribution function from these 25 realizations against the distribution function for the standard normal law. ○

[72] Named after the Ukrainian mathematician Valery Ivanovich Glivenko (1897–1940) and the Italian mathematician Francesco Paolo Cantelli (1875–1966).

5

The Art of Random Sampling

The law of large numbers tells us that expected values can be calculated approximately by averaging over large sets of independently sampled realizations. The accuracy of this procedure can be improved by producing larger and larger samples, but there is a more elegant and more efficient way: reduce the variance in the sample. Furthermore, if direct sampling from the underlying probability law is not available, one can achieve the same result by sampling instead from another law, provided that each realization is rejected with certain probability. These ideas are both powerful and important, and are the subject of the present chapter. Their origins can be traced back to (von Neumann 1951) and the literature devoted to this topic is enormous – see (Glasserman 2004) for a detailed exposition and for a comprehensive list of references.

5.1 Motivation

A particularly common expected value encountered in the domain of finance is

$$\mathsf{E}\Big[\max\Big(e^{\sigma\sqrt{T}\,X+(r-\frac{1}{2}\sigma^2)T} - K\,,\,0\Big)\Big]\,, \quad X \in \mathcal{N}(0,1)\,. \qquad ⓔ_1$$

It gives the price of a European-style call option, based on the Black–Scholes–Merton option pricing model. Since in this model the r.v. X has a standard normal law, there is a closed-form expression for the expectation in $ⓔ_1$ as an explicit function of the parameters r, σ, T, and K – this is the celebrated Black–Scholes–Merton formula.[73] The popularity of this model notwithstanding, it is imperative to have a method at hand that would allow us to compute expected values like the one above even if the choice of the model is such that no closed-form expression is available. What do we do if X is no longer Gaussian, for example? What do we do if we do not have any concrete representation of the distribution law of X, but somehow can sample repeatedly from independent copies of X? More generally, we would like to develop techniques for computing $\mathsf{E}[f(X)]$ for certain deterministic – and, of course, measurable – functions $f\colon \mathbb{R} \mapsto \mathbb{R}$ under the assumption that we can sample repeatedly and independently from the law of X.

---

[73] Some modern computing systems – SageMath or SymPy, for example – are capable of producing the closed-form expression for $ⓔ_1$ with very little human intervention.

(5.1) PLAIN MONTE CARLO SAMPLING: Suppose that $\mathsf{E}[f(X)^2] < \infty$ and let $X_1, \dots, X_n$ be n independent copies of the r.v. X. By the central limit theorem, for a sufficiently large n, a single observation of

$$Y_n \stackrel{\text{def}}{=} \frac{1}{n}\big(f(X_1) + f(X_2) + \cdots + f(X_n)\big) \qquad \dagger$$

would fall into the confidence interval

$$I_n \stackrel{\text{def}}{=} \left[\mathsf{E}[f(X)] - \frac{3}{\sqrt{n}}\sqrt{\mathrm{Var}(f(X))}\,,\ \mathsf{E}[f(X)] + \frac{3}{\sqrt{n}}\sqrt{\mathrm{Var}(f(X))}\right] \qquad \ddagger$$

with probability close to 0.997. ○

One way to make the random sampling from Y_n accurate (despite being random) is to make the interval I_n as narrow as possible. One obvious way to achieve this is to choose a very large sample size n. Unfortunately, random sampling, especially large-scale random sampling, does not come for free. This has led many researchers to actively seek alternative "tricks." But apart from increasing the sample size n, the only other alternative would be to reduce the value of $\sqrt{\mathrm{Var}(f(X))}$. Since f and X are both given, such an objective may at first appear as nothing more than wishful thinking. Yet, as we are about to see, it is perfectly achievable. Before we get to the discussion of some concrete "tricks," we need to look around and discover the features that we can eventually exploit. To this end, notice first that if we were to write e_1 as a generic expression of the form $\mathsf{E}[f(X)]$, then the concrete function f in this case would have the following three properties: (a) it is continuous, (b) it is positive, and (c) it is increasing (being the composition of the increasing functions $x \rightsquigarrow e^{\sigma\sqrt{T}\,x+(r-\frac{1}{2}\sigma^2)T}$ and $x \rightsquigarrow \max(x-K,0)$). These three features may appear innocuous – and certainly unrelated to the central limit theorem – but as we are about to see, they become crucial when trying to reduce the variance of $f(X)$ throughout the sampling process. Before we can do that, we need to develop one particularly useful computational tool.

5.2 Layer Cake Formulas

Let $(\Omega, \mathcal{F}, P)$ be any probability space, let X be any positive r.v. on it ($X \geq 0$ P-a.s.), and let μ be any (not necessarily finite) Borel measure over $\mathbb{R}_+$. A straightforward application of the Tonelli–Fubini theorem gives:

$$\begin{aligned}\int_{\mathbb{R}_+} P(X \geq u)\,\mu(\mathrm{d}u) &= \int_{\mathbb{R}_+}\left(\int_{\Omega} 1_{[u,\infty[}\big(X(\omega)\big)P(\mathrm{d}\omega)\right)\mu(\mathrm{d}u)\\ &= \int_{\Omega}\left(\int_{\mathbb{R}_+} 1_{[0,X(\omega)]}(u)\,\mu(\mathrm{d}u)\right)P(\mathrm{d}\omega) = \mathsf{E}\big[\mu([0,X])\big]\,.\end{aligned}$$

In fact, the last relation admits the following generalization: if Y is yet another (arbitrary) positive r.v., then

$$\int_{\mathbb{R}_+} \mathsf{E}\big[Y 1_{\{X\ge u\}}\big]\,\mu(\mathrm{d}u) = \int_{\mathbb{R}_+}\left(\int_\Omega Y(\omega) 1_{[u,\infty[}\big(X(\omega)\big) P(\mathrm{d}\omega)\right)\mu(\mathrm{d}u)$$

$$= \int_\Omega \left(Y(\omega)\int_{\mathbb{R}_+} 1_{[0,X(\omega)]}(u)\,\mu(\mathrm{d}u)\right) P(\mathrm{d}\omega) = \mathsf{E}\big[Y\mu([0,X])\big]\,.$$

Similarly, if X and Y are two positive random variables on $(\Omega, \mathcal{F}, P)$ and M is any (not necessarily finite) Borel measure on $\mathbb{R}_+ \times \mathbb{R}_+$, then one can write

$$\int_{\mathbb{R}_+\times\mathbb{R}_+} P(X \ge u\,,\ Y \ge v)\, M(\mathrm{d}u, \mathrm{d}v)$$

$$= \int_{\mathbb{R}_+\times\mathbb{R}_+}\left(\int_\Omega 1_{[u,\infty[}\big(X(\omega)\big) 1_{[v,\infty[}\big(Y(\omega)\big)\,\mathrm{d}P(\omega)\right) M(\mathrm{d}u, \mathrm{d}v)$$

$$= \int_\Omega\left(\int_{\mathbb{R}_+\times\mathbb{R}_+} 1_{[0,X(\omega)]}(u) 1_{[0,Y(\omega)]}(v)\, M(\mathrm{d}u, \mathrm{d}v)\right) P(\mathrm{d}\omega)$$

$$= \mathsf{E}\big[M([0,X]\times[0,Y])\big]\,.$$

As a direct application of these relations, we now come to three formulas that will be very useful in a number of contexts, including the derivation of Doob's inequality in (9.86).

(5.2) LAYER CAKE FORMULAS FOR EXPECTED VALUES: In the special case $\mu = \Lambda$ and $M = \Lambda^2$ (the Lebesgue measure on $\mathbb{R}$ and on $\mathbb{R}^2$) we have $\mu([0,X]) = X$ and $M([0,X]\times[0,Y]) = XY$, and thus arrive at the so-called *layer cake formulas* for the expected values $\mathsf{E}[X]$ and $\mathsf{E}[XY]$:

$$\mathsf{E}[X] = \int_{\mathbb{R}_+} P(X \ge u)\,\mathrm{d}u = \int_{\mathbb{R}_+} \mathsf{E}\big[1_{[u,\infty[}(X)\big]\,\mathrm{d}u\,,$$

$$\mathsf{E}[XY] = \int_{\mathbb{R}_+} \mathsf{E}\big[Y 1_{\{X\ge u\}}\big]\,\mathrm{d}u\,,$$

† $$\mathsf{E}[XY] = \int_{\mathbb{R}_+\times\mathbb{R}_+} P(X \ge u\,,\ Y \ge v)\,\mathrm{d}u\mathrm{d}v$$

$$= \int_{\mathbb{R}_+\times\mathbb{R}_+} \mathsf{E}\big[1_{[u,\infty[}(X) 1_{[v,\infty[}(Y)\big]\,\mathrm{d}u\mathrm{d}v\,.$$

We stress that these formulas are only valid for positive random variables X and Y (in which case the expectations are always meaningful). Since the Lebesgue measure is nonatomic, and since the distribution laws of X and Y can have at most countably many atoms, nothing will change in † if we replace $1_{[u,\infty[}$ and/or $1_{[v,\infty[}$, respectively, with $1_{]u,\infty[}$ and $1_{]v,\infty[}$. ○

(5.3) LAYER CAKE FORMULAS FOR THE COVARIANCE: With the previous computation in mind, it is not difficult to obtain layer cake formulas for the covariance $\mathrm{Cov}(X, Y)$ as well (X and Y are still positive). As long as $\mathsf{E}[X] < \infty$ and $\mathsf{E}[Y] < \infty$, $\mathrm{Cov}(X, Y)$ is defined and we have

$$\begin{aligned}
&\int_{\mathbb{R}_+ \times \mathbb{R}_+} \mathrm{Cov}\big(1_{[u,\infty[}(X),\, 1_{[v,\infty[}(Y)\big)\, \mathrm{d}u\,\mathrm{d}v \\
&= \int_{\mathbb{R}_+ \times \mathbb{R}_+} \Big(\mathsf{E}\big[1_{[u,\infty[}(X) 1_{[v,\infty[}(Y)\big] - \mathsf{E}\big[1_{[u,\infty[}(X)\big] \times \mathsf{E}\big[1_{[v,\infty[}(Y)\big]\Big)\, \mathrm{d}u\,\mathrm{d}v \\
&= \int_{\mathbb{R}_+ \times \mathbb{R}_+} \Big(\mathsf{E}\big[1_{[u,\infty[}(X) 1_{[v,\infty[}(Y)\big]\Big)\, \mathrm{d}u\,\mathrm{d}v \\
&\qquad - \int_{\mathbb{R}_+} \Big(\mathsf{E}\big[1_{[u,\infty[}(X)\big]\Big)\, \mathrm{d}u \times \int_{\mathbb{R}_+} \Big(\mathsf{E}\big[1_{[v,\infty[}(Y)\big]\Big)\, \mathrm{d}v \\
&= \mathsf{E}[XY] - \mathsf{E}[X]\,\mathsf{E}[Y] = \mathrm{Cov}(X, Y).
\end{aligned}$$

If X is positive and Y is negative ($X \geq 0$ P-a.s. and $Y \leq 0$ P-a.s.), then

$$\begin{aligned}
\mathrm{Cov}(X, Y) &= -\mathrm{Cov}(X, -Y) = -\int_{\mathbb{R}_+ \times \mathbb{R}_+} \mathrm{Cov}\big(1_{[u,\infty[}(X),\, 1_{]v,\infty[}(-Y)\big)\, \mathrm{d}u\,\mathrm{d}v \\
&= -\int_{\mathbb{R}_+ \times \mathbb{R}_+} \mathrm{Cov}\big(1_{[u,\infty[}(X),\, 1_{]-\infty,-v[}(Y)\big)\, \mathrm{d}u\,\mathrm{d}v \\
&= -\int_{\mathbb{R}_+ \times \mathbb{R}_+} \mathrm{Cov}\big(1_{[u,\infty[}(X),\, 1 - 1_{[-v,\infty[}(Y)\big)\, \mathrm{d}u\,\mathrm{d}v \\
&= \int_{\mathbb{R}_+ \times \mathbb{R}_+} \mathrm{Cov}\big(1_{[u,\infty[}(X),\, 1_{[-v,\infty[}(Y)\big)\, \mathrm{d}u\,\mathrm{d}v \\
&= \int_{\mathbb{R}_+ \times \mathbb{R}_-} \mathrm{Cov}\big(1_{[u,\infty[}(X),\, 1_{[v,\infty[}(Y)\big)\, \mathrm{d}u\,\mathrm{d}v\,.
\end{aligned}$$

Clearly, an analogous chain of relations can be developed in the case where X is negative and Y is positive. The only change that will result in the last integral is that the domain of integration will change from $\mathbb{R}_+ \times \mathbb{R}_-$ to $\mathbb{R}_- \times \mathbb{R}_+$. Thus, as long as X and Y do not change sign and are integrable, we can write

† $$\mathrm{Cov}(X, Y) = \int_{\mathbb{R}_\pm \times \mathbb{R}_\pm} \mathrm{Cov}\big(1_{[u,\infty[}(X),\, 1_{[v,\infty[}(Y)\big)\, \mathrm{d}u\,\mathrm{d}v\,,$$

with the obvious choice for the signs in $\mathbb{R}_\pm \times \mathbb{R}_\pm$ to match the signs of (X, Y). ○

(5.4) EXERCISE: Let X be any (not necessarily positive) random variable and let u and v be any (not necessarily positive) real numbers ($u, v \in \mathbb{R}$). Prove that

$$\mathrm{Cov}\big(1_{[u,\infty[}(X),\, 1_{[v,\infty[}(X)\big) \geq 0\,.$$

HINT: Notice first that the product of the indicator functions $1_{[u,\infty[}$ and $1_{[v,\infty[}$ is either the indicator function $1_{[u,\infty[}$, or the indicator function $1_{[v,\infty[}$. Then observe that $\mathsf{E}[1_{[u,\infty[}(X)] \leq 1$ and $\mathsf{E}[1_{[v,\infty[}(X)] \leq 1$. ○

(5.5) EXERCISE: Let X be any (not necessarily positive) random variable and let $f: \mathbb{R} \mapsto \mathbb{R}$ and $g: \mathbb{R} \mapsto \mathbb{R}$ be any two increasing and càdlàg functions that do not change sign (each function is either positive or negative, but the two functions need not be simultaneously positive or simultaneously negative). Using the formula in (5.3†), prove that if the covariance $\mathrm{Cov}(f(X), g(X))$ is defined, in that $\mathsf{E}[f(X)]$ and $\mathsf{E}[g(X)]$ are both finite, then it must be positive:

$$\mathrm{Cov}(f(X), g(X)) \geq 0\,.$$

HINT: Justify the relations

$$1_{[u,\infty[}(f(x)) = 1_{[f^{-1}(u),\infty[}(x) \quad \text{and} \quad 1_{[v,\infty[}(g(x)) = 1_{[g^{-1}(v),\infty[}(x)\,, \qquad x \in \mathbb{R}\,,$$

where the inverse functions f^{-1} and g^{-1} are understood as in (2.87). ○

(5.6) COROLLARY TO EXERCISE (5.5): There is an interesting corollary to the last exercise: if X is any (not necessarily positive) random variable and if $f: \mathbb{R} \mapsto \mathbb{R}_+$ is any increasing, positive, and *continuous* function, then

$$\mathrm{Cov}(f(X), f(-X)) \leq 0$$

provided that $\mathsf{E}[f(X)]$ and $\mathsf{E}[f(-X)]$ are both finite. This is because the function $\mathbb{R} \ni x \rightsquigarrow -f(-x) \in \mathbb{R}_-$ is increasing, negative, and continuous, so that the statement in exercise (5.5) implies $\mathrm{Cov}(f(X), -f(-X)) \geq 0$. This feature is essential for the method that we are about to develop in the next section. We stress that the requirement for $f(\cdot)$ to be continuous is needed: the functions $x \rightsquigarrow f(x)$ and $x \rightsquigarrow f(-x)$ cannot be simultaneously càdlàg unless they are continuous. ○

5.3 The Antithetic Variates Method

Now we turn to the "magical" variance reduction scheme mentioned in section 5.1. If X is distributed with standard normal law ($X \in \mathcal{N}(0,1)$), then the law of X would be indistinguishable from the law of $-X$. Thus, nothing will change in (5.1†) if we replace the average with

$$\frac{1}{n}\big(f(-X_1) + f(-X_2) + \cdots + f(-X_n)\big)\,.$$

Having gone through the trouble (and the expense) of generating n independent sampling points $(x_1, \ldots, x_n)$ from the law of X (that is, $\mathcal{N}(0,1)$), and also through the trouble of evaluating f at all those sampling points and averaging the values so obtained, we might as well evaluate f at the antithetic sampling points $(-x_1, \ldots, -x_n)$ and average those values, too (typically, the evaluation of a function at a given sampling point requires only a small fraction of the computing resources needed for the simulation of yet another sampling point). On the one hand we want to get as much mileage from the sampling points that we already have, and on the other hand we do not quite know what to do with two *dependent* estimators for $\mathsf{E}[f(X)] = \mathsf{E}[f(-X)]$. With the n sampling points and $2n$ evaluations that are already paid for, we can calculate the average

$$\frac{1}{n}\Big(\frac{f(x_1)+f(-x_1)}{2}+\cdots+\frac{f(x_n)+f(-x_n)}{2}\Big). \qquad ⓔ_1$$

Since the (antithetic) pairs $(x_i, -x_i)$, $1 \le i \le n$, are sampled independently, the law of large numbers is telling us that the average in ⓔ$_1$ is an estimate for

$$\mathsf{E}\Big[\frac{f(X)+f(-X)}{2}\Big] = \mathsf{E}[f(X)],$$

but are we better off with the average ⓔ$_1$ instead of the one in (5.1†)? Given that the expression in ⓔ$_1$ is slightly more expensive to produce, the only meaningful way in which one can determine whether the average ⓔ$_1$ is preferable to the one in (5.1†) is to establish that the confidence interval corresponding to the expression in ⓔ$_1$ is narrower than the one in (5.1‡). But is this so? Since we use the same sample size, n, in both schemes, what we want to know is whether the following relation is in force:

$$\mathrm{Var}\Big(\frac{f(X)+f(-X)}{2}\Big) < \mathrm{Var}\big(f(X)\big).$$

In fact, the following much stronger estimate is in force, provided that f satisfies certain conditions:

(5.7) EXERCISE: Prove that if $f\colon \mathbb{R} \mapsto \mathbb{R}_+$ is any positive, increasing, and continuous function, then

$$\mathrm{Var}\Big(\frac{f(X)+f(-X)}{2}\Big) \le \frac{1}{2}\mathrm{Var}\big(f(X)\big),$$

provided that the variance on the right is finite. ○

As a result, as long as the function f satisfies the properties listed in (5.7), switching from the sampling scheme (5.1†) to the sampling scheme ⓔ$_1$ would shrink the confidence interval by the factor $\frac{1}{\sqrt{2}} \approx 71\%$. We now see that substantially reducing the sampling variance from an already existing sample was not a farfetched objective after all!

(5.8) EXERCISE: Let $X \in \mathcal{N}(0,1)$. Produce 100 independent estimates of the expected value $\mathsf{E}\big[\max\big(e^X - 1, 0\big)\big]$, first by using straight sampling with $n = 50$ data points for each estimate. Then produce another 100 estimates, for the same expected value and from *the same set of sampling points*, but this time by using the antithetic variates method. Finally, calculate the expectation $\mathsf{E}\big[\max\big(e^X - 1, 0\big)\big]$ exactly and build the histogram for each of the two sets of estimates. ○

What if the law of X is not Gaussian? At least in principle, one can sample from the law of X by sampling from uniform law $\mathcal{U}(]0,1[)$ and transforming the sampling points so obtained through the function F_X^{-1}. Observe that if $Y \in \mathcal{U}(]0,1[)$, then $1-Y \in \mathcal{U}(]0,1[)$. Thus, if the data points $(y_1, \dots, y_n)$ are obtained by way of independent sampling from $\mathcal{U}(]0,1[)$, one can work with the antithetic pairs

$$\big(F_X^{-1}(y_i), F_X^{-1}(1-y_i)\big),\ 1 \le i \le n,$$

in much the same way in which we worked with the antithetic pairs $(x_i, -x_i)$ in the Gaussian case. Since the function $y \rightsquigarrow F_X^{-1}(1-y)$ is decreasing, the function $y \rightsquigarrow -f(F_X^{-1}(1-y))$ is increasing, and we stress that $F_X^{-1}(\cdot)$ must be continuous in order to use the result in (5.6).

5.4 The Importance Sampling Method

If one can find a probability measure, Q, that dominates P ($Q \gtrsim P$), then one can write

$$\mathsf{E}^P\big[f(X)\big] = \mathsf{E}^Q\Big[f(X)\frac{\mathrm{d}P}{\mathrm{d}Q}\Big].$$

Thus, instead of averaging over a large set of samples from the law of $f(X)$ under the probability measure P, one may consider averaging over an equally large set of samples from the law of the r.v. $f(X)\frac{\mathrm{d}P}{\mathrm{d}Q}$ under the measure Q – of course, provided that this is somehow feasible. This "change of the paradigm" would be beneficial if the variance of $f(X)$ under P is larger than the variance of $f(X)\frac{\mathrm{d}P}{\mathrm{d}Q}$ under Q. It is not very difficult to think of a situation where this can happen, and, as is often the case, the extreme cases are the easiest to grasp. As an example, suppose that $Q \approx P$ and $f(X) = \frac{\mathrm{d}Q}{\mathrm{d}P}$, in which case $f(X)\frac{\mathrm{d}P}{\mathrm{d}Q} = 1$. This is indeed an extreme case where the distribution law that we have to sample from has the lowest variance possible, that is, 0. Of course, a single observation suffices for estimating – and with infinite accuracy – the expectation of a random variable with vanishing variance (that is, a random variable that is not random at all). Actually, in this case there is no need to carry out any observation procedures, because if it is possible to establish that $f(X)$ is the Radon-Nikodym derivative of some measure Q relative to P, then it would be immediately known that $\mathsf{E}[f(X)] = 1$. Nevertheless, this line of reasoning suggests the following strategy. First, identify a r.v. Z such that $Z > 0$ P-a.s. and $\mathsf{E}^P[Z^{-1}] = 1$. Then define the measure $Q = Z^{-1} \odot P$ so that $\mathsf{E}^P[f(X)] = \mathsf{E}^Q[f(X)Z]$. If $\mathrm{Var}^Q(f(X)Z) < \mathrm{Var}^P(f(X))$, then, at least in principle, it would be preferable to sample from the law of $f(X)Z$ under Q.

(5.9) EXAMPLE: Let $X \in \mathcal{N}(0,1)$ and let

$$f(x) = e^{-\theta x - \frac{1}{2}\theta^2 + \varepsilon x}, \quad x \in \mathbb{R},$$

for some choice of the parameters $\theta, \varepsilon \in \mathbb{R}$. Setting $Z = e^{\theta X + \frac{1}{2}\theta^2}$, recall that $\mathsf{E}[Z^{-1}] = 1$ and that$^{\multimap(3.4)}$ relative to $Q = Z^{-1} \odot P$ we have $X \in \mathcal{N}(-\theta, 1)$. Since for $X \in \mathcal{N}(-\theta, 1)$ the variance of $f(X)Z = e^{\varepsilon X}$ is given by $e^{\varepsilon^2 - 2\varepsilon\theta}\left(e^{\varepsilon^2} - 1\right)$, and the variance of $f(X)$ with $X \in \mathcal{N}(0, 1)$ is given by $e^{\varepsilon^2 - 2\varepsilon\theta}\left(e^{(\varepsilon-\theta)^2} - 1\right)$, we find that the last variance would be larger if (and only if) $\theta^2 - 2\varepsilon\theta > 0$. ○

5.5 The Acceptance–Rejection Method

Generally, the simulation of real-valued random variables is not a major computational challenge. However, there are situations in which one needs to simulate random variables that take values in some less tractable configuration space – say, $\mathbb{R}^n$ or $\mathbb{Z}^n$ – or perhaps even some very large subset of $\mathbb{N}$. This section is a brief review of a technique known as "the acceptance–rejection method," which allows one to emulate the sampling from the distribution of a given r.v. X by actually sampling from the distribution of another r.v. Y. Specifically, suppose that $(\mathcal{X}, \mathcal{S})$ is some measurable space (in particular, this space could be $(\mathbb{R}, \mathcal{B}(\mathbb{R}))$, but, from a practical point of view, this is less interesting), and let μ and ν be any two probability measures on $\mathcal{S}$. The objective is to compute the integral $\int_{\mathcal{X}} f(x)\mu(\mathrm{d}x)$ for some μ-integrable function $f : \mathcal{X} \mapsto \mathbb{R}$, but by way of sampling from the distribution ν – not μ (suppose that, somehow, one knows how to sample from ν, but not from μ). It turns out that this is possible to do, provided that: (a) one can sample from the uniform distribution $\mathcal{U}(]0, 1[)$ and, *independently*, sample from ν; and (b) $\mu \precsim \nu$ (the law μ is absolutely continuous with respect to ν) and the Radon–Nikodym derivative $R = \mathrm{d}\mu/\mathrm{d}\nu$ is globally bounded, in that there is a constant $c \in \mathbb{R}_{++}$ such that $0 \leq R(x) \leq c$ for all $x \in \mathcal{X}$. To see how this might actually work, consider the product space $(\mathcal{X} \times]0, 1[, \mathcal{S} \otimes \mathcal{B}(]0, 1[), \nu \times \Lambda)$ (Λ being the usual Lebesgue measure on $]0, 1[$, so that $\nu \times \Lambda$ is a probability measure on $\mathcal{X} \times]0, 1[$). Now use the Tonelli–Fubini theorem in the following chain of calculations:

$$\begin{aligned}\int_{\mathcal{X} \times]0,1[} & f(x) 1_{[0, R(x)/c[}(u)\, (\nu \times \Lambda)(\mathrm{d}x, \mathrm{d}u) \\ &= \int_{\mathcal{X}} f(x) \left(\int_{]0,1[} 1_{[0, R(x)/c[}(u) \mathrm{d}u \right) \nu(\mathrm{d}x) \\ &= \int_{\mathcal{X}} f(x) \frac{R(x)}{c}\, \nu(\mathrm{d}x) = \frac{1}{c} \int_{\mathcal{X}} f(x)\, \mu(\mathrm{d}x).\end{aligned}$$

If X is any $\mathcal{X}$-valued r.v. with distribution law μ, Y is any $\mathcal{X}$-valued r.v. with distribution law ν, and if U is any $\mathbb{R}$-valued r.v. that is independent of Y and is uniformly distributed in $]0, 1[$, then the last relation says that

$$c\, \mathsf{E}\big[f(Y) 1_{[0, R(Y)/c]}(U)\big] = \mathsf{E}\big[f(X)\big].$$

What is remarkable is that the random variables Y and U do not have to be related to X in any way – the only requirement is that the law of Y dominates the law of X and the respective Radon–Nikodym derivative is bounded.

Now we are ready to describe the acceptance–rejection sampling method. Fix some (sufficiently large) $n \geq 1$ and produce n independent realizations of the r.v. Y (that is, n independent samples from the law ν) $y_i \in \mathbb{X}$, $1 \leq i \leq n$, and also n independent realizations of U, $u_i \in \,]0,1[$, $1 \leq i \leq n$, that are also independent of the realizations of Y. Thanks to the law of large numbers, the average

$$\frac{c}{n}\Big(f(y_1)1_{[0,R(y_1)/c]}(u_1) + \cdots f(y_n)1_{[0,R(y_n)/c]}(u_n)\Big)$$

should be close to the desired expected value $\mathsf{E}[f(X)] = \int_{\mathbb{X}} f(x)\,\mu(\mathrm{d}x)$. The reason why this sampling strategy is called "acceptance–rejection" is now clear. Glivenko–Cantelli's theorem says that$^{\multimap(4.47)}$ one can approximate the distribution μ as follows: first, choose a sufficiently large $n \in \mathbb{N}_{++}$ and create n independent samples from μ, say $x_i \in \mathbb{X}$, $1 \leq i \leq n$; then place probability mass $1/n$ at each of those sampling points; and, finally, use the sum of the point-mass measures so created (obviously, a probability measure on $\mathbb{X}$) as an approximation of μ. The acceptance–rejection method says that one can emulate the same exact distribution μ by sampling from a completely different distribution, ν, provided that every sampling point (from ν), say y, is accepted with probability $R(y)/c$ and is rejected with probability $1 - R(y)/c$, and, when accepted, the probability mass placed at the sampling point is c/n.[74] The importance of this strategy for many practical applications cannot be overstated.

Now we turn to some concrete applications. To keep things simple, we consider the case $\mathbb{X} = \mathbb{R}$, and again note that such a choice is not all that interesting, except for the purpose of illustration. If X and Y are any two $\mathbb{R}$-valued r.v. that are distributed in some (finite or infinite) interval $]\,a,b[\, \subseteq \mathbb{R}$ with densities ϕ_X and ϕ_Y, then, as long as ϕ_Y is strictly positive ($\phi_Y(x) > 0$ for all $x \in \,]\,a,b[$), we can claim that $\mathcal{L}_Y \approxeq \Lambda$. In particular the law of X is absolutely continuous with respect to the law of Y and the Radon–Nikodym derivative $R = \mathrm{d}\mathcal{L}_X/\mathrm{d}\mathcal{L}_Y$ is nothing but the ratio of the two densities $\phi_X(x)/\phi_Y(x)$, $x \in \,]a,b[$. Instead of sampling from $\mathcal{L}_X$, we can use the acceptance–rejection method and sample from the law $\mathcal{L}_Y$ instead, as long as the function $\phi_X(x)/\phi_Y(x)$ is globally bounded on the interval $]a,b[$. As an example, if X is distributed in $]0,1[$ with$^{\multimap(2.28h)}$ beta law of with parameters $\alpha = 1.5$ and $\beta = 3$ and Y is uniformly distributed in $]0,1[$ (that is, $\phi_Y \equiv 1$) then the ratio $\phi_X/\phi_Y = \phi_X$ is bounded on the interval $]0,1[$ by the constant 2 (to be precise, with this choice of the parameters the maximal value of ϕ_X on the interval $]0,1[$ is around 1.87829710109982). In yet another

[74] The total mass placed at all points will be random, and, in general, will not sum to 1, but will still converge to the probability distribution μ as $n \to \infty$.

example, if X has⊶(2.28f) gamma distribution of parameters $c = 11$ and $p = 3.5$ and Y has gamma distribution with parameters $c = 3$ and $p = 1$, then the ratio $\phi_X(x)/\phi_Y(x)$ is again bounded by the constant 2 on the domain $x \in \mathbb{R}_{++}$ (to be precise, the maximal value of of the ratio $\phi_X(x)/\phi_Y(x)$ for $x \in \mathbb{R}_{++}$ is around 1.98411017537704).

(5.10) EXERCISE: Suppose that the r.v. X has beta distribution with parameters $\alpha = 1.5$ and $\beta = 3$. Calculate the approximate value of $\mathsf{E}[\sin(X)]$ by way of sampling from the uniform distribution $\mathcal{U}(]0,1[)$ and using the acceptance–rejection method with $n = 200$. ○

(5.11) EXERCISE: Suppose that the r.v. X has gamma distribution with parameters $c = 11$ and $p = 3.5$. Calculate the approximate value of $\mathsf{E}[\sin(X)]$ by way of sampling from the Gamma distribution with parameters $c = 3$ and $p = 1$, and using the acceptance–rejection method with $n = 1{,}000$. ○

6

Equilibrium Asset Pricing in Finite Economies

In this chapter we take up the topic of "pricing of stochastic payoffs" from where it was left off in section 1.2, and come to our first encounter with the general principles and methods of asset pricing. As we are about to discover, there is a deep connection between those principles and the equivalent change of measure operation that we are already familiar with. In turn, this connection leads to an even deeper connection with martingales and the notion of arbitrage, which will be introduced and studied in subsequent chapters.

The introduction to asset pricing presented here may appear self-contained and brief, but it is barely the scratch of the surface, and the full story of this subject is long and winding. As with any broad subject, its starting point is difficult to identify precisely, but one can certainly point to: Markowitz (1952, 1959), who invented portfolio theory, Arrow and Debreu (1954), who introduced the notions of price equilibrium and state-price density, Modigliani and Miller (1958), who discovered the invariance of the capital structure of the firm, Sharpe (1964) and Lintner (1965), who invented the Capital Asset Pricing Model, Merton (1969a), who introduced the principle that equal risk demands equal expected excess return, and to the immensely influential works (Merton 1970b) and (Black and Scholes 19–), which introduced the ideas of pricing by arbitrage and hedging. This, of course, was only the beginning, and while a detailed review of the history and the development of asset pricing is beyond the scope of this space, we would be remiss to not point to at least some of the groundbreaking publications that followed: (LeRoy 1973), (Ross 1975), (Rubinstein 1976), (Lucas 1978), (Cox et al. 1979), (Cox et al. 1985), (Duffie and Huang 1985), (Duffie 1987), (Dybvig and Ross 1987), (Cox and Huang 1989), (Karatzas et al. 1990a), and (Karatzas 1996).[75]

The brief introduction to asset pricing in discrete time presented in this chapter is meant to be – to the extent to which this is possible – self-contained, and follows closely (Föllmer and Schied 2011) and (Dumas and Lyasoff 2012), which is essentially an elaboration of the Lucas model of the economy.

[75] The Sveriges Riksbank Prize in Economic Sciences in Memory of Alfred Nobel was awarded to Kenneth J. Arrow (1972), Gérard Debreu (1983), Franco Modigliani (1985), Harry M. Markowitz, Merton H. Miller and William F. Sharpe (1990), Robert E. Lucas Jr. (1995), and Robert C. Merton and Myron S. Scholes (1997).

6.1 Information Structure

We are already familiar with one concrete example of information structure, which was developed in the process of constructing the canonical probability measure P_∞ on the coin toss space Ω_∞. The first step in that construction was to define the chain of partitions $(\mathscr{P}_t)_{t\in\mathbb{N}}$ by letting $\mathscr{P}_0$ be the trivial partition, $\{\Omega\}$, and for $t \in \mathbb{N}_{++}$, letting $\mathscr{P}_{t+1}$ be the refinement of $\mathscr{P}_t$ obtained by splitting each atom in $\mathscr{P}_t$ into two atoms from $\mathscr{P}_{t+1}$. The second step was to postulate that all atoms A in any given partition $\mathscr{P}_t$ have the same probability $P_*(A) = 2^{-t}$. This is the same as splitting the probability of any given atom $A \in \mathscr{P}_t$ equally among the two descending atoms from $\mathscr{P}_{t+1}$ that comprise A. Using Caratheodory's extension theorem, we were then able to extend the definition of the probability set function P_* from the algebra $\mathscr{A} = \cup_{t\in\mathbb{N}} \sigma(\mathscr{P}_t)$ to an actual probability measure, denoted by P_∞, on the σ-field $\mathscr{F}_\infty = \sigma(\mathscr{A})$. The same construction can be carried out in a more general setup in which the probability of each atom is not split equally among the two descending atoms. Specifically, if $A \in \mathscr{P}_t$, and $A^+ \in \mathscr{P}_{t+1}$ and $A^- \in \mathscr{P}_{t+1}$ are its two descending atoms (that is, the two child nodes on the tree), so that $A = A^+ \cup A^-$ (e.g., A^+ may correspond to "up movement" and A^- may correspond to "down movement"), then we can postulate that $P_*(A^+) = p\,P_*(A)$ and $P_*(A^-) = q\,P_*(A)$ for some fixed $p, q \in \mathbb{R}_{++}$ with $p + q = 1$. This construction, too, gives rise to a unique probability measure – still denoted P_∞ – on the σ-field $\mathscr{F}_\infty$, except that now $P_\infty(\mathcal{X}_t = +1) = p$ and $P_\infty(\mathcal{X}_t = -1) = q$ (just as before, we again postulate that $\mathcal{X}_0 = 0$). If all that we care about is the random walk $\mathcal{Z} = (\mathcal{Z}_t)_{t\in\mathbb{N}}$, given by

$$\mathcal{Z}_t = \mathcal{X}_0 + \mathcal{X}_1 + \cdots + \mathcal{X}_t\,, \quad t \in \mathbb{N}\,, \qquad e_1$$

as each $\mathcal{Z}_t$ is measurable for $\sigma(\mathscr{P}_t)$ (notation: $\mathcal{Z}_t \prec \sigma(P_t)$), it becomes a convenient simplification to aggregate into a single set those atoms from $\mathscr{P}_t$ on which $\mathcal{Z}_t$ takes the same value. This gives rise to the so-called recombining trees, on which the celebrated binomial asset pricing model of Cox et al. (1979) is based.

As our objective is to eventually connect the binomial model, the general principles of asset pricing, and the theory of martingales, it would be helpful to expand the setup of the coin toss space into a more general information structure. The starting point in this project is to postulate a sufficiently rich probability space $(\Omega, \mathcal{F}, P)$ endowed with a filtration – an object that we introduce next.

(6.1) DEFINITION (FILTRATION): A filtration (aliases: information structure, information arrival dynamics) on some measurable space $(\Omega, \mathcal{F})$ is any sequence of sub-σ-fields $\mathscr{F} \equiv (\mathscr{F}_t \subseteq \mathcal{F})_{t\in\bar{\mathbb{N}}}$ that is increasing, in that $\mathscr{F}_t \subseteq \mathscr{F}_{t+1} \subseteq \mathscr{F}_\infty$ for any $t \in \mathbb{N}$. An increasing sequence of σ-fields $(\mathscr{F}_t)_{t\in\mathbb{N}}$ can always be amended by the σ-field$^{\multimap(1.23)}$ $\mathscr{F}_\infty \overset{\text{def}}{=} \vee_{t\in\mathbb{N}} \mathscr{F}_t$. In what follows the symbol $\mathscr{F}_\infty$ will always be understood as the smallest σ-field that includes $\mathscr{F}_t$ for every $t \in \mathbb{N}$. A filtration may be expressed also as $(\mathscr{F}_t)_{t\in\bar{\mathbb{N}}}$ or as $(\mathscr{F}_t)$, but most often we will write simply $\mathscr{F}$, and the property "$\mathscr{F}$ is a filtration contained in the σ-field $\mathcal{F}$" will be abbreviated as "$\mathscr{F} \setminus\!\subset \mathcal{F}$."

Heuristically, the σ-field $\mathscr{F}_t$ is a metaphor for "all the statements that can be verified from information received up until and including period t." For example, an event $A \in \mathscr{F}_t$ may or may not occur, but information received up until time t (usually expressed as an observed realizations of a particular sequence of random vectors) should be enough to determine which of these two alternatives has been realized, and receiving information after period t cannot change the conclusion that the event A has or has not occurred. It is quite common to suppose that $\mathscr{F}_0$ is the trivial σ-field $\{\emptyset, \Omega\}$, meaning that quantities that are observable during the initial period $t = 0$ are nonrandom, but there are situations where $\mathscr{F}_0$ is chosen to be some nontrivial σ-field. ○

As we have seen already, information structure may be introduced in terms of a refining sequence of partitions $(\mathscr{P}_t)_{t \in \mathbb{N}}$ (every atom in $\mathscr{P}_t$ is a finite union of atoms from $\mathscr{P}_{t+1}$). Indeed, it is straightforward to check that the σ-fields $\mathscr{F}_t \overset{\text{def}}{=} \sigma(\mathscr{P}_t)$, $t \in \mathbb{N}$, form a filtration, but this filtration is just as informative as the refining chain of partitions $(\mathscr{P}_t)_{t \in \mathbb{N}}$.

(6.2) Definition (Adapted Stochastic Process): A stochastic process (alias: random process), defined on a probability space $(\Omega, \mathcal{F}, P)$ and indexed (alias: labeled) by the set $\mathbb{N}$, is any sequence of random variables $\big(X_t \in \mathscr{L}^0(\Omega, \mathcal{F}, P)\big)_{t \in \mathbb{N}}$. A stochastic process may be expressed as $(X_t)_{t \in \mathbb{N}}$ or as (X_t), but most often we will write simply X. In some situations a stochastic process may be labeled by the set $\bar{\mathbb{N}} = \mathbb{N} \cup \{+\infty\}$ if the r.v. $X_\infty \in \mathscr{L}^0(\Omega, \mathcal{F}, P)$ is meaningful. The indexing variable (the index) t denotes "the time" and, eventually, the labeling set $\bar{\mathbb{N}}$ will be replaced with $\bar{\mathbb{R}}_+$. If the nature of the time domain must be emphasized, we will use the phrases "stochastic process in discrete time," or "stochastic process in continuous time."

We say that the process X is *adapted* to the filtration $\mathscr{F}$, and express this feature as $X \prec \mathscr{F}$, if the r.v. X_t is measurable with respect to $\mathscr{F}_t$ (notation: $X_t \prec \mathscr{F}_t$) for every $t \in \mathbb{N}$ (or for every $t \in \bar{\mathbb{N}}$). In practical terms, "adapted" means that a particular variable associated with date t (say, the closing price of TSLA or GOOGL on that date) is not contingent upon events that can be observed after date t. In other words, the closing price for today is forever fixed in that there may be a different closing price tomorrow, but today's closing price will not change tomorrow, or the day after tomorrow, etc. One classical example of an adapted process that we are already familiar with is the random walk in $\mathcal{Z} = (\mathcal{Z}_t)_{t \in \mathbb{N}}$ from e_1, which is adapted to the filtration on the coin toss space generated by the partitions $(\mathscr{P}_t)_{t \in \mathbb{N}}$. ○

6.2 Risk Preferences

Considered in the financial context, a random variable, Π, defined on an appropriate probability space $(\Omega, \mathscr{F}, P)$, often has the meaning of a *contingent payoff* (aliases: *stochastic payoff*, *payoff profile*, or simply *payoff*) that a particular economic agent is entitled to receive. The payoff $\Pi = \Pi(\omega)$ – which in most cases is going to represent consumption or monetary wealth, usually expressed in units of currency ¤ – is contingent upon the state of the economy $\omega \in \Omega$, and is either *risky*, in that ω is sampled from the set of scenarios Ω according to some known (to the agent) probability measure P, or is perhaps *uncertain*, in that the agent has no knowledge of which – if any – probability measure the scenario $\omega \in \Omega$ may be sampled from (from the economic agent's point of view, there is only a measurable space $(\Omega, \mathscr{F})$ and a measurable function Π from Ω into $\mathbb{R}$ that represents their payoff). In what follows we will be concerned exclusively with risky payoffs. It is quite common in the economic literature to call such payoffs *lotteries*, for in a lottery the probabilities for winning and losing are exactly known. The financial decisions that an agent makes can end up in different lotteries that have different payoff profiles and, generally, demand different entry costs. Essentially, asset pricing is the study of all these choices. As an example, is a payoff of either ¤99 or ¤101 with equal probability preferable to a payoff of either ¤90 or ¤110 with equal probability? Are there prices at which an agent would be indifferent to the choice between these two payoff profiles, and, if there are, how do such prices vary in the cross-section of economic agents?

The study of economic behavior lies outside of our main concerns, but we still need to develop a relatively simple and tractable mathematical metaphor for the agents' preferences and attitudes toward various payoff profiles – and unless we do, there will be very little we can say about asset pricing. The classical and widely adopted approach is to follow the expected utility framework. The idea of expected utility goes way back to Bernoulli (1738), but its present-day version is due to von Neumann and Morgenstern (1947). An especially clear, elegant, modern – and perhaps the most satisfactory so far – version of the expected utility approach to expressing risk preferences can be found in (Föllmer and Schied 2011, ch. 2), which we closely follow and refer the reader to for more details, examples, illustrations, and an extensive list of references.

Assuming that the payoffs are modeled as random variables Π, what really matters to the economic agents are the distribution laws $\mathcal{L}_\Pi$. In general, the payoffs Π are distributed in some measurable space $(\mathscr{X}, \mathscr{S})$, which in most practical applications is going to be some subset of the spaces $\mathbb{R}$ or $\mathbb{R}^n$, endowed with the usual Borel structure. Let $\mathcal{M}(\mathscr{X}, \mathscr{S})$ denote the collection of all σ-finite measures on $(\mathscr{X}, \mathscr{S})$ and let $\mathcal{M}_1(\mathscr{X}, \mathscr{S})$ denote the collection of all probability measures on $(\mathscr{X}, \mathscr{S})$, that is, the collection of all possible distribution laws associated with potential payoffs. For an agent, the choice between two different payoff profiles comes down to the choice between two different elements of $\mathcal{M}_1(\mathscr{X},\mathscr{S})$. Let $\Theta \subseteq \mathcal{M}_1(\mathscr{X},\mathscr{S})$ denote the set of choices that are available to a particular agent. We suppose – for now – that Θ can be any subset of $\mathcal{M}_1(\mathscr{X},\mathscr{S})$, but insist that Θ must be convex: we require that $x\mu + (1-x)\nu \in \Theta$ for any $x \in [0,1]$ and any $\mu,\nu \in \Theta$. Suppose that $\mu,\nu \in \Theta$ and the agent is presented with the choice between μ and ν. If they prefer μ to ν we write $\mu \succeq \nu$ and postulate that $\succeq$ is⊶ page xvii a relation on Θ, which we call a *preference relation*. The preference relation $\succeq$ gives rise to the symmetric relation $(\simeq) \overset{\text{def}}{=} (\succeq)\cap(\preceq)$, which we call an *indifference relation*, as $\mu \simeq \nu$ means that $\mu \succeq \nu$ (the agent prefers μ to ν) and $\nu \succeq \mu$ (the agent prefers ν to μ). It also gives rise to the relation $(\succ) \overset{\text{def}}{=} (\succeq)\cap(\not\simeq) = (\succeq)\cap(\not\preceq)$, which we call a *strict preference relation*, as $\mu \succ \nu$ means that the agent prefers μ to ν and is not indifferent to the choice between μ and ν, or, equivalently, prefers μ to ν and does not prefer ν to μ. In order to develop useful economic models, we must impose certain axioms on the preference relation $\succeq$, and this is what we do next. We stress that the relations $\succeq$, $\simeq$, and $\succ$ are agent specific, so that, generally, the economic agents may differ in their risk preferences.

(6.3) PREFERENCE ORDER: A preference relation, $\succeq$, defined on $\Theta \subseteq \mathcal{M}_1(\mathscr{X},\mathscr{S})$, is said to be a *preference order* if the following two axioms are satisfied:

Completeness: For every $\mu,\nu \in \Theta$, either $\mu \succeq \nu$ or $\nu \succeq \mu$, that is, any two available choices can be compared.

Transitivity: $\lambda,\mu,\nu \in \Theta$ and $\lambda \succeq \mu$ and $\mu \succeq \nu$ imply $\lambda \succeq \nu$. ○

(6.4) REMARK: The completeness axiom can be stated as $(\succeq)\cup(\preceq) = \mathscr{X}\times\mathscr{X}$, which is the same property as $(\not\preceq) \subseteq (\succeq)$. As a result, the strict preference relation $\succ$ corresponding to $\succeq$, which is defined as⊶ page xvii $(\succ) \overset{\text{def}}{=} (\succeq)\cap(\not\preceq)$, is such that $(\succ) = (\not\preceq)$, and this is the same property as $(\not\prec) = (\succeq)$; in particular, $(\succ) \subseteq (\succeq) = (\not\prec)$. Thus, one can start with the strict preference relation $\succ$, postulate that $(\succ) \subseteq (\not\prec)$ (the so-called *asymmetry axiom*), and then define the preference relation $\succeq$ simply as $\not\prec$. Without the asymmetry axiom the preference relation $\succeq$ can still be defined in terms of the strict preference relation $\succ$, but one has to set $(\succeq) = (\succ)\cup(\not\prec)$. In any case, "preference order" and "strict preference order" can always be obtained from one another and in a unique way. ○

(6.5) EXERCISE: Prove that if $\succeq$ is a preference order on Θ, then the relation

$$(\simeq) \overset{\text{def}}{=} (\succeq)\cap(\preceq)$$

is an equivalence relation on Θ. ○

(6.6) VON NEUMANN–MORGENSTERN REPRESENTATION: A strict preference order, $\succ$, on $\Theta \subseteq \mathcal{M}_1(\mathscr{X},\mathscr{S})$ is said to admit von Neumann–Morgenstern representation if there is a measurable function $U\colon \mathscr{X} \mapsto \mathbb{R}$ such that all integrals

$$U\bullet\mu \overset{\text{def}}{=} \int_{\mathscr{X}} U(x)\,\mu(\mathrm{d}x), \quad \mu \in \Theta,$$

are defined and $\mu \succ \nu$ if and only if $U{\bullet}\mu > U{\bullet}\nu$ for any $\mu, \nu \in \Theta$, or, which is the same, $\mu \succeq \nu$ if and only if $U{\bullet}\mu \geq U{\bullet}\nu$ for any $\mu, \nu \in \Theta$. A von Neumann–Morgenstern representation, if one exists, is never unique, for, if $\mu \rightsquigarrow U{\bullet}\mu$ is one such representation, then so is also $\mu \rightsquigarrow (aU + b){\bullet}\mu$ for any $a \in \mathbb{R}_{++}$ and $b \in \mathbb{R}$. If $a \in \mathbb{R}_{++}$ and $b \in \mathbb{R}$, we say that the representation $\mu \rightsquigarrow (aU + b){\bullet}\mu$ is a *positive affine transformation* of the representation $\mu \rightsquigarrow U{\bullet}\mu$. ○

An important question in asset pricing is whether von Neumann–Morgenstern representation for a particular preference order exists. Before we can address this matter, we need to introduce a few definitions.

(6.7) THE INDEPENDENCE AND CONTINUITY AXIOMS: A strict preference relation $\succ$ on $\Theta \subseteq \mathcal{M}_1(\mathcal{X}, \mathcal{S})$ is said to satisfy the *independence axiom* if $\mu, \nu \in \Theta$ and $\mu \succ \nu$ imply

$$x\mu + (1-x)\lambda \succ x\nu + (1-x)\lambda \quad \text{for any } \lambda \in \Theta \text{ and any } x \in]0,1] .$$

It is said to satisfy the *continuity axiom* (alias: the *Archimedian axiom*) if $\lambda, \mu, \nu \in \Theta$ and $\lambda \succ \mu \succ \nu$ imply that for some $x, y \in]0, 1[$,

$$x\lambda + (1-x)\nu \succ \mu \succ y\lambda + (1-y)\nu . \quad ○$$

(6.8) EXISTENCE OF REPRESENTATION I: (Föllmer and Schied 2011, 2.21–2.23) A *simple probability measure* on $\mathcal{X}$ is any purely atomic probability measure on $\mathcal{X}$ with a finite number of atoms, that is, any measure of the form

$$\mu = \sum\nolimits_{i \in \mathbb{N}_{|n}} a_i \epsilon_{x_i}$$

for some choice of $n \in \mathbb{N}_{++}$, elements $x_i \in \mathcal{X}$, and scalars $a_i \in]0, 1]$ such that $\sum_{i \in \mathbb{N}_{|n}} a_i = 1$.[76] Let Θ be the collection of all simple probability measures on $(\mathcal{X}, \mathcal{S})$ and suppose that $\succ$ is a strict preference order on Θ such that the independence and continuity axioms hold for $\succ$. Then $\succ$ admits von Neumann–Morgenstern representation $\Theta \ni \mu \rightsquigarrow U{\bullet}\mu$ that is unique up to a positive affine transformation. In particular, if $\mathcal{X}$ is a finite set, then the same claim can be made for the entire collection of probability measures $\Theta = \mathcal{M}_1(\mathcal{X}, \mathcal{S})$, as any measure on a finite set is automatically simple. ○

Restricting the agents' choices to simple distributions is clearly unsatisfactory, for, quantities such as individual income, wealth, asset prices, consumption, and such, typically take values in some continuous range (e.g., an interval), not to mention models in which the continuous-time framework is adoped. For this reason, we now introduce the tools that would allow us to formulate a more general and useful version of the von Neumann–Morgenstern representation theorem. In the rest of this section we suppose that $\mathcal{X}$ is a Polish space and that $\mathcal{S}$ is the Borel

[76] Simple probability distributions on $(\mathcal{X}, \mathcal{S})$ are associated with $\mathcal{X}$-valued random variables that take finitely many values, whence the term "simple."

σ-field on $\mathcal{X}$. All results presented below are essentially a review of (Föllmer and Schied 2011, ch. 2 and sec. A.5).

(6.9) WEAK TOPOLOGY ON $\mathcal{M}_1(\mathcal{X}, \mathcal{S})$: We first fix a continuous function

$$\mathcal{X} \ni x \rightsquigarrow h(x) \in [1, +\infty[$$

and denote by $\mathscr{C}_h(\mathcal{X}; \mathbb{R})$ the collection of those continuous functions $f \in \mathscr{C}(\mathcal{X}; \mathbb{R})$ that are bounded by h in the following sense: there is a constant $C \in \mathbb{R}_{++}$ (possibly depending on f) such that

$$|f(x)| \leq Ch(x) \quad \text{for all } x \in \mathcal{X}$$

(notice the inclusion $\mathscr{C}_b(\mathcal{X}; \mathbb{R}) \subseteq \mathscr{C}_h(\mathcal{X}; \mathbb{R})$, due to the requirement $h \geq 1$). The next step is to define the set of probability measures that integrate h to a finite value:

$$\mathcal{M}_1^h(\mathcal{X}, \mathcal{S}) \overset{\text{def}}{=} \left\{ \mu \in \mathcal{M}_1(\mathcal{X}, \mathcal{S}) : \int_{\mathcal{X}} h \, \mathrm{d}\mu < \infty \right\}.$$

Finally, define a topology on $\mathcal{M}_1^h(\mathcal{X}, \mathcal{S})$: the h-weak topology on $\mathcal{M}_1^h(\mathcal{X}, \mathcal{S})$, denoted by $\mathcal{T}_h$, is the topology generated by the functionals

$$\mathcal{M}_1^h(\mathcal{X}, \mathcal{S}) \ni \mu \rightsquigarrow \int_{\mathcal{X}} f \, \mathrm{d}\mu \quad \text{for all choices of } f \in \mathscr{C}_h(\mathcal{X}; \mathbb{R}).$$

To put it another way, $\mathcal{T}_h$ is the weakest topology for which all functionals above are continuous. Endowed with this topology, $\mathcal{M}_1^h(\mathcal{X}, \mathcal{S})$ becomes a Polish space – due to the fact that $\mathcal{X}$ is a Polish space (Föllmer and Schied 2011, A.28). ○

(6.10) CONTINUOUS PREFERENCES: A strict preference relation $\succ$ on $\mathcal{M}_1^h(\mathcal{X}, \mathcal{S})$ is said to be continuous in the topology $\mathcal{T}_h$ if for any fixed $\mu \in \mathcal{M}_1^h(\mathcal{X}, \mathcal{S})$,

$$\left\{ \nu \in \mathcal{M}_1^h(\mathcal{X}, \mathcal{S}) : \nu \succ \mu \right\} \in \mathcal{T}_h \quad \text{and} \quad \left\{ \nu \in \mathcal{M}_1^h(\mathcal{X}, \mathcal{S}) : \nu \prec \mu \right\} \in \mathcal{T}_h .$$

To state this property more succinctly, for any fixed μ the sets $\{\nu : \nu \succ \mu\}$ and $\{\nu : \nu \prec \mu\}$ are both open in $\mathcal{M}_1^h(\mathcal{X}, \mathcal{S})$. ○

(6.11) EXERCISE: Prove that any strict preference that is continuous with respect to $\mathcal{T}_h$ automatically satisfies the continuity axiom from (6.7).

HINT: The measure-valued function $[0, 1] \ni x \rightsquigarrow x\lambda + (1 - x)\nu$ is continuous in the topology $\mathcal{T}_h$ and converges to ν as $x \searrow 0$ and to λ as $x \nearrow 1$ (Föllmer and Schied 2011, 2.9). ○

(6.12) EXISTENCE OF REPRESENTATION II: (ibid., 2.30) Any strict preference order, $\succ$, on $\Theta = \mathcal{M}_1^h(\mathcal{X}, \mathcal{S})$ that is continuous in the h-weak topology $\mathcal{T}_h$ and satisfies the independence axiom from (6.7) admits von Neumann–Morgenstern representation $\Theta \ni \mu \rightsquigarrow U \bullet \mu$ for some $U \in \mathscr{C}_h(\mathcal{X}; \mathbb{R})$. Up to a positive affine transformation this representation is unique. ○

(6.13) MONOTONICITY AND RISK AVERSION: Because of the natural order on the real line $\mathbb{R}$, the special case where the set $\mathcal{X}$ is chosen to be some (finite or infinite) interval $I \subseteq \mathbb{R}$ is especially interesting in the present context. In the rest of this section we will suppose that the menu of choices for a particular agent is given by the space $\Theta = \mathcal{M}_1^h(I, \mathcal{B}(I))$ for some choice of the interval $I \subseteq \mathbb{R}$ and some choice of the function $h\colon I \mapsto [1, +\infty[$ such that every $\mu \in \Theta$ has a well-defined expectation $\int_I x\,\mu(\mathrm{d}x)$ and every sure payoff ϵ_x, $x \in I$, is an element of Θ. Economic agents are assumed to prefer more (consumption, wealth, etc.) to less. In order to incorporate this feature into the strict preference relation $\succ$ on Θ, we require that $\succ$ must be *monotone*: if $x, y \in I$, then $y > x$ implies $\epsilon_y \succ \epsilon_x$ (if $y > x$, the sure payoff y is strictly preferred to the sure payoff x).

Aversion to risk can be incorporated in the strict preference relation $\succ$ by requiring that

$$\text{if } m \equiv m(\mu) \stackrel{\text{def}}{=} \int_{\mathbb{R}} x\mu(\mathrm{d}x) \quad \text{then} \quad \epsilon_m \succ \mu$$

for any $\mu \in \Theta$ that is not a sure payoff ($\mu \neq \epsilon_x$ for any $x \in \mathbb{R}$). To put it another way, $\succ$ strictly prefers the sure payoff equal to the expected payoff to the actual (risky) payoff, and this property holds for any payoff that is not a sure payoff. ○

(6.14) EXPECTED UTILITY: With the setting and the notation as in (6.13), suppose that $\succ$ is a strict preference order on the space $\Theta = \mathcal{M}_1^h(I, \mathcal{B}(I))$. Suppose further that $\succ$ is continuous in the respective h-weak topology and satisfies the independence axiom, so that ⊸ (6.12) $\succ$ admits von Neumann–Morgenstern representation $\Theta \ni \mu \rightsquigarrow U{\bullet}\mu$ with $U \in \mathscr{C}_h(I; \mathbb{R})$. Then (Föllmer and Schied 2011, 2.35) the property "$\succ$ is monotone" is equivalent to the property "$x \rightsquigarrow U(x)$ is strictly increasing," and the property "$\succ$ is risk averse" is equivalent to the property "$x \rightsquigarrow U(x)$ is strictly concave." Because of this result, we are particularly interested in functions $U : I \mapsto \mathbb{R}$ that are: (a) strictly increasing, (b) strictly concave, and (c) continuous. Any function that satisfies these three properties is called *utility function*, and we remark that any increasing and concave function on I is automatically continuous on any interval $]a, b] \subseteq I$ (see ibid., A.4), so that the continuity requirement (c) is only relevant at the left end point of I, and only if I is left-closed.

In any case, if we postulate that the preferences of any economic agent must be modeled as a preference relation, $\succeq$, on Θ, which satisfies the completeness and the transitivity axioms ($\succeq$ is a preference order), is continuous in the h-weak topology for some h, satisfies the independence axiom, and, finally, is monotone and risk averse, then we end up with models in which all risk preferences admit von Neumann–Morgenstern representations of the form

$$\Theta \ni \mu \rightsquigarrow U{\bullet}\mu = \int_I U(x)\,\mu(\mathrm{d}x)$$

for some choice of the utility function $U : I \mapsto \mathbb{R}$. In other words, a risk preference relation that has the properties just mentioned can always be represented as expected utility.

It is worth recording that the strict concavity of a generic utility function implies that the first derivative is strictly decreasing on its domain, which is to say the marginal increase in satisfaction from any increase in the consumption level (or the wealth, etc.) declines as the consumption level increases. The well-known folklore explanation is this: for a thirsty person the first sip from a glass of water is much more precious than the last. ○

(6.15) IMPORTANT REMARK: The expected utility interpretation of the agents' risk preferences is very widely adopted – deeply rooted, in fact – in the asset pricing literature. We are going to follow this tradition, but we would be remiss if we did not acknowledge that some of the premises of the expected utility framework are inconsistent with a number of empirical studies of human economic behavior, and the entire approach has been the subject of a debate ever since it was first proposed. The axiom that appears to be most questionable is the independence axiom, as illustrated by the Allais paradox – see (Föllmer and Schied 2011, 2.32). The extensive work of the economists Amos Tversky and Daniel Kahneman[77] under the rubric of *prospect theory*, stretching over more than three decades, has concluded, among other things, that if we were to accept that von Neumann–Morgenstern representation exists, then such a representation is likely to involve a function U that is convex in some neighborhood of 0 (in most applications $I = \mathbb{R}_{++}$) and is concave in the rest of its domain, capturing the idea that at very low levels of wealth the agents are not risk averse (Tversky and Kahneman 1992). Unfortunately, the use of such functions instead of standard utility functions is not straightforward and demands the rewriting of many useful and widely adopted results from finance and economics. There are reasons to believe that more useful models and better understanding of how humans react to and perceive risk will be developed in the years to come. Until then, the expected utility paradigm is likely to remain, as the saying goes, "good enough for government work." ○

(6.16) CERTAINTY EQUIVALENT AND RISK PREMIUM: Let $\succeq$ be any preference order on $\Theta = \mathcal{M}_1^h(I, \mathcal{B}(I))$ that has expected utility representation with utility function U. The function $I \ni v \rightsquigarrow U(v)$ is continuous and increasing by definition, and therefore so is also the functions $I \ni v \rightsquigarrow \int_I U(v)\mu(\mathrm{d}x)$ for any finite measure μ. A straightforward application of the intermediate value theorem implies that, given any probability distribution $\mu \in \Theta$, one can find a constant $c \in I$, possibly depending on μ, such that

$$\int_I U(x)\,\mu(\mathrm{d}x) = \int_I U(c)\,\mu(\mathrm{d}x) = U(c) = \int_I U(x)\,\epsilon_c(\mathrm{d}x)\,.$$

In other words, from the point of view of the preference order $\succeq$ one has both $\mu \succeq \epsilon_c$ and $\mu \preceq \epsilon_c$, hence $\mu \simeq \epsilon_c$. An agent whose risk preferences are modeled

[77] In 2002 Kahneman was awarded the Sveriges Riksbank Prize in Economic Sciences in Memory of Alfred Nobel.

by $\succeq$ would be indifferent to the choice between the (risky) payoff μ and the sure payoff ϵ_c. For this reason the scalar c is called the *certainty equivalent of* μ and is expressed $c = c(\mu)$. The certainty equivalent $c(\mu)$ may be viewed as the "price" that the agent is willing to pay (to be precise, the upper bound on the asking prices that the agent would be willing to accept) for the risky payoff μ, or, the "true value" of the risky payoff μ, given the agent's preference order $\succeq$.

The risk premium associated with the payoff $\mu \in \Theta$ is defined as

$$\rho(\mu) = m(\mu) - c(\mu),$$

and we remark that the units dimension of $\rho(\mu)$ is the same as the units dimension of the payoff Π (usually ¤). If the payoff μ is a sure payoff, then $\rho(\mu) = 0$, but, in general, Jensen's inequality (notice that $U(\cdot)$ is concave – not convex) gives

$$U(m(\mu)) = U\Big(\int_I x\,\mu(\mathrm{d}x)\Big) \geq \int_I U(x)\,\mu(\mathrm{d}x) = U(c(\mu)) \equiv U\big(m(\mu) - \rho(\mu)\big),$$

and, since $U(\cdot)$ is increasing, we see that $m(\mu) \geq c(\mu)$. In fact, since $U(\cdot)$ is strictly concave and strictly increasing, all the inequalities above would be strict unless μ is a sure payoff. Thus, we have $\rho(\mu) \geq 0$ with equality possible only if μ is a sure payoff. We also see that the more concave the utility function $U(\cdot)$, the larger the risk premium $\rho(\mu)$. In other words, a very concave utility function will have certainty equivalents associated with the various payoffs that are far below the expected payoffs. At the other extreme, a utility function that is nearly flat will have certainty equivalents that are very close to the expected payoffs. The takeaway from this reasoning is that the expectation of a risky payoff is generally higher than the "market" value of that same payoff, considered from the point of view of a particular set of preferences toward risk – provided that the preferences are risk averse. ○

(6.17) EXERCISE: With the notation as in (6.16), suppose that $I = \mathbb{R}_{++}$ and consider the payoff

$$\mu \stackrel{\text{def}}{=} \sum\nolimits_{i \in \mathbb{N}} 2^{-(i+1)} \epsilon_{2^i}$$

that can be linked with the well-known St. Petersburg paradox.[78] Calculate $m(\mu)$ and $c(\mu)$ with $U(x) = \log(x)$, $x \in \mathbb{R}_{++}$. ○

(6.18) EXAMPLE (MIXTURE OF RISKY AND NON-RISKY PAYOFFS): With the notation as in (6.16), consider a stochastic payoff, expressed as a random variable Π with a nondegenerate distribution law μ, and a sure payoff, expressed as a scalar

[78] A symmetric coin is flipped indefinitely and every H results in the win of ¤2 for every bet of ¤1, while every T results in the loss of ¤1 for every bet of ¤1. There is an entry fee of ¤1 to start betting. Consider the so-called doubling strategy, in which the player begins with an initial bet of ¤1 and doubles the previous bet at every consecutive flip, until the coin lands H, at which point the player stops playing and leaves. The payoff from this strategy, Π, is random, with $\mathsf{P}(\Pi = 2^i) = 2^{-(i+1)}$, $i \in \mathbb{N}$, so that $\mathsf{E}[\Pi] = \sum_{i \in \mathbb{N}} 2^i 2^{-(i+1)} = \infty$, whence the paradox.

value $b \in \mathbb{R}$. Suppose that the random variable Π is bounded, in that $|\Pi| < C$ for some $C \in \mathbb{R}_{++}$ such that $C > |b|$, and consider the family of mixtures of the risky payoff Π and the non-risky (sure) payoff b, defined as

$$Z_\theta = \theta\Pi + (1-\theta)b\,, \quad \theta \in [-A, A]\,,$$

for some fixed $A \in \mathbb{R}_{++}$. Observe that the payoffs Z_θ are distributed inside the interval

$$I = \big]-A(C+b)+b\,,\ A(C-b)+b\big[$$

and let $U(\cdot)$ be some utility function defined on this interval. An economic agent whose preferences are described by $U(\cdot)$ must choose the parameter $\theta \in [-A, A]$ (that is, choose the mixture of the two payoffs) so as to maximize the expected utility

$$[-A, A] \ni \theta \rightsquigarrow f(\theta) \stackrel{\text{def}}{=} \mathsf{E}\big[U(\theta\Pi + (1-\theta)b)\big]\,,$$

where we suppose that the expectation is well-defined and finite for every choice of θ. Since $U(\cdot)$ is strictly concave, so is also $f(\cdot)$. As a result, there is a unique $\theta^* \in [-A, A]$ at which $f(\cdot)$ attains its maximum. It is possible to impose only a one-sided restriction on θ, say $\theta \in [-A, +\infty[$ or $\theta \in]-\infty, A]$, or impose no restriction at all. If the agent is not allowed to take a short position (that is, borrow) in any of the payoffs, then θ is constrained to the interval $[0, 1]$. Depending on the shape of the utility function $U(\cdot)$, if the domain of $f(\cdot)$ is unbounded, concavity alone can no longer guarantee the existence of a solution, since $f(\cdot)$ may increase without a bound. If a solution θ^* exists, then the quantities θ^* and $1-\theta^*$ give the agent's demand for the payoffs Π and b (based on the risk preferences encrypted in the utility function $U(\cdot)$). However, in most practical situations the constraints on the choice of θ are more involved due to a limited budget set and transaction costs. ○

(6.19) EXERCISE: Similar to (6.18), consider the payoff $\Pi \in \mathcal{N}(1, \sigma^2)$ and the sure payoff $b = 1$. Assume risk preferences given by an exponential utility function of the form $\mathbb{R} \ni x \rightsquigarrow U(x) \stackrel{\text{def}}{=} -e^{-ax}$ for some fixed $a \in \mathbb{R}_{++}$. Find the value for $\theta \in \mathbb{R}$ that maximizes the expected utility $\mathsf{E}[U(\theta\Pi + 1 - \theta)]$. Explain the dependence of the optimal solution on the parameter a. ○

(6.20) ABSOLUTE RISK AVERSION: Given any two utility functions $U_i : I \mapsto \mathbb{R}$, $i = 1, 2$, how should the statement "U_2 is more risk averse than U_1" be understood? We know that "risk aversion" means "concavity," and, at least intuitively, the greater risk aversion of U_2 should translate into greater concavity of U_2, compared to U_1. It is easy to check that, given any increasing and concave function $f : \mathrm{Ran}(U_1) \mapsto \mathbb{R}$, the composition $x \rightsquigarrow f(U_1(x))$ is yet another utility function – it is called the *increasing concave transformation of* U_1. It is just as easy to check that increasing concave transformations increase concavity. Thus, the greater risk aversion of U_2 relative to U_1 should be possible to state as "U_2 is an increasing concave transformation of U_1." To justify this claim, notice first that any stochastic payoff, Π, such that

$$\mathsf{E}\big[U_1(v+\Pi)\big] \le U_1(v) \quad \text{for any } v \in I\,,$$

can be interpreted as "a stochastic payoff that U_1 dislikes." If U_2 dislikes (in the same sense) any payoff that U_1 dislikes, then U_2 can be seen as more risk averse. In particular, if $U_2 = f \circ U_1$ for some increasing and concave function $f(\cdot)$ and U_1 dislikes Π, then U_2 also dislikes Π:

$$\mathsf{E}\big[U_2(v+\Pi)\big] = \mathsf{E}\big[f\big(U_1(v+\Pi)\big)\big] \le f\big(\mathsf{E}\big[U_1(v+\Pi)\big]\big) \le f(U_1(v)) = U_2(v).$$

It turns out that the converse statement is also true (LeRoy and Werner 2001, 9.6.1): if U_2 dislikes any payoff that U_1 dislikes, then U_2 is an increasing concave transformation of U_1. In any case, if we were to suppose that $U_2 = f \circ U_1$ and that all three functions U_1, U_2, and f are twice continuously differentiable, then

$$\partial U_2(x) = (\partial f)(U_1(x))\partial U_1(x)$$

and

$$\partial^2 U_2(x) = (\partial^2 f)(U_1(x))\partial U_1(x)^2 + (\partial f)(U_1(x))\partial^2 U_1(x)\,.$$

In particular, since $\partial U_1(x) > 0$, we have

$$\begin{aligned}(\partial^2 f)(U_1(x)) = \frac{\partial^2 U_2(x)}{\partial U_1(x)^2} - \frac{\partial U_2(x)}{\partial U_1(x)}\frac{\partial^2 U_1(x)}{\partial U_1(x)^2} &= \frac{\partial U_2(x)}{\partial U_1(x)^2}\left(\frac{\partial^2 U_2(x)}{\partial U_2(x)} - \frac{\partial^2 U_1(x)}{\partial U_1(x)}\right) \\ &= \frac{\partial U_2(x)}{\partial U_1(x)^2}\big(A_1(x) - A_2(x)\big)\end{aligned}$$

where

$$A_i(x) \stackrel{\text{def}}{=} -\frac{\partial^2 U_i(x)}{\partial U_i(x)}\,, \qquad i = 1, 2\,.$$

Since $\partial U_i(x) > 0$, $i = 1, 2$, we see that the relation $(\partial^2 f)(U_1(x)) \le 0$ is equivalent to $A_2(x) \ge A_1(x)$. Plainly, if $U_2(x)$ is a functional transformation of $U_1(x)$, this functional transformation is concave if and only if $A_2(x) \ge A_1(x)$. Furthermore, the larger the gap between $A_2(x)$ and $A_1(x)$, the more concave the functional transformation. In sum, the risk aversion of a twice continuously differentiable utility function $U(\cdot)$ can be measured by the expression

$$A(x) = -\frac{\partial^2 U(x)}{\partial U(x)}\,, \qquad x \in I\,,$$

which is called the *Arrow–Pratt coefficient of absolute risk aversion.* ○

(6.21) ARROW–PRATT APPROXIMATION: (LeRoy and Werner 2001) A stochastic payoff, ζ, with distribution law $\mu \in \Theta$ $(= \mathcal{M}_1^h(I, \mathcal{B}(I)))$ is said to represent "pure risk" if ζ is not a sure payoff and $m(\mu) = \mathsf{E}[\zeta] = 0$. Assuming that ζ is a pure risk payoff and that $x \in I$ is arbitrarily chosen, consider the payoffs $\Pi_\alpha = x + \alpha\zeta$, $\alpha \in \mathbb{R}$, which can be seen as additive perturbations of the sure payoff x. Let $\rho(\alpha)$

denote the risk premium associated with Π_α. Since $m(\epsilon_\alpha) = \mathsf{E}[\Pi_\alpha] = x$, the certainty equivalent of Π_α is $c(\alpha) = x - \rho(\alpha)$, hence

$$\mathsf{E}\big[U(x+\alpha\zeta)\big] = U\big(x - \rho(\alpha)\big) \quad \text{for all } \alpha \in \mathbb{R}.$$

A formal differentiation in the last identity in α yields

$$\mathsf{E}\big[\zeta \partial U(x+\alpha\zeta)\big] = -\partial\rho(\alpha)\partial U\big(x-\rho(\alpha)\big)$$

and

$$\mathsf{E}\big[\zeta^2\partial^2 U(x+\alpha\zeta)\big] = \partial\rho(\alpha)^2\partial^2 U\big(x-\rho(\alpha)\big) - \partial^2\rho(\alpha)\partial U\big(x-\rho(\alpha)\big).$$

With $\alpha = 0$ we have $\rho(\alpha) = \rho(0) = 0$ and the first identity gives

$$-\partial\rho(0)\partial U(x) = \partial U(x)\mathsf{E}[\zeta] = 0.$$

Since $\partial U(x) > 0$, we must have $\partial\rho(0) = 0$. As a result, setting $\alpha = 0$ in the second identity, we get

$$^\dagger\ \partial^2 U(x)\mathsf{E}[\zeta^2] = -\partial^2\rho(0)\partial U(x) \iff \partial^2\rho(0) = -\frac{\partial^2 U(x)}{\partial U(x)}\mathsf{E}[\zeta^2] = A(x)\mathsf{E}[\zeta^2],$$

which has the effect that

$$\rho(\alpha) = \frac{\alpha^2}{2}A(x)\mathsf{E}[\zeta^2] + O(\alpha^3) = \frac{1}{2}A(x)\mathsf{E}[(\alpha\zeta)^2] + O(\alpha^3).$$

This relation is known as the *Arrow–Pratt approximation* and can be written equivalently as

$$A(x) = \frac{2\rho(\alpha) + O(\alpha^3)}{\mathsf{E}[(\alpha\zeta)^2]} = \frac{2\rho(\alpha)}{\mathsf{E}[(\alpha\zeta)^2]} + O(\alpha).$$

Thus, an alternative equivalent definition of the Arrow–Pratt coefficient of absolute risk aversion can be given in terms of the power series expansion around $\alpha = 0$ of the risk premium, $\rho(\alpha)$, associated with the stochastic payoff $\Pi_\alpha = x + \alpha\zeta$, where ζ is some pure risk. To be precise, since $\partial\rho(0) = 0$ we can write $\rho(\alpha) = \frac{1}{2}C\alpha^2 + O(\alpha^3)$ for some constant $C = C(x,\zeta)$ – and this constant is of course nothing other than $\partial^2\rho(0)$. The coefficient of absolute risk aversion can then be defined as $A(x) = C(x,\zeta)/\mathsf{E}[\zeta^2]$. Note that the utility function $U(\cdot)$ enters this definition through the risk premium $\rho(\alpha)$ and that the units dimension of $A(x)$ is ¤/¤^{2} = 1/¤. ○

(6.22) RELATIVE RISK AVERSION: (LeRoy and Werner 2001) Instead of additive perturbations of sure payoffs $x \in I$, it is often useful to consider multiplicative perturbations of the form $\Pi_\alpha = x(1+\alpha\zeta)$, where ζ is again pure risk as in (6.21). Since x is fixed, this multiplicative perturbation of x can be treated as an additive perturbation with pure risk $x\zeta$. We can again expand the risk premium associated with Π_α near $\alpha = 0$, and write this expansion as $\rho(\alpha) = \frac{1}{2}\bar{C}\alpha^2 + O(\alpha^3)$. Substituting $x\zeta$ for ζ in (6.21), we see⊶(6.21†) that $\bar{C} = x^2 C = x^2 A(x)\mathsf{E}[\zeta^2]$, where $A(x) = -\frac{\partial^2 U(x)}{\partial U(x)}$ is the absolute risk aversion coefficient associated with the additive perturbation from (6.21). Thus, the absolute risk aversion coefficient associated with the multiplicative perturbation must be of the form $A_*(x) = A(x)x^2$. The *Arrow–Pratt coefficient of relative risk aversion* is given by

$$R(x) = \frac{A_*(x)}{x} = A(x)x = -\frac{\partial^2 U(x)}{\partial U(x)}x, \quad x \in I.$$

It is easy to see that the relative risk aversion coefficient is units free ($A(x)$ has units dimension 1/¤, whereas x has units dimension ¤), which is one of its most useful features. ○

(6.23) EXERCISE (CARA UTILITIES): A twice continuously differentiable utility function $U(\cdot)$, defined on some interval $I \subseteq \mathbb{R}$, is said to have constant absolute risk aversion (CARA) if its absolute risk aversion coefficient has the form

$$A(x) = -\frac{\partial^2 U(x)}{\partial U(x)} = \text{constant} > 0 \quad \text{for all } x \in I.$$

Prove that any CARA utility function must be of the form $U(x) = a - be^{-Ax}$, for some choice of $A, b \in \mathbb{R}_{++}$ and $a \in \mathbb{R}$, that is, up to an affine transformation, a CARA utility can be identified as $U(x) = 1 - e^{-Ax}$, $A > 0$. ○

(6.24) EXERCISE (HARA UTILITIES): A twice continuously differentiable utility function $U(\cdot)$, defined on the interval $I = \mathbb{R}_{++}$, is said to have hyperbolic absolute risk aversion (HARA) if its absolute risk aversion coefficient has the form

$$A(x) = -\frac{\partial^2 U(x)}{\partial U(x)} = \frac{1-\gamma}{x}, \quad x \in \mathbb{R}_{++},$$

for some choice of the parameter $\gamma \in [0, 1[$. Prove that up to an affine transformation any HARA utility function can be identified as either $U(x) = x^\gamma/\gamma$, if $\gamma \in]0, 1[$, or with $U(x) = \log(x)$, if $\gamma = 0$. ○

(6.25) EXERCISE (CRRA UTILITIES): A twice continuously differentiable utility function $U(\cdot)$, defined on the interval $I = \mathbb{R}_{++}$, is said to have constant relative risk aversion (CRRA) if its relative risk aversion coefficient is constant, that is,

$$R(x) = -\frac{\partial^2 U(x)}{\partial U(x)}x = \text{constant} > 0 \quad \text{for all } x \in \mathbb{R}_{++}.$$

Prove that up to an affine transformation any CRRA utility function can be identified as either $U(x) = x^{1-R}/(1-R)$, if the CRRA coefficient is $0 < R \neq 1$, or as $U(x) = \log(x)$ if $R = 1$. Prove that $\lim_{R\to 1}(x^{1-R} - 1)/(1-R) = \log(x)$. ○

In what follows we will be working almost exclusively with CRRA utility functions with $R \geq 1$.

6.3 The Multiperiod Endowment Economy

Having developed the most basic tools for quantifying the economic agents' attitudes toward risk, now we turn to the interactions between them. These interactions come down to two things: trading their lots in life (that is, endowments) and sharing the risks that they face. We suppose for now that time is discrete ($t \in \mathbb{N}$) and the economy is finitely lived, with finite time horizon $T \in \mathbb{N}_{++}$. Any risk that the agents face will be modeled on a single probability space $(\Omega, \mathcal{F}, P)$, and the information structure is going to be described as a filtration $\mathscr{F} \equiv (\mathscr{F}_t \subseteq \mathcal{F})_{t \in \mathbb{N}_{|T}}$, chosen so that $\mathscr{F}_0 = \{\emptyset, \Omega\}$. The economy is populated by a finite set of agents, which set we denote by $\mathbb{I}$. There is a single consumption good that comes either in the form of individual income across the agents, or in the form of dividend yield received from a finite collection of tradable assets (alias: securities), which collection we denote by $\mathbb{J}$. The period t individual income we denote by e_t, or by e^i_t, if the association with a particular agent $i \in \mathbb{I}$ must be emphasized. Similarly, the period t dividend payout from security $j \in \mathbb{J}$ we denote by D^j_t, but are going to write simply D_t if the association with security $j \in \mathbb{J}$ is not relevant. We insist that all dividend payoffs be positive, that is, $D^j_t \in \mathbb{R}_+$, $j \in \mathbb{J}$. The period t dividend payoffs from all traded securities we aggregate into the vector $\bar{D}_t = (D^j_t, j \in \mathbb{J})$. All individual incomes and dividend payoffs are stochastic and respect the information structure $\mathscr{F}$; to be precise, $(e^i_t)_{t \in \mathbb{N}_{|T}}$ and $(D^j_t)_{t \in \mathbb{N}_{|T}}$ are assumed to be $\mathscr{F}$-adapted stochastic processes on $(\Omega, \mathcal{F}, P)$ for every $i \in \mathbb{I}$ and every $j \in \mathbb{J}$.

Every agent must consume some strictly positive amount $c_t \in \mathbb{R}_{++}$ (expressed in units of the consumption good) during every period $t \in \mathbb{N}_{|T}$ in order to survive. Because the agents differ in their respective incomes, endowments with securities, and risk preferences (the population of agents is heterogeneous), generally, the consumption levels in the cross-section of agents are going to vary. We will therefore write c^i_t if the association with agent $i \in \mathbb{I}$ is relevant. Every agent has a time additive utility from consumption, and the utility derived during period t for a particular agent is $U_t(c_t)$, for some twice continuously differentiable utility function $U_t \colon \mathbb{R}_{++} \mapsto \mathbb{R}$. This utility function is again agent specific, and we will again write $U^i_t(c^i_t)$ if the association with agent $i \in \mathbb{I}$ is relevant. The consumption good is perishable, so that the aggregate period t economic output (income plus dividends) must exactly match the aggregate period t consumption:

$$\sum\nolimits_{i \in \mathbb{I}} c^i_t = \sum\nolimits_{i \in \mathbb{I}} e^i_t + \sum\nolimits_{j \in \mathbb{J}} D^j_t \quad \text{for all } t \in \mathbb{N}_{|T} .$$

In this expression the consumption levels c^i_t, the incomes e^i_t, and the dividends D^j_t are all expressed in units of the consumption good. Nothing will change if we treat these quantities as nominal amounts, that is, expressed in currency units, as long as the exchange rate between one unit of the consumption good and one unit of currency remains fixed. For simplicity, we set the currency unit ¤1 to be the unit of consumption good.

The next step is to describe the agents' holdings of securities. All securities are considered to be perfectly divisible, meaning, any security can be held in any amount – positive or negative. A holder of θ^j units of security $j \in \mathbb{J}$ – to put it another way, a holder of θ^j shares of security $j \in \mathbb{J}$ – is entitled to the (stochastic) payoff of $\theta^j D^j_t$ units of the consumption good during period t if $\theta^j > 0$, and has a liability of (must deliver) $(-\theta^j)D^j_t$ units of the consumption good during period t if $\theta^j < 0$. The holdings of securities for a particular agent we aggregate into the vector $\bar{\theta} = (\theta^j, j \in \mathbb{J})$ with dimension $\mathrm{len}(\bar{\theta}) = |\mathbb{J}|$,[79] and, as we are already accustomed to, will write $\bar{\theta}^i$ if the association with a particular agent $i \in \mathbb{I}$ is relevant. Thus, $\theta^{i,j}$ denotes the units of security j held by agent i. A negative amount of a particular security may be acquired by short selling that security: an agent may sell a security that they do not have, but is obliged to deliver all future payoffs from it, in effect underwriting (creating) "new" shares from the same security. Notice that such an underwriting of "new" shares does not increase the aggregate amount of shares in the economy (whence the quotation marks), that is, does not increase the shares outstanding for the individual securities, since any shares that are underwritten that way show up as negative amounts in the accounts of those who underwrite them (take a short position) and, at the same time, show up as positive amounts in the accounts of those who acquire them (take a long position).

The agents may use part of their income to purchase consumption, but rational agents anticipate future states of the economy in which their income may not be adequate, and may choose to forfeit part of their present income in exchange for more income in the future. Such agents act as investors. Other agents anticipate more income in the future but have insufficient income in the present period. Such agents act as borrowers: they are willing to forfeit some of their future income in exchange for consumption today, that is, they would like to purchase consumption today with income that will be received in the future. In sum, the agents would like to have the ability to trade their lots in life, so that their intertemporal consumption schedules maximize their expected time additive utility. The channel through which such objectives can be achieved is the securities market. The agents that act as borrowers sell securities and the agents that act as investors buy securities. As a result, the securities holding (alias: portfolio) of any particular agent would be time dependent and therefore expressed as $(\bar{\theta}_t)_{t \in \mathbb{N}_{|T}}$ (or by $(\bar{\theta}^i_t)_{t \in \mathbb{N}_{|T}}$, if the association with agent $i \in \mathbb{I}$ is relevant). Since the agents' portfolio choices are affected by the way in which the stochastic incomes and dividends are revealed over time, the portfolios are going to be $\mathscr{F}$-adapted stochastic processes on $(\Omega, \mathcal{F}, P)$. Every agent $i \in \mathbb{I}$ begins their life with an initial endowment with securities given by the vector $\bar{\theta}^i_{-1}$. The period t spot price at which security $j \in \mathbb{J}$ is traded we denote by

[79] Recall that the token $|\cdot|$ is also used to denote cardinality.

S^j_t (or by S_t if the association with security $j \in \mathscr{J}$ is not relevant), and aggregate all spot prices into the price vector $\bar{S}_t = (S^j_t, j \in \mathscr{J})$. The spot prices, too, are going to be $\mathscr{F}$-adapted stochastic processes on $(\Omega, \mathcal{F}, P)$, but we are yet to establish how these prices come about.

Unless noted otherwise, a security is assumed to be long-lived, that is, assumed to exist for as long as the economy exists. If purchased in period t at the spot price S_t, in period $t+1$ a long-lived security will generate payoff that includes not only the dividend D_{t+1} but also the period $t+1$ spot price S_{t+1}. In contrast, a short-lived security that exists only between period t and period $t+1$ will generate payoff in period $t+1$ equal only to the dividend payoff D_{t+1}. Actually, this distinction becomes relevant only if we need to include a risk-free security in the model. "Risk free" means that the period $t+1$ payoff would be known at time t. To put it another way, treated as an r.v., the period $t+1$ payoff has to be measurable with respect to $\mathscr{F}_t$. At the same time, we would still like to allow the period $t+1$ spot price of the risk-free security to be stochastic relative to the information available in period t, and the only way around this is to impose that the dividend D_{t+1} is $\mathscr{F}_t$-measurable, while a payoff equal to the spot price S_{t+1} is simply not available for purchase at time t (or, if it is available, it is a separate security that is not risk-free); in other words, a risk-free security should be short-lived and its payoff should be just the dividend D_{t+1}, which is known in period t ($D_{t+1} \prec \mathscr{F}_t$). Of course, a market model does not have to include a risk-free security in order to be meaningful.

(6.26) INVESTMENT-CONSUMPTION DECISIONS: An agent enters period t with a portfolio of securities $\bar{\theta}_{t-1}$, and exits period t with a portfolio of securities $\bar{\theta}_t$. Thus, the agent begins period t with (entering) wealth given by (in units of the consumption good, which in our setting is the same as units of currency ¤) $Z_t \overset{\text{def}}{=} e_t + \bar{\theta}_{t-1} \cdot (\bar{S}_t + \bar{D}_t)$. During period t this wealth can be used for two types of transactions: consumption and investment. The rational expectations value of the wealth Z_t – to put it another way, the largest amount of expected utility that the agent can extract during period t and, by way of investing, in all future periods (if any) following period t – we denote by $V_t(Z_t)$, and remark that the function $V_t(\cdot)$ actually depends on $\omega \in \Omega$, and this dependence is consistent with the information arrival dynamics encrypted in $\mathscr{F}$. If we need to make this dependence explicit we must write $V_t(\omega, \cdot)$. In particular, if $\mathscr{F}_t$ is generated by a partition $\mathscr{P}_t$, then there would be one such function $V_t(\cdot)$ attached to every atom in $\mathscr{P}_t$, so that if the filtration $\mathscr{F}$ is represented as an event tree, then there would be one function attached to every node on the tree, which function takes as an argument the wealth with which the agent enters the node (the state of the economy) and returns the rational expectations value of that wealth. During period $t \in \mathbb{N}_{|T}$ the agent chooses their consumption level $x = c_t \in \mathbb{R}_{++}$ and chooses a new portfolio $\bar{y} = \bar{\theta}_t \in \mathbb{R}^{|\mathscr{J}|}$. How much expected utility does the choice of x and $\bar{y}$ create? Clearly, the choice of the consumption level x would generate immediate utility $U_t(x)$, but the choice of the portfolio $\bar{y}$ would affect the utility generated during period $t+1$ and after, which is captured by the function $V_{t+1}(Z_{t+1})$, where $Z_{t+1} = e_{t+1} + \bar{y}\cdot(\bar{S}_{t+1} + \bar{D}_{t+1})$, with the understanding that $V_{T+1} = e_{T+1} = 0$ and $\bar{S}_{T+1} = \bar{D}_{T+1} = \vec{0}$. Naturally, the choice of x and $\bar{y}$ is constrained by the available wealth $z = Z_t$, whence the equation

$$x + \bar{y}\cdot\bar{S}_t = z\,.$$

Another implicit constraint faced by the agent is the random mechanism – in other words, the structure of the event tree – through which the entering wealth in period $t+1$, that is, the variable $Z_{t+1} \prec \mathscr{F}_{t+1}$, depends on the entering wealth $z = Z_t$, the choice of consumption $x = c_t$, and the choice of portfolio $\bar{y} = \bar{\theta}_t$. We will cast this constraint as

$$Z_{t+1} - e_{t+1} - \bar{y}\cdot(\bar{S}_{t+1} + \bar{D}_{t+1}) + x + \bar{y}\cdot\bar{S}_t = z\,, \quad Z_{t+1} \prec \mathscr{F}_{t+1}\,.$$

Notice that if $\mathscr{F}_{t+1}$ is generated by a partition $\mathscr{P}_{t+1}$, then the variables x, $\bar{y}$, and z are going to be attached to an atom $a \in \mathscr{P}_t$, while the variables $Z_{t+1}(\omega)$, $\omega \in \Omega$, and the associated constraints above are going to be attached to all atoms in $\mathscr{P}_{t+1}$ that are contained in a. Thus, in state $a \in \mathscr{P}_t$ the agent will be faced with $1 + |a|$ constraints, where $|a|$ is the number of atoms in $\mathscr{P}_{t+1}$ that are contained in a. Because during period $t \in \mathbb{N}_{|T}$ the agent is maximizing expected utility given information found in $\mathscr{F}_t$, and because their utility is time additive, we can write[80]

$$\dagger \qquad V_t(z) = \max_{x,\bar{y},Z_{t+1} \in \ddagger} \Big(U(x) + \beta\,\mathsf{E}\big[V_{t+1}(Z_{t+1}) \mid \mathscr{F}_t\big]\Big),$$

where the choice variables are constrained as follows:

$$\ddagger \qquad \begin{gathered} Z_{t+1} - e_{t+1} - \bar{y}\cdot(\bar{S}_{t+1} + \bar{D}_{t+1}) + x + \bar{y}\cdot\bar{S}_t = z\,, \quad x + \bar{y}\cdot\bar{S}_t = z\,, \\ x \in \mathbb{R}_{++}\,, \quad y \in \mathbb{R}^{|\mathbb{J}|}\,, \quad Z_{t+1} \prec \mathscr{F}_{t+1}\,. \end{gathered}$$

The scalar value β on the right side is a subjective discount factor, that is, a parameter that captures the agent's impatience toward utilizing their consumption set. This parameter may vary in the cross-section of agents, and therefore will be expressed as $\beta = \beta^i$, if the identification with agent $i \in \mathbb{I}$ is relevant. ○

(6.27) THE FIRST-ORDER CONDITIONS AND THE SHADOW PRICES: Since the rational expectations functions $V_t(\cdot)$ and $V_{t+1}(\cdot)$ can be shown to be concave, the constrained maximization problem (6.26†) is equivalent to the following unconstrained problem:

$$\min_{\varphi \in \mathbb{R},\, \Phi \prec \mathscr{F}_{t+1}} \quad \max_{x \in \mathbb{R}_{++},\, \bar{y} \in \mathbb{R}^{|\mathbb{J}|},\, Z \prec \mathscr{F}_{t+1}} \quad L(x, \bar{y}, Z, \varphi, \Phi)$$

where

[80] This relation can be seen as a consequence of the principle of dynamic programming, which will be introduced later.

$$\begin{aligned}L(x,\bar{y},Z,\varphi,\Phi) \stackrel{\text{def}}{=} \Big(U(x) + \beta\,\mathsf{E}\big[V_{t+1}(Z)\,\big|\,\mathscr{F}_t\big]\Big)\\ + \beta\,\mathsf{E}\Big[\Phi\big(z - Z + e_{t+1} + \bar{y}\cdot(\bar{S}_{t+1} + \bar{D}_{t+1}) - x - \bar{y}\cdot\bar{S}_t\big)\,\Big|\,\mathscr{F}_t\Big]\\ + \varphi\,(z - x - \bar{y}\cdot\bar{S}_t)\,.\end{aligned}$$

Note that the Lagrange multipliers associated with the first set of constraints – there is one such constraint attached to every contingent state of the economy in period $t+1$ that follows the current state in period t – are written formally as $\beta\,\Phi(\omega)P(\mathrm{d}\omega\,|\,\mathscr{F}_t)$ for $\Phi \precsim \mathscr{F}_{t+1}$ (the actual choice is $\Phi(\omega)$). The values for $x \in \mathbb{R}_{++}$, $y \in \mathbb{R}^{|\mathscr{J}|}$, $Z \precsim \mathscr{F}_{t+1}$, $\varphi \in \mathbb{R}$, and $\Phi \precsim \mathscr{F}_{t+1}$, at which the min max above is attained can be computed from the first-order conditions:

$$\begin{aligned}&\partial_\varphi L(x,\bar{y},Z,\varphi,\Phi) = 0\,, \qquad \partial_{\Phi(\omega)} L(x,\bar{y},Z,\varphi,\Phi) = 0\,,\ \ \omega \in \Omega\,,\\ &\partial_x L(x,\bar{y},Z,\varphi,\Phi) = 0\,, \qquad \partial_{Z(\omega)} L(x,\bar{y},Z,\varphi,\Phi) = 0\,,\ \ \omega \in \Omega\,,\\ &\qquad\qquad \partial_{y_j} L(x,\bar{y},Z,\varphi,\Phi) = 0\,,\ \ j \in \mathscr{J}\,.\end{aligned}$$

If this system of conditions has a solution $x = x^*$, $y = y^*$, $Z(\omega) = Z^*(\omega)$, $\varphi = \varphi^*$, and $\Phi(\omega) = \Phi^*(\omega)$, the rational choice for the agent would be to set their consumption level during period t to $c_t = x^*$, and set their new (by the end of period t) portfolio of securities to $\bar{\theta}_t = y^*$. This determines the agent's optimal decision – for a particular moment in time and state of the economy – but in asset pricing we are also interested in, well, pricing. In this regard, the variables φ and $\Phi(\omega)$, that is, the Lagrange multipliers associated with the respective constraints, are instrumental. As is well known, these variables give the shadow prices – in the respective states of economy – of the resource on the right side of the constraints. These shadow prices are nothing but marginal values of nominal wealth available for consumption and investment at the beginning of period t, given the agent's lot in life, lifetime objectives, state of the economy in period t, and given the mechanism through which the event tree allows the agent to transfer wealth from the current state in period t to each of the period $(t+1)$-contingent states (modeled as *child nodes* on the tree).[81] The aggregate shadow price of entering wealth at time t is therefore

$$\phi \stackrel{\text{def}}{=} \varphi + \beta\mathsf{E}\big[\Phi\,\big|\,\mathscr{F}_t\big]\,,$$

and the first-order conditions above can now be transcribed as

$$\begin{aligned}&Z = e_{t+1} + \bar{y}\cdot(\bar{S}_{t+1} + \bar{D}_{t+1})\,,\ \ \omega \in \Omega\,; \quad x + \bar{y}\cdot\bar{S}_t = z\,;\\ \dagger\quad &\phi = (\partial U)(x)\,,\quad \Phi(\omega) = (\partial V_{t+1})(Z(\omega))\,,\ \ \omega \in \Omega\,;\\ &\phi S_t^j = \beta\,\mathsf{E}\big[\Phi\big(S_{t+1}^j + D_{t+1}^j\big)\,\big|\,\mathscr{F}_t\big]\,,\quad j \in \mathscr{J}\,.\end{aligned}$$

[81] The entire event tree – to be specific, the dynamics of the dividend payoffs D_t^j and the individual incomes e_t^i – clearly represent an implicit constraint in the agent's consumption-investment decision. Thus, the inner structure of some trees is more valuable than the inner structure of other trees, or, to put it another way, there are shadow prices associated with the structure of the event tree.

Furthermore, a direct application of the envelope theorem gives

$$(\partial V_t)(z) = \varphi + \beta \mathsf{E}[\Phi \mid \mathscr{F}_t] = \phi = (\partial U)(x) \equiv (\partial U)(c_t)\,.$$

By the same argument, if the agent makes the optimal investment-consumption choice in period $t+1$, having entered that period with wealth $Z(\omega)$ (the optimal entering wealth that they could have designed in period t for that state), then

$$(\partial U)(c_{t+1}(\omega)) = (\partial V_{t+1})(Z(\omega)) = \Phi(\omega)\,,$$

where the last identity is simply the the 4th condition in †. All of this is consistent with the economic interpretation of $\Phi(\omega)$ as the aggregate shadow price of entering wealth for state ω. As a result, the 5th condition in † gives

$$\ddagger \qquad \bar{S}_t = \beta \mathsf{E}\Big[\frac{(\partial U)(c_{t+1})}{(\partial U)(c_t)}\big(\bar{S}_{t+1} + \bar{D}_{t+1}\big) \,\Big|\, \mathscr{F}_t\Big]\,, \quad t \in \mathbb{N}_{|T-1}\,,$$

understood as an identity between vectors of dimension $|\mathscr{J}|$. This relation is fundamental in the realm of asset pricing, as will be evidenced throughout the sequel. It has many important aspects, and, for this reason, in what follows we will rewrite ‡ in several equivalent forms. ○

(6.28) STATE PRICE DENSITIES: The process $\big(\psi_t \overset{\text{def}}{=} \beta^t(\partial U)(c_t)\big)_{t\in\mathbb{N}_{|T}}$ is clearly $\mathscr{F}$-adapted, strictly positive, and, according to (6.27‡), has the property

$$\dagger \qquad \psi_t S_t = \mathsf{E}\big[\psi_{t+1}(S_{t+1} + D_{t+1}) \mid \mathscr{F}_t\big]\,, \quad t \in \mathbb{N}_{|T-1}\,,$$

for every traded security S, with the understanding that $S_{t+1} = 0$ if the security is short-lived. In the final period $t = T$ no security is traded and we set $S_T = 0$. Any positive and $\mathscr{F}$-adapted stochastic process $(\psi_t)_{t\in\mathbb{N}_{|T}}$ that satisfies † simultaneously for all traded securities (S_t) and in all time periods t is called *state price density*, or SPD for short. Since, up to a multiplicative factor, the quantity ψ_t is the shadow price of a unit of nominal wealth for the agent, then $\psi_t S_t$ gives the subjective price (that is, the price from the point of view of the agent's own objectives) of the security S_t. Similarly, $\psi_{t+1}(S_{t+1} + D_{t+1})$ is the agent's subjective value of the payoff $(S_{t+1} + D_{t+1})$. The relation † now says something rather interesting: the price of every security at time t equals the expected payoff at time $t+1$, but only after the market prices and the dividend payoffs are re-priced according to the agent's private pricing rules (which also take into account the agent's impatience, in addition to their risk preferences). We stress that an SPD of the form $\psi_t = \beta^t \partial U(c_t)$ is really agent-specific, that is, $\psi_t = \psi_t^i$ for some $i \in \mathbb{I}$. This is clear: the discount factor β, the utility function U and the consumption level c_t are all agent-specific quantities. Another important observation is that the SPD ψ^i depends on the state of the economy (it is a stochastic process) but does not depend on any particular security. To put it another way, the agent's private pricing system is a pricing system for states of the economy – not securities. ○

(6.29) EXERCISE: Prove that for any SPD (ψ_t), if one exists, and for any long-lived security (S_t),

$$S_t = \sum_{s=t+1}^{T} \mathsf{E}\left[\frac{\psi_s}{\psi_t} D_s \,|\, \mathscr{F}_t\right] \quad \text{for all } t \in \mathbb{N}_{|T-1} . \quad \circ$$

(6.30) STOCHASTIC DISCOUNT FACTORS: Consider the process

$$\xi_{t+1} \stackrel{\text{def}}{=} \frac{\psi_{t+1}}{\psi_t} \equiv \beta \frac{\partial U(c_{t+1})}{\partial U(c_t)}, \quad t \in \mathbb{N}_{|T-1},$$

which is clearly agent-specific, that is, $\xi_{t+1} = \xi^i_{t+1}$ (we again follow our usual convention to suppress the superscripts unless they are needed), and deserves the name "discount factor." Indeed, (6.27‡) gives

$$\dagger \qquad S_t = \mathsf{E}\left[\xi_{t+1}(S_{t+1} + D_{t+1}) \;\middle|\; \mathscr{F}_t\right], \quad t \in \mathbb{N}_{|T-1},$$

which is to say the quantity ξ_{t+1} is telling us how to discount the period $t+1$ payoffs – in fact, how to discount universally all period $t+1$ payoffs, for all securities S_{t+1} – back to period t. We will refer to † as *the fundamental pricing equation.* Any positive and $\mathscr{F}$-adapted stochastic process $(\xi_t)_{t \in \mathbb{N}_{|T}}$ for which † holds simultaneously for all securities (S_t) is called *stochastic discount factor*, or SDF for short. It is remarkable that the payoffs from all securities – very risky and very non-risky alike – are discounted with the same factor, and we stress that ξ_{t+1} is a $\mathscr{F}_{t+1}$-measurable random variable – it is *not*, in general, an $\mathscr{F}_t$-measurable random variable. In other words, ξ_{t+1} is a discount factor for every contingent state of the economy in period $t+1$, given the information available in period t. $\circ$

(6.31) EXERCISE: Prove that for any SDF (ξ_t), if one exists, and for any security S,

$$S_t = \sum_{s=t+1}^{T} \mathsf{E}\left[\left(\prod_{u=t+1}^{s} \xi_u\right) D_s \;|\; \mathscr{F}_t\right] \quad \text{for all } t \in \mathbb{N}_{|T-1} . \quad \circ$$

(6.32) MARGINAL RATES OF SUBSTITUTION: As an illustration of the role that the stochastic discount factors play in asset pricing, suppose that $\mathscr{F}_t = \sigma(\mathscr{P}_t)$ and $\mathscr{F}_{t+1} = \sigma(\mathscr{P}_{t+1})$, where $\mathscr{P}_t$ is some (finite) partition of Ω and $\mathscr{P}_{t+1}$ is another (finite) partition, which is a refinement of $\mathscr{P}_t$. Then the fundamental pricing equation can be cast as

$$\dagger \qquad S_t(a) = \sum_{\hat{a} \in \mathscr{P}_{t+1},\, \hat{a} \subseteq a} \xi_{t+1}(\hat{a}) \left(S_{t+1}(\hat{a}) + D_{t+1}(\hat{a})\right) \frac{P(\hat{a})}{P(a)}, \quad a \in \mathscr{P}_t .$$

Whether such a security exists or not, $\xi_{t+1}(\hat{a})\frac{P(\hat{a})}{P(a)}$ can be thought of as the price – perhaps the hypothetical price – in period t, when the economy is in state a, of a security that promises to pay ¤1 in period $(t+1)$ if the economy happens to be in state $\hat{a}$, and promises to pay ¤0 in any other state that the economy could be in, following state a. As we already know, such an instrument is called *Arrow–Debreu security*.

The reason for the alternative term "marginal rate of substitution" is now clear: since $\xi_{t+1}(\hat{a})$ is nothing but the ratio $\psi_{t+1}(\hat{a})/\psi_t(a)$, it represents the hypothetical gross return for the agent from forfeiting ¤1 during period t in state a (in their private pricing rule this comes down to investing the amount $\psi_t(a)$) in exchange for receiving ¤1 in state $\hat{a}$ in period $(t+1)$ (in their private pricing rule this comes down to receiving the amount $\psi_{t+1}(\hat{a})$); in other words, the discount factor ξ_{t+1} measures the agent's willingness to substitute ¤1 in state a in period t with ¤1 in state $\hat{a}$ in the (future) period $t+1$. ○

(6.33) REMARK: In the economic model described in the present section the aggregate amount of the consumption good is stochastic, but is not affected by the economic activity. The agents can trade the output streams that they are entitled to – and this way re-engineer the (heterogeneous) shocks that they are subjected to and match their (heterogeneous) preferences – but, nevertheless, the individual endowment streams and the dividend streams are as they come along. Whence the name "endowment economy." ○

6.4 General Equilibrium

There is one peculiar aspect of the fundamental pricing equation (6.30†): the stochastic discount factor on the right side is agent specific, whereas the asset price on the left side is not. This phenomenon has the following interpretation: although different agents have different SPDs and SDFs, if the economy is in the state of equilibrium, they all agree on the price, in the sense that their individual heterogeneous pricing rules (the result of their individual investment-consumption choices) produce the same prices for all stochastic payoffs that are traded. How does one arrive at market equilibrium, then? First, we must recognize that in the first-order conditions introduced in (6.27) the price vector $\bar{S}_{t+1}$ and the dividend vector $\bar{D}_{t+1}$ are assumed given. In other words, those first-order conditions are telling us what the optimal investment-consumption plan for a given agent is, but with the stipulation that the price system is fixed in the outset. Clearly, the optimization problems that different agents are faced with are generally different. After all, the agents differ in their risk preferences, income streams, and initial endowments with securities. Generally, such differences are more than enough to force the agents to make different investment-consumption decisions, which then leads to different welfare shocks in the cross-section of agents. In any case, there is no reason to suspect that all agents can solve their (different) first-order conditions (by choosing their consumptions and portfolios over time) from (6.27) under any price system whatsoever. The fundamental premise in the notion of economic equilibrium is that the asset prices and the individual investment-consumption choices (for all agents) are determined *simultaneously* and in such a way that *every* agent's individual choices are optimal (for a price system determined while the individual choices are made – again, simultaneously by all agents). To put it another way, the price process $(\bar{S}_t)_{t \in \mathbb{N}_{|T}}$ is determined endogenously from the agents' concerted effort to be optimal. The fact that all agents agree on the prices of all securities – in other words, the fact that the fundamental pricing equation in (6.30†) holds with $\xi_{t+1} = \xi^i_{t+1}$ for every $i \in \mathbb{I}$ – is simply one of the conditions from which the endogenous price process $(\bar{S}_t)$ is ultimately determined. Figuratively speaking, the agents solve their first-order conditions "in an orchestra," and while doing so they also settle on the equilibrium prices for all traded securities. For this to be possible, one must impose some form of a collective constraint on the economy. In an endowment economy one such requirement is that economic output cannot be stored and that, therefore, the aggregate consumption must equal the aggregate output in every time period and in every state of the world. This requirement is captured by the aggregate resource constraint that we are already familiar with:

$$\sum\nolimits_{i \in \mathbb{I}} c^i_t = \sum\nolimits_{i \in \mathbb{I}} e^i_t + \sum\nolimits_{j \in \mathbb{J}} D^j_t \quad \text{for all } t \in \mathbb{N}_{|T}\,.$$

In addition, we must impose the market-clearing condition:[82]

$$\sum\nolimits_{i \in \mathbb{I}} \bar{\theta}^i_t = (1, 1, \dots, 1) \in \mathbb{R}^{|\mathbb{J}|}\,.$$

The last equation is telling us that every security is in aggregate net supply of one share at all times. Not much will change in the model if some securities are in zero net supply while others are in the net supply of one, that is, if some entries in the last vector are 0 while others are 1.

One obvious consequence of the last two equations is that some agents can consume more only if other agents consume less, and an agent can buy more shares only if it is optimal for other agents to sell shares. In sum, the only exogenous variables in a general equilibrium for an endowment economy are the endowments (individual incomes and dividend streams), while the security prices, the consumption levels, and the portfolio choices for all agents are endogenous. As there is no requirement for such markets to be complete (see below), such models are known under the acronym GEI (general equilibrium for incomplete markets).

(6.34) COMPUTATION: The endogenous quantities in a concrete GEI model are determined from the following four requirements attached to every state of the world in every time period: (a) all individual first-order conditions are satisfied, (b) all agents agree on the price of every traded security, (c) the economy is balanced (the aggregate resource constraint holds), and (d) the market for securities clears. It is remarkable that, at least in the case of finite GEI models (the sample space Ω is finite, or, equivalently, the information arrival dynamics are given by a finite event tree), gathering all these conditions and insisting that they must hold at all times and in all states of the world (i.e., at every node on the event tree) leads

[82] The market-clearing condition can be shown to be just another form of the aggregate resource constraint.

to a giant system, which – see (Dumas and Lyasoff 2012) – has exactly as many equations as there are unknowns (security prices, consumption plans and portfolio choices for all agents in every node on the tree). This claim can be made without any additional simplifying assumptions, and regardless of the number of agents, the number of securities, the size of the event tree, or the level of incompleteness of the market (see below). It is a separate matter whether such a giant system can be solved in practice, and one can certainly ask the question: can the financial markets really function as a giant super computer that allocates risk perfectly, sets asset prices perfectly, and ensures that all agents make perfect consumption-investment choices? Questions of such a nature are beyond the scope of this space, but we stress that one cannot simply round up the usual toolboxes from optimal control (think dynamic programming) and throw them at a real life GEI model. This is because the system of conditions that we just described is intrinsically forward-backward in time: the investment decisions made today depend on today's asset prices, but today's asset prices depend on tomorrow's payoffs, which in turn depend on today's investment choices. Dumas and Lyasoff (2012) proposed a computational method designed specifically for problems of this nature, but so far it has only been implemented in a few rather rudimentary concrete asset pricing models on finite event trees. Among other things, these examples show that multiple solutions are possible (in some cases of extreme disparity in the initial wealth of the agents), even as equations and unknowns match in number. Given the highly nonlinear structure of the GEI (to say the least, the agents' preferences are nonlinear), the possibility for multiple equilibria should not come as a surprise (it is not surprising that a quadratic equation of one unknown may admit two solutions, for example). Generically, such equilibria are isolated (just as in the case of a quadratic equation), which is to say that in equilibrium an agent cannot make small changes in their portfolio allocation without violating the optimality of another agent's portfolio, or the aggregate balance in the economy, or the price-agreement among the agents. ○

(6.35) PARTIAL EQUILIBRIUM: Just as the name suggests, a partial equilibrium market model is similar to the general equilibrium model described above, but without some of the conditions that define GEI. This makes perfect sense: after all, if resolving all the conditions that one is faced with becomes prohibitively complex, why not focus on a subset of conditions that can be realistically dealt with? Typically, one assumes that there is at least one stochastic discount factor, which is generally accepted by all agents, and, in addition, it is postulated that the agents' investment decisions do not affect the prices at which the assets trade (so to speak, all agents are "small investors"). This is a different paradigm, in which one is trying to use various observable quantities as proxies for the SDF, rather than derive the SDF form first principles. To put it another way, one assumes that the proverbial invisible hand has done its job, and instead of trying to understand how, one is merely trying to read the result. This means that some elements of GEI are taken as given, while others are derived from other principles – absence of arbitrage, for example (see below). Generically, a partial equilibrium model would still involve the pricing equation ($6.30^{\dagger}$), but, typically, would ignore the fact that one such equation (associated with a particular security) cannot be solved in isolation, and that the choice of portfolios do affect the spot prices at which the assets are traded. As an example, the classical Black–Scholes–Merton option pricing model is a partial equilibrium model. It assumes that the trading of the stock option has no effect on the spot price of the underlying asset, and takes the latter as exogenously given. All market models discussed in the rest of this space belong to the realm of partial equilibrium models. In particular, all asset prices will be assumed to follow some exogenously given stochastic process. ○

In the rest of this chapter we preserve the notation introduced in the previous section, but are no longer insisting on a finite time horizon, and suppose that all first-order conditions are somehow already solved for in the outset. In addition, unless noted otherwise, we are not going to insist that $\mathscr{F}_0$ is the trivial σ-field $\{\emptyset, \Omega\}$.

(6.36) SELF-FINANCING: Let $(\bar{\theta}_t)_{t \in \mathbb{N}}$ be some individual portfolio choice, treated as an $\mathscr{F}$-adapted stochastic process with values in $\mathbb{R}^{|\mathscr{J}|}$. We say that $(\bar{\theta}_t)_{t \in \mathbb{N}}$ is a *self-financing* trading strategy if $\bar{\theta}_{t-1} \cdot (\bar{S}_t + \bar{D}_t) = \bar{\theta}_t \cdot \bar{S}_t$ for all $t \in \mathbb{N}$. Thus, "self financing" means that the agent exits every period with the same amount of aggregate wealth with which they enter the period. Later in this space we will reintroduce the notion "self-financing" in a way that also takes into account consumption. For now, "self-financing" is just another way of saying that the only quantities that the agents are allowed to change during any particular period are the compositions of the assets they are holding. ○

(6.37) ARBITRAGE: A self-financing trading strategy $(\bar{\theta}_t)_{t \in \mathbb{N}}$ is said to represent arbitrage (or the trading strategy itself is called "arbitrage") if

$$P\Big(\bar{\theta}_t \cdot \bar{S}_t \leq 0 \;\&\; \bar{\theta}_{T-1} \cdot (\bar{S}_T + \bar{D}_T) \geq 0\Big) = 1 \text{ and } P\big(\bar{\theta}_{T-1} \cdot (\bar{S}_T + \bar{D}_T) > 0\big) > 0$$

for some $t, T \in \mathbb{N}$, $t < T$. In other words, *an arbitrage* is a reference to a self-financing trading strategy that leaves a particular time period t either with zero wealth, or with a borrowed amount (that is, negative wealth), and enters some future period $T > t$ with net wealth (assets less liabilities) that is positive with probability 1 and is strictly positive with strictly positive probability. This is the proverbial "free lunch:" the agent has a nontrivial chance to get something for nothing – and without facing any risk for a loss. ○

Now we come to a particularly important result.

(6.38) FIRST FUNDAMENTAL THEOREM OF ASSET PRICING (FTAP): The market is free of arbitrage if and only if there is a state-price density process $(\psi_t)_{t \in \mathbb{N}}$ that is strictly positive (P-a.s.), that is, $P(\psi_t > 0) = 1$ for all $t \in \mathbb{N}$. ○

The qualifier "first" suggests that there is also a "second" fundamental theorem. It will be stated (and proved) in the next section, after introducing the concept of market completeness – see (6.40). Traditionally, a reference to the FTAP without the qualifier "first" or "second" is understood to be a reference to the first fundamental theorem – this convention is justified by the fact that the first FTAP is referenced throughout the literature (and in this space) much more often. The "if" part in the FTAP – that is, in the first FTAP – is completely straightforward and is offered as an exercise next.

(6.39) EXERCISE: Prove that if a strictly positive SPD exists, then no self-financing trading strategy can represent arbitrage.

HINT: Consider proving the statement by contradiction. Using the self-financing condition, show that

$$\psi_t\,(\bar{\theta}_t\cdot\bar{S}_t) = \mathsf{E}\big[\psi_T\,\big(\bar{\theta}_{T-1}\cdot(\bar{S}_T+\bar{D}_T)\big)\;\big|\;\mathscr{F}_t\big] = \mathsf{E}\big[\psi_T\,\big(\bar{\theta}_T\cdot\bar{S}_T\big)\;\big|\;\mathscr{F}_t\big]\,. \quad \circ$$

The proof of the "only if" part in the first FTAP, that is, the claim that nonexistence of arbitrage implies the existence of at least one P-a.s. positive SPD, is somewhat more involved and is the subject of the next section.

(6.40) MARKET COMPLETENESS: In an arbitrage-free market a strictly positive stochastic discount factor always exists, for one can always take $\xi_{t+1} = \psi_{t+1}/\psi_t$, where (ψ_t) is any positive SPD, whose existence the first FTAP guarantees. An arbitrage-free market is complete if, up to identity P-a.s., there is only one strictly positive SDF. This property is essentially equivalent to – and this equivalence is known as the "second fundamental theorem of asset pricing" (see the next section) – the claim that for any fixed $t\in\mathbb{N}$ and for any $\mathscr{F}_t$-measurable r.v. $\Phi\colon\Omega\mapsto\mathbb{R}$, there is a self-financing trading strategy $(\bar{\theta}_t)$ that *replicates* Φ in that

$$\bar{\theta}_{t-1}\cdot(\bar{S}_t+\bar{D}_t) = \Phi \quad P\text{-a.s.}$$

In other words, any conceivable payoff $\Phi \prec \mathscr{F}_t$ whatsoever can be recreated with a particular initial endowment and with a particular self-financing trading strategy that involves only securities that are traded on the market. It would be very instructive to illustrate the meaning of this result in the context of (6.32). One can think of (6.32†) as a linear system of equations for the unknowns $\xi_{t+1}(\hat{a})\frac{P(\hat{a})}{P(a)}$ for all $\hat{a}\in\mathscr{P}_{t+1}$, $\hat{a}\subseteq a$ – there is one equation for every security (S_t), which is to say, there are $|\mathscr{J}|$ equations in total. The matrix of this linear system has one row for every security and has one column, namely $\bar{S}_{t+1}(\hat{a})+\bar{D}_{t+1}(\hat{a})\in\mathbb{R}^{|\mathscr{J}|}$, for every contingent state $\hat{a}\in\mathscr{P}_{t+1}$, $\hat{a}\subseteq a$. This matrix plays an important role in asset pricing and has a special name: it is called the period $t+1$ payoff matrix. Interpreted as a list of vector-columns, it can be expressed as

$$\dagger \qquad (\bar{S}_{t+1}+\bar{D}_{t+1}) \stackrel{\text{def}}{=} \big\{\bar{S}_{t+1}(\hat{a})+\bar{D}_{t+1}(\hat{a})\in\mathbb{R}^{|\mathscr{J}|} : \hat{a}\in\mathscr{P}_{t+1}\,,\ \hat{a}\subseteq a\big\}\,.$$

Every column in $(\bar{S}_{t+1} + \bar{D}_{t+1})$ gives the period $(t+1)$ payoffs from all securities ($|\mathscr{J}|$ in number) in a particular contingent state $\hat{a}$, and we note that the right side in † depends on $a \in \mathscr{P}_t$, and so does the left side – the entire payoff matrix is a quantity associated with a particular state (node) $a \in \mathscr{P}_t$. Without further notice we are going to assume that: (a) no security is traded at the price of 0, that is, in order to be traded, a security must pay a strictly positive dividend amount in at least one future state of the world, and (b) there are no redundant securities, in that there is no row in $(\bar{S}_{t+1} + \bar{D}_{t+1})$ that can be expressed as a linear combination of other rows. In conjunction with the first assumption, according to which no row can be identified as the zero vector, the second assumption implies that the row rank of the payoff matrix must equal $|\mathscr{J}|$. In particular, the column rank must equal $|\mathscr{J}|$ and this is only possible if there are at least as many contingent states $\hat{a} \in \mathscr{P}_{t+1}$, $\hat{a} \subseteq a$, as there are securities $|\mathscr{J}|$. As a result, the payoff matrix in every state is of full rank and has at least as many columns as rows. In this setting, market completeness comes down to the claim that the (full-rank) payoff matrix is square, that is, the number of assets exactly matches the number of contingent states – this feature will be clarified in the next section. ○

(6.41) RETURNS: In asset pricing one is concerned almost exclusively with returns (in some sense, the actual asset prices are merely an accounting convention). "Returns" usually means "gross returns," and the *net returns* are understood to be gross returns minus 1. The period $t+1$ return from security (S_t) is the $\mathscr{F}_{t+1}$-measurable random variable

$$R_{t+1} = \frac{S_{t+1} + D_{t+1}}{S_t}\,.$$

This is nothing but the value at time $t+1$ of an investment in the security of ¤1 made at time t. Written in terms of the returns, the fundamental pricing equation from (6.30†) becomes

$$\dagger \qquad 1 = \mathsf{E}\big[\xi_{t+1} R_{t+1} \mid \mathscr{F}_t\big]\,.$$

This last form is somewhat more insightful than (6.30†). Notice that, in the context of GEI, even if the SDF differs across the agents, the above relation will still hold for any agent-specific SDF ξ_{t+1} and for any return R_{t+1} from any traded security. For a risk-free security, assuming that one exists, the return R_{t+1} is $\mathscr{F}_t$-measurable (another way of saying risk-free). It is customary to express the risk-free return as $R_{t+1} = 1 + r_t$, where (r_t) is an $\mathscr{F}$-adapted process known as "the interest rate process." Plainly, r_t is the net return on a risk-free investment from period t to period $t+1$, and we stress that this return is generally stochastic, except that it depends on information available in period t, not period $t+1$ at which the actual return is realized. The following connection is now straightforward:

$$1 + r_t = \frac{1}{\mathsf{E}\big[\xi_{t+1} \mid \mathscr{F}_t\big]}\,.$$

Here we again see that, generally, the right side is agent specific, while the left side is not. This is yet another way of saying that in equilibrium all agents agree on the price of the risk-free security. In particular, as long as the market includes a risk-free security, all SDF must have the same expected value. Even if the market does not include a risk-free security, we may still think of $\mathsf{E}[\xi_{t+1}|\mathscr{F}_t]^{-1} - 1$ as some sort of a hypothetical risk-free rate, but unless the market happens to be complete, one can no longer claim that this quantity would be the same across all agents, for, if a security with a particular payoff is not traded, there is nothing to enforce agreement on its price. ○

6.5 The Two Fundamental Theorems of Asset Pricing

The phrase "fundamental theorem of asset pricing" was most likely introduced by Dybvig and Ross (1987), but this result has been in the center of attention practically since the dawn of contemporary finance, as is evident from (Merton 1973), (Merton 1977), and (Ross 1978). The theorem itself has evolved through several stages: from (Harrison and Kreps 1979), followed by (Dybvig and Ross 1987) and (Dalang et al. 1990), until it was more or less finalized by (Delbaen and Schachermayer 1994). Newer variants and simplified proofs are trickling in, still – see (Kabanov and Stricker 2001) and (Lyasoff 2014), for example – and, generally, the matter is still not completely settled, mainly because a result that is as simple to formulate as the FTAP should also have a simple and intuitive justification, which so far has not been found in full generality. From a practitioner's point of view, however, the FTAP appears to be not all that important, and for a number of reasons. Perhaps the most important reason is that the FTAP explicitly posits that any (long or short) market position is available for trade and, furthermore, trading – and, in particular, arbitrage trading – has no effect on the prices at which the assets are traded. Thus, the argument of the FTAP ignores the fact that, in order to take advantage of an arbitrage opportunity, an agent must find a counterparty willing to take the opposite side of the arbitrage trade; to put it another way, the agent must find a set of traders willing to pay for the "free lunch." Nevertheless, understanding the nature of the FTAP remains crucial for understanding the principles of asset pricing. Our only objective in this section is to give a reasonably complete proof of the first and the second fundamental theorems of asset pricing in the case where the proofs are least technical and most intuitive, namely in the case where the filtration $\mathscr{F}$ is finitely generated, that is, $\mathscr{F}_t = \sigma(\mathscr{P}_t)$ for some refining chain of partitions $(\mathscr{P}_t)_{t\in\mathbb{N}}$, so that $\mathscr{P}_t \subseteq \mathscr{F}_{t+1}$ and $\mathscr{P}_0 = \{\Omega\}$. The proof given below follows the argument of (Dybvig and Ross 1987), but uses a somewhat different set of prerequisites from convex analysis. For the general discrete-time case, we refer to (Dalang et al. 1990), while the discussion of the continuous-time case we postpone for chapter 13.

For simplicity, we now specialize our setup to a two-period economy ($t \in \mathbb{N}_{|T} = \mathbb{N}_{|1} = \{0, 1\}$) – the proof of the first FTAP for a multiperiod economy follows by essentially the same argument, as will be detailed below. There is only one state in period $t = 0$, interpreted as the entire sample space Ω, while the period $t = 1$ states (the atoms in the partition $\mathscr{P}_1$) are simply the elements $\omega \in \Omega$, assumed to be finitely many with $2 \leq |\Omega| < \infty$. In particular, we can write ω instead of $\hat{a}$. There is only one payoff matrix of dimensions $|\mathscr{J}| \times |\Omega|$, which we will express as $\bar{S}_1 + \bar{D}_1$. The columns in this matrix, associated with the various states $\omega \in \Omega$, we denote by $\bar{S}_1(\omega) + \bar{D}_1(\omega)$, and the period $t = 0$ price-vector (column) we denote by $\bar{S}_0$. As a first step in the proof of the first FTAP, define the $|\mathscr{J}|$-by-$(1 + |\Omega|)$-matrix

$$M = \left\{-\bar{S}_0, \bar{S}_1 + \bar{D}_1\right\},$$

which is simply the payoff matrix prepended by the vector column $(-\bar{S}_0)$, and then define the linear subspace of $\mathbb{R}^{1+|\Omega|}$

$$L \stackrel{\text{def}}{=} \left\{x^\intercal M : x \in \mathbb{R}^{|\mathscr{J}|}\right\},$$

which is nothing but the span of all possible payoffs in the cross-section of all period $t = 1$ contingent states of the economy, prepended by the payoff in (the only state of) period $t = 0$, that can be attained with the various choices (during period $t = 0$) of the portfolio $x \in \mathbb{R}^{|\mathscr{J}|}$.

(6.42) EXERCISE: Verify that the absence of arbitrage in the two period economy can be stated as the condition

$$L \cap \mathbb{R}_+^{1+|\Omega|} = \{\vec{0}\}.$$

HINT: This condition reads: the only payoff that is positive (that is, nonnegative) in all states of the economy (in all periods) is the payoff of 0 in all states of the economy. ○

Recall that a set $A \subseteq \mathbb{R}^{1+|\Omega|}$ is said to be convex if

$$\alpha x + (1 - \alpha) y \in S \quad \text{for all } x, y \in A \text{ and all } \alpha \in [0, 1],$$

and is said to be a cone if

$$\alpha x \in A \quad \text{for all } x \in A \text{ and all } \alpha \in \mathbb{R}_{++}.$$

A convex cone is a set that is both convex and a cone, and a pointed cone is a cone that contains the origin $\vec{0}$, but does not contain an entire line passing through the origin, that is, $x \in A$ and $x \neq \vec{0} \in A$ implies $(-x) \notin A$. Given any two subsets $A, B \subseteq \mathbb{R}^{1+|\Omega|}$, define the set $A - B$ as

$$A - B = \left\{x - y \in \mathbb{R}^{1+|\Omega|} : x \in A,\ y \in B\right\}.$$

(6.43) EXERCISE: Prove that if the set A is a convex cone, then A is closed under addition of vectors: $x + y \in A$ for every choice of $x, y \in A$. ○

(6.44) EXERCISE: Prove that: (a) if the sets A and B are both closed, then the set $A - B$ must be closed; (b) if the sets A and B are both cones, then the set $A - B$ must be a cone; (c) if A and B are both convex cones, then $A - B$ must be a convex cone, too. ○

(6.45) EXERCISE: Prove that if the set A is a cone and the set B is a pointed convex cone, then the condition $A \cap B = \{\vec{0}\}$ is equivalent to $(A - B) \cap B = \{\vec{0}\}$.

HINT: Given any $x \in A$ and any $y, z \in B$, we have $x - y = z$ if and only if $x = y + z$, and $x = z$ if and only if $x - \vec{0} = z$, where $\vec{0} \in B$. ○

By (6.42) and (6.45), the absence of arbitrage is equivalent to the condition

$$\left(L - \mathbb{R}_+^{1+|\Omega|}\right) \cap \mathbb{R}_+^{1+|\Omega|} = \{\vec{0}\}\,.$$

Let e_i, $i \in \mathbb{N}_{|\Omega|}$, denote the standard orthonormal basis of vectors in $\mathbb{R}^{1+|\Omega|}$. Since $e_i \in \mathbb{R}_+^{1+|\Omega|}$ and since $e_i \neq \vec{0}$, the last condition implies $e_i \notin L - \mathbb{R}_+^{1+|\Omega|}$ for any i. As L and $\mathbb{R}_+^{1+|\Omega|}$ are both closed convex cones, so is also the set $L - \mathbb{R}_+^{1+|\Omega|}$. As a result, the (non-intersecting) closed convex sets $L - \mathbb{R}_+^{1+|\Omega|}$ and $\{e_i\}$ can be separated strictly, that is, for every i, there is a vector $h_i \in \mathbb{R}^{1+|\Omega|}$ such that

$$h_i \cdot (x - y) < h_i \cdot e_i \quad \text{for all } \ x \in L \ \text{ and } \ y \in \mathbb{R}_+^{1+|\Omega|}\,.$$

Setting above $x = \vec{0} \in L$ and $y = e_i \in \mathbb{R}_+^{1+|\Omega|}$ we get $-h_i \cdot e_i < h_i \cdot e_i$, which is the same as $h_i \cdot e_i > 0$. Furthermore, since L and $\mathbb{R}_+^{1+|\Omega|}$ are cones, we can replace above x with αx and y with αy for some fixed $\alpha \in \mathbb{R}_{++}$. As a result, the last relation is actually

$$\alpha h_i \cdot (x - y) < h_i \cdot e_i \quad \text{for all } \ \alpha \in \mathbb{R}_{++}\,, \quad x \in L\,, \ \text{ and } \ y \in \mathbb{R}_+^{1+|\Omega|}\,,$$

and, since $h_i \cdot e_i > 0$, this is possible only if $h_i \cdot (x - y) \leq 0$ for all $x \in L$ and $y \in \mathbb{R}_+^{1+|\Omega|}$. In particular, with $y = \vec{0} \in \mathbb{R}_+^{1+|\Omega|}$ we get $h_i \cdot x \leq 0$ for all $x \in L$, and since L is a liner space ($-x \in L$ whenever $x \in L$) this is only possible if $h_i \cdot x = 0$ for all $x \in L$, that is, $h_i \perp L$. With $x = \vec{0} \in L$ and with $y = e_k \in \mathbb{R}_+^{1+|\Omega|}$, $h_i \cdot (x - y) \leq 0$ becomes $h_i \cdot e_k \geq 0$ for all $k \in \mathbb{N}_{|\Omega|}$. It is now easy to see that the vector

$$\upsilon = h_0 + h_1 + \cdots + h_{|\Omega|} \in \mathbb{R}^{1+|\Omega|}$$

has the property $\upsilon \cdot e_i > 0$ for any i (all coordinates of υ are strictly positive), in addition to the property $\upsilon \perp L$. Let υ_0 denote first coordinate of υ and let υ_+ denote the list of all remaining coordinates of υ, so that $\upsilon = (\upsilon_0, \upsilon_+)$. Since the entries in the list υ_+ can be associated with states $\omega \in \Omega$ (the columns in the payoff matrix), we can think of this list as a function of the form $\upsilon_+ \colon \Omega \mapsto \mathbb{R}_{++}$. For every portfolio choice $\bar{\theta} \in \mathbb{R}^{|\mathcal{J}|}$ we have by definition $\bar{\theta}^\intercal M \in L$, so that (below we treat υ as a vector column)

$$0 = (\bar{\theta}^\intercal M)\upsilon = -\bar{\theta} . \bar{S}_0\, \upsilon_0 + \sum\nolimits_{\omega \in \Omega} \bar{\theta} . \big(S(\omega) + D(\omega)\big)\, \upsilon_+(\omega)\,.$$

Setting $\psi_0 = v_0$ and $\psi_1(\omega) = v_+(\omega)/P(\omega)$, the last equation can be rearranged as

$$\bar{\theta} \cdot \bar{S}_0 \, \psi_0 = \mathsf{E}\big[\bar{\theta} \cdot (S_1 + D_1)\, \psi_1\big] ,$$

and, as this relation holds for any $\bar{\theta} \in \mathbb{R}^{|\mathscr{J}|}$, by selecting $\bar{\theta}$ to be any of the standard orthonormal basis vectors in $\mathbb{R}^{|\mathscr{J}|}$, we get

$$S_0^j \, \psi_0 = \mathsf{E}\big[\big(S_1^j + D_1^j\big)\, \psi_1\big] \quad \text{for any } j \in \mathscr{J} . \qquad \mathrm{e}_1$$

Thus, $\{\psi_0, \psi_1\}$ can be claimed to be a strictly positive SPD.

It is not difficult to guess how the same argument can be carried out in the multiperiod case, as long as the information arrival dynamics $(\mathscr{F}_t)$ are generated by a refining sequence of (finite) partitions. In fact, the only adjustment to be made is to interpret the space L as the span of all period T payoffs in the cross-section of all period T states of the economy, prepended by the period $t = 0$ payoff, that can be attained with all self-financing trading strategies $(\bar{\theta}_t)_{t \in \mathbb{N}_{|T}}$. It is not difficult to see that L is again a linear space, and this is the only feature that we used at the previous step. The reason why L is a linear space is that for any two self-financing strategies $(\bar{\theta}_t')_{t \in \mathbb{N}_{|T}}$ and $(\bar{\theta}_t'')_{t \in \mathbb{N}_{|T}}$ and for any $a, b \in \mathbb{R}$, one can claim that $(a\bar{\theta}_t' + b\bar{\theta}_t'')_{t \in \mathbb{N}_{|T}}$ is a self-financing strategy, as is immediate from the definition. Given any self-financing strategy $(\bar{\theta}_t)_{t \in \mathbb{N}_{|T}}$ we have $\big\{-\bar{\theta}_0 \cdot \bar{S}_0, \bar{\theta}_{T-1}^{\intercal}(\bar{S}_T + \bar{D}_T)\big\} \in L$, and we can again write

$$0 = \big\{-\bar{\theta}_0 \cdot \bar{S}_0, \bar{\theta}_{T-1}^{\intercal}(\bar{S}_T + \bar{D}_T)\big\}\, v .$$

Note that any vector $\bar{\theta}_0 \in \mathbb{R}^{|\mathscr{J}|}$ can be identified as the initial portfolio in some self-financing strategy. This property follows from the fact that the payoff matrix in every state is of full rank and has at least as many columns as rows, which ⊶ (6.40) we automatically assume. We have thus completely proved the following result.

(6.46) THE "ONLY IF" PART IN THE FIRST FTAP FOR A FINITELY GENERATED INFORMATION STRUCTURE: Suppose that the information structure $\mathscr{F}$ is generated by a refining sequence of (finite) partitions of Ω. Then, under the standard assumptions in the model, the absence of arbitrage from period t to period $T > t$, in any period t state of the economy, implies that there is an $\mathscr{F}_t$-measurable random variable $\psi_t \colon \Omega \mapsto \mathbb{R}_{++}$ and an $\mathscr{F}_T$-measurable random variable $\psi_T \colon \Omega \mapsto \mathbb{R}_{++}$ such that

$$\bar{\theta}_t \cdot \bar{S}_t \, \psi_t = \mathsf{E}\big[\bar{\theta}_{T-1} \cdot \big(\bar{S}_T + \bar{D}_T\big)\psi_T \,|\, \mathscr{F}_t\big]$$

for every self-financing trading strategy $(\bar{\theta}_t)_{t \in \mathbb{N}_{|T}}$. In particular, if the trading strategy holds a single stock $j \in \mathscr{J}$, reinvests immediately all dividends (into the same stock), and closes the position in period T, then $S_t^j \psi_t = \mathsf{E}[A_T^j \psi_T \,|\, \mathscr{F}_t]$, where A_T^j is the random amount collected at closing. ○

Now we turn to the notion of market completeness and the second FTAP. Back to the two-period setting introduced at the beginning of this section, suppose that the market is free of arbitrage in any state and in any period. This means that a strictly positive SPD $\{\psi_0, \psi_1\}$ exists, and e_1 can be re-stated as

$$\bar{S}_0 = \sum\nolimits_{\omega\in\Omega}\big(\bar{S}_1(\omega) + \bar{D}_1(\omega)\big)\,\frac{\psi_1(\omega)}{\psi_0}P(\omega) = \big(\bar{S}_1 + \bar{D}_1\big)\Big(\frac{1}{\upsilon_0}\upsilon_+\Big),$$

where the last expression is understood as the (matrix) product between the payoff matrix and the vector column $\upsilon_+/\upsilon_0 \in \mathbb{R}^{|\Omega|}_{++}$. Suppose that all spot prices and all dividend payouts are somehow given in any period and any state of the economy. The payoff matrix $(\bar{S}_1 + \bar{D}_1)$ is of full rank by assumption and has at least as many columns as rows. If it happens to be a square matrix ($|\mathcal{J}| = |\Omega|$), then the vector υ_+/υ_0 must be unique, and this is the same as the claim that the stochastic discount factor $\xi_1(\omega) = \psi_1(\omega)/\psi_0$ is unique. To connect the uniqueness of the SDF with the notion of replication, consider a generic stochastic payoff $\Pi : \Omega \mapsto \mathbb{R}$ in period $t = 1$, that is, Π is some (arbitrary) $\mathbb{R}$-valued r.v. that is measurable for $\mathscr{F}_1$. Any such payoff Π can be treated as a vector from the space $\mathbb{R}^{|\Omega|}$ in the obvious way, and the replication of Π comes down to the choice of a vector $\bar{\theta} \in \mathbb{R}^{|\mathcal{J}|}$, a portfolio purchased in period $t = 0$, such that

$$\bar{\theta}^{\intercal}\big(\bar{S}_1 + \bar{D}_1\big) = \Pi^{\intercal}. \qquad e_2$$

Clearly, this relation is only possible if the vector row $\Pi^{\intercal}$ belongs to the linear span of the rows of the matrix $(\bar{S}_1 + \bar{D}_1)$. As a result, to be able to claim that any payoff whatsoever can be replicated (the definition of completeness), the rows of $(\bar{S}_1 + \bar{D}_1)$ must span $\mathbb{R}^{|\Omega|}$, and since$^{\multimap (6.40)}$ $|\mathcal{J}| \leq |\Omega|$, this is only possible if $|\mathcal{J}| = |\Omega|$. Thus, market completeness implies that the payoff matrix is square (it is of full rank by assumption). Conversely, if $|\mathcal{J}| = |\Omega|$ (and $(\bar{S}_1 + \bar{D}_1)$ is again of full rank by assumption), then the system e_2 has a unique solution $\bar{\theta} \in \mathbb{R}^{|\mathcal{J}|}$ for every choice of the payoff vector Π. Thus, under the assumptions in our model, the properties "all period 1 payoffs are replicable" and "the period 1 payoff matrix is square" are the same and come down to what we call "market completeness."

It is easy to see from the argument above that in the absence of arbitrage market completeness implies that the stochastic discount factor is unique. To prove this implication in the opposite direction, suppose that the market is free of arbitrage and the strictly positive stochastic discount factor is unique but the market is not complete, that is, $|\mathcal{J}| < |\Omega|$. Since the row rank of the payoff matrix equals $|\mathcal{J}|$, the column rank also must equal $|\mathcal{J}|$. As the columns are of dimension $|\mathcal{J}|$ and are at least $|\mathcal{J}| + 1$ in number, at least one column must be a nontrivial linear combination of the remaining columns. Thus, there is a vector $k \in \mathbb{R}^{|\Omega|}$, $k \neq \vec{0}$, such that $(\bar{S}_1 + \bar{D}_1)\,k = \vec{0}$. If a vector k has this property, then so does also the vector εk for any $\varepsilon \in \mathbb{R}_{++}$, and if $\upsilon_{++} \in (\mathbb{R}_{++})^{|\Omega|}$ is any vector with strictly positive entries, one can always choose ε so that $\upsilon_{++} + \varepsilon k$ also has strictly positive entries. If v_{++} is such that $\bar{S}_0 = (\bar{S}_1 + \bar{D}_1)v_{++}$, then

$$\bar{S}_0 = (\bar{S}_1 + \bar{D}_1)v_{++} = (\bar{S}_1 + \bar{D}_1)(v_{++} + \varepsilon k),$$

for all sufficiently small $\varepsilon \in \mathbb{R}_{++}$. As a result, a strictly positive SDF cannot be unique, which is a contradiction. We have proved completely the following result.

(6.47) THE SECOND FTAP FOR A FINITELY GENERATED INFORMATION STRUCTURE: Suppose that the information structure $\mathscr{F}$ is generated by a refining sequence of (finite) partitions of Ω, and that the financial market is free of arbitrage. Then, under the standard assumptions in the model, the market is complete if and only if there is precisely one strictly positive stochastic discount factor (the existence of which the First FTAP guarantees), or, equivalently, the strictly positive SPD is unique up to a multiplicative factor (assuming that $\mathscr{F}_0 = \{\emptyset, \Omega\}$ – otherwise the SPD is unique up to a multiplication by an $\mathscr{F}_0$-measurable random variable). ○

6.6 From Stochastic Discount Factors to Equivalent Measures and Local Martingales

Adopt the setting and the notation introduced in section 6.3, and suppose that the market is free of arbitrage and admits a risk-free security in all periods and all states of the economy. The first FTAP guarantees the existence of an SDF process, that is, an $\mathscr{F}$-adapted and strictly positive stochastic process $\xi \equiv (\xi_t)_{t\in\mathbb{N}}$ that gives the stochastic discount factors for all states and all time periods. Since a risk-free security is assumed to be traded (at a finite price), we must have $\mathsf{E}[\xi_{t+1} \mid \mathscr{F}_t] < \infty$. Consider the process $(X_t)_{t\in\mathbb{N}}$ defined recursively by the recipe: $X_0 = 1$ and

$$X_t = X_{t-1}\frac{\xi_t}{\mathsf{E}[\xi_t \mid \mathscr{F}_{t-1}]} \equiv X_{t-1}(1 + r_{t-1})\xi_t \quad \text{for } t \in \mathbb{N}_{++}.$$

(6.48) EXERCISE: Prove that the process $X \equiv (X_t)_{t\in\mathbb{N}}$ defined above has the following properties: (a) X is adapted to the filtration $\mathscr{F}$; (b) $\mathsf{E}[|X_t|] < \infty$ for any $t \in \mathbb{N}$; (c) $X_t = \mathsf{E}[X_{t+1} \mid \mathscr{F}_t]$ for any $t \in \mathbb{N}$; (d) $X_t > 0$ P-a.s. for any $t \in \mathbb{N}$; (e) $\mathsf{E}[X_t] = 1$ for any $t \in \mathbb{N}$. ○

(6.49) LOCAL MARTINGALES AND MARTINGALES: A stochastic process that satisfies conditions (a), (b), and (c) in (6.48) is called a *martingale*, but we postpone the official definition for the next chapter. At this point we note that among those conditions only (a) and (c) are truly interesting and important. The integrability condition (b) is merely a guarantee that the conditional expectation is defined, but conditional expectations and expectations may still be meaningful⊸ (3.39) even without the integrability requirement. A stochastic process that satisfies only conditions (a) and (c) in (6.48) – that is, an adapted process for which the conditional expectation in (c) is somehow meaningful (not necessarily because of (b)) and the identity in (c) holds – is called *local martingale*, but we again postpone the precise definition for the next chapter, where it will become clear in what sense such objects have the martingale property only "locally." ○

Thanks to (d) and (e) from (6.48), given any fixed $t \in \mathbb{N}$, we can define the probability measure Q_t on the σ-field $\mathscr{F}_t$ so that

$$\frac{\mathrm{d}Q_t}{\mathrm{d}P} = X_t \quad \Leftrightarrow \quad Q_t = X_t \odot P\,.$$

In particular, (d) implies that $Q_t \approx P$ on $\mathscr{F}_t$. This feature will play an important role in what follows.

(6.50) EXERCISE: Prove that the probability measures $(Q_t)_{t \in \mathbb{N}}$ are compatible in that $Q_{t+1} \restriction \mathscr{F}_t = Q_t$ for every $t \in \mathbb{N}$, that is, if $A \in \mathscr{F}_t$, then $Q_t(A) = Q_{t+1}(A)$. ○

(6.51) REMARK: One may be tempted to declare that the probability measures $(Q_t)_{t \in \mathbb{N}}$ give rise to a probability measure on the σ-field $\mathscr{F}_\infty \overset{\text{def}}{=} \bigvee_{t \in \mathbb{N}} \mathscr{F}_t$, but in order to be able to justify this claim we need to develop some understanding of the intrinsic nature of martingales first. Everything that we need is already encrypted in the connection between the measures Q_t and the stochastic discount factors ξ_t, namely,

† $$Q_t = X_t \odot P \equiv \big((1 + r_0)\xi_1(1 + r_1)\xi_2 \,\cdots\, (1 + r_{t-1})\xi_t\big) \odot P\,. \quad ○$$

(6.52) EXERCISE: Explain where and how the no-arbitrage condition is used in the construction of the measures Q_t. Is the existence of a risk-free security used? In particular, explain the connection between the no-arbitrage property and the absolute continuity of the measure Q_t with respect to P. ○

(6.53) EXCESS RETURNS: Given a risky security (S_t), consider initiating during period t the following chain of transactions that does not involve any initial capital: (a) borrow ¤1 at the risk-free rate and invest ¤1 in security S; (b) during period $(t + 1)$ close the position and receive the amount $R_{t+1} - (1 + r_t)$. This amount is called period-$(t + 1)$ *excess return* (from security S). Generally, the excess return is the difference between the one period return on an investment in a particular risky security and the one period return on an investment in the risk-free security. We aggregate the excess returns into the process $(\mathcal{E}_t)_{t \in \mathbb{N}}$ defined by

$$\mathcal{E}_0 = 0 \quad \text{and} \quad \mathcal{E}_t = \sum\nolimits_{s \in \mathbb{N}_{|t} \setminus \{0\}} \big(R_s - (1 + r_{s-1})\big) \quad \text{for } t \in \mathbb{N}_{++}\,.$$

As usual, we make the convention to write $(\mathcal{E}_t^j)_{t \in \mathbb{N}}$, $j \in \mathscr{J}$, if the identification of the security is relevant. ○

(6.54) EXERCISE: Prove that for any finite time horizon $T \in \mathbb{N}_{++}$ the measure Q_T makes the process $(\mathcal{E}_t)_{t \in \mathbb{N}_{|T}}$ into a local martingale, that is,

$$\mathcal{E}_t = \mathsf{E}^{Q_T}\big[\mathcal{E}_{t+1} | \mathscr{F}_t\big] \quad \Leftrightarrow \quad \mathsf{E}^{Q_T}\big[R_{t+1} - (1 + r_t) | \mathscr{F}_t\big] = 0 \ \text{ for all } t \in \mathbb{N}_{|T-1}\,. \quad ○$$

Now we want to reverse our line of reasoning and understand whether the existence of an equivalent measure Q_T that makes the aggregate excess returns $(\mathcal{E}_t)_{t \in \mathbb{N}_{|T}}$ into a local martingale implies the no-arbitrage condition. The next exercise contains an interesting converse of the statement in (6.50).

(6.55) EXERCISE: Suppose that for any $t \in \mathbb{N}$ there is a probability measure Q_t, which is defined on $\mathscr{F}_t$ and is such that: (a) Q_t is absolutely continuous relative to P ($Q_t \lesssim P$ on $\mathscr{F}_t$), and (b) the measures $(Q_t)_{t \in \mathbb{N}}$ are compatible, in that $Q_{t+1} \upharpoonright \mathscr{F}_t = Q_t$ for every $t \in \mathbb{N}$. Prove that the process $X_t \overset{\text{def}}{=} (\mathrm{d}Q_t/\mathrm{d}P)\upharpoonright \mathscr{F}_t$, $t \in \mathbb{N}$, is a martingale: $X_t = \mathsf{E}[X_{t+1} \mid \mathscr{F}_t]$ for every $t \in \mathbb{N}$. ○

One important consequence of the last exercise is that⊸ (6.51) if Q happens to be another probability measure on $\mathscr{F}_\infty$ such that $Q \lesssim P$ (on the entire $\mathscr{F}_\infty$), and if X_t stands for the Radon–Nikodym derivative $\mathrm{d}Q/\mathrm{d}P$ on $\mathscr{F}_t$, then the process (X_t) must be a martingale (relative to the measure P). The first FTAP can now be restated as follows:

(6.56) THE FIRST FTAP REVISITED: Suppose that the financial market is finitely lived, with time horizon $T \in \mathbb{N}_{++}$, and admits a risk-free security in every period and in every state of the economy. Then the market is free of arbitrage if and only if there is a probability measure Q_T on $\mathscr{F}_T$, which is equivalent to P ($Q_T \approx P$) and is such that any excess-returns process, associated with any of the traded securities, is a local martingale relative to Q_T. ○

We are already familiar with the way in which the no-arbitrage condition implies the existence of an equivalent measure Q_T that shares the properties stated in (6.56). In lieu with our previous formulation of the FTAP in (6.38), proving that the existence of such a measure precludes the possibility for arbitrage is not difficult – see the next exercise. More important, the proof reveals the meaning of the change of measure operation, which is instrumental in asset pricing. A measure Q_T that has the properties stated in (6.56) is often called a *risk-neutral measure*, or an *equivalent martingale measure*, or, somewhat more accurately, an *equivalent local martingale measure*. If any of these terms is used, it is implicitly assumed that it is the excess returns process that has the local martingale property.

(6.57) EXERCISE: Prove that the existence of a measure Q_T that has the properties listed in (6.56) precludes the possibility for arbitrage.

HINT: By the FTAP in (6.38), it is enough to establish that a strictly positive SPD, or, which is the same, a strictly positive SDF, exists. To see how to construct a positive SDF, observe that the properties of Q_T imply that the conditional expectation under Q_T of any (one-period) excess return is 0. Then transform the conditional expectation with respect to Q_T into a conditional expectation relative to P in terms of the Radon–Nikodym derivative $X_t \overset{\text{def}}{=} (\mathrm{d}Q_T \upharpoonright \mathscr{F}_t)/(\mathrm{d}P \upharpoonright \mathscr{F}_t)$. Conclude that $\xi_{t+1} \overset{\text{def}}{=} \frac{X_{t+1}}{(1+r_t)X_t}$ has all the features of a strictly positive SDF. ○

(6.58) REMARK: The message encoded in the last exercise is that any measure that is equivalent to P, and has the additional property that with respect to it all excess returns processes are local martingales, can be linked to a unique SDF process through its Radon–Nikodym derivatives relative to P on $\mathscr{F}_t$. This remark will be clarified further below. ○

(6.59) ON THE MEANING OF "RISK-NEUTRAL": Let Q_T denote any probability measure on $\mathscr{F}_T$ that has the properties listed in (6.56). Then we have $Q_T \approx P$ and, for any risky security with return $R_{t+1} = (S_{t+1} + D_{t+1})/S_t$,

† $$1 + r_t = \mathsf{E}^{Q_T}[R_{t+1} \mid \mathscr{F}_t] \text{ for all } t \in \mathbb{N}_{|T-1} .$$

In other words, under Q_T the expected conditional return from any risky security is identical to the return from the risk-free security. In general, such a property cannot hold under the real-world probability measure P. Indeed, rational investors expect to be rewarded for the risk that they take by receiving *on average* a higher return from securities that are risky. To put it another way, the investors may be willing to accept returns that are randomly distributed, but provided that the mean value in the associated distribution law is larger than the (fixed) return from the risk-free asset. In this respect, the risk-neutral measure may appear as something purely theoretical, but it is important to keep in mind where it comes from: it is built$^{\multimap (6.51^{\dagger})}$ out of the stochastic discount factors and the risk-free rate, and can be viewed as a mathematical metaphor for the way in which uncertain payoffs in the future are priced by the market (and also for the way in which investors are rewarded). Indeed, we can rewrite † in the form

‡ $$S_t = \frac{1}{1+r_t}\mathsf{E}^{Q_T}\left[S_{t+1} + D_{t+1} \mid \mathscr{F}_t\right], \quad t \in \mathbb{N}_{|T-1}$$

and observe that the risk-neutral measure gives us something very useful and elegant: the price of every uncertain payoff can be expressed as the discounted (at the risk-free rate) conditional expectation (with respect to the risk-neutral measure) of the uncertain payoff. It is remarkable that absence of arbitrage is equivalent to the existence of a risk-neutral measure that prices simultaneously all risky payoffs in the market. ○

(6.60) CHANGE OF NUMÉRAIRE: When security prices are quoted in yens, then we use 1¥ as a numéraire. If security prices are measured in euros, then we use 1€ as a numéraire. Of course, the yen and the euro are securities (traded more or less in the way other securities are traded), and, in principle, any security can be used as a numéraire. Switching from one numéraire to another is often useful because the change of the numéraire changes the properties of the price process under consideration (to wit: the share price of TSLA expressed in dollars will have different dynamics compared to the share price of TSLA expressed in yens). When we think of the period-t payoff, $S_t + D_t$, we think of a particular currency amount that this security will pay to its holder in period t. Since an investment of ¤1 placed in the risk-free asset at time $t = 0$ becomes ¤B_t by the beginning of period $t > 0$ (assuming that any interest is reinvested), where

$$B_t = (1 + r_0)(1 + r_1) \cdots (1 + r_{t-1}) \prec \mathscr{F}_{t-1}\,, \quad t \in \mathbb{N},$$

it would seem more natural to use the process (B_t) as a numéraire, and thus work with the discounted price process $\tilde{S}_t = (B_t)^{-1} S_t$, $t \in \mathbb{N}$, rather than the process (S_t). Accordingly, the dividend process (D_t) is to be replaced with the process $\tilde{D}_t = (B_t)^{-1} D_t$, $t \in \mathbb{N}$. We can now rewrite (6.59‡) even more elegantly in the form

$$\tilde{S}_t = \mathsf{E}^{Q_T}\left[\tilde{S}_{t+1} + \tilde{D}_{t+1} \,|\, \mathscr{F}_t\right]. \quad \circ$$

(6.61) EXERCISE: Suppose that the financial market is finitely lived with time horizon $T \in \mathbb{N}_{++}$ and consider a security (S_t) that pays dividends only in the final period T. Assuming that the market is free of arbitrage, prove that for any equivalent risk-neutral probability measure $Q_T \approx P$ on $\mathscr{F}_T$, the price process $(\tilde{S}_t)_{t \in \mathbb{N}_{|T-1}}$ (expressed in terms of the risk-free asset as a numéraire as above) is a local martingale with respect to the measure Q_T. Explain under what condition the price process can be claimed to be a martingale.

HINT: Try to answer the more general question: when is a positive local martingale a martingale? ○

(6.62) FINAL REMARK: The main reason behind the change of numéraire step described above is to remove the risk-free rate from the picture entirely. In the new numéraire the risk-free rate is 0, so that "excess returns" and "net returns" come down to the same quantity: returns minus 1. Of course, in order to implement such a change of numéraire in practice one must first come up with a concrete model for the dynamics of the risk-free rate. Thus, the simplification of one aspect of the model leads to an increased complexity in another aspect. ○

7

Crash Course on Discrete-Time Martingales

The intuition behind many important concepts from the theory of stochastic processes is easier to develop in discrete time first. The notions that play a central role in this subject are (sub/super)martingales, the convergence of such objects and the related maximal inequalities, predictable compensation, and stopping times and the events that precede them. In essence, this chapter is a brief synopsis of the key results from the theory of martingales in discrete time, and in most part it follows closely (Shiryaev 2004, VII).

7.1 Basic Concepts and Definitions

Our first step is to introduce the classes of processes around which much of what follows in this space is going to be be built. For now, time is discrete and the range of the time-parameter t is the set, $\mathbb{N}$ (or $\bar{\mathbb{N}}$).

(7.1) DEFINITION: Let $(\Omega, \mathcal{F}, P)$ be a probability space, let $\mathscr{F} = (\mathscr{F}_t \subseteq \mathcal{F})_{t \in \bar{\mathbb{N}}}$ be some$^{\multimap (6.1)}$ filtration inside $\mathcal{F}$ (notation: $\mathscr{F} \setminus\!\subset \mathcal{F}$), and let $X \equiv (X_t)_{t \in \mathbb{N}}$ be any stochastic process on Ω that is adapted to $\mathscr{F}$ (notation: $X \prec \mathscr{F}$). We say that X is a *martingale*, or a *submartingale*, or a *supermartingale*, if $\mathsf{E}[|X_t|] < \infty$ for every $t \in \mathbb{N}$ and, respectively,

$$X_t = \mathsf{E}[X_{t+1} \,|\, \mathscr{F}_t] \ P\text{-a.s. for all } t \geq 0 \text{ (martingale);}$$

† $$X_t \leq \mathsf{E}[X_{t+1} \,|\, \mathscr{F}_t] \ P\text{-a.s. for all } t \geq 0 \text{ (submartingale);}$$

$$X_t \geq \mathsf{E}[X_{t+1} \,|\, \mathscr{F}_t] \ P\text{-a.s. for all } t \geq 0 \text{ (supermartingale).}$$

Thus, the attributes martingale, submartingale, and supermartingale depend on both the filtration $\mathscr{F}$ and the probability measure P. In situations where this dependence is not immediately transparent and must be emphasized we will say that X is an $(\mathscr{F}, P)$-martingale (respectively, $(\mathscr{F}, P)$-submartingale, or $(\mathscr{F}, P)$-supermartingale). If no filtration is specified, the notion (sub/super)martingale will be understood relative to the natural filtration generated by X, which we denote by $\mathscr{F}^X$ and define as

$$\mathscr{F}_t^X \overset{\text{def}}{=} \sigma\{X_0, X_1, \dots, X_t\}, \quad t \in \mathbb{N},$$

(this is the smallest filtration to which X can be adapted). It is straightforward to check that the relations † can be restated in the following form:

‡ $$X_s = (\leq, \geq)\ \mathsf{E}[X_t \,|\, \mathscr{F}_s] \quad (P\text{-a.s.}) \text{ for any } s, t \in \mathbb{N},\ s < t\,.$$

The notions of (sub/super)martingale can be extended in the obvious way for processes indexed by $\bar{\mathbb{N}}$ by requiring that $X_t \prec \mathscr{F}_t$ for any $t \in \bar{\mathbb{N}}$ and that the respective relation in ‡ is satisfied for any $t \in \bar{\mathbb{N}}$ and any $s \in \mathbb{N}$, $s < t$. But unless X_∞ is explicitly defined, we will suppose that X is indexed by $\mathbb{N}$.

A process X, defined on $(\Omega, \mathcal{F}, P)$, is said to be *predictable*, or $\mathscr{F}$*-predictable* (notation: $X \prec \mathscr{F}_-$), if $X_0 \prec \mathscr{F}_0$ and $X_t \prec \mathscr{F}_{t-1}$ for any $t \in \mathbb{N}_{++}$. ○

It is clear from the last definition that X is a martingale if and only if it is simultaneously a submartingale and a supermartingale. In addition, X is a submartingale if and only if $-X$ is a supermartingale; or, which is the same, X is a supermartingale if and only if $-X$ is a submartingale. The next exercise points to a particularly simple and intuitive way to construct a martingale. We will discover later that most "reasonably behaved" martingales arise that way, but, unfortunately, not all martingales that we will have to work with are going to be "reasonably behaved" in that sense.

(7.2) EXERCISE: Given a probability space $(\Omega, \mathcal{F}, P)$, let $\mathscr{F} = (\mathscr{F}_t \subseteq \mathcal{F})_{t \in \bar{\mathbb{N}}}$ be any filtration inside $\mathcal{F}$, and let $\xi \in \mathscr{L}^1(\Omega, \mathcal{F}, P)$ be any integrable random variable. Prove that the process

$$\left(X_t \overset{\text{def}}{=} \mathsf{E}[\xi \,|\, \mathscr{F}_t]\right)_{t \in \bar{\mathbb{N}}}$$

is an $(\mathscr{F}, P)$-martingale. ○

(7.3) EXERCISE: Prove that the canonical random walk on coin toss space is a martingale with respect to the probability measure P_∞ and the filtration on that space generated by the partitions $(\mathscr{P}_t)_{t \in \mathbb{N}}$. ○

(7.4) EXERCISE: Let $(Z_i)_{i \in \mathbb{N}}$ be any sequence of P-a.s. positive i.i.d. random variables on some probability space $(\Omega, \mathcal{F}, P)$, and suppose that $\mathsf{E}[Z_i] = 1$. Prove that the process $\left(X_t = Z_0 Z_1 \cdots Z_t\right)_{t \in \mathbb{N}}$ is a martingale with respect to its natural filtration $\mathscr{F}^X$. ○

The next exercise describes a particularly common construction of a submartingale, as a convex transformation of a martingale or a submartingale.

(7.5) EXERCISE: Let X be any martingale (any submartingale) and let $f : \mathbb{R} \mapsto \mathbb{R}$ be any convex function (any increasing convex function) such that $\mathsf{E}[|f(X_t)|] < \infty$ for every $t \in \mathbb{R}_+$. Prove that in either case the process $(f(X_t))_{t \in \mathbb{N}}$ is a submartingale (for the same filtration and probability measure for which X is a martingale or a submartingale).

HINT: This is a straightforward application of Jensen's inequality for conditional expectations. ○

(7.6) EXAMPLES: To illustrate the connection between submartingales and asset pricing, suppose that the price process (S_t), associated with a particular risky security, happens to be a positive martingale, say, under the risk-neutral measure Q and with the risk-free security used as a numéraire. Then consider an American-style derivative with termination payoff $\Phi_t \stackrel{\text{def}}{=} (S_t - K)^+$, or $\Phi_t \stackrel{\text{def}}{=} (K - S_t)^+$, for some fixed constant $K > 0$. As the functions $x \rightsquigarrow (x - K)^+$ and $x \rightsquigarrow (K - x)^+$ are both convex, the termination payoff process (Φ_t) is a submartingale. However, the price of the derivative, (X_t), must be a supermartingale with respect to Q. To see this, observe that as the risk-free rate is 0, $\mathsf{E}^Q[X_{t+1} \,|\, \mathcal{F}_t]$ gives the continuation value of the option at time t, that is, the value attached to the decision not to exercise in period t and hold the option until period $(t+1)$. Then observe that during period t the holder of the option is faced with the following three choices: (a) sell the option at the spot price X_t, (b) exercise the option and collect the termination payoff Φ_t, or (c) do not take any action and hold the option until the next period $t+1$. Clearly, the spot price X_t at which the contract trades must be at least as large as the value generated by any of the decisions that its holder can choose from, that is, $X_t \geq \Phi_t$ and $X_t \geq \mathsf{E}^Q[X_{t+1} \,|\, \mathcal{F}_t]$ (everywhere in Ω). As a result, the process (X_t) is a supermartingale. If taking no action is optimal, $X_t = \mathsf{E}^Q[X_{t+1} \,|\, \mathcal{F}_t] \geq \Phi_t$, and if termination is optimal, $\Phi_t = X_t \geq \mathsf{E}^Q[X_{t+1} \,|\, \mathcal{F}_t]$. Heuristically, for as long as taking no action is optimal (X_t) must be a martingale, and for as long as termination is optimal (X_t) (a supermartingale) must coincide with (Φ_t) (a submartingale), so that both must be martingales. We will see later on that this amounts to the claim that (X_t) is a martingale and that termination is never optimal before the expiration date. It is indeed intuitive that it is never optimal to terminate a reward process that is a submartingale, but there is a long way in front of us until we can make such heuristic statements precise. Our first step is to develop the tools that would allow us to make phrases as "for as long" and "until" mathematically meaningful. The obvious difficulty is that, generally, such "time domains" are going to be random and must be coordinated with the flow of information. ○

We come now to an important concept.

(7.7) STOPPING TIMES: The notion of *stopping time* captures a particularly useful and important concept in the theory of stochastic processes and asset pricing. Intuitively, a stopping time may be interpreted as a random moment in time that does not depend on the future that follows it, in that events that occur after that (random) time cannot change its (random) value. To put it another way, if the outcome from a particular experiment can be thought of as a random sequence of observations received sequentially over time, then a stopping time is a function of the observed sequence, such that its value depends only on the portion of the sequence from its beginning until the moment on the time scale marked by the stopping time's value. Thus, having observed the portion of the sequence from the beginning of time until some fixed time $t \in \mathbb{N}$, if the value of the stopping time attached to that particular realization of the sequence occurs before time t, then the observer should be able to read that value from the observed portion of the sequence until time t – and without the need to observe the sequence after time t. ○

A classical example of a stopping time, taken from the realm of finance, is the time of exercise for an American-style derivative instrument. It depends on the realization of the underlying price process, but only on the portion of that realization from the beginning of time until the moment of exercise. Thus, if an agent monitors the realization of the stock price sequentially over time, and if the time of exercise (for that particular realization of the stock price) is before time t (fixed), then by time t the agent must know that the time of exercise has passed. One can also think of the moment of exercise as the first moment when the spot price enters a particular domain (and choosing that domain amounts to prescribing a concrete exercise rule). Naturally, such ideas must be made precise.

(7.8) DEFINITION: Let $(\Omega, \mathcal{F}, P)$ be any probability space and let $\mathscr{F} \equiv (\mathscr{F}_t)_{t\in\bar{\mathbb{N}}}$ be any filtration inside $\mathcal{F}$ ($\mathscr{F} \setminus\subset \mathcal{F}$). A stopping time relative to $\mathscr{F}$ (or an $\mathscr{F}$-stopping time, or simply a stopping time, if the filtration $\mathscr{F}$ is understood) is any function $\mathcal{t}\colon \Omega \mapsto \bar{\mathbb{N}}$ such that $\{\mathcal{t} = t\} \in \mathscr{F}_t$ for any $t \in \bar{\mathbb{N}}$. ○

(7.9) REMARK: Any stopping time $\mathcal{t}$ is automatically a random variable (that is, an $\mathcal{F}$-measurable function on Ω). Indeed, for every set $B \subseteq \bar{\mathbb{R}}_+$ the event

$$\{\mathcal{t} \in B\} = \cup_{t\in\bar{\mathbb{N}}\cap B}\{\mathcal{t} = t\}$$

belongs to $\mathcal{F}$ (in fact, belongs $\mathscr{F}_\infty \subseteq \mathcal{F}$), as the union of at most countably many events that belong to $\mathscr{F}_\infty$. ○

(7.10) EXERCISE: Let $\mathcal{t}$ be any $\mathscr{F}$-stopping time. Prove that the events $\{\mathcal{t} < t\}$ and $\{\mathcal{t} \geq t\}$ belong to $\mathscr{F}_{t-1}$ for every $t \in \mathbb{N}_{++}$. In addition, prove that the events $\{\mathcal{t} > t\}$ and $\{\mathcal{t} \leq t\}$ belong to $\mathscr{F}_t$ for every $t \in \mathbb{N}$. ○

(7.11) IMPORTANT EXAMPLE: Let $X = (X_t)_{t\in\mathbb{N}}$ be any $\mathscr{F}$-adapted stochastic process and let $B \in \mathcal{B}(\mathbb{R})$ be any Borel set. Then, setting $\inf(\emptyset) = \infty$ by convention, the random variable

$$\mathcal{t}(\omega) \stackrel{\text{def}}{=} \inf\{t \in \mathbb{N}\colon X_t(\omega) \in B\}$$

is a stopping time. Indeed, for any $t \in \mathbb{N}$,

$$\{\mathcal{t} = t\} = \{X_0 \notin B\} \cap \{X_1 \notin B\} \cap \cdots \cap \{X_{t-1} \notin B\} \cap \{X_t \in B\},$$

while $\{\mathcal{t} = +\infty\} = \cap_{t\in\mathbb{N}}\{X_t \notin B\}$, and these events are clearly in $\mathscr{F}_t$ and in $\mathscr{F}_\infty$, respectively. ○

(7.12) EXERCISE: Let $(\Omega, \mathcal{F}, P)$ be any probability space, let $\mathscr{F} = (\mathscr{F}_t)_{t\in\bar{\mathbb{N}}}$ be any filtration inside $\mathcal{F}$, let $X = (X_t)_{t\in\mathbb{N}}$ be any $\mathscr{F}$-adapted stochastic process, and let $\mathcal{t}$ be any $\mathscr{F}$-stopping time. Prove that the mapping $\omega \rightsquigarrow X_{\mathcal{t}(\omega)}(\omega)$ is $\mathcal{F}$-measurable (in fact, $\mathscr{F}_\infty$-measurable) on the set $\{\mathcal{t} < \infty\}$, that is, $X_{\mathcal{t}}$ is a legitimate random variable on $\{t < \infty\}$, or, equivalently, $1_{\{t<\infty\}}X_t$ is a legitimate random variable on $(\Omega, \mathcal{F}, P)$. ○

It is quite common to think of martingales as metaphors for fair games. It is natural to ask whether the game would still be fair if the players are allowed to bail-out at certain (possibly random) moments of their choosing. In other words, is a martingale stopped at random time still a martingale? The answer is yes, provided that the choice of time to stop is only based on information available up until that time. We now make this claim precise.

(7.13) PROCESS STOPPED AT RANDOM TIME: Let t be any $\mathscr{F}$-stopping time and let $X = (X_t)_{t \in \mathbb{N}}$ be any stochastic process (defined on the same probability space). The "process X stopped at time t" is the process

$$X^t = (X_t^t \overset{\text{def}}{=} X_{t \wedge t})_{t \in \mathbb{N}},$$

where $X_{t\wedge t}$ is understood as the r.v. $\omega \rightsquigarrow X_{t\wedge t(\omega)}(\omega)$. If $X \prec \mathscr{F}$ (X is adapted to $\mathscr{F}$), then $X^t \prec \mathscr{F}$ (X^t is adapted to $\mathscr{F}$). Indeed, $X_0^t = X_0$ by definition, and for any $t \in \mathbb{N}_{++}$,

$$X_t^t = X_{t\wedge t} = \sum\nolimits_{s \in \mathbb{N}_{|t-1}} X_s 1_{\{t=s\}} + X_t 1_{\{t \geq t\}} .$$

As all r.v. on the right side are $\mathscr{F}_t$-measurable, X_t^t must be $\mathscr{F}_t$-measurable, too. ○

(7.14) EXERCISE: Prove that if $X = (X_t)_{t \in \mathbb{N}}$ is a (sub/super)martingale relative to some filtration $\mathscr{F}$, then for any $\mathscr{F}$-stopping time t, the stopped process X^t is a (sub/super)martingale relative to $\mathscr{F}$.

HINT: Prove the identity

† $$X_{(t+1)\wedge t} - X_{t\wedge t} = 1_{\{t>t\}}(X_{t+1} - X_t)$$

and use the basic properties of the conditional expectation. ○

(7.15) REMARK: Since $1_{\{t>t\}}1_{\{t>0\}} = 1_{\{t>t\}}$ for every $t \in \mathbb{N}$, the relation (7.14†) will not change if both sides are multiplied by $1_{\{t>0\}}$. Thus, whether the stopped process X^t is a martingale or not depends only on the process $(1_{\{t>0\}}X_t)$. ○

(7.16) EVENTS THAT OCCUR BEFORE A STOPPING TIME: Let $(\Omega, \mathcal{F}, P)$ be a probability space, let $\mathscr{F} = (\mathscr{F}_t)_{t \in \bar{\mathbb{N}}}$ be any filtration inside $\mathcal{F}$, and let $X = (X_t)_{t \in \mathbb{N}}$ be any stochastic process on Ω that is adapted to $\mathscr{F}$. By definition, this means that the r.v. X_t is measurable with respect to $\mathscr{F}_t$ for any $t \in \mathbb{N}$. Following the apparent analogy, one may feel prompted to declare that X_t, that is, the r.v. that gives the position of X at the random time t, is measurable for $\mathscr{F}_t$, but $\mathscr{F}_t$ must be defined first. The σ-field of events that occur before the stopping time t is given by

$$\mathscr{F}_t \overset{\text{def}}{=} \left\{A \in \mathcal{F} : A \cap \{t = t\} \in \mathscr{F}_t \text{ for all } t \in \mathbb{N}\right\}.$$

The intuition behind this definition is clear: $\mathscr{F}_t$ consists of those events $A \in \mathcal{F}$ for which one can determine whether A has occurred or not only in terms of the history up until the particular realization of the r.v. t. For example, if $A \in \mathscr{F}_t$ and the event $\{\mathscr{t} = 10\} \in \mathscr{F}_{10}$ has been observed, the information contained in the σ-field $\mathscr{F}_{10}$ – to be precise, the information contained in $\{\mathscr{t} = 10\} \cap \mathscr{F}_{10}$ – should be enough to determine whether the event A has occurred or not, and adding more information, say, by observing events from $\mathscr{F}_{11}$, cannot change the conclusion. ○

(7.17) EXERCISE: With the notation as in (7.16), prove that $\mathscr{F}_{\mathscr{t}}$ is a σ-field and that $X_{\mathscr{t}}$ is an $\mathscr{F}_{\mathscr{t}}$-measurable r.v.

HINT: For any $B \in \mathcal{B}(\mathbb{R})$,

$$\{X_{\mathscr{t}} \in B\} \cap \{\mathscr{t} = t\} = \{X_t \in B\} \cap \{\mathscr{t} = t\}. \quad ○$$

It turns out that the characterization of (sub/super)martingales in terms of conditional expectations from (7.1‡) can be extended by replacing the deterministic times s and t with random stopping times $\mathscr{s}$ and $\mathscr{t}$ – this was the main reason for introducing the σ-fields of events that occur before a stopping time. The precise statement is the following:

(7.18) DOOB'S OPTIONAL STOPPING THEOREM:[83] (Shiryaev 2004, VII sec. 2) Let $X = (X_t)_{t\in\mathbb{N}}$ be any $(\mathscr{F}, P)$-(sub/super)martingale and let $\mathscr{s}$ and $\mathscr{t}$ be any two $\mathscr{F}$-stopping times such that the following conditions are satisfied:

$$\mathsf{E}[|X_{\mathscr{s}}|] < \infty, \qquad \mathsf{E}[|X_{\mathscr{t}}|] < \infty,$$

$$\limsup_{t\to\infty} \mathsf{E}\big[|X_t| 1_{\{\mathscr{s}>t\}}\big] = 0 \quad \text{and} \quad \limsup_{t\to\infty} \mathsf{E}\big[|X_t| 1_{\{\mathscr{t}>t\}}\big] = 0.$$

Then P-a.s. on the set $\{\mathscr{s} \leq \mathscr{t}\}$ one has

$$X_{\mathscr{s}} = (\leq, \geq) \mathsf{E}\big[X_{\mathscr{t}} | \mathscr{F}_{\mathscr{s}}\big].$$

In particular, if $\mathscr{s} \leq \mathscr{t}$ P-a.s., then the last relation holds P-a.s. ○

The practical implications of the last result are rather important: if martingales are to be thought of as models of fair games, then the optional stopping theorem says that one cannot "cheat" by bailing out of the game (the proverbial "take the money and run" scenario) at random times – as long as one cannot peek into the future following those random times. However, we stress that merely not being able to "peek into the future" is not enough: there are integrability conditions that must be met as well – see above – and without those conditions the optional stopping result may not hold. One situation in which the conditions from (7.18) are rather easy to check is described in the next exercise.

(7.19) EXERCISE: Just as in (7.18), let $X = (X_t)_{t\in\mathbb{N}}$ be any $(\mathscr{F}, P)$-(sub/super) martingale and let $\mathscr{s}$ and $\mathscr{t}$ be any two $\mathscr{F}$-stopping times. Prove that if there is a (finite) integer $N \in \mathbb{N}_{++}$ such that $\mathscr{s} \leq N$ P-a.s. and $\mathscr{t} \leq N$ P-a.s., then all of the assumptions in the optional stopping theorem are met.

HINT: Observe that the r.v. $|X_{\mathscr{s}}|$ and $|X_{\mathscr{t}}|$ are both dominated P-a.s. by the r.v. $|X_0| + \cdots + |X_N|$. Furthermore, for any $t \geq N$ the r.v. $|X_t| 1_{\{\mathscr{s}>t\}}$ and $|X_t| 1_{\{\mathscr{t}>t\}}$ are both equal to 0 P-a.s. ○

[83] Named after the American mathematician Joseph Leo "Joe" Doob (1910–2004).

Now we turn to yet another use of stopping times: the description of the "fair game" property without the need to impose the integrability requirement.

(7.20) LOCAL MARTINGALES: Let $(\Omega, \mathcal{F}, P)$ be any probability space, let $\mathscr{F} = (\mathscr{F}_t)_{t \in \mathbb{N}}$ be any filtration inside $\mathcal{F}$, and let $X = (X_t)_{t \in \mathbb{N}}$ be any stochastic process on Ω that is adapted to $\mathscr{F}$. We say that X is a *local martingale* (or $(\mathscr{F}, P)$*-local martingale*) if there is a sequence of $\mathscr{F}$-stopping times $(t_i)_{i \in \mathbb{N}}$, called a *localizing sequence for* X, such that:

(a) $t_i \leq t_{i+1}$ P-a.s. for all $i \in \mathbb{N}$;

(b) $\lim_{i \to \infty} t_i = \infty$ P-a.s.;

(c) for any fixed $i \in \mathbb{N}$ the stopped process $\left(1_{\{t_i > 0\}} X_t^{t_i} \equiv 1_{\{t_i>0\}} X_{t \wedge t_i}\right)_{t \in \mathbb{N}}$ is an $(\mathscr{F}, P)$-martingale. ○

(7.21) REMARK: If we introduce the process $Y_t = X_t - X_0$, $t \in \mathbb{N}$, then condition (c) above comes down to two things: the stopped process $(Y_t^{t_i})_{t \in \mathbb{N}}$ is a martingale and X_0 is integrable on the set $\{t_i > 0\}$ (notice that $X_0^{t_i} = X_0$) for every $i \in \mathbb{N}$. Without the indicator $1_{\{t_i>0\}}$ condition (c) would imply that $\mathsf{E}[|X_0|] < \infty$, which is precisely the kind of condition that the notion of local martingale is trying to circumvent. This is only relevant if the σ-field $\mathscr{F}_0$ is not trivial, for if it is (that is, if $\mathscr{F}_0 = \{\emptyset, \Omega\}$), then X_0 is always a constant and there are no integrability issues to address. ○

(7.22) FIRST STEP TOWARD STOCHASTIC INTEGRATION: Let $\mathscr{F}$ be any filtration in the probability space $(\Omega, \mathcal{F}, P)$, let $X \prec \mathscr{F}_-$ be any $\mathscr{F}$-predictable process, and let $Y \prec \mathscr{F}$ be any $\mathscr{F}$-adapted process on $(\Omega, \mathcal{F}, P)$. We can then define the process

$$(X \cdot Y)_t \stackrel{\text{def}}{=} X_0 Y_0 + \sum_{s \in \mathbb{N}_{|t-1}} X_{s+1} (Y_{s+1} - Y_s), \quad t \in \mathbb{N},$$

with the understanding that the summation vanishes if $t = 0$. In finance this construction typically arises in connection with modeling the dynamics of the wealth created by a particular trading strategy $(A_t)_{t \in \mathbb{N}_{++}}$: with no initial investment whatsoever, at time $t \in \mathbb{N}$ one borrows the amount $A_{t+1} \prec \mathscr{F}_t$ and invests this amount in a particular risky asset with⊶ (6.53) excess returns process $(\mathcal{E}_t)_{t \in \mathbb{N}}$. At time $t+1$ the position is closed, the amount $A_{t+2} \prec \mathscr{F}_{t+1}$ is borrowed and invested in the same way until the next period – and so on. The amount accumulated as a result of this trading strategy by time $t \in \mathbb{N}_{++}$ can be expressed as $(A \cdot \mathcal{E})_t$ (recall that $\mathcal{E}_0 = 0$ by definition).

In any case, the process $X \cdot Y$ above is a precursor of what we will soon call a stochastic integral, but here we call this process *the transform of* Y *by* X, and if Y happens to be a martingale, we call $X \cdot Y$ a *martingale transform*. ○

(7.23) EXERCISE: Prove that any stochastic process that can be expressed as a martingale transform of some martingale Y by some predictable process X is a local martingale.

HINT: Consider using a localizing sequence of stopping times of the form $t_i = \inf\{t \in \mathbb{N} : X_{t+1} \geq i\}$, $i \in \mathbb{N}$ ($t_i = +\infty$ if $X_t < i$ for all $t \in \mathbb{N}$) and argue that

$(X \bullet Y)_t^{t_i} = (X^{t_i} \bullet Y^{t_i})_t$. Why is the process $\left(1_{\{t_i>0\}}(X^{t_i} \bullet Y^{t_i})_t\right)_{t \in \mathbb{N}}$ a martingale for every $i \in \mathbb{N}$? ○

(7.24) THEOREM: CHARACTERIZATION OF LOCAL MARTINGALES: (Shiryaev 2004, p. 651) Let $\mathscr{F}$ be any filtration in $(\Omega, \mathcal{F}, P)$ and let $X \prec \mathscr{F}$ be some $\mathscr{F}$-adapted stochastic process on Ω. The following statements are then equivalent:

(a) X is a local $(\mathscr{F}, P)$-martingale;

(b) the conditional expectation $\mathsf{E}[X_{t+1} | \mathscr{F}_t]$ is defined (not necessarily from the condition $\mathsf{E}[|X_{t+1}|] < \infty$) and $X_t = \mathsf{E}[X_{t+1} | \mathscr{F}_t]$ for every $t \in \mathbb{N}$;

(c) the process X can be expressed as a martingale transform (of some appropriately chosen martingale by some appropriately chosen predictable process). ○

7.2 Predictable Compensators

(7.25) DOOB'S DECOMPOSITION THEOREM: Given any $(\mathscr{F}, P)$-submartingale X, there is a $\mathscr{F}$-predictable process $A \prec \mathscr{F}_-$, called the *predictable compensator of* X, which is increasing, in that $A_t \leq A_{t+1}$ (P-a.s.) for every $t \in \mathbb{N}$, starts from 0, in that $A_0 = 0$ (P-a.s.), and is such that the process $(M_t \overset{\text{def}}{=} X_t - A_t)_{t \in \mathbb{N}}$ is an $(\mathscr{F}, P)$-martingale. A process A with these properties is P-a.s. unique: if $A^* \prec \mathscr{F}_-$ is another predictable and increasing process that starts from 0 and has the property that $X - A^*$ is a martingale, then $A_t^* = A_t$ (P-a.s.) for every $t \in \mathbb{N}$. ○

The predictable process A from (7.25) can be interpreted as a "correction" that makes the submartingale X into a martingale. What is remarkable is that such a correction can be chosen to be predictable, even though, generally, the innovations in a martingale are not predictable. Equivalently, Doob's theorem can be restated as the claim that every $(\mathscr{F}, P)$-submartingale X can be decomposed into the sum

$$X_t = M_t + A_t \quad (P\text{-a.s.}) \text{ for every } t \in \mathbb{N},$$

where M is an $(\mathscr{F}, P)$-martingale and A is a predictable and increasing process that starts from 0. The next exercise illustrates these features.

(7.26) EXERCISE: Prove Doob's decomposition theorem.

HINT: Show that the processes M and A, given by $M_0 = X_0$, $A_0 = 0$, and

$$A_t = X_0 + \sum\nolimits_{s \in \mathbb{N}_{|t-1}} \mathsf{E}\left[X_{s+1} - X_s \,|\, \mathscr{F}_s\right], \quad t \in \mathbb{N}_{++},$$

and

$$M_t = \sum\nolimits_{s \in \mathbb{N}_{|t-1}} \Big(\left(X_{s+1} - X_s\right) - \mathsf{E}\left[X_{s+1} - X_s \,|\, \mathscr{F}_s\right] \Big), \quad t \in \mathbb{N}_{++},$$

have the desired properties. To prove the uniqueness, suppose that $X = M^* + A^*$ is another such decomposition and show that

$$A_{t+1}^* - A_t^* = (A_{t+1} - A_t) + (M_{t+1} - M_t) - (M_{t+1}^* - M_t^*), \quad t \in \mathbb{N}.$$

By taking conditional expectations and using the fact that A and A^* are both predictable, conclude that (P-a.s.) $A^*_{t+1} - A^*_t = A_{t+1} - A_t$. ○

As we are already aware, if M is any martingale and $f : \mathbb{R} \mapsto \mathbb{R}$ is any convex function such that $\mathsf{E}[|f(M_t)|] < \infty$ for all $t \in \mathbb{N}$, then the process $\big(f(M_t)\big)_{t\in\mathbb{N}}$ is a submartingale. In particular, if M is a square integrable martingale, in that $\mathsf{E}[M_t^2] < \infty$ for all $t \in \mathbb{N}$, then the process $M^2 = (M_t^2)_{t\in\mathbb{N}}$ is an integrable submartingale.

(7.27) PREDICTABLE QUADRATIC VARIATION: If M is any square integrable martingale, then the predictable compensator, A, of the submartingale M^2, that is, the unique predictable and increasing process A that starts from 0 ($A_0 = 0$) and is such that $M^2 - A$ is a martingale, is called the *quadratic characteristic*, or the *predictable quadratic variation*, of M. As we are about to realize, the predictable quadratic variation encodes certain intrinsic features of the square integrable martingales and its role in the theory of stochastic processes is instrumental. As a first illustration, observe that since $M^2 - A$ is a martingale that starts from M_0, we have for every $t \in \mathbb{N}$,

$$\mathsf{E}\big[M_t^2 - A_t\big] = \mathsf{E}[M_0^2] \iff \mathsf{E}[M_t^2] = \mathsf{E}[A_t] + \mathsf{E}[M_0^2]. \quad ○$$

(7.28) EXERCISE: Suppose that M is a square integrable martingale and that A is the predictable compensator of M^2. Prove that

$$\mathsf{E}\big[(M_t - M_s)^2 \,|\, \mathscr{F}_s\big] = \mathsf{E}\big[M_t^2 - M_s^2 \,|\, \mathscr{F}_s\big] = \mathsf{E}\big[A_t - A_s \,|\, \mathscr{F}_s\big],$$

for all $s, t \in \mathbb{N}$, $s \leq t$. ○

The property established in the last exercise can be rephrased as

$$\mathsf{E}\big[\big(M_t - \mathsf{E}[M_t \mid \mathscr{F}_s]\big)^2 \mid \mathscr{F}_s\big] = \mathsf{E}\big[A_t - A_s \mid \mathscr{F}_s\big],$$

where the expression on the left side is nothing but the conditional variance of M_t given $\mathscr{F}_s$. In particular, since A is predictable,

$$\mathsf{E}\big[\big(M_{t+1} - \mathsf{E}[M_{t+1} \mid \mathscr{F}_t]\big)^2 \mid \mathscr{F}_t\big] = A_{t+1} - A_t \quad \text{for every } t \geq 0.$$

Thus, the quadratic characteristic of M is nothing but the process given as the sum of all one-period conditional variances of M in the first $t \in \mathbb{N}_{++}$ periods:

$$\begin{aligned} A_t &= \sum\nolimits_{s\in\mathbb{N}_{|t-1}} \mathsf{E}\big[\big(M_{s+1} - \mathsf{E}[M_{s+1} \mid \mathscr{F}_s]\big)^2 \mid \mathscr{F}_s\big] \\ &= \sum\nolimits_{s\in\mathbb{N}_{|t-1}} \mathsf{E}\big[\big(M_{s+1} - M_s\big)^2 \mid \mathscr{F}_s\big]. \end{aligned}$$

It is clear now why the quadratic characteristic of M (the predictable compensator of M^2) deserves to be called the *predictable quadratic variation* of M and deserves to be denoted by $\langle M, M\rangle$.

(7.29) PREDICTABLE QUADRATIC COVARIATION: If M and N are any two square integrable $(\mathscr{F}, P)$-martingales, the *predictable quadratic covariation*, or the *angled bracket*, of the pair (M, N) is the process $\langle M, N\rangle$ given by

$$\langle M, N\rangle \stackrel{\text{def}}{=} \frac{1}{4}\big(\langle M+N\rangle - \langle M-N\rangle\big)\,. \quad \circ$$

(7.30) EXERCISE: Let M and N be any two square integrable $(\mathscr{F}, P)$-martingales. Prove that the process $(M_t N_t - \langle M, N\rangle_t)_{t\in\mathbb{N}}$ is a martingale. Conclude that for any $s, t \in \mathbb{N}$, $s \le t$,

$$\begin{aligned}\mathsf{E}\big[(M_t - M_s)(N_t - N_s)\,|\,\mathscr{F}_s\big] &= \mathsf{E}\big[M_t N_t - M_s N_s\,|\,\mathscr{F}_s\big] \\ &= \mathsf{E}\big[\langle M, N\rangle_t - \langle M, N\rangle_s\,|\,\mathscr{F}_s\big]\,,\end{aligned}$$

and observe that the expectation on the left is nothing but the conditional covariance

$$\mathrm{Cov}(M_t, N_t\,|\,\mathscr{F}_s) \stackrel{\text{def}}{=} \mathsf{E}\big[\big(M_t - \mathsf{E}[M_t\,|\,\mathscr{F}_s]\big)\big(N_t - \mathsf{E}[N_t\,|\,\mathscr{F}_s]\big)\;\big|\;\mathscr{F}_s\big]\,.$$

Conclude that for any $t \in \mathbb{N}_{++}$,

$$\dagger \qquad \begin{aligned}\langle M, N\rangle_t &= \sum\nolimits_{s\in\mathbb{N}_{|t-1}} \mathrm{Cov}(M_{s+1}, N_{s+1}\,|\,\mathscr{F}_s) \\ &= \sum\nolimits_{s\in\mathbb{N}_{|t-1}} \mathsf{E}\big[(M_{s+1} - M_s)(N_{s+1} - N_s)\;\big|\;\mathscr{F}_s\big]\,. \quad \circ\end{aligned}$$

In addition to the angled bracket $\langle\cdot,\cdot\rangle$, we are also going to need the square bracket $[\cdot,\cdot]$, which is defined⊸(7.30†) in the same way as $\langle\cdot,\cdot\rangle$ but with the conditional expectation removed, that is, assuming as before that M and N are two square integrable martingales, we set $[M, N]_0 = 0$ and

$$[M, N]_t = \sum\nolimits_{s\in\mathbb{N}_{|t-1}} (M_{s+1} - M_s)(N_{s+1} - N_s) \quad \text{for all } t \in \mathbb{N}_{++}\,.$$

It would not be an exaggeration to say that in the most part stochastic calculus is essentially calculus with the brackets $\langle\cdot,\cdot\rangle$ and $[\cdot,\cdot]$ – these objects are much more interesting in the continuous-time setting, which will be introduced later in this space.

7.3 Fundamental Inequalities and Convergence

The running maximum of a process $X = (X_t)_{t\in\mathbb{N}}$ is the process

$$X_t^* \stackrel{\text{def}}{=} \max\nolimits_{s\in\mathbb{N}_{|t}} X_s\,, \quad t \in \mathbb{N}\,.$$

The following result is instrumental for both the theory and the practice of stochastic processes:

(7.31) DOOB'S MAXIMAL INEQUALITY: If $X = (X_t)_{t\in\mathbb{N}}$ is any $(\mathscr{F}, P)$-submartingale and $\lambda \in \mathbb{R}_{++}$ is arbitrarily chosen, then for any $t \in \mathbb{N}$,

$$\lambda\, P\big(X_t^* \ge \lambda\big) \le \mathsf{E}\big[X_t\, 1_{\{X_t^* \ge \lambda\}}\big] \le \mathsf{E}[X_t^+]\,,$$

where $X_t^+ \overset{\text{def}}{=} X_t \vee 0 \equiv \max\{X_t, 0\}$. ○

PROOF OF (7.31): If $t = 0$, then $X_t^* = X_0^* = X_0$, the event $\{X_0 \geq \lambda\}$ is the same as $\{X_0^+ \geq \lambda\}$, and the maximal inequality follows from⊸ (2.77e) Chebyshev inequality. Given any $t \in \mathbb{N}_{++}$, let $t^* = \inf\{s \in \mathbb{N}_{|t} : X_s \geq \lambda\}$, with the understanding that $t^* = t$ if $X_s < \lambda$ for all $s \in \mathbb{N}_{|t}$. It is easy to check that t^* is a stopping time with $0 \leq t^* \leq t$. Furthermore, the event $\{X_t^* \geq \lambda\}$ occurs before time t^*, that is, $\{X_t^* \geq \lambda\} \in \mathscr{F}_{t^*}$. To see why this claim can be made, it is enough to notice that

$$\{X_t^* \geq \lambda\} \cap \{t^* = s\} = \{X_0 < \lambda\} \cap \cdots \cap \{X_{s-1} < \lambda\} \cap \{X_s \geq \lambda\} \in \mathscr{F}_s$$

for any $s \in \mathbb{N}_{|t}$. If $s > t$, then, from the very definition of t^*, $\{t^* = s\} = \emptyset$, so that the relation $\{X_t^* \geq \lambda\} \cap \{t^* = s\} \in \mathscr{F}_s$ becomes trivial, namely $\emptyset \in \mathscr{F}_s$. Next, using (7.19), we can apply the optional stopping theorem from (7.18) with the stopping times t and $t^* \leq t$, and this will give us

$$X_{t^*} \leq \mathsf{E}\big[X_t \,|\, \mathscr{F}_{t^*}\big].$$

Since, as was established earlier, $1_{\{X_t^* \geq \lambda\}}$ is $\mathscr{F}_{t^*}$-measurable,

$$X_{t^*} 1_{\{X_t^* \geq \lambda\}} \leq 1_{\{X_t^* \geq \lambda\}} \mathsf{E}[X_t \,|\, \mathscr{F}_{t^*}] = \mathsf{E}[X_t 1_{\{X_t^* \geq \lambda\}} \,|\, \mathscr{F}_{t^*}],$$

which gives

$$\begin{aligned} \lambda\, P\big(X_t^* \geq \lambda\big) &\leq \mathsf{E}[X_{t^*} 1_{\{X_t^* \geq \lambda\}}] \leq \mathsf{E}[X_t 1_{\{X_t^* \geq \lambda\}}] \\ &\leq \mathsf{E}[X_t^+ 1_{\{X_t^* \geq \lambda\}}] \leq \mathsf{E}[X_t^+]. \quad ○ \end{aligned}$$

(7.32) EXERCISE: Let $\mathcal{Z} = (\mathcal{Z}_t)_{t \in \mathbb{N}}$ denote the canonical (symmetric) random walk on the coin toss space. Using Doob's maximal inequality, estimate the probabilities of the following events:

(a) $\{\max_{0 \leq t \leq 10} \mathcal{Z}_t \geq 5\}$;

(b) $\{\max_{0 \leq t \leq 10} |\mathcal{Z}_t| \geq 5\}$. ○

One of the reasons as to why the (sub/super)martingales are as interesting and useful as mathematical objects as they indeed are, is that, under certain conditions, processes with these attributes have a limit as the time variable t converges to ∞, and in any such case the time domain can be extended to $\bar{\mathbb{N}} = \mathbb{N} \cup \{\infty\}$, as we detail next. This is the very reason for using the term "martingale," which is the name of a device that equestrians often use to allow horses to wander around, but prevents them from running too far away. The precise convergence statement is the following:

(7.33) DOOB'S CONVERGENCE THEOREM: (Meyer 1966 V 17) Let X be any $(\mathscr{F}, P)$-supermartingale such that

$$\sup_{t \in \mathbb{N}} \mathsf{E}[X_t^-] < \infty.$$

Then there is an r.v. $X_\infty \in \mathscr{L}^1(\Omega, \mathscr{F}_\infty, P)$ such that $X_t \to X_\infty$ as $t \to \infty$ P-a.s. ○

(7.34) RESTATEMENT OF (7.33): The last theorem can be restated for submartingales in the obvious way, and in what follows we will pretend that (7.33) is stated for both submartingales and supermartingales. Indeed, let X be any $(\mathscr{F}, P)$-submartingale. Then the convergence of X is the same as the convergence of $-X$, which is a supermartingale. Thus, by the last theorem the convergence of X is guaranteed by the condition

$$\sup_{t\in\mathbb{N}} \mathsf{E}[X_t^+] < \infty .$$

In particular, if X is any $(\mathscr{F}, P)$-(sub/super)martingale, then the condition

$$\sup_{t\in\mathbb{N}} \mathsf{E}[|X_t|] < \infty$$

guarantees the existence of an $\mathscr{F}_\infty$-measurable and integrable r.v. X_∞ such that $X_t \to X_\infty$ as $t \to \infty$ P-a.s. We stress that although Doob's convergence theorem guarantees that $X_\infty \in \mathscr{L}^1$, it does not guarantee that the convergence $X_t \to X_\infty$ takes place in L^1. This point is important and is illustrated in exercise (7.35). ○

To illustrate the power of the last result, suppose that X is some negative $(\mathscr{F}, P)$-submartingale. Since $X_t^+ = 0$ P-a.s., the assumptions in Doob's convergence theorem are satisfied. By Fatou's lemma[84] for conditional expectations,

$$\mathsf{E}[X_\infty | \mathscr{F}_t] = \mathsf{E}\big[\lim_{s\to\infty} X_s | \mathscr{F}_t\big] \geq \limsup_{s\to\infty} \mathsf{E}\big[X_s | \mathscr{F}_t\big] \geq X_t , \quad t \in \mathbb{N} .$$

In other words, X continues to be a submartingale even if the range of the time variable t is expanded to include $+\infty$, with $X_\infty \overset{\text{def}}{=} \lim_{s\to\infty} X_s$.

Another interesting and extremely useful application of Doob's convergence theorem is in the case of a positive $(\mathscr{F}, P)$-martingale X. Since $X_t^- = 0$, the assumptions of (7.33) are again met and $X_\infty = \lim_{t\to\infty} X_t$ exists P-a.s. Furthermore, X_∞ is integrable, but can one claim that $\mathsf{E}[X_\infty | \mathscr{F}_t] = X_t$ for every $t \in \mathbb{N}$? The next exercise provides an answer.

(7.35) EXERCISE: Let $(X_t)_{t\in\mathbb{N}}$ be any sequence of i.i.d. r.v. with $X_0 \in \mathcal{N}(0, 1)$ (all r.v. are standard normal and independent). Define the process

$$M_t = e^{-X_0 - \frac{1}{2}} e^{-X_1 - \frac{1}{2}} \cdots e^{-X_t - \frac{1}{2}} , \quad t \in \mathbb{N} ,$$

and observe that⊸(7.4) M is a martingale with respect to its natural filtration $\mathscr{F}^M$. In fact, M is a positive martingale with $\mathsf{E}[M_t] = 1$ for every $t \in \mathbb{N}$. As was shown above, Doob's convergence theorem implies that $M_\infty = \lim_{t\to\infty} M_t$ exists a.s. Prove that the only possibility is $M_\infty = 0$ P-a.s. Verify this claim independently, showing that $\lim_{t\to\infty} M_t = 0$ without relying on Doob's convergence theorem.

HINT: Use the fact that convergence a.s. implies convergence in distribution, and argue that $P(\log(M_t) > a) \to 0$ as $t \to \infty$ for any $a \in \mathbb{R}$. What can you say

[84] Fatou's lemma was stated previously for positive random variables. There is an obvious restatement of this result for negative random variables in which lim inf is to be replaced with lim sup and the direction of the inequality is reversed. However, one must keep in mind that, in general, Fatou's lemma cannot be stated for random variables that are neither positive nor negative.

about the law of $\log(M_t)$? As for the second proof, using the law of large numbers and the central limit theorem, show that $\log(M_t) \to -\infty$ P-a.s. ○

It is clear from the last exercise that without imposing additional assumptions on the martingale X we will not be able to claim that $\mathsf{E}[X_\infty \,|\, \mathscr{F}_t] = X_t$. These additional assumptions come down to the requirement that the family of random variables $X = (X_t)_{t \in \mathbb{N}}$ is uniformly integrable, that is,

$$\lim_{C\to\infty}\Big(\sup_{t\in\mathbb{N}} \mathsf{E}\big[|X_t| 1_{\{|X_t|>C\}}\big]\Big) = 0\,.$$

(7.36) EXERCISE: Let X be any positive $(\mathscr{F}, P)$-martingale and let X_∞ denote the limit $\lim_{t\to\infty} X_t$, which, as was shown above, always exists. Show that the martingale X is uniformly integrable and $\mathsf{E}[X_\infty \,|\, \mathscr{F}_t] = X_t$ for all $t \in \mathbb{N}$ if and only if $\mathsf{E}[X_\infty] = \mathsf{E}[X_0]$ $(= \mathsf{E}[X_t]$ for every $t \in \mathbb{N})$. Explain why this last condition implies that $X_t \to X_\infty$ also in L^1-norm.

HINT: This statement is essentially a rephrasing of (4.23). Since $X_\infty \geq 0$ P-a.s., the expectation $\mathsf{E}[X_\infty]$ is always meaningful. The relation $\mathsf{E}[X_t] \to \mathsf{E}[X_\infty]$, which is equivalent to uniform integrability, actually reads $\mathsf{E}[X_0] = \mathsf{E}[X_\infty]$. The L^1-convergence, too, is a consequence of (4.23). ○

Looking back at the martingale described in exercise (7.2), it would be natural to ask whether $\lim_{t\to\infty} \mathsf{E}[\xi \,|\, \mathscr{F}_t]$ exists, and, if it does exist, whether it can be identified with the r.v. $\mathsf{E}[\xi \,|\, \mathscr{F}_\infty]$. By the result in (4.24), the martingale $(\mathsf{E}[\xi \,|\, \mathscr{F}_t])_{t\in\mathbb{N}}$ is uniformly integrable for any $\xi \in \mathscr{L}^1(\Omega, \mathcal{F}, P)$, and by the last exercise $\xi^* \stackrel{\text{def}}{=} \lim_{t\to\infty} \mathsf{E}[\xi \,|\, \mathscr{F}_t]$ exists both a.s. and in the L^1-norm. However intuitive it may seem that ξ^* can be nothing other than $\mathsf{E}[\xi \,|\, \mathscr{F}_\infty]$, the matter is somewhat delicate and hinges on⊶(8.76) the monotone class theorem. This is a classical result and the detailed argument can be found, for example, in (Meyer 1966, V 18). It is essential that (as we assume by default) $\mathscr{F}_\infty \stackrel{\text{def}}{=} \vee_{t\in\mathbb{N}} \mathscr{F}_t$, which then allows one to conclude (using the monotone class theorem) that the identity $\mathsf{E}[\xi \,|\, \mathscr{F}_t] = \mathsf{E}[\xi^* \,|\, \mathscr{F}_t]$ for every $t \in \mathbb{N}$ is only possible if $\xi^* \equiv \mathsf{E}[\xi^* \,|\, \mathscr{F}_\infty] = \mathsf{E}[\xi \,|\, \mathscr{F}_\infty]$ (all understood a.e.). In any case, we have the following classical result (Meyer 1966, V 18):

(7.37) LÉVY'S CONVERGENCE THEOREM FOR MARTINGALES: Let $\mathscr{F}$ be any filtration in $(\Omega, \mathcal{F}, P)$ and let $\xi \in \mathscr{L}^1(\Omega, \mathcal{F}, P)$. Then both P-a.s. and in L^1-sense

$$\mathsf{E}[\xi \,|\, \mathscr{F}_t] \;\to\; \mathsf{E}[\xi \,|\, \mathscr{F}_\infty] \quad \text{as } t \to \infty\,. \quad ○$$

8

Stochastic Processes and Brownian Motion

The very first known description of a stochastic process in continuous time can be found in the work of the Danish astronomer, statistician, and actuary Thorvald Nicolai Thiele (1838–1910). He wrote the continuous-time dynamics of the premium reserve in a life insurance policy in terms of a random differential equation, which now carries his name. However, Thiele did not publish his work on this equation, and his continuous-time model of life insurance policy was made public only after his death in the publication (Gram 1910), which, ironically, was Thiele's academic obituary. Nevertheless, among his published works, (Thiele 1880) contains the very first known formulation of a process with stationary, independent, and Gaussian increments, which, as is now well known, is nothing other than Brownian motion. Thiele arrived at the idea for such a process from his interpretation of the way random errors accumulate over time. In any case, if one is to point to the very first description – in probabilistic language – of what we call today Brownian motion, that would most likely be (Thiele 1880). Unfortunately, this work was way ahead of its time. So much so that it even contains a prototype of the widely used (today) Kálmán–Bucy linear filtering algorithm – see (Kálmán and Bucy 1961). The paper (Lauritzen 1981) provides a detailed review of (Thiele 1880), from the point of view of where probability theory and statistics stood a century later.

A better known pioneering work on the process that we call today Brownian motion is (Bachelier 1900), which is generally considered to be the starting point of continuous-time finance. This work, too, was ahead of its time, and was more or less forgotten for nearly half a century, until it was brought back to the center of attention by Paul Samuelson in the 1950s.[85] Nevertheless, the research on the phenomenon of Brownian motion literally exploded shortly after Louis Bachelier (1870–1946) presented his doctoral dissertation in 1900, although the general direction of this research took a slight turn – or, perhaps, not so slight, depending on the point of view. The turn was in the direction of studying physically observed random motion of particles bombarded by other particles – as,

[85] In 1970 the American economist Paul Samuelson (1915–2009) was awarded the Sveriges Riksbank Prize in Economic Sciences in Memory of Alfred Nobel for his contribution to raising "the level of scientific analysis in economic theory."

say, the motion of pollen particles suspended in water, which was first documented by the Scottish botanist Robert Brown (1773–1858) in his 16-page summary (Brown 1828). Interestingly, the idea – as opposed to an actually observed phenomenon – for such a motion can be traced more than two millennia back to the work (Lucretius 50 BC), which was rediscovered at the end of the fifteenth century and is still in print today. Among other things, we learn from this work that the Roman poet and (Epicurean) philosopher Titus Lucretius (99 BC–55 BC) believed that unless the elementary particles in the universe move chaotically, they will not be able to "meet" and will thus dissipate in "empty space," and, as a result, the objects around us would not exist – talk about someone being ahead of their time! In any case, a more or less rigorous model of such a chaotic motion of particles, based on the principles of modern physics, was developed – independently from the works of Thiele and Bachelier – in the seminal papers (Einstein 1905), (Smoluchowski 1906), (Perrin 1909), and (Wiener 1923).[86] This, of course, was only the beginning, and the mathematical construct used to model the random walk of a particle in space-time continuum was eventually named after Robert Brown.[87] Our main objective in the present and in all subsequent chapters is to briefly outline the theory and the methods that were developed as a follow up to the groundbreaking publications mentioned above. A detailed and up-to-date account of the most interesting and useful features of Brownian motion can be found in the recent work (Mörters and Peres 2010) – among many other sources.

8.1 General Properties and Definitions

The starting point in our exploration of stochastic processes in continuous time is still the random walk $\mathcal{Z} = (\mathcal{Z}_t)_{t \in \mathbb{N}}$, which we were able to construct on the coin toss space $(\Omega_\infty, \mathscr{F}_\infty, P_\infty)$. Recall that this probability space was endowed with the filtration $\left(\mathscr{F}_t = \sigma(\mathscr{P}_t)\right)_{t \in \mathbb{N}}$, and the random walk was defined as

$$\mathcal{Z}_t = \sum\nolimits_{i \in \mathbb{N}_{|t}} \mathcal{X}_i \quad \text{for all } t \in \mathbb{N},$$

where $\mathcal{X}_i$, $i \in \mathbb{N}$, are the canonical coordinate mappings on Ω_∞, with $\mathcal{X}_0 = 0$ by convention. The random variables $\mathcal{X}_i$, $i \in \mathbb{N}_{++}$, defined on $(\Omega_\infty, \mathscr{F}_\infty, P_\infty)$, are i.i.d. with $P(\mathcal{X}_i = 1) = P(\mathcal{X}_i = -1) = \frac{1}{2}$. Starting with the discrete-time process $\mathcal{Z}$, labeled by the set $\mathbb{N}$, we can now construct a process labeled by the set $\mathbb{R}_+$.

(8.1) First encounter with a continuous-time process: Given any $t \in \mathbb{R}_+$, let $W_t^1 \overset{\text{def}}{=} \mathcal{Z}_{\lfloor t \rfloor}$, or equivalently, $W_t \overset{\text{def}}{=} \mathcal{Z}_i$ for any $t \in [i, i+1[$ for any $i \in \mathbb{N}$.

[86] Albert Einstein (1879–1955) and Jean Baptiste Perrin (1870–1942) were both awarded the Nobel Prize in Physics, respectively, in 1921 and 1926.

[87] The term "Wiener process" is still in use today as an alias for "Brownian motion." This use was quite common until the early 1990s, which explains why a particularly common notation for the Brownian motion process is still the token W.

Now we have a family of random variables, $(W^1_t)_{t\in\mathbb{R}_+}$, that is labeled by the set $\mathbb{R}_+$ instead of $\mathbb{N}$. ○

(8.2) STOCHASTIC PROCESSES IN CONTINUOUS TIME: A *stochastic process in continuous time*, often called *stochastic process*, is any family of random variables with values in some metric space $\mathbb{S}$, all defined on the same probability space $(\Omega, \mathcal{F}, P)$ and labeled by the time parameter $t \in \mathbb{R}_+$. Such a family is usually expressed as $(X_t)_{t\in\mathbb{R}_+}$. The range space $\mathbb{S}$ is called the *state space of* X, and we will often say that "X is a stochastic process on $(\Omega, \mathcal{F}, P)$ with values in $\mathbb{S}$," or that "X is an $\mathbb{S}$-valued process on $(\Omega, \mathcal{F}, P)$." In most situations the state space will be either $\mathbb{R}$ or $\bar{\mathbb{R}}$, or perhaps a particular subset of $\mathbb{R}^n$ (e.g., $\mathbb{R}_{++}$, or $\mathbb{R}^n_+$). However, most of the features that we are going to develop will be applicable to state spaces $\mathbb{S}$ that have the structure of a$^{\circ\!\!\rightarrow (A.42)}$ Polish space. In fact, technically, stochastic processes with values in a generic Polish space are no different from stochastic process with values in the real line $\mathbb{R}$ – this has to do with the notion of "Borel equivalence," which lies outside of our main concerns. If the state space of a particular process is not understood from the context, the default assumption is that it is some Polish space, but not much will be lost if the reader understands "Polish space" to mean "the real line $\mathbb{R}$." In any case, if the state space of X is $\bar{\mathbb{R}}$, we will say that X is a real process, or a scalar process; if it is $\mathbb{R}$, we will say that X is a finite real process; if it is $\bar{\mathbb{R}}_+$ we will say that X is a positive process; if it is $\mathbb{R}_+$, we will say that X is a finite positive process – and so on.

Any stochastic process X with state space $\mathbb{S}$ can be treated as a function in three different ways: (a) a function of the time parameter t, that is, a function defined on $\mathbb{R}_+$ that takes values in the space$^{\circ\!\!\rightarrow (2.41)}$ of $\mathbb{S}$-valued random variables $\mathscr{L}^0(\Omega, \mathcal{F}, P; \mathbb{S})$; (b) a function of the pair (ω, t) that maps the product $\Omega\times\mathbb{R}_+$ into $\mathbb{S}$; and (c) a function of ω that maps the sample space Ω into the space of all $\mathbb{S}$-valued functions on $\mathbb{R}_+$. These three points of view can be expressed as:

$$\mathbb{R}_+ \ni t \rightsquigarrow X_t \in \mathscr{L}^0(\Omega, \mathcal{F}, P; \mathbb{S}), \quad \Omega\times\mathbb{R}_+ \ni (\omega, t) \rightsquigarrow X(\omega, t) \in \mathbb{S},$$
$$\Omega \ni \omega \rightsquigarrow X_\omega \equiv X(\omega, \cdot) \in \mathbb{S}^{\mathbb{R}_+}.$$

For any fixed $\omega \in \Omega$, the function $\mathbb{R}_+ \ni t \rightsquigarrow X(\omega, t) \in \mathbb{S}$ is called a *sample path, trajectory*, or *realization* of X (associated with the outcome ω). In particular, according to the third point of view, a continuous-time stochastic process can be viewed as a trajectory-valued random variable defined on the probability space $(\Omega, \mathcal{F}, P)$. In most cases we are going to be working with interpretations (b) and (c), and remark that if the $\mathbb{S}$-valued process X is treated as a mapping of the form $X: \Omega\times\mathbb{R}_+ \mapsto \mathbb{S}$, then some joint measurability with respect to the pair of arguments (ω, t) must be imposed. At the very minimum, we would like X to be $\mathcal{B}(\mathbb{S})/\big(\mathcal{F}\otimes\mathcal{B}(\mathbb{R}_+)\big)$-measurable. If this condition is satisfied, we will say that X is *jointly measurable*, or simply *measurable*. Other – and much more important – types of measurability will be introduced in (8.53). In any case, in all circumstances X_t will be assumed $\mathcal{B}(\mathbb{S})/\mathcal{F}$-measurable as a function from Ω into $\mathbb{S}$ for any $t \in \mathbb{R}_+$. ○

(8.3) NOTATION: As we have already seen, stochastic processes in continuous time are commonly expressed as $X = (X_t)_{t \in \mathbb{R}_+}$, or as $X = (X_t)_{t \in [0,T]}$, in the case of a finite time domain. For brevity, we will often refer to such objects only by their "first name," X. The random variable associated with the position of the process X at time $t \in \mathbb{R}_+$ we will express as X_t, but may also write $X_t(\omega)$, $X(\omega, t)$, or $X_\omega(t)$ if the dependence on $\omega \in \Omega$ needs to be emphasized. Given any (fixed) $\omega \in \Omega$, the sample path $\mathbb{R}_+ \ni t \rightsquigarrow X(\omega, t) \in \mathbb{S}$ may be expressed as X_ω, that is, X_ω stands for the function $X_\omega : \mathbb{R}_+ \mapsto \mathbb{S}$ (an element of $\mathbb{S}^{\mathbb{R}_+}$) given by $X_\omega(t) = X(\omega, t) \equiv X_t(\omega)$ for any $t \in \mathbb{R}_+$.

A deterministic function on $\mathbb{R}_+$ is of course a special case of a stochastic process X with sample paths X_ω that do not depend on $\omega \in \Omega$. One particularly simple stochastic process, which we are going to encounter quite often, is the identity mapping $\iota : \mathbb{R}_+ \mapsto \mathbb{R}_+$, which we will interpret as the stochastic process $\iota(\omega, t) = \iota(t) = t$. ○

(8.4) THE FAMILY OF FINITE DIMENSIONAL DISTRIBUTIONS: Let $X = (X_t)_{t \in \mathbb{R}_+}$ be any stochastic process on $(\Omega, \mathcal{F}, P)$ with state space $\mathbb{S}$ (Polish space); in particular, X_t is $\mathcal{B}(\mathbb{S})/\mathcal{F}$-measurable r.v. on Ω for every $t \in \mathbb{R}_+$. Thus, for every choice of $n \in \mathbb{N}_{++}$ and every choice of $t_1, \dots, t_n \in \mathbb{R}_+$, we can define the mapping

$$\Omega \ni \omega \rightsquigarrow (X_{t_1}(\omega), \dots, X_{t_n}(\omega)) \in \mathbb{S}^n,$$

which is clearly measurable (for the product σ-field on $\mathbb{S}^n$) and maps the probability measure P from Ω into a (Borel) probability measure over $\mathbb{S}^n$. We denote this measure over $\mathbb{S}^n$ by $P_{t_1, \dots, t_n}$, and call the family of all such probability measures, for all possible choices of the finite sets $\{t_1, \dots, t_n\} \subset \mathbb{R}_+$ (of all possible sizes $n \in \mathbb{N}_{++}$), "family of finite dimensional distributions of X." This family is *consistent* in the following sense: if $\{s_1, \dots, s_k\}$, $k \le n$, is another finite set inside $\mathbb{R}_+$, which is included into the finite set $\{t_1, \dots, t_n\}$, then there is a canonical projection, π, from $\mathbb{S} \times \cdots \times \mathbb{S}$ (n-times) onto $\mathbb{S} \times \cdots \times \mathbb{S}$ (k-times), corresponding to the selection of the coordinates $\{s_1, \dots, s_k\}$ from the list of coordinates $\{t_1, \dots, t_n\}$, which is such that

$$\dagger \qquad\qquad P_{s_1, \dots, s_k} = P_{t_1, \dots, t_n} \circ \pi^{-1}.$$

That is, $P_{s_1, \dots, s_k}$ is the image of $P_{t_1, \dots, t_n}$ taken by the projection $\pi : \mathbb{S}^n \mapsto \mathbb{S}^k$. This is nothing but a description of the joint distribution law of $(X_{s_1}, \dots, X_{s_k})$ as a marginal distribution law from the joint law of $(X_{t_1}, \dots, X_{t_n})$. ○

It can be argued that practical experiments can only lead to knowing the family of finite dimensional distributions of physically observed processes. We are therefore free to choose any mathematical model of a stochastic process (probability space and a particular collection of random variables on that space with values in the state space $\mathbb{S}$) that we like – as long as it produces a family of finite dimensional distributions that matches the reality reasonably well. The question then arises as to whether *any* given collection of probability distributions over $\mathbb{S}^n$, attached to the various choices of the finite subsets $\{t_1, \dots, t_n\} \subset \mathbb{R}_+$, is constructable, meaning one can define a mathematical model of a stochastic process whose family of finite dimensional distributions is precisely the given collection of distributions. It is clear that a given collection of probability measures, $P_{t_1,\dots,t_n}$, attached to the various choices for $\{t_1, \dots, t_n\} \subset \mathbb{R}_+$, must be consistent in the sense of ($8.4^{\dagger}$) in order to be constructable. It turns out that the attribute "consistent" is also sufficient; to be precise, we have the following classical result, which we borrow from (Revuz and Yor 1999, I (3.2)):

(8.5) KOLMOGOROV'S EXTENSION THEOREM: Let $\mathbb{S}$ be any Polish space, let $\mathbb{S}^{\mathbb{R}_+}$ denote the space of all functions $\omega\colon \mathbb{R}_+ \mapsto \mathbb{S}$, and let $\mathcal{B}(\mathbb{S})^{\mathbb{R}_+}$ denote the smallest σ-field on $\mathbb{S}^{\mathbb{R}_+}$ for which every coordinate mapping $\mathbb{S}^{\mathbb{R}_+} \ni \omega \rightsquigarrow \epsilon_t(\omega) = \omega(t) \in \mathbb{S}$, $t \in \mathbb{R}_+$, is measurable for the Borel σ-field on $\mathbb{S}$, that is, $\mathcal{B}(\mathbb{S})^{\mathbb{R}_+} = \vee_{t\in\mathbb{R}_+}\epsilon_t^{-1}(\mathcal{B}(\mathbb{S}))$. Let $P_{t_1,\dots,t_n}$ be a (Borel) probability measure over $\mathbb{S}^n$ uniquely assigned to any finite subset $\{t_1, \dots, t_n\} \subset \mathbb{R}_+$, and suppose that all these probability measures are consistent in the sense of ($8.4^{\dagger}$). Then there is a unique probability measure $\mathcal{P}$ on the σ-field $\mathcal{B}(\mathbb{S})^{\mathbb{R}_+}$ such that the probability measure over $\mathbb{S}^n$ that arises as the image of $\mathcal{P}$ by the mapping

$$\mathbb{S}^{\mathbb{R}_+} \ni \omega \rightsquigarrow (\epsilon_{t_1}(\omega), \dots, \epsilon_{t_n}(\omega)) \in \mathbb{S}^n ,$$

is precisely $P_{t_1,\dots,t_n}$ for every choice of the finite subset $\{t_1, \dots, t_n\} \subset \mathbb{R}_+$. To put it another way, the canonical coordinate mappings $(\epsilon_t)_{t\in\mathbb{R}_+}$ give a stochastic process on $\left(\mathbb{S}^{\mathbb{R}_+}, \mathcal{B}(\mathbb{S})^{\mathbb{R}_+}, \mathcal{P}\right)$, whose family of finite dimensional distributions is precisely the family of all $P_{t_1,\dots,t_n}$ for all choices of $\{t_1, \dots, t_n\} \subset \mathbb{R}_+$. ○

Powerful as the last theorem certainly is, it leaves a lot to be desired since the space of functions $\mathbb{S}^{\mathbb{R}_+}$ is simply too general to be practical. Thus, our search for more tractable mathematical models that would be compatible with our particular choices of finite dimensional distributions must go on. The first step in this direction is to introduce the following concepts.

(8.6) MODIFICATIONS AND EQUIVALENCE: Given any two stochastic processes, X and X^*, possibly defined on two different probability spaces but sharing the same state space, we say that X and X^* are *equivalent*, or that X and X^* are *versions* of one another, if their respective families of finite dimensional distributions are identical, that is,

$$(X_{t_1}, \dots, X_{t_n}) \overset{\text{law}}{=} (X^*_{t_1}, \dots, X^*_{t_n})$$

for every $n \in \mathbb{N}_{++}$ and every finite subset $\{t_1, \dots, t_n\} \subset \mathbb{R}_+$.

If X and X^* are defined on the same probability space $(\Omega, \mathcal{F}, P)$, we say that X and X^* are *modifications of one another* if $X_t = X^*_t$ P-a.s. for every fixed $t \in \mathbb{R}_+$.

Finally, we say that X and X^* are *indistinguishable* if the sample paths X_ω and X^*_ω coincide (as elements of $\mathbb{S}^{\mathbb{R}_+}$) for P-a.e. $\omega \in \Omega$. ○

(8.7) CONTINUITY: If X is a stochastic process on $(\Omega, \mathcal{F}, P)$, then we say that X is *continuous* if the sample path X_ω is a continuous function on $\mathbb{R}_+$ for every $\omega \in \Omega$. If the sample path X_ω can be claimed to be continuous for all ω outside some P-negligible set, then we say that X is continuous a.s., or that X is continuous P-a.s., or that X is continuous with probability 1. This definition can be repeated with "continuous" replaced by "left-continuous," or "right-continuous," or "càdlàg." Since most of the time we will be working not with stochastic processes but, rather, with classes of indistinguishable processes, we will often drop the qualifier "P-a.s.," and will say that a particular process is continuous (or right-continuous, or càdlàg) in situations where this property actually holds only P-a.s. ○

Although the process W^1 that we constructed in (8.1) is labeled by $\mathbb{R}_+$, it is essentially the standard random walk in discrete time, except that it is expressed in continuous-time notation. The obvious next step for us is to write down processes labeled by $\mathbb{R}_+$ that are truly continuous-time – not just discrete-time processes in continuous-time dressing. Heuristically, such structures should arise as the limits of discrete-time processes. To gain some insight into the way such limits may come about, consider the process W^1 from (8.1), but change the unit of time from 1 to $\Delta \in \mathbb{R}_{++}$, change the unit of distance in the state space from 1 to $h = h(\Delta) > 0$, and set

$$W_t^\Delta \stackrel{\text{def}}{=} h(\Delta)\, \mathcal{Z}_{\lfloor \frac{t}{\Delta} \rfloor} \equiv h(\Delta) \left(\mathcal{X}_0 + \mathcal{X}_1 + \cdots + \mathcal{X}_{\lfloor \frac{t}{\Delta} \rfloor} \right) \quad \text{for every } \; t \in \mathbb{R}_+ .$$

This definition is the same as $W_t = h(\Delta)\, \mathcal{Z}_i$ for $t \in [i\Delta, (i+1)\Delta[\,,\; i \in \mathbb{N}$. Now we can pass to the limit as $\Delta \searrow 0$. By the central limit theorem,

$$\frac{1}{\sqrt{\lfloor \frac{t}{\Delta} \rfloor}} \mathcal{Z}_{\lfloor \frac{t}{\Delta} \rfloor} = \frac{1}{\sqrt{\lfloor \frac{t}{\Delta} \rfloor}} \left(\mathcal{X}_1 + \cdots + \mathcal{X}_{\lfloor \frac{t}{\Delta} \rfloor} \right) \underset{\text{as } \Delta \to 0}{\overset{\text{in law}}{\longrightarrow}} \mathcal{N}(0,1) .$$

Thus, we would like to choose $h(\Delta)$ in such a way that $h(\Delta)\sqrt{\lfloor \frac{t}{\Delta} \rfloor} \equiv \frac{h(\Delta)}{\sqrt{\Delta}} \sqrt{\Delta \lfloor \frac{t}{\Delta} \rfloor}$ has a limit as $\Delta \searrow 0$.

(8.8) EXERCISE: Prove that $\lim_{\Delta \searrow 0} \Delta \lfloor \frac{t}{\Delta} \rfloor = t$ for any $t \in \mathbb{R}_{++}$. ○

Since $\sqrt{\Delta \lfloor \frac{t}{\Delta} \rfloor} \to \sqrt{t}$ we only need to ensure that $\frac{h(\Delta)}{\sqrt{\Delta}}$ converges to a finite limit as $\Delta \searrow 0$. In particular, if $h(\Delta) \stackrel{\text{def}}{=} \sigma\sqrt{\Delta}$ for some fixed $\sigma \in \mathbb{R}\backslash\{0\}$ and for all sufficiently small values of Δ, then the law of W_t^Δ will be approximately $\mathcal{N}(0, \sigma^2 t)$. Without any loss of generality, we set for now $\sigma = 1$, and stress that all that we can claim at this point is convergence in distribution as $\Delta \to 0$ for fixed $t \in \mathbb{R}_+$.

(8.9) EXERCISE: Let $\mathcal{Z}$ denote the canonical symmetric random walk labeled by the set $\mathbb{N}$. For some fixed $\Delta \in \mathbb{R}_{++}$, define the process $W_t^\Delta = \sqrt{\Delta}\, \mathcal{Z}_{\lfloor \frac{t}{\Delta} \rfloor}$, $t \in \mathbb{R}_+$, and prove that:

(a) $W_0^\Delta = 0$ P-a.s.;
(b) if $0 \le s < t$, then

$$(W_s^\Delta, W_t^\Delta - W_s^\Delta) \underset{\text{as } \Delta \searrow 0}{\overset{\text{in law}}{\longrightarrow}} (\xi_1, \xi_2),$$

where $\xi_1 \in \mathcal{N}(0, s)$ and $\xi_2 \in \mathcal{N}(0, t - s)$ are independent r.v.;
(c) if $0 = t_0 < t_1 < t_2 < \cdots < t_n$, then

$$(W_{t_1}^\Delta - W_{t_0}^\Delta, W_{t_2}^\Delta - W_{t_1}^\Delta, \ \dots \ , W_{t_n}^\Delta - W_{t_{n-1}}^\Delta) \underset{\text{as } \Delta \searrow 0}{\overset{\text{in law}}{\longrightarrow}} (\xi_1, \xi_2, \ \dots \ , \xi_n),$$

where $\xi_i \in \mathcal{N}(0, t_i - t_{i-1})$, $1 \le i \le n$, are independent Gaussian r.v.
HINT: Consider using the theorems from (4.17), as in exercise (4.19). ○

If we were to believe, contrary to fact, that we have established that $\lim_{\Delta \searrow 0} W_t^\Delta$ exists P-a.s. for all $t \in \mathbb{R}_+$, and not just in distribution, then we could call this limit W_t (treated as a r.v.) and end up with a stochastic process $W = (W_t)_{t \in \mathbb{R}_+}$, which would be defined on the same probability space on which the random walk $\mathcal{Z}$ is defined. The process W would then have the following properties:

(a) $W_0 = 0$ P-a.s.;
(b) $W_t - W_s \in \mathcal{N}(0, t - s)$ for every choice of $0 \le s < t < \infty$;
(c) for any $0 = t_0 < t_1 < \cdots < t_n < \infty$, $n \ge 2$, the random variables

$$W_{t_1} - W_{t_0}, W_{t_2} - W_{t_1}, \ \dots \ , W_{t_n} - W_{t_{n-1}}$$

are independent.

We stress that, at this point, the existence of a mathematical model of a process that shares properties (a), (b), and (c) is purely hypothetical. Nevertheless, we introduce the following definitions:

(8.10) DEFINITION: If W is any finite real stochastic process that satisfies condition (c) above, then we say that W is a *process of independent increments*. ○

(8.11) TEMPORARY DEFINITION OF BROWNIAN MOTION: A finite real stochastic process, W, labeled by $\mathbb{R}_+$ or by $[0, T]$ for some $T \in \mathbb{R}_{++}$, and defined on some probability space $(\Omega, \mathcal{F}, P)$, is called *Brownian motion* if it shares properties (a), (b), and (c) above. The additional requirement that a Brownian motion process has a.s. continuous sample paths will be incorporated into the definition as soon as we develop the tools that would allow us to make such a requirement practical. ○

(8.12) DONSKER'S FUNCTIONAL CENTRAL LIMIT THEOREM:[88] In the foregoing we used the classical central limit theorem (CLT) for every fixed $t \in \mathbb{R}_+$, in effect relying on infinitely many calls to the CLT. It is possible to arrive at the same construction with a single call to the CLT, but for this purpose we need a CLT of a different kind. To see how such a result may look like, consider the finite time domain $[0, T]$ and observe that for any fixed $\omega \in \Omega$ and $n \in \mathbb{N}_{++}$ the function (as

[88] Named after the American mathematician Monroe David Donsker (1924–1991).

before, $\mathcal{X}$ is the canonical coordinate process on the coin toss space)

$$[0,T] \ni t \rightsquigarrow W_t^{1/n}(\omega) \equiv \frac{\sigma}{\sqrt{n}}\left(\mathcal{X}_0(\omega) + \mathcal{X}_1(\omega) + \cdots + \mathcal{X}_{\lfloor nt \rfloor}(\omega)\right) \in \mathbb{R}$$

is an element of the Skorokhod space $\mathbb{D}([0,T];\mathbb{R})$. Thus, $W_\cdot^{1/n}$ can be treated as a r.v. with values in $\mathbb{D}([0,T];\mathbb{R})$. Its distribution law is a probability measure on $\mathbb{D}([0,T];\mathbb{R})$ and as $n \to \infty$ this probability measure converges weakly to a probability measure, such that the coordinate process on $\mathbb{D}([0,T];\mathbb{R})$ is a Brownian motion with respect to it. This result is known as "functional CLT" and is due to Donsker (1951). We encountered a CLT of this type back in (4.47), and these two results are indeed intimately connected – see (Donsker 1952).

We have yet to establish that a Brownian motion, and therefore a probability distribution on $\mathbb{R}^{\mathbb{R}_+}$ associated with it, exists. Furthermore, a careful reader must have noticed that so far we have only introduced weak convergence of probability measures on $\mathbb{R}^n$. Nevertheless, the above formulation of the functional CLT makes it clear that weak convergence of probability measures can be introduced on more general spaces – see (Dudley 2014). The importance of such results notwithstanding, in what follows we are not going to rely on the functional CLT, and this result was stated here only as an aside. ○

(8.13) EXERCISE: Let $W = (W_t)_{t\in\mathbb{R}_+}$ be a Brownian motion, assuming that one exists, and let $0 \le t_1 < t_2 < \cdots < t_n < \infty$ be arbitrarily chosen. Write down the distribution density in $\mathbb{R}^n$ of the random vector $(W_{t_1}, \ldots, W_{t_n})$. In other words, describe explicitly the family of finite dimensional distributions of the Brownian motion. In addition, show that this family is consistent in the sense of $(8.4^\dagger)$, and conclude that a Brownian motion, as defined in (8.11), indeed exists. ○

8.2 Limit of the Binomial Asset Pricing Model

Our objective in this section is to determine the limiting behavior of the celebrated model of (Cox et al. 1979), colloquially known as "the CRR model." The limit will be taken with the time step Δ decreasing to 0. The CRR model postulates that time is discrete, every time period is Δ units of time long, and the price process associated with the risky security is $S_t^\Delta = S_0^\Delta R_1^\Delta \cdots R_t^\Delta$, $t \in \mathbb{N}$. The initial price S_0^Δ is fixed and the returns R_t^Δ, $t \in \mathbb{N}_{++}$, are i.i.d. binary random variables (random variables that take only two distinct values), representing two possible realizations of the return during any given period $t \in \mathbb{N}_{++}$. To be precise, every return R_t^Δ takes the values $1+b\Delta\pm\sigma\sqrt{\Delta}$, where $b \in \mathbb{R}$ and $\sigma \in \mathbb{R}_{++}$ are model parameters. In addition to the risky security, the market also admits a short-lived risk-free security with a constant one period return equal to $1 + r\Delta$. Looking at the price dynamics from one period to the next, we find ourselves in the setting of the one period model that we encountered in section 1.2, except that here we postulate that $u = 1+b\Delta+\sigma\sqrt{\Delta}$

and $d = 1 + b\Delta - \sigma\sqrt{\Delta}$, and therefore require that

$$b - \frac{\sigma}{\sqrt{\Delta}} \le r \le b + \frac{\sigma}{\sqrt{\Delta}} \quad \Leftrightarrow \quad -1 \le \frac{r-b}{\sigma}\sqrt{\Delta} \le 1\,,$$

which comes from the condition $0 < d \le 1 + r\Delta \le u$. The compelling intuition behind this modeling choice is that if the two possible outcomes for the return R_t^Δ are equally likely in the physical (real) world, then we can write $\mathsf{E}[R_t^\Delta] = 1+b\Delta$ and $\mathrm{Var}[R_t^\Delta] = \sigma^2\Delta$, where both expectation and variance are understood with respect to the real-world probability measure P. However, as we are already aware, in this model the market is complete, and the only probabilities that really matter from the point of view of asset pricing are the risk-neutral probabilities, which we discovered back in (1.8†). With the current choice for u and d, we have

$$\tilde{p} = \frac{1}{2}\Big(1 + \frac{r-b}{\sigma}\sqrt{\Delta}\Big) \quad \text{and} \quad \tilde{q} = \frac{1}{2}\Big(1 - \frac{r-b}{\sigma}\sqrt{\Delta}\Big).$$

It is convenient to express the returns R_t^Δ as fixed (nonrandom) returns subjected to unbiased multiplicative random i.i.d. shocks (in the real-world probability measure); to be precise $R_t^\Delta = (1 + b\Delta)(1 + X_t^\Delta)$, where $(X_t^\Delta)_{t\in\mathbb{N}}$ are i.i.d. binary r.v. with $\mathsf{E}[X_t^\Delta] = 0$ and $\mathrm{Var}[X_t^\Delta] = \frac{\sigma^2\Delta}{(1+b\Delta)^2}$ (again, in the real-world probability measure).

(8.14) EXERCISE: Let $\widetilde{\mathsf{E}}$ and $\widetilde{\mathrm{Var}}$ denote, respectively, the expectation and the variance with respect to the risk-neutral probabilities $(\tilde{p}, \tilde{q})$. Prove that

$$\widetilde{\mathsf{E}}[X_t^\Delta] = \frac{r-b}{1+b\Delta}\Delta\,, \quad \widetilde{\mathsf{E}}\big[(X_t^\Delta)^2\big] = \frac{\sigma^2}{(1+b\Delta)^2}\Delta\,,$$

$$\text{and} \quad \widetilde{\mathrm{Var}}[X_t^\Delta] = \frac{\sigma^2\Delta - (r-b)^2\Delta^2}{(1+b\Delta)^2}\,. \quad \circ$$

A straightforward power series expansion at $\Delta = 0$ gives

$$\widetilde{\mathsf{E}}[X_t^\Delta] = (r-b)\Delta + \mathcal{O}(\Delta^2)\,, \quad \widetilde{\mathsf{E}}\big[(X_t^\Delta)^2\big] = \sigma^2\Delta + \mathcal{O}(\Delta^2)\,,$$

$$\text{and} \quad \widetilde{\mathrm{Var}}[X_t^\Delta] = \sigma^2\Delta + \mathcal{O}(\Delta^2)\,,$$

and those relations easily translate to

$$\widetilde{\mathsf{E}}[R_t^\Delta] = 1 + r\Delta + \mathcal{O}(\Delta^2) \quad \text{and} \quad \widetilde{\mathrm{Var}}[R_t^\Delta] = \sigma^2\Delta + \mathcal{O}(\Delta^2)\,.$$

Thus, if we were to neglect quantities of order $\mathcal{O}(\Delta^2)$, the switch from the real-world probabilities to the risk-neutral probabilities would not alter the variance in the returns, but it would make the expected returns equal to the risk-free returns, whence the name "risk-neutral."

Next, consider the continuous-time interpretation of the price, namely the process $\big(S_{\lfloor t/\Delta\rfloor}^\Delta\big)_{t\in\mathbb{R}_+}$, and its logarithm

$$\log(S_{\lfloor t/\Delta\rfloor}^\Delta) = \log(S_0^\Delta) + \sum\nolimits_{i=1}^{\lfloor t/\Delta\rfloor} \log(R_i^\Delta)\,, \qquad \mathrm{e}_1$$

where the summation is assumed to vanish if $\lfloor t/\Delta \rfloor = 0$. Since

$$\log(1+x) = x - \frac{1}{2}x^2 + \mathcal{O}(x^3) \quad \text{and} \quad \big(\log(1+x)\big)^2 = x^2 + \mathcal{O}(x^3),$$

neglecting the terms of order $\mathcal{O}(\Delta^2)$ – *this operation requires attention!* – we can write

$$\log(1+b\Delta) \approx b\Delta, \quad \mathsf{E}[\log(1+X_i^\Delta)] \approx -\frac{1}{2}\sigma^2\Delta, \quad \mathrm{Var}\big[\log(1+X_i^\Delta)\big] \approx \sigma^2\Delta,$$

$$\widetilde{\mathsf{E}}[\log(1+X_i^\Delta)] \approx (r-b)\Delta - \frac{1}{2}\sigma^2\Delta, \quad \widetilde{\mathrm{Var}}\big[\log(1+X_i^\Delta)\big] \approx \sigma^2\Delta,$$

and rearrange $\log(R_i^\Delta)$ as

$$\begin{aligned}\log(R_i^\Delta) &= \log(1+b\Delta) + \log(1+X_i^\Delta) \approx b\Delta + \log(1+X_i^\Delta)\\ &= \Big(b - \frac{1}{2}\sigma^2\Big)\Delta + \sigma\sqrt{\Delta}\,\frac{\log(1+X_i^\Delta) + \frac{1}{2}\sigma^2\Delta}{\sigma\sqrt{\Delta}}\\ &= \Big(r - \frac{1}{2}\sigma^2\Big)\Delta + \sigma\sqrt{\Delta}\,\frac{\log(1+X_i^\Delta) + \frac{1}{2}\sigma^2\Delta + (b-r)\Delta}{\sigma\sqrt{\Delta}}.\end{aligned}$$

The rationale behind these two forms of the same expression, $b\Delta + \log(1+X_i^\Delta)$, is clear: we would like to express the stochastic shocks as normalized random variables, that is, as fractions with a centered r.v. in the numerator and with its standard deviation in the denominator. In the first fraction this is achieved relative to the real-world probability measure and in the second fraction this is achieved relative to the risk-neutral measure. Consequently, using equation e_1, we can cast the logarithmic price as

$$\begin{aligned}&\log\big(S^\Delta_{\lfloor t/\Delta\rfloor}\big) - \log(S_0^\Delta)\\ &\overset{!}{\approx} \Big(b - \frac{1}{2}\sigma^2\Big)\Delta\Big\lfloor\frac{t}{\Delta}\Big\rfloor + \sigma\sqrt{\Delta\Big\lfloor\frac{t}{\Delta}\Big\rfloor}\,\frac{\sum_{i=1}^{\lfloor t/\Delta\rfloor}\Big(\log(1+X_i^\Delta) + \frac{1}{2}\sigma^2\Delta\Big)}{\sigma\sqrt{\Delta\lfloor t/\Delta\rfloor}} \qquad \mathrm{e}_2\\ &\equiv \Big(r - \frac{1}{2}\sigma^2\Big)\Delta\Big\lfloor\frac{t}{\Delta}\Big\rfloor + \sigma\sqrt{\Delta\Big\lfloor\frac{t}{\Delta}\Big\rfloor}\,\frac{\sum_{i=1}^{\lfloor t/\Delta\rfloor}\Big(\log(1+X_i^\Delta) + \frac{1}{2}\sigma^2\Delta + (b-r)\Delta\Big)}{\sigma\sqrt{\Delta\lfloor t/\Delta\rfloor}}.\end{aligned}$$

A straightforward application of the CLT leads to the conclusion that with respect to the real-world probabilities we have

$$\log(S_t^\Delta) - \log(S_t^\Delta) \underset{\text{as } \Delta\to 0}{\overset{\text{in law}}{\longrightarrow}} \Big(b - \frac{1}{2}\sigma^2\Big)t + \sigma W_t, \quad t \in \mathbb{R}_+, \qquad \mathrm{e}_3$$

and with respect to the risk-neutral probabilities we have

$$\log(S_t^\Delta) - \log(S_t^\Delta) \underset{\text{as } \Delta\to 0}{\overset{\text{in law}}{\longrightarrow}} \Big(r - \frac{1}{2}\sigma^2\Big)t + \sigma \tilde{W}_t, \quad t \in \mathbb{R}_+, \qquad \mathrm{e}_4$$

where W and $\tilde{W}$ are Brownian motions. Since what we claim here is only convergence in distribution, we do not need to be specific about the probability space on which W and $\tilde{W}$ live, nor does it make sense to insist on the distinction between W and $\tilde{W}$.

At this point a careful reader may be dismayed at the steps just taken, and for a good reason: the terms $\mathcal{O}(\Delta^2)$ that we neglected are actually random variables, and, furthermore, thanks to the summation operation $\sum_{i=1}^{\lfloor t/\Delta \rfloor}$, the number of these terms becomes arbitrarily large as $\Delta \to 0$. Thus, taking it for granted that the effect of the neglected terms vanishes in the limit does require some suspension of disbelief (whence the exclamation mark in e_2). Nevertheless, while a bit technical, a more pedantic account of the terms involved would show that in the limit the effect of the error terms does indeed go away, and the relations in e_3 and e_4 do indeed hold. While we prefer not to be sidetracked into technical matters of such nature, keeping in mind all the steps in the trek from the binomial asset pricing model to the right sides of e_3 and e_4 will be very instructive. What is especially important is that replacing the expected growth rate b in e_3 with the risk-free interest rate r in e_4 is ultimately a consequence of the completeness of the market and the absence of arbitrage.

To fully appreciate the connection between discrete- and continuous-time asset pricing models that we have discovered so far in this section, it would be very instructive to suspend our disbelief one more time and imagine that all (in effect, discrete-time) price processes $\big(S^{\Delta}_{\lfloor t/\Delta \rfloor}\big)_{t\in\mathbb{R}_+}$, for all choices of $\Delta \in \mathbb{R}_{++}$, live in one and the same probability space $(\Omega, \mathcal{F}, P)$, in which P is the real-world probability measure (the one that makes the states "up" and "down" equally likely). Imagine further that the limit in e_3 is not just in distribution but takes place uniformly for all $t \in [0, T]$ P-a.s., for any fixed $T \in \mathbb{R}_{++}$. The Brownian motion on the right side of e_3 is then a well-defined process on $(\Omega, \mathcal{F}, P)$, and we can think of the limiting process $\lim_{\Delta\to 0} S^{\Delta}$ as the (genuinely continuous-time) price process $(S_t)_{t\in\mathbb{R}_+}$ given by

$$\log(S_t) - \log(S_0) = \big(b - \frac{1}{2}\sigma^2\big)t + \sigma W_t \quad \Leftrightarrow \quad S_t = S_0 e^{\sigma W_t + (b - \frac{1}{2}\sigma^2)t}\,.$$

Next, imagine that the period-by-period risk-neutral probabilities that we used above all obtain from a single probability measure, Q, which is defined on $\mathcal{F}$ and is such that $Q \approx P$. The speculation here is that Q provides the one-period risk-neutral probabilities $(\tilde{p}, \tilde{q})$ for all periods, all states, and all timesteps Δ. One can then imagine – contrary to what we know – that the convergence in e_4 takes place Q-a.s. Since $Q \approx P$, convergence Q-a.s. is no different from convergence P-a.s., and this means that the process on the right side of e_4 is a.s. the same as the process on the right side of e_3. As a result,

$$\big(b - \frac{1}{2}\sigma^2\big)t + \sigma W_t = \big(r - \frac{1}{2}\sigma^2\big)t + \sigma \tilde{W}_t \quad \Leftrightarrow \quad \tilde{W}_t = W_t + \frac{b-r}{\sigma}t\,.$$

Thus, formally, the measure Q can be characterized by the fact that, while W is a Brownian motion under P, the process $\tilde{W} = W + \frac{b-r}{\sigma}\iota$ must be a Brownian motion under Q. To put it another way, switching from the measure P to the measure Q eliminates the effect of the drift $\frac{b-r}{\sigma}\iota$. The mathematics of this transformation will be developed later (and the term "drift" needs an explanation, too), but let us always keep in mind the underlying intuition: because asset prices reflect agents' preferences and attitudes toward risk, in equilibrium, future risky payoffs will be priced by way of averaging under a different probability measure, Q, which is equivalent to P, and the Radon–Nikodym derivative $\mathrm{d}Q/\mathrm{d}P$ is intimately related to the stochastic discount factors (which, in turn, are intimately related to the agents' risk preferences). It just so happens that in complete markets this connection has yet another interpretation: the measure $Q \approx P$ removes the drift $\frac{b-r}{\sigma}\iota$, and this has the effect of making the average (under Q) of all risky returns identical to the risk-free return.

8.3 Construction of Brownian Motion and First Properties

Although at this point we are aware$^{\multimap (8.13)}$ that a mathematical model of Brownian motion exists, so far we have only been able to deduce the existence of such a model from$^{\multimap (8.5)}$ the fairly abstract setting of Kolmogorov's extension theorem, and, as was already noted, such a construction is somewhat unsatisfactory. It should be clear by now that the explicit construction of a mathematical model of Brownian motion, especially a construction that can facilitate a deeper study of its features, is a highly nontrivial project. This project was completed – to the extent to which it can be completed within the paradigm of contemporary mathematics – in the course of many years in the works of Norbert Wiener, Albert Einstein, Paul Lévy, Andrei Kolmogorov, Richard Feynman, and many of their followers. What follows in this section is a cursory look at some of the key steps in their work.

(8.15) GAUSSIAN PROCESS: A Gaussian stochastic process is any $\mathbb{R}$-valued stochastic process the finite dimensional distributions of which are$^{\multimap (3.50)}$ multivariate Gaussian. Equivalently, $X = (X_t)_{t\in\mathbb{R}_+}$ is a Gaussian process, if, treated as a family of random variables, it is a Gaussian family in the sense of (3.50). We say that X is *centered* if $\mathsf{E}[X_t] = 0$ for all $t \in \mathbb{R}_+$. ○

(8.16) COVARIANCE FUNCTION: Let $X = (X_t)_{t\in\mathbb{R}_+}$ be any real stochastic process such that $\mathsf{E}[X_t^2] < \infty$ and $\mathsf{E}[X_t] = 0$ for every $t \in \mathbb{R}_+$. The *covariance function* of X is the function

$$\mathbb{R}_+ \times \mathbb{R}_+ \ni (s,t) \rightsquigarrow C(s,t) \overset{\text{def}}{=} \mathsf{E}[X_s X_t]. \quad ○$$

(8.17) EXERCISE: Prove that, as defined in (8.16), the covariance function $C(\cdot,\cdot)$ has the following two attributes: (a) it is symmetric, in that $C(s,t) = C(t,s)$ for any $s,t \in \mathbb{R}_+$, and (b) it is positive definite, in that for every finite subset $\{t_1,\dots,t_n\} \subset$

$\mathbb{R}_+$ the matrix with entries $C(t_i,t_j)$, $1 \leq i,j \leq n$, is a positive definite (that is, nonnegative definite) matrix with dimensions (n,n). ○

(8.18) EXERCISE: Let $\mathbb{R}_+ \times \mathbb{R}_+ \ni (s,t) \rightsquigarrow C(s,t) \in \mathbb{R}$ be any function that is symmetric and positive definite (satisfies (a) and (b) from (8.17)). Prove that a *Gaussian* process $X = (X_t)_{t\in\mathbb{R}_+}$, such that $\mathsf{E}[X_t] = 0$ and $\mathsf{E}[X_s X_t] = C(s,t)$ for every $s,t \in \mathbb{R}_+$, exists.

HINT: The function $C(\cdot,\cdot)$ completely determines the finite dimensional distributions of X. One only has to check that those finite dimensional distributions are consistent. ○

(8.19) EXERCISE: Prove that the Brownian motion W is a Gaussian process with $\mathsf{E}[W_t] = 0$ and with $\mathsf{E}[W_s W_t] = s \wedge t$ for any $s,t \in \mathbb{R}_+$ (or $s,t \in [0,T]$). ○

(8.20) EXERCISE: Prove that the converse to the statement in exercise (8.19) is also true: if X is a Gaussian process such that $\mathsf{E}[X_t] = 0$ and $\mathsf{E}[X_s X_t] = s \wedge t$ for any $s,t \in \mathbb{R}_+$, then X must be a Brownian motion. ○

The next exercise offers an interesting alternative construction of a Brownian motion.

(8.21) EXERCISE: Let $(\xi_i)_{i\in\mathbb{N}}$ be any infinite sequence of i.i.d. $\mathcal{N}(0,1)$ random variables and let $(\varphi_i)_{i\in\mathbb{N}}$ be any$^{\multimap(A.65)}$ orthonormal basis in the Hilbert space $L^2([0,T]) \equiv L^2([0,T],\mathcal{B}([0,T]),\Lambda)$, for some fixed $T \in \mathbb{R}_{++}$. Prove that

$$W_t \overset{\text{def}}{=} \sum_{i\in\mathbb{N}} \xi_i \left(\int_0^t \varphi_i(s)\,\mathrm{d}s \right), \quad t \in [0,T],$$

is a Brownian motion.

HINT: The fact that $(\varphi_i)_{i\in\mathbb{N}}$ is an orthonormal basis in $L^2([0,T])$ means that:

(a) $\int_0^T \varphi_i(s)^2\,\mathrm{d}s = 1$;

(b) $\int_0^T \varphi_i(s)\varphi_j(s)\,\mathrm{d}s = 0$ for $i \neq j$;

(c) any $f \in L^2([0,T])$ can be expressed as $f = \sum_{i\in\mathbb{N}} a_i\varphi_i$, the convergence of the series being understood in the norm of $L^2([0,T])$, where

$$a_i = \int_0^T f(s)\,\varphi_i(s)\,\mathrm{d}s\,, \quad i \in \mathbb{N}\,.$$

It would be enough to show that $\mathsf{E}[W_s W_t] = \int_0^T 1_{[0,s]}(u)1_{[0,t]}(u)\,\mathrm{d}u$. ○

According to the last exercise, a Brownian motion on $[0,T]$ can be constructed on any probability space that is rich enough to support an i.i.d. sequence of standard normal random variables. One such probability space, among many other choices, is the product of countably infinitely many copies of the real line $\mathbb{R}$, endowed with the infinite product of identical probability measures on $\mathbb{R}$, every one of which can be identified as $\mathcal{N}(0,1)$.

(8.22) EXERCISE: Let $W = (W_t)_{t \in \mathbb{R}_+}$ be the Brownian motion. Prove that:

(a) for any fixed $t \in \mathbb{R}_+$, $(W_{t+s} - W_t)_{s \in \mathbb{R}_+}$ is a Brownian motion that is independent of the Brownian motion $(W_s)_{s \in [0,t]}$;

(b) $(-W_t)_{t \in \mathbb{R}_+}$ is a Brownian motion (symmetry);

(c) $(a\, W_{t/a^2})_{t \in \mathbb{R}_+}$ is a Brownian motion for any $a \in \mathbb{R}_{++}$ (the *scaling property* of Brownian motion). ○

(8.23) EXERCISE: Prove that if $(W_t)_{t \in [0,1]}$ is a Brownian motion, then the process $(W_{1-t} - W_1)_{t \in [0,1]}$ also must be a Brownian motion. ○

(8.24) KOLMOGOROV'S CONTINUITY CRITERION:[89] (Revuz and Yor 1999, I (1.9)) Let $(\Omega, \mathcal{F}, P)$ be any probability space and let X be any stochastic process defined on it. If one can find constants $\alpha, \beta, C \in \mathbb{R}_{++}$ such that

$$\mathsf{E}\left[|X_{t+h} - X_t|^{\alpha}\right] \leq C h^{1+\beta} \quad \text{for all } t, h \in \mathbb{R}_{++},$$

then there is a P-a.s. continuous process $\tilde{X}$, defined on $(\Omega, \mathcal{F}, P)$, which is a modification of X ($\tilde{X}_t = X_t$ P-a.s. for every $t \in \mathbb{R}_+$). ○

(8.25) EXERCISE: Using the continuity theorem above, prove that every Brownian motion process admits an a.s. continuous modification (on the same probability space on which it is defined). ○

(8.26) BROWNIAN MOTION (FINAL DEFINITION): A stochastic process, W, on the time domain $\mathbb{R}_+$, or on $[0, T]$ for some $T \in \mathbb{R}_{++}$, is said to be a *Brownian motion* if it is a Brownian motion in the sense of definition (8.11) and, in addition, has a.s. continuous samples paths.

REMARK: If the sample paths are required to be a.s. continuous, then the requirement in condition (b) from (8.11) for the increments to be Gaussian becomes redundant; specifically, it would be enough to replace (b) with the requirement that the law of $W_t - W_s$ (whatever its nature) depends only on the difference $(t - s)$ (conditions (a) and (c) are still needed). This will become clear later, in the study of Lévy processes in chapter 16. ○

(8.27) LINEAR BROWNIAN MOTION: A stochastic process W, labeled by $\mathbb{R}_+$, or by $[0, T]$ for some $T \in \mathbb{R}_{++}$, is said to be a linear Brownian motion if the r.v. W_0 is equal to some fixed scalar value from $\mathbb{R}$ and the process $W - W_0 = (W_t - W_0)_{t \in \mathbb{R}_+}$ is a Brownian motion. For example, if B is a Brownian motion in the sense of (8.26) and $x \in \mathbb{R}$, then the process $W = x + B$ is a linear Brownian motion starting from x. ○

(8.28) EXERCISE: Using Monte Carlo simulation, find (empirically) the distribution of the r.v. $\int_0^1 W_s \, \mathrm{d}s$, where W is the Brownian motion. Explain why this integral is a well-defined r.v. ○

---

[89] Named after the Russian mathematician Andrei Nikolaevich Kolmogorov (1903–1987).

(8.29) EXERCISE: Given a Brownian motion W, let $B_t \overset{\text{def}}{=} tW_{1/t}$ for any $t \in \mathbb{R}_{++}$ and set $B_0 \overset{\text{def}}{=} 0$. Prove that $B = (B_t)_{t \in \mathbb{R}_+}$ is a Brownian motion. This is the so-called *time reversal property* of Brownian motion.

HINT: The only nontrivial part is to show that the sample paths of B are a.s. continuous at 0. Consider using the following line of reasoning. First, show that for a function $f : \mathbb{R}_{++} \mapsto \mathbb{R}$ the convergence $\lim_{x \searrow 0} f(x) = 0$ is equivalent to $\lim_{n \to \infty} \left(\sup_{x \in]0,1/n]} |f(x)| \right) = 0$ and observe that if the function is continuous on $]0, 1]$, then the sequence

$$a_n \overset{\text{def}}{=} \left(\sup_{x \in]0,1/n]} |f(x)| \right) \equiv \left(\sup_{x \in]0,1/n] \cap \mathbb{Q}} |f(x)| \right), \quad n \in \mathbb{N},$$

is monotonically decreasing as $n \to \infty$. Then use the fact that $(B_t)_{t \in \mathbb{Q}_{++}}$ and $(W_t)_{t \in \mathbb{Q}_{++}}$ are equivalent families of r.v. and also the fact that $\lim_{t \searrow 0} W_t = 0$ a.s. Consult (4.7) and (4.11). ○

8.4 The Wiener Measure

Since the space of continuous sample paths will crop up regularly in our study, it deserves to be given a simpler name: we will denote it by $\mathbb{W} \overset{\text{def}}{=} \mathscr{C}(\mathbb{R}_+; \mathbb{R})$. Every a.s. continuous real-valued process X, defined on $(\Omega, \mathcal{F}, P)$, can be treated as a random variable that takes values in $\mathbb{W}$, that is, as a mapping of the form

$$\Omega \ni \omega \rightsquigarrow X_\omega \in \mathbb{W}, \qquad e_1$$

defined P-a.e. in Ω. If we can point to a natural measurable structure on the path space $\mathbb{W}$ and if the mapping in e_1 respects that structure, then X would transfer the probability measure P from Ω onto $\mathbb{W}$. The resulting measure $P \circ X^{-1}$ would be nothing but the distribution law of X, treated as a $\mathbb{W}$-valued r.v. In particular, if W is the Brownian motion, then the measure $\mathscr{W} \overset{\text{def}}{=} P \circ W^{-1}$ (it will be made meaningful shortly) is the celebrated Wiener measure on the path space $\mathbb{W}$. This measure is supported by the set of sample paths that start from 0, namely,

$$\mathbb{W}_0 \overset{\text{def}}{=} \left\{ f \in \mathbb{W} : f(0) = 0 \right\}.$$

To be able to develop the structures that we just described, we must first endow $\mathbb{W}$ with an appropriate measurable structure. Since $\mathbb{W}$ is a Polish space with respect to$^{\multimap (0.8)}$ the locally uniform distance, a choice that immediately comes to mind is the Borel structure associated with that distance (notice that $\mathbb{W}_0$ is closed, and therefore Borel, set). With a little bit of work, the Borel σ-field on $\mathbb{W}$ – which we denote, as usual, by $\mathcal{B}(\mathbb{W})$ – can be identified also as the σ-field generated by the canonical coordinate mappings

$$\mathbb{W} \ni f \rightsquigarrow \epsilon_t(f) = f(t) \in \mathbb{R}, \quad t \in \mathbb{R}_+ .$$

Unfortunately, the explicit construction of measures on this σ-field is not easy and involves tools that would be too much of a detour for us to start developing.

The reason for this difficulty is that although $\mathbb{W}$ has a "natural" distance, namely $d_{\text{loc},\infty}(\cdot,\cdot)$, this distance does not have the invariance properties that the Euclidean distance has, and, as a result, it is not possible to construct a measure on $\mathbb{W}$ that would be translation invariant, as the Lebesgue measure on $\mathbb{R}^n$ is. To put it another way, there is simply no measure on $\mathbb{W}$ that is immediately obvious and intuitive, as is the case in $\mathbb{R}^n$, which could allow us to build other measures in terms of densities, as we did in $\mathbb{R}^n$. At the same time, as we have already discovered, the Brownian motion has a number of interesting and useful symmetry properties that its distribution must reflect.

Perhaps the main stumbling block in constructing the Brownian distribution law $\mathscr{W}$ (the Wiener measure) was that many mathematicians and physicists did not give up on the pursuit of some sort of a canonical "uniform measure" on $\mathscr{W}$, relative to which the distribution of the Brownian motion can be defined – more or less in the way the Gaussian distribution is defined in terms of its density with respect to the Lebesgue measure. After completing exercise (8.13) we find that the density of $(W_{t_1}, \ldots, W_{t_n})$ is given by

$$\phi(x_1, \ldots, x_n) = \frac{1}{\sqrt{(2\pi)^n t_1(t_2-t_1)\ldots(t_n-t_{n-1})}} e^{-\frac{x_1^2}{2t_1}} e^{-\frac{(x_2-x_1)^2}{2(t_2-t_1)}} \cdots e^{-\frac{(x_n-x_{n-1})^2}{2(t_n-t_{n-1})}} \qquad \mathrm{e}_2$$

and now we see why R. Feynman wrote – formally – the distribution law of W as some sort of an "infinite dimensional Gaussian density" relative to some sort of a "uniform measure" $\mathrm{d}f$, understood to be the analogue of $\mathrm{d}x_1 \ldots \mathrm{d}x_n$ (though, of course, no such analogue exists), namely:

$$\mathrm{d}\mathscr{W} = \left(\prod\nolimits_{t\in\mathbb{R}_+}(2\pi\,\mathrm{d}t)\right)^{-\frac{1}{2}} e^{-\int_0^\infty |\partial f(t)|\,\mathrm{d}t} \text{“}\mathrm{d}f\text{”}. \qquad \mathrm{e}_3$$

It is remarkable that an expression that is as meaningless as the one above can still be full of insight: neither the product of uncountably many copies of the Lebesgue measure $\mathrm{d}t$, nor the measure on the space of functions "$\mathrm{d}f$" can be understood in any meaningful way, and, as we will soon realize, with probability 1 the sample paths of the Brownian motion do not admit a derivative at any point, which renders the quantity $\partial f(t)$ in e_3 meaningless. Yet the expression in e_3 still captures the heuristic view that the distribution law of the Brownian motion arises as the limit – of some sort – of well-understood finite dimensional distributions, namely the distributions in e_2, more or less in the way the common notation for the Riemann integral captures the heuristic view that this integral gives the sum of the areas of "infinitely thin" rectangles. Somehow, the expression in e_3 is still meaningful as a limit, even if most components of that expression have no meaning on their own. To put it another way, the failures of the various terms in e_3 to exist magically cancel out, and the whole expression is still meaningful, as a measure that does exist and can indeed be constructed.

In any case, the matter of existence of Brownian motion was never fully settled until it was established that there is, after all, a probability measure, called $\mathscr{W}$, on the Borel σ-field $\mathcal{B}(\mathbb{W})$, such that the coordinate mappings $(\epsilon_t)_{t\in\mathbb{R}_+}$ constitute a Brownian motion (alias: Wiener process) on the probability space

$$\big(\mathbb{W}, \mathcal{B}(\mathbb{W}), \mathscr{W}\big).$$

This was achieved by Wiener (1923), which is why the above probability space is called the *Wiener space*, and $\mathscr{W}$ is called the *Wiener measure*.[90] However, Wiener's work was somewhat ahead of its time, and preceded by at least one whole decade the invention of modern probability theory. Thus, at the time it was discovered his result was yet to be restated in the probabilistic terminology that we use today. The search for such a restatement became the subject of an intense research, which was essentially concluded by Nelson (1967).

Before we continue further, we must note that in many applications one needs to work with a multivariate Brownian motion. Given some $d \in \mathbb{N}_{++}$, a d-dimensional Brownian motion (alias: Brownian motion in $\mathbb{R}^d$), is any vector process of the form

$$(W_t^1, \dots, W_t^n) \in \mathbb{R}^d, \quad t \in \mathbb{R}_+,$$

the components of which are independent (scalar) Brownian motions. Its sample paths are distributed in the space $\mathbb{W}^d \stackrel{\text{def}}{=} \mathscr{C}(\mathbb{R}_+;\mathbb{R}^d)$; in fact, they are distributed in

$$\mathbb{W}_0^d \stackrel{\text{def}}{=} \{f \in \mathbb{W}^d : f(0) = 0\}.$$

The distribution law of the d-dimensional Brownian motion in the space $\mathbb{W}^d$, treated as a probability measure on $\mathcal{B}(\mathbb{W}^d)$, is the d-dimensional Wiener measure $\mathscr{W}^d$. In order to keep the notation simple, we will often omit the superscript and write $\mathbb{W}$ and $\mathscr{W}$ instead of $\mathbb{W}^d$ and $\mathscr{W}^d$, since in most cases the dimension of the Brownian motion will be implied by the context.

8.5 Filtrations, Stopping Times, and Such

Our main objective in this section is to introduce the continuous-time counterparts of the notions "filtration," "process," "adapted," "stopping time," and all other related structures that were developed earlier in the context of stochastic processes in discrete time. Despite the continuous-time setting that we are about to introduce, the underlying intuition and motivation remain the same: all practically meaningful random moments in time, or collections of random variables labeled by time, must be coordinated with the information arrival dynamics.[91]

[90] Named after the American mathematician Norbert Wiener (1894–1964).

[91] In the present context, the term "information," which we never formally introduced, is understood as a particular collection of random events, for every one of which there is sufficient knowledge to determine whether it has been realized or not. The arrival of new information can then be thought of as the addition of new events to those already categorized as "verifiable."

(8.30) FILTRATIONS: A filtration (aliases: information structure, information arrival dynamics) on the measurable space $(\Omega, \mathcal{F})$ is any family of sub-σ-fields $\mathscr{F} = \{\mathscr{F}_t \subseteq \mathcal{F} : t \in \bar{\mathbb{R}}_+\}$ that is increasing, in that $\mathscr{F}_s \subseteq \mathscr{F}_t \subseteq \mathscr{F}_\infty$ for every choice of $s, t \in \mathbb{R}_+$ with $s \leq t$. An increasing family of σ-fields, $(\mathscr{F}_t)_{t \in \mathbb{R}_+}$, can always be amended by the σ-field$^{\multimap(1.23)}$ $\mathscr{F}_\infty \overset{\text{def}}{=} \vee_{t \in \mathbb{R}_+} \mathscr{F}_t$, or by $\mathscr{F}_\infty \overset{\text{def}}{=} \mathcal{F}$. The default assumption is that $\mathscr{F}_\infty = \vee_{t \in \mathbb{R}_+} \mathscr{F}_t$. The fact that $\mathscr{F}$ is a filtration contained in the σ-field $\mathcal{F}$ will be expressed as $\mathscr{F} \setminus\!\subset \mathcal{F}$. Given a filtration $\mathscr{F}$, the left and right filtrations of $\mathscr{F}$, denoted, respectively, by $\mathscr{F}_-$ and $\mathscr{F}_+$, are given by

$$\mathscr{F}_{t-} \overset{\text{def}}{=} \vee_{s \in [0,t[} \mathscr{F}_s \quad \text{and} \quad \mathscr{F}_{t+} \overset{\text{def}}{=} \cap_{s \in]t,\infty[} \mathscr{F}_s \quad t \in \mathbb{R}_+ ,$$

with the understanding that $\mathscr{F}_{0-} = \mathscr{F}_0$.

Heuristically speaking, the σ-field $\mathscr{F}_t$ consists of the events that precede the moment t, in that whether an event $A \in \mathscr{F}_t$ has occurred or not can be determined from everything observed up until and including time t, whereas $\mathscr{F}_{t-}$ consists of the events that precede t strictly. Thus, a filtration, $\mathscr{F}$, can be thought of as a model of information arrival dynamics in real time, or, to put it another way, as a mathematical metaphor for the "flow of time" (associated with a particular random environment). If $\mathscr{F}_t = \mathscr{F}_{t-}$ for every $t \in \mathbb{R}_+$ we say that $\mathscr{F}$ is left-continuous. If $\mathscr{F}_t = \mathscr{F}_{t+}$ for every $t \in \mathbb{R}_+$ we say that $\mathscr{F}$ is right-continuous. Figuratively speaking, right-continuity of the filtration means that the information arrival dynamics are such that "peeking a little bit into the future" following time t would not produce any new information that is not already available at time t. This is an assumption that we will often impose on the filtrations that show up in our study (and there are also some technical reasons for imposing this assumption, as will be detailed soon). ○

(8.31) NATURAL FILTRATIONS: A filtration that immediately comes to mind is the one generated by a stochastic process $(X_t)_{t \in \mathbb{R}_+}$. Since every r.v. X_t represents an observation received at time t, the random events that can be detected by observing the process X up until (and including) time t form a σ-field, which is the smallest σ-field that includes all random events associated with all random variables X_s for all $s \in [0, t]$. This σ-field can be expressed as

$$\mathscr{F}_t^X = \sigma\{X_s : s \in [0, t]\} .$$

It is the smallest σ-field that contains all random events $X_s^{-1}(B)$ for all choices of $s \in [0, t]$ and all choices of the set B from the Borel σ-field in the state space of X. Clearly, $\mathscr{F}^X \overset{\text{def}}{=} (\mathscr{F}_t^X)_{t \in \mathbb{R}_+}$ is a filtration. It represents the minimal "past" of X and is called the *natural filtration* of X. ○

Usually, decisions made in a random environment – the exercise of an American-style call option, for example – are made at random moments in time. Random time events are often attached to the realizations – gradually over time – of certain time-dependent configurations (trajectories), and now we need to formalize the requirement for a given time event to be "coordinated with the flow of time," in the sense that the event depends on the realized configuration, but it depends only on the portion of it that has been revealed until the time of the event. Heuristically speaking, if a particular configuration leads to a time event that occurs before time t, then all other configurations that are identical to it until time t must lead to the same exact time event. To put it another way, a time event that is "coordinated with the flow of time" depends on the (entire) observed configuration, but if one only wants to determine whether the time event occurs before time t or not, then one would only need to know the portion of the realized configuration revealed until time t (knowing the history until time t is enough to determine whether the time event has occurred before time t or not). We thus come to the following definition.

(8.32) STOPPING TIMES: Let $(\Omega, \mathcal{F}, P)$ be a probability space and let $\mathscr{F} \setminus\!\subset \mathcal{F}$ be some filtration inside $\mathcal{F}$. A *stopping time relative to* $\mathscr{F}$, or an $\mathscr{F}$*-stopping time*, or simply a *stopping time*, is any function $\mathcal{t}: \Omega \mapsto \bar{\mathbb{R}}_+$ such that $\{\mathcal{t} \leq t\} \in \mathscr{F}_t$ for every $t \in \mathbb{R}_+$. This condition can be expressed equivalently as $\mathcal{t}^{-1}([0,t]) \in \mathscr{F}_t$, or even as $\mathcal{t}^{-1}(\mathcal{B}([0,t])) \subseteq \mathscr{F}_t$ for every $t \in \mathbb{R}_+$. If the range space of the stopping time $\mathcal{t}$ is $\mathbb{R}_+$ instead of $\bar{\mathbb{R}}_+$, we say that $\mathcal{t}$ is a *finite stopping time*, and if the range space is some finite time interval $[0, T]$, we say that $\mathcal{t}$ is a *bounded stopping time*. ○

By definition, every stopping time is an $\mathscr{F}_\infty$-measurable random variable with values in $\bar{\mathbb{R}}_+$, but, as is clear from the preceding definition, there is more to the structure of a stopping time than the structure of a random variable: we insist that the measurability of $\mathcal{t}$ must be coordinated with the flow of time in both the range and the domain, in the sense that the random event $\{\mathcal{t} \in A\} \equiv \mathcal{t}^{-1}(A)$ is required to belong to $\mathscr{F}_t$ – not just to $\mathscr{F}_\infty$ or $\mathcal{F}$ – for every choice of $t \in \bar{\mathbb{R}}_+$ and every choice of $A \in \mathcal{B}([0,t])$.

(8.33) EXAMPLES OF STOPPING TIMES: Quite often stopping times arise as the "hitting time" of a particular set in the state space of a particular stochastic process. We are going to revisit shortly this aspect of stopping times in a more systematic manner, after introducing several new tools and concepts, but at this point, just for the purpose of illustration, it would suffice for us to consider a finite real process and its hitting time of the interval $[a, \infty[$, or of the interval $]\,a, \infty[$, for some fixed $a \in \mathbb{R}$ (nothing will change in what follows if we replace $[a, \infty[$ with a generic closed subset of $\mathbb{R}$ and replace $]\,a, \infty[$ with a generic open subset of $\mathbb{R}$). As we are about to see, there are some subtle differences between the hitting times of open sets and the hitting times of closed sets, and also between the hitting times for processes with continuous sample paths and for processes with right-continuous sample paths.

Let X be some continuous $\mathbb{R}$-valued process with natural filtration $\mathscr{F}^X$, and for some fixed $a \in \mathbb{R}$ consider the hitting time of $[a, \infty[$ by X, namely,

$$\mathcal{t}_a(\omega) \stackrel{\text{def}}{=} \inf\{t \in \mathbb{R}_+ : X_t(\omega) \geq a\} \equiv \inf\{t \in \mathbb{R}_+ : X_t(\omega) \in [a, \infty[\},$$

with the understanding that $\inf(\emptyset) = \infty$ (a convention that we will often follow), that is, $\mathscr{t}_a(\omega) = \infty$ if $X_t(\omega) < a$ for all $t \in \mathbb{R}_+$. Since $t \rightsquigarrow X_t(\omega)$ is a continuous function, it is not difficult to check that the event $\mathscr{t}_a(\omega) \leq t$ is the same as the event "$X_s(\omega) = a$ for some $s \in [0, t]$." Therefore (using again the continuity of the sample paths),

$$\{\mathscr{t}_a \leq t\} = \{\omega \in \Omega : \inf_{s \in \mathbb{Q} \cap [0,t]} |X_s - a| = 0\}$$

for every $t \in \mathbb{R}_+$. Since the r.v. $\inf_{s \in \mathbb{Q} \cap [0,t]} |X_s - a|$ is $\mathscr{F}_t^X$-measurable, we see that $\{\mathscr{t}_a \leq t\} \in \mathscr{F}_t^X$. Thus, the hitting time $\mathscr{t}_a$ is a stopping time relative to $\mathscr{F}^X$. We note that due to the continuity of the function $t \rightsquigarrow X_t(\omega)$, if $X_0 < a$, then

$$\inf\{t \in \mathbb{R}_+ : X_t(\omega) \geq a\} = \inf\{t \in \mathbb{R}_+ : X_t(\omega) = a\} .$$

In addition, if $\mathscr{t}_a(\omega) < \infty$, then $X_{\mathscr{t}_a(\omega)}(\omega) = a$.

Now suppose that X is only right-continuous and this time consider the hitting time of the open set $]a, \infty[\subset \mathbb{R}$ by X, namely,

$$\mathscr{t}_a(\omega) \stackrel{\text{def}}{=} \inf\{t \in \mathbb{R}_+ : X_t(\omega) > a\} \equiv \inf\{t \in \mathbb{R}_+ : X_t(\omega) \in]a, \infty[\} .$$

Clearly, $X_s(\omega) \leq a$ for any $s \in [0, \mathscr{t}_a(\omega)[$, and the event $\{\mathscr{t}_a(\omega) \geq t\}$ is the same as the event "$X_s(\omega) \leq a$ for any $s \in [0, t[$." Due to the right-continuity of the sample paths,

$$\{\mathscr{t}_a \geq t\} = \bigcap_{s \in \mathbb{Q} \cap [0,t[} \{X_s \leq a\} \in \mathscr{F}_t^X ,$$

which shows that $\{\mathscr{t}_a < t\} = \{\mathscr{t}_a \geq t\}^{\complement} \in \mathscr{F}_t^X$. Since this relation holds for any $t \in \mathbb{R}_+$, it is easy to see that $\{\mathscr{t}_a \leq t\} \in \mathscr{F}_{t+}^X$ for any $t \in \mathbb{R}_+$, that is, $\mathscr{t}_a$ is an $\mathscr{F}_+^X$-stopping time. Typically, the hitting times for right-continuous processes are going to be stopping times only if the filtration is right-continuous. This is also true for the hitting time of an open interval by a continuous process. In most cases, we are going to need to work with right-continuous filtrations in order to ensure that the hitting times are stopping times.

Apart from the hitting time of a continuous process to a closed interval, the only other interesting example of a hitting time that is a stopping time with respect to the natural filtration of the process is the following: Suppose that the sample paths of X are right-continuous and *increasing*, and again consider the hitting time of the closed set $[a, \infty[\subset \mathbb{R}$ by X, namely,

$$\mathscr{t}_a(\omega) \stackrel{\text{def}}{=} \inf\{t \in \mathbb{R}_+ : X_t(\omega) \geq a\} .$$

In this case the event $\{\mathscr{t}_a \leq t\}$ is the same as the event $\{X_t \geq a\}$, which obviously belongs to $\mathscr{F}_t^X$, so that $\mathscr{t}_a$ is an $\mathscr{F}_t^X$-stopping time (no need to go to $\mathscr{F}_+^X$).

We stress that most of the interesting and useful hitting times that we are going to encounter will not be stopping times automatically, without imposing any restrictions on the sample paths, the filtration, and the topology of the set. ○

(8.34) EXERCISE: Prove that if $\mathscr{t}\colon \Omega \mapsto [0,T]$ is any bounded stopping time for the filtration $\mathscr{F}$, then the mapping $\Omega \ni \omega \rightsquigarrow (\omega, \mathscr{t}(\omega)) \in \Omega\times[0,T]$ can be claimed to be $\big(\mathscr{F}_T \otimes \mathcal{B}([0,T])\big)/\mathscr{F}_T$-measurable. ○

(8.35) STOPPED STOCHASTIC PROCESS: If $\mathscr{t}$ is any stopping time and X is any stochastic process, then one can imagine the process obtained by freezing the sample paths of X at the random time $\mathscr{t}$. This stopped version of X is denoted by $X^{\mathscr{t}}$ and is given by

$$\mathbb{R}_+ \ni t \rightsquigarrow X^{\mathscr{t}}_\omega(t) = \begin{cases} X_\omega(t) & \text{if } t \leq \mathscr{t}(\omega)\,, \\ X_\omega(\mathscr{t}(\omega)) & \text{if } t > \mathscr{t}(\omega)\,, \end{cases} \qquad \omega \in \Omega\,.$$

We can express the process $X^{\mathscr{t}}$ more succinctly as $X^{\mathscr{t}}_t = X_{t\wedge\mathscr{t}}$, $t \in \mathbb{R}_+$, where the random variable $X_{t\wedge\mathscr{t}}$ is understood as $\omega \rightsquigarrow X(\omega, t\wedge\mathscr{t}(\omega))$. ○

(8.36) EVENTS THAT OCCUR BEFORE A STOPPING TIME: A random time event can be thought of as a collection of possible realizations – revealed gradually over time – of a certain time-dependent and random configuration (a random sample path, say). Heuristically speaking, the claim that a particular random event, A, occurs before the stopping time $\mathscr{t}$ boils down to the claim that whether a particular realization belongs to A or not depends only on the portion of it revealed until time $\mathscr{t}$. In particular, looking a particular realization that leads to the time-event $\mathscr{t}$ occurring before time t, whether or not such a realization also belongs to A should be possible to determine only from the portion of it revealed before time t. Motivated by this observation, we now define the collection of random events that occur before the stopping time $\mathscr{t}$ as

$$\mathscr{F}_{\mathscr{t}} \overset{\text{def}}{=} \big\{A \in \mathscr{F}_\infty\colon A \cap \{\mathscr{t} \leq t\} \in \mathscr{F}_t\,,\ \text{for all } t \in \mathbb{R}_+\big\}\,.$$

The collection of random events that occur strictly before the stopping time $\mathscr{t}$ we define as the σ-field

$$\mathscr{F}_{\mathscr{t}-} \overset{\text{def}}{=} \mathscr{F}_0 \vee \sigma\big\{A \cap \{\mathscr{t} > t\}\colon t \in \mathbb{R}_+\,,\ A \in \mathscr{F}_t\big\}\,.$$

The interpretation of the events $A\cap\{\mathscr{t} > t\}$, $A \in \mathscr{F}_t$, is this: among the realizations leading to the time-event $\mathscr{t}$ occurring strictly after time t, we select subsets membership to which depends only on the portion of the realization revealed before time $t < \mathscr{t}$. In addition, we are going to need the following collection of events:

$$\mathscr{F}_{\mathscr{t}+} \overset{\text{def}}{=} \big\{A \in \mathscr{F}_\infty\colon A \cap \{\mathscr{t} \leq t\} \in \mathscr{F}_{t+}\,,\ \text{for all } t \in \mathbb{R}_+\big\}\,.$$

This definition immediately gives the inclusion $\mathscr{F}_{\mathscr{t}} \subseteq \mathscr{F}_{\mathscr{t}+}$. ○

(8.37) EXAMPLE: Let W be the Brownian motion and let $X \overset{\text{def}}{=} e^{W-\frac{1}{2}\iota}$. Consider the stopping times

$$\mathscr{s} \overset{\text{def}}{=} \inf\big\{t \in \mathbb{R}_+\colon X_t \geq 1\big\} \quad \text{and} \quad \mathscr{t} \overset{\text{def}}{=} \inf\big\{t \in \mathbb{R}_+\colon X_t \geq 2\big\}\,.$$

It is intuitive that for any fixed $T \in \mathbb{R}_{++}$, the event $\{\mathscr{s} \leq T\}$ must occur before the stopping time $\mathscr{t}$, that is, must belong to $\mathscr{F}^W_{\mathscr{t}}$. This is indeed the case, as we now verify. We need to show that

$$A_t \overset{\text{def}}{=} \{\mathcal{s} \le T\} \cap \{\mathcal{t} \le t\} \in \mathcal{F}_t^W \quad \text{for every } t \in \mathbb{R}_+ .$$

If $t \le T$, then $\{\mathcal{t} \le t\} \subseteq \{\mathcal{s} \le T\}$, so that $A_t = \{\mathcal{t} \le t\} \in \mathcal{F}_t^W$. If $t > T$, then $\{\mathcal{s} \le T\} \in \mathcal{F}_T^W \subset \mathcal{F}_t^W$ and it is again the case that $A_t \in \mathcal{F}_t^W$. ○

(8.38) EXERCISE: Prove that for any stopping time $\mathcal{t}$ the families $\mathcal{F}_{\mathcal{t}}$ and $\mathcal{F}_{\mathcal{t}+}$ defined above are both σ-fields. ○

(8.39) EXERCISE: Prove that for any fixed $t \in \mathbb{R}_+$ the deterministic time $\mathcal{t}(\omega) = t$, $\omega \in \Omega$, is a stopping time and that, with this special choice of $\mathcal{t}$, $\mathcal{F}_{\mathcal{t}} = \mathcal{F}_t$. ○

(8.40) EXERCISE: Prove that $\mathcal{F}_{\mathcal{t}-} \subseteq \mathcal{F}_{\mathcal{t}}$ for any stopping time $\mathcal{t}$.

HINT: It is enough to show that $A \cap \{\mathcal{t} > t\} \in \mathcal{F}_{\mathcal{t}}$ for any $t \in \mathbb{R}_+$ and any $A \in \mathcal{F}_t$. If $s \in \mathbb{R}_{++}$ is chosen so that $s \le t$, then $A \cap \{\mathcal{t} > t\} \cap \{\mathcal{t} \le s\} = \emptyset$. If $s \in \mathbb{R}_{++}$ is chosen so that $s > t$, then $A \in \mathcal{F}_t \subseteq \mathcal{F}_s$, $\{\mathcal{t} > t\} \in \mathcal{F}_t \subseteq \mathcal{F}_s$, and $\{\mathcal{t} \le s\} \in \mathcal{F}_s$ Thus, $A \cap \{\mathcal{t} > t\} \cap \{\mathcal{t} \le s\} \in \mathcal{F}_s$ again. ○

(8.41) EXERCISE: Prove that if $\mathcal{t}$ is a stopping time, then the random variable $\omega \rightsquigarrow \mathcal{t}(\omega)$ is $\mathcal{F}_{\mathcal{t}-}$-measurable, and therefore also $\mathcal{F}_{\mathcal{t}}$-measurable.

HINT: By the definition of stopping time, the event $A \overset{\text{def}}{=} \{\mathcal{t} > t\}$ is in $\mathcal{F}_t$. By the definition of $\mathcal{F}_{\mathcal{t}-}$, the event $A \cap \{\mathcal{t} > t\} = \{\mathcal{t} > t\}$ must be in $\mathcal{F}_{\mathcal{t}-}$ ○

(8.42) EXERCISE: Prove that if $\mathcal{t}$ is a stopping time, then $\{\mathcal{t} = \infty\} \in \mathcal{F}_{\mathcal{t}-}$. Conclude that $\{\mathcal{t} < \infty\} \in \mathcal{F}_{\mathcal{t}-}$. ○

(8.43) EXERCISE: Let $\mathcal{s}$ and $\mathcal{t}$ be any two $\mathcal{F}$-stopping times such that $\mathcal{s} \le \mathcal{t}$. Prove that $\mathcal{F}_{\mathcal{s}} \subseteq \mathcal{F}_{\mathcal{t}}$ and $\mathcal{F}_{\mathcal{s}-} \subseteq \mathcal{F}_{\mathcal{t}-}$.

HINT: If $A \in \mathcal{F}_{\mathcal{s}}$ and $\mathcal{s} \le \mathcal{t}$, then $A \cap \{\mathcal{t} \le t\} = A \cap \{\mathcal{s} \le t\} \cap \{\mathcal{t} \le t\}$. Also, if $A \in \mathcal{F}_t$ and $\mathcal{s} \le \mathcal{t}$, then $A \cap \{\mathcal{s} > t\} = A \cap \{\mathcal{s} > t\} \cap \{\mathcal{t} > t\}$, where $\{\mathcal{s} > t\} \in \mathcal{F}_t$. ○

(8.44) EXERCISE: Prove that if $\mathcal{s}$ and $\mathcal{t}$ are stopping times, then $\mathcal{s} \wedge \mathcal{t}$ and $\mathcal{s} \vee \mathcal{t}$ are stopping times.

HINT: $\{\mathcal{s} \wedge \mathcal{t} \le t\} = \{\mathcal{s} \le t\} \cup \{\mathcal{t} \le t\}$, $\{\mathcal{s} \vee \mathcal{t} \le t\} = \{\mathcal{s} \le t\} \cap \{\mathcal{t} \le t\}$. ○

(8.45) EXERCISE: Prove that if $\mathcal{s}$ is a stopping time and the r.v. $\mathcal{t} \prec \mathcal{F}_{\mathcal{s}}$ is such that $\mathcal{t} \ge \mathcal{s}$, then $\mathcal{t}$ is a stopping time.

HINT: For $t \in \mathbb{R}_+$, $\{\mathcal{t} \le t\} \subseteq \{\mathcal{s} \le t\}$, so that $\{\mathcal{t} \le t\} = \{\mathcal{t} \le t\} \cap \{\mathcal{s} \le t\}$, where $\{\mathcal{t} \le t\} \in \mathcal{F}_{\mathcal{s}}$ and $\{\mathcal{s} \le t\} \in \mathcal{F}_t$. ○

(8.46) EXERCISE: Prove that if $\mathcal{s}$ and $\mathcal{t}$ are stopping times and $A \in \mathcal{F}_{\mathcal{s}}$, then

$$A \cap \{\mathcal{s} \le \mathcal{t}\} \in \mathcal{F}_{\mathcal{t}}, \quad A \cap \{\mathcal{s} < \mathcal{t}\} \in \mathcal{F}_{\mathcal{t}}, \quad \text{and} \quad A \cap \{\mathcal{s} = \mathcal{t}\} \in \mathcal{F}_{\mathcal{t}} .$$

Conclude that the events $\{\mathcal{s} \le \mathcal{t}\}$, $\{\mathcal{s} < \mathcal{t}\}$, and $\{\mathcal{s} = \mathcal{t}\}$, all belong to $\mathcal{F}_{\mathcal{s}} \cap \mathcal{F}_{\mathcal{t}}$ (for any two stopping times $\mathcal{s}$ and $\mathcal{t}$).

HINT: $A \cap \{\mathcal{s} \le \mathcal{t}\} \cap \{\mathcal{t} \le t\} = \big(A \cap \{\mathcal{s} \le t\}\big) \cap \{\mathcal{s} \wedge t \le \mathcal{t} \wedge t\} \cap \{\mathcal{t} \le t\}$ and, similarly, $A \cap \{\mathcal{s} < \mathcal{t}\} \cap \{\mathcal{t} \le t\} = \big(A \cap \{\mathcal{s} \le t\}\big) \cap \{\mathcal{s} \wedge t < \mathcal{t} \wedge t\} \cap \{\mathcal{t} \le t\}$. Verify the relations $\mathcal{s} \wedge t \prec \mathcal{F}_{\mathcal{s} \wedge t} \subseteq \mathcal{F}_t$ and $\mathcal{t} \wedge t \prec \mathcal{F}_{\mathcal{t} \wedge t} \subseteq \mathcal{F}_t$. ○

(8.47) EXERCISE: Let $\mathscr{t}$ be any stopping time and let $A \in \mathscr{F}_{\mathscr{t}}$. Prove that the mapping

$$\Omega \ni \omega \rightsquigarrow \mathscr{t}_A(\omega) \stackrel{\text{def}}{=} \begin{cases} \mathscr{t}(\omega) & \text{if } \omega \in A\,, \\ +\infty & \text{if } \omega \notin A\,, \end{cases}$$

is a stopping time. In particular, if $t \in \mathbb{R}_+$ is any deterministic time and $A \in \mathscr{F}_t$, then t_A is a (generally random) stopping time.

HINT: Explain why for any finite $t \in \mathbb{R}_+$, $\{\mathscr{t}_A \leq t\} = A \cap \{\mathscr{t} \leq t\}$. ○

(8.48) STOPPED FILTRATION: Given any filtration $\mathscr{F}$ and any $\mathscr{F}$-stopping time $\mathscr{t}$, one can always define the stopped filtration $\mathscr{F}^{\mathscr{t}} = (\mathscr{F}^{\mathscr{t}}_t \stackrel{\text{def}}{=} \mathscr{F}_{t \wedge \mathscr{t}})_{t \in \mathbb{R}_+}$. Notice that since $t \wedge \mathscr{t}$ is a stopping time dominated by t, we have⊸(8.43) $\mathscr{F}_{t\wedge\mathscr{t}} \subseteq \mathscr{F}_t$, so that the stopped filtration $\mathscr{F}^{\mathscr{t}}$ is always included in $\mathscr{F}$. ○

(8.49) EXERCISE: Prove that if $\mathscr{s}$ and $\mathscr{t}$ are any two stopping times for the filtration $\mathscr{F}$, then $\mathscr{s} \wedge \mathscr{t}$ is a stopping time for the filtration $\mathscr{F}^{\mathscr{t}}$.

HINT: You must show that

$$A \stackrel{\text{def}}{=} \{\mathscr{s} \wedge \mathscr{t} \leq t\} \cap \{t \wedge \mathscr{t} \leq s\} \in \mathscr{F}_s \quad \text{for every } \ s, t \in \mathbb{R}_+\,.$$

If $t \leq s$, then $A = \{\mathscr{s} \wedge \mathscr{t} \leq t\}$, and if $t > s$, then $A = \{\mathscr{t} \leq s\}$. ○

(8.50) EXERCISE: Let $\mathscr{t}$ be a stopping time for the filtration $\mathscr{F}$. Prove that if $\mathscr{F}$ is right-continuous, then the stopped filtration $\mathscr{F}^{\mathscr{t}}$ is right-continuous, and if $\mathscr{F}$ is complete, in the sense that $\mathscr{F}_0$ contains all P-null sets, then $\mathscr{F}^{\mathscr{t}}$ is complete in the same sense.

HINT: If $A \in \mathscr{F}^{\mathscr{t}}_{t+1/n}$ for any $n \in \mathbb{N}_{++}$, then $A \cap \{(t+1/n) \wedge \mathscr{t} \leq s\} \in \mathscr{F}_s$ for any $s \in \mathbb{R}_{++}$. Thus, for all $s \in \mathbb{R}_+$,

$$A \cap \{t \wedge \mathscr{t} \leq s\} = \bigcap\nolimits_{n \in \mathbb{N}_{++}} \big(A \cap \{(t + 1/n) \wedge \mathscr{t} \leq s\}\big) \in \mathscr{F}_s\,,$$

so that $A \in \mathscr{F}^{\mathscr{t}}_t$. If $\mathscr{F}_0$ contains all null sets, then so does $\mathscr{F}^{\mathscr{t}}_t \supseteq \mathscr{F}^{\mathscr{t}}_0 = \mathscr{F}_0$. ○

(8.51) SEQUENCES OF STOPPING TIMES:[92] If $(\mathscr{s}_i)_{i \in \mathbb{N}}$ is any sequence of stopping times for the filtration $\mathscr{F}$, then:

(a) $\sup_{i\in\mathbb{N}} \mathscr{s}_i$ is a stopping time for the filtration $\mathscr{F}$ and $\inf_{i\in\mathbb{N}} \mathscr{s}_i$ is a stopping time for the filtration $\mathscr{F}_+$;

(b) if the filtration $\mathscr{F}$ is right-continuous (as is going to be the case most often), then $\inf_{i\in\mathbb{N}} \mathscr{s}_i$, $\liminf_{i\in\mathbb{N}} \mathscr{s}_i$, and $\limsup_{i\in\mathbb{N}} \mathscr{s}_i$ are all stopping times for $\mathscr{F}$. If, in addition, the sequence $(\mathscr{s}_i)_{i\in\mathbb{N}}$ is decreasing, then the stopping time $\mathscr{t} \stackrel{\text{def}}{=} \inf_{i\in\mathbb{N}} \mathscr{s}_i$ is such that $\mathscr{F}_{\mathscr{t}} = \cap_{i\in\mathbb{N}} \mathscr{F}_{\mathscr{s}_i}$.[93] ○

(8.52) EXERCISE: Let $\mathscr{t}$ be any $\mathscr{F}$-stopping time. Prove that $\{\mathscr{t} < t\} \in \mathscr{F}_{t-} \subseteq \mathscr{F}_t$ for any fixed $t \in \mathbb{R}_+$, and, therefore, $\{\mathscr{t} = t\} \in \mathscr{F}_t$. In addition, prove that if

---

[92] These properties are not very difficult. Consider proving as an exercise. If needed, consult, e.g., (Dellacherie 1972).

[93] This property is one of the many reasons why in most situations we need to work with right-continuous filtrations.

the filtration $\mathscr{F}$ is right-continuous, then the condition $\{\mathcal{t} < t\} \in \mathscr{F}_t$ (which is equivalent to $\{t \leq \mathcal{t}\} \in \mathscr{F}_t$) for all $t \in \mathbb{R}_+$ implies $\{\mathcal{t} \leq t\} \in \mathscr{F}_t$ for all $t \in \mathbb{R}_+$.

HINT: $\{\mathcal{t} < t\} = \cup_{n \in \mathbb{N}_{++}} \{\mathcal{t} \leq t - \frac{1}{n}\}$ and $\{\mathcal{t} \leq t\} = \cap_{n \in \mathbb{N}_{++}} \{\mathcal{t} < t + \frac{1}{n}\}$. ○

One immediate corollary from the last exercise is that for a right-continuous filtration $\mathscr{F}$ the function $\mathcal{t} : \Omega \mapsto \bar{\mathbb{R}}_+$ is an $\mathscr{F}$-stopping time if and only if the process $\big(X_t = 1_{[0,\mathcal{t}]}(t) = 1_{\{t \leq \mathcal{t}\}}\big)_{t \in \mathbb{R}_+}$ is coordinated with $\mathscr{F}$, in the sense that $X_t \lessdot \mathscr{F}_t$ (X_t is measurable for $\mathscr{F}_t$) for every $t \in \mathbb{R}_+$. This is our first encounter with the notion of an "adapted process" in continuous time that we are about to introduce. There are other measurability matters that we need to address. We have already introduced⊶(8.2) one rather rudimentary type of measurability, which we called "joint measurability." However, this notion is not coordinated with the flow of time in any way. To get some insight into the issues that motivate our next definition, observe that the value of the process X at the random time $\mathcal{t}$ can be identified as the random variable

$$\Omega \ni \omega \rightsquigarrow X_{\mathcal{t}}(\omega) \stackrel{\text{def}}{=} X(\omega, \mathcal{t}(\omega)) \equiv X_{\mathcal{t}(\omega)}(\omega) \in \mathbb{S},$$

which may be viewed as the composition of the mappings

$$\Omega \ni \omega \rightsquigarrow (\omega, \mathcal{t}(\omega)) \ni \Omega \times \mathbb{R}_+ \quad \text{and} \quad X : \Omega \times \mathbb{R}_+ \mapsto \mathbb{S}.$$

We insist that $X_{\mathcal{t}}$ must be $\mathcal{F}$-measurable – and thus a legitimate r.v. on $(\Omega, \mathcal{F}, P)$ – but this measurability alone is not going to be enough. For example, it is not too difficult to foresee situations where we would like to be able to claim that the events of the r.v. $X_{\mathcal{t}}$ occur before (not necessarily strictly before) the random moment $\mathcal{t}$, that is, $X_{\mathcal{t}} \lessdot \mathscr{F}_{\mathcal{t}}$. It is clear that without any additional measurability conditions for X – other than joint measurability – this property may not hold in general. We introduce such conditions next.

(8.53) PROGRESSIVELY MEASURABLE SETS AND FUNCTIONS: Let $(\Omega, \mathcal{F}, P)$ be a probability space. If one is to give a σ-field to the product space $\Omega \times \mathbb{R}_+$, the σ-field that immediately comes to mind is the product σ-field $\mathcal{F} \otimes \mathcal{B}(\mathbb{R}_+)$. When a stochastic process $(\omega, t) \rightsquigarrow X(\omega, t)$ happens to be measurable for $\mathcal{F} \otimes \mathcal{B}(\mathbb{R}_+)$, we say⊶(8.2) that X is jointly measurable, but as was already noted, this type of measurability does not take into account the flow of time. If the probability space $(\Omega, \mathcal{F}, P)$ is given a filtration $\mathscr{F} \setminus\!\subset \mathcal{F}$, then a subset $B \subseteq \Omega \times \mathbb{R}_+$ is said to be *progressively measurable* (relative to $\mathscr{F}$) if $B \cap (\Omega \times [0, t]) \in \mathscr{F}_t \otimes \mathcal{B}([0, t])$ for every $t \in \mathbb{R}_+$. The collection of all progressively measurable subsets of $\Omega \times \mathbb{R}_+$ is easily seen to be a σ-field, which we denote by $\mathrm{Prog}(\mathscr{F})$ and call the *progressive σ-field* (for $\mathscr{F}$). A stochastic process $(\omega, t) \rightsquigarrow X(\omega, t)$ that is measurable for $\mathrm{Prog}(\mathscr{F})$ is said to be *$\mathscr{F}$-progressively measurable*, or simply *progressively measurable*, if the filtration $\mathscr{F}$ is understood. It is not difficult to check that the progressive measurability of X is equivalent to the claim that the restriction $X \restriction (\Omega \times [0, t])$ is $\mathscr{F}_t \otimes \mathcal{B}([0, t])$-measurable for every $t \in \mathbb{R}_+$. ○

In addition to progressive measurability, there is another – seemingly more straightforward and intuitive but, as we will soon realize, less useful – way to coordinate a stochastic process with the flow of time.

(8.54) ADAPTED PROCESSES: Let $X \equiv (X_t)_{t\in\mathbb{R}_+}$ be any stochastic process defined on $(\Omega, \mathcal{F}, P)$ with values in some Polish space $\mathbb{S}$ and let $\mathscr{F} \equiv (\mathscr{F}_t)_{t\in\mathbb{R}_+}$ be any filtration inside $\mathcal{F}$. We say "X is adapted to $\mathscr{F}$," "X is $\mathscr{F}$-adapted," or simply "X is adapted" if the filtration is understood, if $X_t \prec \mathscr{F}_t$ (the r.v. X_t is $\mathscr{F}_t$-measurable) for every $t \in \mathbb{R}_+$. To put it another way, "X is adapted to $\mathscr{F}$" means that $X_t^{-1}(B) \in \mathscr{F}_t$ for every $t \in \mathbb{R}_+$ and every $B \in \mathcal{B}(\mathbb{S})$. We will express this feature more succinctly as $X \prec \mathscr{F}$. ○

(8.55) EXERCISE: Prove that every $\mathscr{F}$-progressively measurable process is automatically $\mathscr{F}$-adapted and jointly measurable.

HINT: Consult exercise (2.8). ○

The main reason for inventing the notion "progressively measurable" is to be able to establish the following feature:

(8.56) EXERCISE: Prove that if X is any progressively measurable process for the filtration $\mathscr{F}$ and if $\mathcal{t}$ is any stopping time for the same filtration, then the random variable $\{\mathcal{t} < \infty\} \ni \omega \rightsquigarrow X_{\mathcal{t}}(\omega)$ is measurable for the σ-field $\mathscr{F}_{\mathcal{t}} \cap \{\mathcal{t} < \infty\}$.

HINT: You must show that the event

$$\{X_{\mathcal{t}} \in B\} \cap \{\mathcal{t} \le t\} = \{X_{t\wedge\mathcal{t}} \in B\} \cap \{\mathcal{t} \le t\}$$

is in $\mathscr{F}_t$ for every choice of $t \in \mathbb{R}_+$ and $B \in \mathcal{B}(\mathbb{S})$. Notice that $\mathcal{s} \stackrel{\text{def}}{=} t\wedge\mathcal{t}$ is a bounded stopping time with values in $[0,t]$ and that the r.v. $\omega \rightsquigarrow X_{\mathcal{s}(\omega)}(\omega)$ is the composition of the mappings $\Omega\times[0,t] \ni (\omega,s) \rightsquigarrow X(\omega,s)$ and $\Omega \ni \omega \rightsquigarrow (\omega, \mathcal{s}(\omega)) \in \Omega\times[0,t]$. Use exercise (8.34) and the definition of "progressively measurable." ○

(8.57) EXERCISE: Prove that if X is any progressively measurable process for the filtration $\mathscr{F}$ and if $\mathcal{t}$ is any $\mathscr{F}$-stopping time, then the stopped process $X^{\mathcal{t}}$ is progressively measurable as well.

HINT: The restriction $X^{\mathcal{t}} \restriction (\Omega\times[0,t])$ is the composition of the mappings

$$\Omega\times[0,t] \ni (\omega,s) \rightsquigarrow X(\omega,s) \quad \text{and} \quad \Omega\times[0,t] \ni (\omega,s) \rightsquigarrow (\omega, s\wedge\mathcal{t}(\omega)).$$

It is enough to establish that these mappings are $\mathscr{F}_t \otimes \mathcal{B}([0,t])$-measurable. For the first mapping, this property is clear. As for the second mapping, observe that

$$\begin{aligned}&\big\{(\omega,s) \in \Omega\times[0,t] : \omega \in A,\ s\wedge\mathcal{t}(\omega) \in [0,u]\big\}\\ &\quad = \Big(\big(A \cap \{\mathcal{t} \le u\}\big)\times[0,t]\Big) \cup \Big(\big(A \cap \{\mathcal{t} > u\}\big)\times[0,u]\Big)\end{aligned}$$

for any $u \in [0,t]$ and any $A \in \mathscr{F}_t$. Consult theorem (2.4) and exercise (2.5). ○

(8.58) EXERCISE: Explain why in the special case where the filtration $\mathscr{F}$ is such that $\mathscr{F}_t = \mathscr{F}_0$ for every $t \in \mathbb{R}_+$ the property "progressively measurable" is the same as "jointly measurable." ○

(8.59) PROGRESSIVE VS. ADAPTED: Although every progressively measurable process is automatically adapted, it is not very difficult to construct an adapted process that is not progressively measurable. Specifically, (Dellacherie 1972, III 2) provides the following construction, which we slightly modify. Let $\Omega \stackrel{\text{def}}{=} \mathbb{R}_+$ and let $\mathscr{F}_t = \mathscr{F}_0$ for every t (so that "progressively measurable" and "jointly measurable" become the same property) be the σ-field generated by all singletons $\{\omega\}, \omega \in \Omega \equiv \mathbb{R}_+$, and completed for the Lebesgue measure Λ. Notice that any nonempty finite sub-intervals of $\mathbb{R}_+$ would not be in $\mathscr{F}_0$; in fact, argue that the Lebesgue measure of any set that belongs to $\mathscr{F}_0$ must be either 0 or ∞. To see this, argue that the collection of all Borel sets that have measure either 0 or ∞ form a σ-field, and that all singletons belong to that collection. Now consider the process $X(\omega, t) \stackrel{\text{def}}{=} 1_{\{t\wedge 1\}}(\omega)$ and observe that the state space for this process is the discrete set $\{0, 1\} \subset \mathbb{R}_+$. Since for every fixed t the set $\{\omega \in \Omega\colon X_t(\omega) = 1\}$ is a singleton, this set must belong to $\mathscr{F}_0 = \mathscr{F}_t$. Thus, the process X is adapted. However, the set $\{(\omega, t) \in \mathbb{R}^2\colon X(\omega, t) = 1\}$ is nothing but

$$\big\{(\omega, t) \in \mathbb{R}^2\colon \omega \in [0, 1],\ t = \omega\big\} \cup \big\{(\omega, t) \in \mathbb{R}^2\colon \omega = 1\,,\ t \in\,]1, \infty[\big\}$$

and this set is not in $\mathscr{F}_0 \otimes \mathcal{B}(\mathbb{R})$, for if it were, by theorem (2.12) its projection onto Ω, which is nothing but the interval $[0, 1]$, should belong to $\mathscr{F}_0$, and it clearly does not, since the Lebesgue measure of this interval is neither 0 nor ∞. This is yet another situation in which the completion of the σ-fields involved is instrumental. ○

(8.60) "ADAPTED" IS NOT ENOUGH: We have already seen the importance of having the most elementary operations on measurable functions preserve measurability. However, the only measurability that can be preserved is relative to the σ-field for which all functions involved can be declared measurable. For example, if the $\mathbb{R}$-valued process X is jointly measurable, then, by Tonelli–Fubini's theorem, the random variable $\omega \rightsquigarrow \int_0^\infty X(\omega, s)^2\,\mathrm{d}s$ is going to be $\mathscr{F}_\infty$-measurable P-a.e. (that is, $\mathscr{F}_\infty$-measurable on some set $A \in \mathscr{F}_\infty$ such that $\mathsf{P}(A) = 1$). However, there would be no reason for the random variable $\omega \rightsquigarrow \int_0^t X(\omega, s)^2\,\mathrm{d}s$ to be $\mathscr{F}_t$-measurable P-a.e., even if the process X is required to be adapted, simply because Tonelli–Fubini's theorem cannot be applied on the product space $\Omega\times[0, t]$ – unless the restriction $X\restriction(\Omega\times[0, t])$ happens to be $\mathscr{F}_t \otimes \mathcal{B}([0, t])$-measurable, which is not guaranteed by the requirement for X to be adapted. We must have $X \in \mathrm{Prog}(\mathscr{F})$ to claim that the process $\int_0^t X(\omega, s)^2\,\mathrm{d}s$, $t \in \mathbb{R}_+$, is $\mathscr{F}$-adapted. ○

Nevertheless we have the following remarkable result.

(8.61) THEOREM: (Meyer 1966, IV 46) Every jointly measurable and adapted process with values in some Polish space $\mathbb{S}$ has a progressively measurable modification. ○

There are situations where, in conjunction with certain type of regularity of the sample paths, "jointly measurable" plus "adapted" implies "progressively measurable," as we now detail.

(8.62) EXERCISE: Prove that every adapted process, with values in some Polish space $\mathbb{S}$, that has either right-continuous sample paths, or has left-continuous sample paths,[94] is progressively measurable (for the filtration to which it is adapted).

HINT: For a right-continuous X, the restriction $X\upharpoonright(\Omega\times[0,t])$ can be identified as the limit of the mappings $(\omega,s)\rightsquigarrow X(\omega,\lceil s2^n/t\rceil 2^{-n}t)$, $n\in\mathbb{N}_{++}$, every one of which is measurable for $\mathscr{F}_t\otimes\mathcal{B}([0,t])$ (explain why). For a left-continuous X, the same restriction is the limit of the mappings $(\omega,s)\rightsquigarrow X(\omega,\lfloor s2^n/t\rfloor 2^{-n}t)$, $n\in\mathbb{N}_{++}$. You must again explain why these mappings are measurable for $\mathscr{F}_t\otimes\mathcal{B}([0,t])$. ○

(8.63) EXERCISE: Let X be any stochastic process with left-continuous sample paths (in particular, a process with continuous sample paths) and let $\mathscr{F}^X$ be the natural filtration of X. Prove that $X_t \prec \mathscr{F}^X_{t-}$ (X_t is measurable for $\mathscr{F}^X_{t-}$) for any $t\in\mathbb{R}_+$. Conclude that the natural filtration $\mathscr{F}^X$ is left-continuous. ○

Completing the filtration with the the null sets of the respective measure is a standard, and more or less "given," step in stochastic analysis and asset pricing. The fundamental reason for this is that many interesting and important processes arise as limits (usually, limits in probability) of adapted processes. It is therefore desirable to work in a setting where the notion "adapted" survives such limits. In other words, we would like to work in a setting where a process that admits an adapted modification should be adapted as well. Yet another – and perhaps even more important – reason to work with complete filtrations can be found in theorem (2.12), which allows one to conclude that certain hitting times are stopping times, as will be evidenced in (8.66).

(8.64) COMPLETION AND USUAL AUGMENTATION: Given a probability space $(\Omega,\mathcal{F},P)$, let $\mathscr{F}\equiv(\mathscr{F}_t)_{t\in\bar{\mathbb{R}}_+}$ be some filtration inside $\mathcal{F}$. Following the procedure for completion of the σ-field described in (1.48), we can now complete the filtration $\mathscr{F}$ by amending the σ-field $\mathscr{F}_0$ – and therefore $\mathscr{F}_t\supseteq\mathscr{F}_0$ for every t – with the collection, $\mathscr{N}_P(\mathscr{F}_\infty)$, of all P-null sets for $\mathscr{F}_\infty$. To put this another way, we can replace every $\mathscr{F}_t$ with$^{\multimap(1.23)}$ the larger σ-field $\tilde{\mathscr{F}}^P_t \overset{\text{def}}{=} \mathscr{F}_t\vee\mathscr{N}_P(\mathscr{F}_\infty)$ for any $t\in\bar{\mathbb{R}}_+$. The larger filtration $\tilde{\mathscr{F}}^P\equiv(\tilde{\mathscr{F}}^P_t)_{t\in\bar{\mathbb{R}}_+}$ is called the *completion* of $\mathscr{F}$. In what follows we will be working almost exclusively with complete filtrations. It is not very difficult to see that $\tilde{\mathscr{F}}^P$ cannot be enlarged further by way of augmenting it with null sets for $\mathscr{F}_\infty$. In other words, the completion of $\tilde{\mathscr{F}}^P$ is $\tilde{\mathscr{F}}^P$ itself. A filtration that coincides with its completion is said to be complete.

For reasons that we have already noted (and others that will become clear later in the sequel), in addition to completeness, the filtrations that we will be working with will need to be right-continuous as well. The nearest right-continuous and complete filtration for a given filtration $\mathscr{F}$ is the filtration $\tilde{\mathscr{F}}^P_+\equiv(\tilde{\mathscr{F}}^P_{t+})_{t\in\bar{\mathbb{R}}_+}$, where the σ-field $\tilde{\mathscr{F}}^P_{t+}$ is understood as the completion of $\mathscr{F}_{t+}$, or, equivalently,

[94] The left continuity of a function $f:\mathbb{R}_+\mapsto\mathbb{S}$ is understood to mean left continuity of $f\upharpoonright\mathbb{R}_{++}$; in other words, left continuity at 0 is automatically assumed.

as $\bigcap_{\varepsilon \in \mathbb{R}_{++}} \tilde{\mathscr{F}}^P_{t+\varepsilon}$.[95] This filtration is called *the usual augmentation* of $\mathscr{F}$, and it is again easy to see that the usual augmentation of $\tilde{\mathscr{F}}^P_+$ is $\tilde{\mathscr{F}}^P_+$ itself. A filtration that coincides with its usual augmentation is said to *satisfy the usual conditions*. Another property that is also easy to check is that if $\mathscr{F}$ is right-continuous, then so is also its completion $\tilde{\mathscr{F}}^P$, so that for a right-continuous filtration the completion and the usual augmentation amount to the same operation. ○

(8.65) REMARK: Switching from the filtration $\mathscr{F}$ to its completion $\tilde{\mathscr{F}}^P$ is not as innocuous as it may seem. This will be illustrated in the next section, where we are going to discover that the usual augmentation of the Brownian filtration can alter significantly its structure and its connection to the Brownian motion. ○

Now we come to the main reasons for working with right-continuous and complete filtrations, and also for introducing the notion "progressively measurable."

(8.66) DEBUTS AND HITTING TIMES: Let A be any subset of $\Omega \times \mathbb{R}_+$. The debut of the set A is defined as the mapping (recall that $\inf\{\emptyset\} = \infty$ by convention)

$$\Omega \ni \omega \rightsquigarrow d_A(\omega) \stackrel{\text{def}}{=} \inf \left\{ t \in \mathbb{R}_+ : (\omega, t) \in A \right\},$$

and we have the following fundamental result (Meyer 1966, IV 52):

THEOREM: Let $\mathscr{F}$ be any right-continuous and complete filtration in some probability space $(\Omega, \mathcal{F}, P)$, and suppose that the set $A \subseteq \Omega \times \mathbb{R}_+$ is progressively measurable for the filtration $\mathscr{F}$. Then the debut, d_A, of the set A is a stopping time for the filtration $\mathscr{F}$.

This result can be seen as an immediate corollary from the theorem in (2.12): the set $A_t \stackrel{\text{def}}{=} A \cap (\Omega \times [0, t[)$ is an element of the σ-field $\mathscr{F}_t \otimes \mathcal{B}([0, t])$, and its projection onto Ω in the product space $\Omega \times [0, t]$ is precisely the event $\{d_A < t\}$. Consequently, $\{d_A < t\} \in \mathscr{F}_t$ for all $t \in \mathbb{R}_+$.

If X is a stochastic process with values in some Polish space $\mathbb{S}$, and if B is any subset of $\mathbb{S}$, then the function

$$\Omega \ni \omega \rightsquigarrow h_B(\omega) \stackrel{\text{def}}{=} \inf \left\{ t \in \mathbb{R}_+ : X(\omega, t) \in B \right\}$$

is called *the hitting time of* B (for the process X). The hitting time of B is nothing but the debut to the set $X^{-1}(B)$. Thus, if X is progressively measurable (for a complete and right-continuous filtration) and $B \in \mathcal{B}(\mathbb{S})$, then the hitting time of B is a stopping time. We have seen in (8.33) that there are situations where a hitting time can be a stopping time even if the filtration is neither right-continuous nor complete, but such situations are special and rare. Fundamentally, this phenomenon has to do with the subtle measurability issues that we noted in (2.12). ○

[95] That the completion of $\mathscr{F}_{t+}$ is included in $\bigcap_{\varepsilon \in \mathbb{R}_{++}} \tilde{\mathscr{F}}^P_{t+\varepsilon} = \bigcap_{n \in \mathbb{N}_{++}} \tilde{\mathscr{F}}^P_{t+2^{-n}}$ is clear. If $B \in \bigcap_{n \in \mathbb{N}_{++}} \tilde{\mathscr{F}}^P_{t+2^{-n}}$, then for every n there is a set $A_n \in \mathscr{F}_{t+2^{-n}}$ such that $B \cap A_n^{\complement} \in \mathscr{N}_P(\mathscr{F}_\infty)$ and $B^{\complement} \cap A_n \in \mathscr{N}_P(\mathscr{F}_\infty)$. The set $R = \bigcap_{n \in \mathbb{N}_{++}} \bigcup_{m \geq n} A_n$ is clearly in $\mathscr{F}_{t+}$ and it is not very difficult to check that $B \cap R^{\complement} \in \mathscr{N}_P(\mathscr{F}_\infty)$ and $B^{\complement} \cap R \in \mathscr{N}_P(\mathscr{F}_\infty)$, so that B belongs to the completion of $\mathscr{F}_{t+}$.

(8.67) EXERCISE: (Jacod and Shiryaev 1987, I 1.19) Suppose that the probability space $(\Omega, \mathcal{F}, P)$ is given a right-continuous filtration $\mathscr{F}$. Prove that if t is a stopping time for the completion $\mathscr{F}^P$, then there is a stopping time, t', for the filtration $\mathscr{F}$ such that $t = t'$ P-a.s.

HINT: Observe that there is a one-to-one correspondence between positive random variables (in particular, stopping times) t and increasing families of events U_r, $r \in \mathbb{Q}_+$, connected to t through the relation $U_r = \{t < r\}$. For any $r \in \mathbb{Q}_+$ there is a set $A_r \in \mathscr{F}_r$ such that $P(A_r \,\Delta\, \{t < r\}) = 0$. Thus, given any $r \in \mathbb{Q}_+$, define $U'_r = \cup_{q \in \mathbb{Q}_+ \cap [0,r]} A_r$, consider the r.v. t' defined through the relations $U'_r = \{t' < r\}$ for every $r \in \mathbb{Q}_+$, and prove that t' is the stopping time that you are looking for. ○

(8.68) PREDICTABLE σ-FIELDS: The σ-field on $\Omega \times \mathbb{R}_+$ generated by all left-continuous and $\mathscr{F}$-adapted finite real processes $X : \Omega \times \mathbb{R}_+ \mapsto \mathbb{R}$ is called *the predictable σ-field for $\mathscr{F}$* and is denoted by $\mathscr{P}(\mathscr{F})$. In other words, $\mathscr{P}(\mathscr{F})$ is the intersection of all σ-fields every one of which contains all $X^{-1}(\mathcal{B}(\mathbb{R}))$, for all possible choices of the left-continuous and $\mathscr{F}$-adapted process $X : \Omega \times \mathbb{R}_+ \mapsto \mathbb{R}$ (the smallest σ-field that contains all σ-fields $X^{-1}(\mathcal{B}(\mathbb{R}))$). The elements of $\mathscr{P}(\mathscr{F})$ are called *predictable sets*, and a process $X : \Omega \times \mathbb{R}_+ \mapsto \mathbb{S}$, with values in some Polish space $\mathbb{S}$, is said to be *predictable* if it is $\mathcal{B}(\mathbb{S})/\mathscr{P}(\mathscr{F})$-measurable. Of course, a process can be predictable without being left-continuous, as is evidenced in exercise (8.69). The intuition behind the attribute "predictable" is that the events of a predictable process that occur before time t are the same as the events that occur strictly before time t. Heuristically speaking, there are no events associated with a predictable process that can be observed at time t, but cannot be inferred from the events that occur strictly before time t.

It is not very difficult to see that the predictable σ-field $\mathscr{P}(\mathscr{F})$ is actually generated by the smaller class of continuous and adapted processes. Indeed, a process $X : \Omega \times \mathbb{R}_+ \mapsto \mathbb{R}$ with left-continuous sample paths is the pointwise limit of the processes $(\omega, t) \rightsquigarrow X(\omega, 2^{-n}(\lceil t2^n \rceil - 1)^+)$, $n \in \mathbb{N}_{++}$, and each of these processes can be approximated with continuous ones, since$^{\multimap(0.11)}$ every indicator of the form $1_{]u,v]}$ is the pointwise limit of *continuous* (even smooth) functions $f_k : \mathbb{R} \mapsto [0, 1]$, chosen so that $f_k(t) = 1$ for $t \in [u+1/k, v+1/k]$ and $f_k(t) = 0$ for $t \notin [u, v+2/k]$.

Finally, we note that$^{\multimap(8.62)}$ since every left-continuous process is automatically progressively measurable, so are also all predictable processes, that is, we have the inclusion $\mathscr{P}(\mathscr{F}) \subseteq \mathrm{Prog}(\mathscr{F})$. ○

(8.69) EXERCISE: Give an example of an adapted process which is not left-continuous, but is nevertheless predictable. Give an example of an adapted process which is right-continuous, and is therefore progressively measurable, but is not predictable. ○

Our next step is to introduce one special tool which, among other things, will be instrumental in the construction of the stochastic integral.

(8.70) STOCHASTIC INTERVALS: Given any two functions $\mathscr{s}, \mathscr{t} \colon \Omega \mapsto \bar{\mathbb{R}}_+$ (in particular, $\mathscr{s}$ and $\mathscr{t}$ could be stopping times), the subset of $\Omega \times \mathbb{R}_+$ given by

$$[\![\mathscr{s}, \mathscr{t}]\!] \overset{\text{def}}{=} \left\{(\omega, t) \in \Omega \times \mathbb{R}_+ : \mathscr{s}(\omega) \le t \le \mathscr{t}(w)\right\}$$

is called the *stochastic interval* with end-points $\mathscr{s}$ and $\mathscr{t}$. Note that the stochastic interval $[\![\mathscr{s}, \mathscr{t}]\!]$ is meaningful for every choice of the functions $\mathscr{s}$ and $\mathscr{t}$. For example, if $\mathscr{s}(\omega) > \mathscr{t}(\omega)$ for every $\omega \in \Omega$, then the stochastic interval $[\![\mathscr{s}, \mathscr{t}]\!]$ is the empty set $\emptyset \subset \Omega \times \mathbb{R}_+$. The stochastic intervals $]\!]\mathscr{s}, \mathscr{t}]\!]$, $[\![\mathscr{s}, \mathscr{t}[\![$, and $]\!]\mathscr{s}, \mathscr{t}[\![$, will be understood in the obvious way by replacing in the definition above $\le$ with $<$ accordingly. The stochastic interval $[\![\mathscr{t}, \mathscr{t}]\!]$ will be expressed equivalently as $[\![\mathscr{t}]\!]$, and we note that $[\![\mathscr{t}]\!] \subset \Omega \times \mathbb{R}_+$ is nothing but the intersection between the graph of the mapping $\mathscr{t} \colon \Omega \mapsto \bar{\mathbb{R}}_+$ (a subset of $\Omega \times \bar{\mathbb{R}}_+$) and the set $\Omega \times \mathbb{R}_+$. ○

(8.71) EXERCISE: Prove that the intersection of any two left-open-right-closed stochastic intervals, say $]\!]\mathscr{s}, \mathscr{t}]\!]$ and $]\!]\mathscr{s}', \mathscr{t}']\!]$ is another stochastic interval, which is also left-open-right-closed. Furthermore, if $\mathscr{s}$, $\mathscr{s}'$, $\mathscr{t}$, and $\mathscr{t}'$ are all stopping times, then the end-points of the intersection are stopping times as well. ○

(8.72) EXERCISE: Prove that if $\mathscr{t}$ is any $\mathscr{F}$-stopping time, then the indicator $1_{[\![0, \mathscr{t}]\!]}$, treated as a stochastic process of the form $\left(1_{[\![0, \mathscr{t}]\!]}(\omega, t)\right)_{t \in \mathbb{R}_+}$, is left-continuous and adapted, and thus predictable.

HINT: If $t \in \mathbb{R}_+$ is fixed, then the set $\left\{\omega \in \Omega \colon 1_{[\![0, \mathscr{t}]\!]}(\omega, t) = 0\right\}$ is the same as the set $\{\mathscr{t} < t\}$. Consult exercise (8.52). ○

(8.73) THEOREM: (Jacod and Shiryaev 1987, I 2.2) The predictable σ-field $\mathscr{P}(\mathscr{F})$ is generated by any of the following two families of subsets of $\Omega \times \mathbb{R}_+$:

(a) $\left\{A \times \{0\} \colon A \in \mathscr{F}_0\right\} \cup \left\{[\![0, \mathscr{t}]\!] : \mathscr{t} \text{ is a finite } \mathscr{F}\text{-stopping time}\right\}$;

(b) $\left\{A \times \{0\} \colon A \in \mathscr{F}_0\right\} \cup \left\{A \times]s, t] \colon s, t \in \mathbb{R}_+,\ s < t,\ A \in \mathscr{F}_s\right\}$. ○

Although rigorous proofs are not of our primary concern, the proof of the last result is especially instructive and is the subject of the next exercise.

(8.74) EXERCISE: Give a complete proof to theorem (8.73) without consulting the original source.

HINT: Argue first that the σ-field generated by the sets in (a) – call it $\sigma(\text{a})$ – must be included in $\mathscr{P}(\mathscr{F})$. Then show that $\sigma(\text{b}) \subseteq \sigma(\text{a})$ by using the following argument: if $A \in \mathscr{F}_s$ and $s < t$ then the set $A \times]s, t] \subseteq \Omega \times \mathbb{R}_+$ is nothing but the stochastic interval $]\!]s_A, t_A]\!] \subseteq \Omega \times \mathbb{R}_+$, with s_A and t_A are understood as in (8.47), and $]\!]s_A, t_A]\!] = [\![0, t_A]\!] \setminus [\![0, s_A]\!] \in \sigma(\text{a})$. Finally, show that $\mathscr{P}(\mathscr{F}) \subseteq \sigma(\text{b})$ by arguing that any left-continuous and adapted finite real process X is the pointwise limit as $n \to \infty$ of the processes $(\omega, t) \rightsquigarrow X(\omega, 2^{-n}(\lceil t2^n \rceil - 1)^+)$, $n \in \mathbb{N}_{++}$, and that every such process is measurable for $\sigma(\text{b})$. Note that any indicator of the form $1_{\Omega \times]s,t]}$, for $s < t$, is no different from the indicator $1_{]\!]s,t]\!]}$, in which s and t are treated⊸ (8.39) as deterministic stopping times. ○

Our interest in the predictable σ-field and the collection of stochastic intervals was motivated mostly by the need to arrive at the following result, which will be instrumental in the construction of the stochastic integral.

(8.75) PROPOSITION: Let $\mathfrak{S} = \mathfrak{S}(\mathscr{F})$ denote the collection (linear space, in fact) of all finite sums of functions on $\Omega \times \mathbb{R}_+$ that are either of the form

$$\Omega \times \mathbb{R}_+ \ni (\omega, t) \rightsquigarrow \xi(\omega) 1_{[\![0]\!]}(\omega, t) \in \mathbb{R},$$

or of the form

$$\Omega \times \mathbb{R}_+ \ni (\omega, t) \rightsquigarrow \eta(\omega) 1_{]\!]u,v]\!]}(\omega, t) \in \mathbb{R},$$

for all possible choices of $u, v \in \mathbb{R}_+$ with $u < v$, and all choices of the bounded real random variables $\xi \prec \mathscr{F}_0$ and $\eta \prec \mathscr{F}_u$. Suppose that m is some measure on $(\Omega \times \mathbb{R}_+, \mathscr{P}(\mathscr{F}))$, such that $m([\![0]\!]) < \infty$ and $m(]\!]u, v]\!]) < \infty$ for any finite $u, v \in \mathbb{R}_+$ with $u < v$. Then the space $\mathfrak{S}$ is dense in the Hilbert space $L^2(\Omega \times \mathbb{R}_+, \mathscr{P}(\mathscr{F}), m)$ (strictly speaking, the m-equivalence classes associated with the elements of $\mathfrak{S}$ are dense). ○

Somewhat surprisingly, the justification of the last proposition appears to be harder than one might think, and involves the following highly nontrivial result, which is due to Dynkin (1959) and can be found in several variations throughout most texts on stochastic processes. The particular version that we are going to use is the one from (Meyer 1966, I 20):

(8.76) MONOTONE CLASS THEOREM: Let $\mathscr{H}$ be any vector space of bounded real functions on Ω, which is closed with respect to the uniform convergence, contains the constant 1, and is such that for any uniformly bounded and increasing sequence, $(f_i \in \mathscr{H})_{i \in \mathbb{N}}$, of positive elements of $\mathscr{H}$ one has $\lim_{i \to \infty} f_i \in \mathscr{H}$ ($\mathscr{H}$ is closed under pointwise monotone convergence of positive elements). Then for any subset $\mathscr{C} \subset \mathscr{H}$ that is closed under pointwise multiplication, one can claim that $\mathscr{H}$ contains all bounded functions that are measurable with respect to $\sigma(\mathscr{C})$, the σ-field generated by $\mathscr{C}$. ○

Here is an illustration of how this powerful result is typically used.

(8.77) JUSTIFICATION OF (8.75): Apart from the Monotone Class Theorem, the key step was already taken in exercise (8.71), which allows us to conclude (see the next exercise) that the family $\mathfrak{S}$ is closed under pointwise multiplication. Let $\mathscr{H}$ denote the vector space of all bounded real functions on $\Omega \times \mathbb{R}_+$ that belong to the closure of $\mathfrak{S}$ in $L^2(\Omega \times \mathbb{R}_+, \mathscr{P}(\mathscr{F}), m)$. It is straightforward to check that $\mathscr{H}$ has the properties listed in (8.76), and that, therefore, the space $\mathscr{H}$ must contain all bounded functions on $\Omega \times \mathbb{R}_+$ that are measurable with respect to the σ-field generated by $\mathfrak{S}$, and, by the result in (8.73), this σ-field is nothing but $\mathscr{P}(\mathscr{F})$. As all bounded and $\mathscr{P}(\mathscr{F})$-measurable functions are dense in $L^2(\Omega \times \mathbb{R}_+, \mathscr{P}(\mathscr{F}), m)$, so is also $\mathscr{H}$, and therefore also $\mathfrak{S}$. ○

(8.78) EXERCISE: Verify that the family $\mathfrak{S}$ is closed under pointwise multiplication and the space $\mathscr{H}$, as defined in (8.77), does indeed satisfy the conditions required by the Monotone Class Theorem (8.76). ○

Another extremely useful application of the Monotone Class Theorem, in conjunction with theorem (8.73), is the following result:

(8.79) PROPOSITION: (Jacod and Shiryaev 1987, I 2.4) Let X be any predictable process and let t be any stopping time. Then the stopped process X^t is also predictable and the r.v. $X_t 1_{\{t<\infty\}}$ is $\mathscr{F}_{t-}$-measurable. ○

8.6 Brownian Filtrations

A Brownian filtration is simply the natural filtration, $\mathscr{F}^W$, generated by a particular Brownian motion W. By definition, the sample paths of W are continuous a.s., but it is easy to imagine a Brownian motion with continuous sample paths – take the canonical coordinate process on the Wiener space, for example. If the sample paths of W are continuous everywhere, then, by the result from exercise (8.63), the natural filtration $\mathscr{F}^W$ is left-continuous. Can we claim that $\mathscr{F}^W$ is also right-continuous – after all, why should the left-continuity of the sample paths be any different from the right-continuity when it comes to influencing the continuity of the generated filtration? To see that there is actually a subtle (or, perhaps not so subtle, depending on the point of view) difference, fix some $t \in \mathbb{R}_{++}$ and consider the event[96]

$$A_t = \{\omega \in \Omega\colon \text{the sample-path } W_\omega \text{ has a local right-maximum at } t\}.$$

(8.80) EXERCISE: Prove that $A_t \in \mathscr{F}^W_{t+}$. ○

If the property $\mathscr{F}^W_{t+} = \mathscr{F}^W_t$ were to hold, then the last exercise would imply $A_t \in \mathscr{F}^W_t$, which is counterintuitive: an observer who has complete information about the history of the Brownian path until time t would know for certain whether the next move of the Brownian path, immediately following time t, would be "up" (A_t does not occur) or "down" (A_t occurs).

In some sense the situation that we just described is not all that interesting, since, typically, the Brownian motions that we will be working with are going to be continuous only a.s. and all filtrations are going to be complete. However, a remarkable feature of a generic Brownian filtration $\mathscr{F}^W$ is that, once completed, it becomes also right-continuous, as we now detail.

(8.81) $\mathscr{F}^W$ IS LEFT- AND RIGHT-CONTINUOUS: The completion, denoted $\tilde{\mathscr{F}}^W$, of any Brownian filtration $\mathscr{F}^W$ is right-continuous. To put it another way, the completion of $\mathscr{F}^W$ and its usual augmentation give the same result. This is a general

[96] A function $f\colon \mathbb{R}_+ \mapsto \mathbb{R}$ has a local right-maximum at $t \in \mathbb{R}_+$ if there is an $\varepsilon \in \mathbb{R}_{++}$ such that $f(t) \geq f(s)$ for every $s \in [t, t+\varepsilon]$.

property of the class of the so-called Feller processes, to which the Brownian motion belongs – see (Revuz and Yor 1999, III (2.10)). At the same time $\tilde{\mathscr{F}}^W$ is also left-continuous, but this property is a special case of exercise (8.63). Indeed, we have $W_t(\omega) = \lim_n W_{t-\frac{1}{n}}(\omega)$ for every ω outside some P-negligible set, with the implication that W_t is measurable for $\tilde{\mathscr{F}}^W_{t-}$. Thus, $\mathscr{F}^W_t \subseteq \tilde{\mathscr{F}}^W_{t-}$. ○

(8.82) BROWNIAN MOTION AS A MARKOV PROCESS:[97] Just as any process of independent increments is, the Brownian motion W is a Markov process: for every bounded Borel function $f : \mathbb{R} \mapsto \mathbb{R}$ and every $t, h \in \mathbb{R}_+$,

$$\mathsf{E}\big[f(W_{t+h}) \mid \mathscr{F}^W_t\big] = \mathsf{E}\big[f(W_{t+h}) \mid W_t\big].$$

Indeed, since $W_{t+h} - W_t$ is independent of $\mathscr{F}^W_t$, and since W_t is $\mathscr{F}^W_t$-measurable, we have

$$\begin{aligned}\mathsf{E}\big[f(W_{t+h})|\mathscr{F}^W_t\big] &= \mathsf{E}\big[f\big(W_t + (W_{t+h} - W_t)\big) \mid \mathscr{F}^W_t\big] \\ &= \frac{1}{\sqrt{2\pi h}} \int_{\mathbb{R}} f(W_t + x)e^{-\frac{x^2}{2h}}\,\mathrm{d}x = \mathsf{E}\big[f(W_{t+h}) \mid W_t\big].\end{aligned}$$ ○

(8.83) STRONG MARKOV PROPERTY:[98] The Markov property of Brownian motion can be stated as: if W is Brownian motion and $t \in \mathbb{R}_+$ is fixed, then the process $(W_{t+s} - W_t)_{s \in \mathbb{R}_+}$ is a Brownian motion that is independent of $\mathscr{F}^W_t$. What is much more interesting, however, is that every Brownian motion shares the so-called strong Markov property: given any $\mathscr{F}^W_+$-stopping time $\mathscr{t}$ such that $P(\mathscr{t} < \infty) = 1$, the process $(W_{\mathscr{t}+t} - W_{\mathscr{t}})_{t \in \mathbb{R}_+}$ is a Brownian motion that is independent of the σ-field $\mathscr{F}^W_{\mathscr{t}+}$. This remarkable feature is often stated as: "Brownian motion starts afresh at stopping times" – see (Grünbaum et al. 2015, p. 318), (Itô and McKean 1974, sec. 1.6), or (Hunt 1956).[99]

In a more modern parlance, the strong Markov property of Brownian motion can be seen as a corollary from the fact that any diffusion with sufficiently regular coefficients is a Feller process, and every Feller process is known to have the strong Markov property – see (11.11) for a brief discussion of these results. ○

Most of the intrinsic nature of the Brownian motion W is encrypted in its *infinitesimal generator*, defined as the family of linear operators that are labeled by the time parameter $t \in \mathbb{R}_+$ and act on test functions $\varphi \in \mathscr{C}^\infty_K(\mathbb{R};\mathbb{R})$ according

[97] The class of processes known as "Markov" is named after the Russian mathematician Andrei Andreievich Markov (1856–1922). More details about this important class will be given in section 11.2 in connection with our study of diffusion processes.

[98] The discovery of this feature is attributed to the American mathematicians Eugene B. Dynkin and Gilbert A. Hunt.

[99] In (Itô and McKean 1974) and other earlier sources, stopping times are defined through the relation $\{\mathscr{t} < t\} \in \mathscr{F}_t$. In today's nomenclature, this corresponds to stopping times relative to the filtration $\mathscr{F}_+$. Stopping times relative to $\mathscr{F}_+$ are sometimes called optional stopping times.

to the prescription

$$\mathcal{A}_t\varphi(x) \stackrel{\text{def}}{=} \lim_{\varepsilon\searrow 0} \frac{1}{\varepsilon}\mathsf{E}\big[\varphi\big(W_{t+\varepsilon}\big) - \varphi\big(W_t\big) \mid W_t = x\big]\,, \quad x \in \mathbb{R}\,.$$

Since the Brownian motion starts afresh at time t, the operation $\mathcal{A}_t$ does not depend on t and we can write simply $\mathcal{A}$.

(8.84) EXERCISE: Prove that the infinitesimal generator of the Brownian motion is given by

$$\mathcal{A}\varphi(x) = \frac{1}{2}\partial^2\varphi(x)\,, \quad \varphi \in \mathscr{C}_K^\infty(\mathbb{R};\mathbb{R})\,. \quad \circ$$

(8.85) LARGER FILTRATIONS: The Markov property and the strong Markov property can be understood with respect to a filtration that is larger than $\tilde{\mathscr{F}}^W$. Specifically, if W is Brownian motion and $\mathscr{F}$ is any filtration to which W happens to be adapted, then the Markov property relative to $\mathscr{F}$ is understood as the claim that the Brownian motion $(W_{t+s} - W_t)_{s\in\mathbb{R}_+}$ is independent of $\mathscr{F}_t$ for every t, while the strong Markov property is understood as the claim that for any finite $\mathscr{F}$-stopping time $\mathcal{t}$ the process $(W_{\mathcal{t}+s} - W_{\mathcal{t}})_{s\in\mathbb{R}_+}$ is Brownian motion that is independent of the σ-field $\mathscr{F}_{\mathcal{t}}$. If this property holds, then we say that W is an $\mathscr{F}$-Brownian motion, or that $(W,\mathscr{F})$ is a Brownian motion, or that W is an $(\mathscr{F},P)$-Brownian motion. It is well known – see (Revuz and Yor 1999, III (3.1)), for example – that any Brownian motion W is also an $\tilde{\mathscr{F}}^W$-Brownian motion, that is, any Brownian motion is still a Brownian motion, and retains the strong Markov property, with respect to⊸ (8.81) the usual augmentation (the completion) of its natural filtration. $\circ$

8.7 Total Variation

(8.86) SUBDIVISIONS: Let $[a,b]$ be any finite interval in $\mathbb{R}$. A subdivision (alias: partition) of $[a,b]$ is any finite set $\mathcal{p} = \{t_0, t_1, \ldots, t_n\} \subset [a,b]$ such that $a = t_0 < t_1 < \ldots < t_n = b$. The number $|\mathcal{p}| \stackrel{\text{def}}{=} \max_{i\in\mathbb{N}_{|n-1}} |t_{i+1} - t_i|$ is called the *modulus* (aliases: mesh, size, diameter) of the subdivision $\mathcal{p}$. We say that the subdivision $\mathcal{p}'$ is a refinement of $\mathcal{p}$ if $\mathcal{p}' \supseteq \mathcal{p}$. A dyadic subdivision of order n is the subdivision given by $t_i = a + i\,(b-a)\,2^{-n}$, $0 \le i \le 2^n$. If $\mathcal{p}$ is any subdivision of a larger interval $[a',b'] \supseteq [a,b]$, then $\mathcal{p}$ can be treated as a subdivision of $[a,b]$ in the obvious way, that is, as the subdivision $\{a\} \cup \{b\} \cup ([a,b] \cap \mathcal{p})$.

A subdivision of $\mathbb{R}_+ = [0,\infty[$ is any countable collection $\mathcal{p} = \{t_0, t_1, \ldots\}$ such that the sequence $(t_i)_{i\in\mathbb{N}}$ is strictly increasing with $t_0 = 0$ and $\lim_i t_i = \infty$. $\circ$

(8.87) TOTAL VARIATION: Let $[a,b]$ be any interval and let $\mathcal{p} = \{t_0, t_1, \ldots, t_n\}$ be any subdivision of $[a,b]$. The $\mathcal{p}$-variation of a function $f\colon [a,b] \mapsto \mathbb{R}$ (on the interval $[a,b]$) is defined as

$$\mathcal{p}\text{-}|\mathrm{var}|_{[a,b]}(f) \stackrel{\text{def}}{=} \sum\nolimits_{i\in\mathbb{N}_{|n-1}} |f(t_{i+1}) - f(t_i)|\,,$$

and the total variation of f (on the interval $[a,b]$) is defined as

$$|\mathrm{var}|_{[a,b]}(f) \stackrel{\text{def}}{=} \sup_{\mathcal{p}} \left(\mathcal{p}\text{-}|\mathrm{var}|_{[a,b]}(f)\right),$$

where the sup is taken over all possible subdivisions of $[a,b]$. If $a = b$, then we set, formally, $|\mathrm{var}|_{[a,b]}(f) = 0$. Notice that $\mathcal{p}\text{-}|\mathrm{var}|_{[a,b]}(f)$ is an increasing function of $\mathcal{p}$ in the sense that if $\mathcal{p}' \supseteq \mathcal{p}$ is a refinement of $\mathcal{p}$ then

$$\mathcal{p}\text{-}|\mathrm{var}|_{[a,b]}(f) \le \mathcal{p}'\text{-}|\mathrm{var}|_{[a,b]}(f).$$

The function f is said to be of finite variation on $[a,b]$ if $|\mathrm{var}|_{[a,b]}(f) < \infty$. If defined on $\mathbb{R}_+$, we say that f has finite variation if $|\mathrm{var}|_{[0,t]}(f) < \infty$ for every $t \in \mathbb{R}_+$, and if $\lim_{t\to\infty} |\mathrm{var}|_{[0,t]}(f) < \infty$, then we say that f has bounded variation. ○

(8.88) EXERCISE: Suppose that the function $f\colon [a,b] \mapsto \mathbb{R}$ is continuous and is also continuously differentiable on $]a,b[$ (that is, $\partial f(t)$ exists for every $t \in\,]a,b[$ and the function ∂f is continuous on $]a,b[$). Prove that

$$|\mathrm{var}|_{[a,b]}(f) = \int_a^b |\partial f(t)|\, dt. \quad ○$$

(8.89) EXERCISE: Suppose that W is a Brownian motion on $(\Omega, \mathcal{F}, P)$ and let

† $$X(\omega,t) \stackrel{\text{def}}{=} \int_0^t W_\omega(s)\, ds \equiv \int_0^t W_\omega\, d\iota \quad \text{for all } t \in \mathbb{R}_+ .$$

Is the process X well defined? Is it well defined a.s.? If it is well defined a.s., is it continuous a.s.? Can you define a process that is indistinguishable from X and is continuous? Can you define a process that is indistinguishable from X and has sample paths that are of finite variation? ○

(8.90) SOME USEFUL FEATURES: (a) Every$^{\multimap(1.57)}$ increasing function has finite variation on every finite interval on which it is defined (follows immediately from the definition).

(b) If μ is any finite Borel measure on some finite interval $[a,b]$ and if the set $B \in \mathcal{B}([a,b])$ is such that $\mu(B) = 0$, then the function $[a,b] \ni t \rightsquigarrow A(t) \stackrel{\text{def}}{=} \mu([a,t])$ is such that the derivative $\partial A(t)$ exists and equals 0 for Λ-a.e. $t \in B$ (Dudley 2002, 7.2.3).

(c) A function of finite variation on $[a,b]$ is continuous everywhere in $[a,b]$, except on an at most countable set of points (ibid., 7.2.5).

(d) A function of finite variation on $[a,b]$ is differentiable Λ-a.e. in $]\,a,b\,[$ and its derivative belongs to $\mathscr{L}^1(]\,a,b\,[, \Lambda)$ (ibid., 7.2.7).

(e) Any function, f, that has finite variation on $[a,b]$ can be written as the difference of two increasing functions. As an example, the functions

† $$f^+(x) \stackrel{\text{def}}{=} \frac{1}{2}\left(|\mathrm{var}|_{[a,x]}(f) + f(x)\right) \quad \text{and} \quad f^-(x) \stackrel{\text{def}}{=} \frac{1}{2}\left(|\mathrm{var}|_{[a,x]}(f) - f(x)\right)$$

are both increasing and such that $f(x) = f^+(x) - f^-(x)$).

(f) For every increasing and right-continuous function $\mathbb{R}_+ \ni t \rightsquigarrow A(t) \in \mathbb{R}$ there is a unique Borel measure $\mu = \mu_A$ on $\mathbb{R}_{++}$ such that $A(t) - A(0) = \mu_A(]0,t])$ for every $t \in \mathbb{R}_{++}$ (see (1.59)). ○

With the notation as in (8.90f), given a Borel function $g\colon \mathbb{R}_+ \mapsto \mathbb{R}$, whether it actually exists or not, the integral $\int_0^t g(s)\mathrm{d}A(s)$, $t \in \mathbb{R}_{++}$, will be understood⊸ (2.39) as a Lebesgue–Stieltjes integral in the usual way, and this integral is nothing but the Lebesgue integral $\int_{]0,t]} g(s)\,\mu_A(\mathrm{d}s)$ (heuristically, $\mathrm{d}A(s)$ is just another notation for $\mu_A(\mathrm{d}s)$). If the function A is right-continuous and of finite variation (but is not necessarily increasing) then the integral $\int_0^t g(s)\mathrm{d}A(s)$, $t \in \mathbb{R}_{++}$, will be understood as the difference of two integrals with respect to two increasing functions A^+ and A^-, following (8.90e).

(8.91) REMARK: In general, a function of finite variation can be written as the difference of two increasing functions in infinitely many ways. It is therefore necessary to verify that understanding the integral $\int_0^t g(s)\mathrm{d}A(s)$ as the difference

$$\int_0^t g(s)\mathrm{d}A^+(s) - \int_0^t g(s)\mathrm{d}A^-(s)$$

is consistent, in the sense that the difference above depends on A, but not on the particular choice of A^+ and A^- such that $A^+ - A^- = A$. Alternatively, one can work only with the A^+ and the A^- defined in $(8.90^\dagger)$, in which case no ambiguity exists. Without further notice we will always suppose that the decomposition of a right-continuous function of finite variation into the difference of two increasing functions is given by $(8.90^\dagger)$. This choice can be seen as canonical, in the sense that $\mathrm{d}A^+(t)\times\mathrm{d}A^-(t) = 0$, that is, the functions A^+ and A^- cannot increase simultaneously, or, to say it another way, the measures $\mathrm{d}A^+$ and $\mathrm{d}A^-$ are singular ($\mathrm{d}A^+ \perp \mathrm{d}A^-$). It is not difficult to check that there is only one choice for the increasing functions A^+ and A^- such that $A = A^+ - A^-$, $\mathrm{d}A^+ \perp \mathrm{d}A^-$, and $A_0^+ = -A_0^- = \frac{1}{2}A_0$. This is the choice that will be automatically assumed from now on. ○

(8.92) NOTATION FOR INTEGRALS: For simplicity, if it is defined, we will express the function $\mathbb{R}_+ \ni t \rightsquigarrow \int_0^t g(s)\mathrm{d}A(s)$ as $g{\bullet}A$, so that we can write simply $(g{\bullet}A)_t$ instead of $\int_0^t g(s)\mathrm{d}A(s)$. It will be permissible for us to write $g{\bullet}A_t$ instead of $(g{\bullet}A)_t$ and write $\int_0^t g\,\mathrm{d}A$ instead of $\int_0^t g(s)\mathrm{d}A(s)$. In particular, the Lebesgue–Stieltjes integral $\int_0^t g(s)\mathrm{d}s \equiv \int_{]0,t]} g\,\mathrm{d}\Lambda$ can be expressed as $g{\bullet}\iota_t$. ○

Of course, there may be situations where $\mathrm{d}A^+ \precsim \Lambda$ and $\mathrm{d}A^- \precsim \Lambda$. In any such case, setting $\rho_A = \mathrm{d}A^+/\mathrm{d}\Lambda - \mathrm{d}A^-/\mathrm{d}\Lambda$, we can write

$$\int_0^t g(s)\mathrm{d}A(s) = \int_0^t g(s)\rho_A(s)\mathrm{d}s\,,$$

so that, formally, $\mathrm{d}A(s) = \rho_A(s)\mathrm{d}s$.[100]

(8.93) EXERCISE: Give an example of an increasing function $A\colon [0,1] \mapsto [0,1]$, which is continuous and is such that the associated measure μ_A is singular with respect to Λ. Verify that nevertheless the derivative $\frac{\mathrm{d}}{\mathrm{d}s}A(s)$ exists for Λ-a.e. $s \in [0,1]$, as it should, and belongs to $\mathscr{L}^1([0,1],\Lambda)$, as it should.

HINT: Consider using (8.90b) and some familiar features of the probability measure on the coin toss space. ○

(8.94) EXERCISE: Let X be the process defined in (8.89†) and let Y be any continuous process. How would you give meaning to the stochastic integral

$$(Y \bullet X)_t \equiv \int_0^t Y_s \mathrm{d}X_s\,, \quad t \in \mathbb{R}_{++}\,? \quad ○$$

(8.95) EXERCISE: Let X be the process defined in (8.89†). Prove that X is a centered Gaussian process and compute its covariance function $C(s,t) = \mathrm{Cov}(X_s, X_t)$, $s,t \in \mathbb{R}_+$. ○

(8.96) EXERCISE: Let X be the process defined in (8.89†). Prove that if f is any deterministic function from the space $\mathscr{L}^2([0,1],\Lambda)$, then the process

$$Z_s \stackrel{\text{def}}{=} (f \bullet X)_t \equiv \int_0^t f(s)\mathrm{d}X_s\,, \quad t \in [0,1]\,,$$

is still a centered Gaussian process. ○

8.8 Quadratic Variation

It was noted earlier that the quadratic variation is one crucially important tool in the theory of stochastic processes. Now we turn to the study of this tool.

Let $\wp = \{t_0, t_1, \ldots, t_n\}$ be any$^{\multimap (8.86)}$ subdivision of $[0,t]$. The quadratic $\wp$-variation of the function $f\colon [0,t] \mapsto \mathbb{R}$ (on the interval $[0,t]$) is defined as

$$\mathcal{Q}^{\wp}(f,f)_t \stackrel{\text{def}}{=} \sum\nolimits_{i \in \mathbb{N}_{|n-1}} \big(f(t_{i+1}) - f(t_i)\big)^2\,.$$

By analogy with the total variation, one may be tempted to define the total quadratic variation of f as $\sup_{\wp} \mathcal{Q}^{\wp}(f,f)_t$. Such a definition would be perfectly meaningful, but, unfortunately, would not be all that useful. The reason is that unlike the $\wp$-variation, the quadratic $\wp$-variation does not increase monotonically as the subdivision $\wp$ increases. A somewhat less ambitious project would be to consider $\lim_n \mathcal{Q}^{\wp_n}(f,f)_t$ for a sequence of subdivisions chosen so that $|\wp_n| \to 0$ as $n \to \infty$,

[100] This relation is not just formal: Lebesgue's version of the fundamental theorem of calculus says that the function $t \rightsquigarrow A(t) - A(0) = \mu(]0,t]) = \int_0^t \rho_A(s)\mathrm{d}s$ is differentiable and its derivative is $\rho_A(t)$ for Λ-a.e. t.

and if the function f is some random sample path, then we must also decide what type of convergence to impose.

(8.97) EXERCISE: Suppose that $f: [0,t] \mapsto \mathbb{R}$ is continuous and has finite variation on $[0,t]$. Let $(\mathcal{P}_n)_{n \in \mathbb{N}}$ be any sequence of subdivisions of $[0,t]$ such that $|\mathcal{P}_n| \to 0$ as $n \to \infty$. Prove that $\lim_{n\to\infty} \mathcal{Q}^{\mathcal{P}_n}(f,f)_t = 0$.

HINT: $\mathcal{Q}^{\mathcal{P}_n}(f,f)_t \leq |\mathrm{var}|_{[0,t]}(f) \times \varepsilon$ where ε could be arbitrarily small. ○

(8.98) DEFINITION: Let $X = (X_t)_{t \in \mathbb{R}_+}$ be any stochastic process defined on probability space $(\Omega, \mathcal{F}, P)$. We say that X *has finite quadratic variation*, or that X *is of finite quadratic variation*, if there is a (finite) $\mathbb{R}_+$-valued process on $(\Omega, \mathcal{F}, P)$, denoted by $[X,X]$ and called the *quadratic variation of* X, such that for any $t \in \mathbb{R}_{++}$ and for any sequence, $(\mathcal{P}_n)_{n \in \mathbb{N}}$, of subdivisions of $[0,t]$, chosen so that $|\mathcal{P}_n| \to 0$ as $n \to \infty$,

$$[X,X]_t(\omega) = P\text{-}\lim_n \mathcal{Q}^{\mathcal{P}_n}(X_\omega, X_\omega)_t ,$$

the limit being understood in probability P. ○

(8.99) EXERCISE: If the process X has finite quadratic variation, then the process $[X,X]$ is increasing in probability, in that $P([X,X]_t \geq [X,X]_s) = 1$ for every choice of $s,t \in \mathbb{R}_+$, $0 \leq s < t$. Prove this statement. In addition, prove that if $[X,X]$ exists and has a right-continuous modification, then such a modification, still denoted by $[X,X]$, can be chosen to be also increasing (in addition to right-continuous), that is, a.e. sample path $[X,X]_\omega$ is càdlàg and increasing. ○

(8.100) EXERCISE: Prove that the Brownian motion W has finite quadratic variation given by $[W,W] = \iota$.

HINT: If $\mathcal{P} = \{t_0, t_1, \dots, t_n\}$ is a subdivision of $[0,t]$, show that the quantity

$$\mathsf{E}\left[\left(\sum_{i \in \mathbb{N}_{|n-1}} \left((W_{t_{i+1}} - W_{t_i})^2 - (t_{i+1} - t_i)\right)\right)^2\right]$$

can be made arbitrarily small, if the size of the subdivision, $|\mathcal{P}|$, is chosen to be sufficiently small. Explain why this is more than enough. ○

In conjunction with exercise (8.97), the next exercise shows that with probability 1 the Brownian sample paths have infinite variation on any finite interval.

(8.101) EXERCISE: Let W be the Brownian motion and let $\mathcal{P}_n$ be the nth dyadic subdivision of $[0,t]$. Prove that

† $$\lim_n \mathcal{Q}^{\mathcal{P}_n}(W_\omega, W_\omega)_t = t \quad \text{for } P\text{-a.e. } \omega \in \Omega .$$

HINT: Prove that

$$\sum_{n \in \mathbb{N}_{++}} \mathsf{E}\left[\left|\mathcal{Q}^{\mathcal{P}_n}(W_\omega, W_\omega)_t - t\right|\right] \leq \sum_{n \in \mathbb{N}_{++}} \mathsf{E}\left[\left|\mathcal{Q}^{\mathcal{P}_n}(W_\omega, W_\omega)_t - t\right|^2\right]^{\frac{1}{2}} < \infty$$

and use exercise (4.31). ○

(8.102) REMARK: The following more general result is well known: if, instead of choosing dyadic subdivisions, in exercise (8.101) one takes any sequence of subdivisions $\wp_1 \subset \wp_2 \subset \ldots$ such that $|\wp_n| \to 0$, then the relation (8.101†) still holds (Revuz and Yor 1999, II (2.12)). ○

(8.103) QUADRATIC COVARIATION: Let X and Y be any two stochastic processes. Given a subdivision $\wp\colon 0 = t_0 < t_1 < \ldots < t_n = t$, let

† $$\mathcal{Q}^{\wp}(X_\omega, Y_\omega)_t \stackrel{\text{def}}{=} \sum_{i \in \mathcal{N}_{|n-1}} \big(X_{t_{i+1}}(\omega) - X_{t_i}(\omega)\big)\big(Y_{t_{i+1}}(\omega) - Y_{t_i}(\omega)\big)\,.$$

Using the simple algebraic relation (the so-called polarization identity)

$$a\,b = \frac{1}{4}\Big[(a+b)^2 - (a-b)^2\Big]\,, \qquad a, b \in \mathbb{R}\,,$$

it is straightforward to check that

‡ $$\mathcal{Q}^{\wp}(X_\omega, Y_\omega)_t = \frac{1}{4}\Big[\mathcal{Q}^{\wp}(X_\omega + Y_\omega, X_\omega + Y_\omega)_t - \mathcal{Q}^{\wp}(X_\omega - Y_\omega, X_\omega - Y_\omega)_t\Big]\,.$$

We say that X *and* Y *have finite quadratic covariation* if there is a finite process, denoted by $[X,Y]$ and called the *quadratic covariation of* X *and* Y, such that

$$\mathcal{Q}^{\wp_n}(X_\omega, Y_\omega)_t \to [X,Y]_t(\omega) \quad \text{as } n \to \infty \text{ in probability } P$$

for any $t \in \mathbb{R}_+$ and for any sequence of subdivisions of $[0,t]$, $(\wp_n)_{n \in \mathcal{N}}$, chosen so that $|\wp_n| \to 0$ as $n \to \infty$.

It is straightforward from the relation ‡ that if $X+Y$ and $X-Y$ are both of finite quadratic variation, then X and Y have finite quadratic covariation given, by the polarization identity,

$$[X,Y] = \frac{1}{4}\Big[[X+Y, X+Y] - [X-Y, X-Y]\Big]\,. \qquad ○$$

(8.104) EXERCISE: Let X and Y be any two stochastic processes such that the sample paths of X are a.s. continuous and the sample paths of Y are a.s. of finite variation on every finite interval. Prove that X and Y have finite quadratic covariation given by $[X,Y] = 0$ P-a.s.

HINT: This exercise is tautological with (8.97). ○

8.9 Brownian Sample Paths Are Nowhere Differentiable

It would not be an exaggeration to say that the phenomenon that we are about to describe in this section is the very reason for inventing stochastic calculus. This is one situation where a detailed proof provides more intuition and takes less space and effort than trying to explain and justify the feature in some other way. We break the proof into several exercises, following (Revuz and Yor 1999, I (2.9)).

(8.105) EXERCISE: Let f be any function from $\mathbb{R}_+$ into $\mathbb{R}$. Suppose that the right derivative of f exists and is finite at some fixed $t \in \mathbb{R}_+$. Prove that there is a positive real $\varepsilon \in \mathbb{R}_{++}$ and a positive integer $L \in \mathbb{N}_{++}$ such that

$$|f(t+h) - f(t)| \leq Lh \quad \text{for all } h \in]0, \varepsilon[.$$

HINT: Simply write down the definition of right derivative. ○

(8.106) EXERCISE: With the notation as in (8.105), prove that for all sufficiently large integers $n \in \mathbb{N}_{++}$,

$$\left|f(s+1/n) - f(s)\right| \leq \frac{11L}{n},$$

for every

$$s \in \left\{ \frac{\lfloor tn \rfloor + 1}{n}, \frac{\lfloor tn \rfloor + 2}{n}, \frac{\lfloor tn \rfloor + 3}{n}, \frac{\lfloor tn \rfloor + 4}{n}, \frac{\lfloor tn \rfloor + 5}{n} \right\}.$$

HINT: You have $\lfloor tn \rfloor / n \leq t < (\lfloor tn \rfloor + 1)/n$. Then notice that

$$\left|f(s+1/n) - f(s)\right| \leq \left|f(s+1/n) - f(t)\right| + \left|f(s) - f(t)\right| . \quad ○$$

Our next step is to develop some estimates involving the Gaussian distribution that are instrumental in the current context, and are also useful in a broad range of applications.

Suppose that the r.v. ξ has Gaussian law of mean 0 and variance σ^2, that is, $\xi \in \mathcal{N}(0, \sigma^2)$. Then

$$P\left(|\xi| \leq k\sigma^2\right) = \frac{1}{\sigma\sqrt{2\pi}} \int_{-k\sigma^2}^{k\sigma^2} e^{-\frac{x^2}{2\sigma^2}}\, dx = \operatorname{erf}(k\sigma/\sqrt{2}) \quad \text{for any } k \in \mathbb{R}_{++} .$$

Since $\partial \operatorname{erf}(x) = 2e^{-x^2}/\sqrt{\pi}$ and $\partial^2 \operatorname{erf}(x) = -4x e^{-x^2}/\sqrt{\pi}$, we see that the function $\operatorname{erf}(\cdot)$ is concave on $\mathbb{R}_+$, and, since $\partial \operatorname{erf}(0) = 2/\sqrt{\pi}$, it is straightforward to conclude that $\operatorname{erf}(x) \leq 2x/\sqrt{\pi}$ for all $x \in \mathbb{R}_+$ (due to the concavity, the graph of $\operatorname{erf}(\cdot)$ must be below its tangent at 0). We thus arrive at the following important estimate:

$$P\left(|\xi| \leq k\sigma^2\right) \leq \sqrt{\frac{2}{\pi}}\, k\sigma \quad \text{for any } k \in \mathbb{R}_{++} \quad \text{for any } \xi \in \mathcal{N}(0, \sigma^2) .$$

Now suppose that W is a Brownian motion on $(\Omega, \mathcal{F}, P)$.

(8.107) EXERCISE: Prove that for any $s \in \mathbb{R}_+$,

$$P\left(\left|W_{s+1/n} - W_s\right| \leq \frac{11L}{n}\right) \leq \sqrt{\frac{2}{\pi}}\, \frac{11L}{\sqrt{n}} .$$

Conclude that

$$P\Big(\bigcap_{1 \leq j \leq 5} \Big\{ \left|W_{s+(j+1)/n} - W_{s+j/n}\right| \leq \frac{11L}{n} \Big\} \Big) \leq \frac{2^{5/2}}{\pi^{5/2}}\, \frac{(11L)^5}{n^{5/2}} . \quad ○$$

For some fixed $T \in \mathbb{R}_{++}$ and fixed $L \in \mathbb{N}_{++}$ consider the events

$$A(T,L\,;\,n) \stackrel{\text{def}}{=} \bigcup_{0 \le i \le \lfloor Tn \rfloor} \Big(\bigcap_{1 \le j \le 5} \Big\{ \big|W_{i/n+(j+1)/n} - W_{i/n+j/n}\big| \le \frac{11L}{n} \Big\} \Big), \quad n \in \mathbb{N}_{++},$$

and let

$$B(T,L) \stackrel{\text{def}}{=} \limsup_{n\to\infty} A(T,L\,;\,n) = \bigcap_{n=1}^{\infty} \bigcup_{m=n}^{\infty} A(T,L\,;\,m)\,;$$

in other words, $B(T,L)$ is the event that infinitely many of $A(T,L\,;\,n)$, $n \in \mathbb{N}_{++}$, occur.

(8.108) EXERCISE: Prove that for every fixed $L \in \mathbb{N}_{++}$, $P\big(B(T,L)\big) = 0$, and conclude that

$$P\Big(\bigcup_{L\in\mathbb{N}} B(T,L)\Big) = 0\,.$$

HINT: First use the Borel–Cantelli lemma and then use the fact that $\mathbb{N}_{++}$ is a countable set. ○

(8.109) EXERCISE: Prove that the event

$$\Big\{\text{the sample path } W_\omega \text{ has a right derivative at at least one } t \in [0,T]\Big\}$$

is included in the event $\cup_{L\in\mathbb{N}_{++}} B(T,L)$, and is therefore P-negligible. Conclude that the event

$$\Big\{\text{the sample path } W_\omega \text{ has a right derivative at at least one } t \in \mathbb{R}_{+}\Big\}$$

must be P-negligible as well. ○

With a completely analogous line of reasoning, one can establish that the event

$$\Big\{\text{the sample path } W_\omega \text{ has a left derivative at at least one } t \in \mathbb{R}_{++}\Big\}$$

is P-negligible. As a result, for P-a.e. $\omega \in \Omega$, the sample path W_ω has no derivative (neither left, nor right) at any point in $\mathbb{R}_{+}$.

8.10 Some Special Features of Brownian Sample Paths

Let W be any Brownian motion process defined on $(\Omega, \mathcal{F}, P)$.

(8.110) LAW OF THE ITERATED LOGARITHM: (Revuz and Yor 1999, II (1.9)) We have the following identity, known as the *law of the iterated logarithm*:[101]

$$P\Big[\limsup_{t\searrow 0} \frac{W_t}{\sqrt{2t\log(\log(1/t))}} = 1\Big] = 1\,. \quad ○$$

[101] The discovery of the law of the iterated logarithm is attributed to the Russian mathematician Aleksandr Yakovlevich Khinchin (1894–1959).

(8.111) EXERCISE: Using the preceding result, prove that

$$P\left[\liminf_{t\searrow 0}\frac{W_t}{\sqrt{2t\log(\log(1/t))}}=-1\right]=1\,.$$

HINT: What can you say about the process $(-W_t)$? ○

Since the event

$$\left\{\limsup_{t\searrow 0}\frac{W_t}{\sqrt{2t\log(\log(1/t))}}=1\right\}\bigcap\left\{\liminf_{t\searrow 0}\frac{W_t}{\sqrt{2t\log(\log(1/t))}}=-1\right\}$$

occurs with probability 1, and since the Brownian motion starts afresh at any moment in time, we see that once at 0, with probability 1 the Brownian motion will cross level 0 infinitely many times in the immediate future, that is, during any – however short – time period that begins at the moment when level 0 has been reached. By translation, this property holds for any level $a \in \mathbb{R}$: if $W_t = a$, then with probability 1 there will be infinitely many points $s \in \,]t, t+\varepsilon[$ such that $W_s = a$, and this claim can be made for every $\varepsilon \in \mathbb{R}_{++}$.

Now we turn to the recurrence property (see bellow) of Brownian motion:

(8.112) EXERCISE: Prove that

$$P\left[\limsup_{t\to\infty}\frac{W_t}{\sqrt{2t\log(\log(t))}}=1\right]=1=P\left[\liminf_{t\to\infty}\frac{W_t}{\sqrt{2t\log(\log(t))}}=-1\right].$$

HINT: Use the time reversal property of Brownian motion. ○

Since the event

$$\left\{\limsup_{t\to\infty}\frac{W_t}{\sqrt{2t\log(\log(t))}}=1\right\}\bigcap\left\{\liminf_{t\to\infty}\frac{W_t}{\sqrt{2t\log(\log(t))}}=-1\right\}$$

occurs with probability 1, we see that once at 0, with probability 1 the Brownian motion will be back to 0 after any finite time horizon. By translation, this property holds for any level $a \in \mathbb{R}$: if $W_t = a$, then with probability 1 the Brownian motion will be back to level a at some moment in time after time T, and this claim can be made for every finite $T > t$. Since the Brownian motion starts afresh at stopping times, this also means that the Brownian motion will visit a infinitely many times with probability 1. In fact, the result in (8.112) shows that the linear Brownian motion will visit infinitely many times any $a \in \mathbb{R}$ with probability 1 – and regardless of its starting position $W_0 = x \in \mathbb{R}$.

(8.113) BROWNIAN MOTION OF DIMENSIONS 1, 2, AND ≥ 3 (TRANSIENCE AND RECURRENCE): We have seen above that for every choice of the starting point $\mathbb{R} \ni x = W_0$, for every choice of $a \in \mathbb{R}$, and for a.e. $\omega \in \Omega$, one can find a sequence $(\tau_i(\omega) \in \mathbb{R}_+)_{i\in\mathbb{N}}$ such that $\tau_i(\omega) \nearrow \infty$ and $W_{\tau_i(\omega)}(\omega) = a$ for all $i \in \mathbb{N}$. This property is known as *point recurrence*. The 1-dimensional Brownian motion is therefore point recurrent. It turns out that Brownian motion of any dimension strictly larger than 1 is not point recurrent – see (Mörters and Peres 2010).[102] To be precise, if W is the linear Brownian motion in $\mathbb{R}^d$ with $d \geq 2$, then for every starting point $\mathbb{R}^d \ni x = W_0$ and for every $a \in \mathbb{R}^d$, the event

$$\{\omega \in \Omega \colon W_t(\omega) = a \text{ for some } t \in \mathbb{R}_+\}$$

has probability 0. This property is quite intuitive: the event that a Brownian motion with dimension $d \geq 2$ revisits 0, after starting from 0, means that several independent 1-dimensional Brownian motions that start from 0 revisit 0 simultaneously, and this should be a 0-probability event, although a rigorous proof is still needed. Nevertheless – see ibid.[102] – the linear Brownian motion in $\mathbb{R}^2$, W, can be claimed to be ε-recurrent in the following sense: for every starting point $\mathbb{R}^2 \ni x = W_0$, for every $a \in \mathbb{R}^2$, for every $\varepsilon \in \mathbb{R}_{++}$, and for a.e. $\omega \in \Omega$, there is a sequence $(\tau_i(\omega) \in \mathbb{R}_+)_{i \in \mathbb{N}}$ such that $\tau_i(\omega) \nearrow \infty$ and $|W_{\tau_i(\omega)}(\omega) - a| < \varepsilon$.

A Brownian motion W in $\mathbb{R}^d$ is said to be *transient* if $\lim_{t \nearrow \infty} |W_t| = \infty$. Clearly, being ε-recurrent, a Brownian motion with dimension $d \leq 2$ cannot be transient. It turns out – see ibid.[102] – that a Brownian motion with dimension $d \geq 3$ is always transient, which, in particular, implies that a Brownian motion with dimension $d \geq 3$ cannot be ε-recurrent. ○

The running maximum of the Brownian motion W is defined as the process

$$\mathcal{S}_t \equiv \mathcal{S}_t(W) \stackrel{\text{def}}{=} \sup_{s \in [0,t]} W_s\,, \quad t \in \mathbb{R}_+\,.$$

This process is continuous, increasing, and starts from 0 ($\mathcal{S}_0 = 0$). Furthermore, an immediate application of the law of the iterated logarithm gives $\mathcal{S}_\infty \stackrel{\text{def}}{=} \lim_{t \to \infty} \mathcal{S}_t = \infty$ with probability 1. It would not be a restriction for us to suppose that $\mathcal{S}_\infty(\omega) = \infty$ for every $\omega \in \Omega$. Given any $a \in \mathbb{R}_+$, define the $\tilde{\mathscr{F}}^W$-stopping time

$$\mathscr{t}_a = \inf\{t \in \mathbb{R}_+ \colon W_t \geq a\} \equiv \inf\{t \in \mathbb{R}_+ \colon \mathcal{S}_t \geq a\}\,,$$

which is to say, $\mathscr{t}_a$ is the moment of the first visit of W to level $a \in \mathbb{R}_{++}$ (starting from level 0 at time $t = 0$), and this is the same as the moment of the first visit of its running maximum $\mathcal{S}$ to level a. The stopping time $\mathscr{t}_a$ is P-a.s. finite (it is understood to be ∞ if the running maximum never reaches a in finite time).

(8.114) EXERCISE: Prove that the function $\mathbb{R}_+ \ni a \rightsquigarrow \mathscr{t}_a(\omega)$ is left-continuous and increasing (thus càglàd) for P-a.e. $\omega \in \Omega$. In addition, prove that

$$\mathcal{S}_t = \inf\{a \in \mathbb{R}_{++} \colon \mathscr{t}_a \geq t\} \quad \text{a.s.}\,,$$

and argue that the events $\{\mathscr{t}_a \leq t\}$ and $\{\mathcal{S}_t \geq a\}$ are a.s. the same.

HINT: Consult exercises (2.89), (2.90), and (2.91). ○

Now we come to a remarkable feature that is important for many applications.

[102] These results are classical and can be found in many texts – old or new.

(8.115) REFLECTION PRINCIPLE:[103] Let $b \in \mathbb{R}_{++}$ and $a \in\,]-\infty, b]$ be arbitrarily chosen. Then

$$P(S_t \geq b, W_t < a) = P(W_t > 2b - a)\,. \quad \circ$$

The reason for calling the last relation the *reflection principle* is that it can be justified by the following heuristic argument: It is clear that $\{\mathscr{t}_b \leq t\}$ on the set $\{S_t \geq b\}$. The Brownian motion being at level b at time $\mathscr{t}_b \leq t$, its chances of falling below level $a = b - (b - a)$ by time $t \geq \mathscr{t}_b$ are the same as its chances of rising above level $b + (b - a) = 2b - a$. In other words, statistically, reaching level b before time t and then falling below level a at time t should be just as likely as reaching level b before time t and then rising above level $2b-a$ at time t. Of course, being above level $2b-a$ at time t implies that level b must have been reached some time before time t.

The more rigorous argument is based on the strong Markov property and is not very difficult to establish: First, notice that, since $W_{\mathscr{t}_b} = b$,

$$\begin{aligned} P\big(S_t \geq b, W_t < a\big) &= P\big(\mathscr{t}_b \leq t, W_{\mathscr{t}_b} + (W_{\mathscr{t}_b+(t-\mathscr{t}_b)} - W_{\mathscr{t}_b}) < a\big) \\ &= P\big(\mathscr{t}_b \leq t, W_{\mathscr{t}_b+(t-\mathscr{t}_b)} - W_{\mathscr{t}_b} < a - b\big)\,. \end{aligned} \qquad \mathrm{e}_1$$

Next, observe that $\{\mathscr{t}_b \leq t\} \in \tilde{\mathscr{F}}^W_{\mathscr{t}_b}$ and, due to the strong Markov property, $(W_{\mathscr{t}_b+s} - W_{\mathscr{t}_b})_{s\in\mathbb{R}_+}$ is a Brownian motion that is independent of $\tilde{\mathscr{F}}^W_{\mathscr{t}_b}$. Therefore, the probability in e_1 will not change if we were to replace $(W_{\mathscr{t}_b+s} - W_{\mathscr{t}_b})_{s\in\mathbb{R}_+}$ with another Brownian motion, which is also independent of $\tilde{\mathscr{F}}^W_{\mathscr{t}_b}$. In particular, we can use the Brownian motion $(-W_{\mathscr{t}_b+s} + W_{\mathscr{t}_b})_{s\in\mathbb{R}_+}$ instead. As a result, the probability in e_1 must be the same as

$$\begin{aligned} &P\big(\mathscr{t}_b \leq t, W_{\mathscr{t}_b+(t-\mathscr{t}_b)} - W_{\mathscr{t}_b} > b - a\big) \\ &\quad = P\big(\mathscr{t}_b \leq t, W_{\mathscr{t}_b} + (W_{\mathscr{t}_b+(t-\mathscr{t}_b)} - W_{\mathscr{t}_b}) > 2b - a\big) \\ &\quad = P\big(\mathscr{t}_b \leq t, W_t > 2b - a\big) = P\big(S_t \geq b, W_t > 2b - a\big) \\ &\quad = P\big(W_t > 2b - a\big)\,. \end{aligned}$$

Since W_t has Gaussian density of mean 0 and variance t, we arrive at the following important formula:

$$\begin{aligned} P\big(S_t \geq b, W_t < a\big) &= P\big(2b - W_t < a\big) = P\big(2b + W_t < a\big) \\ &= \frac{1}{\sqrt{2\pi t}} \int_{-\infty}^{a} e^{-\frac{(x-2b)^2}{2t}}\, \mathrm{d}x\,. \end{aligned}$$

Deriving the distribution density of the bi-variate r.v. (W_t, S_t) – that is, the joint density of the Brownian motion and its running maximum at time t – in the domain $\big\{(a, b) \in \mathbb{R}^2 : b > 0\,,\ a \leq b\big\}$, is now a matter of straightforward calculation.

[103] The discovery of the reflection principle is attributed to the French mathematician Désiré André [André Antoine Désiré] (1840–1918?).

(8.116) EXERCISE: Prove that the joint density of the Brownian motion and its running maximum at time t in the domain $\mathscr{D} = \{(a,b) \in \mathbb{R}^2 : b > 0\,,\ a \leq b\}$ is given by

$$\phi(a,b) = \sqrt{\frac{2}{\pi t^3}}\,(2b-a)\,e^{-\frac{(2b-a)^2}{2t}}\,, \quad (a,b) \in \mathscr{D}\,. \quad \circ$$

(8.117) EXERCISE: Using the result in the previous exercise, prove that for every fixed $t \in \mathbb{R}_{++}$, the joint laws of $(S_t - W_t, S_t)$ and $(S_t, S_t - W_t)$ coincide. $\circ$

Another interesting – and crucial for many applications in the domain of finance – corollary of the reflection principle is the following:

(8.118) EXERCISE: Prove that for every fixed $t \in \mathbb{R}_{++}$,

$$\dagger \qquad P\big(S_t \geq b\big) = P\big(|W_t| \geq b\big) \text{ for all } b \in \mathbb{R}_{++}\,.$$

HINT: Observe that $P\big(S_t \geq b\big) = P\big(W_t \geq b\big) + P\big(S_t \geq b, W_t < b\big)$. $\circ$

The relation established in ($8.118^{\dagger}$) is truly remarkable: for fixed t, the distribution of the r.v. S_t is identical to the distribution of the r.v. $|W_t|$, which is straightforward to write down. We stress, however, that the distribution law of $(S)_{t\in\mathbb{R}_+}$ as a stochastic process is very different from the distribution law of $(|W_t|)_{t\in\mathbb{R}_+}$ as a stochastic process, for to say the least, the sample paths S_ω are a.s. increasing, whereas the sample paths $|W_\omega|$ are not.

As a direct application of ($8.118^{\dagger}$) we now obtain yet another important formula:

$$P(\mathfrak{t}_b \leq t) = \sqrt{\frac{2}{\pi\, t}} \int_b^\infty e^{-\frac{x^2}{2t}}\,\mathrm{d}x \quad \text{for any } t \in \mathbb{R}_{++}\,. \qquad \mathbb{e}_2$$

(8.119) EXERCISE: Using the formula in $\mathbb{e}_2$, prove that for fixed $b \in \mathbb{R}_{++}$, the distribution law of the stopping time $\mathfrak{t}_b$ in $\mathbb{R}_{++}$ admits density given by

$$\varphi_b(t) = \frac{b}{\sqrt{2\pi t^3}}\,e^{-\frac{b^2}{2t}}\,, \quad t \in \mathbb{R}_{++}\,.$$

HINT: Consult exercises (8.114) and (8.118), and use the basic differentiation rules from calculus. $\circ$

9

Crash Course on Continuous-Time Martingales

We saw in chapter 6 that, at least in the context of some basic discrete-time models, there is an intrinsic connection between the principles of asset pricing and the notion of a martingale. Since we are already aware of this connection, and since most of the intuition behind continuous-time models builds from discrete-time considerations, in this chapter we will focus almost exclusively on the mathematics of continuous-time martingales, local martingales, and semimartingales.

9.1 Definitions and First Properties

To begin with, we fix a probability space $(\Omega, \mathcal{F}, P)$ and a filtration $\mathscr{F} \setminus\subset \mathcal{F}$, which will have the usual meaning of "flow of time," or "information arrival dynamics." We suppose that these structures are sufficiently rich and can support all stochastic processes and random variables that crop up in our study. Having already adopted the continuous-time paradigm in the previous chapter, in what follows the range of the time parameter t will be $\mathbb{R}_+$, or $\bar{\mathbb{R}}_+$, or some finite interval $[0, T] \subseteq \mathbb{R}_+$. In those situations where we are indifferent to the concrete form of the time domain, we will use the token $\mathbb{T}$ as a generic substitute for any of symbols $\mathbb{R}_+$, $\bar{\mathbb{R}}_+$, or $[0, T]$.

(9.1) SUBMARTINGALES: An $\mathbb{R}$-valued process $X = (X_t)_{t \in \mathbb{T}}$ is said to be an $(\mathscr{F}, P)$-*submartingale*, or simply a *submartingale* if both $\mathscr{F}$ and P are understood, if the following conditions are satisfied:

(a) $X \lesssim \mathscr{F}$ (X is adapted to $\mathscr{F}$);

(b) $\mathsf{E}[X_t^+] < \infty$ for all $t \in \mathbb{T}$;

(c) $X_s \leq \mathsf{E}[X_t | \mathscr{F}_s]$ (a.s.) for all $s, t \in \mathbb{T}$, $s \leq t$.

Note that as long as $\mathsf{E}[X_t^+] < \infty$, any conditional expectation of X_t is well-defined without the requirement for X_t to be integrable ($\mathsf{E}[|X_t|] < \infty$). ○

(9.2) SUPERMARTINGALES AND MARTINGALES: The process X is said to be an $(\mathscr{F}, P)$-*supermartingale*, or simply a *supermartingale* if both $\mathscr{F}$ and P are understood, if the process $(-X)$ is an $(\mathscr{F}, P)$-submartingale. The attribute "submartingale" is equivalent to $X \lesssim \mathscr{F}$, $\mathsf{E}[X_t^-] < \infty$, and $X_s \geq \mathsf{E}[X_t | \mathscr{F}_s]$ (a.s.) for all $s, t \in \mathbb{T}$, $s \leq t$.

We say that X is an $(\mathscr{F}, P)$-*martingale*, or simply a *martingale* if both $\mathscr{F}$ and P are understood, if X is simultaneously a submartingale and a supermartingale.

The attribute "martingale" is therefore equivalent to $X \prec \mathscr{F}$, $\mathsf{E}[|X_t|] < \infty$, and $X_s = \mathsf{E}[X_t \,|\, \mathscr{F}_s]$ (a.s.) for all $s, t \in \mathbb{T}$, $s \leq t$. ○

(9.3) REMARK: (a) Martingales are required to be integrable ($\mathsf{E}[|X_t|] < \infty$), but submartingales and supermartingales are required to be integrable only half way, that is, $\mathsf{E}[X_t^+] < \infty$ and $\mathsf{E}[X_t^-] < \infty$, respectively. In some situations submartingales or supermartingales will be required to be integrable, and this would mean imposing the additional requirements $\mathsf{E}[X_t^-] < \infty$ for submartingales and $\mathsf{E}[X_t^+] < \infty$ for supermartingales.

(b) The result from exercise (7.5) remains valid for (sub)martingales in continuous time, and the argument is the same. Thus, if X is any martingale (any submartingale) and $f : \mathbb{R} \mapsto \mathbb{R}$ is any convex function (any increasing convex function) such that $\mathsf{E}[|f(X_t)|] < \infty$ for every $t \in \mathbb{R}_+$, then in either case the process $(f(X_t))_{t \in \mathbb{N}}$ is a submartingale. A parallel statement involving supermartingales mapped through functions that are concave and decreasing can be formulated in the obvious way. ○

(9.4) EXERCISE: Suppose that $X = (X_t)_{t \in \mathbb{T}}$ is a martingale (or submartingale, or supermartingale) with respect to $\mathscr{F} = (\mathscr{F}_t)_{t \in \mathbb{T}}$. Then suppose that another filtration, $\mathscr{G} = (\mathscr{G}_t)_{t \in \mathbb{T}}$, is such that $\mathscr{G}_t \subseteq \mathscr{F}_t$ for every $t \in \mathbb{T}$. Prove that if X happens to be adapted to $\mathscr{G}$, then X must be a martingale (or submartingale, or supermartingale) relative to $\mathscr{G}$ as well. ○

(9.5) EXERCISE AND FIRST EXAMPLES: Any adapted, integrable, and increasing process is a submartingale. Similarly, any adapted, integrable, and decreasing process is a supermartingale. Although trivial, these examples illustrate the main idea: heuristically speaking, martingales are the stochastic analogues of constant functions, submartingales are the stochastic analogues of increasing functions, and supermartingales are the stochastic analogues of decreasing functions.

Suppose now that the integrable r.v. ξ represents a particular quantity of interest and the filtration $\mathscr{F} \equiv (\mathscr{F}_t)_{t \in \mathbb{R}_+}$ represents the flow of time. The conditional expectation $\mathsf{E}[\xi \,|\, \mathscr{F}_t]$ is then a mathematical metaphor for "the best forecast of ξ given the information available at time t." Prove that the process $\big(\mathsf{E}[\xi \,|\, \mathscr{F}_t]\big)_{t \in \mathbb{R}_+}$ is an $(\mathscr{F}, \mathsf{P})$-martingale. Ideally, we would like all martingales to be of this form, as it allows an entire process to be described in terms of a single r.v. As we are about to discover, although such a representation would not be possible for all martingales that are of interest to us, it is still possible to characterize the class of martingales that can be written in this form, and this characterization will play an important role in our study.

In some situations supermartingales come about through the following procedure: given an adapted process $A = (A_t)_{t \in \mathbb{R}_+}$, the sample paths of which are increasing and are such that $A_\infty(\omega) \stackrel{\text{def}}{=} \lim_{t \to \infty} A_t(\omega)$ is an integrable r.v., consider the process

$$X_t \stackrel{\text{def}}{=} \mathsf{E}[A_\infty \,|\, \mathscr{F}_t] - A_t\,, \quad t \in \bar{\mathbb{R}}_+\,,$$

which gives the time evolution of the forecast of the distance between the current (at time t) level of A and the level of A at time $t = \infty$. Prove that the process X is a supermartingale[104] (heuristically, a "decreasing function" in a stochastic sense, as the error in the prediction of A_∞ should decrease over time). ○

As supermartingales and submartingales are mirror images of one another, one of these two notions is mathematically redundant. However, the main interest in such objects is not about the mathematics, and the established tradition is to work with both.

(9.6) EXERCISE: Prove that if X is a supermartingale such that $\mathsf{E}[X_T] = \mathsf{E}[X_0]$ for some $T \in \mathbb{R}_{++}$, then $(X_t)_{t \in [0,T]}$ must be a martingale.

HINT: Using the relations $\mathsf{E}[X_0 - X_T \,|\, \mathscr{F}_0] \geq 0$ and $\mathsf{E}[X_T] = \mathsf{E}[X_0]$, conclude that $\mathsf{E}[X_0 - X_T \,|\, \mathscr{F}_0] = 0$, that is, $X_0 = \mathsf{E}[X_T \,|\, \mathscr{F}_0]$, which then implies

$$X_0 \geq \mathsf{E}[X_t \,|\, \mathscr{F}_0] \geq \mathsf{E}\big[\mathsf{E}[X_T \,|\, \mathscr{F}_t] \,|\, \mathscr{F}_0\big] = \mathsf{E}[X_T \,|\, \mathscr{F}_0] = X_0 \,.$$

Conclude that $X_t = \mathsf{E}[X_T \,|\, \mathscr{F}_t]$ for every $t \in [0, T]$. ○

(9.7) EXERCISE (FROM CONTINUOUS TO DISCRETE TIME): Prove that if X is an $(\mathscr{F}, P)$-(sub/super)martingale, then for any discrete sequence of time values $0 \leq t_0 < t_1 < t_2 < \ldots$, the sequence $(X_{t_i})_{i \in \mathbb{N}}$ is a discrete-time (sub/super)martingale for the discrete-time filtration $(\mathscr{F}_{t_i})_{i \in \mathbb{N}}$. ○

One particularly important and insightful example of a continuous-time martingale is the Brownian motion, as we now detail.

(9.8) EXERCISE: Let W be the Brownian motion and suppose that the base filtration $\mathscr{F}$ is taken to be its natural filtration $\mathscr{F}^W$. Prove that for every $\sigma \in \mathbb{R}_{++}$:

(a) σW is a martingale;

(b) $\sigma W \pm \sigma \iota$ is either a submartingale or a supermartingale, depending on whether the sign is $+$ or $-$ (try to understand which is which and why);

(c) $\sigma^2 W^2$ and $e^{\sigma W}$ are submartingales;

(d) $\sigma^2 W^2 - \sigma^2 \iota$ is a martingale;

(e) $e^{\sigma W - \frac{1}{2}\sigma^2 \iota}$ is a martingale. ○

(9.9) REMARK: Although completely elementary, the statement in (d) above provides an important insight and a guiding principle: it is telling us that the process $W^2 - [W, W]$ is a martingale; in other words, what one needs to subtract from the submartingale W^2 in order to make it into a martingale is precisely the quadratic variation process $[W, W]$. As we are about to find out, this property is a special case of a much more general result known as the Doob–Meyer decomposition. To see how powerful and useful such a result can be, suppose that we have somehow established that the process that "corrects" a submartingale and turns it into a martingale exists and is unique, and, furthermore, when the submartingale at hand

[104] Supermartingales of this form are known as "potentials" – see (Meyer 1966) and (Dellacherie and Meyer 1975) for details. Such objects are outside of our main concerns.

happens to be the square of a particular martingale, then that (unique) correction is given by the quadratic variation of the martingale. Since it is not very difficult to guess that subtracting the process ι from the submartingale W^2 yields a martingale, we would then know that $[W, W] = \iota$ – and without the need to compute the limit of $Q^{\mathcal{P}_n}(W_\omega, W_\omega)_t$ as $|\mathcal{P}_n| \to 0$. This phenomenon is one of the cornerstones of stochastic analysis, and is one of the reasons for the need to replace the classical Leibniz–Newton calculus in the context of deterministic functions with the Itô calculus in the context of (sub/super)martingales.

The insight provided by part (e) is just as important. It is telling us that $e^{\sigma W - \frac{\sigma^2}{2}[W,W]}$ is a martingale. This property, too, transcends the Brownian motion and, as we will soon realize, has implications that are very broad, deep, and crucially important for many applications. ○

The next two exercises illustrate the extent to which the martingale property depends on the filtration – and in a crucial way. This is consistent with the intuition: enlarging the filtration by including information about the future of the martingale should destroy the martingale property.

(9.10) EXERCISE: Let $(W_t)_{t \in [0,1]}$ be the Brownian motion and let

$$\mathcal{G}_t = \sigma(W_1) \vee \sigma\big(W_s : s \in [0,t]\big), \quad t \in [0,1],$$

that is, for any $t \in [0, 1]$, in addition to the information about the random variables $(W_s)_{s \in [0,t]}$, the σ-field $\mathcal{G}_t$ contains information also about the r.v. W_1. Calculate $\mathsf{E}\big[W_t \,|\, \mathcal{G}_s\big]$ for $0 \le s < t \le 1$, and show that $(W_t)_{t \in [0,1]}$ is not a $\mathcal{G}$-martingale.

HINT: Since $W_1 = W_s + (W_1 - W_s)$ and since $W_1 - W_s$ is independent of $\sigma(W_u : u \in [0, s])$ for $s \in [0, 1]$, the only additional information found in $\mathcal{G}_s$ that is not in $\mathcal{F}^W_s$ is the information about $W_1 - W_s$. What can you say about $\mathsf{E}[W_t \,|\, W_1 - W_s]$ for $0 \le s < t \le 1$? Consult exercise (3.66). ○

In some situations it is possible to enlarge the filtration without breaking the martingale property. Indeed, if the new information added to the filtration happens to be unrelated to the martingale, then the enlargement cannot affect the martingale property.

(9.11) EXERCISE: Let B and W be two independent Brownian motions and let

$$\mathcal{G}_t = \sigma\big(W_s : s \in [0,t]\big) \vee \sigma\big(B_u : u \in \mathbb{R}_+\big), \quad t \in \mathbb{R}_+,$$

that is, $\mathcal{G}_t$ contains information about the history of W up until time t and all the information about B, including information about its entire future after time t. Prove that W is a $\mathcal{G}$-Brownian motion; in particular, a $\mathcal{G}$-martingale. ○

9.2 Poisson Process and First Encounter with Lévy Processes

The Brownian motion was our first canonical example of a martingale. However, this important process is more than a martingale: it is also a process with stationary and independent increments, and, as we saw in the previous chapter, while nowhere differentiable, its sample paths are still continuous with probability 1. We are yet to find other interesting examples of martingales, and, while looking for such examples, it would be useful and insightful to also look for concrete examples of martingales that are not continuous – perhaps even find examples of processes with stationary and independent increments that are not continuous. To gain some insight into this matter, we now introduce an important class of processes, the main study of which will be taken up in chapter 16.

(9.12) LÉVY PROCESSES : A Lévy process is any process with stationary and independent increments that starts from 0. To be precise, the process $X \equiv (X_t)_{t \in \mathbb{R}_+}$ is said to be a *Lévy process* if it has the following attributes: (a) the sample paths X_ω are a.s. càdlàg; (b) given any $s, t \in \mathbb{R}_+$, $s \leq t$, the r.v. $X_t - X_s$ is independent of the σ-field $\mathscr{F}_s^X$ and has the same distribution law as the r.v. X_{t-s} (in particular, $X_0 = 0$ a.s.). ○

It is clear that the Brownian motion is a Lévy process, but, thanks to the continuity of its sample paths, it is not particularly interesting as a Lévy process, in the sense that its structure of a Lévy process is not any more revealing than its structure of an a.s. continuous martingale with quadratic variation equal to the deterministic process $\imath$. We have some compelling reasons not to insist on continuous sample paths. Indeed, heuristically speaking, being constrained to be continuous, the Brownian motion manages to be an interesting martingale by forcing its sample paths to wiggle so much that they cannot be differentiated even at a single point. However, we would like to allow the uncertainty in our models to be manifested not just in terms of highly irregular but otherwise continuous sample paths, but also in terms of sample paths that are quite regular (of finite variation) but discontinuous (later we will learn that these are the only choices, unless we want to give up on the class of martingales altogether). The canonical example of a process of the latter catagory is the Poisson counting process N, the construction of which we now outline. The Poisson process is also our first truly interesting example of a Lévy process. After an appropriate compensation, it will become our second canonical example of a martingale, and our first truly interesting example of a martingale with discontinuous sample paths.

(9.13) POISSON PROCESS: Let $(\xi_i)_{i \in \mathbb{N}_{++}}$ be any sequence of i.i.d. random variables on $(\Omega, \mathcal{F}, P)$ with common distribution law identified as the exponential law of parameter $c \in \mathbb{R}_{++}$, that is, $P(\xi_i > t) = e^{-c\,t}$, $t \in \mathbb{R}_+$. Consider the sequence of events that occur (one at the time) at the random moments $t_1 = \xi_1$, $t_2 = \xi_1 + \xi_2$, $t_3 = \xi_1 + \xi_2 + \xi_3$, … . The Poisson process of parameter $c \in \mathbb{R}_{++}$, $N = (N_t)_{t \in \mathbb{R}_+}$, counts the events that occur before time t; to be precise: $N_t = \sum_{n \in \mathbb{N}_{++}} 1_{\{t_n \le t\}}$, $t \in \mathbb{R}_+$, where the summation can be shown[9.16] to be a.s. finite. ○

It is clear that the sample paths $t \rightsquigarrow N_\omega(t)$ are càdlàg functions, with jumps of size 1 occurring at times $t_1(\omega), t_2(\omega), \dots$, that are flat between the jumps. Indeed, the r.v. $1_{\{t_n \le t\}}$ can be written equivalently as $1_{[t_n(\omega),\infty[}(t)$, and this is certainly a càdlàg function of t. One can show[9.15] that t_n is distributed with gamma law of parameters (c, n), that is, distributed in $\mathbb{R}_+$ with density $t \rightsquigarrow \frac{c^n}{(n-1)!} t^{n-1} e^{-c\,t}$. As a result,[9.17] for any fixed $t \in \mathbb{R}_{++}$, the r.v. N_t is distributed with a Poisson law of parameter ct; in particular, $\mathsf{E}[N_t] = c\,t$. Since the exponential distribution lacks memory, in that[9.14] $P(\xi_1 > t \,|\, \xi_1 > s) = P(\xi_1 > t - s)$ for every choice of $0 \le s < t$, the increment $N_t - N_s$ must be independent of the history $(N_u)_{u \in [0,s]}$. Indeed, this history comes down to information about the number of jumps before time s, the exact times of those jumps, and also information about the time elapsed since the last jump – clearly, the time until the next jump would be independent of this information. Thus, if at time s the clock and the events-count are both reset to 0, the world would look the same as it looked at time 0. In particular, the r.v. $N_t - N_s$ would be distributed (independently from $(N_u)_{u \in [0,s]}$) according to the Poisson law of parameter $c(t - s)$. It is not difficult to show[9.18] that the process $N - c\,t \equiv (N_t - c\,t)_{t \in \mathbb{R}_+}$ is a martingale relative to the filtration $\mathscr{F}^N$, and so is also the process $(N - c\,t)^2 - c\,t$.

TO SUMMARIZE: a *Poisson process* is any a.s. càdlàg stochastic process N that has independent increments, starts from 0, and is such that for every choice of $s, t \in \mathbb{R}_+$, $s \le t$, the r.v. $N_t - N_s$ is distributed with Poisson law of parameter $c(t - s)$, for some fixed $c \in \mathbb{R}_{++}$. The parameter c is called the intensity of N (in addition to being called the parameter of N). Every Poisson process is thus a Lévy process, as is immediate from the definition.

(9.14) EXERCISE: Prove that the exponential distribution lacks memory: if the r.v. ξ is such that $P(\xi > t) = e^{-ct}$ for every $t \in \mathbb{R}_{++}$, then $P(\xi > t \,|\, \xi > s) = P(\xi > t - s)$ for every choice of $s, t \in \mathbb{R}_+$, $s \le t$. ○

(9.15) EXERCISE: Prove that if ξ and η are two *independent* r.v. distributed[2.28f] with gamma laws of parameters, respectively, (c, m) and (c, n), then $\xi + \eta$ is distributed with gamma law of parameters $(c, m+n)$. Prove by induction that for every $n \in \mathbb{N}_{++}$, the r.v. $t_n = \xi_1 + \cdots + \xi_n$ from (9.13) is distributed with gamma law of parameters (c, n).

HINT: The gamma density of parameters $(c, 1)$ is the same as the exponential density of parameter c. Use the fact that the density of two independent r.v.[3.27] is given by the convolution of their respective densities. ○

(9.16) EXERCISE: Consider the random variables $t_1 \le t_2 \le \dots$ that were introduced in (9.13) and prove that for any fixed $t \in \mathbb{R}_{++}$ the probability for infinitely many of the events $\{t_n \le t\}$, $n \in \mathbb{N}_{++}$, to occur is 0.

HINT: Use the first Borel–Cantelli lemma in conjunction with the result from the previous exercise. ○

(9.17) EXERCISE: Using the result from exercise (9.15), prove that if N is a Poisson process of parameter $c \in \mathbb{R}_{++}$, then for every fixed $t \in \mathbb{R}_{++}$ the r.v. N_t is distributed with Poisson law of parameter ct.

HINT: Observe that

$$P(N_t = k) = P(t_k \leq t, t_{k+1} > t) = P(t_k \leq t, \xi_{k+1} > t - t_k),$$

where t_k and ξ_{k+1} are independent. ○

(9.18) EXERCISE: Let N be the Poisson process of parameter $c \in \mathbb{R}_{++}$. Using the fact that N is a process of independent increments and that $N_t - N_s$ has a Poisson law of parameter $c\,(t-s)$, prove that the process $N - c\iota$ is a martingale with respect to the filtration $\mathscr{F}^N$, and that therefore $(N - c\iota)^2$ is an $\mathscr{F}^N$-submartingale. Finally, prove that $(N - c\iota)^2 - c\iota$ is an $\mathscr{F}^N$-martingale. ○

9.3 Regularity of Paths, Optional Stopping, and Convergence

We begin this section with the continuous-time counterpart of Doob's maximal inequality. This result captures one of the fundamental features of submartingales and illustrates one of the reasons why submartingales (martingales) are such a useful modeling tool. Since most relations between random variables that we are ever going to encounter will be understood to hold only outside of some set of measure 0, from now on we will drop the qualifiers "a.s." and "P-a.s." for brevity. In addition, unless noted otherwise, limits of random variables will be understood as limits a.s.

(9.19) DOOB'S MAXIMAL INEQUALITY REVISITED: If X is any right-continuous submartingale and $X^* \equiv (X_t^* \stackrel{\text{def}}{=} \sup_{s\in[0,t]} X_s)_{t\in\mathbb{R}_+}$ is its running maximum, then (recall the notation $X_t^+ \stackrel{\text{def}}{=} X_t \vee 0$)

† $$\lambda\, P\big(X_t^* \geq \lambda\big) \leq \mathsf{E}\big[X_t\, 1_{\{X_t^* \geq \lambda\}}\big] \leq \mathsf{E}[X_t^+],$$

for every $\lambda \in \mathbb{R}_{++}$ and $t \in \mathbb{R}_+$. ○

A careful reader might have noticed that the above formulation of Doob's maximal inequality is very similar to its discrete-time counterpart from (7.31). As the next exercise makes it clear, this is not a coincidence: the continuous-time version of Doob's inequality is a more or less straightforward corollary of the discrete-time version.

(9.20) EXERCISE: Prove Doob's maximal inequality in (9.19†).

HINT: Argue that the right-continuity implies

$$X_t^* = \lim_{n\to\infty} \left(\max_{i\in\mathbb{N}_{|2^n}} X_{i\,2^{-n}t}\right),$$

and notice that the max above is an increasing function of n. Then observe that

$$\{X_t^* \geq \lambda\} = \bigcap_{k \in \mathbb{N}_{++}} \bigcup_{n \in \mathbb{N}_{++}} \left\{ \max_{i \in \mathbb{N}_{|2^n}} X_{i\,2^{-n}t} \geq \lambda - \frac{1}{k} \right\},$$

notice that the events under the union are increasing, and, using the continuity of the probability set function, conclude that

$$P\left(X_t^* \geq \lambda\right) = \lim_{k\to\infty}\left(\lim_{n\to\infty} P\left(\max_{i \in \mathbb{N}_{|2^n}} X_{i\,2^{-n}t} \geq \lambda - \frac{1}{k}\right)\right).$$

Finally, use⊸(7.31) the discrete version of Doob's maximal inequality applied to every (discrete-time) submartingale $(X_{i\,2^{-n}t})_{i \in \mathbb{N}_{|2^n}}$, for every $n \in \mathbb{N}_{++}$. ○

(9.21) AN ILLUSTRATION: Applied to the Brownian motion W, in conjunction with the relation established in (8.118†), Doob's maximal inequality gives:

$$\frac{2\lambda}{\sqrt{2\pi t}} \int_\lambda^\infty e^{-\frac{x^2}{2t}}\,\mathrm{d}x \leq \frac{1}{\sqrt{2\pi t}} \int_0^\infty x\,e^{-\frac{x^2}{2t}}\,\mathrm{d}x,$$

and this comes down to

$$\lambda\left(1 - \operatorname{erf}\left(\frac{\lambda}{\sqrt{2t}}\right)\right) \leq \sqrt{\frac{t}{2\pi}} \quad \text{for all} \quad \lambda \in \mathbb{R}_{++} \text{ and } t \in \mathbb{R}_{++}.$$

This is one example of a relatively straightforward probabilistic argument that leads to a functional relation that is not immediately obvious to guess or verify with other means. ○

Now we turn to the classical regularity results for submartingales. These results can be restated in terms of supermartingales in the obvious way, and can be traced back to the groundbreaking works of (Doob 1953) and (Doob 1984) – perhaps two of the most influential books in the theory of stochastic processes. In a first step, we fix the probability space $(\Omega, \mathcal{F}, P)$ and the filtration $\mathscr{F} \setminus\!\!\subset \mathcal{F}$, but insist, for now, that $\mathscr{F}$ may not be right-continuous and may not be P-complete. The theorem that we state first is telling us that the right limit X_{t+} and the left limit X_{t-} are always meaningful and finite for any submartingale X (and therefore also for any supermartingale or martingale), and without the need to impose any additional regularity conditions on X or the filtration $\mathscr{F}$; in particular, (sub/super)martingales *cannot have oscillating discontinuities* – this is important.

(9.22) REGULARITY THEOREM I: (Revuz and Yor 1999, II (2.5), (2.6), (2.7)) Let X be any $(\mathscr{F}, P)$-submartingale. Then the following claim can be made for P-a.e. $\omega \in \Omega$: the limit $X_\omega(t-) \stackrel{\text{def}}{=} \lim_{\mathbb{Q} \ni s \nearrow t} X_\omega(s)$ exists for every $t \in \mathbb{R}_{++}$ and the limit $X_\omega(t+) \stackrel{\text{def}}{=} \lim_{\mathbb{Q} \ni s \searrow t} X_\omega(s)$ exists for every $t \in \mathbb{R}_{+}$. In addition, with probability 1

$$X_t \leq \mathsf{E}\left[X_{t+} \mid \mathscr{F}_t\right] \quad \text{for every } t \in \mathbb{R}_+,$$

and

$$X_{t-} \leq \mathsf{E}\big[X_t \mid \mathscr{F}_{t-}\big] \quad \text{for every } t \in \mathbb{R}_+ .$$

If $\mathsf{E}[X_t] > -\infty$ for every $t \in \mathbb{R}_+$, then the first inequality becomes equality if the real function $t \rightsquigarrow \mathsf{E}[X_t]$ is right-continuous, and the second inequality becomes equality if $t \rightsquigarrow \mathsf{E}[X_t]$ is left-continuous. The processes $X_+ \equiv (X_{t+})_{t \in \mathbb{R}_+}$ and $X_- \equiv (X_{t-})_{t \in \mathbb{R}_{++}}$ are both submartingales for the filtrations, respectively, $\mathscr{F}_+$ and $\mathscr{F}_-$, and become martingales (for $\mathscr{F}_+$ and $\mathscr{F}_-$) if X is a martingale. ○

The next theorem says something rather interesting: in conjunction with the submartingale property, a.s. right-continuity implies that the sample paths admit left limits a.s., that is, it implies that the sample paths are a.s. càdlàg. In such situations, too, we will omit "a.s." with the understanding that all qualifiers attached to sample paths apply only outside of some set of measure 0; for example "right continuous" will be understood as "a.s. right continuous," "admits left limits," will be understood as "admits left limits a.s." – and so on.

(9.23) REGULARITY THEOREM II: (ibid., II (2.8)) Let X be any right-continuous $(\mathscr{F}, P)$-submartingale. Then:

(a) X is a submartingale also with respect to $\mathscr{F}_+$ and with respect to its completion $\tilde{\mathscr{F}}_+$;

(b) for P-a.e. $\omega \in \Omega$ the sample path X_ω is càdlàg and bounded on every finite interval (in particular, X_ω does not explode in finite time).[105] ○

The next theorem gives the martingale analogue – of a sort – of Kolmogorov's continuity theorem. One should not take this analogy too far, however, as the two statements are very different in nature.

(9.24) REGULARITY THEOREM III: (Meyer 1966, VI 4) Suppose that the filtration $\mathscr{F}$ is right-continuous and P-complete, and let X be any $(\mathscr{F}, P)$-submartingale, such that $\mathsf{E}[X_t] > -\infty$ for every $t \in \mathbb{R}_+$.[106] Then X admits a right-continuous – and, consequently, according to (9.23), càdlàg – modification if and only if the function $\mathbb{R}_+ \ni t \rightsquigarrow \mathsf{E}[X_t] \in \mathbb{R}$ is right-continuous. ○

(9.25) EXERCISE: Suppose that the filtration $\mathscr{F}$ is right-continuous and P-complete, and let X be any $(\mathscr{F}, P)$-martingale. Prove that X always admits a càdlàg modification. Prove that if, in addition, X happens to be L^1-continuous, that is, $\lim_{s \to t} \mathsf{E}[|X_t - X_s|] = 0$, then X admits a left-continuous modification as well.

HINT: The first part is an immediate corollary from (9.24). For the second part, argue that X_- is a modification of X by following these steps: The family $\mathsf{E}[|X_t| \,|\, \mathscr{F}_s]$, $s \in [0, t[$, plus $\mathsf{E}[|X_t| \,|\, \mathscr{F}_{t-}]$ is u.i., and therefore$^{\multimap (4.25)}$ so is also the family $|X_s|$, $s \in [0, t[$, plus $|X_{t-}|$ (observe that$^{\multimap (9.22)}$ $X_{t-} = \mathsf{E}[X_t \,|\, \mathscr{F}_{t-}]$), and therefore so is also the family $|X_s - X_{t-}|$, $s \in [0, t[$, which converges to 0 as $s \nearrow t$. Thus$^{\multimap (4.23)}$ $\lim_{s \nearrow t} \mathsf{E}[|X_s - X_{t-}|] = 0$. Notice that this last property holds

105 The claim that X_ω must be bounded on every finite interval is from (Meyer 1966, VI 3).

106 This condition is the same $\mathsf{E}[|X_t|] < \infty$ (X is integrable – not just half-integrable).

without the L^1-continuity of X, and, in fact, claiming that X is L^1-continuous is the same as claiming that $\mathsf{E}[|X_t - X_{t-}|] = 0$. ○

(9.26) REMARK: According to the last theorem and exercise, the fact that any Brownian motion admits both left-continuous and right-continuous modifications could have been established without the use of Kolmogorov's continuity theorem, as long as one can show that the Brownian motion retains the martingale property after completing the filtration, which, as we know, makes the filtration right-continuous. On the other hand, if we use Kolmogorov's theorem to establish that the Brownian motion admits a continuous modification (actually, we only need to establish the existence of a right-continuous modification), then theorem (9.23) would guarantee that the Brownian motion retains the martingale property with respect to the complete Brownian filtration. ○

(9.27) THE USUAL CONDITIONS WARNING: Unless otherwise noted, and without further warning, from now on all filtrations that crop up in our study will be assumed to obey the usual condition introduced in (8.64). In particular, the natural filtration of a given process X, which we introduced in (8.31) and denoted by $\mathscr{F}^X$, will always be assumed (unless noted otherwise) to have been passed through the usual augmentation procedure (the symbol $\mathscr{F}^X$ will remain unchanged, even though we just augmented its meaning). The regularity theorems stated above make it clear why this convention is being made: if the usual conditions are imposed, then the most common regularity issues involving the sample paths of martingales will be automatically resolved; in particular, a càdlàg modification will always exist. By way of an example, if $\mathscr{F}$ is a filtration and ξ is an integrable r.v., then the process $\mathsf{E}[\xi|\mathscr{F}] \equiv (\mathsf{E}[\xi|\mathscr{F}_t])_{t\in\mathbb{R}_+}$ is certainly a martingale. Ideally, we would like to be able to write any martingale in this form. While, just in general, this may be too much to ask for, as we will find out soon, under certain not too restrictive conditions, this wish does come true. However, without imposing any conditions on the filtration $\mathscr{F}$, we cannot expect the sample paths of $\mathsf{E}[\xi|\mathscr{F}]$ to have any useful regularity features, in which case working with processes of the form $\mathsf{E}[\xi|\mathscr{F}]$ would be rather impractical. At the same time, if we were to insist that $\mathscr{F}$ obeys the usual conditions, then $\mathsf{E}[\xi|\mathscr{F}]$ would always admit a càdlàg modification – and regardless of the choice of ξ. Another fundamental reason for insisting on complete and right-continuous filtrations is the crucially important theorem in (8.66). ○

(9.28) WHY CÀDLÀG: With the usual conditions now in place, it is clear from theorem (9.24) and exercise (9.25) that nothing essential will be lost if we were to remain within the confines of models in which only processes with càdlàg paths are allowed – we will see more evidence in section 9.4. For this reason, after the present section all (sub/super)martingales will be assumed to be càdlàg by default. It is interesting that if we insists on continuity – and in many cases we will do so – then we must give up on finite variation paths. This will be evidenced in (9.60) and in (15.15) where we will show that a predictable martingale with finite variation paths must be trivial, in that its sample paths must be constant. To put it another way, a nontrivial martingale can be left-continuous – or, more generally, predictable – only if its paths "wiggle wildly" (have infinite total variation on finite intervals). The Brownian motion is a prime example of this phenomenon. However, càdlàg paths without predictability can be had with essentially perfect regularity (i.e., finite variation), and the Poisson process is a prime example: except for a finite number of jumps, its sample paths are perfectly flat on finite intervals. The essence of the result in (15.15) is that, heuristically speaking, if combined with finite variation, predictability removes the truly stochastic part in the instantaneous innovation and makes the martingale uninteresting. By way of an example, the (continuous) sample paths of the process $\left(\int_0^t W_s \, ds\right)_{t \in \mathbb{R}_+}$ are "too nice" to make it possible for this process to be a martingale. Not only is this process predictable, but its sample paths have finite variation. The instantaneous change during an infinitesimally short time period $[t, t + dt]$ is $W_t \, dt + \mathcal{O}(dt^2)$ and can be considered (at infinitesimal level) known at time t up to first-order. Thus, the instantaneous stochastic innovation is of second-order. In contrast, the instantaneous stochastic innovation in the Brownian motion, which we express formally as dW_t, is of first-order (and the price for this feature is the non-differentiability of the paths). We also see that for a continuous process the order of stochastic innovation is captured by the quadratic variation: vanishing quadratic variation means that the innovation is of second-order, whereas a positive quadratic variation means that the innovation is of first-order, and nontrivial quadratic variation is not possible with continuous sample paths of finite variation. ○

Our next goal is to understand whether – and under what conditions – it may be possible to express a martingale X in the form $X_t = \mathsf{E}[X_\infty \,|\, \mathcal{F}_t]$, $t \in \mathbb{R}_+$, for some appropriately chosen $\mathcal{F}_\infty$-measurable r.v. X_∞, with $\mathcal{F}_\infty$ understood to be the σ-field $\vee_{t \in \mathbb{R}_+} \mathcal{F}_t$. In other words, we would like to know whether we can extend the range of the time variable from $\mathbb{R}_+$ to $\bar{\mathbb{R}}_+ = \mathbb{R}_+ \cup \{\infty\}$ and still maintain the martingale property of X. In an ideal world, this extension should be provided by $\lim_{t \to \infty} X_t$ – at least in those situations where this limit exists. We will see in exercise (9.31) that our world is not ideal – at least not in this respect – but even so, after imposing certain reasonable conditions on X the representation $X = \mathsf{E}[X_\infty \,|\, \mathcal{F}]$ will become possible. As we are about to see, generally, such results hinge on uniform integrability. The first step in this program is to state the continuous-time analogue of Doob's convergence theorem.

(9.29) DOOB'S CONVERGENCE THEOREM: (Meyer 1966, VI 6) Let X be any right-continuous $(\mathcal{F}, P)$-submartingale, such that $\sup_{t \in \mathbb{R}_+} \mathsf{E}[X_t^+] < \infty$ (in particular, X could be uniformly integrable).[107] Then there is an $\mathcal{F}_\infty$-measurable r.v.

[107] In (Meyer 1966) all (sub/super)martingales are integrable by definition. However, see (Revuz and Yor 1999, II (2.1), (2.2), (2.10)), the actual argument involves only one-sided integrability, and this is

$X_\infty < +\infty$, such that $\lim_{t\to\infty} X_t = X_\infty$. The last convergence takes place also in L^1-sense if the submartingale $(X_t)_{t\in\mathbb{R}_+}$ is uniformly integrable, in which case $(X_t)_{t\in\bar{\mathbb{R}}_+}$ is also a uniformly integrable submartingale. ○

(9.30) REMARK: According to the last theorem, any right-continuous and negative submartingale converges as $t \to \infty$. Consequently, the same must be true for any positive and right-continuous supermartingale. ○

(9.31) EXERCISE: Let W be the Brownian motion. Prove that $\lim_{t\nearrow\infty} e^{W_t - \frac{1}{2}t} = 0$, even though $(e^{W_t-\frac{1}{2}t})_{t\in\mathbb{R}_+}$ is a positive martingale with $\mathsf{E}[e^{W_t-\frac{1}{2}t}] = 1$ for every $t \in \mathbb{R}_+$. Conclude that $(W_t - \frac{1}{2}t) \to -\infty$.

HINT: Explain why $\lim_{t\nearrow\infty} e^{W_t-\frac{1}{2}t}$ always exists. Then explain why, if it exists, this limit cannot be anything other than 0. Can the law of the iterated logarithm be of any help? ○

(9.32) EXERCISE: Let W be the Brownian motion. Explain why the convergence theorem from (9.29) does not apply to the submartingales W and W^2. ○

Since theorem (9.29) can be re-stated in terms of supermartingales in the obvious way, and since a martingale is both a submartingale and a supermartingale, we have the following corollary from (9.29).

(9.33) THEOREM$^{\multimap(9.89)}$: (Revuz and Yor 1999, II (3.1)) Let X be any $(\mathscr{F}, P)$-martingale. Then the following conditions are equivalent:

(a) the family of random variables $(X_t)_{t\in\mathbb{R}_+}$ is uniformly integrable;

(b) L^1-$\lim_{t\to\infty} X_t$ exists;

(c) there is an integrable r.v. $X_\infty \prec \mathscr{F}_\infty$ such that $X_t = \mathsf{E}\big[X_\infty \,|\, \mathscr{F}_t\big]$ for any $t \in \mathbb{R}_+$.

If these conditions are satisfied, then $\lim_{t\to\infty} X_t$ exists also a.s. and coincides (a.s.) with the r.v. X_∞ from part (c). ○

(9.34) REMARKS: (a) The last theorem is essentially a rephrasing of Lévy's convergence theorem in (7.37) and the result in (7.36). If the r.v. $\xi \prec \mathscr{F}_\infty$ is integrable, then by the result in (4.24) the martingale $X_t \overset{\text{def}}{=} \mathsf{E}[\xi \,|\, \mathscr{F}_t]$, $t \in \mathbb{R}_+$, is uniformly integrable, and, just as in section 7.3, it is again possible to claim that $X_\infty = \lim_{t\to\infty} \mathsf{E}[\xi \,|\, \mathscr{F}_t] = \xi$, but we stress one more time that the matter is delicate and hinges on the monotone class theorem, which allows one to conclude that $\mathsf{E}[\xi \,|\, \mathscr{F}_t] = \mathsf{E}[X_\infty \,|\, \mathscr{F}_t]$ for every $t \in \mathbb{R}_+$ is only possible if $\xi = X_\infty$ (it is essential that $\mathscr{F}_\infty = \vee_{t\in\mathbb{R}_+} \mathscr{F}_t$ and that $\xi \prec \mathscr{F}_\infty$ and $X_\infty \prec \mathscr{F}_\infty$ for this claim to be made).

(b) We saw in (9.30) that every right-continuous and negative submartingale must converge, but now we see that "negative" can be replaced with the condition $X_t \le \mathsf{E}[\xi \,|\, \mathscr{F}_t]$, $t \in \mathbb{R}_+$, for some integrable r.v. $\xi \prec \mathscr{F}_\infty$. Indeed, with this choice

---

the main reason why the definitions found in more recent texts require supermartingales or submartingales to be integrable only half-way. In many cases results obtained for integrable submartingales can be extended for half-integrable ones by considering the process $X \vee a$, for a fixed $a \in \mathbb{R}$, which is always an integrable submartingale as long as X is a submartingale.

$(X_t - \mathsf{E}[\xi\,|\,\mathscr{F}_t])_{t\in\mathbb{R}_+}$ is a negative submartingale, so that $\lim_{t\to\infty}(X_t - \mathsf{E}[\xi\,|\,\mathscr{F}_t])$ exists, and since $\lim_{t\to\infty}\mathsf{E}[\xi\,|\,\mathscr{F}_t] = \xi$ exists, it follows that $\lim_{t\to\infty} X_t$ must exist. Furthermore, we have $X_\infty \stackrel{\text{def}}{=} \lim_{t\to\infty} X_t \leq \xi$. Similarly, if X is some right-continuous supermartingale and there is an integrable r.v. $\xi \prec \mathscr{F}_\infty$ such that $X_t \geq \mathsf{E}[\xi\,|\,\mathscr{F}_t]$ for every $t \in \mathbb{R}_+$, then $\lim_{t\to\infty} X_t$ exists and $X_\infty \stackrel{\text{def}}{=} \lim_{t\to\infty} X_t \geq \xi$.

(c) Any martingale is automatically uniformly integrable if restricted to a finite time interval. Indeed, in that case condition (9.33b) is satisfied in a trivial way, but the feature also obtains directly from (4.24).

(d) We stress that, as we saw in exercise (9.31), the conclusions from theorem (9.33) may not hold for martingales that are bounded only in L^1-sense, since such martingales may not be u.i. However, martingales that are L^p-bounded for some $p > 1$ are$^{\multimap(4.22)}$ u.i., and therefore satisfy the conditions in (9.33). We will establish in (9.89) that for such martingales the convergence in (9.33b) actually takes place in the L^p-norm. ○

The next exercise provides a particularly interesting and useful illustration of the situation described in (9.33).

(9.35) EXERCISE: Let $(\Omega, \mathcal{F}, P)$ be a probability space, let Q be another probability measure on $\mathcal{F}$ such that $Q \lesssim P$, let $\mathscr{F} \subset \mathcal{F}$ be some (any) filtration inside $\mathcal{F}$, and let $R_t \stackrel{\text{def}}{=} (\mathrm{d}Q/\mathrm{d}P)\!\restriction\!\mathscr{F}_t$, $t \in \mathbb{R}_+$. Show that R is a uniformly integrable $(\mathscr{F}, P)$-martingale. What is $\lim_{t\to\infty} R_t$ in this case? ○

(9.36) EXERCISE: Show that the martingale $(e^{W_t - \frac{1}{2}t})_{t\in\mathbb{R}_+}$ is not bounded in L^p-norm for any $p > 1$. Does this matter and why? ○

We are now ready to address the important matter of *optional stopping*. The intuition and the motivation for this result are the same as in the discrete-time case.

(9.37) THEOREM (OPTIONAL STOPPING FOR SUBMARTINGALES): (Meyer 1966, VI 13) Let X be any right-continuous and integrable $(\mathscr{F}, P)$-submartingale, such that, for some integrable r.v. ξ,

$$X_t \leq \mathsf{E}[\xi\,|\,\mathscr{F}_t] \quad \text{for all } t \in \mathbb{R}_+ .$$

Then, for any two stopping times, $\mathcal{s}$ and $\mathcal{t}$, that take values in $\bar{\mathbb{R}}_+$ and are such that $P(\mathcal{s} \leq \mathcal{t}) = 1$, the r.v. $X_{\mathcal{s}}$ and $X_{\mathcal{t}}$ are both integrable and

$$X_{\mathcal{s}} \leq \mathsf{E}\big[X_{\mathcal{t}}\,|\,\mathscr{F}_{\mathcal{s}}\big],$$

with the understanding that$^{\multimap(9.34b)}$ $X_{\mathcal{s}} = X_\infty$ on the set $\{\mathcal{s} = \infty\}$ and $X_{\mathcal{t}} = X_\infty$ on the set $\{\mathcal{t} = \infty\}$.

By symmetry, an analogous optional stopping result can be formulated for supermartingales in the obvious way. ○

(9.38) REMARKS: (a) The requirement in the last theorem for the submartingale X to be integrable can be removed by the following standard argument: If X is a submartingale (not necessarily integrable), then for any $a \in \mathbb{R}$ the process $X \vee a =$

$(a \vee X_t)_{t \in \mathbb{R}_+}$ is an integrable submartingale and the condition $X_t \leq \mathsf{E}[\xi \,|\, \mathscr{F}_t]$ for all $t \in \mathbb{R}_+$, for some integrable r.v. ξ, implies[108]

$$a \vee X_t \leq a \vee \mathsf{E}[\xi \,|\, \mathscr{F}_t] \leq \mathsf{E}[a \vee \xi \,|\, \mathscr{F}_t],$$

where $a \vee \xi$ is integrable (since ξ is). The last theorem gives

$$a \vee X_{\mathcal{s}} \leq \mathsf{E}\big[a \vee X_{\mathcal{t}} \,|\, \mathscr{F}_{\mathcal{s}}\big],$$

for any two stopping times $\mathcal{s}$ and $\mathcal{t}$ such that $\mathcal{s} \leq \mathcal{t}$. Passing to the limit as $a \to -\infty$ we get $X_{\mathcal{s}} \leq \mathsf{E}\big[X_{\mathcal{t}} \,|\, \mathscr{F}_{\mathcal{s}}\big]$. We can no longer claim that $X_{\mathcal{s}}$ and $X_{\mathcal{t}}$ are integrable, but, since $a \vee X_{\mathcal{s}}$ and $a \vee X_{\mathcal{t}}$ are integrable for $a < 0$, we see that $\mathsf{E}[X^+_{\mathcal{s}}] < \infty$ and $\mathsf{E}[X^+_{\mathcal{t}}] < \infty$.

(b) If X is any negative and right-continuous submartingale, then we can repeat the argument above with $\xi = 0$. Nothing changes if "negative" is replaced with "bounded from above."

(c) Optional stopping with bounded stopping times always holds for right-continuous submartingales that are integrable. Indeed, if X is one such submartingale and the stopping times involved take values in $[0, T]$, then the optional stopping theorem can be applied with $\xi = X_T$ (effectively, we are replacing X with X^T). The integrability requirement for X can again be removed as in (a) above: as we just realized, for any finite $a < 0$ and any two bounded stopping times $\mathcal{s}$ and $\mathcal{t}$ such that $\mathcal{s} \leq \mathcal{t}$, we have $a \vee X_{\mathcal{s}} \leq \mathsf{E}\big[a \vee X_{\mathcal{t}} \,|\, \mathscr{F}_{\mathcal{s}}\big]$.

(d) By symmetry, any right-continuous supermartingale can be stopped optionally if it is positive (that is, bounded from below), or stopped optionally with bounded stopping times. ○

(9.39) POSITIVE SUPERMARTINGALES: (Dellacherie and Meyer 1980, VI 17) This class of processes plays an important role in the theory of stochastic processes, as will become evident from our encounter with the general version of Girsanov's theorem in (15.44), for example. It is also instrumental in the domain of asset pricing. In a first step toward decrypting the nature of a positive supermartingale X, suppose$^{\multimap (9.28)}$ that X is càdlàg with $P(X_0 > 0) = 1$, and extend X to the time domain $\bar{\mathbb{R}}_+$ by postulating $X_\infty = 0$. Given any $n \in \mathbb{N}_{++}$, let

$$\mathcal{t}_n(\omega) \stackrel{\text{def}}{=} \inf\Big\{t \in \bar{\mathbb{R}}_+ : X_t(\omega) < \frac{1}{n}\Big\},$$

and observe that $\mathcal{t}_n$ is a stopping time.[109] By definition $\mathcal{t}_{n+1} \geq \mathcal{t}_n$, so that $\mathcal{t}_\infty \stackrel{\text{def}}{=} \lim_n \mathcal{t}_n$ is an $\bar{\mathbb{R}}_+$-valued stopping time. Since X is càdlàg, $X_{\mathcal{t}_n} \leq \frac{1}{n}$, and by optional stopping,

$$\frac{1}{n} \geq X_{\mathcal{t}_n} \geq \mathsf{E}[X_{\mathcal{t}_\infty} \,|\, \mathscr{F}_{\mathcal{t}_n}] \quad \text{for every } n \in \mathbb{N}_{++}.$$

---

[108] Note that $x \rightsquigarrow a \vee x$ is an increasing convex function.

[109] Note that the filtration is right continuous.

This means that $\mathsf{E}[X_{t_\infty}] = 0$, and, since $X_{t_\infty} \geq 0$, it means that $X_{t_\infty} = 0$. Using optional stopping one more time, it is easy to see that $X_{t_\infty + q} = 0$ for every $q \in \mathbb{Q}_{++}$. Since X is càdlàg, the sample paths of $X 1_{[\![t_\infty, \infty]\!]}$ must be identically 0 on $\mathbb{R}_+$,[110] and, clearly, the sample paths of X must be strictly positive on $[0, t_\infty[$; in fact, they must be uniformly bounded away from 0 on $[0, s]$ for any stopping time s such that $P(s < t_\infty) = 1$. If it is known at the outset that $P(X_t > 0) = 1$ for every $t \in \mathbb{R}_+$, then $P(t_\infty \leq t) = 0$ for every $t \in \mathbb{R}_+$, which is to say, $P(t_\infty = \infty) = 1$, or, to put it another way, the sample paths of X are strictly positive on $\mathbb{R}_+$; in fact, they are uniformly bounded away from 0 on any finite time interval. ○

(9.40) EXERCISE: Using theorems (9.33) and (9.37), in conjunction with theorem (4.24), prove the following important result. ○

(9.41) THEOREM (OPTIONAL STOPPING FOR MARTINGALES): (Revuz and Yor 1999, II (3.2), (3.3)) Let X be any uniformly integrable and right-continuous $(\mathscr{F}, P)$-martingale. Then the entire collection of all r.v. X_t, for all choices of the $\mathscr{F}$-stopping time t, is uniformly integrable (in particular, X_t is an integrable r.v. for every stopping time t). Furthermore,

$$X_s = \mathsf{E}\big[X_t \,|\, \mathscr{F}_s\big] = \mathsf{E}\big[X_\infty \,|\, \mathscr{F}_s\big]$$

for every two $\mathscr{F}$-stopping times s and t, such that $P(s \leq t) = 1$. ○

(9.42) REMARK: Since martingales constrained to finite time intervals are uniformly integrable, obviously, optional stopping on finite time-intervals (or, equivalently, with bounded stopping times) for right-continuous martingales always holds. However, we stress that, in general, the optional stopping theorem does not hold for unbounded stopping times without the uniform integrability requirement, as is evident from the following example, which we borrow from (ibid., p. 70). Given a Brownian motion W, consider the martingale $(X_t \stackrel{\text{def}}{=} e^{W_t - \frac{1}{2}t})_{t \in \mathbb{R}_+}$, which starts from $X_0 = 1$ and, as we already know, is not uniformly integrable. Then define the stopping times $s = 0$ and $t = \inf\{t > 0 \colon X_t \geq 2\}$ and observe that, since X is continuous, $X_t = 2$ and therefore

$$\mathsf{E}[X_t \,|\, \mathscr{F}_s] = \mathsf{E}[2 \,|\, \mathscr{F}_0] = 2 \neq 1 = X_s \equiv X_0 .$$

Optional stopping fails in this case because (a) the martingale X is not uniformly integrable, and (b) the stopping time t is not bounded. ○

Our next result is a simple but important application of the optional stopping theorem. What is interesting is that it provides an alternative characterization of the attribute "martingale," which, quite surprisingly, does not involve conditional expectations.

(9.43) THEOREM: (ibid., II (3.5)) Suppose that the process X is $\mathscr{F}$-adapted and càdlàg. Then X is an $(\mathscr{F}, P)$-martingale if and only if for every bounded $\mathscr{F}$-stopping time t, $\mathsf{E}[|X_t|] < \infty$ and $\mathsf{E}[X_t] = \mathsf{E}[X_0]$. ○

110 Any such property is understood to hold with probability 1.

PROOF OF (9.43): If X is any $\mathscr{F}$-martingale and t is any bounded stopping time, then X can be stopped optionally at time t. In particular, X_t is integrable and $\mathsf{E}[X_t \,|\, \mathscr{F}_0] = X_0$.

To prove the statement in the other direction, let $A \in \mathscr{F}_s$ for some $s \in \mathbb{R}_+$, and let $t = s1_A + t1_{A^\complement}$ for some $t \in [s, +\infty[$. It is easy to check that t is a stopping time (obviously bounded). Thus

$$\mathsf{E}[X_0] = \mathsf{E}[X_t] = \mathsf{E}[X_t 1_A] + \mathsf{E}[X_t 1_{A^\complement}] = \mathsf{E}[X_s 1_A] + \mathsf{E}[X_t 1_{A^\complement}].$$

Furthermore, since every deterministic time is bounded stopping time,

$$\mathsf{E}[X_0] = \mathsf{E}[X_t] = \mathsf{E}[X_t 1_A] + \mathsf{E}[X_t 1_{A^\complement}].$$

Comparing the two identities, we find that $\mathsf{E}[X_t 1_A] = \mathsf{E}[X_s 1_A]$. Since this relation is satisfied for every $A \in \mathscr{F}_s$, and since $X_s \prec \mathscr{F}_s$ (X is $\mathscr{F}$-adapted by assumption), by the definition of conditional expectation, we have $X_s = \mathsf{E}[X_t \,|\, \mathscr{F}_s]$. ○

Apart from plain curiosity, the main reason for including the last result in the present chapter is the application outlined in the next two exercises. We remind the reader that at this point in the exposition we assume without warning that all filtrations satisfy the usual conditions all processes are càdlàg.

(9.44) EXERCISE: Let X be any $(\mathscr{F}, P)$-martingale and let t be any $\mathscr{F}$-stopping time. Using the optional stopping theorem and the result from (8.56), conclude that the stopped process X^t is a martingale relative to⊸(8.48) the stopped filtration $\mathscr{F}^t \overset{\text{def}}{=} (\mathscr{F}_{t\wedge t})_{t\in\mathbb{R}_+}$.

HINT: Since⊸(8.43) $\mathscr{F}^t_t \equiv \mathscr{F}_{t\wedge t} \subseteq \mathscr{F}_t$, any $\mathscr{F}^t$-stopping time is automatically an $\mathscr{F}$-stopping time. Argue that the stopped process X^t is adapted to the filtration $\mathscr{F}^t$ and that $s \wedge t$ is a bounded $\mathscr{F}$-stopping time for every choice of the bounded $\mathscr{F}^t$-stopping time s.[111] ○

(9.45) EXERCISE: Given a filtration $\mathscr{F}$ and an $\mathscr{F}$-stopping time t, prove that any $\mathscr{F}^t$-martingale is automatically a $\mathscr{F}$-martingale. In particular, the stopped process X^t from the previous exercise is also an $\mathscr{F}$-martingale.

HINT: The filtration $\mathscr{F}^t$ is⊸(8.50) right-continuous and complete, and any $\mathscr{F}^t$-adapted process is also $\mathscr{F}$-adapted. Since⊸(8.49) $t \wedge t$ is a bounded $\mathscr{F}^t$-stopping time, if X is an $\mathscr{F}^t$-martingale, optional stopping gives $X_{t\wedge t} = \mathsf{E}[X_t \,|\, \mathscr{F}^t_{t\wedge t}] = X_t$. Thus, the right-continuous process X^t is a modification of the right-continuous process X, which is to say X^t and X are indistinguishable (due to the right continuity). If s is any bounded $\mathscr{F}$-stopping time, then $X_s = X_{s\wedge t}$ and⊸(8.49) $s \wedge t$ is a bounded $\mathscr{F}^t$-stopping time, so that⊸(9.43) $\mathsf{E}[X_s] = \mathsf{E}[X_{s\wedge t}] = \mathsf{E}[X_0]$. ○

---

[111] Using the same argument, one can show that X^t is an $\mathscr{F}$-martingale and conclude that X^t must be an $\mathscr{F}^t$-martingale by the argument of (9.4).

9.4 Doob–Meyer Decomposition

If X is any uniformly integrable (u.i.) and càdlàg martingale, then the family of random variables of the form X_t, for all choices of the stopping time t, is u.i. This follows from the result in (4.24) in conjunction with the optional stopping theorem for u.i. martingales that gives $X_t = \mathsf{E}[X_\infty \,|\, \mathscr{F}_t]$. However, if X fails to be a martingale – say, if X is only a submartingale, or a supermartingale – u.i. of X will not imply u.i. of the family of all X_t. For this reason we need to identify the general class of processes that preserve the u.i. of its values at stopping times.

(9.46) PROCESSES OF CLASS (D) AND CLASS (DL): Given any (finite or infinite, open or closed) interval $I \subseteq \mathbb{R}_+$, by $\mathscr{T}_I$ we denote the collection of all stopping times t (for a filtration that is understood from the context) that take values in I. A right-continuous and adapted process X is said to be of *class* (D) if the family of random variables $\{X_t 1_{\{t<\infty\}} : t \in \mathscr{T}_{[0,\infty]}\}$ is uniformly integrable. If one can claim only that $\{X_t : t \in \mathscr{T}_{[0,T]}\}$ is uniformly integrable for every $T \in \mathbb{R}_{++}$, then X is said to be of *class* (DL) (understood as "locally in class (D)"). Thus, any bounded, right-continuous and adapted process is automatically of class (D), and therefore of class (DL). Also, since any deterministic time is automatically a stopping time, any process that is of class (D) is uniformly integrable, and any process that is of class (DL) is integrable (uniformly on finite intervals). ○

(9.47) INCREASING PROCESSES: A finite real-valued process A is said to be an *increasing process* if it is right-continuous, adapted, starts from 0 ($A_0 = 0$), and its sample paths A_ω are increasing functions on $\mathbb{R}_+$.[112] By convention, the left limit of an increasing process at 0 is set to $A_{0-} = A_0 = 0$, and this is another way of saying that the measure $\mathrm{d}A_\omega$ has no mass at $\{0\}$. If the increasing process A is also such that $\mathsf{E}[A_t] < \infty$ for every $t \in \mathbb{R}_+$, we say that A is an *integrable increasing process*. Note that for a process with increasing sample paths the properties "right-continuous" and "càdlàg" are the same. If A is any increasing process, then $A_\infty \overset{\text{def}}{=} \lim_{t\to\infty} A_t$ is a well-defined random variable with $0 \equiv A_0 \leq A_\infty \leq \infty$. If $\mathsf{E}[A_\infty] < \infty$, we say that the increasing process A is *integrable at* ∞, and this property is easily seen to be equivalent to the uniform integrability of A. Any deterministic, càdlàg, and increasing function $A : \mathbb{R}_+ \mapsto \mathbb{R}$ that vanishes at 0 is a special case of an increasing process. ○

The concept that we are about to develop is crucial for the general theory of stochastic processes. It is also instrumental for many applications.

(9.48) DOOB–MEYER DECOMPOSITION:[113] A right-continuous submartingale X is said to admit Doob–Meyer decomposition (henceforth D-M-d for short) if there is a *predictable* and⊶(9.47) increasing process A with the property that $M \overset{\text{def}}{=} X - A$

[112] We remind the reader that all such features are understood to hold with probability 1.

[113] Named after the American mathematician Joseph Leo "Joe" Doob (1910–2004), and the French mathematician Paul-André Meyer (1934–2003).

is a martingale. In other words, if the submartingale property of X is viewed as a "deficiency" of a sort, in the sense that it represents a deviation from the martingale property, then the D-M-d for X (if it exists) says that the "deficiency" of X is caused by a predictable and increasing additive component A, so that

$$X = M + A\,.$$

To put it another way, "X admits D-M-d" means that X can be corrected into a martingale by subtracting from it an increasing process A, and we stress that this correction must be predictable, and also integrable if X is integrable. Note that since an increasing process A is right-continuous by definition, and since we assume that X is right-continuous, if it exists, the martingale component in the decomposition above, M, must be right-continuous as well.

As the submartingale property of X is the same as the supermartingale property of $(-X)$, and the martingale property of M is the same as the martingale property of $(-M)$, the D-M-d for supermartingales is simply the mirror image of the D-M-d for submartingales. To be precise, a right-continuous supermartingale X is said to admit D-M-d if there is a predictable (for the same filtration) and increasing process A with the property that $M \stackrel{\text{def}}{=} X + A$ is a (right-continuous) martingale. ○

Clearly, for the concept that we just introduced to be useful, an existence result must be available. We state this result next, essentially rephrasing (Meyer 1966, VII 31), after taking into account the clarification from (Dellacherie 1972, V 27).

(9.49) THEOREM (EXISTENCE OF D-M-D): An integrable and right-continuous (sub/super)martingale X admits D-M-d if and only if it is of class (DL), and if this condition is satisfied, then, up to indistinguishability, the predictable and increasing component, A, in the D-M-d of X is unique, and therefore so is also the martingale component M. ○

(9.50) EXERCISE: Prove that any càdlàg submartingale which is bounded from below (in particular, any positive submartingale) is integrable and of class (DL), and therefore admits D-M-d. Prove that any uniformly integrable submartingale which is bounded from below must be of class (D).

HINT: A submartingale that is bounded from below is integrable and⊸ (9.38) optional stopping on finite intervals holds. Thus, if $T \in \mathbb{R}_{++}$ and $t \in \mathscr{T}_{[0,T]}$ is any stopping time bounded by T, then $C \leq X_t \leq \mathsf{E}[X_T | \mathscr{F}_t]$. If X is uniformly integrable, then the convergence theorem in (9.29) gives $X_t \leq \mathsf{E}[X_\infty | \mathscr{F}_t]$ for every $t \in \mathbb{R}_{++}$. In addition, optional stopping gives $C \leq X_t \leq \mathsf{E}[X_\infty | \mathscr{F}_t]$ for every stopping time t. ○

(9.51) PREDICTABLE COMPENSATORS: Given any integrable and right-continuous submartingale, or a supermartingale, X, that is of class (DL), the unique integrable, predictable, and increasing process from the D-M-d of X is called the *predictable compensator*, or simply the *compensator of* X, and is denoted by $\langle X \rangle$.

Note that the process $\langle X \rangle$ is automatically càdlàg and starts from 0, since these features are incorporated in the definition of "increasing process." If the process X is a square integrable martingale (that is, $\mathsf{E}[X_t^2] < \infty$ for every $t \in \mathbb{R}_+$), then the process $X^2 = (X_t^2)_{t \in \mathbb{R}_+}$ is a positive submartingale and its predictable compensator $\langle X^2 \rangle$ is well defined. For reasons that will become clear very soon, this predictable compensator is also denoted by $\langle X, X \rangle$ and is also called the *angled bracket of* X, or the *predictable quadratic variation of* X (it may or may not coincide with the quadratic variation of X – see below). Without a qualifier, the term "bracket" will always be understood to mean "angled bracket." ○

(9.52) EXERCISE: Let W be the Brownian motion (for the underlying filtration $\mathscr{F}$ and probability P). Prove that

$$\langle W, W \rangle = \iota = [W, W]. \quad ○$$

(9.53) EXERCISE: Let N be the Poisson process with parameter $c \in \mathbb{R}_{++}$ and suppose that $\mathscr{F} = \mathscr{F}^N$ is the natural filtration of N, so that⊸ (9.18) $N - c\iota$ is an $(\mathscr{F}, P)$-martingale. Give the predictable quadratic variation $\langle N - c\iota, N - c\iota \rangle$. ○

Given the instrumental role that the predictable quadratic variation (angled bracket) plays in both theory and practice, it would be desirable to establish certain general regularity features for this process. For now, we are going to recall the results from (Dellacherie 1972, V 50 and 52), which we state next, but are going to revisit the matter later – see (15.18).

(9.54) THEOREM (UNIFORM INTEGRABILITY OF THE ANGLED BRACKET): Suppose that the right-continuous submartingale, or supermartingale, X is of class (D) (in particular, X could be bounded). Then its predictable compensator $\langle X \rangle$ is uniformly integrable, that is, $\mathsf{E}[\langle X \rangle_\infty] < \infty$. ○

The continuity property of the predictable compensator is somewhat more involved. In order to formulate this result we need to introduce a few definitions.

(9.55) DEFINITION (REGULAR PROCESS): (a) A stopping time t, with values in $\bar{\mathbb{R}}_+$, is said to be *predictable* if there is a sequence, $(s_i)_{i \in \mathbb{N}}$, of $\bar{\mathbb{R}}_+$-valued stopping times, which is increasing in that $s_i \leq s_{i+1}$, $i \in \mathbb{N}$, converges to t in that $\lim_{i \to \infty} s_i = t$, and is such that $s_i < t$ on $\{t > 0\}$. If these conditions are met, then the sequence $(s_i)_{i \in \mathbb{N}}$ is said to be an *announcing sequence* for the (predictable) stopping time t.

(b) A right-continuous and uniformly integrable (sub/super)martingale X is said to be *regular* if the following equivalent conditions are satisfied for every predictable stopping time t:

$$\mathsf{E}[X_{t-}] = \mathsf{E}[X_t] \quad \Leftrightarrow \quad X_{t-} = \mathsf{E}[X_t \,|\, \mathscr{F}_{t-}].$$

That these two conditions are equivalent follows from the martingale convergence theorem, in conjunction with the fact that the r.v. X_{t-} is integrable and, in the case of a supermartingale, $X_{t-} \geq \mathsf{E}[X_t \,|\, \mathscr{F}_{t-}]$ for every predictable stopping time t –

see (Dellacherie 1972, V 50–51). Furthermore, it can be shown – see (Dellacherie and Meyer 1980, VI 50) – that if the sequence of stopping times $(\mathscr{s}_i)_{i\in\mathbb{N}}$ announces t, then the relation $X_{t-} = \mathsf{E}[X_t \,|\, \mathscr{F}_{t-}]$ follows from the relation

$$\mathsf{E}[X_{t-}] \equiv \mathsf{E}[X_{\lim_i \mathscr{s}_i}] = \lim_i \mathsf{E}[X_{\mathscr{s}_i}]\,.$$

In particular, the last relation is automatically satisfied if the u.i. (sub/super)martingale X is continuous.

(c) (Dellacherie and Meyer 1980, VI 50) A right-continuous and adapted process X that is of class (D) is said to be *regular* if

$$\mathsf{E}[X_{\lim_i \mathscr{s}_i}] = \lim_i \mathsf{E}[X_{\mathscr{s}_i}]$$

for every increasing and uniformly bounded sequence of stopping times $(\mathscr{s}_i)_{i\in\mathbb{N}}$. Any continuous and adapted process that is of class (D) is easily seen to be regular in this sense. ○

(9.56) REMARK: As we have seen above, the attribute "regular" from (9.55c) is compatible with the one from (9.55b): a right-continuous and uniformly integrable (sub/super)martingale X that is regular in the sense of (9.55c) – in particular, that is continuous, adapted, and of class (D) – is automatically regular in the sense of (9.55b). As it stands the notion of "regular" from (9.55c) is more stringent than the one from (9.55b) – and it should be, as it is applicable to a broader class of processes. ○

(9.57) THEOREM (CONTINUITY OF THE ANGLED BRACKET): (Dellacherie 1972, V 52) Suppose that the right-continuous (sub/super)martingale X is of class (D) (in particular, X could be bounded). Then its u.i. predictable compensator $\langle X\rangle$ is continuous if and only if X is regular in the sense of (9.55b). ○

In addition, we have the following general result.

(9.58) THEOREM: (ibid., V 10) Every uniformly integrable and right-continuous martingale X is regular: if t is any predictable stopping time, then (P-a.s.)

$$X_{t-} = \mathsf{E}[X_t \,|\, \mathscr{F}_{t-}] = \mathsf{E}[X_\infty \,|\, \mathscr{F}_{t-}]\,. \quad ○$$

9.5 Local Martingales and Semimartingales

Our goal in this section is to connect the angled bracket introduced in the previous section with the important notion of quadratic variation, which we encountered earlier in our study of Brownian motion. In addition, we are going to broaden the class of (sub/super)martingales, and will introduce the largest class of processes for which a useful form of "calculus" can be developed – this is a project that we will undertake in the next chapter. Such tools are instrumental for the models that will be developed in later chapters. We will gradually narrow the focus of our study to continuous processes until we return to the study of processes with discontinuities in chapter 15.

Following the setup introduced at the beginning of this chapter, the probability space $(\Omega, \mathcal{F}, P)$ and the filtration $\mathscr{F} \subset \mathcal{F}$ are assumed given (with the usual conditions automatically assumed). All attributes such as "adapted," "(sub/super)martingale," and so on, are understood relative to $\mathscr{F}$ and P.

(9.59) EXERCISE: Let M be any L^2-martingale (alias: square integrable martingale), that is, let $\mathsf{E}[M_t{}^2] < \infty$ for every $t \in \mathbb{R}_+$. Prove that for $0 \le s \le u < t$,

$$\mathsf{E}\left[(M_t - M_s)^2 \mid \mathscr{F}_u\right] = \mathsf{E}\left[(M_t - M_u)^2 \mid \mathscr{F}_u\right] + (M_u - M_s)^2$$

and

$$\mathsf{E}\left[M_t{}^2 \mid \mathscr{F}_s\right] = M_s{}^2 + \mathsf{E}\left[(M_t - M_s)^2 \mid \mathscr{F}_s\right],$$

which is equivalent to

$$\mathsf{E}\left[(M_t - M_s)^2 \mid \mathscr{F}_s\right] = \mathsf{E}\left[M_t{}^2 - M_s{}^2 \mid \mathscr{F}_s\right]. \quad \circ$$

The next result, which is also the subject of exercise (15.15), will be instrumental for many important steps that we will need to take later.

(9.60) THEOREM: (Dellacherie 1972, V 39) If M is a martingale and its sample paths are continuous and of finite variation on finite intervals, then the sample paths of M are constants. To put it another way, if satisfied simultaneously, the properties "M is a martingale" and "M_ω is a continuous function of finite variation for P-a.e. $\omega \in \Omega$" make the martingale M trivial. $\circ$

The result that we state next is both interesting and important. It tells us that the same procedure$^{\multimap (8.98)}$ that gave us the quadratic variation $[W, W]$ of the Brownian motion W can be repeated with any bounded and continuous martingale M in place W. In this general case, too, one can obtain convergence in probability of the discretized (subdivision based) quadratic variations by way of establishing L^2-convergence first – just as we did in the Brownian motion case, except that with a generic martingale one has to extract the L^2 convergence from the general properties of martingales, rather than the concrete features of Brownian motion, which is somewhat easier. Much of what we call "stochastic calculus" hinges on the following fundamental result.

(9.61) THEOREM: (Revuz and Yor 1999, IV (1.3)) Let M be any continuous and bounded ($|M_t| < C < \infty$ for every t) martingale. Then M has finite quadratic variation, denoted by $[M, M]$, that can be obtained with the procedure described in (8.98). Furthermore, $[M, M]$ is the unique (up to a modification) process that satisfies simultaneously the following three conditions:

(a) $[M, M]$ is adapted and starts from 0;
(b) $[M, M]$ is continuous and increasing;
(c) $M^2 - [M, M]$ is a martingale (obviously continuous). $\circ$

(9.62) CONNECTION WITH THE ANGLED BRACKET: In the setting of (9.61), the process M^2 is a continuous (in particular, right-continuous) and bounded (in particular, of class (DL)) submartingale. Therefore$^{\multimap (9.49)}$ M^2 admits D-M-d, and the last theorem comes down to the claim that the quadratic variation process $[M, M]$ is nothing but the predictable compensator of M^2, that is, $[M, M]$ can be identified as the angled bracket $\langle M^2 \rangle \equiv \langle M, M \rangle$ from (9.51). If the continuity of the bounded martingale M is relaxed to right-continuity, that is, if M is only càdlàg, then, even if the quadratic variation $[M, M]$ is defined, and even if $M^2 - [M, M]$ can be shown to be a martingale – all of which is indeed true, as we will see later – in general, $[M, M]$ is going to be adapted, but, typically, will not be predictable. Thus, in general, the identification of the quadratic variation $[M, M]$ as the predictable compensator $\langle M, M \rangle$ is possible only if M is continuous. There would have been no need for us to ever introduce the angled bracket if throughout this space we were to remain confined to the realm of continuous processes. Since later in the sequel we are going to work with processes with jumps, the angled bracket and the square bracket are both needed. In what follows we will adopt the following convention: the quadratic variation of a continuous process is always going to be written as an angled bracket, while the square bracket is going to be used to denote the quadratic variation only for processes with discontinuities. This way the angled bracket will always denote a predictable process (its original meaning), while the square bracket will show up as a "red flag" of a sort, reminding us that the process may not be predictable. ○

It is clear that the angled bracket (predictable quadratic variation) is much simpler and easier to work with, compared to the martingale itself. For this reason, it would be useful to develop some form of calculus based on the angled bracket alone. Our first step in this direction is to broaden the notion of angled bracket for a class of processes that is larger than the class of bounded and continuous martingales (e.g., the Brownian motion is not in this class, yet, as we know, has finite quadratic variation). In fact, we are going to extend the notion of martingale by relaxing the martingale property. Another important reason for this extension is to avoid the integrability requirement in the definition of martingale – this would be the only way for us to eventually arrive at constructions that are invariant under equivalent changes of the measure.

(9.63) LEMMA: (Revuz and Yor 1999, IV (1.4)) Let M be any bounded and continuous martingale, let $\mathcal{t}$ be any stopping time, and let $M^{\mathcal{t}}$ denote the martingale M stopped at time $\mathcal{t}$. Then the quadratic variation of $M^{\mathcal{t}}$ (recall that $M^{\mathcal{t}}$ is a martingale – obviously, bounded and continuous) coincides with the quadratic variation of M stopped at time $\mathcal{t}$: for P-a.e. $\omega \in \Omega$ the sample paths $\langle M^{\mathcal{t}}, M^{\mathcal{t}} \rangle_\omega$ and $\langle M, M \rangle_\omega^{\mathcal{t}}$ coincide, that is,

$$\langle M^{\mathcal{t}}, M^{\mathcal{t}} \rangle_\omega(t) = \langle M, M \rangle_\omega^{\mathcal{t}}(t) \quad \text{for all } t \in \mathbb{R}_+ . \quad ○$$

(9.64) REDUCTION TO A U.I. MARTINGALE: Let X be any adapted and right-continuous process and let $\mathcal{t}$ be any stopping time. We say that $\mathcal{t}$ reduces X (to a u.i. martingale) if the stopped process $(1_{\{\mathcal{t}>0\}} X_t^{\mathcal{t}})_{t \in \mathbb{R}_+}$ is a u.i. martingale. ○

Note that the property "t reduces X" is equivalent to: $\mathsf{E}[1_{\{t>0\}}|X_0|] < \infty$ (that is, X_0 is integrable on $\{t > 0\}$) and $(X_t^t - X_0^t)_{t\in\mathbb{R}_+}$ is a u.i. martingale (obviously starting from 0).[114] The multiplication by the indicator $1_{\{t>0\}}$ is simply meant to remove the requirement for X_0 to be integrable without stopping. However, in many concrete applications X_0 is going to be integrable, or even constant, and in any such case the factor $1_{\{t>0\}}$ becomes superfluous.

(9.65) EXERCISE: Let X be any adapted and right-continuous process and let t be any stopping time that reduces X. Prove that any stopping time that is dominated by t also reduces X: if s be a stopping time such that $s \leq t$ then s reduces X. ○

Now we come to an important definition.

(9.66) DEFINITION (LOCAL MARTINGALES): Let M be any adapted and right-continuous process. We say that M is a *local martingale* if there is a sequence of stopping times, $(t_i)_{i\in\mathbb{N}}$, called *localizing sequence*, such that: (a) $t_i \leq t_{i+1}$ for every $i \in \mathbb{N}$, (b) $\lim_i t_i = \infty$, and (c) t_i reduces M for every $i \in \mathbb{N}$. ○

(9.67) REMARK: The last definition will not change if condition (c) is replaced with the requirement that $1_{\{t_i>0\}}M^{t_i}$ is a martingale (not necessarily u.i.) for every $i \in \mathbb{N}$. Indeed, martingales stopped at finite times are automatically uniformly integrable. Thus, if $1_{\{t_i>0\}}M_t^{t_i}$ is only a martingale, then $1_{\{t_i\wedge i>0\}}M_t^{t_i\wedge i}$ is a uniformly integrable martingale, so that $(t_i \wedge i)_{i\in\mathbb{N}}$ becomes a localizing sequence that satisfies (a), (b), and (c) in (9.66). More important, if M is a continuous local martingale, as defined in (9.66), then, without any loss of generality, the associated localizing sequence can always be chosen so that M^{t_i} is a bounded (thus u.i.) martingale. Indeed, setting

$$s_i = \inf\{t \in \mathbb{R}_+ : |M_t| \geq i\}, \quad i \in \mathbb{N},$$

and assuming that $(t_i)_{i\in\mathbb{N}}$ is any localizing sequence that reduces M, we have[115] $t_i \wedge s_i \to \infty$, so that $(t_i \wedge s_i)_{i\in\mathbb{N}}$ is a localizing sequence that reduces M to a bounded – and not just u.i. – martingale. ○

The next exercise lists some of the most basic and common features of local martingales that will be used throughout the sequel. Part (d) is especially important in the context of mathematical finance and will be called upon in a number of occasions.

(9.68) EXERCISE: Prove that: (a) every right-continuous martingale is a local martingale; (b) the sum, or the difference, of two local martingales is a local mar-

[114] Notice that $1_{\{t>0\}}X_t^t - 1_{\{t>0\}}X_0^t = X_t^t - X_0^t$.

[115] As always, in this and in other similar expressions we suppose that $\inf\emptyset = \infty$. Notice that the relation $\lim_{i\to\infty} s_i < \infty$ could only mean that the sample paths of M explode in finite time. Such explosions cannot occur if M is continuous on $\mathbb{R}_+$; in particular, $\lim_{i\to\infty} s_i = \infty$.

tingale; (c) a local martingale stopped at any stopping time is still a local martingale; (d) any positive local martingale is a supermartingale.[116]

HINT: For (a), take $t_i = i$; for (d), consider using$^{\multimap (3.42)}$ Fatou's lemma. ○

(9.69) IMPORTANT REMARK: We have seen$^{\multimap (7.24)}$ that discrete-time local martingales are just martingales that lack full integrability, while the conditional expectations are still meaningful (say, in terms of one-sided integrability). However, there is no counterpart to this statement in the context of continuous-time local martingales, as is clarified in (Revuz and Yor 1999, p. 123) with the following example: if W^i, $1 \leq i \leq 3$, are 3 independent Brownian motions, then, given any $x = (x_1, x_2, x_3) \in \mathbb{R}^3$ with $|x| \neq 0$, the process$^{\multimap (10.30)}$ $1/\sqrt{\sum_{i=1}^{3}(x_i + W_t^i)^2}$ is L^2-bounded (and therefore uniformly integrable) local martingale, but is not a martingale (ibid., V (2.13)). We see that even uniform integrability may not be enough to turn a local martingale into a martingale! ○

It turns out, however, that a stronger version of uniform integrability can force a local martingale to become a martingale. The precise result is the following:

(9.70) THEOREM: (ibid., IV (1.7)) The local martingale M is a martingale if and only if it is$^{\multimap (9.46)}$ of class (DL). ○

According to the last theorem, one particularly straightforward step to establish that a local martingale M is actually a martingale is to find a positive random variable, ξ, such that $\mathsf{E}[\xi] < \infty$ and $|M_t| \leq \xi$ for every $t \in \mathbb{R}_+$ P-a.s. Indeed, if this last condition is satisfied, then for any stopping time t, $|M_t| 1_{\{t<\infty\}} \leq \xi$, so that the entire family

$$\left\{M_t 1_{\{t<\infty\}} : t \text{ is a stopping time}\right\}$$

becomes uniformly integrable. Unfortunately, in most practical situations this property would be too much to ask for. Nevertheless, as we will soon realize, for at least one important class of local martingales this condition holds.

The main reason for inventing local martingales is the following result.

(9.71) THEOREM: (ibid., IV (1.8)) Let M be any continuous local martingale. Then M has finite quadratic variation, denoted by $[M, M]$, that can be obtained with the procedure described in (8.98). Furthermore, $[M, M]$ is the unique (up to a modification) process that satisfies simultaneously the following three conditions:

(a) $[M, M]$ is $\mathscr{F}$-adapted and starts from 0;

(b) $[M, M]$ is continuous and increasing;

(c) $M^2 - [M, M]$ is a continuous local martingale.

---

[116] In (Revuz and Yor 1999, IV (1.46)) this property is stated in the following more general form: if M is a local martingale such that the stopped process $(M^-)^i \equiv (-(M \wedge 0))^i$ is of class (D) for any integer $i \in \mathbb{N}$, then M is a supermartingale.

Thus, if N is another continuous local martingale, then$^{\circ\!\!\to (8.103)}$ the *quadratic covariation process*

$$[M,N] = \frac{1}{4}\Big([M+N,M+N] - [M-N,M-N]\Big)$$

has a continuous modification with finite variation sample paths, as the difference between two continuous and increasing processes. In addition, one has the following approximation result (ibid., IV (1.8)). For a given partition of $\mathbb{R}_+$, $\mathcal{p} = \{t_0, t_1, \ldots\}$, let

$$\mathcal{Q}^{\mathcal{p}}(M,N)_t \stackrel{\text{def}}{=} \sum\nolimits_{i \in \mathbb{N}} (M_{t\wedge t_{i+1}} - M_{t\wedge t_i})(N_{t\wedge t_{i+1}} - N_{t\wedge t_i}), \quad t \in \mathbb{R}_+ .$$

Then for every sequence of partitions of $\mathbb{R}_+$, $(\mathcal{p}_n)_{n\in\mathbb{N}}$, such that $|\mathcal{p}_n| \to 0$,

$$\mathsf{P}\text{-}\lim\nolimits_n \Big(\sup\nolimits_{t\in[0,T]} \big|\mathcal{Q}^{\mathcal{p}_n}(M,N)_t - [M,N]_t\big|\Big) = 0, \quad T \in \mathbb{R}_{++} ;$$

in particular,

$$\mathsf{P}\text{-}\lim\nolimits_n \Big(\sup\nolimits_{t\in[0,T]} \big|\mathcal{Q}^{\mathcal{p}_n}(M,M)_t - [M,M]_t\big|\Big) = 0, \quad T \in \mathbb{R}_{++} . \quad \circ$$

(9.72) CONVENTION AND REMARK: Since the quadratic variation associated with any continuous local martingale is continuous, it is automatically predictable, and we see from (9.71c) that it can be identified as a "predictable compensator" in the localized sense. That in the absence of jumps the quadratic variation indeed acts as a predictable compensator will be further evidenced in the next section, where we are going to establish that if M is any continuous L^2-bounded martingale, in that $\sup_{t\in\mathbb{R}_+} \mathsf{E}[M_t^2] < \infty$, then the predictable compensator of the submartingales M^2, as defined in (9.51), is precisely the quadratic variation $[M,M]$. For this reason, following our earlier convention – see (9.62) – in the rest of the sequel we will write $\langle M,M\rangle$ and $\langle M,N\rangle$ instead of $[M,M]$ and $[M,N]$ in any situation where the local martingales involved happen to be continuous – and only in situations where the local martingales involved are continuous. $\circ$

(9.73) IMPORTANT COMMENT: We have introduced – see (8.98) – the quadratic variation as a particular limit in probability for every $t \in \mathbb{R}_+$, and such a construction depends on the measure only up to equivalence. Furthermore, limit in probability P implies limit in probability Q for any $\mathsf{Q} \lesssim \mathsf{P}$. In particular, the quadratic variation will not change if the underlying measure is replaced with another measure that is absolutely continuous with respect to it, even though under the new measure the process may no longer be a local martingale. $\circ$

The following property is used – usually without a notice or a reference – in many calculations involving quadratic variation.

(9.74) PROPOSITION: (ibid., IV (1.11)) If M and N are any two continuous local martingales and if $\mathcal{t}$ is any stopping time, then

$$\langle M^{\mathcal{t}}, N^{\mathcal{t}}\rangle = \langle M, N^{\mathcal{t}}\rangle = \langle M^{\mathcal{t}}, N\rangle = \langle M, N\rangle^{\mathcal{t}} . \quad \circ$$

The next exercise is a special case of exercise (15.15).

(9.75) EXERCISE: Prove that a continuous local martingale M is constant, that is, the sample path M_ω is a constant function (on $\mathbb{R}_+$) for P-a.e. $\omega \in \Omega$, if and only if $\langle M, M\rangle_\omega$ is the constant function 0 for P-a.e. $\omega \in \Omega$.

HINT: Since $M - M_0$ has the same quadratic variation as M, without loss of generality set $M_0 = 0$. Then consider any localizing sequence that reduces $M^2 - \langle M, M\rangle$ to a sequence of bounded martingales (this is always possible since $M^2 - \langle M, M\rangle$ is a continuous local martingale). For each of these bounded martingales use the result from (9.60), and, finally, use the last proposition. ○

In fact, the conclusion from the last exercise admits the following extremely useful extension (often taken for granted).

(9.76) PROPOSITION: (ibid., IV (1.13)) Let M be any continuous local martingale. Then for P-a.e. $\omega \in \Omega$ one can claim that the intervals on which M_ω is constant are the same as the intervals on which $\langle M, M\rangle_\omega$ is constant, that is, for any interval $[a, b] \subset \mathbb{R}_+$ the claim that the restriction $M_\omega \restriction [a, b]$ is a constant is equivalent to the claim that the restriction $\langle M, M\rangle_\omega \restriction [a, b]$ is constant. ○

The next result is useful in a number of applications.

(9.77) THEOREM: (ibid., IV (1.15)) Let $X : \Omega\times\mathbb{R}_+ \mapsto \mathbb{R}$ and $Y : \Omega\times\mathbb{R}_+ \mapsto \mathbb{R}$ be any two stochastic processes that are $\mathcal{F}\otimes\mathcal{B}(\mathbb{R}_+)$-measurable (in particular, X and Y could be prog. measurable) and let M and N be any two continuous local martingales. Then the following inequality$^{\multimap (8.92)}$ holds (P-a.s.) for all $t \in \bar{\mathbb{R}}_+$:

$$\left|XY\bullet\langle M, N\rangle_t\right| \le \left(X^2\bullet\langle M, M\rangle_t\right)^{\frac{1}{2}} \left(Y^2\bullet\langle N, N\rangle_t\right)^{\frac{1}{2}} . \quad ○$$

In many applications – especially applications in the domain of asset pricing – one must work with more than one probability measure. How do the attributes "martingale" and "local martingale" change when the underlying measure changes? This question is addressed in the next exercise.

(9.78) EXERCISE: Let Q be another probability measure on $\mathcal{F}$ such that $Q \lesssim P$, and let $R_t \overset{\text{def}}{=} \frac{dQ}{dP}\restriction\mathscr{F}_t$ for any $t \in \mathbb{R}_+$. Prove that a process $M \prec \mathscr{F}$ is an $(\mathscr{F}, Q)$-martingale if and only if the process MR is an $(\mathscr{F}, P)$-martingale. Conclude that a process $M \prec \mathscr{F}$ is a local $(\mathscr{F}, Q)$-martingale if and only if the process MR is a local $(\mathscr{F}, P)$-martingale.

HINT: Use the relation$^{\multimap (3.45^{\dagger})}$

$$\mathsf{E}^Q[M_t \,|\, \mathscr{F}_s] = \frac{1}{\mathsf{E}^P[R_t \,|\, \mathscr{F}_s]} \mathsf{E}^P[M_t R_t \,|\, \mathscr{F}_s], \quad 0 \le s < t < \infty .$$

Argue that R is a uniformly integrable martingale and explain why $\frac{dQ}{dP}\restriction\mathscr{F}_{t\wedge\mathcal{t}} = R_{t\wedge\mathcal{t}}$ for any stopping time $\mathcal{t}$. Consult Exercises (9.44) and (9.45). ○

We conclude this section with the introduction of the largest class of continuous stochastic processes that can be used as integrators in stochastic integrals – an object that we will encounter shortly.

(9.79) CONTINUOUS SEMIMARTINGALES: A *continuous semimartingale* – or, if there is a risk of mistake, *continuous* $(\mathscr{F}, P)$*-semimartingale* – is any continuous process X that can be decomposed into the sum $X = M + A$, in which M is some continuous local martingale and A is some continuous and adapted (thus predictable) process that starts from 0 and has sample paths of finite variation on finite intervals.

A crucially important – for both the theory of stochastic processes and the methods of asset pricing – feature of the class of semimartingales is that this class is invariant under equivalent changes of the probability measure: a semimartingale with respect to P is also a semimartingale with respect to any probability measure $Q \approx P$. This result was established by Jacod and Memin (1976) – see also (Dellacherie and Meyer 1980, VII 45). It is valid for semimartingales that may not be continuous, but we postpone the introduction of such objects for chapter 15. ○

(9.80) EXERCISE: Prove that the decomposition of a continuous semimartingale into the sum of a continuous local martingale and a continuous and adapted process of finite variation that starts from 0 is unique up to indistinguishability. ○

(9.81) EXERCISE: Prove that any two continuous semimartingales, X and Y, have finite quadratic covariation $[X, Y]$, which, in fact, coincides with the quadratic covariation between their respective local martingale components. For this reason, following our earlier convention, if the semimartingales X and Y are continuous – and only if they are continuous – we will express their quadratic covariation as an angled bracket, rather than square bracket. Thus, if $X = M + A$ and $Y = M' + A'$, then $\langle X, Y\rangle = \langle M, M'\rangle$ will be just another notation for $[X, Y] = [M, M']$. ○

(9.82) QUADRATIC VARIATION AND REALIZED VOLATILITY: In many continuous-time market models the asset prices are expressed as exponential semimartingales, that is, as stochastic processes of the form $S = e^Z$, for a given semimartingale Z.[117] As an example, in the classical Black–Scholes–Merton model the stock price is postulated to be of this form with $Z = \sigma W + (b - \frac{1}{2}\sigma^2)\iota$, for some choice of the Brownian motion W and the parameters $\sigma, b \in \mathbb{R}$.

The term "volatility" is an alias for "standard deviation of the logarithm of the daily gross returns" – or perhaps gross returns over even shorter periods – divided by the length of the respective time period (in the adopted units of time). Volatility squared is thus called "variance." The realized variance and volatility are closely related to the sum of the $\log^2$ of those returns. To clarify this connection, suppose that the price process S is discretely monitored until some finite moment $t > 0$,

[117] The fact that e^Z is a semimartingale as long as Z is a semimartingale follows from Itô's formula, which will be developed in the next chapter.

and let n denote the number of observations. If S is a continuous exponential semimartingale of the form $S = e^Z$, we can write

$$\sum_{i \in \mathbb{N}_{|n-1}} \log^2\left(\frac{S_{(i+1)t/n}}{S_{it/n}}\right) = \sum_{i \in \mathbb{N}_{|n-1}} \left(Z_{(i+1)t/n} - Z_{it/n}\right)^2 .$$

In our usual notation this is just the discrete quadratic variation $\mathcal{Q}^{p_n}(Z,Z)_t$ for a particular subdivision p_n of $[0,t]$. By choosing $|p_n| = \frac{1}{n}$ to be sufficiently small, the expression $\mathcal{Q}^{p_n}(Z,Z)_t$ can be made close to $\langle Z, Z\rangle_t$ with probability close to 1. In the case of the Black–Scholes–Merton model we have $\langle Z, Z\rangle_t = \sigma^2 t$, so that the quantity

$$\sqrt{\frac{1}{t}\mathcal{Q}^{p_n}(Z,Z)_t}$$

must be close to the volatility parameter σ, if the number of observations n is sufficiently large. The above quantity is often called "realized volatility" since it obtains from a particular realization, S_ω, of the price process S – to be precise from a particular realization of the process $Z = \log(S)$.

"Realized volatility" is an important component of the market micro-structure and has been studied extensively throughout the literature. As this topic lies outside of our main concerns, here we will only point to these sources: (Hansen and Lunde 2006), (Shephard et al. 2012), and (Aït-Sahalia and Jacod 2014). ○

9.6 The Space of L^2-Bounded Martingales

In this section we turn to the study of the important space of L^2-bounded martingales, that is, the space of martingales M such that $\sup_{t \in \mathbb{R}_+} \mathsf{E}[M_t^2] < \infty$. The probability space $(\Omega, \mathcal{F}, P)$ and the filtration $\mathscr{F} \subset \mathcal{F}$ (with the usual conditions automatically assumed) are again fixed throughout, and we again remind the default assumption $\mathscr{F}_\infty = \vee_{t \in \mathbb{R}_+} \mathscr{F}_t$.

(9.83) THE SPACES $\mathbb{H}^2$ AND H^2: The space of all L^2-bounded and càdlàg martingales (for a filtration and probability measure understood from the context) we denote by $\mathbb{H}^2$, and the space of all continuous elements of $\mathbb{H}^2$ we denote by H^2. The space of all elements of $\mathbb{H}^2$ (all elements of H^2) that start from 0 we denote by $\mathbb{H}^2_0$ (denote by H^2_0). ○

(9.84) REMARK: As a straightforward application of the result in (4.22), every $M \in \mathbb{H}^2$ is a uniformly integrable martingale. By the convergence theorem in (9.33), $M_\infty \overset{\text{def}}{=} \lim_{t\to\infty} M_t$ exists and $M_t = \mathsf{E}\left[M_\infty \,|\, \mathscr{F}_t\right]$ for every $t \in \mathbb{R}_+$. What is important for us, however, is to establish that in fact $M_\infty = L^2\text{-}\lim_{t\to\infty} M_t$ – notice that (9.33) only guarantees the convergence of M_t as $t \to \infty$ in L^1 norm. ○

(9.85) DOOB'S L^2-INEQUALITY: Let X be any right-continuous and positive submartingale (such as, say, $X = |M|$, for some right-continuous martingale M) and let $X_t^* \stackrel{\text{def}}{=} \sup_{s\in[0,t]} X_s$. Then for any $t \in \mathbb{R}_{++}$,

$$\mathsf{E}\big[(X_t^*)^2\big] \le 4\,\mathsf{E}\big[(X_t)^2\big]\,.$$

In particular, if $X_t \in L^2$, then $X_t^* \in L^2$. ○

(9.86) EXERCISE: Prove Doob's L^2-inequality.

HINT: Using$^{\multimap(5.2)}$ the layer cake formula, show that

$$\mathsf{E}\big[(X_t^*)^2\big] = \int_{\mathbb{R}_+} P\big(X_t^* \ge \sqrt{v}\big)\,\mathrm{d}v = 2\int_{\mathbb{R}_+} u\,P(X_t^* \ge u)\,\mathrm{d}u\,.$$

Next, use$^{\multimap(9.19)}$ Doob's maximal inequality to estimate the quantity $uP(X_t^* \ge u)$, and, finally, use the layer cake formula (the second formula in $(5.2^\dagger)$) one more time to estimate the integral

$$2\int_{\mathbb{R}_+} \mathsf{E}\big[X_t\,1_{\{X_t^*\ge u\}}\big]\,\mathrm{d}u = 2\,\mathsf{E}\big[X_tX_t^*\big] \le 2\,\mathsf{E}\big[X_t^2\big]^{\frac{1}{2}}\,\mathsf{E}\big[(X_t^*)^2\big]^{\frac{1}{2}}\,. \quad ○$$

With some minor adjustments of the algebra in the steps outlined above, one can establish the following more general result.

(9.87) DOOB'S L^p-INEQUALITY: Let X be any right-continuous and positive submartingale and let $X_t^* \stackrel{\text{def}}{=} \sup_{s\in[0,t]} X_s$. Then for any $t \in \mathbb{R}_{++}$ and any $p > 1$,

$$\mathsf{E}\big[(X_t^*)^p\big] \le \Big(\frac{p}{p-1}\Big)^p\,\mathsf{E}\big[(X_t)^p\big]\,.$$

In particular, if $X_t \in L^p$, then $X_t^* \in L^p$. ○

(9.88) COROLLARY: With the notation as in (9.85), suppose that the (positive and right-continuous) submartingale X is also L^2-bounded, that is, $\mathsf{E}[X_t{}^2] < C < \infty$ for every $t \in \mathbb{R}_+$. Doob's L^2-inequality gives $\mathsf{E}[(X_t^*)^2] < 4C$ for every $t \in \mathbb{R}_+$, and, since the sample paths of X^* are increasing functions with $\lim_{t\to\infty} X_t^* = X_\infty^* = \sup_{t\in\mathbb{R}_+} X_t$, the monotone convergence theorem would give us $\mathsf{E}\big[(X_\infty^*)^2\big] \le 4C < \infty$, so that $X_\infty^* \in L^2$. Similarly, if X is L^p-bounded for some $p > 1$, then $X_\infty^* \in L^p$.

Given any $M \in \mathbb{H}^2$, observe that $|M|$ is an L^2-bounded and positive submartingale, and set $|M|_\infty^* \stackrel{\text{def}}{=} \sup_{t\in\mathbb{R}_+} |M_t|$. By the foregoing, $\mathsf{E}[(|M|_\infty^*)^2] < \infty$, and since $(M_t)^2 \le (|M|_\infty^*)^2$, setting $M_\infty \stackrel{\text{def}}{=} \lim_{t\to\infty} M_t$ (the existence of this limit is guaranteed by the convergence theorem in (9.33)) we see that $(M_\infty)^2 \le (|M|_\infty^*)^2$. As a result,

$$(M_t - M_\infty)^2 \le 2(M_t{}^2 + M_\infty{}^2) \le 4\,(|M|_\infty^*)^2\,,$$

which shows that $\big((M_t - M_\infty)^2\big)_{t\in\mathbb{R}_+}$ is a uniformly integrable family, and therefore

$$\lim_{t\to\infty} \mathsf{E}\big[(M_t - M_\infty)^2\big] = \mathsf{E}\big[\lim_{t\to\infty}(M_t - M_\infty)^2\big] = 0\,.$$

Thus, we can now claim that $M_\infty = L^2\text{-}\lim_{t\to\infty} M_t$, whereas theorem (9.33) only guarantees convergence in the L^1-norm. In particular, $\lim_{t\to\infty}\mathsf{E}[M_t{}^2] = \mathsf{E}[M_\infty{}^2]$.

Passing to the limit in Doob's L^2-inequality, we arrive at the following useful and important inequality:

$$\mathsf{E}\big[(|M|^*_\infty)^2\big] \le 4\,\mathsf{E}\big[(M_\infty)^2\big]\,.$$

One immediate use of this estimate is to measure how "far apart" two martingales $M, N \in \mathbb{H}^2$ are, namely, by measuring their distance in L^2-norm at ∞: if applied to the martingale $(M - N) \in \mathbb{H}^2$, the last inequality gives

† $$\mathsf{E}\Big[\Big(\sup_{t\in\mathbb{R}_+} |M_t - N_t|\Big)^2\Big] \le 4\,\mathsf{E}\big[\big(M_\infty - N_\infty\big)^2\big]\,. \quad \circ$$

We have the following general result, which we just established in the special case $p = 2$ (the proof in the general case $p > 1$ follows a similar argument).

(9.89) ADDENDUM TO THEOREM (9.33): Let X be any L^p-bounded martingale, that is, $\sup_{t\in\mathbb{R}_+} \mathsf{E}[|X_t|^p] < \infty$ for some $p > 1$, in which case X is automatically u.i. and the equivalent conditions (a), (b), and (c) from (9.33) hold. Then the convergence in (9.33b) takes place also in L^p-norm. ○

(9.90) EXERCISE: Since any $M \in \mathbb{H}^2$ is a u.i. martingale, theorem (9.33) guarantees that

$$M_t = \mathsf{E}\big[M_\infty \,|\, \mathscr{F}_t\big] \quad \text{for all } t \in \mathbb{R}_+\,.$$

Give an independent proof of this relation by repeating some of the steps in (9.88).

HINT: This relation is the same as $\mathsf{E}[M_\infty - M_t \,|\, \mathscr{F}_t] = 0$, which is the same as $\mathsf{E}\big[\big(\mathsf{E}[M_\infty - M_t \,|\, \mathscr{F}_t]\big)^2\big] = 0$. Recall that⊸(3.41c) the conditional expectation operator is continuous in the L^2-norm. ○

The results established so far in this section are quite remarkable: according to (9.88†), comparing the sample paths of two martingales from $\mathbb{H}^2$ comes down to comparing their values at infinity, treated as elements of $L^2(\Omega, \mathscr{F}_\infty, P)$. Since every random variable $\xi \in L^2(\Omega, \mathscr{F}_\infty, P)$ gives rise to the (L^2-bounded and càdlàg) martingale

$$M_t \overset{\text{def}}{=} \mathsf{E}[\xi \,|\, \mathscr{F}_t]\,, \quad t \in \mathbb{R}_+\,,$$

we can identify the space $\mathbb{H}^2$ as the much simpler space of square integrable random variables $L^2(\Omega, \mathscr{F}_\infty, P)$. Consequently, the space H^2 can be identified as the subspace of those random variables $\xi \in L^2(\Omega, \mathscr{F}_\infty, P)$ for which the martingale $\mathsf{E}[\xi \,|\, \mathscr{F}]$ admits a continuous modification. It is not very difficult to conclude from (9.88†) that, under this identification, H^2 can be treated as a closed subspace of the Hilbert space $L^2(\Omega, \mathscr{F}_\infty, P)$.

Since every $M \in H^2$ is a continuous martingale, it is also⊸(9.68) a continuous local martingale, and the result in (9.71) is telling us that $M^2 - \langle M, M\rangle$ is a continuous local martingale as well (here⊸(9.72) $\langle M, M\rangle$ stands for the quadratic variation of M). Nevertheless, the process $M^2 - \langle M, M\rangle$ can be claimed to be a martingale, which would then allow us to identify the (continuous) quadratic variation $\langle M, M\rangle$ as a predictable compensator, and thus justify the use of angled brackets instead of square ones. In fact such a result is not entirely new to us, for, at least in the case where M is globally bounded, we know from (9.61) that $M^2 - \langle M, M\rangle$ must be a martingale. To see why this property actually holds for every $M \in H^2$, let $(t_i)_{i\in\mathbb{N}}$ be some (any) localizing sequence of stopping times for the local martingale $M^2 - \langle M, M\rangle$, chosen so that $(M^{t_i})^2 - \langle M, M\rangle^{t_i} - {M_0}^2$ is a bounded martingale starting from 0 for every $i \in \mathbb{N}$. In particular,

$$\mathsf{E}\big[\langle M, M\rangle_{t\wedge t_i}\big] = \mathsf{E}\big[{M_{t\wedge t_i}}^2\big] - \mathsf{E}\big[{M_0}^2\big], \quad t \in \mathbb{R}_+ . \qquad e_1$$

Since $(M_{t\wedge t_i})^2 \le (M^*_\infty)^2$ for every $t \ge 0$, a straightforward application of the dominated and monotone convergence theorems gives, after passing to the limit in the above relation as $i \to \infty$,

$$\mathsf{E}\big[\langle M, M\rangle_t\big] = \mathsf{E}\big[{M_t}^2\big] - \mathsf{E}\big[{M_0}^2\big].$$

We can pass to the limit one more time with $t \to \infty$, using the monotone convergence on the left, to get

$$\mathsf{E}\big[\langle M, M\rangle_\infty\big] = \mathsf{E}\big[{M_\infty}^2\big] - \mathsf{E}\big[{M_0}^2\big] < \infty .$$

For every stopping time t we have ${M_t}^2 \le (M^*_\infty)^2$ and $\langle M, M\rangle_t \le \langle M, M\rangle_\infty$, so that the family

$$\big\{{M_t}^2 - \langle M, M\rangle_t : t \text{ is a stopping time}\big\}$$

is uniformly integrable, that is, $M^2 - \langle M, M\rangle$ is of class (D). The result in (9.70) then guarantees that local martingale $M^2 - \langle M, M\rangle$ is a martingale – in fact, u.i. martingale as we just showed.

Now suppose that M is only known to be a continuous local martingale and let the sequence $(t_i)_{i\in\mathbb{N}}$ be chosen as above, so that e_1 still holds for every $i \in \mathbb{N}$. In fact, since optional stopping for bounded martingales always holds, e_1 would still hold with t replaced by an arbitrary stopping time s. We can still pass to the limit in e_1 (with s in place of t), but without the condition $M \in H^2$ must rely on the less precise Fatou's lemma and the monotone convergence theorem, which would give us

$$\mathsf{E}[M_s^2] \le \mathsf{E}[M_0^2] + \mathsf{E}[\langle M, M\rangle_\infty] \quad \text{for every stopping time } s .$$

Thus, if the local M is such that $\mathsf{E}[M_0^2] + \mathsf{E}[\langle M, M\rangle_\infty] < \infty$, then all expectations $\mathsf{E}[M_s^2]$, for all choices of the stopping time s, would be uniformly bounded. The local martingale M is then of class D and is therefore a martingale; in fact a martingale with uniformly bounded second moments $\mathsf{E}[M_t^2]$, which is to say, an element of H^2. We have thus established the following result:

(9.91) THEOREM: (Revuz and Yor 1999, IV (1.23)) A continuous local martingale M is an element of H^2 if and only $\mathsf{E}[{M_0}^2] < \infty$ and $\mathsf{E}\big[\langle M, M\rangle_\infty\big] < \infty$. When these conditions hold, the local martingale $M^2 - \langle M, M\rangle$ is a uniformly integrable martingale. More generally, if M and N are elements of H^2, then the local martingale $MN - \langle M, N\rangle$ is a uniformly integrable martingale. ○

TO SUMMARIZE: The spaces H^2 and H_0^2 can be identified as Hilbert sub-spaces of $L^2(\Omega, \mathcal{F}, P)$. Under this identification, the (Hilbert) L^2-norm of any $M \in H^2$ can be expressed as $\mathsf{E}[{M_0}^2] + \mathsf{E}[\langle M, M\rangle_\infty]$. In addition, we can now define a "dot-product" between two elements $M, N \in H_0^2$ as

$$M \cdot N \overset{\text{def}}{=} \mathsf{E}\big[\langle M, N\rangle_\infty\big] = \mathsf{E}\big[M_\infty N_\infty\big].$$

If the mapping $\ell \colon H_0^2 \mapsto \mathbb{R}$ is linear and continuous, in the sense that for some constant $C > 0$,

$$|\ell(M)| \leq C\sqrt{M \cdot M} \equiv C\sqrt{\mathsf{E}\big[\langle M, M\rangle_\infty\big]} \quad \text{for all } M \in H_0^2, \qquad \mathrm{e}_2$$

then,⊶(A.64) Riesz lemma implies that there is a unique element $N \in H_0^2$ such that $\ell(M) = M \cdot N$ for every $M \in H_0^2$. Furthermore, by the same lemma, $\sqrt{N \cdot N}$ can be identified as the smallest constant $C > 0$ for which e_2 holds. This feature is sometimes used in the construction of stochastic integrals for integrators $M \in H^2$ – see (Revuz and Yor 1999), for example – but it has other important applications as well.

(9.92) EXERCISE: Using the result from theorem (9.77), prove that

$$\mathsf{E}\big[|\langle M, N\rangle_t|\big] \leq \sqrt{\mathsf{E}\big[\langle M, M\rangle_t\big]}\sqrt{\mathsf{E}\big[\langle N, N\rangle_t\big]}$$

for every $M, N \in H_0^2$. Conclude that (taking $t \to \infty$) $|M \cdot N| \leq \sqrt{M \cdot M}\sqrt{N \cdot N}$ (as it should be, given that H_0^2 is a Hilbert space). ○

One immediate conclusion from the last exercise – or from the fact that H_0^2 is a Hilbert space – is that if, for some fixed $N \in H_0^2$, one defines the linear mapping

$$H_0^2 \ni M \rightsquigarrow \ell(M) \overset{\text{def}}{=} M \cdot N \equiv \mathsf{E}\big[\langle M, N\rangle_\infty\big] \equiv \mathsf{E}\big[M_\infty N_\infty\big],$$

then this mapping is continuous in the sense of e_2, and the smallest constant with which e_2 holds is indeed $\sqrt{N \cdot N}$.

9.7 The Binomial Asset Pricing Model Revisited

In this section we revisit the continuous-time counterpart of the binomial asset pricing model that we introduced in section 8.2. From where we stood back then the mere existence of Brownian motion – which naturally showed up in the limit – required certain suspension of disbelief, but now this is no longer the case. Since, from practical point of view, what ultimately matters in financial modeling are the statistical properties of those quantities that one can associate with actual physical observations, we are really indifferent to the probability space on which a particular market model is built. Nevertheless, it is important to know – and, as we have already seen, this matter is far from trivial – that there is a probability space on which we can define all the random variables that the model involves, and in such a way that their joint distributions are consistent with those statistical features that we have a reason to insist on. Once the existence of at least one such probability space is established, we would be indifferent to its choice.

At this point we are aware that it is always possible to construct a probability space $(\Omega, \mathcal{F}, P)$ that can support a Brownian motion W. On any such space the continuous-time limit (as a probability distribution) of the price of the risky asset in the binomial model can be identified with

$$S_t = S_0\, e^{\sigma W_t + (b - \frac{1}{2}\sigma^2)t}, \quad t \in \mathbb{R}_+,$$

where σ and b are the parameters inherited from the discrete CRR model. Back in section 8.2 we realized that if we were to describe the statistical behavior of the price process under the risk-neutral measure, then it would be enough for us to replace – assuming that such an operation is feasible – the probability measure P with another probability measure $Q \approx P$, such that under Q the process $\tilde{W} \overset{\text{def}}{=} W + \frac{b-r}{\sigma}\iota$ has all the features of a Brownian motion. In particular, given any fixed $t \in \mathbb{R}_+$, the new measure that we are looking for must preserve the normality of the r.v. $W_t + \frac{b-r}{\sigma}t$ – in fact, it must preserve also the variance – but must annihilate the expected value $\frac{b-r}{\sigma}t$. We saw in section 3.1 that such a change in the distribution law can be achieved with an equivalent transformation of P provided by the Radon–Nikodym derivative

$$R_t = e^{-\frac{b-r}{\sigma} t \frac{W_t}{t} - \frac{1}{2}\frac{1}{t}\frac{t^2(b-r)^2}{\sigma^2}} = e^{-\frac{b-r}{\sigma} W_t - \frac{1}{2}\frac{(b-r)^2}{\sigma^2} t}. \qquad \mathrm{e}_1$$

However, on the one hand this Radon–Nikodym derivative depends on t, and on the other hand we would like to change the measure from P to Q just once. We thus need to understand how a single change of the underlying measure can have the effect of infinitely many such changes, for all choices of the time parameter t. This is where the martingale tools that we developed in this chapter become useful.

(9.93) EXERCISE: Assuming that the initial price S_0 at time $t = 0$ is a given constant, prove that the filtration generated by the process S is the same as the Brownian filtration $\mathscr{F}^W$, generated by the same Brownian motion, W, that drives the stock price. ○

Let $\mathscr{F} \equiv \mathscr{F}^W$ denote the filtration introduced in the last exercise (after being passed through$^{\multimap\,(8.64)}$ the usual augmentation procedure, which, at this point we automatically assume). As is straightforward to check – which we have done in a number of occasions already – the process R from e_1 is an $(\mathscr{F}, P)$-martingale. Consider a finite time horizon $T \in \mathbb{R}_{++}$ and the probability measure $Q_T \approx P$ given by $Q_T \overset{\text{def}}{=} R_T \circ P$. Given any $t \in [0, T]$, let ξ denote any bounded and $\mathscr{F}_t$-

measurable random variable. Since R is a martingale, we have

$$\mathsf{E}^{Q_T}[\xi] = \mathsf{E}^P[\xi\, R_T] = \mathsf{E}^P\big[\xi\,\mathsf{E}^P[R_T \,|\, \mathscr{F}_t]\big] = \mathsf{E}^P[\xi\, R_t]\,.$$

This relation shows that the restriction $Q_T \restriction \mathscr{F}_t$ can be identified as the measure $R_t \odot P$ – precisely the equivalent change of P that annihilates the mean of $\tilde{W}_t$. Thus, the measure $Q_T = R_T \odot P$ indeed provides simultaneously all the changes that we need, for all choices of $t \in [0, T]$. The fundamental reason why this became possible is that the infinitely many Radon–Nikodym derivatives that we needed can be expressed succinctly as $R_t = \mathsf{E}[R_T \,|\, \mathscr{F}_t]$, $t \in [0, T]$. Plainly, there really is just one Radon–Nikodym derivative in our collection. Furthermore, using$^{\multimap\,(3.45^\dagger)}$ the Bayes' rule for conditional expectations, for every $0 \le u < t \le T$ and every positive r.v. ξ,

$$\mathsf{E}^{Q_T}\big[\xi \mid \mathscr{F}_u\big] = \frac{1}{\mathsf{E}^P\big[R_T \,|\, \mathscr{F}_u\big]}\,\mathsf{E}^P\big[\xi\, R_T \mid \mathscr{F}_u\big] = \mathsf{E}^P\Big[\xi\,\frac{R_t}{R_u} \mid \mathscr{F}_u\Big].$$

As a result, since $S_t = S_0\, e^{\sigma \tilde{W}_t + (r - \frac{1}{2}\sigma^2)t}$, we have

$$\mathsf{E}^{Q_T}\big[S_t \mid \mathscr{F}_u\big] = S_u\, e^{r\,(t-u)}\,\mathsf{E}^P\Big[e^{\sigma\,(\tilde{W}_t - \tilde{W}_s) - \frac{1}{2}\,\sigma^2\,(t-u)}\,\frac{R_t}{R_u} \mid \mathscr{F}_u\Big] = S_u\, e^{r\,(t-u)}\,.$$

The last relation gives the familiar risk-neutral pricing equation

$$S_u = e^{-r\,(t-u)}\,\mathsf{E}^{Q_T}\big[S_t \,|\, \mathscr{F}_u\big]\,, \quad u \in [0, t]\,,$$

but it also gives the even more important relation

$$S_u = \mathsf{E}^P\Big[S_t\, e^{-r\,(t-u)}\,\frac{R_t}{R_u} \mid \mathscr{F}_u\Big], \quad u \in [0, t]$$

that says that the stochastic discount factor between times u and $t > u$ is

$$e^{-r\,(t-u)}\,\frac{R_t}{R_u} = e^{-r\,(t-u)}\, e^{-\frac{b-r}{\sigma}\,(W_t - W_u) - \frac{1}{2}\,\frac{(b-r)^2}{\sigma^2}\,(t-u)}\,.$$

Before we get to some concrete examples, it is important to clarify whether it is possible to push in the above the time horizon to $T = \infty$. In particular, that would mean that the martingale R admits the representation $R_t = \mathsf{E}[R_\infty \,|\, \mathscr{F}_t]$, $t \in \mathbb{R}_+$, which, as we are well aware, is not possible since R is not a uniformly integrable martingale. In fact,$^{\multimap\,(9.31)}$ we have seen that $\lim_{t\to\infty} R_t = 0$ P-a.s. It is simply not possible to define a risk-neutral measure in this model if the time horizon is pushed to infinity!

Another important question for us is whether the risk-neutral measure Q_T – or, equivalently, the stochastic discount factor – in this asset pricing model is unique. We are yet to develop the tools that would allow us to address such matters. We are going to establish later on that in this particular model the risk-neutral measure Q_T is indeed unique for any finite (and fixed) time horizon $T \in \mathbb{R}_{++}$.

As long as we know the risk-neutral measure – or, equivalently, we know the stochastic discount factor – associated with any finite time-horizon T, we are able to price *any* risky payoff $\Pi \preccurlyeq \mathscr{F}_T$ delivered at time $t \in [0, T]$. The equilibrium price of this payoff at time $s < t$ must be given by

$$e^{-r(t-s)}\mathsf{E}^{Q_T}\left[\Pi \mid \mathscr{F}_s\right] = \mathsf{E}^P\left[\Pi \, e^{-r(t-s)} \frac{R_t}{R_s} \mid \mathscr{F}_s\right].$$

In the special case where $\Pi \stackrel{\text{def}}{=} f(S_T)$, that is, in the case of a contingent claim with payoff function $f : \mathbb{R}_+ \mapsto \mathbb{R}$, spot asset S, and closing date T, the price of the claim at time $t = 0$ must be given by

$$e^{-rT}\mathsf{E}^{Q_T}\left[f\left(S_0 \, e^{\sigma \tilde{W}_T + (r - \frac{1}{2}\sigma^2)T}\right)\right] = \frac{e^{-rT}}{\sigma\sqrt{2\pi T}} \int_{-\infty}^{\infty} f\left(S_0 \, e^{x + (r - \frac{1}{2}\sigma^2)T}\right) e^{-\frac{x^2}{2T\sigma^2}} \, \mathrm{d}x \,.$$

(9.94) BLACK–SCHOLES–MERTON (BSM) FORMULA:[118] With $f(x) \stackrel{\text{def}}{=} (x - K)^+$ (call option with strike price $K > 0$), the last integral can be computed in closed-form. This computation – which can be done rather easily using computer algebra on systems such as SymPy or SageMath – gives the renowned Black–Scholes–Merton formula that reads:

$$BSM(S_0, K, \sigma^2, r, T) = \frac{1}{2}\, e^{-rT} K \operatorname{erf}\left(\frac{T\,(\sigma^2/2 - r) + \log\left(\frac{K}{S_0}\right)}{\sigma\sqrt{2T}}\right)$$
$$+ \frac{1}{2}\, S_0 \operatorname{erf}\left(\frac{T\,(\sigma^2/2 + r) - \log\left(\frac{K}{S_0}\right)}{\sigma\sqrt{2T}}\right) - \frac{K}{2}\, e^{-rT} + \frac{1}{2}\, S_0 \,. \quad \circ$$

[118] Named after the American economists Fischer Sheffey Black (1938–1995), Myron Samuel Scholes, and Robert C. Merton. In 1997 Merton and Scholes received the Sveriges Riksbank Prize in Economic Sciences in Memory of Alfred Nobel for developing “a new method to determine the value of derivatives.”

10

Stochastic Integration

Integration theory has many interesting and important applications. One such application is modeling the motion of a particle submerged in a velocity vector field $V = (V(x) \in \mathbb{R})_{x \in \mathbb{R}}$. Speaking figuratively, this can be pictured as the motion of a pollen particle blown by wind V over the line $\mathbb{R}$. If the scalar function $\mathbb{R}_+ \ni t \rightsquigarrow f(t) \in \mathbb{R}$ models the magnitude of the wind as a function of time, the trajectory of the pollen particle, $\mathbb{R}_+ \ni t \rightsquigarrow x(t) \in \mathbb{R}$, starting from position $x(0)$ at time $t = 0$, would be governed by the equation

$$\frac{\mathrm{d}x(t)}{\mathrm{d}t} = V(x(t))\, f(t) \quad \Longleftrightarrow \quad \mathrm{d}x(t) = V(x(t))\, f(t)\mathrm{d}t\,,$$

which can also be cast as

$$x(t) - x(0) = \int_0^t V(x(s))\, f(s)\mathrm{d}s\,.$$

If we want to model the movement of the pollen particle as a random motion, then we must introduce an appropriate probability space $(\Omega, \mathcal{F}, P)$, and allow the velocity field V and the amplitude f to depend on the outcome $\omega \in \Omega$, that is, set $V = V(\omega, x)$ and $f = f(\omega, t)$. With this choice, the trajectory of the particle can be expressed as

$$\mathbb{R}_+ \ni t \rightsquigarrow X(\omega, t) \equiv x(0) + \int_0^t V(\omega, X(\omega, s))\, f(\omega, s)\mathrm{d}s \in \mathbb{R}\,,$$

and can thus be treated as a sample path of a particular stochastic process X. However, sample paths of this form are still differentiable, and, as we are well aware, continuous stochastic processes with sample paths of this type are not particularly interesting. To say the least, it would not be possible to identify a random motion of this type as Brownian motion – a type of motion discovered precisely by observing the movements of a grain of pollen suspended in water.[119] Thus, to develop truly interesting and useful models of random motion, we must replace the differential expression $V(\omega, X(\omega, t))\, f(\omega, t)\mathrm{d}t = \mathrm{d}X(\omega, t)$ with something more general that would include $\mathrm{d}W(\omega, t)$ (the formal differential of a Brownian trajectory W_ω) as a special case. If the derivative $\partial W(\omega, t)$ were to exist, we could set $V \equiv 1$

[119] This discovery was made in 1827 by the Scottish botanist Robert Brown (1773–1858), after whom the Brownian motion is named.

and $f(\omega,t) = \partial W(\omega,t)$, which would then give us $\mathrm{d}X(\omega,t) = 1\partial W(\omega,t)\mathrm{d}t = \mathrm{d}W(\omega,t)$, and, consequently, allow us to write

$$X(\omega,t) = x(0) + \int_0^t \mathrm{d}W(\omega,t) = x(0) + W(\omega,t)\,.$$

We see that $x(0)+W$ is a perfectly meaningful stochastic process (a linear Brownian motion starting from $x(0)$), even though the derivative $\partial W(\omega,t)$ is not. As we can see, despite the fact that $\partial W(\omega,t)$ does not exist, there is hope that formal expressions of the form

$$\int_0^t V(\omega, X(\omega,s))\big(\partial W(\omega,s)\big)\,\mathrm{d}s = \int_0^t V(\omega, X(\omega,s))\mathrm{d}W(\omega,s)$$

can still be made meaningful. Such expressions are called *stochastic integrals* and are the subject of the present chapter.

As a conceptual and practical tool, stochastic integrals are instrumental in continuous-time finance, since these objects allow one to model the price dynamics of an actively managed portfolio of financial assets with randomly fluctuating returns. In this context, we can think of the (hypothetical) derivative $\partial W(\omega,t)$ as the instantaneous rate of capital gains from ¤1 invested in a particular risky asset. Thus, if the process $t \rightsquigarrow V(\omega,t)$ represents the time evolution of asset holdings (expressed in units of currency ¤), as prescribed by a particular trading strategy, then the integral $\int_0^t V(\omega,s)\partial W(\omega,s)\mathrm{d}s$ would give us the aggregate capital gains from the strategy accumulated by time t. This connection will be the subject of chapter 13, but it is important to recognize at this point that the truly interesting continuous-time models would be the ones in which the instantaneous rate of capital gains $\partial W(\omega,t)$ does not exist as a derivative in the usual sense, while the accumulated capital gains can still be understood as an integral of the form $\int_0^t V(\omega,s)\mathrm{d}W(\omega,s)$, whose construction we now detail.

10.1 Basic Examples and Intuition

Our first step is to introduce stochastic integrals that can be interpreted as random Lebesgue–Stieltjes integrals. Strictly speaking, such integrals do not deserve to be called "stochastic," but, as we will soon realize, it is convenient to have at hand a single universal integration machine designed to work in every situation in which integration of one form or another is going to be needed. With this purpose in mind, we fix a probability space $(\Omega, \mathcal{F}, P)$ and filtration $\mathscr{F} \equiv (\mathscr{F}_t \subseteq \mathcal{F})_{t\in\mathbb{R}_+}$ (the usual conditions are again automatically assumed), and suppose that these objects are rich enough to support all random variables and processes that will show up in our study. Let A stand for a stochastic process with sample paths A_ω, $\omega \in \Omega$, that have finite variation for P-a.e. $\omega \in \Omega$. If X is another stochastic process, the integral

$$\mathcal{I}_\omega(t) \stackrel{\text{def}}{=} \int_{0+}^{t} X_\omega(s)\,\mathrm{d}A_\omega(s) \equiv \int_{]0,t]} X_\omega\,\mathrm{d}A_\omega\,, \quad t \in \mathbb{R}_+\,, \qquad \mathrm{e}_1$$

can be understood⊸(8.91) as a usual Lebesgue–Stieltjes integral for P-a.e. (fixed) $\omega \in \Omega$. The only requirement for the sample paths X_ω is to be Borel-measurable and $\mathrm{d}A_\omega$-integrable on any finite interval $]0,t]$, for P-a.e. $\omega \in \Omega$. As an example, if X is $\mathscr{F}$-progressive – in particular, if it is $\mathscr{F}$-adapted and is either left- or right-continuous – then X_ω will be Borel-measurable on every interval $]0,t]$, and, if X_ω is also bounded on every interval $[0,t]$, the integral in e_1 would be perfectly meaningful as a path-by-path Lebesgue–Stieltjes integral. If these conditions hold, the sample path $\mathcal{I}_\omega$ is well defined P-a.e., and the symbol $\mathcal{I}$ can be interpreted as a stochastic process in the usual way.

(10.1) DEFINITION (LEBESGUE–STIELTJES STOCHASTIC INTEGRALS): The process $\mathcal{I}$ from e_1 we call the *stochastic integral* of (the integrand) X with respect to (the integrator) A. We may prepend "stochastic integral" with "Lebesgue–Stieltjes," if the distinction with the true stochastic integrals (yet to be introduced) has to be made. Following the French tradition, we will denote⊸(8.92) the process $\mathcal{I}$ by the symbol $X \bullet A$, though we will also use the standard integral notation

$$X \bullet A_t = \int^t X\,\mathrm{d}A \equiv \int^t X_s\,\mathrm{d}A_s\,, \quad t \in \mathbb{R}_+\,.$$

The common convention is to include the upper limit t in the integration domain, but not the lower limit 0, and to treat the stochastic integral as a process that starts from 0 by definition. Such a "fine point" becomes relevant only if the measure $\mathrm{d}A_\omega$ contains atoms. A more pedantic notation may be $\int_{0+}^{t}$, but it is often convenient to simply drop the lower bound from the notation if it happens to be 0+. More generally, the difference $X \bullet A_t - X \bullet A_s$, for $s,t \in \mathbb{R}_+$, $s < t$, can be expressed as $\int_{s+}^{t} X_u\,\mathrm{d}A_u$, or as $\int_{]s,t]} X\,\mathrm{d}A$, but in all cases in which the measure $\mathrm{d}A_\omega$ is nonatomic it is permissible to write $\int_s^t$ instead of $\int_{s+}^t$.

Naturally, the integrator A could be some deterministic function of finite variation. As an example, if $A_t = \imath(t) \equiv t$, then $X \bullet \imath$ can be interpreted as the process

$$X \bullet \imath(\omega,t) \equiv X \bullet \imath_\omega(t) \equiv X \bullet \imath_t(\omega) = \int_0^t X(\omega,s)\,\mathrm{d}s\,, \quad t \in \mathbb{R}_+\,, \quad \omega \in \Omega\,,$$

the last integral being understood as a Lebesgue integral for fixed ω. In particular, treated as a function of t, the Lebesgue integral $t \rightsquigarrow \int_0^t f(s)\,\mathrm{d}s$ can be viewed as a "stochastic integral" of the deterministic integrand f with respect to the deterministic finite variation process $\imath$. Thus, with the nomenclature just introduced, we can write $f \bullet \imath_t$ instead of $\int_0^t f(s)\,\mathrm{d}s$. It is instructive to keep in mind that, unlike in calculus, where integrals are introduced as scalars, a stochastic integral is understood to be a process. ○

Clearly, the path-by-path construction of the integral is only possible for integrators that are of finite variation, and, as was noted earlier, this is inadequate for many important applications. The first step in the construction of the stochastic integral for more general integrators, which may not be of finite variation, is to define such integrals for integrands that are simple enough to still allow for path-by-path interpretation, and, more important, allow for an interpretation that is consistent with the intuition. Recall that in (8.75) we introduced the space $\mathfrak{S} = \mathfrak{S}(\mathscr{F})$, and defined this space to be the collection of all finite sums of functions on $\Omega\times\mathbb{R}_+$ that are either of the form⊸ (8.70)

$$\Omega\times\mathbb{R}_+ \ni (\omega, t) \rightsquigarrow \xi(\omega)1_{[\![0]\!]}(\omega, t) \in \mathbb{R}$$

or of the form e_2

$$\Omega\times\mathbb{R}_+ \ni (\omega, t) \rightsquigarrow \eta(\omega)1_{]\!]u,v]\!]}(\omega, t) \in \mathbb{R}$$

for some choice of $u, v \in \mathbb{R}_+$ with $u < v$, and some choice of the bounded random variables $\xi \prec \mathscr{F}_0$ and $\eta \prec \mathscr{F}_u$. We call the elements of the space $\mathfrak{S}$ *simple integrands*. If Z is any finite stochastic process[120] on $(\Omega, \mathcal{F}, P)$, and X is any stochastic process of the form e_2, we set

$$X\bullet Z_t \stackrel{\text{def}}{=} \begin{cases} 0 & \text{if } X = \xi 1_{[\![0]\!]}, \\ \eta(Z_{t\wedge v} - Z_{t\wedge u}) & \text{if } X = \eta 1_{]\!]u,v]\!]}, \end{cases} \qquad t \in \mathbb{R}_+. \qquad \mathrm{e}_3$$

(10.2) DEFINITION (SIMPLE STOCHASTIC INTEGRALS): The definition of $X\bullet Z$ above extends by linearity for all simple integrands $X \in \mathfrak{S}$. To be precise, every process $X \in \mathfrak{S}$ can be expressed as a finite sum, $X = \sum_{i\in\mathbb{N}_{|n}} X^i$, where each X^i is of the form e_2, and we set

$$X\bullet Z = \sum\nolimits_{i\in\mathbb{N}_{|n}} X^i\bullet Z .$$

Although there are infinitely many ways in which a simple integrand $X \in \mathfrak{S}$ can be written as the sum of processes of the form e_2, it is easy to check that if $\sum_{i\in\mathbb{N}_{|n}} X^i = \sum_{j\in\mathbb{N}_{|m}} Y^j$, where all X^i's and all Y^j's are of the form e_2, then

$$\sum\nolimits_{i\in\mathbb{N}_{|n}} X^i\bullet Z = \sum\nolimits_{j\in\mathbb{N}_{|m}} Y^j\bullet Z .$$

As a result, the stochastic integral $X\bullet Z$ is well defined for every simple integrand $X \in \mathfrak{S}$. We call such integrals *simple stochastic integrals*. The process $X\bullet Z$ may be written also as $\int^t X_s \mathrm{d}Z_s$, $t \in \mathbb{R}_+$, and we note that this definition of the stochastic integral is consistent with the definition given in (10.1) for the case where the sample paths Z_ω are of finite variation. In particular, the integral again starts from 0 and is right-continuous at 0 if Z is right-continuous at 0. ○

[120] A "finite stochastic process" is a process with sample paths that are finite on finite intervals, or, to put it another way, a process with non-exploding sample paths.

(10.3) REMARK: The definition of $X \bullet Z$ would be perfectly meaningful for a slightly broader class of simple integrands X, that are again finite linear combinations of processes of the form e_2, but with u and v chosen to be finite stopping times with $u < v$. In fact, except for the change in the meaning of the symbols u and v, nothing else in the construction of $X \bullet Z$ needs to change. ○

Note that so far neither the measurability requirements for the r.v. ξ and η from e_2, nor the requirement for these r.v. to be bounded, was ever used. These requirements are needed in the next phase of our project, in which we are going to extend the construction of $X \bullet Z$ for more general (and interesting) integrands X, that may not be elements of $\mathfrak{S}$. Since in this extension we are going to insist that the measure $\mathrm{d}Z_\omega$ should not be required to exist for fixed $\omega \in \Omega$, we must restrict the integrator Z in some other way. To gain some insight into this matter, we now turn to the special case where Z is replaced by an $(\mathscr{F}, P)$-Brownian motion W, and observe the following:

(10.4) EXERCISE: Prove that for any $X \in \mathfrak{S}$ the process $X \bullet W$ is a continuous $(\mathscr{F}, P)$-martingale. In addition, prove that

$$\mathsf{E}\big[(X \bullet W_t)^2\big] = \int_0^t \mathsf{E}[X_s{}^2]\,\mathrm{d}s, \quad t \in \mathbb{R}_+ . \quad ○$$

(10.5) EXERCISE: Prove that for any $X, Y \in \mathfrak{S}$ and $s, t \in \mathbb{R}_+$,

$$\mathsf{E}\big[X \bullet W_s \times Y \bullet W_t\big] = \int_0^{s \wedge t} \mathsf{E}[X_r Y_r]\,\mathrm{d}r .$$

Conclude that

$$\mathsf{E}\big[\big(X \bullet W_t - Y \bullet W_t\big)^2\big] = \int_0^t \mathsf{E}\big[(X_r - Y_r)^2\big]\,\mathrm{d}r . \quad ○$$

The conclusions from the last two exercises are quite insightful. We know from (8.75) that $\mathfrak{S}$ is a dense subset of $L^2(\Omega \times \mathbb{R}_+, \mathscr{P}(\mathscr{F}), P \times \Lambda)$. Thus, given any predictable process $X \in \mathscr{L}^2(\Omega \times \mathbb{R}_+, \mathscr{P}(\mathscr{F}), P \times \Lambda)$, there is a sequence of simple processes, $(S^i \in \mathfrak{S})_{i \in \mathbb{N}}$, that converges to X in the L^2-norm for $P \times \Lambda$, that is,

$$\lim_{i \to \infty} \int_{\mathbb{R}_+} \mathsf{E}\big[(X_r - S_r^i)^2\big]\,\mathrm{d}r = 0 .$$

Since $(S^i)_{i \in \mathbb{N}}$ is a Cauchy sequence in the L^2-norm,

$$\lim_{i \to \infty} \sup_{j \ge i} \mathsf{E}\big[\big(S^j \bullet W_t - S^i \bullet W_t\big)^2\big] = \lim_{i \to \infty} \sup_{j \ge i} \int_{\mathbb{R}_+} \mathsf{E}\big[(S_r^i - S_r^j)^2\big]\,\mathrm{d}r = 0$$

for every $t \in \mathbb{R}_+$. As a result, for every fixed $t \in \mathbb{R}_+$ the sequence of random variables $(S^i \bullet W_t)_{i \in \mathbb{N}}$ is a Cauchy sequence in $L^2(\Omega, \mathscr{F}_t, P)$, and we can denote its limit by $X \bullet W_t$. Strictly speaking, this limit is a P-equivalence class, but we can always think of the r.v. $\Omega \ni \omega \rightsquigarrow X \bullet W_t(\omega) \in \mathbb{R}$ as one particular member of that class (here is an example of a situation where the axiom of choice is needed). Now we have a well-defined process, namely $X \cdot W \equiv (X \cdot W_t)_{t \in \mathbb{R}_+}$, that deserves to be called the *stochastic integral of X with respect to W*. It is not difficult to show that the conclusions from (10.4) and (10.5) remain valid not just for $X, Y \in \mathfrak{S}$, but also for $X, Y \in \mathscr{L}^2(\Omega \times \mathbb{R}_+, \mathscr{P}(\mathscr{F}), P \times \Lambda)$ – except for the continuity property of the martingale $X \cdot W$, which can still be established with some additional work.

In many introductory expositions the construction of the stochastic integral usually concludes not very far from where we now stand. However, this standpoint leaves a lot to be desired, and for a number of reasons. In the first place, we would like to be able to work with integrators that are more general than Brownian motion – to say the least, we would like to allow the stochastic integral itself to become an integrator. And even if we remain confined to stochastic integration with respect to Brownian motion only, we still need to establish some measurability and regularity properties for the process $X \cdot W$. By way of an example, at this point we cannot even claim that $(\omega, t) \rightsquigarrow X \cdot W(\omega, t)$ is jointly measurable as a function on $\Omega \times \mathbb{R}_+$. Another limitation – especially relevant for the domain of mathematical finance – is the square integrability of the integrand X. This is a limitation because such a property is not invariant under an equivalent change of the probability measure. In other words, even if the process $X \cdot W$ is still meaningful (as a stochastic process) under any other probability measure $Q \approx P$, from the point of view of the construction that we have at the present, there would be no reason to continue to interpret this process as a stochastic integral if treated under some other equivalent measure Q, since $X \cdot W$ may no longer be the L^2-limit of simple integrals under Q. To wit: a process that has the meaning of "accumulated capital gains" should not lose this meaning if the underlying measure is to be replaced with an equivalent one. For all these reasons, in the rest of this chapter we are going to develop a more general construction of the stochastic integral, which depends on the probability measure only up to equivalence. For now, we are going to limit the scope of this construction only to continuous integrators. The case of càdlàg integrators will be taken up later in section 15.2.

10.2 Stochastic Integrals with Respect to Continuous Local Martingales

Our goal in this section is to outline the construction of the stochastic integral for integrators that are generic continuous local martingales. The probability space $(\Omega, \mathcal{F}, P)$ and the filtration $\mathscr{F} \equiv (\mathscr{F}_t \subseteq \mathcal{F})_{t \in \mathbb{R}_+}$ are again fixed throughout (and the usual conditions are assumed). Suppose that M is any continuous local martingale (relative to $\mathscr{F}$ and P), and recall the definition of the class of simple processes $\mathfrak{S}$ introduced in (8.75), and also the construction of the simple stochastic integral in (10.2) and its amendment in (10.3). Given any $X \in \mathfrak{S}$, the (simple) integral $X \cdot M$ is a well-defined stochastic process. The immediate problem in front of us is how to extend the definition of $X \cdot M$ for the broadest possible class of integrands X. It is straightforward to check that if $X \in \mathfrak{S}$, then the sample path $X \cdot M_\omega$ is continuous for any $\omega \in \Omega$ for which the sample path M_ω is continuous, that is, for P-a.e. $\omega \in \Omega$. As a result, if $X \in \mathfrak{S}$, then $X \cdot M$ can be treated as a random variable with values in the Polish space $\mathscr{C}(\mathbb{R}_+;\mathbb{R})$, equipped with⊶ (0.8) the locally uniform distance $d_{\text{loc},\infty}$. Thus, to define $X \cdot M$ for more general integrands X, we must be able to approximate X with a sequence $(S^i \in \mathfrak{S})_{i \in \mathbb{N}}$ in some yet to be determined topology, but topology chosen so that the simple integrals⊶ (2.41)

$$S^i \cdot M \in \mathscr{L}^0\big(\Omega, \mathscr{F}_\infty, P; \mathscr{C}(\mathbb{R}_+;\mathbb{R})\big), \quad i \in \mathbb{N},$$

must converge as long as $S^i \to X$ (in the yet to be determined topology). There is a choice to be made at this juncture: in what sense should the simple integrals $S^i \cdot M$ be required to converge? From what we know about convergence of r.v., the least demanding convergence would be the convergence in probability P. Coincidentally, this convergence is also the one that is most interesting and useful for us, since it depends on the measure only up to equivalence. Given our general objectives in this space, this is a particularly desirable feature, since we would like any objects defined as stochastic integrals to retain their "stochastic integral" interpretation even if the measure is replaced with an absolutely continuous (or equivalent) one. Before we continue, recall that, since $\mathscr{C}(\mathbb{R}_+;\mathbb{R})$ is a Polish space, the above space $\mathscr{L}^0$ is⊶ (4.9), (4.18) complete for the Ky Fan metric, the convergence in which is nothing but the convergence in probability P. Recall also that the convergence in $\mathscr{C}(\mathbb{R}_+;\mathbb{R})$ in the locally uniform distance is equivalent to convergence in the uniform distance on every finite interval $[0,T]$, and that, therefore, convergence in probability for random variables taking values in $\mathscr{C}(\mathbb{R}_+;\mathbb{R})$ is equivalent to convergence in probability for the restrictions of such random variables to $[0,T]$ for every $T \in \mathbb{R}_{++}$, the restrictions being treated as random variables from $\mathscr{L}^0\big(\Omega, \mathscr{F}_\infty, P; \mathscr{C}([0,T];\mathbb{R})\big)$ and the space $\mathscr{C}([0,T];\mathbb{R})$ being treated as a Banach space equipped with the uniform distance. To get started with this program, we first derive the basic features of the integral $X \cdot M$ for simple integrands $X \in \mathfrak{S}$. These features are just as important as they are straightforward.

(10.6) EXERCISE: Let M be any continuous local martingale, and let $X, Y \in \mathfrak{S}$ be arbitrarily chosen. Prove the following:

(a) If $a, b \in \mathbb{R}$, then $(aX + bY) \cdot M = aX \cdot M + bY \cdot M$.

(b) If t is a stopping time, then⊶ (8.56), (8.70), (10.3)

$$(X\,1_{[\![0,t]\!]}) \cdot M = X \cdot (M^t) = (X \cdot M)^t\,,$$

or, equivalently,

$$(X\,1_{[\![0,t]\!]}) \cdot M_t = X \cdot (M^t)_t = X \cdot M_{t \wedge t} \quad \text{for any } t \in \mathbb{R}_+\,.$$

(c) If M is a bounded and continuous martingale, then so is also $X \cdot M$.

(d) If M is an L^2-bounded and continuous martingale, then so is also $X{\bullet}M$, and

$$\mathsf{E}\big[(X{\bullet}M_t)^2\big] = \mathsf{E}\big[X^2{\bullet}\langle M,M\rangle_t\big]\,, \quad t \in \mathbb{R}_+ .$$

(e) If M is a continuous local martingale (the initial meaning of M), then so is also $X{\bullet}M$ and

$$\langle X{\bullet}M, X{\bullet}M\rangle = X^2{\bullet}\langle M,M\rangle\,.$$

If N is another continuous local martingale, then

$$\langle X{\bullet}M, N\rangle = X{\bullet}\langle M,N\rangle\,.$$

(f) We have $XY \in \mathfrak{S}$ and $X{\bullet}(Y{\bullet}M) = (XY){\bullet}M$. ○

(10.7) REMARK: When the class of simple processes $\mathfrak{S}$ was introduced back in (8.75) it was not immediately clear why its elements $\eta\, 1_{]\!]u,v]\!]}$ had to be chosen so that $\eta \prec \mathscr{F}_u$. In fact, this property was never used until (10.6) above. Now we see that although $X{\bullet}M$ is well defined without the requirement for X to be predictable (which comes down to the conditions $\eta \prec \mathscr{F}_u$ and $\xi \prec \mathscr{F}_0$ in (8.75)), if X lacks predictability, in general, $X{\bullet}M$ will not inherit the (local)martingale property from M and, as a result, our ability to extend the definition of $X{\bullet}M$ for a broader class of integrands X will be severely limited. This is because the extension that we are about to outline hinges in a crucial way on the properties of martingales and local martingales. ○

The key step in the construction of the stochastic integral with respect to continuous local martingales is the estimate that we are about to formulate. This estimate is essentially a variation of the one found in (McKean 1969) for stochastic integrals with respect to Brownian motion. The key insight in this crucial result is that the size of the stochastic integral $X{\bullet}M$, treated as a $\mathscr{C}(\mathbb{R}_+;\mathbb{R})$-valued random variable, can be controlled by the size of the integral $X^2{\bullet}\langle M,M\rangle$, which is$^{\multimap(10.1)}$ of Lebesgue–Stieltjes type. Notice that since the sample paths $X^2{\bullet}\langle M,M\rangle_\omega$ are continuous for P-a.e. $\omega \in \Omega$, the process $X^2{\bullet}\langle M,M\rangle$ can be treated as a $\mathscr{C}(\mathbb{R}_+;\mathbb{R})$-valued random variable as well. Furthermore, the sample paths $X^2{\bullet}\langle M,M\rangle_\omega$ are increasing, and this means that the uniform norm of any such path on any finite interval $[0,t]$ is simply its value at the right end-point, namely $X^2{\bullet}\langle M,M\rangle_\omega(t)$. As a result, the convergence in probability to the constant function 0 for $\mathscr{C}(\mathbb{R}_+;\mathbb{R})$-valued random variables of the form $X^2{\bullet}\langle M,M\rangle_\omega$ comes down to convergence to $0 \in \mathbb{R}$ in probability for the $\mathbb{R}$-valued random variables $X^2{\bullet}\langle M,M\rangle_t$ for every fixed $t \in \mathbb{R}_+$.

(10.8) ONE USEFUL ESTIMATE: (McKean 1969) If M is any continuous local martingale, then

† $$P\Big(\sup\nolimits_{t\in[0,T]} \big|X{\bullet}M_t\big| > \delta\Big) \le \frac{\varepsilon}{\delta^2} + P\Big(X^2{\bullet}\langle M,M\rangle_T > \varepsilon\Big)$$

for every choice of the simple process $X \in \mathfrak{S}$ and the scalars $T, \varepsilon, \delta \in \mathbb{R}_{++}$. ○

The complete proof of ($10.8^\dagger$) is included at the end of this section.

(10.9) COMMENTS: If we choose X and ε in such a way that the probability in the right side of ($10.8^\dagger$) would be bounded by ε, then we would have

$$P\Big(\sup_{t\in[0,T]} \big|X\bullet M_t\big| > \delta\Big) \le \frac{\varepsilon}{\delta^2} + \varepsilon\,.$$

If we take the infimum over all such ε (keeping X fixed), we arrive at the relation

$$P\Big(\sup_{t\in[0,T]} \big|X\bullet M_t\big| > \delta\Big) \le \Big(\frac{1}{\delta^2} + 1\Big) d(X^2\bullet\langle M,M\rangle_T\,,\,0)\,,$$

where$^{\multimap(4.8)}$ d is the Ky Fan distance in the space $\mathscr{L}^0(\Omega,\mathcal{F},P;\mathbb{R})$. In particular, if $\delta\in\,]0,1]$ and $d(X^2\bullet\langle M,M\rangle_T\,,\,0)\le\frac{1}{2}\delta^3$, then

$$d\Big(\sup_{t\in[0,T]} \big|X\bullet M_t\big|\,,\,0\Big) \le \delta\,. \quad \circ$$

The steps that we now have to take suggest themselves. We would like to extend the definition of the stochastic integral for integrands X that are predictable (why predictable will become clear very soon) and such that $X^2\bullet\langle M,M\rangle_t<\infty$ a.s. for any $t\in\mathbb{R}_{++}$, by using the extension result from (A.41), but we have to make this idea precise.

(10.10) THE CLASS OF STOCHASTIC INTEGRANDS $\mathfrak{L}^2_{\mathrm{loc}}(M)$: Given a continuous local $(\mathscr{F},P)$-martingale M, we use the symbol $\mathfrak{L}^2_{\mathrm{loc}}(M)$ to denote the collection of all predictable processes, that is, $\mathscr{P}(\mathscr{F})$-measurable mappings, of the form

$$\Omega\times\mathbb{R}_+ \ni (\omega,t) \rightsquigarrow X(\omega,t)\in\mathbb{R}\,,$$

such that for P-a.e. $\omega\in\Omega$ the sample path X_ω belongs to the space

$$\mathscr{L}^2_{\mathrm{loc}}(\mathbb{R}_+,\mathcal{B}(\mathbb{R}_+),\mathrm{d}\langle M,M\rangle_\omega)\,,$$

thereby, for P-a.e. $\omega\in\Omega$, $\int_{[0,t]} {X_\omega}^2\,\mathrm{d}\langle M,M\rangle_\omega<\infty$ for every $t\in\mathbb{R}_+$. Any process X that has these properties can be associated with the $\mathscr{C}(\mathbb{R}_+;\mathbb{R})$-valued r.v. $X^2\bullet\langle M,M\rangle$, treated as an element of $\mathscr{L}^0\big(\Omega,\mathscr{F}_\infty,P;\mathscr{C}(\mathbb{R}_+;\mathbb{R})\big)$. As we are already aware, this space can be endowed with the topology of convergence in probability, and this topology can be metrized by the respective Ky Fan distance, which we denote by $\rho(\cdot,\cdot)$.[121] Thus, we can define "distance to 0" in the space $\mathfrak{L}^2_{\mathrm{loc}}(M)$ in the following way:

$$\rho^*(X,0)=\rho\big(X^2\bullet\langle M,M\rangle\,,\,0\big)\,,\quad X\in\mathfrak{L}^2_{\mathrm{loc}}(M)\,.$$

Since $\mathfrak{L}^2_{\mathrm{loc}}(M)$ is a linear space, the distance to 0 defines topology (convergence) in that space, and since the stochastic integral is a linear operation, it would be enough for us to show that it is continuous at 0 in that topology. As an aside, we

[121] Recall that the distance $\rho(\cdot,\cdot)$ turns the space of equivalence classes $L^0\big(\Omega,\mathscr{F}_\infty,P;\mathscr{C}(\mathbb{R}_+;\mathbb{R})\big)$ into a complete metric space.

note that the class $\mathfrak{L}^2_{\text{loc}}(M)$ is actually quite broad: at the very least it includes all adapted and continuous processes. In addition, the space $\mathfrak{L}^2_{\text{loc}}(M)$ depends on the local martingale M only through the increasing process $\langle M, M\rangle$, so that there is no ambiguity if we were to use the symbol $\mathfrak{L}^2_{\text{loc}}(\langle M, M\rangle)$ as an alias for $\mathfrak{L}^2_{\text{loc}}(M)$. ○

(10.11) COMMENTS: If one insists on working only with the Ky Fan distance d in the simpler space $\mathscr{L}^0(\Omega, \mathcal{F}, P; \mathbb{R})$ that consists of real random variables, then the distance (in probability) in the space $\mathscr{L}^0\big(\Omega, \mathscr{F}_\infty, P; \mathscr{C}(\mathbb{R}_+; \mathbb{R})\big)$ can be defined (equivalently) as

$$\rho(X,Y) = \sum\nolimits_{n \in \mathbb{N}_{++}} 2^{-n} d\big(\sup\nolimits_{t \in [0,n]} |X_t - Y_t|\,,\, 0\big)\,,$$

in which case, given any $X \in \mathfrak{L}^2_{\text{loc}}(M)$, we can write

$$\rho^*(X,0) = \sum\nolimits_{n \in \mathbb{N}_{++}} 2^{-n} d\big(X^2 \bullet \langle M, M\rangle_n\,,\, 0\big)\,. \quad \circ$$

Getting back to the estimate in (10.8), we see that the stochastic integral, treated as a mapping of the form

$$\mathfrak{L}^2_{\text{loc}}(M) \ni X \rightsquigarrow X \bullet M \in \mathscr{L}^0\big(\Omega, \mathscr{F}_\infty, P; \mathscr{C}(\mathbb{R}_+; \mathbb{R})\big)\,,$$

is continuous (that is, continuous at 0) when restricted to the space of simple integrands $\mathfrak{S} \subset \mathfrak{L}^2_{\text{loc}}(M)$, the continuity being understood with respect to the distance ρ^* to 0 in the domain and the Ky Fan distance ρ in the range. Actually, we need uniform continuity in order to utilize the extension argument from (A.41), and (10.8) does indeed provide this feature, as we have seen in (10.9). To state the conclusion from (10.8) and (10.9) in another way, the stochastic integral maps any Cauchy sequence (for the distance ρ^*) of simple integrands into a Cauchy sequence (for the distance ρ) of simple stochastic integrals, and this is all that we need.

In any case, now we know that the stochastic integral can be extended from the space of simple integrands $\mathfrak{S}$ to the closure of $\mathfrak{S}$ inside the space of candidate integrands $\mathfrak{L}^2_{\text{loc}}(M)$, the closure being understood with respect to the distance ρ^*. However, the closure of $\mathfrak{S}$ is the entire space $\mathfrak{L}^2_{\text{loc}}(M)$, or, to put it another way, the space $\mathfrak{S}$ is dense in $\mathfrak{L}^2_{\text{loc}}(M)$. In fact, this is a property that we have essentially established in (8.75). Indeed, if we define the measure

$$m(\mathrm{d}\omega, \mathrm{d}t) = P(\mathrm{d}\omega)\mathrm{d}\langle M, M\rangle_\omega(t)\,,$$

where $\mathrm{d}\langle M, M\rangle_\omega(t)$ is understood as the Borel measure on $\mathbb{R}_+$ associated with the increasing function $t \rightsquigarrow \langle M, M\rangle_\omega(t)$, then we can claim$^{\multimap (8.75)}$ that the closure of the space $\mathfrak{S}$ with respect to the norm

$$\|X\|_2 \overset{\text{def}}{=} \Big(\int_{\Omega\times\mathbb{R}_+} X(\omega,t)^2 m(\mathrm{d}\omega, \mathrm{d}t)\Big)^{\frac{1}{2}} = \big(\mathsf{E}\big[X^2 \bullet \langle M, M\rangle_\infty\big]\big)^{\frac{1}{2}}$$

is the entire space $L^2(\Omega\times\mathbb{R}_+, \mathscr{P}(\mathscr{F}), m)$, and, clearly, $\mathscr{L}^2(\Omega\times\mathbb{R}_+, \mathscr{P}(\mathscr{F}), m)$ is contained inside $\mathfrak{L}^2_{\text{loc}}(M)$. Furthermore, the topology associated with the norm

$\|\cdot\|_2$ is stronger than the topology generated by the distance (to 0) ρ^*, as is immediate from the relation

$$P\Big(X^2{\bullet}\langle M,M\rangle_t > \varepsilon\Big) \le \frac{1}{\varepsilon}\mathsf{E}\big[X^2{\bullet}\langle M,M\rangle_\infty\big] \quad \text{for any } t \in \mathbb{R}_{++},$$

which is telling us that for any $\varepsilon \in]0,1]$ the relation $\mathsf{E}\big[X^2{\bullet}\langle M,M\rangle_\infty\big] \le \varepsilon^2$ implies

$$\rho^*\big(X^2{\bullet}\langle M,M\rangle, 0\big) \le \varepsilon.$$

Thus, $\mathfrak{S}$ is also dense in $\mathscr{L}^2(\Omega{\times}\mathbb{R}_+, \mathscr{P}(\mathscr{F}), m)$ relative to the topology of ρ^*. However, $\mathscr{L}^2(\Omega{\times}\mathbb{R}_+, \mathscr{P}(\mathscr{F}), m)$ is dense in $\mathfrak{L}^2_{\mathrm{loc}}(M)$ in that same topology. To see this, given any $X \in \mathfrak{L}^2_{\mathrm{loc}}(M)$, define the stopping times (recall that $\inf\emptyset \overset{\mathrm{def}}{=} \infty$)

$$t_n = \inf\big\{t \in \mathbb{R}_{++} : X^2{\bullet}\langle M,M\rangle_t \ge n\big\}, \quad n \in \mathbb{N}_{++},$$

and observe that the process $X 1_{[\![0,t_n]\!]}$ belongs to $\mathscr{L}^2(\Omega{\times}\mathbb{R}_+, \mathscr{P}(\mathscr{F}), m)$ since, by its very definition,

$$(X 1_{[\![0,t_n]\!]})^2{\bullet}\langle M,M\rangle_\infty = X^2{\bullet}\langle M,M\rangle_{t_n} \le n.$$

Since $X \in \mathfrak{L}^2_{\mathrm{loc}}(M)$, the increasing function $t \rightsquigarrow X^2{\bullet}\langle M,M\rangle_\omega(t)$ can explode to ∞ in finite time only on a set of measure 0, and this means that $t_n(\omega) \to +\infty$ for P-a.e. $\omega \in \Omega$. In particular, P-a.s.

$$(X - X 1_{[\![0,t_n]\!]})^2{\bullet}\langle M,M\rangle_t \to 0 \ \text{ for every } t \in \mathbb{R}_+.$$

Since the a.s. convergence implies convergence in probability (governed by the Ky Fan distance), in conjunction with the result from (10.11) the last relation implies

$$\rho^*(X - X 1_{[\![0,t_n]\!]}, 0) = \rho\big((X - X 1_{[\![0,t_n]\!]})^2{\bullet}\langle M,M\rangle, 0\big) \to 0 \quad \text{as } n \to \infty.$$

In particular, $\mathscr{L}^2(\Omega{\times}\mathbb{R}_+, \mathscr{P}(\mathscr{F}), m)$ is dense in $\mathfrak{L}^2_{\mathrm{loc}}(M)$ in the topology of ρ^* as claimed, and since $\mathfrak{S}$ is dense in $\mathscr{L}^2(\Omega{\times}\mathbb{R}_+, \mathscr{P}(\mathscr{F}), m)$, we conclude that $\mathfrak{S}$ is dense in $\mathfrak{L}^2_{\mathrm{loc}}(M)$ as well (in the topology of ρ^*).

Now we come to an important definition.

(10.12) DEFINITION (STOCHASTIC INTEGRALS I): The stochastic integral with respect to a continuous local martingale M is the unique linear mapping

$$\mathfrak{L}^2_{\mathrm{loc}}(M) \ni X \rightsquigarrow X{\bullet}M \in \mathscr{L}^0\big(\Omega, \mathscr{F}_\infty, P; \mathscr{C}(\mathbb{R}_+;\mathbb{R})\big)$$

that coincides with the simple stochastic integral on the space $\mathfrak{S} \subseteq \mathfrak{L}^2_{\mathrm{loc}}(M)$ and is continuous with respect to the distance ρ^* in the domain and the Ky Fan distance ρ in the range. There are situations where the usual integral notation would be preferable, and in any such situation we will write

$$X{\bullet}M_t = \int_0^t X_s \,\mathrm{d}M_s \equiv \int_0^t X\,\mathrm{d}M, \quad t \in \mathbb{R}_+, \quad X \in \mathfrak{L}^2_{\mathrm{loc}}(M).$$

When using the integral notation one must keep in mind that the integral is understood as $\int_{0+}^{t}$ (it includes the upper limit t but not the lower limit 0) – such a "fine point" is immaterial in the present setting, since the measures $\mathrm{d}\langle M,M\rangle_\omega$, $\omega\in\Omega$, are nonatomic, but it will become important later, when we expand the construction of the integral to include integrators with jumps. For this reason, if the integral notation is desirable, the difference $X\bullet M_t - X\bullet M_s$, for $s<t$, should be written as $\int_{s+}^{t} X_u\,\mathrm{d}M_u$, or, better yet, $\int_{]s,t]} X\,\mathrm{d}M$. For simplicity, the lower bound can be omitted from the notation if it happens to be 0+, and in situations where $\mathrm{d}\langle M,M\rangle_\omega$ is nonatomic it is permissible to write $\int_s^t$ instead of $\int_{s+}^t$. ○

(10.13) COMMENTS: (a) Since the construction of the stochastic integral involves only convergence in probability, its definition depends only on the null sets for the measure P. Thus, if the $\mathscr{C}(\mathbb{R}_+;\mathbb{R})$-valued random variable $X\bullet M$ is treated under another measure $Q\approx P$, then $X\bullet M$ is not only a $\mathscr{C}(\mathbb{R}_+;\mathbb{R})$-valued random variable under Q, but is actually a stochastic integral under Q, understood as a limit in probability Q of simple integrals, even if M may no longer be a local martingale under Q. This observation is especially important in the domain of asset pricing. The same claim can be made if Q is only absolutely continuous with respect to P, since convergence in probability P automatically implies convergence in probability $Q\lesssim P$ (enlarging the collection of null sets makes it easier for a sequence to converge in probability).

(b) It should be clear by now why we insist that the integrand X must be predictable (however, see the next item). The reason is that it is not possible to approximate non-predictable integrands with simple ones, and in such a way that the approximating sequence of simple integrals would converge – in probability, or in some other meaningful way. Indeed, the estimate in (10.8) rests in a crucial way on the fact that $X\bullet M$ is a local martingale if M is a local martingale, and the predictability of X (the requirement for ξ and η from e_2 in section 10.1 to be measurable, respectively, for $\mathscr{F}_0$ and $\mathscr{F}_u$) is instrumental for this feature to hold.

(c) Strictly speaking, the available integrands for the stochastic integral are the equivalence classes with respect to the measure $P(\mathrm{d}\omega)\mathrm{d}\langle M,M\rangle_\omega(t)$ associated with the processes from $\mathfrak{L}^2_{\mathrm{loc}}(M)$. We will denote the space of these equivalence classes by $\mathbb{L}^2_{\mathrm{loc}}(M)$, or, equivalently, by $\mathbb{L}^2_{\mathrm{loc}}(\langle M,M\rangle)$, which is more cumbersome but more accurate. Clearly, a stochastic process may belong to some equivalence class from the space $\mathbb{L}^2_{\mathrm{loc}}(M)$ without being predictable. As an example, if $Y:\Omega\times\mathbb{R}_+\mapsto\mathbb{R}$ is any $\mathcal{F}\otimes\mathcal{B}(\mathbb{R}_+)$-measurable mapping (not necessarily predictable, and not necessarily adapted even) such that for P-a.e. $\omega\in\Omega$, $Y(\omega,t)=X(\omega,t)$ for $\mathrm{d}\langle M,M\rangle_\omega$-a.e. $t\in\mathbb{R}_{++}$, for some $X\in\mathfrak{L}^2_{\mathrm{loc}}(M)$, then the integral $Y\bullet M$ would be well defined, and the (continuous) sample paths $Y\bullet M_\omega$ and $X\bullet M_\omega$ would be identical for P-a.e. $\omega\in\Omega$. ○

(10.14) EXERCISE: Prove that the crucial estimate in (10.8†) remains valid for any $X\in\mathfrak{L}^2_{\mathrm{loc}}(M)$.

HINT: The property extends by continuity for the respective types of convergence. Alternatively, the proof included at the end of this section does not really use the fact that X is a simple integrand from $\mathfrak{S}$ and can be repeated with a generic $X \in \mathfrak{L}^2_{\text{loc}}(M)$ instead. ○

(10.15) ISOMETRY: Suppose that the continuous local martingale M is such that $\mathsf{E}[\langle M, M\rangle_\infty] < \infty$, that is, $M \in H^2$ (M is an L^2-bounded and continuous martingale). By (d) and (e) in (10.6), for every $X \in \mathfrak{S}$, we have $X \bullet M \in H^2_0$ and

$$\mathsf{E}\big[(X \bullet M_\infty)^2\big] = \mathsf{E}\big[X^2 \bullet \langle M, M\rangle_\infty\big] < \infty .$$

Recall that H^2_0 can be identified with the (Hilbert) subspace of $L^2(\Omega, \mathscr{F}_\infty, P)$ that contains all elements with a vanishing mean, consider again the measure

$$m(\mathrm{d}\omega, \mathrm{d}t) = P(\mathrm{d}\omega)\mathrm{d}\langle M, M\rangle_\omega(t) ,$$

and introduce the linear mapping

$$L^2(\Omega\times\mathbb{R}_+, \mathscr{P}(\mathscr{F}), m) \ni X \rightsquigarrow X \bullet M_\infty \in L^2(\Omega, \mathscr{F}_\infty, P) ,$$

which is no different from

$$L^2(\Omega\times\mathbb{R}_+, \mathscr{P}(\mathscr{F}), m) \ni X \rightsquigarrow X \bullet M \in H^2_0 .$$

If restricted to simple $X \in \mathfrak{S}$, this mapping is an isometry (defined on $\mathfrak{S}$) with respect to the Hilbert norms in the domain and the range, as we just showed. Since, as we showed earlier, $\mathfrak{S}$ is dense in the domain, we can extend $X \rightsquigarrow X \bullet M$ to an isometry with values in H^2_0 defined on the entire Hilbert space $L^2(\Omega\times\mathbb{R}_+, \mathscr{P}(\mathscr{F}), m)$, the norm in which is stronger than the distance in probability ρ^*. As a result, the isometry that we just introduced is no different from the stochastic integral that we introduced earlier, since clearly $L^2(\Omega\times\mathbb{R}_+, \mathscr{P}(\mathscr{F}), m) \subset \mathbb{L}^2_{\text{loc}}(M)$. This observation has the effect that if $X \in \mathfrak{L}^2_{\text{loc}}(M)$ is such that $\mathsf{E}[X^2 \bullet \langle M, M\rangle_\infty] < \infty$, then $X \bullet M$ is not only well defined as a continuous process, but can be claimed to be a continuous and L^2-bounded martingale that starts from 0 (an element of H^2_0). In particular, $X \bullet M_t$ is a square integrable r.v. with

$$\mathsf{E}\big[(X \bullet M_t)^2\big] = \mathsf{E}\big[X^2 \bullet \langle M, M\rangle_t\big] \text{ and } \mathsf{E}\big[X \bullet M_t\big] = 0 \quad \text{for every } t \in \mathbb{R}_+ . \quad ○$$

(10.16) EXERCISE: Prove that properties (a), (b), and (e) in exercise (10.6) remain valid if the token $\mathfrak{S}$ is replaced everywhere with $\mathfrak{L}^2_{\text{loc}}(M)$. In addition, prove the following extended version of (f): if $Y \in \mathfrak{L}^2_{\text{loc}}(M)$ and X is predictable and such that $XY \in \mathfrak{L}^2_{\text{loc}}(M)$, then $X \bullet (Y \bullet M) = (XY) \bullet M$. The extension of property (d) was developed in (10.15).

HINT: All properties extend by continuity for the respective types of convergence. ○

We see from the last exercise that the stochastic integral $X \bullet M$ is a continuous local martingale as long as M is a continuous local martingale and $X \in \mathfrak{L}^2_{\text{loc}}(M)$.

In general, the random variables $X \bullet M_t$, $t \in \mathbb{R}_+$, have no reason to have well-defined expected values unless some additional assumptions are imposed on the integrator M and the integrand X – see (10.15).

(10.17) APPROXIMATION WITH RIEMANN SUMS: We know from calculus that any continuous real-valued function is integrable in the sense of Riemann on every closed and finite interval on which it is defined. We have also seen that if the Riemann integral is defined, then so is also the Lebesgue integral and both integrals coincide. The stochastic integral that we defined in this section has a similar feature: for continuous integrands and integrators it can be defined equivalently as the limit of Riemann sums. To see this, observe that any simple process $X \in \mathfrak{S}$ can be written in the form

$$\xi 1_{[\![0]\!]} + \sum_{i \in \mathbb{N}_{|n}} \eta_i 1_{]\!]t_i\,,\,t_{i+1}]\!]}$$

for some choice of $\xi \prec \mathscr{F}_0$, $0 \leq t_0 < t_1 \ldots < t_n < t_{n+1}$, and $\eta_i \prec \mathscr{F}_{t_i}$ for every $i \in \mathbb{N}_{|n}$. Since any $X \in \mathfrak{L}^2_{\text{loc}}(M)$ can be approximated in the distance ρ^* with simple integrands of the form above, the stochastic integral $X \bullet M$ can be approximated in the Ky Fan distance ρ with processes of the form

$$\sum_{i \in \mathbb{N}_{|n}} \eta_i (M_{t_{i+1} \wedge t} - M_{t_i \wedge t}) = \Big(\sum_{i \in \mathbb{N}_{|n}} \eta_i (M^{t_{i+1}} - M^{t_i})\Big)_t \quad t \in \mathbb{R}_+ .$$

In particular, if X is any P-a.s. continuous and $\mathscr{F}$-adapted process, then it is $\mathscr{P}(\mathscr{F})$-measurable and its sample paths X_ω are a.s. bounded on any finite interval with probability 1, so that $X \in \mathfrak{L}^2_{\text{loc}}(M)$. It is straightforward to check that

$$X = \rho^*\text{-}\lim_n \sum_{i=0}^{2^n} X_{i/n} 1_{]\!]i/n\,,\,(i+1)/n]\!]}$$

and that therefore

$$X \bullet M = \rho\text{-}\lim_n \sum_{i=0}^{2^n} X_{i/n}\big(M^{(i+1)/n} - M^{i/n}\big) .$$

More generally, if $0 = t^n_0 < t^n_1 \ldots < t^n_i < t^n_{i+1} < \ldots \to \infty$, $n \in \mathbb{N}$, is any sequence of partitions of $\mathbb{R}_+$ such that $\lim_n \sup_{i \in \mathbb{N}} |t^n_{i+1} - t^n_i| = 0$, then

† $$X \bullet M = \rho\text{-}\lim_n \sum_{i \in \mathbb{N}} X_{t^n_i}\big(M^{t^n_{i+1}} - M^{t^n_i}\big) . \quad \circ$$

We conclude this section with the proof of the key step on which our construction of the stochastic integral was based.

(10.18) PROOF OF (10.8): Let $(\mathscr{t}_n)_{n \in \mathbb{N}}$ be any localizing sequence for the local martingale M, so that $M^{\mathscr{t}_n}$ is a bounded martingale for every $n \in \mathbb{N}$. Let $\varepsilon \in \mathbb{R}_{++}$ be arbitrarily chosen, let

$$\mathscr{s}_\varepsilon \overset{\text{def}}{=} \inf\big\{t \in \mathbb{R}_+ : X^2 \bullet \langle M, M\rangle_t \geq \varepsilon\big\} ,$$

and let $t_{n,\varepsilon} = t_n \wedge s_\varepsilon$, $n \in \mathbb{N}$. Clearly, $M^{t_{n,\varepsilon}}$ is a bounded martingale (as a bounded martingale stopped at time s_ε), and for every $t \in \mathbb{R}_{++}$ one has

$$\mathsf{E}\big[\big(X\bullet(M^{t_{n,\varepsilon}})_t\big)^2\big] = \mathsf{E}\big[X^2\bullet\langle M^{t_{n,\varepsilon}}, M^{t_{n,\varepsilon}}\rangle_t\big] = \mathsf{E}\big[X^2\bullet\langle M, M\rangle_{t\wedge t_{n,\varepsilon}}\big] \le \varepsilon\,.$$

Next, define the processes

$$a_t^{n,\varepsilon} = X\bullet M_{t\wedge t_{n,\varepsilon}} \equiv X\bullet(M^{t_{n,\varepsilon}})_t \quad \text{and} \quad b_t^{n,\varepsilon} = X\bullet M_t - a_t^{n,\varepsilon} \quad t \in \mathbb{R}_+\,,$$

and observe that for every fixed $T > 0$ and every fixed $\delta > 0$, (notice that $b^{n,\varepsilon}(\omega, t) = 0$ if $(\omega, t) \in [\![0, t_{n,\varepsilon}]\!]$)

$$\begin{aligned}
P\Big(\sup_{t\in[0,T]} |X\bullet M_t| > \delta\Big) &= P\Big(\sup_{t\in[0,T]}|a_t^{n,\varepsilon} + b_t^{n,\varepsilon}| > \delta\Big) \\
&= P\Big(\sup_{t\in[0,T]}|a_t^{n,\varepsilon} + b_t^{n,\varepsilon}| > \delta, t_{n,\varepsilon} < T\Big) \\
&\qquad + P\Big(\sup_{t\in[0,T]}|a_t^{n,\varepsilon} + b_t^{n,\varepsilon}| > \delta, t_{n,\varepsilon} \ge T\Big) \\
&\le P\big(t_{n,\varepsilon} < T\big) + P\Big(\sup_{t\in[0,T]}|a_t^{n,\varepsilon}| > \delta\Big)\,.
\end{aligned}$$

Now observe that $(a_t^{n,\varepsilon})$ is an L^2-bounded martingale and Doob's maximal inequality gives

$$P\Big(\sup_{t\in[0,T]}|a_t^{n,\varepsilon}| > \delta\Big) = P\Big(\sup_{t\in[0,T]}|a_t^{n,\varepsilon}|^2 > \delta^2\Big) \le \frac{\mathsf{E}[|a_T^{n,\varepsilon}|^2]}{\delta^2} \le \frac{\varepsilon}{\delta^2}\,,$$

which results in the estimate

$$\dagger \qquad P\Big(\sup_{t\in[0,T]} |X\bullet M_t| > \delta\Big) \le P\big(t_{n,\varepsilon} < T\big) + \frac{\varepsilon}{\delta^2} \quad \text{for every } n \in \mathbb{N}\,.$$

As the sets $\big(t_{n,\varepsilon} < T\big)$ decrease monotonically (with $n \nearrow +\infty$) to the set

$$(s_\varepsilon < T) = \Big(X^2\bullet\langle M, M\rangle_T > \varepsilon\Big),$$

passing to the limit in † with $n \to \infty$ we arrive at the estimate

$$P\Big(\sup_{t\in[0,T]} |X\bullet M_t| > \delta\Big) \le \frac{\varepsilon}{\delta^2} + P\Big(X^2\bullet\langle M, M\rangle_T > \varepsilon\Big)\,. \quad \circ$$

10.3 Stochastic Integrals with Respect to Continuous Semimartingales

From where we now stand, constructing stochastic integrals with respect to continuous semimartingales is entirely straightforward. The first step in this program is the following definition:

(10.19) DEFINITION: A progressively measurable process Y is said to be locally bounded if there is a localizing sequence of stopping times $(t_n)_{n\in\mathbb{N}}$ that increases to $+\infty$ monotonically and is such that for any $n \in \mathbb{N}$ the stopped process Y^{t_n}

is bounded, that is, there is a constant $C_n \in \mathbb{R}_{++}$ such that for P-a.e. $\omega \in \Omega$, $|Y_{t\wedge t_n}(\omega)| \le C_n$ for all $t \in \mathbb{R}_+$. ○

(10.20) EXERCISE: Prove that every adapted and continuous process is locally bounded. Prove that if M is any continuous local martingale and Y is any locally bounded predictable process, then P-a.s.

$$Y^2 \cdot \langle M, M\rangle_t < \infty \quad \text{for every } t \in \mathbb{R}_+ . \quad ○$$

(10.21) DEFINITION (STOCHASTIC INTEGRALS II): Given any⊸• (9.79) continuous semimartingale $X = M + A$ and any locally bounded and predictable process Y, we define the integral $Y \cdot X$ in the obvious way as the process

$$Y \cdot X \stackrel{\text{def}}{=} Y \cdot M + Y \cdot A .$$

In particular, $Y \cdot X$ is itself a semimartingale. The difference $Y \cdot X_t - Y \cdot X_s$, for $s < t$, can again be expressed as $\int_{s+}^t Y_u \, dX_u$, and $\int^t Y_u \, dX_u$ will be understood as $\int_{0+}^t Y_u \, dX_u = Y \cdot X_t$, although the + sign in the lower limits of the integrals will remain superfluous until we introduce semimartingales with jumps. ○

(10.22) BASIC PROPERTIES: Let X be any continuous semimartingale and let Y and Z be any two locally bounded predictable processes. Then:

(a) $Z \cdot (Y \cdot X) = (ZY) \cdot X$;

(b) $Y \cdot (X^t) = (Y 1_{[\![0,t]\!]}) \cdot X = (Y \cdot X)^t$;

(c) with probability 1 the sample path $Y \cdot X_\omega$ is constant on any interval $[a, b]$ on which either the sample path Y_ω vanishes or the sample path X_ω is constant;

(d) the Riemann sums approximation from (10.17†) also holds for stochastic integrals with respect to continuous semimartingales, provided that the integrand is continuous and adapted. ○

(10.23) MULTIVARIATE INTEGRANDS AND INTEGRATORS: A multivariate semimartingale is simply a matrix whose entries are all semimartingales (for the same filtration and probability measure). For a multivariate integrator X and a multivariate integrand Y the integral $Y \cdot X$ will be understood in the obvious way by following the convention for products between multivariate objects introduced on page xvi. For example, if X is a semimartingale vector column with dimension n and Y is a semimartingale matrix with dimensions (m, n), then $Y \cdot X$ is understood as a semimartingale vector column with entries $\sum_j Y^{i,j} \cdot X^j$, $1 \le i \le n$. If X and Y are both semimartingale vectors of the same dimension n and the same position (both are columns, or both are rows), then $Y \cdot X$ is understood as the semimartingale vector (of the same position) with entries $Y^i \cdot X^i$, $1 \le i \le n$. If Y is a 1-dimensional semimartingale and X is an n-dimensional semimartingale, then $Y \cdot X$ is understood as the vector semimartingale with entries $Y \cdot X^i$, $1 \le i \le n$. If Y is n-dimensional and X is one-dimensional, then $Y \cdot X$ is understood as the vector semimartingale with entries $Y^i \bullet X$, $1 \le i \le n$, and so on. Our general strategy is to use notation that looks one-dimensional but allows for multidimensional interpretation as well. ○

10.4 Itô's Formula

(10.24) INTEGRATION BY PARTS: Let X and Y be any two continuous semimartingales. Then

$$XY = X_0 Y_0 + X \bullet Y + Y \bullet X + \langle X, Y \rangle ,$$

understood as an identity between two processes up to indistinguishability. This relation is often abbreviated as $\mathrm{d}(X_t Y_t) = X_t \,\mathrm{d}Y_t + Y_t \,\mathrm{d}X_t + \mathrm{d}\langle X, Y \rangle_t$. ○

To see where the preceding result is coming from, take any subdivision of the interval $[0,t]$, $\mathcal{P} = \{t_0, t_1, \dots, t_n\}$, and observe that

$$\begin{aligned} X_t Y_t - X_0 Y_0 &= \sum\nolimits_{i \in \mathbb{N}_{|n-1}} \big(X_{t_{i+1}} Y_{t_{i+1}} - X_{t_i} Y_{t_i} \big) \\ &= \sum\nolimits_{i \in \mathbb{N}_{|n-1}} \big(X_{t_{i+1}} Y_{t_{i+1}} - X_{t_i} Y_{t_{i+1}} + X_{t_i} Y_{t_{i+1}} - X_{t_i} Y_{t_i} \big) \\ &= \sum\nolimits_{i \in \mathbb{N}_{|n-1}} \big((X_{t_{i+1}} - X_{t_i}) Y_{t_{i+1}} + X_{t_i} (Y_{t_{i+1}} - Y_{t_i}) \big) \\ &= \sum\nolimits_{i \in \mathbb{N}_{|n-1}} \big((X_{t_{i+1}} - X_{t_i}) Y_{t_i} + X_{t_i} (Y_{t_{i+1}} - Y_{t_i}) \\ &\qquad\qquad + (X_{t_{i+1}} - X_{t_i})(Y_{t_{i+1}} - Y_{t_i}) \big) . \end{aligned}$$

The integration by parts rule is now a straightforward application of the Riemann sums approximation from (10.22d), in conjunction with the same type of approximation for the quadratic variation. One immediate application of this result is stated in the next exercise – it will be put to use later in the derivation of the Feynman–Kac formula.

(10.25) EXERCISE: Let M be any continuous local martingale and let A be any adapted and càdlàg process of finite variation that starts from 0. Then the process $AM - M \bullet A$ is a local martingale.

HINT: If A is continuous, then (10.24) gives $AM - M \bullet A = A \bullet M$, which is a local martingale since M is (stochastic integrals with respect to local martingales are local martingales – see (10.16)). If A is only càdlàg (and of finite variation), then use the preceding argument to show that $AM - M \bullet A = A_- \bullet M$, where A_- is the process given by the left limits of A, that is, the process $(A_{t-})_{t \in \mathbb{R}_+}$ with $A_{t-}(\omega) \overset{\text{def}}{=} \lim_{s \nearrow t} A_s(\omega)$. ○

The steps toward expanding the scope of the integration by parts discovered in (10.24) now suggest themselves.

(10.26) EXERCISE: Prove the following generalization of the integration by parts formula: if X, Y, and Z are any three (not necessarily different) continuous semimartingales, then

$$XYZ = X_0 Y_0 Z_0 + (XY)\cdot Z + (XZ)\cdot Y + (YZ)\cdot X \\ + X\cdot\langle Y, Z\rangle + Y\cdot\langle X, Z\rangle + Z\cdot\langle X, Y\rangle . \quad \circ$$

(10.27) EXERCISE: Let X be any continuous semimartingale and let X^n denote the process $(X_t{}^n)_{t \in \mathbb{R}_+}$ (the nth power of X). Prove by induction that

$$X^n = X_0^n + n\,(X^{n-1})\cdot X + \frac{1}{2}\,n\,(n-1)\,(X^{n-2})\cdot\langle X, X\rangle$$

for every $n \in \mathbb{N}$, $n \geq 2$. $\circ$

The following fundamental result is essentially an elaboration and generalization of the foregoing:

(10.28) ITÔ'S FORMULA: (Revuz and Yor 1999, IV (3.3)) Let X^i, $i \in \mathbb{N}_{|n}$, be any $n+1$ continuous semimartingales, let $f : \mathbb{R}^{n+1} \mapsto \mathbb{R}$ be any continuous function with continuous partial derivatives $\partial_i f$ and $\partial_i \partial_j f$, and let $f(X^0, \ldots, X^n)$ denote the process $\big(f(X_t^0, \ldots, X_t^n)\big)_{t \in \mathbb{R}_+}$. Then $f(X^0, \ldots, X^n)$ is a continuous semimartingale and

$$f(X^0, \ldots, X^n) = f(X_0^0, \ldots, X_0^n) + \sum\nolimits_{i \in \mathbb{N}_{|n}} (\partial_i f)(X^0, \ldots, X^n)\cdot X^i \\ + \frac{1}{2} \sum\nolimits_{i \in \mathbb{N}_{|n}} \sum\nolimits_{j \in \mathbb{N}_{|n}} (\partial_i \partial_j f)(X^0, \ldots, X^n)\cdot\langle X^i, X^j\rangle ,$$

understood as an identity up to indistinguishability. $\circ$

(10.29) EXERCISE: Explain how the integration by parts rule stated in (10.24) obtains formally from Itô's formula. $\circ$

(10.30) EXERCISE: Let W be the linear Brownian motion in $\mathbb{R}^3$ that starts from $W_0 = x \in \mathbb{R}^3$, $x \neq \vec{0}$. Using Itô's formula, show that the process$^{\circ\!\!\to (9.69)}$ $|W|^{-1}$ is a local martingale. $\circ$

(10.31) EXERCISE: Given any two continuous semimartingales, X and Y, argue that $f(X, Y)$ is also a continuous semimartingale for every sufficiently smooth function $f : \mathbb{R}^2 \mapsto \mathbb{R}$. Using Itô's formula, give the local martingale part and the finite variation part of the semimartingale $f(X, Y)$. Do the same for the continuous semimartingale $e^{X+Y} = (e^{X_t + Y_t})_{t \in \mathbb{R}_+}$. $\circ$

The following result gives a particularly interesting and useful application of Itô's formula.

(10.32) LÉVY'S CHARACTERIZATION THEOREM: (Revuz and Yor 1999, IV (3.6)) Let M^i, $i \in \mathbb{N}_{|n}$, be continuous local $(\mathscr{F}, P)$-martingales, all starting from 0 and such that $\langle M^i, M^j\rangle = 0$ for $i \neq j$ and $\langle M^i, M^i\rangle = \imath$ for every $i \in \mathbb{N}_{|n}$. Then M^i, $i \in \mathbb{N}_{|n}$, are independent $(\mathscr{F}, P)$-Brownian motions. In particular, any continuous local martingale M such that $M_0 = 0$ and $\langle M, M\rangle = \imath$ is a Brownian motion. $\circ$

(10.33) EXERCISE: Let M be any continuous local martingale such that

$$\mathrm{d}\langle M, M\rangle_\omega \approx \Lambda \text{ for } P\text{-a.e. } \omega \in \Omega ,$$

that is, there is a strictly positive process γ such that $\langle M, M\rangle = \gamma \bullet \iota$. Prove that the process γ can always be chosen to be predictable, and that, with any such choice, the process $B \overset{\text{def}}{=} (\gamma^{-1/2}) \bullet M$ is a well-defined local martingale, which, in fact, is a Brownian motion.

HINT: A reader who may be stuck with the predictable choice for γ may consult the next section. ○

(10.34) EXERCISE: Prove that if X is any continuous semimartingale, then the process $\mathscr{E}(X) \overset{\text{def}}{=} e^{X - \frac{1}{2}\langle X, X\rangle}$ is also a continuous semimartingale that satisfies the relation $\mathscr{E}(X) = e^{X_0} + \mathscr{E}(X) \bullet X$. Conclude that $\mathscr{E}(X)$ must be a local martingale if X is a local martingale. A process of the form $\mathscr{E}(X)$ is often called the *Doléan–Dade exponent of X* – see (11.38) and (16.39). ○

(10.35) EXERCISE: Prove that every continuous and strictly positive local martingale M can be expressed in the form $M = e^{L - \frac{1}{2}\langle L, L\rangle}$ for some continuous local martingale L.

HINT: Apply Itô's formula to the process $\log(M)$. ○

10.5 Stochastic Integrals with Respect to Brownian Motion

Every $(\mathscr{F}, P)$-Brownian motion W is a continuous martingale, thus a local martingale. Since for P-a.e. $\omega \in \Omega$ we have $\langle W, W\rangle_\omega = \iota$ and the measure $\mathrm{d}\langle W, W\rangle_\omega$ is nothing but the Lebesgue measure Λ, the associated space of stochastic integrands $\mathfrak{L}^2_{\text{loc}}(W) \equiv \mathfrak{L}^2_{\text{loc}}(\langle W, W\rangle) \equiv \mathfrak{L}^2_{\text{loc}}(\iota)$ consists of all predictable processes X with sample paths$^{\multimap (2.76)}$ $X_\omega \in \mathscr{L}^2_{\text{loc}}(\mathbb{R}_+)$ for a.e. $\omega \in \Omega$. However, as was noted in (10.13c), the actual space of admissible integrands is the space of all equivalence classes for the measure $P \times \Lambda$ that contain a predictable process with sample paths from $\mathscr{L}^2_{\text{loc}}(\mathbb{R}_+)$. One immediate consequence of this observation is that every adapted process X that is jointly measurable (that is, $\mathcal{F} \otimes \mathcal{B}(\mathbb{R}_+)$-measurable – a prog. measurable process would have this property, for example) and is such that $X_\omega \in \mathscr{L}^2_{\text{loc}}(\mathbb{R}_+)$ for a.e. $\omega \in \Omega$, can be integrated against the Brownian motion W. In order to verify this claim, we must point to at least one predictable process $\tilde{X}$ such that for P-a.e. $\omega \in \Omega$ the sample paths $\tilde{X}_\omega$ and X_ω coincide Λ-a.e. in $\mathbb{R}_+$. Following the argument of Ikeda and Watanabe (1989), observe first that by theorem (8.61) any jointly measurable and adapted (to $\mathscr{F}$) process X admits a $\operatorname{Prog}(\mathscr{F})$-measurable modification. If X' is one such modification, then, by Tonelli–Fubini's theorem, $\int_{\mathbb{R}_+} \mathrm{d}t \int_\Omega |X(\omega, t) - X'(\omega, t)| P(\mathrm{d}\omega) = 0$, so that X' belongs to the $P \times \Lambda$-equivalence class of X. Thus, it would be enough to show that the equivalence class of every $\operatorname{Prog}(\mathscr{F})$-measurable process with P-a.s. locally square integrable sample paths contains a predictable process from the space $\mathfrak{L}^2_{\text{loc}}(\imath)$. To see why this is the case, suppose that X is some (any) $\text{Prog}(\mathscr{F})$-measurable process such that $X_\omega \in \mathscr{L}^2_{\text{loc}}(\mathbb{R}_+)$ for P-a.e. $\omega \in \Omega$, and define the sequence of predictable processes

$$\tilde{X}^n(\omega,t) \overset{\text{def}}{=} n \int_{(t-1/n)\vee 0}^{t} X(\omega,s)\,\mathrm{d}s\,, \quad n \in \mathbb{N}_{++}\,.$$

The reason why these processes are predictable is that they are simultaneously continuous and adapted.[122] As a result, the process $\tilde{X}(\omega,t) \overset{\text{def}}{=} \limsup_{n\in\mathbb{N}_{++}} \tilde{X}^n(\omega,t)$ is predictable. Furthermore, since $\tilde{X}^n(\omega,t) \to X(\omega,t)$ for Λ-a.e. $t \in \mathbb{R}_+$, we have $\tilde{X}(\omega,t) = X(\omega,t)$ for Λ-a.e. $t \in \mathbb{R}_+$, so that $\tilde{X} = X$ $P\times\Lambda$-a.e.

Because of the preceding argument, the space of integrands $\mathfrak{L}^2_{\text{loc}}(W) \equiv \mathfrak{L}^2_{\text{loc}}(\imath)$ will be understood to include not just predictable, but also jointly measurable and adapted processes with a.s. locally square integrable sample paths. This change in the meaning of the symbol $\mathfrak{L}^2_{\text{loc}}(W)$ is innocuous, since, as we just realized, it does not change the actual space of integrands $\mathbb{L}^2_{\text{loc}}(W) \equiv \mathbb{L}^2_{\text{loc}}(\imath)$.

To summarize: the space of integrands $\mathfrak{L}^2_{\text{loc}}(\imath)$ can be described as the collection of all jointly measurable and adapted processes $X : \Omega\times\mathbb{R}_+ \mapsto \mathbb{R}$ such that the (continuous) sample paths of the process $X^2 \bullet \imath$ are a.s. non-exploding (remain finite on finite intervals with probability 1). For any such process X the stochastic integral $X \bullet W$ is a well-defined local martingale with quadratic variation

$$\langle X\bullet W, X\bullet W\rangle = X^2\bullet\imath\,.$$

In the special case where X satisfies the stronger condition $\mathsf{E}[X^2\bullet\imath_t] < \infty$ for every $t \in \mathbb{R}_+$, one can claim$^{\multimap(9.91)}$ that $X\bullet W$ is a square integrable martingale and the following isometry property holds:

$$\mathsf{E}\big[(X\bullet W_t)^2\big] = \mathsf{E}\big[X^2\bullet\imath_t\big] \equiv \mathsf{E}\Big[\int_0^t X_s^2\,\mathrm{d}s\Big] < \infty \quad \text{for every } t \in \mathbb{R}_+\,.$$

In particular $X\bullet W_t \in \mathscr{L}^2(\Omega,\mathscr{F}_t,P)$, so that $X\bullet W_t$ is an integrable r.v. with

$$\mathsf{E}[X\bullet W_t] = \mathsf{E}[X\bullet W_0] = 0 \ \text{ for any } t \in \mathbb{R}_+\,.$$

We emphasize once again that the expectation of $X\bullet W_t$ may not be defined for a generic integrand $X \in \mathfrak{L}^2_{\text{loc}}(\imath)$.

(10.36) DETERMINISTIC INTEGRANDS: As every deterministic function is a special case of a stochastic process, stochastic integrals with deterministic integrands are perfectly meaningful. By the foregoing, given any deterministic function $f \in \mathscr{L}^2_{\text{loc}}(\mathbb{R}_+)$, the integral $f\bullet W$ is a well-defined square integrable martingale with

$$\mathsf{E}[f\bullet W_t] = 0 \quad\text{and}\quad \mathsf{E}\big[\big(f\bullet W_t\big)^2\big] = (f^2\bullet\imath)_t \equiv \int_0^t f(s)^2\,\mathrm{d}s\,, \quad t \in \mathbb{R}_+\,.$$

[122] One cannot claim that $\tilde{X}^n$ is adapted without the requirement for X to be progressively measurable – see (8.60).

Intuitively, $f{\bullet}W$ should be a Gaussian process with vanishing mean and covariance function

$$\mathsf{E}\big[f{\bullet}W_s\, f{\bullet}W_t\big] = (f^2{\bullet}\imath)_{s\wedge t} \equiv \int_0^{s\wedge t} f(u)^2\,\mathrm{d}u\,, \quad s,t \in [0,T]\,,$$

but this claim must be justified. The argument is both simple and intuitive. Given any $n \in \mathbb{N}_{++}$ and any $i \in \mathbb{N}$, let $a_i^n \overset{\text{def}}{=} 2^n \int_{i2^{-n}}^{(i+1)2^{-n}} f(t)\,\mathrm{d}t$ and let $f_n\colon \mathbb{R}_+ \mapsto \mathbb{R}$ be the simple function given by

$$f_n = \sum\nolimits_{i\in\mathbb{N}} a_i^n 1_{[i2^{-n},(i+1)2^{-n}[}\,.$$

For every fixed dyadic number $T = k2^{-N}$ we have $f_n{\restriction}[0,T] \mapsto f{\restriction}[0,T]$, in the norm of $L^2([0,T])$. This feature can be seen as a consequence of the martingale convergence theorem. Indeed, the functions $\frac{1}{T}f_n{\restriction}[0,T]$ can be treated as conditional expectations of f with respect to a refining sequence of dyadic partitions of $[0,T]$. By the very definition of the simple stochastic integral, $f_n{\bullet}W$ is a centered Gaussian process for every n, and

$$\mathsf{E}\big[\big(f{\bullet}W_t - f_n{\bullet}W_t\big)^2\big] = \int_0^t \big(f(s) - f_n(s)\big)^2\,\mathrm{d}s \;\to\; 0 \quad \text{as } n\to\infty\,.$$

In conjunction with the properties of the Gaussian distribution listed in section 3.4, it is not very difficult to conclude from the last relation that $f{\bullet}W$ is a Gaussian process with $\mathsf{E}[f{\bullet}W_t] = 0$ and $\mathsf{E}\big[f{\bullet}W_s\, f{\bullet}W_t\big] = (f^2{\bullet}\imath)_{s\wedge t}$. ○

We now introduce a special class of processes that are instrumental for many applications – especially for applications in continuous-time finance.

(10.37) ITÔ PROCESSES: A stochastic process X, defined on probability space $(\Omega, \mathcal{F}, P)$ that is endowed with filtration $\mathscr{F} \subset \mathcal{F}$, is said to be an *Itô process* if there is an $(\mathscr{F}, P)$-Brownian motion W and two processes $\alpha, \beta\colon \Omega{\times}\mathbb{R} \mapsto \mathbb{R}$ that are: jointly measurable, $\mathscr{F}$-adapted ($\alpha, \beta \prec \mathscr{F}$), with sample paths $\alpha_\omega \in \mathscr{L}^2_{\text{loc}}(\mathbb{R}_+)$ and $\beta_\omega \in \mathscr{L}^1_{\text{loc}}(\mathbb{R}_+)$ (P-a.s.), and such that

† $$X = X_0 + \alpha{\bullet}W + \beta{\bullet}\imath\,,$$

understood as an identity up to indistinguishability between two stochastic processes. The last relation is often abbreviated as $\mathrm{d}X_t = \alpha_t\,\mathrm{d}W_t + \beta_t\,\mathrm{d}t$. Equivalently, an Itô process can be described as a continuous semimartingale X such that its quadratic variation $\langle X, X\rangle$ and its finite variation predictable component (from the semimartingale representation of X) are both absolutely continuous with respect to the Lebesgue measure Λ.

Describing a multivariate Itô process is essentially a matter of interpreting the symbols in † as multivariate objects with the conventions from (10.23) in mind. Specifically, X can be taken to be a vector[123] process of dimension n, the elements

[123] Recall that vectors are understood to be vector-columns by default.

of which are scalar-valued semimartingales (we say that X is an $\mathbb{R}^n$-valued semimartingale, or a vector semimartingale of dimension n). The symbol W can be understood as a Brownian motion of dimension d, that is, a vector process with components W^i that are d independent scalar Brownian motions. The symbol α can be understood as a matrix process of dimensions (n, d) with entries $\alpha^{i,j}$ that are jointly measurable and adapted scalar processes with sample paths $\alpha^{i,j}_\omega \in \mathscr{L}^2_{\text{loc}}(\mathbb{R}_+)$. Finally, the symbol β can be understood as a vector process of dimension n (same dimension as X) with components β^i that are jointly measurable and adapted scalar processes with sample paths $\beta^i_\omega \in \mathscr{L}^1_{\text{loc}}(\mathbb{R}_+)$. With this new meaning of the symbols involved, the relation † is still meaningful as it stands, but if one insists on a more detailed form, it can be cast as

$$X^i = \sum_{j=1}^{d} \alpha^{i,j} \boldsymbol{\cdot} W^j + \beta^i \boldsymbol{\cdot} \iota\,, \quad 1 \le i \le n. \quad \circ$$

The application of Itô's formula for the class of processes that we just introduced requires the following result.

(10.38) EXERCISE: Suppose that B and W are any two independent Brownian motions. Prove that their joint quadratic variation is null, that is, the sample paths of the process $\langle W, B\rangle$ are a.s. equal to the constant function 0.

HINT: Consider showing that $\mathsf{E}\big[\big(\mathcal{Q}^{\mathcal{P}_n}(W, B)_t\big)^2\big] \to 0$ as $|\mathcal{P}_n| \to 0$. $\circ$

The calculus with brackets involving Itô processes is now entirely straightforward: if X is the n-dimensional Itô process from (10.37†), then

$$\langle X^i, X^j\rangle_t = \int_0^t \Big(\sum_{k=1}^{d} \alpha_s^{i,k}\alpha_s^{j,k}\Big)\,\mathrm{d}s\,, \quad t \ge 0\,, \quad 1 \le i, j \le n\,,$$

which is often written in differential form as

$$\mathrm{d}\langle X^i, X^j\rangle_t = \Big(\sum_{k=1}^{d} \alpha_t^{i,k}\alpha_t^{j,k}\Big)\,\mathrm{d}t\,, \quad 1 \le i, j \le n\,.$$

We note that the last sum is nothing but the dot-product between the ith row and the jth row in the n-by-d-matrix α_t, which is nothing but the entry $(\alpha_t\alpha_t^{\intercal})^{i,j}$ in the n-by-n (symmetric and nonnegative definite) matrix $\alpha_t\alpha_t^{\intercal}$. If we now define $\langle X, X\rangle$ to be the symmetric-matrix-valued process with entries

$$\langle X, X\rangle^{i,j} \stackrel{\text{def}}{=} \langle X^i, X^j\rangle\,, \quad 1 \le i \le n,\ 1 \le j \le n\,,$$

then we can simply write $\mathrm{d}\langle X, X\rangle_t = (\alpha_t\alpha_t^{\intercal})\mathrm{d}t$, or $\langle X, X\rangle = (\alpha\alpha^{\intercal})\boldsymbol{\cdot}\iota$.

Finally, we turn to some examples.

(10.39) EXERCISE: Assuming that W is the scalar Brownian motion, develop a closed-form expression for the Itô integral $W \boldsymbol{\cdot} W$. $\circ$

(10.40) EXERCISE: We have encountered the integral $W{\bullet}\iota$ before and realized that this integral is a Gaussian process. Prove this claim one more time by verifying the identity $W{\bullet}\iota = \iota W - \iota{\bullet}W$, that is,

$$\int_0^t W_s \,\mathrm{d}s = t\,W_t - \int_0^t s\,\mathrm{d}W_s\,, \quad t \in \mathbb{R}_+\,.$$

In addition, calculate $\mathsf{E}\big[W_t\,(\iota{\bullet}W_t)\big]$. ○

(10.41) EXERCISE: Let $X = e^{\sigma W + (b - \frac{1}{2}\sigma^2)\iota}$ for some fixed parameters $\sigma, b \in \mathbb{R}$. Prove that

$$X = 1 + \sigma X{\bullet}W + bX{\bullet}\iota\,,$$

understood as an identity up to indistinguishability. ○

The next statement is much more than a mere exercise and will be used in many occasions (often without a warning or a reference) throughout the sequel.

(10.42) EXERCISE: Let W be the scalar Brownian motion, let $\alpha \in \mathfrak{L}^2_{\mathrm{loc}}(\iota)$, let $M = \alpha{\bullet}W$, and let $Y = e^{M - \frac{1}{2}\langle M,M\rangle}$. Prove that Y is a local martingale and conclude that Y is also a supermartingale such that $\mathsf{E}[Y_t] \leq 1$ for any $t \in \mathbb{R}_+$. Argue that the same claim can be made for any continuous local martingale M that starts from 0 (not necessarily written as a stochastic integral). ○

(10.43) EXERCISE: Let $W = (W^1, W^2)$ be the two-dimensional Brownian motion, let α be any scalar process that is jointly measurable and adapted (for the same filtration), and let A denote the matrix-valued process

$$A_t = \begin{Bmatrix} \cos(\alpha_t) & -\sin(\alpha_t) \\ \sin(\alpha_t) & \cos(\alpha_t) \end{Bmatrix}, \quad t \in \mathbb{R}_+\,.$$

Prove that the process $Z = A{\bullet}W$ is also a two-dimensional Brownian motion, that is, its components, Z^1 and Z^2, are two independent scalar Brownian motions. ○

(10.44) EXERCISE: Let everything be as in the previous exercise, except that now the matrix process A is given by

$$A_t = \begin{Bmatrix} \cos(\alpha_t) & -\sin(\alpha_t) \\ \sin(\beta_t) & \cos(\beta_t) \end{Bmatrix}, \quad t \in \mathbb{R}_+\,,$$

where β is another jointly measurable and adapted scalar process. Prove that the components of $Z \stackrel{\text{def}}{=} A{\bullet}W$, Z^1 and Z^2, are still Brownian motions, though not necessarily independent (this is a convenient way for constructing correlated Brownian motions). Give the semimartingale decomposition of the process $X \stackrel{\text{def}}{=} e^{Z^1 + Z^2 + \iota}$. ○

(10.45) EXERCISE: Consider the three-dimensional Itô process

$$X = X_0 + \sigma{\bullet}W + b{\bullet}\iota$$

with the deterministic function $b(t) \stackrel{\text{def}}{=} (t, -t, 1) \in \mathbb{R}^3$, $t \in \mathbb{R}_+$, and the constant matrix

$$\sigma = \begin{pmatrix} 1 & 2 & 0 \\ 3 & 0 & 1 \\ 0 & 1 & 4 \end{pmatrix}.$$

Prove that for any function $f \in \mathscr{C}^{2,1}(\mathbb{R}^3 \times \mathbb{R}_+; \mathbb{R})$ the process $f(X, \iota)$, that is, $f(X^1, X^2, X^3, \iota)$, is a continuous semimartingale, in fact, an Itô process, and give its semimartingale decomposition. ○

Now we come to a particularly important feature of the Brownian motion, which, among many other important applications, is instrumental for the domain of mathematical finance, in connection with the notions of market completeness, hedging, and replication.

(10.46) PREDICTABLE REPRESENTATION PROPERTY (PRP): (Revuz and Yor 1999, V (3.5)) Let $W = (W_t^1, \dots, W_t^d)$ be the d-dimensional Brownian motion, let $\mathscr{F}^W$ be the right-continuous and complete filtration that W generates, and let M be any $\mathbb{R}$-valued local $\mathscr{F}^W$-martingale (not necessarily continuous). Then M has a continuous version that can be expressed as a stochastic integral with respect to W, that is, a continuous version of the form[124]

$$\dagger \qquad M = M_0 + h^\intercal \bullet W$$

for some choice of the $\mathscr{F}^W$-predictable $\mathbb{R}^d$-valued process h, with sample paths $h_\omega \in \mathscr{L}^2_{\text{loc}}(\mathbb{R}_+; \mathbb{R}^d)$. If more detailed notation is desirable, the predictable representation of M can be expressed as

$$M_t = M_0 + \sum\nolimits_{j=1}^{d} \int^t h_s^j \, \mathrm{d}W_s^j \quad \text{for all } t \in \mathbb{R}_+, \quad P\text{-a.s.} \quad ○$$

(10.47) REMARK: The calculation of the predictable integrands h^j, which give the predictable representation in (10.46†), is important for a number of applications, especially in connection with pricing by arbitrage and the construction of replicating strategies. The computational aspect of the PRP will be taken up later in section 14.1, but one particularly simple connection that we can immediately point to is to identify h_ω^j as the Radon–Nikodym derivative of the measure $\mathrm{d}\langle M, W^j \rangle_\omega$ with respect to the Lebesgue measure Λ. ○

(10.48) EXERCISE: Given a local $\mathscr{F}^W$-martingale M, prove that, if it exists, the predictable process h from (10.46†) is unique $P(\mathrm{d}\omega) \times \mathrm{d}t$-a.e.

HINT: Use the previous remark, or, alternatively, observe that if $M = 0$, then $\langle M, M \rangle = 0$. ○

[124] Since M is $\mathscr{F}^W$-adapted, M_0 must be $\mathscr{F}_0^W$-measurable and this property is the same as $M_0 =$ constant a.s.

(10.49) EXERCISE: Let W be the (scalar) Brownian motion, suppose that $T > 0$ is fixed, and let

$$M_t \stackrel{\text{def}}{=} \mathsf{E}\big[W_T{}^2 \,|\, \mathscr{F}_t^W\big], \quad t \in [0,T].$$

Find the predictable representation of the martingale M on $[0,T]$. ○

Another interesting and useful for many applications representation type result for local martingales is the following:

(10.50) DAMBIS–DUBINS–SCHWARTZ THEOREM: (Revuz and Yor 1999, V (1.6)) Let M be any continuous local $(\mathscr{F}, P)$-martingale that starts from 0 and has the property $\langle M, M\rangle_\infty = \infty$ (P-a.s.), and let

$$\jmath_t \stackrel{\text{def}}{=} \left\{s \in \mathbb{R}_{++} : \langle M, M\rangle_s > t\right\}, \quad t \in \mathbb{R}_+ .$$

Then the process $(B_t \stackrel{\text{def}}{=} M_{\jmath_t})_{t\in\mathbb{R}_+}$ is a Brownian motion relative to the filtration $(\mathscr{F}_{\jmath_t})_{t\in\mathbb{R}_+}$ and the sample path M_ω coincides with the sample path $B_\omega \circ \langle M, M\rangle_\omega$, that is, coincides with $\mathbb{R}_+ \ni t \rightsquigarrow M(\omega,t) = B(\omega, \langle M, M\rangle_t(\omega))$ for P-a.e. $\omega \in \Omega$. To put it another way, a continuous local martingale that starts from 0 is nothing but a Brownian motion, relative to a particular filtration, that runs on a particular random clock, and the bracket of the local martingale is that random clock. ○

10.6 Girsanov's Theorem

Back in section 3.1 we realized that any shift in the distribution of a Gaussian r.v. by a fixed scalar value can be achieved by changing the probability measure on the sample space Ω accordingly – and without the need to modify the r.v. as a function on Ω in any way. Since the Wiener measure can be seen as an infinite dimensional analogue of the multivariate Gaussian law, it is reasonable to suspect that certain shifts of the distribution law of the Brownian motion may be possible to attain by way of changing the underlying probability measure, and without the need to modify the Brownian motion as a path-valued function on Ω. This suspicion is correct, and the precise general statement, of which the quasi-invariance of the Brownian motion under translation is a special case, is the following (see (Ikeda and Watanabe 1989, IV thm. 4.1)):

(10.51) GIRSANOV'S THEOREM[125] (FIRST VERSION):[126] Suppose that the domain of the time parameter t is reduced to the finite interval $[0,T]$. Given a probability space $(\Omega, \mathcal{F}, P)$ and a filtration $\mathscr{F} \subset \mathcal{F}$ (with the usual conditions automatically assumed), suppose that N is any continuous local $(\mathscr{F}, P)$-martingale that starts

[125] Named after the Russian mathematician Igor Vladimirovich Girsanov (1934–1967). An earlier prototype of this result was established by the American mathematicians Robert H. Cameron (1908–1989) and William T. Martin (1911–2004). Girsanov's theorem is sometimes referred to as Cameron–Martin–Girsanov's theorem.

[126] A second version of this result, concerned with local martingales that are not required to be continuous, will be given in (15.44).

from 0 and satisfies the condition

† $$\mathsf{E}^P\left[e^{N_T-\frac{1}{2}\langle N,N\rangle_T}\right]=1\,,$$

and let $Q \approx P$ denote the probability measure on $\mathscr{F}_T$ with Radon–Nikodym derivative

‡ $$\frac{\mathrm{d}Q}{\mathrm{d}P}=e^{N_T-\frac{1}{2}\langle N,N\rangle_T}\,.$$

Then the following claim can be made: for every continuous local $(\mathscr{F},P)$-martingale M, the process $M-\langle M,N\rangle$ is a continuous $(\mathscr{F},Q)$-local martingale. Thus, if one simultaneously changes the measure P to $Q=e^{N_T-\frac{1}{2}\langle N,N\rangle_T}\odot P$ and translates M by the continuous finite variation process $\langle M,-N\rangle$, then the local martingale property will remain intact: if M is a continuous local martingale relative to P, then $M+\langle M,-N\rangle$ is a continuous local martingale relative to Q. The quadratic variation will remain intact as well, since it depends on the probability measure only up to equivalence, and is unaffected by the addition of a continuous finite variation component. In particular, if M is a Brownian motion under P, then $M-\langle M,N\rangle$ is a Brownian motion under Q.

The same result can be rephrased as follows: Let $X=M+A$ be any continuous $(\mathscr{F},P)$-semimartingale, and suppose that the finite variation component of X can be expressed as $A=\langle M,-N\rangle$, for a particular continuous local martingale N that starts from 0 and satisfies condition †. If the measure Q is given by ‡, then the process X is a local martingale with respect to $Q \approx P$. In other words, one can remove the statistical effects of the finite variation component A – that is, make the semimartingale X behave statistically as a local martingale – simply by changing the probability measure from P to Q, and we again stress that such an equivalent change of the probability measure, and the resulting transformation of X into a local martingale, have no effect on the quadratic variation $\langle X,X\rangle$. ○

(10.52) EXERCISE: Let N be any continuous local martingale that starts from 0. Prove that condition $(10.51^{\dagger})$ is equivalent to the claim that the process

$$\mathscr{E}(N) \overset{\text{def}}{=} e^{N-\frac{1}{2}\langle N,N\rangle}$$

is a martingale on $[0,T]$, which is equivalent to the claim that $\mathsf{E}[\mathscr{E}(N)_t]=1$ for every $t\in[0,T]$.

HINT: The process $\mathscr{E}(N)$ is a positive local martingale, hence⊸ (9.68d) a supermartingale that starts from $\mathscr{E}(N)_0=e^0=1$. Consult exercise (9.6). ○

(10.53) EXERCISE: Using the result from exercise (9.78), give a complete proof of Girsanov's theorem stated in (10.51).

HINT: Use the integration by parts rule from (10.24) to show that the process $(M-\langle M,N\rangle)\mathscr{E}(N)$ can be written as the sum of two stochastic integrals with respect to the local martingales M and N. ○

(10.54) REMARK: Even if condition (10.51†) were to hold for every choice of $T \in \mathbb{R}_{++}$ – in which case the process $\mathscr{E}(N)$ would be a martingale on the entire time domain $t \in \mathbb{R}_+$ – in general, without imposing any additional conditions, it may not be possible to define a single probability measure, Q, on $\mathscr{F}_\infty$ so that the process $M - \langle M, N\rangle$ would be a local martingale under Q on entire time domain $\mathbb{R}_+$. One particular set of conditions that guarantees that such a universal Q exists are conditions about the concrete structure of the sample space Ω. To be specific, if the sample space Ω is the path space $\mathbb{W} \stackrel{\text{def}}{=} \mathscr{C}(\mathbb{R}_+; \mathbb{R}^n)$ and $\mathscr{F}$ is the natural filtration on $\mathbb{W}$, that is, the right-continuous and complete filtration generated by the coordinate mappings on $\mathbb{W}$, then the condition $\mathsf{E}[\mathscr{E}(N)_t] = 1$ for every $t \in \mathbb{R}_+$ is enough to ensure that there is a measure, Q, on $\mathscr{F}_\infty$ such that $Q\restriction\mathscr{F}_t \approx P\restriction\mathscr{F}_t$ and $\mathrm{d}(Q\restriction\mathscr{F}_t)/\mathrm{d}(P\restriction\mathscr{F}_t) = \mathscr{E}(N)_t$ for every $t \in \mathbb{R}_+$ (Stroock and Varadhan 1979, p. 34). Another approach is to require that $\mathscr{E}(N)$ can be extended to a martingale on $\bar{\mathbb{R}}_+$, so that $\mathsf{E}[\mathscr{E}(N)_\infty] = 1$. This is the same$^{\multimap (9.33)}$ as the requirement that $\mathscr{E}(N)$ is a uniformly integrable martingale. In this later case the measure $Q \stackrel{\text{def}}{=} \mathscr{E}(N)_\infty \odot P$ will be equivalent to P not just on $\mathscr{F}_t$ for every $t \in \mathbb{R}_+$, but also on $\mathscr{F}_\infty$. ○

In general, the matter of verifying condition (10.51†) for a given local martingale N is far from trivial. We state next two of the most widely used criteria that can guarantee that (10.51†) holds. We saw in exercise (10.52) that what needs to be verified is the claim that the positive local martingale $\mathscr{E}(N)$ is not only a supermartingale, which is automatic, but is also a martingale.

(10.55) KAZAMAKI'S CRITERION: (Revuz and Yor 1999, VIII (1.14)) Let N be any continuous local martingale. If the submartingale $e^{\frac{1}{2}N}$ is uniformly integrable, then the local martingale $\mathscr{E}(N)$ is a uniformly integrable martingale. ○

In order to apply Kazamaki's criterion in the context of Girsanov's theorem it would suffice$^{\multimap (4.22)}$ to find two constants $C, \varepsilon \in \mathbb{R}_{++}$ such that $\mathsf{E}\big[e^{\frac{1+\varepsilon}{2}N_t}\big] \le C$ for all $t \in \mathbb{R}_+$. Alternatively, it would suffice to find a r.v. ξ such that $\mathsf{E}[e^\xi] < \infty$ and $\frac{1}{2}N_t \le \xi$ a.s. for every $t \in \mathbb{R}_+$. One scenario in which this last condition would be satisfied in a trivial way is when the local martingale N is globally bounded, but such a condition is not all that interesting, as it is too restrictive for most situations encountered in practical applications.

(10.56) NOVIKOV'S CRITERION: (ibid., VIII (1.15)) If N is any continuous local martingale, then

$$\mathsf{E}\big[e^{\frac{1}{2}\langle N,N\rangle_\infty}\big] < \infty$$

implies that $\mathscr{E}(N) \stackrel{\text{def}}{=} e^{N-\frac{1}{2}\langle N,N\rangle}$ is a uniformly integrable martingale, and, in addition, implies that $\mathsf{E}[e^{\frac{1}{2}N^*_\infty}] < \infty$, where $N^*_\infty = \sup_{t\in\mathbb{R}_+} N_t$. ○

(10.57) REMARKS: Since $\frac{1}{2}N_t \le \frac{1}{2}N^*_\infty$ P-a.s. for every $t \in \mathbb{R}_+$, we see that if Novikov's criterion is satisfied, then so is also Kazamaki's criterion. That Novikov's criterion applies to a narrower range of cases is not surprising, since it can be satisfied only simultaneously by both N and $(-N)$. Nevertheless, in many situations applying Novikov's criterion is more straightforward. One such situation is when the bracket of N happens to be bounded, that is, $\langle N, N\rangle_\infty \leq C$ for some constant C, although under this condition the property $\mathsf{E}[\mathscr{E}(N)_t] = 1$ is easy to establish directly from the relation

$$\mathscr{E}(N) = 1 + \mathscr{E}(N)\bullet N\,,$$

which says that $\mathscr{E}(N)$ is a local martingale, and thus a supermartingale (as it is positive). If $\langle N, N\rangle_\infty \leq C$, then for every bounded stopping time t,

$$\mathscr{E}(N)^2_t = e^{2N_t-\langle N,N\rangle_t} = e^{2N_t-2\langle N,N\rangle_t}e^{\langle N,N\rangle_t} \leq e^{2N_t-2\langle N,N\rangle_t}e^C\,.$$

The optional stopping theorem, applied to the positive supermartingale $e^{2N-2\langle N,N\rangle}$, gives $\mathsf{E}\big[e^{2N_t-2\langle N,N\rangle_t}\big] \leq 1$, with the implication that $\mathsf{E}\big[\mathscr{E}(N)^2_t\big] \leq e^C$, which then guarantees that $\mathscr{E}(N)$ is of class (DL), and⊶(9.70) is therefore a martingale. In conjunction with Hölder's inequality, and after a minor adjustment in the algebra,[127] the same argument can be used also if the condition $\langle N, N\rangle_\infty \leq C$ is relaxed to $\mathsf{E}\big[e^{\frac{1+\varepsilon}{2}\langle N,N\rangle_\infty}\big] < \infty$ for some $\varepsilon > 0$. Thus, the truly remarkable aspect of Novikov's result is that one can actually get away with $\varepsilon = 0$. ○

(10.58) AN APPLICATION: One common application of Girsanov's theorem, especially in the realm of continuous-time finance, is the following. Let W be the d-dimensional Brownian motion for some filtration $\mathscr{F}$ and probability measure P, and let θ be any jointly measurable and $\mathscr{F}$-adapted stochastic process with values in $\mathbb{R}^d$, such that for some $T \in \mathbb{R}_{++}$,

$$\int_0^T |\theta_\omega(t)|^2\,\mathrm{d}t < +\infty \quad \text{for } P\text{-a.e. } \omega \in \Omega\,.$$

Define the local martingale $N = \theta^\intercal \bullet W$, that is,

$$N_t = \int^t \theta^\intercal_s\,\mathrm{d}W_s \equiv \sum_{j=1}^d \int^t \theta^j_s\,\mathrm{d}W^j_s\,, \quad t \in [0,T]\,,$$

and suppose that $\mathsf{E}\big[\mathscr{E}(-N)_T\big] = 1$. Novikov's criterion is telling us that this last condition is satisfied if the process θ satisfies also the condition

$$\mathsf{E}\Big[e^{\frac{1}{2}\int^T |\theta_s|^2\,\mathrm{d}s}\Big] < +\infty\,.$$

Now consider the probability measure $Q \approx P$ on $\mathscr{F}_T$ given by

$$\frac{\mathrm{d}Q}{\mathrm{d}P} = \mathscr{E}(-N)_T \equiv e^{-\int^T \theta^\intercal_s\,\mathrm{d}W_s - \frac{1}{2}\int^T |\theta_s|^2\,\mathrm{d}s}\,.$$

Girsanov's theorem now implies that the process $X = W + \langle W, N\rangle$, that is,

† $$X_t = W_t + \int^t \theta_s\,\mathrm{d}s\,, \quad t \in [0,T]\,,$$

[127] Instead of $\mathscr{E}(N)^2_t$ consider $\mathscr{E}(N)_t^{\frac{1+2\delta}{1+\delta}}$ with $\delta = \varepsilon/2$.

is an $\mathbb{R}^d$-valued Brownian motion with respect to $\mathcal{F}$ and Q. ○

(10.59) EXERCISE: Prove that the process in (10.58†) is indeed a d-dimensional Brownian motion with respect to the measure Q, as claimed.

HINT: What can you say about the brackets $\langle X^i, X^j \rangle$? ○

10.7 Local Times and Tanaka's Formula

The object that we turn to now is a somewhat peculiar, but, as we will soon realize, rather important set: the set of zeros of the Brownian sample path W_ω, that is,

$$\mathcal{Z}_0(W_\omega) \stackrel{\text{def}}{=} \{t \in \mathbb{R}_+ : W_\omega(t) = 0\} .$$

Had the Brownian sample paths been differentiable functions, the set $\mathcal{Z}_0(W_\omega)$ would not have been all that interesting. However, heuristically speaking, the Brownian paths wiggle wildly before deciding which way to go, and, as a result, the amount of time that any particular path spends in 0 – or, for that matter, in any other level $a \in \mathbb{R}$ – becomes nontrivial, interesting, and, as we are about to see, quite important and useful.

In what follows the probability space $(\Omega, \mathcal{F}, P)$ and the filtration $\mathcal{F} \setminus\subset \mathcal{F}$ are fixed, W is an $(\mathcal{F}, P)$-Brownian motion.

(10.60) EXERCISE: Prove that for P-a.e. $\omega \in \Omega$ the set of zeros for W_ω is a negligible set for the Lebesgue measure, that is, $\Lambda(\mathcal{Z}_0(W_\omega)) = 0$. Conclude that if B is another $(\mathcal{F}, P)$-Brownian motion (possibly identical to W), then the process $1_{\{0\}}(W)\bullet B$ is indistinguishable from 0.

HINT: Due to the continuity of the sample paths, $\mathcal{Z}_0(W_\omega)$ is a Borel set. It is enough to prove that

$$\mathsf{E}\big[\Lambda\big(\mathcal{Z}_0(W)\big)\big] \equiv \mathsf{E}\Big[\int_{\mathbb{R}_+} 1_{\{0\}}(W_s)\,\mathrm{d}s\Big] = 0 .$$

Consult (2.59), and compute the quadratic variation of $1_{\{0\}}(W)\bullet B$. ○

Nothing would have changed in the last exercise if the set $\mathcal{Z}_0(W_\omega)$ were to be replaced with the times of visits to level $a \in \mathbb{R}$, namely

$$\mathcal{Z}_a(W_\omega) \stackrel{\text{def}}{=} \{t \in \mathbb{R}_+ : W_\omega(t) = a\} .$$

The sets $\mathcal{Z}_a(W_\omega)$, $\omega \in \Omega$, are similar in nature to the Cantor set, in that $\mathcal{Z}_a(W_\omega)$ is a.s. of Lebesgue measure 0, and, at the same time, one can construct on that set a "uniform measure" (obviously singular to Λ). This measure plays an important role in the study of the Brownian motion. The cumulative function associated with any such measure is very similar in nature to the distribution function of the Cantor measure, except that the points of increase are now random. This cumulative function can be defined for P-a.e. $\omega \in \Omega$ as

$$L_t^a(\omega) = \lim_{\varepsilon \searrow 0} \int^t \frac{1}{2\varepsilon} 1_{]a-\varepsilon, a+\varepsilon[}(W_s(\omega))\,\mathrm{d}s \quad \text{for all } t \in \mathbb{R}_+ .$$

The process L^a is called *Brownian local time in level a*. The heuristic reason for this definition can be illustrated by attempting to write down Itô's formula for the process $f(W)$ with $f(x) = |x - a|$. Of course, Itô's formula cannot be applied directly in this case, since the function f is not differentiable at $x = a$. However, since formal expressions for the first and the second derivative of f at $x = a$ still exist, we can at least speculate as to how an "Itô formula" might look like. Proceeding formally, we get

$$f(W) - f(W_0) = |W - a| - |a| = \partial f(W) \bullet W + \frac{1}{2}\partial^2 f(W) \bullet \iota \qquad e_1$$

but now must find a way to make the last two integrals meaningful. Since $\partial f(x)$ does not exist only at $x = a$, since the set $\{s \in \mathbb{R}_+ : W_s = a\}$ is Λ-negligible, and since $\partial f(x) = \text{sign}(x - a)$ for $x \neq a$, by the argument in the last exercise it is not too difficult to see that the nonexistence of $\partial f(x)$ at $x = a$ is really inessential, and the first integral is perfectly meaningful if we replace in that integral the function $\partial f(\cdot)$ with the function $\text{sign}(\cdot - a)$. However, it is not immediately obvious how to interpret $\partial^2 f(x) = \partial_x \,\text{sign}(x - a)$. The notion of generalized derivatives that we introduced in (0.12) now comes to the rescue: given any test function $\varphi(\cdot)$, a formal integration by parts would give us

$$\begin{aligned}
&\int_{\mathbb{R}} \varphi(x)\partial_x \,\text{sign}(x - a)\,dx \\
&\qquad = -\int_{\mathbb{R}} \partial\varphi(x)\,\text{sign}(x - a)\,dx = \int_{]-\infty,a[} \partial\varphi(x)\,dx - \int_{[a,\infty[} \partial\varphi(x)\,dx \\
&\qquad = 2\,\varphi(a) = 2\int_{\mathbb{R}} \varphi(x)\epsilon_a(dx) \equiv 2\int_{\mathbb{R}} \varphi(x)\delta_a(x)\,dx\,,
\end{aligned}$$

where ϵ_a is the unit point-mass measure concentrated at $a \in \mathbb{R}$, and $\delta_a(x)$ is Dirac's delta (the formal density $\epsilon_a(dx)/dx$). Thus, at least formally, we can write $\partial^2 f(x) = \partial_x \,\text{sign}(x - a) = 2\delta_a(x)$, and then speculate that, if it is at all possible to extend Itô's formula for this case, the extension should have the form

$$|W - a| - |a| = \text{sign}(W - a) \bullet W + \delta_a(W) \bullet \iota\,,$$

provided that the second integral above can be made meaningful. Since as $\varepsilon \searrow 0$ the integral $\int_{\mathbb{R}} \varphi(x)\frac{1}{2\varepsilon}1_{]a-\varepsilon,a+\varepsilon[}(x)\,dx$ approximates $\varphi(a) = \int_{\mathbb{R}} \varphi(x)\delta_a(x)\,dx$, at least intuitively, the function $x \rightsquigarrow \frac{1}{2\varepsilon}1_{]a-\varepsilon,a+\varepsilon[}(x)$ can be seen as an approximation of Dirac's delta function $x \rightsquigarrow \delta_a(x)$. Heuristically at least, the second integral in e_1 can be identified as

$$\int^t \delta_a(W_s(\omega))\,ds = \lim_{\varepsilon \searrow 0} \int^t \frac{1}{2\varepsilon} 1_{]a-\varepsilon,a+\varepsilon[}(W_s(\omega))\,ds = L_t^a(W_\omega)\,.$$

In any case, the local time L^a is a well-defined increasing process that starts from 0, and we have the following result, due to Tanaka (1963).

(10.61) TANAKA'S FORMULA: The process $|W-a|-a$ is indistinguishable from $\text{sign}(W-a)\bullet W + L^a$, that is, with probability 1,

$$\dagger \qquad |W_t - a| - |a| = \int^t \text{sign}(W_s - a)\mathrm{d}W_s + L_t^a \quad \text{for all } t \in \mathbb{R}_+ . \quad \circ$$

More generally, the local time process $L^a(X) = (L_t^a(X))_{t\in\mathbb{R}_+}$ can be defined for any continuous semimartingale X. To do this, we need the following result.

(10.62) EXTENSION OF ITÔ'S FORMULA: (Revuz and Yor 1999, VI (1.1)) Let X be any continuous semimartingale and let $f\colon \mathbb{R} \mapsto \mathbb{R}$ be any finite convex function. Then there is a continuous, increasing, and adapted process, A^f, that depends on both f and X, such that (up to indistinguishability)

$$\dagger \qquad f(X) = f(X_0) + \partial_- f(X)\bullet X + \frac{1}{2}A^f ,$$

where $\partial_- f$ is the left derivative of f. $\circ$

Clearly, ($10.61^\dagger$) is a special case[128] of ($10.62^\dagger$) with $X = W$, $f(x) = |x-a|$, and $A^f = 2L^a(W)$. We note that the extended Itô formula has an obvious counterpart for concave functions $f\colon \mathbb{R} \mapsto \mathbb{R}$: ($10.62^\dagger$) remains the same except that the process A^f on the right side has to be replaced with $(-A^{-f})$. One immediate corollary from (10.62) is that if X is any continuous semimartingale any $a \in \mathbb{R}$, then there are continuous, increasing, and adapted processes A, A^+, and A^-, that depend on both X and a, such that (up to indistinguishability):

$$\begin{aligned} |X-a| &= |X_0 - a| + \text{sign}_-(X-a)\bullet X + \frac{1}{2}A , \\ (X-a)^+ &= (X_0-a)^+ + 1_{]a,\infty[}(X)\bullet X + \frac{1}{2}A^+ , \qquad \mathrm{e}_2 \\ (X-a)^- &= (X_0-a)^- - 1_{]-\infty,a]}(X)\bullet X + \frac{1}{2}A^- . \end{aligned}$$

(10.63) EXERCISE: Prove that the processes A, A^+, and A^- above are connected through the relations

$$A^+ = A^- \qquad \text{and} \qquad A = A^+ + A^- = 2A^+ = 2A^- .$$

HINT: What can you say about the semimartingale $(X-a)^+ + (X-a)^-$ and about the semimartingale $(X-a)^+ - (X-a)^-$? $\circ$

(10.64) LOCAL TIMES AND TANAKA'S FORMULA FOR CONTINUOUS SEMIMARTINGALES: Given a continuous semimartingale X and a scalar $a \in \mathbb{R}$, the unique continuous and increasing process $\frac{1}{2}A = A^+ = A^-$ from the last exercise is called *local time of X in a* and is denoted by $L^a(X)$, or simply by L^a if the semimartingale X is understood from the context. In its most general version, Tanaka's

[128] Nothing will change in (10.61) if we replace $\text{sign}(x-a)$ with the left derivative of $x \rightsquigarrow |x-a|$, which we denote by $\text{sign}_-(x-a)$, understood to be 1 for $x > a$ and -1 for $x \le a$.

formula can be stated as

$$\dagger \qquad |X-a| = |X_0 - a| + \operatorname{sign}_-(X-a)\bullet X + L^a(X)\,. \quad \circ$$

If the semimartingale X happens to be a continuous local martingale, then L^a is bicontinuous in the variables (a,t) (up to a modification), but, in general, L^a can only be claimed to be continuous in t and càdlàg in a (again, up to a modification). The lack of full continuity with respect to a is due to the fact that there is nothing to prevent the measure associated with the finite variation component of X to have a singular component with respect to the Lebesgue measure – even if the finite variation component is continuous. To be precise (Revuz and Yor 1999, VI (1.7)), if $X = M + V$ (the canonical decomposition of a semimartingale), then (notice that $V=0$ if X is a local martingale)

$$L^a - L^{a-} = 2 1_{\{a\}}(X)\bullet V = 2 1_{\{a\}}(X)\bullet X\,.$$

This feature is revealed through the following identities, in which X is a continuous semimartingale and M is a continuous local martingale (ibid., VI (1.9)): for any $a \in \mathbb{R}$ and any $t \in \mathbb{R}_+$, (P-a.s.)

$$L_t^a(X) = \lim_{\varepsilon\searrow 0}\int^t \frac{1}{\varepsilon} 1_{[a,a+\varepsilon[}(X_s(\omega))\,\mathrm{d}\langle X,X\rangle_s\,,$$

and e_3

$$L_t^a(M) = \lim_{\varepsilon\searrow 0}\int^t \frac{1}{2\varepsilon} 1_{]a-\varepsilon,a+\varepsilon[}(M_s(\omega))\,\mathrm{d}\langle M,M\rangle_s\,.$$

Since every local martingale is a semimartingale, the right sides in e_3 must coincide if $X = M$, that is, for continuous local martingales we have the choice of treating the local time as a two-sided limit or as a one-sided limit. In particular, if W is the Brownian motion, then

$$L_t^a(W) = \lim_{\varepsilon\searrow 0}\int^t \frac{1}{\varepsilon} 1_{[a,a+\varepsilon[}(W_s)\,\mathrm{d}s = \lim_{\varepsilon\searrow 0}\int^t \frac{1}{2\varepsilon} 1_{]a-\varepsilon,a+\varepsilon[}(W_s)\,\mathrm{d}s\,.$$

(10.65) IMPORTANT OBSERVATION: One immediate corollary from the first equation in e_3 is that the process $L^a(X)$ is adapted to the natural filtration $\mathscr{F}^X$ – and this is true for any continuous semimartingale X and any $a \in \mathbb{R}$. In the case of the Brownian motion W, since the event $\{-\varepsilon < W_s < \varepsilon\}$ is the same as the event $\{0 \le |W_s| < \varepsilon\}$, we see that $L^0(W)$ is adapted to the narrower filtration $\mathscr{F}^{|W|}$ – this observation has a number of important implications that will be taken up later in the sequel. $\circ$

(10.66) EXERCISE: If W is the Brownian motion, prove that $L^0(|W|) = 2L^0(W)$.

HINT: Write down the canonical semimartingale decomposition of $|W|$ and use the appropriate formula from e_3. Observe that $1_{[0,\varepsilon[}(|W|) = 1_{]-\varepsilon,\varepsilon[}(W)$. $\circ$

(10.67) EXERCISE: If W is the Brownian motion, prove that with probability 1 $L^0(W) = 1_{\{0\}}(W)\boldsymbol{\cdot}L^0(W)$. Conclude that with probability 1 $|W|\boldsymbol{\cdot}L^0(W) = 0$ and the measure $\mathrm{d}L^0(W)$ is supported by the set $\mathcal{Z}_0(W)$.

HINT: Use the result from the previous exercise and the third formula in e_2 with $a = 0$ and with $X = |W| = \mathrm{sign}\_(W)\boldsymbol{\cdot}W + L^0(W)$. ○

(10.68) GENERALIZED DERIVATIVES AND LOCAL TIMES: Getting back to the setting of (10.62), if the second derivative of f happens to exist in the generalized sense, and can be identified with the positive measure $m(\mathrm{d}x)$, in that for every test function $\varphi(\cdot)$,

$$\int_{\mathbb{R}} \partial^2\varphi(x) f(x)\,\mathrm{d}x = \int_{\mathbb{R}} \varphi(x)\, m(\mathrm{d}x),$$

then (Jeanblanc et al. 2009, 4.1.9.1) the process A^f in (10.62†) can be expressed as $A^f = \int_{\mathbb{R}} L^a(X) m(\mathrm{d}a)$, in that $A^f_t = \int_{\mathbb{R}} L^a_t(X) m(\mathrm{d}a)$ P-a.s. for every $t \in \mathbb{R}_+$. In particular, the process $\int_{\mathbb{R}} L^a_t(X) m(\mathrm{d}a)$, $t \in \mathbb{R}_+$, has a continuous modification, which we denote by $\int_{\mathbb{R}} L^a(X) m(\mathrm{d}a)$, and can therefore cast the extended Itô formula as:

† $$f(X) = f(X_0) + \partial_- f(X)\boldsymbol{\cdot}X + \frac{1}{2}\int_{\mathbb{R}} L^a(X)\, m(\mathrm{d}a)\,. \quad ○$$

The common use of (10.68†) in financial applications is with $f(\cdot) = (\cdot - K)^+$ or with $f(\cdot) = (\cdot - K)^-$, for some fixed $K > 0$. The way in which (10.68†) specializes to these two cases is the subject of the next exercise.

(10.69) EXERCISE: Let X be any continuous semimartingale and let $K \in \mathbb{R}$ be arbitrarily chosen. Prove the relations

$$(X-K)^+ = (X_0 - K)^+ + 1_{]K,+\infty[}(X)\boldsymbol{\cdot}X + \frac{1}{2}L^K(X),$$

and

$$(X-K)^- = (X_0 - K)^- - 1_{]-\infty,K]}(X)\boldsymbol{\cdot}X + \frac{1}{2}L^K(X),$$

and verify that they are compatible with the following obvious ones:

$$(X-K)^+ - (X-K)^+ = X - K \quad \text{and} \quad (X-K)^+ + (X-K)^+ = |X-K|\,.$$

HINT: What are the first and the second derivatives (in the generalized sense) of the functions $(\cdot - K)^+$ and $(\cdot - K)^-$? ○

We conclude this section with the following important result (Revuz and Yor 1999, VI (1.6)), which reveals the connection between local times and brackets:

(10.70) OCCUPATION TIMES FORMULA: For any continuous semimartingale X, any positive Borel function $f : \mathbb{R} \mapsto \mathbb{R}_+$, and any $t \in \mathbb{R}_+$,

$$f(X)\boldsymbol{\cdot}\langle X, X\rangle_t = \int_{\mathbb{R}} L^a_t(X) f(a)\,\mathrm{d}a \quad P\text{-a.s.}$$

In particular, back in the setting of (10.62) and (10.68), if the finite convex function f admits a generalized derivative of order 2, if this derivative can be identified as the measure $m(\mathrm{d}x)$, and, finally, if this measure is absolutely continuous with respect to the Lebesgue measure Λ, so that we can set $\partial^2 f(x) \overset{\text{def}}{=} m_{\partial^2 f}(\mathrm{d}x)/\mathrm{d}x$ – in other words, if the generalized second derivative $\partial^2 f$ exists and can be identified as a positive function defined Λ-a.e. – then the extended Itô formula in (10.68†) can be cast in the usual Itô form

$$f(X) = f(X_0) + \partial_- f(X)\bullet X + \frac{1}{2}\partial^2 f(X)\bullet\langle X, X\rangle\,, \qquad \dagger$$

except that $\partial^2 f$ is now understood in the generalized sense. ○

(10.71) EXERCISE: (Jeanblanc et al. 2009, 4.1.9.3) Let X be any continuous semimartingale such that $\langle X, X\rangle = h(X, \iota)^2 \bullet \iota$, for some Borel function $h(\cdot, \cdot)$, and suppose that the r.v. X_t admits probability density $\psi_t(\cdot)$ for every t. Given any $\alpha \in \mathbb{R}$, prove that

$$\mathsf{E}[L_t^\alpha(X)] = \int^t h(\alpha, s)^2 \psi_s(\alpha)\mathrm{d}s \quad \text{for any } t \in \mathbb{R}_+\,.$$

Conclude that the measure $\mathrm{d}\mathsf{E}[L_t^\alpha(X)]$ must be absolutely continuous with respect to the Lebesgue measure, with derivative

$$\partial_t \mathsf{E}[L_t^\alpha(X)] = h(\alpha, t)^2 \psi_t(\alpha)\,.$$

HINT: In the occupation times formula set $f(\cdot) \equiv \delta_\alpha(\cdot)$ and observe that, at least formally, $\mathsf{E}[\delta_\alpha(X_s)h(X_s, s)^2] = h(\alpha, s)^2 \psi_s(\alpha)$. ○

10.8 Reflected Brownian Motion

Since $(-W)$ is a Brownian motion whenever W is a Brownian motion, the reflection across 0 of any piece of the sample path W_ω that is contained in the domain $]-\infty, 0[$ would be statistically indistinguishable from a piece contained in the domain $\mathbb{R}_{++} =]0, \infty[$. In particular, this means that, locally, the process $|W|$ should be statistically indistinguishable from the Brownian motion W – except when W is in 0! Because $|W|$ and W are quite different, we see to what extent the local behavior of the Brownian motion in 0 matters, and how much the behavior of the Brownian motion can be altered by merely altering its behavior in 0. The object of study in this section is the reflected Brownian motion, which is defined as a process that has the law of $|W|$. Due to Tanaka's formula, such a process is a continuous semimartingale; to be precise (all identities are understood P-a.s. and in terms of indistinguishability)

$$|W| = B + L^0(W)\,, \qquad \mathrm{e}_1$$

where $B \overset{\text{def}}{=} \mathrm{sign}(W)\bullet W$ is a Brownian motion.

(10.72) EXERCISE: Explain why the process $B \stackrel{\text{def}}{=} \text{sign}(W)\boldsymbol{\cdot}W$ is a Brownian motion (for the same filtration and probability measure for which W is a Brownian motion). ○

Equation e_1 can be seen from yet another angle, namely as a way to "steer" the Brownian motion B into a reflected Brownian motion by way of adding an appropriately chosen increasing and continuous process. This is a remarkable feature, and a natural question to ask is this: can any Brownian motion be steered into a reflected Brownian motion with the addition of an increasing and continuous process that is adapted to the Brownian filtration (does not foresee the future)? The answer is positive and obtains from the following important result (see (Revuz and Yor 1999, VI (2.1))):

(10.73) SKOROKHOD'S LEMMA: Let f be any continuous function from $\mathbb{R}_+$ into $\mathbb{R}$ such that $f(0) \geq 0$. Then there is a unique pair of functions, (g, a), from $\mathbb{R}_+$ into $\mathbb{R}$, such that: (a) $g(t) = f(t) + a(t)$ for every $t \in \mathbb{R}_+$; (b) $g(t) \geq 0$ for every $t \in \mathbb{R}_+$; (c) $\mathbb{R}_+ \ni t \rightsquigarrow a(t)$ is increasing, continuous, vanishing at 0 ($a(0) = 0$) and such that the measure $\mathrm{d}a$ is supported by the set $\{s \in \mathbb{R}_+ : g(s) = 0\}$. In addition, the function a is given by

† $$a(t) = \sup_{s\in[0,t]} f(s)^- \equiv \sup_{s\in[0,t]}\big(\max\{-f(s), 0\}\big), \quad t \in \mathbb{R}_+ . \quad ○$$

The proof of Skorokhod's lemma becomes straightforward (see ibid., p. 240), as soon as one realizes that if the function $a(\cdot)$ is defined as in (10.73†), and the function g is defined as $g = f + a$, then the pair (g, a) has the desired properties. The reason is that $a(t) = \sup_{s\in[0,t]} f(s)^-$ can increase only at those t's in which $f(\cdot)$ is at its lowest point on $[0, t]$ that is not above 0, that is, only if $f(t) = -a(t) \leq 0$. Clearly, $g(t) = 0$ at any such t, so that the points of increase for $a(\cdot)$ must be included in the set of zeros for $g(\cdot)$. The uniqueness statement in Skorokhod's lemma now allows us to conclude that the local time in 0, $L^0(W)$, can be expressed in terms of the Brownian motion $B = \text{sign}(W)\boldsymbol{\cdot}W$ as

$$L^0_t(W) = \sup_{s\in[0,t]} B_s^- \equiv \sup_{s\in[0,t]}\big(-B_s\big) \quad \text{for all } t \in \mathbb{R}_+ .$$

In fact, if B is any Brownian motion, and the procedure with which the function g obtains from f in (10.73) is applied to the sample path B_ω instead of f, then the result would be a sample path distributed as a reflected Brownian motion. To put it another way, if B is the Brownian motion, then the process

$$X_t \stackrel{\text{def}}{=} B_t + \sup_{s\in[0,t]}\big(-B_s\big), \quad t \in \mathbb{R}_+ , \qquad e_2$$

is (has the distribution of) a reflected Brownian motion, and its local time in 0 is nothing but

$$L^0_t(X) = \sup_{s\in[0,t]}\big(-B_s\big), \quad t \in \mathbb{R}_+ .$$

Alternatively, we can think of X as the solution to the equation

$$X = B + L^0(X), \qquad e_3$$

for a given Brownian motion B (the solution is required to be adapted to the filtration of B). The solution, given by e_2, is a reflected Brownian motion, and is indeed adapted to the filtration of B, so that $\mathscr{F}_t^X \subseteq \mathscr{F}_t^B$ for every $t \in \mathbb{R}_+$. At the same time, since the local time $L^0(X)$ is⊸(10.65) adapted to the filtration of X, it is also clear that $B = X - L^0(X)$ must be adapted to the filtration of X, so that $\mathscr{F}_t^B \subseteq \mathscr{F}_t^X$ for every $t \in \mathbb{R}_+$. It is important to recognize that claiming that the solution to e_3 is a reflected Brownian motion is not the same as claiming that $X = |B|$, which is certainly false – one can only claim that the law of X is the same as the law of $|B|$. We can now summarize these results as follows:

(10.74) PROPOSITION: If B is the Brownian motion, then the solution, X, to equation e_3 is given by e_2. Furthermore, the solution is a reflected Brownian motion that generates the filtration of B, that is, $\mathscr{F}^X = \mathscr{F}^B$. ○

In particular, we see that the filtration of the Brownian motion

$$B \stackrel{\text{def}}{=} \mathrm{sign}\_(W) \bullet W \equiv \mathrm{sign}(W) \bullet W ,$$

that showed up earlier in this section, is the same as the filtration $\mathscr{F}^{|W|}$, which is strictly smaller than $\mathscr{F}^W$. This is due to the relations⊸(10.66)

$$L^0(W) = \frac{1}{2} L^0(|W|) \quad \text{and} \quad |W| = B + L^0(W) = B + \frac{1}{2} L^0(|W|) .$$

Indeed, by the uniqueness part of Skorokhod's lemma, the last relation implies that the sample path $\frac{1}{2}L^0(|W_\omega|)$, and therefore also $|W_\omega|$, can be constructed from the sample path B_ω by using a completely deterministic procedure, so that $\mathscr{F}^{|W|}$ must be contained in $\mathscr{F}^B$. At the same time, $\mathscr{F}^B$ must be contained in $\mathscr{F}^{|W|}$ since $B = |W| - \frac{1}{2}L^0(|W|)$.

Since in the relations e_2 and e_3 the Brownian motion B is arbitrarily chosen, we can replace in those relations B with $(-B)$. This allows us to bring into the picture the running maximum

$$S(\omega, t) \stackrel{\text{def}}{=} \sup\nolimits_{s \in [0,t]} B(\omega, s) , \quad t \in \mathbb{R}_+ ,$$

and conclude that the process $\tilde{X} = -B + S$ is a reflected Brownian motion with local time in 0 given by the process S (the running maximum). Furthermore, the sample paths S_ω and $\tilde{X}_\omega$ can again be constructed from the Brownian sample path $(-B_\omega)$ in terms of the fully deterministic mechanism of Skorokhod's lemma. This brings us to the following remarkable result.

(10.75) LÉVY'S EQUIVALENCE-IN-LAW THEOREM: If W is the Brownian motion and S its running maximum, then the two-dimensional processes

$$(S_t - W_t, S_t)_{t \in \mathbb{R}_+} \quad \text{and} \quad (|W_t|, L_t^0(W))_{t \in \mathbb{R}_+}$$

share the same distribution law (as processes). In particular, for fixed $t \in \mathbb{R}_+$, $\mathrm{Law}\big(L_t^0(W)\big) = \mathrm{Law}(S_t) = \mathrm{Law}(|W_t|)$. ○

(10.76) REFLECTED BROWNIAN MOTION AS A MARKOV PROCESS: When away from 0 a reflected Brownian motion behaves locally as the Brownian motion. When in 0 its history cannot affect its local behavior. As a result, the reflected Brownian motion is a Markov process. However, its infinitesimal generator is not defined at 0, as we now detail. By Itô's formula, for any test function $\varphi \in \mathscr{C}_K^\infty(\mathbb{R};\mathbb{R})$,

$$\frac{1}{\varepsilon}\mathsf{E}\big[\varphi(|W_\varepsilon|) - \varphi(0)\big] = \frac{1}{\varepsilon}\int_0^\varepsilon \mathsf{E}\Big[\frac{1}{2}\partial^2\varphi(|W_s|)\Big]\,\mathrm{d}s + \frac{1}{\varepsilon}\mathsf{E}\Big[\int_0^\varepsilon \partial\varphi(|W_s|)\,\mathrm{d}L_s^0(W)\Big].$$

As $\varepsilon \to 0$ the first term on the right would converge to $\frac{1}{2}\partial^2\varphi(0)$, but the second term would not have a finite limit unless $\partial\varphi(0) = 0$. Indeed, since the measure $\mathrm{d}L^0(W)$ is supported only on the set $\{s\colon W_s = 0\}$, we have

$$\begin{aligned}\int_0^\varepsilon \partial\varphi(|W_s|)\,\mathrm{d}L_s^0(W) &= \partial\varphi(0)L_\varepsilon^0(W) + \int_0^\varepsilon \big(\partial\varphi(|W_s|) - \partial\varphi(0)\big)\,\mathrm{d}L_s^0(W)\\ &= \partial\varphi(0)L_\varepsilon^0(W),\end{aligned}$$

so that

$$\begin{aligned}\frac{1}{\varepsilon}\mathsf{E}\Big[\int_0^\varepsilon \partial\varphi(|W_s|)\,\mathrm{d}L_s^0(W)\Big] &= \partial\varphi(0)\frac{1}{\varepsilon}\mathsf{E}\big[L_\varepsilon^0(W)\big]\\ &= \partial\varphi(0)\frac{1}{\varepsilon}\mathsf{E}\big[|W_\varepsilon|\big] = \partial\varphi(0)\frac{1}{\sqrt{\varepsilon}}\sqrt{\frac{2}{\pi}}.\end{aligned}$$

As a result, the infinitesimal generator

$$\mathcal{A}_t^{|W|}\varphi(x) \stackrel{\text{def}}{=} \lim_{\varepsilon\searrow 0}\frac{1}{\varepsilon}\mathsf{E}\Big[\varphi\big(|W_{t+\varepsilon}|\big) - \varphi\big(|W_t|\big) \;\big|\; |W_t| = x\Big], \quad x \in \mathbb{R}_+,$$

can be defined – everywhere in $\mathbb{R}_+ = [0, \infty[$ – only for those $\varphi \in \mathscr{C}_K^\infty(\mathbb{R}_+;\mathbb{R})$ that have the additional property $\partial\varphi(0) = 0$, and, for any such φ, $\mathcal{A}_t^{|W|}\varphi = \frac{1}{2}\partial^2\varphi$. ○

(10.77) APPROXIMATION WITH RANDOM WALKS: We have seen earlier that the Brownian motion arises as the limit of appropriately constructed random walks. How should these random walks be modified in order to produce a reflected Brownian motion in the limit? The correct answer is the intuitive one: when the random walk is in 0 and attempts to move "down," it should be kept in 0, but if it attempts to move "up" it should be let go – see (Chung and Williams 1990, sec. 8.4). ○

11

Stochastic Differential Equations

We saw in the introduction to chapter 10 that the motion of a particle immersed in a velocity vector field – so to speak, the motion of a particle blown by wind – can be modeled as the solution $t \rightsquigarrow x(t)$ to an equation of the form

$$x(t) - x(0) = \int_0^t V(x(s), s)\mathrm{d}s\,, \quad t \in \mathbb{R}_+\,,$$

which is familiar to the reader from their very first course on differential equations. In that chapter we recognized that interesting and useful models of a genuinely stochastic motion – so to speak, motion caused by a "chaotic wind" – must involve stochastic integration. While the object of study in the previous chapter was the construction of the stochastic integral and its basic features, in the present chapter we study the actual motion described in terms of such integrals, that is, the motion of a particle immersed in a "stochastic velocity field." Our plan is to introduce and study equations that are analogous to the one above, with the integral on the right side replaced by a stochastic integral with respect to a particular semimartingale.

11.1 An Example

In one way or another, the meaning of the word "equation" invariably comes down to a question of the form "Which objects from class such and such satisfy relation(s) such and such?" As an example, we can rephrase exercise (10.41) as the question: given a Brownian motion W, which processes $X \in \mathfrak{L}^2_{\mathrm{loc}}(\imath)$ satisfy the relation[129]

$$X = 1 + \sigma X {\scriptstyle\bullet} W + b X {\scriptstyle\bullet} \imath\,?$$

To answer this question – that is, to find a solution – one must construct a process X (on some probability space, relative to a particular filtration) such that both integrals above are meaningful (as stochastic processes) and the identity holds, that is, the process X is indistinguishable from $1 + \sigma X {\scriptstyle\bullet} W + b X {\scriptstyle\bullet} \imath$. It is also customary to write relations like the one above in differential form:

$$\mathrm{d}X_t = \sigma X_t \mathrm{d}W_t + b X_t \mathrm{d}t \quad \Leftrightarrow \quad \frac{\mathrm{d}X_t}{X_t} = \sigma\,\mathrm{d}W_t + b\,\mathrm{d}t\,, \quad X_0 = 1\,.$$

[129] Unless noted otherwise, an identity between two stochastic processes is understood as an identity up to indistinguishability – see (8.6).

In this particular case we already know that $X = e^{\sigma W + (b - \frac{1}{2}\sigma^2)\imath}$ is one solution, but can we find another one? To address the uniqueness question, we first observe that, since the integrals $X{\bullet}W$ and $X{\bullet}\imath$ are continuous by definition, any solution, X, must be continuous as well.

(11.1) EXERCISE: Given a probability space $(\Omega, \mathcal{F}, P)$, filtration $\mathscr{F} \subset \mathcal{F}$, and an $(\mathscr{F}, P)$-Brownian motion W, suppose that X and Y are any two $\mathscr{F}$-adapted processes on Ω, that have continuous sample paths and are such that

$$X_0 \neq 0, \quad Y_0 \neq 0, \quad X = X_0 + \sigma X{\bullet}W + bX{\bullet}\imath\,, \quad \text{and} \quad Y = Y_0 + \sigma Y{\bullet}W + bY{\bullet}\imath\,.$$

Prove that the sample paths of the process X/Y are P-a.s. constant. In particular, if $X_0 = Y_0$ P-a.s., then $X_\omega = Y_\omega$ (understood as an identity between two continuous functions on $\mathbb{R}_+$) for P-a.e. $\omega \in \Omega$.

HINT: Using Itô's formula, show that $X/Y = X_0/Y_0$. ○

11.2 Strong and Weak Solutions

Back in section 8.4 we introduced the path space $\mathbb{W} = \mathscr{C}(\mathbb{R}_+; \mathbb{R}^n)$ in connection with the canonical construction of the Wiener process and measure. This space will play a central role in the present chapter, and our first step is to equip it with an appropriate filtration. With that purpose in mind, recall that the canonical coordinate process (the collection of coordinate mappings) on $\mathbb{W}$ is given by

$$\mathbb{W} \ni w \rightsquigarrow \epsilon_t(w) = w(t) \in \mathbb{R}^n\,, \quad t \in \mathbb{R}_+\,.$$

The unique filtration on $\mathbb{W}$ generated by this coordinate process (the natural filtration of ϵ) we denote by $\mathscr{B}$, that is, $\mathscr{B}_t = \sigma\{\epsilon_s : s \in [0,t]\}$, $t \in \mathbb{R}_+$, and note that $\mathscr{B} \subset \mathcal{B}(\mathbb{W})$; in fact, it is not difficult to see that $\mathscr{B}_\infty \overset{\text{def}}{=} \vee_{t \in \mathbb{R}_+} \mathscr{B}_t = \mathcal{B}(\mathbb{W})$. When attached to functions defined on $\mathbb{W}$, the attributes "progressively measurable" and "predictable" can now be understood relative to $\mathscr{B}$.

Next, consider a probability space $(\Omega, \mathcal{F}, P)$ with filtration $\mathscr{F} \subset \mathcal{F}$, and suppose that $X : \Omega{\times}\mathbb{R}_+ \mapsto \mathbb{R}^n$ is some $\mathscr{F}$-adapted process with continuous sample paths $X_\omega \in \mathbb{W}$. Any such process can be treated as a function of the form $\Omega \ni \omega \rightsquigarrow X_\omega \in \mathbb{W}$, which we are going to denote by the same symbol X. Since, treated as a process on Ω, X is $\mathscr{F}$-adapted, we have $X^{-1}(\mathscr{B}_t) \subseteq \mathscr{F}_t$, that is, the mapping $X : \Omega \mapsto \mathbb{W}$ is $\mathscr{B}_t/\mathscr{F}_t$-measurable for any $t \in \mathbb{R}_+$. Thus, if we now define the mapping

$$(X, \imath) : \Omega{\times}\mathbb{R}_+ \mapsto \mathbb{W}{\times}\mathbb{R}_+$$

so that $(X, \imath)(\omega, t) = (X_\omega, t)$, then it would not be difficult to see that this mapping is $\text{Prog}(\mathscr{B})/\text{Prog}(\mathscr{F})$-measurable.[130] The same mapping can also be shown

[130] If $B \in \text{Prog}(\mathscr{B})$, then $B \cap (\mathbb{W}{\times}[0,t]) \in \mathscr{B}_t \otimes \mathcal{B}([0,t])$ and it is not difficult to check by inspection that $(X,\imath)^{-1}(B) \cap (\Omega{\times}[0,t]) = (X,\imath)^{-1}\big(B \cap (\mathbb{W}{\times}[0,t])\big)$, where $B \cap (\mathbb{W}{\times}[0,t]) \in \mathscr{B}_t \otimes \mathcal{B}([0,t])$. Since $X^{-1}(\mathscr{B}_t) \subseteq \mathscr{F}_t$ and $\imath^{-1}(\mathcal{B}([0,t])) = \mathcal{B}([0,t])$, $(X,\imath)^{-1}\big(\mathscr{B}_t \otimes \mathcal{B}([0,t])\big) \subseteq \mathscr{F}_t \otimes \mathcal{B}([0,t])$.

to be $\mathscr{P}(\mathscr{B})/\mathscr{P}(\mathscr{F})$-measurable.[131] If E stands for some Euclidean space and if $\varphi\colon \mathbb{W}\times\mathbb{R}_+ \mapsto E$ is some Prog($\mathscr{B}$)-measurable function (some $\mathscr{P}(\mathscr{B})$-measurable function), then the composition $\varphi \circ (X, \imath)$, which we denote by $\varphi(X, \imath)$, is going to be a Prog($\mathscr{F}$)-measurable function ($\mathscr{P}(\mathscr{F})$-measurable function) from $\Omega\times\mathbb{R}_+$ into E. Note that the symbol $\varphi(X, \imath)$ is understood as the name of a function, and the result from the evaluation of this function at the point (ω, t) will be expressed as $\varphi(X, \imath)(\omega, t) = \varphi(X_\omega, t)$. In particular, if $\varphi(w, .)$ is continuous on $\mathbb{R}_+$ for every $w \in \mathbb{W}$, and if $\varphi(\cdot, t) \prec \mathscr{B}_t$ for every $t \in \mathbb{R}_+$, then $\varphi(X, \imath)$ can be treated as a continuous and $\mathscr{F}$-adapted process on $\Omega\times\mathbb{R}_+$.

One special, but important, class of functions from $\mathbb{W}\times\mathbb{R}_+$ into E is given by

$$\mathbb{W}\times\mathbb{R}_+ \ni (w, t) \rightsquigarrow f(\varepsilon_\imath, \imath)(w, t) = f(w(t), t) \in E$$

for some Borel-measurable function $f\colon \mathbb{R}^n\times\mathbb{R}_+ \mapsto E$. Since the $\mathbb{R}^n\times\mathbb{R}_+$-valued process

$$\mathbb{W}\times\mathbb{R}_+ \ni (w, t) \rightsquigarrow (\varepsilon_\imath, \imath)(w, t) = (w(t), t) \in \mathbb{R}^n\times\mathbb{R}_+$$

is continuous and $\mathscr{B}$-adapted by definition, it is automatically $\mathscr{P}(\mathscr{B})$-measurable (predictable for $\mathscr{B}$). As a result, the composition $f(\varepsilon_\imath, \imath) = f \circ (\varepsilon_\imath, \imath)$ is $\mathscr{P}(\mathscr{B})$-measurable as well. For simplicity, the composition of the function $f(\varepsilon_\imath, \imath)$ and the mapping $(X, \imath)$ we will express as $f(X, \imath)$, and write

$$f(X, \imath)(\omega, t) = f(X_\omega(t), t) = f(X_t(\omega), t),$$

or write simply $f(X_t, t)$, if suppressing the dependence on ω is preferable.

Given any two Prog($\mathscr{B}$)-measurable functions

$$\alpha\colon \mathbb{W}\times\mathbb{R}_+ \mapsto \mathbb{R}^{n\otimes d} \quad \text{and} \quad \beta\colon \mathbb{W}\times\mathbb{R}_+ \mapsto \mathbb{R}^n,$$

and a probability distribution law μ over $\mathbb{R}^n$, one can formulate the stochastic differential equation[132] (SDE for short)$^{\multimap(10.23)}$

$$\mathrm{Law}(X_0) = \mu, \quad X = X_0 + \alpha(X, \imath){\scriptstyle\bullet} W + \beta(X, \imath){\scriptstyle\bullet}\imath, \qquad \mathbb{e}_\mu(\alpha, \beta)$$

in which W stands for a d-dimensional Brownian motion. The same equation is often written in differential form as

$$\mathrm{Law}(X_0) = \mu, \quad \mathrm{d}X_t = \alpha(X, t)\mathrm{d}W_t + \beta(X, t)\mathrm{d}t, \quad t \in \mathbb{R}_+,$$

or, in integral form as

$$\mathrm{Law}(X_0) = \mu, \quad X_t = X_0 + \int_0^t \alpha(X, s)\mathrm{d}W_s + \int_0^t \beta(X, s)\mathrm{d}s, \quad t \in \mathbb{R}_+.$$

131 The σ-field $\mathscr{P}(\mathscr{B})$ is generated by functions $f\colon \mathbb{W}\times\mathbb{R}_+ \mapsto \mathbb{R}$ such that $f(\cdot, t) \prec \mathscr{B}_t$ for any $t \in \mathbb{R}_+$, and $f(w, \cdot)$ is continuous for every $w \in \mathbb{W}$. For any such f the process $(\omega, t) \rightsquigarrow f(X_\omega, t)$ is clearly $\mathscr{F}$-adapted and continuous. Thus, $(X, \imath)^{-1}(\mathscr{P}(\mathscr{B})) \subseteq \mathscr{P}(\mathscr{F})$.

132 Apart from a minor typographical variation, the token $\mathbb{e}_\mu(\alpha, \beta)$, which we use extensively throughout the sequel, is borrowed from (Revuz and Yor 1999), which this chapter closely follows.

However, the first form of $\mathbb{e}_\mu(\alpha,\beta)$ is preferable, since an SDE is understood as an identity between two stochastic processes up to indistinguishability – not as an identity between random variables for every fixed $t \in \mathbb{R}_+$.

(11.2) DEFINITION: A solution to $\mathbb{e}_\mu(\alpha,\beta)$ is a collection of:

(a) probability space $(\Omega, \mathcal{F}, P)$;

(b) filtration $\mathscr{F} \subset \mathcal{F}$;

(c) d-dimensional $(\mathscr{F}, P)$-Brownian motion W;

(d) $\mathscr{F}$-adapted and continuous $\mathbb{R}^n$-valued process X such that $\mathrm{Law}(X_0) = \mu$, the integrals in $\mathbb{e}_\mu(\alpha,\beta)$ are meaningful, and the process on the right side of $\mathbb{e}_\mu(\alpha,\beta)$ is indistinguishable from X.

In most cases a solution to $\mathbb{e}_\mu(\alpha,\beta)$ will be expressed as the pair (X, W), the probability space and the filtration being understood from the context, or even just as the symbol X, if the Brownian motion W is understood from the context as well. The functions α and β are called *coefficients of* $\mathbb{e}_\mu(\alpha,\beta)$. In the special case where these coefficients have the form $\alpha(w,t) = a(w_t,t)$ and $\beta(w,t) = b(w_t,t)$, for some choice of the Borel functions $a\colon \mathbb{R}^n \times \mathbb{R}_+ \mapsto \mathbb{R}^{n\otimes d}$ and $b\colon \mathbb{R}^n \times \mathbb{R}_+ \mapsto \mathbb{R}^n$, equation $\mathbb{e}_\mu(\alpha,\beta)$ is said to be a *diffusion SDE*. A diffusion SDE will be expressed as $\mathbb{e}_\mu(a,b)$ and will be often written as[133]

$$X = X_0 + a(X,\iota)\bullet W + b(X,\iota)\bullet\iota\,.$$

The functions a and b, which define the diffusion SDE $\mathbb{e}_\mu(a,b)$, are called, respectively, *diffusion coefficient* (alias: *covariance coefficient*) and *drift coefficient*, or simply *drift*. If (X, W) is a solution to a diffusion SDE, then the process X is said to be a *diffusion process*, or simply a *diffusion*. If the coefficients a and b do not depend on the time-variable (have the form $a(w_t)$ and $b(w_t)$), we say that the diffusion SDE (the diffusion process) is *time-homogeneous*, or simply *homogeneous*, and if we want to stress that the coefficients do depend on the time variable, we say that the SDE (the diffusion process) is *non-homogeneous*.

There are situations where the choice of the initial distribution $\mu = \mathrm{Law}(X_0)$ is generic, in the sense that $\mathbb{e}_\mu(\alpha,\beta)$ is formulated for an arbitrary choice of the initial distribution μ. In such situations we will write $\mathbb{e}(\alpha,\beta)$ instead of $\mathbb{e}_\mu(\alpha,\beta)$. On the other extreme, there will be situations where we will need to work with SDEs that have a nonrandom initial value $X_0 = x \in \mathbb{R}^n$ – those are simply $\mathbb{e}_{\epsilon_x}(\alpha,\beta)$. ○

(11.3) REMARK: Equation $\mathbb{e}_\mu(\alpha,\beta)$ can be formulated with some local martingale, or semimartingale, in place of the Brownian motion W. In this more general setup the coefficients α and β must be required to be predictable and not merely progressive, in order for the stochastic integral in $\mathbb{e}_\mu(\alpha,\beta)$ to be meaningful. ○

---

133 This convention assumes that the context makes it clear – say, from the definition of the symbols a and b – whether $a(X,\iota)$ and $b(X,\iota)$ depend only on the location of the process X at a given moment in time, or depend on its entire history leading to that moment.

(11.4) INFINITESIMAL GENERATORS: Given any two Borel-measurable functions $a\colon \mathbb{R}^n \times \mathbb{R}_+ \mapsto \mathbb{R}^{n\otimes d}$ and $b\colon \mathbb{R}^n \times \mathbb{R}_+ \mapsto \mathbb{R}^n$ (in other words, some time-dependent fields of matrices and vectors distributed over the state-space $\mathbb{R}^n$) we can formulate the diffusion SDE $\mathbb{e}(a,b)$ – and we already did. But we can also define the mapping

$$\mathbb{R}_+ \ni t \rightsquigarrow \mathcal{A}_t = \frac{1}{2} \sum_{i,j=1}^{n} \left(\sum_{k=1}^{d} a_{i,k}(x,t) a_{j,k}(x,t) \right) \partial_{x_i,x_j} + \sum_{i=1}^{n} b_i(x,t) \partial_{x_i}$$

that sends the time domain $\mathbb{R}_+$ into the space of second-order differential operators that operate on functions from the space $\mathscr{C}^2(\mathbb{R}^n;\mathbb{R})$, or on functions from the space $\mathscr{C}^{2,1}(\mathbb{R}^n \times \mathbb{R}_+;\mathbb{R})$, with the understanding that for any fixed $t \in \mathbb{R}_+$, $\mathcal{A}_t$ operates only on the first n variables in the list $(x_1,\ldots,x_n,t) \in \mathbb{R}^n \times \mathbb{R}_+$. We can cast $\mathcal{A}_t$ more succinctly in terms of the gradient operator $\nabla \overset{\text{def}}{=} (\partial_{x_1},\ldots,\partial_{x_n})$ (understood to be a column) as

$$\mathcal{A}_t = \frac{1}{2} \left(a(x,t)\, a(x,t)^{\intercal} \nabla \right) \cdot \nabla + b(x,t) \cdot \nabla ,$$

and if we want to stress the dependence of $\mathcal{A}_t$ on the coefficients a and b, we will write $\mathcal{A}_t^{aa^{\intercal},b}$. We are going to realize very soon that, in a certain sense, the mapping $t \rightsquigarrow \mathcal{A}_t^{aa^{\intercal},b}$ is just another encryption of the most useful features of $\mathbb{e}(a,b)$ – the next exercise is our first step toward revealing this connection. If the process X happens to be a solution to $\mathbb{e}(a,b)$, together with some Brownian motion W, we will say that $t \rightsquigarrow \mathcal{A}_t^{aa^{\intercal},b}$ is the *infinitesimal generator* of X – the reason for this nomenclature will become clear shortly. ○

(11.5) EXERCISE: With the notation as in (11.4), prove that if X is a solution to $\mathbb{e}(a,b)$, then for any function $f \in \mathscr{C}^{2,1}(\mathbb{R}^n \times \mathbb{R}_+;\mathbb{R})$, the process[134]

$$M^f \overset{\text{def}}{=} f(X,\iota) - \big(\partial f(X,\iota) + \mathcal{A} f(X,\iota)\big) \bullet \iota ,$$

that is,

$$M_t^f \overset{\text{def}}{=} f(X_t,t) - \int_0^t \big(\partial f(X_s,s) + \mathcal{A}_s f(X_s,s)\big)\, \mathrm{d}s , \quad t \in \mathbb{R}_+ ,$$

is a local martingale (for the probability measure and the filtration that are implicit from the formulation of $\mathbb{e}(a,b)$). The relevance of this statement to our discussion in the present and the following chapters will become clear in (11.7) and later in section 12.1. ○

(11.6) REMARK: Given the applications that are of primary concern to us, stochastic equations of diffusion type are especially interesting. Heuristically speaking, a solution X to the diffusion equation $\mathbb{e}_\mu(a,b)$, can be thought of as a model of a motion such that, neglecting terms of order $o(\varepsilon)$, the displacement from time t to time $t+\varepsilon$, $X_{t+\varepsilon} - X_t$, can be expressed as a multivariate Gaussian vector, the realization

---

[134] Here we suppose that the symbol ∇, and therefore $\mathcal{A} = \mathcal{A}^{aa^{\intercal},b}$, operates only on the first n-variables, while the differentiation symbol ∂ operates only on the last variable in the list $(x_1,\ldots,x_n,t)$.

of which is independent of X_t and the entire history up to time t, which has covariance matrix $a(X_t,t)a(X_t,t)^\intercal\varepsilon$ and has mean $b(X_t,t)\varepsilon$. In many practical situations, however, one would be more interested in the dynamics of $f(X)$ for various choices of the function f, rather than tracking X directly. Indeed, in practice one can only see the interactions between the available measuring instruments (functions f) and the various phenomena (processes X) in the real world. If one can read the measurements from a very large number of instruments f, one would then have a reasonably complete knowledge about the actual phenomenon X. We see now that the infinitesimal generator is just another way of encrypting the mechanics of the instantaneous changes in a particular system – in much the same way in which the first derivative of a function is an encryption of the way the value changes as a result of infinitesimally small changes in the argument. There is one crucial difference, however: the infinitesimal generator encrypts only the "local average" in the infinitesimal change of $f(X)$ – not the actual infinitesimal change. Or, to put it another way, it gives the infinitesimal change only in the predictable component of the semimartingale $f(X)$. ○

It is clear from the preceding remark that it would be desirable to somehow conceptualize the solution X to a particular diffusion SDE without the need to bring into the picture the sample space Ω or the Brownian motion W. After all, if the sample paths of X are all that there is to see, one should be able to come up with a reasonably good description of X only in terms of the probability distribution of its sample paths in the space $\mathbb{W}$. Motivated by this observation, Stroock and Varadhan (1979) proposed the following alternative tack:

(11.7) THE MARTINGALE PROBLEM OF STROOCK AND VARADHAN: Given a function $f(\cdot)$, defined on $\mathbb{R}_+$ and taking values in some (any) space, define the family of *translations*, $\theta_s f(\cdot)$, $s \in \mathbb{R}_+$, in the obvious way as $\theta_s f(\cdot) = f(\cdot + s)$, that is, $\theta_s f(t) = f(t+s)$ for any $t \in \mathbb{R}_+$. In particular, the application of θ_s to the infinitesimal generator introduced in (11.4) would give

$$(\theta_s\mathcal{A})_t = \mathcal{A}_{t+s} = \frac{1}{2}\left(a(x,t+s)\,a(x,t+s)^\intercal \nabla\right)\cdot\nabla + b(x,t+s)\cdot\nabla\,.$$

Instead of $\mathbb{e}_{\epsilon_x}(a,b)$ for a fixed $x \in \mathbb{R}^n$, with a and b as in (11.4), Stroock and Varadhan formulated the following problem, which is known as *the martingale problem* (or the *martingale formulation*), and will be expressed as $\mathbb{M}_{x,s}(aa^\intercal, b)$: given a starting point $(x,s) \in \mathbb{R}^n \times \mathbb{R}_+$, construct a probability measure, $\mathcal{Q}_{x,s}$, defined on $(\mathbb{W}, \mathcal{B}(\mathbb{W}))$, such that $\mathcal{Q}_{x,s}(w(0) = x) = 1$ and for every test function $\varphi \in \mathscr{C}_K^\infty(\mathbb{R}^n;\mathbb{R})$ the process (here the symbol $\mathcal{A}$ is a shorthand for $\mathcal{A}^{aa^\intercal,b}$)

$$M_w^\varphi \stackrel{\text{def}}{=} \varphi(w) - (\theta_s\mathcal{A})\varphi(w)\bullet\iota\,,\quad w \in \mathbb{W}\,,$$

that is

$$M^\varphi(w,t) \stackrel{\text{def}}{=} \varphi(w(t)) - \int_0^t (\mathcal{A}_{u+s}\varphi)(w(u))\mathrm{d}u\,,\quad t \in \mathbb{R}_+\,,\ \ w \in \mathbb{W}\,,$$

is a $(Q_{x,s}, \mathscr{B})$-*martingale* on the path space $\mathbb{W}$. The idea is that, if it exists, $Q_{x,s}$ would be able to capture all essential features of the law of the process $(X_{t+s})_{t\in\mathbb{R}_+}$, conditioned to the realization $X_s = x$. Notice that M^{φ} gives rise to a process that starts from position $\varphi(x)$ at time $t = s$ in the obvious way:

$$M^{\varphi}(\theta_s w, t-s) = \varphi(w(t-s)) - \int_s^t (\mathcal{A}_v\varphi)(w(v))\,\mathrm{d}v\,, \quad t \in [s,\infty[\,, \quad w \in \mathbb{W}\,.$$

The benefits from the martingale formulation will become evident in the next chapter, where we study the connection between stochastic differential equations of diffusion type with coefficients a and b and parabolic partial differential equations with coefficients $\frac{1}{2}aa^{\intercal}$ and b. In this regard, we stress that, unlike the stochastic equation $\mathbb{e}(a,b)$, the martingale problem depends on the coefficient a only through the positive definite matrix $aa^{\intercal}$. ○

Another interesting aspect of the heuristic description in (11.6) is that the instantaneous displacement in a diffusion process appears to depend only on the current state of the process, and not on the history leading to that state. Basic intuition then suggests that all future events associated with a diffusion process depend on its history only through the present state; in other words, being part of the history, the present state is all the history that would be relevant for predicting the future. In the theory of stochastic processes this feature is known as "the Markov property," but we must formulate it precisely to be able to use it later on. In a first step, we would like to see the Markov property of a process X, with some Polish space $\mathbb{S}$ as its state space, as the claim that for every Borel set $B \in \mathcal{B}(\mathbb{S})$ and every $s,t \in \mathbb{R}_+$, $s < t$, the conditional expectation $\mathsf{E}\big[1_B(X_t)\,|\,\mathscr{F}_s^X\big]$ is the same as $\mathsf{E}\big[1_B(X_t)\,|\,X_s\big]$, which is a (deterministic) function of X_s. Writing this function as $P_{s,t}(\cdot, B)$, we now realize that for every fixed $x \in \mathbb{S}$, the mapping

$$\mathcal{B}(\mathbb{S}) \ni B \rightsquigarrow P_{s,t}(x,B) = \mathsf{E}\big[1_B(X_t)\,|\,X_s = x\big] \equiv \mathsf{P}(X_t \in B\,|\,X_s = x)$$

gives a probability measure on $\mathbb{S}$. This measure is nothing but the conditional distribution of X_t given $X_s = x$. It is also called *transition probability*. Clearly, the probability measures $P_{s,t}(x,\cdot)$, with the various choices of s, t, and x, must be coordinated. In particular, given $0 \le s < t < u$, we would like to be able to write

$$\begin{aligned}\mathsf{E}\big[1_B(X_u)\,|\,X_s\big] &= \mathsf{E}\big[1_B(X_u)\,|\,\mathscr{F}_s^X\big]\\ &= \mathsf{E}\big[\mathsf{E}[1_B(X_u)\,|\,\mathscr{F}_t^X]\,|\,\mathscr{F}_s^X\big] = \mathsf{E}\big[\mathsf{E}[1_B(X_u)\,|\,X_t]\,|\,X_s\big]\,,\end{aligned}$$

and this amounts to the following relation, which is known as *the Chapman–Kolmogorov equation*:[135]

$$P_{s,u}(x,B) = \int_{\mathbb{S}} P_{t,u}(y,B)P_{s,t}(x,\mathrm{d}y)\,. \qquad \mathrm{e}_1$$

[135] Named after the British mathematician Sydney Chapman (1988–1970) and the Russian mathematician Andrei Nikolaevich Kolmogorov (1903–1987).

At this point, such a relation is merely a desirable feature, and, for this reason, it will now be incorporated into the following definition, which is standard in the theory of Markov processes:

(11.8) TRANSITION FUNCTIONS AND MARKOV PROCESSES: Let $\mathbb{S}$ be some fixed Polish space. A family of transition probabilities, $P_{s,t}(x,\cdot)$, $x \in \mathbb{S}$, $s,t \in \mathbb{R}_+$, $s < t$, defined on $\mathcal{B}(\mathbb{S})$, is said to be a *transition function* on $\mathbb{S}$ if the Chapman–Kolmogorov equation e_1 is satisfied for every choice of $x \in \mathbb{S}$, $B \in \mathcal{B}(\mathbb{S})$, and $0 \leq s < t < u$. If the transition function is *homogeneous*, in that the transition probabilities $P_{s,t}(\cdot,\cdot)$ depend on s and t only through the difference $t - s$, we will write P_{t-s} instead of $P_{s,t}$.

A stochastic process X, with state space $\mathbb{S}$, defined on a probability space $(\Omega, \mathcal{F}, \mathsf{P})$, and adapted to filtration $\mathscr{F} \setminus\!\subset \mathcal{F}$, is said to be a *Markov process* (relative to $\mathscr{F}$ and P) if there is a transition function P on the state space $\mathbb{S}$, such that

$$\mathsf{E}[f(X_t)\,|\,\mathscr{F}_s] = \int_{\mathbb{S}} f(y) P_{s,t}(X_s, \mathrm{d}y)$$

for every positive Borel function $f : \mathbb{S} \mapsto \mathbb{R}_+$ and every $s,t \in \mathbb{R}_+$, $s < t$. If the transition function P happens to be homogeneous, we say that X is a *homogeneous Markov process.* ○

Since our interest in the Markov property was motivated by our first encounter with diffusions, at this point we must ask: is the solution to a diffusion SDE indeed a Markov process, as our intuition told us that it should be? With the martingale formulation in mind, after some reflection one quickly realizes that there must be an intrinsic connection between the Markov property and the existence of a unique solution to the associated martingale problem for every choice of the starting point (x, s). The precise statement is the following, and its proof is straightforward:

(11.9) THEOREM: (Revuz and Yor 1999, IX (1.9))[136] Suppose that for every given $(x, s) \in \mathbb{R}^n \times \mathbb{R}_+$ there is precisely one probability measure $\mathcal{Q}_{x,s}$ on the path space $\mathbb{W}$ that solves $\mathbb{M}_{x,s}(aa^\intercal, b)$. In addition, suppose that this unique measure is such that for every $B \in \mathcal{B}(\mathbb{R}^n)$ and every $t \in \mathbb{R}_+$, the function $x \rightsquigarrow \mathcal{Q}_{x,s}\big(w(t) \in B\big)$ is Borel-measurable. Then, for every fixed $x \in \mathbb{R}^n$, the canonical coordinate process on $\mathbb{W}$ is Markov relative to the probability measure $\mathcal{Q}_{x,0}$. This Markov process starts from x (that is, $\mathcal{Q}_{x,0}(w(0) = x) = 1$) and its transition function is given by $P_{s,t}(x, B) = \mathcal{Q}_{x,s}\big(w(t - s) \in B\big)$, $t \geq s$. ○

(11.10) REMARK: Suppose that $\mathcal{P}$ is some probability measure on $(\mathbb{W}, \mathscr{B}_\infty)$ (recall that $\mathscr{B}_\infty = \mathcal{B}(\mathbb{W})$) such that the coordinate process on $\mathbb{W}$ is Markov relative to $\mathcal{P}$ and relative to its natural filtration $\mathscr{B}$. In this setting, the Markov structure can be described in terms of transition probabilities that are more general than

[136] In (Revuz and Yor 1999) this result is established only for the homogeneous case, but we will see in (11.28) that this is not a restriction. Alternatively, the direct proof of this statement in the nonhomogeneous case would differ only in the notation.

those we used earlier, as we now detail. Given any $t \in \mathbb{R}_+$ and any $B \in \mathscr{B}_\infty$, the function⊶(11.7) $w \rightsquigarrow 1_B(\theta_t w)$ depends only on the future of the path w after time t. The Markov property is then the claim that, for every choice of t and B,

$$\mathsf{E}^{\mathcal{P}}\left[1_B \circ \theta_t \,|\, \mathscr{B}_t\right] = \mathsf{E}^{\mathcal{P}}\left[1_B \circ \theta_t \,|\, \epsilon_t\right] \quad \mathcal{P}\text{-a.s.}$$

The conditional expectation on the right side is a function of $w(t) \in \mathbb{R}^n$ (for fixed t and B) and we will express this function as $\mathcal{P}_{\cdot,t}(B)$. In particular, $B \rightsquigarrow \mathcal{P}_{x,t}(B)$ is a probability measure on $\mathscr{B}_\infty$ for every $x \in \mathbb{R}^n$ and every $t \in \mathbb{R}_+$. If we denote the integral (that is, the expectation) with respect to $\mathcal{P}_{x,t}$ as $\mathsf{E}_{x,t}$, then the Markov property of the coordinate process can be expressed as

$$\mathsf{E}^{\mathcal{P}}\left[F \circ \theta_t \,|\, \mathscr{B}_t\right] = \mathsf{E}_{w(t),t}[F] \quad \mathcal{P}\text{-a.s.}$$

for every $t \in \mathbb{R}_+$ and every positive Borel function $F\colon \mathbb{W} \mapsto \mathbb{R}_+$. Thus, if μ_t stands for the distribution in $\mathbb{R}^n$ of $w(t)$ (relative to $\mathcal{P}$), then for every such function

$$\mathsf{E}^{\mathcal{P}}[F \circ \theta_t] = \int_{\mathbb{R}^n} \mathsf{E}_{x,t}[F]\, \mu_t(\mathrm{d}x)\,.$$

In particular, this relation shows that the measure $\mathcal{P}$ can be restored from the distribution of $w \rightsquigarrow w(0)$ and the collection of probability measures $\mathcal{P}_{x,0}$, $x \in \mathbb{R}^n$.

The homogeneity of the coordinate process (as a Markov process) is simply the claim that the probability measures $\mathcal{P}_{x,t}(\cdot)$ depend on $x \in \mathbb{R}^n$ but not on $t \in \mathbb{R}_+$. If this is the case, then we will write simply E_x instead of $\mathsf{E}_{x,t}$.

Finally, we note that if the measure $\mathcal{Q}_{x,s}$ is a solution to the martingale problem with starting point (x, s), and if the coordinate process on $\mathbb{W}$ is Markov with respect to $\mathcal{Q}_{x,s}$, then one can define the transition probabilities $\mathcal{P}_{y,t}$, $y \in \mathbb{R}^n$, $t \geq s$, in the obvious way through the relation

$$\mathcal{P}_{w(t),t}(B) = \mathsf{E}^{\mathcal{Q}_{x,s}}\left[1_B \circ \theta_{t-s} \,|\, \mathscr{B}_{t-s}\right], \quad B \in \mathscr{B}_\infty\,.$$

In particular, we have $\mathcal{P}_{x,s}(B) = \mathcal{Q}_{x,s}(B)$. The symbol $\mathsf{E}_{y,t}$ can now be understood as the integral with respect to $\mathcal{P}_{y,t}$, but one must be aware that this operation depends also on $\mathcal{Q}_{x,s}$ (using the symbol $\mathsf{E}^{\mathcal{Q}_{x,s}}_{y,t}$ instead is less ambiguous but more cumbersome and should be avoided). Consistent with this and our previous convention, if there is no possibility for mistake, we will write $\mathsf{E}_{x,s}$ for the expectation with respect to $\mathcal{Q}_{x,s}$. ○

(11.11) Strong Markov property: We saw in (8.83) that the Brownian motion is not only a Markov process, but is also a strong Markov process. Back then this feature was stated as "Brownian motion starts afresh at stopping times." How can we formulate such a feature for the coordinate process on $\mathbb{W}$ under some generic probability measure $\mathcal{P}$ on $\mathcal{B}(\mathbb{W})$? With the notation as in (11.10), the answer is the obvious one: we would like to be able to claim that

$$\mathsf{E}^{\mathcal{P}}\left[1_{\{t<\infty\}} F \circ \theta_t \,|\, \mathscr{B}_t\right] = 1_{\{t<\infty\}} \mathsf{E}_{w(t),t}[F] \quad (\mathcal{P}\text{-a.s.})$$

for every choice of the stopping time $t\colon \mathbb{W} \mapsto [0,\infty]$ and every choice of the positive Borel function $F\colon \mathbb{W} \mapsto \mathbb{R}_+$. This feature is instrumental for many important applications in optimal control and other applied areas – especially for optimal stopping problems, which will be studied in later chapters. The mere use of the adjective "strong" suggests that not all Markov processes are Markov in the strong sense. Revuz and Yor (1999) note[137] that if, in addition to the assumptions imposed in theorem (11.9), the coefficients a and b are required to be continuous, then the coordinate process can be claimed to be a Feller process (an important special class that is outside of our scope of interests), and thus (ibid., III (3.1)) has the strong Markov property. Other conditions that guarantee the strong Markov property will be discussed later in this section. ○

As we just realized, the uniqueness of the solution to the martingale problem is rather easy to formulate, and, at the same time, has important and far-reaching consequences. From where we now stand, we see no reason to suspect that the matter of uniqueness is as delicate as we are about to discover.

(11.12) UNIQUENESS OF SOLUTIONS: The martingale formulation aside, the notion of a "unique solution to $\mathbb{e}_\mu(\alpha,\beta)$" can be understood in two different ways:

Uniqueness in distribution law: if (X,W) and (X',W') are two solutions (possibly defined on different probability spaces, and relative to different filtrations), then the distribution laws of X and X' in the space $\mathbb{W}$ are identical.

Uniqueness in pathwise sense: if (X,W) and (X',W') are two solutions defined on the same probability space, with respect to the same filtration, and such that the Brownian motions W and W' are indistinguishable, then X and X' are indistinguishable as well, that is, there is a set $\mathcal{N} \in \mathcal{F}$ with $P(\mathcal{N}) = 0$ such that the sample paths X_ω and X'_ω are identical for every $\omega \in \Omega\backslash\mathcal{N}$.

We say that uniqueness in distribution law, or uniqueness in pathwise sense, holds for $\mathbb{e}(\alpha,\beta)$ if the respective property holds for $\mathbb{e}_\mu(\alpha,\beta)$ for every choice of the initial distribution μ, including unit point-mass measures of the form $\mu = \epsilon_x$ for fixed $x \in \mathbb{R}$. ○

(11.13) PROPOSITION: (Revuz and Yor 1999, IX (1.4)) Uniqueness in distribution law holds for $\mathbb{e}(\alpha,\beta)$ if and only if uniqueness in distribution law holds for $\mathbb{e}_{\epsilon_x}(\alpha,\beta)$ for every choice of the starting point $x \in \mathbb{R}^n$. ○

(11.14) TANAKA'S EXAMPLE: Let B stand for the Brownian motion on $(\Omega,\mathcal{F},P)$, and let $a(x) \stackrel{\text{def}}{=} \operatorname{sign}(x)$, $x \in \mathbb{R}$. Given any $x \in \mathbb{R}$, define the processes

$$X \stackrel{\text{def}}{=} x + B \quad \text{and} \quad W \stackrel{\text{def}}{=} a(X)\boldsymbol{\cdot} B\,.$$

Due to⊶(10.32) Lévy's characterization theorem, W is a Brownian motion and

$$a(X)\boldsymbol{\cdot} W = a(X)a(X)\boldsymbol{\cdot} B = B \quad \Rightarrow \quad X = x + B = x + a(X)\boldsymbol{\cdot} W\,.$$

[137] See the remark following (Revuz and Yor 1999, IX (1.9)).

Thus, (X, W) is a solution to $\mathbb{e}_{\epsilon_x}(a, 0)$. It is easy to check that if (X, W) is any solution to $\mathbb{e}_{\epsilon_x}(a, 0)$, then the law of X can be nothing other than the distribution law of a linear Brownian motion starting from x, so that uniqueness in distribution law holds for $\mathbb{e}_{\epsilon_x}(a, 0)$.

Now suppose that (X, W) is any solution to $\mathbb{e}_{\epsilon_0}(a, 0)$. This means that $X = a(X)\bullet W$, which is the same as $-X = -a(X)\bullet W$; in particular, X is a Brownian motion that starts from 0. Since

$$a(x) = \operatorname{sign}(x) = 1_{\{0\}}(x) - 1_{]-\infty,0[}(x) + 1_{]0,\infty[}(x)\,,$$

it is easy to check that $-a(x) = a(-x) - 2\,1_{\{0\}}(x)$. As a result,

$$-X = a(-X)\bullet W - 2\,1_{\{0\}}(X)\bullet W\,, \quad t \in \mathbb{R}_+\,.$$

Since X is a Brownian motion, and since the aggregate time that a Brownian path spends in 0, as measured by the Lebesgue measure, is 0 – see exercise (10.60) – it follows that the second stochastic integral above must be 0. However, this implies that $(-X, W)$ is also a solution to $\mathbb{e}_{\epsilon_0}(a, 0)$, and, clearly, since X is a Brownian motion, X and $-X$ cannot be indistinguishable. Therefore, pathwise uniqueness does not hold for $\mathbb{e}_{\epsilon_0}(a, 0)$, while uniqueness in distribution law holds. ○

(11.15) STRONG SOLUTIONS AND WEAK SOLUTIONS: We say that $\mathbb{e}(\alpha, \beta)$ admits a strong solution if, given any Brownian motion W, defined on any probability space on which a Brownian motion can be defined, and given any choice for the initial value $x \in \mathbb{R}^n$, one can construct a process X, which is a non-anticipating function of W, and is such that (X, W) is a solution to $\mathbb{e}_{\epsilon_x}(\alpha, \beta)$. The attribute "non-anticipating function of W" is an alias for "process adapted to the natural filtration of W." Plainly, a strong solution is any solution (X, W) that has the additional property that X is adapted to $\mathscr{F}^W$, but the key point here is that such a solution must exist for every choice of W, and can be constructed on any probability space on which a Brownian motion can be constructed.

The precise description of *strong solution to* $\mathbb{e}_\mu(\alpha, \beta)$ is the following: There is a measurable mapping (thought of as a deterministic recipe for constructing the solution) of the form

$$\mathbb{W} \times \mathbb{R}^n \ni (w, x) \rightsquigarrow G(w, x) \in \mathbb{W}$$

that depends on the coefficients α and β, and is *non-anticipating*, in that for every $x \in \mathbb{R}^n$ and every $t \in \mathbb{R}_+$ the function $\mathbb{W} \ni w \rightsquigarrow G(w, x)_t \equiv \epsilon_t(G(w, x)) \in \mathbb{R}^n$ is $\mathscr{B}_t$-measurable, and, furthermore, given any probability space $(\Omega, \mathcal{F}, P)$ that can support a filtration $\mathscr{F} \subset \mathcal{F}$, an $(\mathscr{F}, P)$-Brownian motion W, and an $\mathscr{F}_0$-measurable r.v. X_0 with $\operatorname{Law}(X_0) = \mu$, the process $X \overset{\text{def}}{=} G(W, X_0)$ and the Brownian motion W give a solution, (X, W), to $\mathbb{e}_\mu(\alpha, \beta)$. A solution (X, W) that cannot be expressed in the form $X = G(W, X_0)$, for some deterministic function $G(\cdot, \cdot)$, is said to be a *weak solution*. ○

(11.16) REMARK: Since $G\colon \mathbb{W}\times\mathbb{R}^n \mapsto \mathbb{W}$ is a deterministic function, ultimately, a strong solution can be viewed as a nonrandom "black box" that yields random results only because the input, W, is random. Thus, practical models of risk based on strong solutions cannot incorporate one important aspect of systems driven by stochastic dynamics: the generation of risk from within the dynamics – a feature that will be detailed in (11.19). ○

(11.17) THEOREM: (Revuz and Yor 1999, IX (1.7)) The pathwise uniqueness for $\mathbb{e}(\alpha,\beta)$ implies uniqueness in distribution law, and implies the existence of a strong solution for $\mathbb{e}(\alpha,\beta)$. ○

With the last theorem in mind, it is not surprising that Tanaka's equation is an example of an equation that does not have a strong solution, as we now detail.

(11.18) EXERCISE: Prove that Tanaka's equation $\mathbb{e}_{\epsilon_0}(a,0)$, with $a(x) = \operatorname{sign}(x)$, $x \in \mathbb{R}$, that is, the equation $X = \operatorname{sign}(X)\cdot W$, cannot have a strong solution.

HINT: Assuming that (X, W) is any solution to $\mathbb{e}_{\epsilon_0}(a,0)$, prove that X must be a Brownian motion. Using Tanaka's formula, show that $W = |X| - L^0(X)$ and conclude that W must be adapted to the filtration $\mathscr{F}^{|X|}$ (consult (10.65) and (10.66)). Argue that this last observation is incompatible with the assumption that X is adapted to $\mathscr{F}^W$ (strong solution exists), which entails that $\mathscr{F}_t^X \subseteq \mathscr{F}_t^W \subseteq \mathscr{F}_t^{|X|} \subset \mathscr{F}_t^X$. Explain why the information generated by the absolute value of a Brownian motion is included strictly in the information generated by the same Brownian motion; in particular, the relation $\mathscr{F}_t^X = \mathscr{F}_t^{|X|}$ is not possible. ○

(11.19) STRIKING AND IMPORTANT OBSERVATION: The main message encoded in Tanaka's example is that the filtration generated by the solution, X, is *strictly larger* than the filtration generated by the Brownian motion that drives the equation! In other words, a stochastic differential equation may be able to generate more information (and risk) from within, by simply manipulating the information that is provided as an input, and without any other external sources of randomness. What is particularly striking is that this phenomenon can crop up in diffusion equations with rather elementary coefficients. ○

We conclude this section with a straightforward but instructive example.

(11.20) EXAMPLE OF MULTIPLE STRONG SOLUTIONS: Consider the equation

$$X = 1_{\mathbb{R}\setminus\{0\}}(X)\cdot W\,.$$

Clearly, $(0, W)$ is a solution, and, as is established in the next exercise, (W, W), is a solution as well. ○

(11.21) EXERCISE: Let W be the Brownian motion. Prove that

$$1_{\mathbb{R}\setminus\{0\}}(W)\cdot W = W$$

and complete the argument in the previous example. ○

11.3 Existence of Solutions

Having introduced the notions "solution" and "unique solution," the next step for us is to address the problem of existence. With that purpose in mind, we start with an example of nonexistence.

(11.22) EXAMPLE OF NONEXISTENCE: (A. Shiryaev) Consider the equation

$$\dagger \qquad X = -\frac{1}{2X} 1_{\mathbb{R}\setminus\{0\}}(X)\bullet\iota + W\,,$$

in which $\frac{1}{2x} 1_{\mathbb{R}\setminus\{0\}}(x)$ is understood to be 0 if $x = 0$. As the next exercise demonstrates, if we were to assume that (X, W) is a solution, then we will arrive at the conclusion that $X = 0$, which is a contradiction, because with $X = 0$ the right side of † becomes W (which is not 0). As a result, † cannot have a solution – whether weak or strong. The main reason for this phenomenon is the explosion of the coefficient b at 0. ○

(11.23) EXERCISE: Prove that if (X, W) is a solution to (11.22†), then one must have $P(X_\omega \equiv 0) = 1$.

HINT: Using Itô's formula, show that

$$X^2 = 1_{\{0\}}(X)\bullet\iota + 2\,X\bullet W\,.$$

Then use the occupation times formula from (10.70) to show that the first integral above is 0. Conclude that X^2 is a positive supermartingale that starts from 0. ○

(11.24) TRANSFORMATION OF THE DRIFT: Consider $\mathbb{e}_{\epsilon_x}(\alpha, \beta)$ with progressively measurable coefficients $\alpha\colon \mathbb{W}\times\mathbb{R}_+ \mapsto \mathbb{R}^{n\otimes d}$ and $\beta\colon \mathbb{W}\times\mathbb{R}_+ \mapsto \mathbb{R}^n$, and suppose that (X, W) is a solution (W is a d-dimensional Brownian motion and X is of dimension n). Then suppose that, for some progressive $\theta\colon \mathbb{W}\times\mathbb{R}_+ \mapsto \mathbb{R}^d$ and for some $T \in \mathbb{R}_{++}$,

$$\dagger \qquad |\theta(X, \iota)|^2\bullet\iota_T < \infty \text{ (a.s.)} \quad \text{and} \quad \mathsf{E}\big[e^{-\theta(X,\iota)^\intercal\bullet W_T - \frac{1}{2}|\theta(X,\iota)|^2\bullet\iota_T}\big] = 1\,.$$

In particular, the last relation would be guaranteed by Novikov's condition

$$\mathsf{E}\big[e^{\frac{1}{2}|\theta(X,\iota)|^2\bullet\iota_T}\big] < \infty\,.$$

Consider the measure $Q \approx P$ on $\mathscr{F}_T$ given by

$$\frac{\mathrm{d}Q}{\mathrm{d}P} = e^{-\theta(X,\iota)^\intercal\bullet W_T - \frac{1}{2}|\theta(X,\iota)|^2\bullet\iota_T}$$

and observe that$^{\multimap (10.58)}$ $B \overset{\text{def}}{=} W + \theta(X, \iota)\bullet\iota$ is a d-dimensional Brownian motion on $[0, T]$ relative to Q. The process X can now be written (with respect to the either measure P or Q) as

$$\begin{aligned} X &= X_0 + \alpha(X, \iota)\bullet W + \beta(X, \iota)\bullet\iota \\ &= X_0 + \alpha(X, \iota)\bullet B + \big(\beta(X, \iota) - \alpha(X, \iota)\theta(X, \iota)\big)\bullet\iota\,. \end{aligned}$$

Thus, if X solves $\mathbb{e}_{\epsilon_x}(\alpha, \beta)$ under P, then, treated under Q, the same process solves $\mathbb{e}_{\epsilon_x}(\alpha, \beta - \alpha\theta)$. If, in particular, $n \leq d$, the d-by-d-matrix $\alpha(w,t)^\intercal\alpha(w,t)$ is invertible, and the process

$$\theta(w,t) \stackrel{\text{def}}{=} \left(\alpha(w,t)^\intercal\alpha(w,t)\right)^{-1}\alpha(w,t)^\intercal\beta(w,t)$$

is globally bounded – or somehow condition † holds – then under the measure Q the process X would be a solution to $\mathbb{e}_{\epsilon_x}(\alpha, 0)$. We see that in some situations the effect of the drift can be removed by changing the measure accordingly. In any such situation the existence of a solution would be governed by the diffusion matrix alone. One benefit from this observation is that the existence of a solution for a particular drift can be deduced from the existence of a solution for a different – and possibly easier to deal with – drift. ○

(11.25) CONSTRUCTION OF A SOLUTION: One possible program for constructing a strong solution to $\mathbb{e}_\mu(\alpha, \beta)$, if one exists, is the following. Choose a probability space $(\Omega, \mathcal{F}, P)$ that is rich enough to support a filtration $\mathscr{F}$, a d-dimensional $(\mathscr{F}, P)$-Brownian motion W, and a r.v. $X_0 \prec \mathscr{F}_0$ such that $\text{Law}(X_0) = \mu$. Then choose a partition of $\mathbb{R}_+$, $\mathcal{P} = \{t_0, t_1, \ldots, t_i, \ldots\}$, such that $t_0 = 0$ and $\lim_i t_i = \infty$, and define the process $X^{\mathcal{P}}$ inductively, so that $X^{\mathcal{P}}_{t_0} = X_{t_0} \equiv X_0$, and

$$X^{\mathcal{P}}_t = X^{\mathcal{P}}_{t_i} + \alpha(X^{\mathcal{P}}, t_i)(W_t - W_{t_i}) + \beta(X^{\mathcal{P}}, t_i)(t - t_i) \quad \text{for } t \in \left]t_i, t_{i+1}\right], \ i \in \mathbb{N}.$$

Since the sample paths of $X^{\mathcal{P}}$ are continuous by construction, the process $X^{\mathcal{P}}$ can be treated as an element of the space $\mathscr{L}^0\big(\Omega, \mathscr{F}_\infty, P; \mathscr{C}(\mathbb{R}_+; \mathbb{R}^n)\big)$, which, as we already know, is a complete metric space for the Ky Fan distance (associated with the convergence in probability in the locally uniform distance in $\mathscr{C}(\mathbb{R}_+; \mathbb{R}^n)$, that is, the uniform convergence on any finite interval in probability P). The idea now is to impose conditions on the coefficients α and β that ensure the convergence of $X^{\mathcal{P}}$ in the Ky Fan distance as $|\mathcal{P}| \stackrel{\text{def}}{=} \sup_{i \in \mathbb{N}}|t_{i+1} - t_i| \searrow 0$. To be precise, we would like to be able to claim that

$$\lim_{|\mathcal{P}|\searrow 0,\, |\mathcal{P}'|\searrow 0} P\Big[\sup_{t\in[0,T]}\big|X^{\mathcal{P}}_t - X^{\mathcal{P}'}_t\big| \geq \varepsilon\Big] = 0 \quad \text{for any } \varepsilon, T \in \mathbb{R}_{++},$$

which would guarantee the existence of a limit, $X \in \mathscr{L}^0\big(\Omega, \mathscr{F}_\infty, P; \mathscr{C}(\mathbb{R}_+; \mathbb{R}^n)\big)$, characterized by

$$\lim_{|\mathcal{P}|\searrow 0} P\Big[\sup_{t\in[0,T]}\big|X_t - X^{\mathcal{P}}_t\big| \geq \varepsilon\Big] = 0 \quad \text{for any } \varepsilon, T \in \mathbb{R}_{++}.$$

The requirements imposed on the coefficients α and β must be such that (X, W) can be shown to be a solution to $\mathbb{e}_\mu(\alpha, \beta)$. ○

(11.26) LIPSCHITZ COEFFICIENTS: Suppose that the coefficients α and β satisfy the so-called Lipschitz condition: there is a universal constant $C > 0$ such that[138]

---

[138] Recall that for a matrix A the symbol $|A|$ is understood to be the Frobenius norm of A, that is, the Euclidean norm of A treated as a vector in the obvious way.

$$\dagger \qquad |\alpha(w,t) - \alpha(w',t)| + |\beta(w,t) - \beta(w',t)| \leq C \sup_{s\in[0,t]} |w(s) - w'(s)|\,,$$
$$\text{for all } \; w, w' \in \mathbb{W}\,, \quad \text{for all } \; t \in \mathbb{R}_+\,.$$

If, in addition to the Lipschitz condition, the coefficients $\alpha(w,t)$ and $\beta(w,t)$ are locally bounded with respect to the time variable, in the sense that for any constant function $\bar{w} \in \mathbb{W}$ the functions $\mathbb{R}_+ \ni t \rightsquigarrow a(\bar{w},t)$ and $\mathbb{R}_+ \ni t \rightsquigarrow b(\bar{w},t)$ are bounded on every finite interval, then (Revuz and Yor 1999, IX (2.1)) equation $\mathbb{e}_\mu(\alpha,\beta)$ admits a unique strong solution for every choice of $\mu = \mathrm{Law}(X_0)$. This is essentially Itô's original result.

It is useful to note that the result in (ibid., IX (2.1)) is stated in the following more general form. Let $\tilde{\alpha}(w,t) \in \mathbb{R}^{n\otimes(d+1)}$ denote the matrix obtained by appending $\alpha(w,t)$ (treated as a list of vector columns) with the vector-column $\beta(w,t)$, and let Z be the $\mathbb{R}^{d+1}$-valued continuous semimartingale obtained by appending W (treated as a list of d independent Brownian motions) with the semimartingale ι. Equation $\mathbb{e}_\mu(\alpha,\beta)$ can be cast as

$$X = X_0 + \tilde{\alpha}(X,\iota){\boldsymbol\cdot}Z\,,$$

and the following more general claim can be made. Suppose that the coefficient $\tilde{\alpha}\colon \mathbb{W}\times\mathbb{R}_+ \mapsto \mathbb{R}^{n\otimes d}$ is predictable, satisfies the Lipschitz condition, and is locally bounded in the time variable t. Then, given *any* continuous $\mathbb{R}^d$-valued semimartingale Z (defined on some, that is, any, probability space, relative to some, that is, any, filtration $\mathscr{F}$), and given any starting point $x \in \mathbb{R}^n$, there is a unique (up to indistinguishability) continuous $\mathbb{R}^n$-valued process $X \prec \mathscr{F}^Z$ that is indistinguishable from the process $x + \tilde{\alpha}(X,\iota){\boldsymbol\cdot}Z$. ○

(11.27) Dependence on the initial value: We have seen that a strong solution to $\mathbb{e}_{\epsilon_x}(a,b)$ can be treated as a function of the starting point x. To be precise, we can express such a solution as a function of the form

$$\mathbb{R}^n \times \Omega \times \mathbb{R}_+ \ni (x,\omega,t) \rightsquigarrow X^x(\omega,t) \in \mathbb{R}^n\,.$$

By the very definition of a strong solution, for fixed $x \in \mathbb{R}^n$, the sample paths X^x_ω must be continuous a.s. (as functions of the time variable $t \in \mathbb{R}_+$). However, even if a strong solution is known to exist for every $x \in \mathbb{R}^n$, there would be no reason for the sample paths X^x_ω to be continuous as functions of both variables x and t. The classical conditions that guarantee this continuity are the Lipschitz condition in ($11.26^\dagger$) plus the requirement for the coefficients α and β to be bounded – see (Revuz and Yor 1999, IX (2.4)) for a detailed proof. For a much more detailed exposition on this topic we refer the curious reader to the monograph (Kunita 1990). ○

The case of diffusion coefficients $\alpha(w,t) = a(w_t,t)$ and $\beta(w,t) = b(w_t,t)$ is rather important and deserves special attention. We first note that a nonhomogeneous diffusion SDE can be reduced to a homogeneous one with the following standard manipulation:

(11.28) REDUCTION TO TIME-INVARIANT COEFFICIENTS: Suppose that the coefficients

$$\mathbb{R}^n \times \mathbb{R}_+ \ni (x,t) \rightsquigarrow a(x,t) \in \mathbb{R}^{n\otimes d} \quad \text{and} \quad \mathbb{R}^n \times \mathbb{R}_+ \ni (x,t) \rightsquigarrow b(x,t) \in \mathbb{R}^n$$

are given, and construct the matrix $\tilde{a}(x,t) \in \mathbb{R}^{(n+1)\otimes d}$ by appending to $a(x,t)$ a row of zeros and construct the vector $\tilde{b}(x,t) \in \mathbb{R}^{n+1}$ by appending to the vector $b(x,t)$ the constant function 1. Any vector $\tilde{x} \in \mathbb{R}^{n+1}$ can be expressed as the pair (x,u), $x \in \mathbb{R}^n$, $u \in \mathbb{R}$, in the obvious way, and $\tilde{a}$ and $\tilde{b}$ can be treated as functions on $\mathbb{R}^{n+1}$ via the relations

$$\mathbb{R}^{n+1} \ni \tilde{x} \equiv (x,u) \rightsquigarrow \tilde{a}(x,|u|) \quad \text{and} \quad \mathbb{R}^{n+1} \ni \tilde{x} \equiv (x,u) \rightsquigarrow \tilde{b}(x,|u|).$$

Now suppose that $\tilde{X}$ is a solution to $\tilde{X} = \tilde{X}_0 + \tilde{a}(\tilde{X}){\bullet}W + \tilde{b}(\tilde{X}){\bullet}\imath$, such that the last coordinate of $\tilde{X}_0$ equals 0, while the first n-coordinates of $\tilde{X}_0$ are distributed in $\mathbb{R}^n$ with law μ. If we write the solution as $\tilde{X} = (X,U)$, where X is the $\mathbb{R}^n$-valued process given by the first n components of $\tilde{X}$ and U is the $\mathbb{R}$-valued process given by the last component of $\tilde{X}$, then, just by construction, we would have $U = \imath$, and, consequently, X would satisfy the relation

$$X = X_0 + a(X,\imath){\bullet}W + b(X,\imath){\bullet}\imath\,,$$

with $\mathrm{Law}(X_0) = \mu$. Conversely, if X solves $X = X_0 + a(X,\imath){\bullet}W + b(X,\imath){\bullet}\imath$ with $\mathrm{Law}(X_0) = \mu$, then $\tilde{X} = (X,\imath)$ solves $\tilde{X} = \tilde{X}_0 + \tilde{a}(\tilde{X}){\bullet}W + \tilde{b}(\tilde{X}){\bullet}\imath$. ○

The next result is a more or less straightforward consequence of what we have accomplished so far in this section.

(11.29) LOCALLY LIPSCHITZ COEFFICIENTS: (Revuz and Yor 1999, IX (2.10)) In the diffusion case – and, as we have just realized, it would suffice for us to consider only the case of time-invariant diffusion coefficients – the Lipschitz condition can be localized in the following sense: for every $R \in \mathbb{R}_{++}$, there is a constant $C_R \in \mathbb{R}_{++}$ such that

$$|a(x) - a(y)| + |b(x) - b(y)| \le C_R|x-y| \quad \text{for all } x,y \in \mathbb{R}^n \text{ with } |x|,|y| \le R\,.$$

If this condition is satisfied, then the following claim can be made: given any d-dimensional Brownian motion W, defined on any probability space $(\Omega,\mathcal{F},P)$ on which a d-dimensional Brownian motion can be defined, and given any $x \in \mathbb{R}^n$, one can construct a continuous and $\mathscr{F}^W$-adapted process X and an $\mathscr{F}^W$-stopping time ζ, so that on the set $[\![0,\zeta[\![$

$$X = x + a(X){\bullet}W + b(X){\bullet}\imath\,,$$

which identity we interpret in the obvious way as an identity between two $\mathbb{R}^n$-valued functions on a particular subset of $\Omega\times\mathbb{R}_+$. The stopping time ζ gives the time of explosion for X, that is,

$$\zeta(\omega) = \inf\{t \in \mathbb{R}_{++} : \lim_{s\nearrow t}|X_\omega(s)| = \infty\}$$

($\zeta = +\infty$ means that X does not explode in finite time). Moreover, the process X and the stopping time ζ are unique. Note that the local Lipschitz condition always holds if, for example, $x \rightsquigarrow a(x)$ and $x \rightsquigarrow b(x)$ are continuous and have finite and continuous partial derivatives.

If, in addition to the local Lipschitz property, a and b satisfy also the growth condition:

$$|a(x)|^2 + |b(x)|^2 \le C\,(1 + |x|^2) \ \text{ for all } \ x \in \mathbb{R}^n \ \text{ for some } \ C \in \mathbb{R}_{++}\,,$$

then one can show that $\mathsf{E}[|X_t|^2] < \infty$ for all $t \in \mathbb{R}_+$, so that $P(\zeta = +\infty) = 1$.

The reason why the case of locally Lipschitz coefficients is a more or less straightforward corollary from the case of globally Lipschitz coefficients is that for every $i \in \mathbb{N}_{++}$ one can construct (e.g., by using the mollification step outlined in (0.11)) globally Lipschitz coefficients a^i and b^i that coincide with a and b on the domain $\mathscr{D}_i = \{x \in \mathbb{R}^n : |x| < i\}$. The corresponding solutions X^i, $i = 1, 2, \ldots$, are then compatible in the sense that X^i and X^{i+1} are indistinguishable up until the entry time of X^i to the closed domain $\mathscr{D}_i^{\complement}$. If we denote this entry time by ζ_i, then $\zeta \overset{\text{def}}{=} \lim_i \zeta_i$ is the time of explosion. ○

(11.30) THE STRONG MARKOV PROPERTY REVISITED: In the case of diffusion coefficients that are bounded and globally Lipschitz (the constant C_R in (11.29) is universal and does not depend on R), the classical result from (Stroock and Varadhan 1979, 6.3.4) states that the associated martingale problem has a unique solution given by a continuous family of probability measures on the path space $\mathbb{W}$. As we are already aware, under these conditions the coordinate process on $\mathbb{W}$ is strong Markov relative to the distribution law of the solution. For a more recent treatment of this result we can point the reader to (Revuz and Yor 1999, IX (2.5)), among several other sources. ○

(11.31) REMARK (EXIT FROM A DOMAIN): In some concrete applications the diffusion coefficients may be defined only in some open domain $\mathscr{D} \subset \mathbb{R}^n$. Unless these coefficients are specifically designed to keep the diffusion inside $\mathscr{D}$, in general, the diffusion may exit from $\mathscr{D}$. This is yet another situation where a stochastic equation would be satisfied only until a particular stopping time, and, clearly, such a situation may arise even if the coefficients satisfy the Lipschitz property globally (that is, with some universal Lipschitz constants) on the entire domain $\mathscr{D}$. ○

There are several – by now classical – results that allow one to establish existence of strong solutions without the Lipschitz condition. These results provide valuable insight about the intrinsic nature of SDEs and are briefly reviewed next.

(11.32) ZVONKIN'S METHOD: Consider the diffusion case with a square diffusion matrix of dimensions (n, n). Zvonkin's assumptions about the coefficients

$$a\colon \mathbb{R}^n \times \mathbb{R}_+ \mapsto \mathbb{R}^{n \otimes n} \quad \text{and} \quad b\colon \mathbb{R}^n \times \mathbb{R}_+ \mapsto \mathbb{R}^n$$

are the following (Borel measurability for a and b is understood): (a) for some $\varepsilon > 0$, $|a(x,t)\,y| \ge \varepsilon |y|$ for all $x, y \in \mathbb{R}^n$ and all $t \in \mathbb{R}_+$; (b) for some constant

$C > 0$, $|a(x,t)| + |b(x,t)| \le C$ for all $t \in \mathbb{R}_+$ and all $x \in \mathbb{R}^n$; (c) one of the following two conditions holds:

(c_1) (global Lipschitz condition for a) there is a constant $C \in \mathbb{R}_{++}$ such that

$$|a(x,t) - a(y,t)| \le C\,|x-y| \quad \text{for all } x, y \in \mathbb{R}^n \text{ and all } t \in \mathbb{R}_+ ,$$

and (Dini condition for $x \rightsquigarrow a(x,t)$) the modulus of continuity

$$\varpi_t(h) \stackrel{\text{def}}{=} \sup_{|x-y|\le h} |a(x,t) - a(y,t)|$$

satisfies $\int_0^1 \frac{\varpi_t(h)}{h}\,\mathrm{d}h < \infty$;

(c_2) $n = 1$ and for some $C \in \mathbb{R}_{++}$ and some $p \in [½, \infty[$,

$$|a(t,x) - a(t,y)| \le C\,|x-y|^p \quad \text{for all } x, y \in \mathbb{R}^n \text{ and all } t \in \mathbb{R}_+ .$$

It is remarkable that the only requirement for the drift b is to be bounded and Borel-measurable. Nevertheless, by using a clever transformation of the state-space, Zvonkin (1974) was able to show that conditions (a), (b), and (c) imply the existence of a unique strong solution. ○

An immediate corollary from Zvonkin's method is that the one-dimensional diffusion SDE (here we have $n = d = 1$)

$$X = x + W + b(X, \imath)^{\bullet}\imath ,$$

always has a unique strong solution, as long as the function $b\colon \mathbb{R}\times\mathbb{R}_+ \mapsto \mathbb{R}$ is measurable and bounded. In particular, under these conditions the filtration $\mathscr{F}^X$ is always included in the Brownian filtration $\mathscr{F}^W$. But only a year after Zvonkin discovered his method, Tsirel'son published a particularly striking counterexample, showing that if the drift b is of non-diffusion type, that is, if one is to replace in the preceding equation $b(X, \imath)$ with $\beta(X, \imath)$, for some appropriately constructed predictable – and globally bounded, still – function $\beta\colon \mathbb{W}\times\mathbb{R}_+ \mapsto \mathbb{R}$, then it becomes possible that only a weak solution would exist. The reason why this is striking is that the existence of a weak solution without the existence of a strong one implies that the filtration generated by the solution X is strictly larger than the filtration of the Brownian motion that drives the equation. We have seen that this phenomenon can be caused by the lack of continuity in the diffusion coefficient, but the fact that an increase in the flow of information can be caused solely by the drift, which is to say, by actions that are fully predictable, in conjunction with the smoothest possible choice for the diffusion coefficient, is indeed striking.

(11.33) TSIREL'SON'S EXAMPLE: (Tsirel'son 1975) Choose a strictly decreasing sequence $(t_i)_{i\in\mathbb{N}}$ such that $1 = t_0 > t_1 > t_2 > \ldots \to 0$, and define the function (we still suppose $n = d = 1$) $\beta\colon \mathbb{W}\times\mathbb{R}_+ \mapsto \mathbb{R}$ so that

$$\beta(w,t) = \begin{cases} \dfrac{w_{t_i} - w_{t_{i+1}}}{t_i - t_{i+1}} - \left\lfloor \dfrac{w_{t_i} - w_{t_{i+1}}}{t_i - t_{i+1}} \right\rfloor & \text{if } t \in\,]t_i, t_{i-1}],\ i \in \mathbb{N}_{++} \\ 0 & \text{if } t = 0 \text{ or } t > 1 = t_0 . \end{cases}$$

Clearly, so defined, $\beta(w,t)$ is predictable with $0 \le \beta(w,t) < 1$, and the equation

† $$X = W + \beta(X, \imath)\bullet\imath$$

can be written as

$$X_t - X_{t_i} = W_t - W_{t_i} + \left(\frac{X_{t_i} - X_{t_{i+1}}}{t_i - t_{i+1}} - \left\lfloor \frac{X_{t_i} - X_{t_{i+1}}}{t_i - t_{i+1}} \right\rfloor \right)(t - t_i),$$
$$\text{for } t \in \,]t_i, t_{i-1}] \text{ for } i \in \mathbb{N}_{++}.$$

Given any $t \in \,]0,1]$, let

$$\eta_t = \frac{X_t - X_{t_i}}{t - t_i} \quad \text{and} \quad \varepsilon_t = \frac{W_t - W_{t_i}}{t - t_i} \quad \text{for all } t \in \,]t_i, t_{i-1}] \text{ for } i \in \mathbb{N}_{++}.$$

In terms of the notation just introduced, the SDE in † can be put in the form

$$\eta_t = \varepsilon_t + (\eta_{t_i} - \lfloor \eta_{t_i} \rfloor) \quad \text{for } t \in \,]t_i, t_{i-1}] \text{ for } i \in \mathbb{N}_{++}.$$

The key step in this example is to show that: (a) for any $t \in [0,1]$ the r.v. $(\eta_t - \lfloor \eta_t \rfloor)$ is independent of $\mathscr{F}_1^W$ and is uniformly distributed in $[0,1]$; and (b) for any $t \in \,]0,1]$ and any $s \in \,]0,t]$, $\mathscr{F}_t^X = \sigma\{\eta_s - \lfloor \eta_s \rfloor, W_u : 0 \le u \le t\}$ – we refer the reader to (Revuz and Yor 1999, IX (3.6)), or to (Tsirel'son 1975) for details. As a result, the filtration $\mathscr{F}^X$ is strictly larger than the filtration $\mathscr{F}^W$. ○

Another cornerstone in the theory of stochastic differential equations is the following:

(11.34) THE METHOD OF STROOCK AND VARADHAN: The method developed by Stroock and Varadhan (1979)[139] allows one to work with conditions that are weaker than Zvonkin's, in the sense that there is no need to impose a Lipschitz-type condition for the diffusion coefficient, but the price is the loss of strong solutions. Their conditions can be stated as:

(a) the diffusion matrix is time-invariant ($a(x,t) \equiv a(x)$) and is bounded, continuous, and strictly positive, in that $|a(x)y| > 0$ for any $x \in \mathbb{R}^n$ and any $0 \ne y \in \mathbb{R}^d$;

(b) the drift coefficient $b(x,t)$ is only required to be bounded and measurable.

Under these conditions Stroock and Varadhan were able to show that a weak solution (to be precise: a solution to their martingale problem) exists, and uniqueness in distribution law holds. With some mild additional conditions their original result is actually stated for a time-dependent diffusion matrix $a(x,t)$. But perhaps the most curious aspect of their work is that in the 1-dimensional case $n = d = 1$ continuity of $a(x,t)$ is no longer necessary, as long as $a(x,t)$ is uniformly positive on all finite domains in $\mathbb{R}\times\mathbb{R}_+$, that is, for every finite domain $\mathscr{D} \subset \mathbb{R}\times\mathbb{R}_+$ there is an $\varepsilon > 0$ such that $a(x,t) \ge \varepsilon$ for all $(x,t) \in \mathscr{D}$. ○

[139] Stroock (1987) provides a particularly compact and streamlined version.

(11.35) ONE-DIMENSIONAL DIFFUSIONS: It is clear from the above that one-dimensional diffusions are somewhat simpler, and existence and uniqueness can be established with steps that cannot be carried out in higher dimensions. In particular, the result in (Revuz and Yor 1999, IX (3.5)) states that if the diffusion coefficients $(x,t) \rightsquigarrow a(x,t)$ and $(x,t) \rightsquigarrow b(x,t)$ are scalar functions, then pathwise uniqueness for $\mathbb{e}(a,b)$ holds, provided that one of the following three scenarios is in force (notice that in cases 1 and 2 the diffusion coefficient a is required to be time-invariant).

Case 1: $a(x) \geq \varepsilon$ for some fixed $\varepsilon \in \mathbb{R}_{++}$, a and b are bounded, and there is a function $\rho\colon \mathbb{R}_{++} \mapsto \mathbb{R}_{++}$ such that $\int_{\mathbb{R}_{++}} \rho(u)^{-1}\,du = +\infty$ and $|a(x) - a(y)|^2 \leq \rho(|x-y|)$ for all $x, y \in \mathbb{R}$, $x \neq y$.

Case 2: $a(x) \geq \varepsilon > 0$, b is bounded, and there is an increasing and bounded function $f : \mathbb{R} \mapsto \mathbb{R}$ such that $|a(x) - a(y)|^2 \leq |f(x) - f(y)|$.

Case 3: For some function ρ as in case 1,

$$\left|a(x,t) - a(y,t)\right|^2 \leq \rho(|x-y|)$$

for every $t \in \mathbb{R}_+$, and, in addition, b is Lipschitz continuous.

If in case 3 the coefficient $b(x,t)$ is only locally Lipschitz continuous, then the solution can be claimed to exist only until a stopping time ζ, which is the time of explosion. If the coefficients are defined only in some open domain, then the solution may again be defined only until some stopping time, the time of exit from the domain. It is remarkable that the pathwise uniqueness is controlled by the quantity $|a(x) - a(y)|^2$ – not by the quantity $|a(x) - a(y)|$. ○

(11.36) COMPARISON THEOREM: In the case of scalar coefficients, as above, basic intuition would suggest that if $b^1(x) \geq b^2(x)$, and X^1 and X^2 satisfy (we stress that both equations are defined on the same probability space and use the same Brownian motion)

$$X^i = X^i_0 + a(X^i){\bullet}W + b^i(X^i){\bullet}\iota\,, \quad i = 1,2\,,$$

then with probability 1 the sample path X^1_ω must dominate the sample path X^2_ω (everywhere in $\mathbb{R}_+$) if the starting points are such that $X^1_0 \geq X^2_0$ with probability 1. Generally, this intuition is correct, but we must formulate a precise statement. Results of this nature are known as "comparison theorems," and one such result that we can point to is (ibid., IX (3.7)). It states:

Suppose that the diffusion coefficient a satisfies the conditions of case 2 above, and suppose that the drift coefficients b^1 and b^2 are bounded, at least one of them satisfies the Lipschitz condition, and $b^1(x) \geq b^2(x)$ for every $x \in \mathbb{R}$. Then

$$\mathsf{P}\big(X^1_0 \geq X^2_0\big) = 1 \quad \text{implies} \quad \mathsf{P}\big(X^1_t \geq X^2_t \text{ for all } t \in \mathbb{R}_{++}\big) = 1\,.$$

The same relation between X^1 and X^2 also holds in the case of time-dependent diffusion coefficient a, provided that, instead of the conditions from case 2, the coefficient a satisfies the condition $\left|a(x,t) - a(y,t)\right|^2 \leq \rho(|x-y|)$ for every $t \in \mathbb{R}_{++}$, where the function ρ is as in case 1. ○

11.4 Linear Stochastic Differential Equations

We turn now to study of the special class of diffusion equations with linear coefficients, or linear SDEs for short. The nature of such equations is easier to illustrate first in dimensions $n = d = 1$ and in terms of general continuous semimartingales, rather than the $\mathbb{e}(a,b)$ framework that we introduced earlier. Given a continuous process of finite variation, A, and a continuous semimartingale, Z, both real-valued and starting from 0, consider the equation

$$X = X_0 + X \bullet A + Z\,, \qquad \mathrm{e}_1$$

which can be cast in differential form as $\mathrm{d}X = X\,\mathrm{d}A + \mathrm{d}Z$. Its solution is understood to be a continuous semimartingale X for which the process on the right side of e_1 is indistinguishable from X. Since A and Z are given, the solution is understood to be strong.

(11.37) EXERCISE: Prove that e_1 has a unique strong solution given by

$$X = e^A\left(X_0 + e^{-A} \bullet Z\right),$$

or, equivalently, $X_t = e^{A_t}\left(X_0 + \int^t e^{-A_s}\,\mathrm{d}Z_s\right)$, $t \in \mathbb{R}_+$.

HINT: Apply$^{\multimap\,(10.24)}$ integration by parts to the product $e^{-A}X$, assuming that X is a semimartingale that satisfies e_1 with $Z_0 = A_0 = 0$. ○

Now consider the following more general version of e_1, in which the continuous finite variation process A is replaced by a generic continuous semimartingale Y that starts from 0:

$$X = X_0 + X \bullet Y + Z\,. \qquad \mathrm{e}_2$$

We are again looking for a continuous semimartingale X for which the right side of e_2 is indistinguishable from X, and consider the case $Z = 0$ first.

(11.38) EXERCISE (DOLÉAN–DADE EXPONENT): Prove that with $Z = 0$ equation e_2 has a unique strong solution given by $X = X_0 e^{Y - \frac{1}{2}\langle Y,Y\rangle}$. The positive semimartingale $\mathscr{E}(Y) = e^{Y - \frac{1}{2}\langle Y,Y\rangle}$, which is a local martingale if Y is a local martingale, is often called the *Doléan–Dade exponent* of Y – see (16.39).

HINT: Assuming that $X = X_0 + X \bullet Y$, apply$^{\multimap\,(10.24)}$ the integration by parts formula to the product $X\mathscr{E}(Y)^{-1}$. ○

The formula that we are about to establish in the next exercise provides a universal recipe for constructing strong solutions to linear SDEs in the scalar case.

(11.39) EXERCISE: Prove that e_2 has a unique strong solution given by

† $$X = \mathscr{E}(Y)\left(X_0 + \mathscr{E}(Y)^{-1} \bullet Z - \mathscr{E}(Y)^{-1} \bullet \langle Y, Z\rangle\right).$$

HINT: Assuming that X is a semimartingale for which e_2 holds, apply the integration by parts formula to the product $X\mathscr{E}(Y)^{-1} = X\, e^{-Y+\frac{1}{2}\langle Y,Y\rangle}$. ○

(11.40) EXPONENTIAL (GEOMETRIC) BROWNIAN MOTION WITH DRIFT: As a first application of (11.39†), consider the familiar SDE

$$X = X_0 + X\bullet(\sigma W + b\iota) \equiv X_0 + \sigma X\bullet W + bX\bullet\iota\,,$$

in which $\sigma, b \in \mathbb{R}$ are given parameters, W is the Brownian motion, and ι is the deterministic semimartingale $\iota_t = t$, $t \in \mathbb{R}_+$. This is a special case of e_2 with $Y = \sigma W + b\iota$ and with $Z = 0$, and, by the formula in (11.39†),

$$X = X_0\mathscr{E}(Y) = X_0\, e^{\sigma W + b\iota - \frac{\sigma^2}{2}\langle W,W\rangle} = X_0\, e^{\sigma W + (b-\frac{\sigma^2}{2})\iota}\,. \quad ○$$

(11.41) ORNSTEIN-UHLENBECK (OU) PROCESS: For reasons that will become clear below, this process is also called *mean-reverting Brownian motion*. The scalar OU process with parameters $m \in \mathbb{R}$, $\sigma, b \in \mathbb{R}_{++}$, driven by the one-dimensional Brownian motion W, is the unique strong solution, X, to

$$X = X_0 + \sigma W + b(m - X)\bullet\iota\,.$$

The d-dimensional OU process, X, can be described by the same exact equation except that W is the d-dimensional Brownian motion, $m \in \mathbb{R}^d$ is a vector, and $\sigma, b \in \mathbb{R}_{++}$ are scalar parameters, as in the one-dimensional case. ○

(11.42) EXERCISE: Prove that the OU process (of any dimension) can be expressed as the following explicit function of the Brownian motion W:

$$X = e^{-b\iota}\left(X_0 + e^{b\iota}\bullet(\sigma W + bm\iota)\right).$$

Then suppose that X_0 is some Gaussian r.v. that is independent of W, and prove that with this choice the solution X is a Gaussian process. Calculate $\mathsf{E}[X_t]$ and $\mathrm{Var}(X_t)$. Finally, choose the parameters of the Gaussian r.v. X_0 (expectation and variance) in such a way that $t \rightsquigarrow \mathsf{E}[X_t]$ and $t \rightsquigarrow \mathrm{Var}(X_t)$ are constant functions, in which case X becomes a stationary Gaussian process. Explain the mean-reversion aspect of the drift coefficient $b(m - X)$. ○

(11.43) EXERCISE: Given some fixed $T \in \mathbb{R}_{++}$ and any two (deterministic) continuous functions $f, g\colon [0,T[\, \mapsto \mathbb{R}$, consider the SDE

$$X = X_0 + W + (fX + g)\bullet\iota\,. \qquad (\dagger)$$

Prove that this SDE has a unique strong solution given by

$$X = e^{f\bullet\iota}\Big(X_0 + (e^{-f\bullet\iota})\bullet W + \big(ge^{-f\bullet\iota}\big)\bullet\iota\Big) \quad \text{on } [0,T[\,,$$

or, if one insists on full integral form,

$$X_t = e^{\int_0^t f(s)\mathrm{d}s}\left(X_0 + \int_0^t e^{-\int_0^s f(u)\mathrm{d}s}\mathrm{d}W_s + \int_0^t e^{-\int_0^s f(u)\mathrm{d}u}g(s)\mathrm{d}s\right),\quad t \in [0,T[\,. \quad ○$$

(11.44) EXERCISE (PINNED BROWNIAN MOTION): As a special case of (11.43†), for some fixed scalars $x, \xi \in \mathbb{R}$, consider the equation

$$\dagger \quad X = x + W + \frac{\xi - X}{T - \iota} \bullet \iota = x + W + \Big(\frac{\xi}{T-\iota} - \frac{1}{T-\iota} X\Big) \bullet \iota \quad \text{on } [0, T[\,.$$

Prove that the unique strong solution to this equation is given by

$$\ddagger \quad X = (T - \iota)\Big(\frac{1}{T-\iota} \bullet W\Big) + \frac{T-\iota}{T} x + \frac{\iota}{T}\xi \quad \text{on } [0, T[\,. \quad \circ$$

The process in (11.44‡) is obviously Gaussian. It is called *pinned Brownian motion* with pinning points $X_0 = x$ at $t = 0$ and $X_T = \xi$ at $t = T$ – the reason for the name "pinned" becomes clear below. Notice that (11.44†) is only meaningful in the open interval $[0, T[$ and, as it stands, the right side of (11.44‡) is not meaningful with $t = T$. Pinned Brownian motion with pinning points $X_0 = X_T = 0$ is called the *Brownian bridge*. Such a process can be expressed as $(T - \iota)\big(\frac{1}{T-\iota} \bullet W\big)$ on $[0, T[$, or, if one insists on a more detailed form,

$$X_t = (T - t) \int_0^t \frac{1}{T-s} \mathrm{d}W_s\,, \quad 0 \le t < T\,.$$

(11.45) EXERCISE (BROWNIAN BRIDGE): Prove that $\big(\int_0^t \frac{1}{T-s} \mathrm{d}W_s\big)_{t \in [0,T[}$ is a well-defined process that has the same law as the process $\big(W_{\frac{t}{T(T-t)}}\big)_{t \in [0,T[}$. In addition, prove that $\big(W_{\frac{1}{t}}\big)_{t \in \mathbb{R}_+}$ and $\big(\frac{1}{t} W_t\big)_{t \in \mathbb{R}_+}$ share the same distribution law, and conclude that $\big(W_{\frac{t}{T(T-t)}}\big)_{t \in [0,T[}$, has the same distribution as $\big(\frac{t}{T(T-t)} W_{\frac{T(T-t)}{t}}\big)_{t \in [0,T[}$. Finally, show that the Brownian bridge process X has the same distribution as $\big(\frac{t}{T} W_{\frac{T(T-t)}{t}}\big)$ and that, as a result, $\lim_{t \nearrow T} X_t = 0$ with probability 1, which justifies the name "bridge."

HINT: To show that $\frac{1}{T-\iota} \bullet W$ and $\big(W_{\frac{t}{T(T-t)}}\big)_{t \in [0,T[}$ have the same distribution, compare the two covariance functions and use the fact that the distribution of any Gaussian process with vanishing mean is completely determined by its covariance function. $\circ$

As the last exercise demonstrates, although the solution to equation (11.44†) is meaningful only on $[0, T[$, with probability 1 its sample paths have a limit as $t \nearrow T$, and this limit is the (nonrandom) scalar value $\xi \in \mathbb{R}$ (whence the name "pinned"). As a result, the solution X admits a unique continuous extension to the entire interval $[0, T]$, and this extension is such that $X_T = \xi$. It is remarkable that the resulting process has a predetermined and nonrandom value at time T, that is, the dynamics of equation (11.44†) steer the solution X toward the fixed scalar value ξ. The heuristic explanation is this: if X is above the value ξ, the drift is pushing the process in the negative direction, and if X is below the value ξ, the drift is pushing the process in the positive direction. Moreover, the push becomes overwhelming as the time parameter t approaches T.

We conclude our discussion of the Brownian bridge process with the following useful feature: a Brownian motion process on the time interval $[0, T]$ can be interpreted as a pinned Brownian motion with randomized pinning point. The precise statement is the following:

(11.46) EXERCISE: Assuming that $T \in \mathbb{R}_{++}$ is given and the Gaussian r.v. $\xi \in \mathcal{N}(0, T)$ is independent of the Brownian motion W, prove that the process

$$B_t = (T - t) \int_0^t \frac{1}{T - s} \mathrm{d}W_s + \frac{t}{T} \xi, \quad t \in [0, T],$$

is a standard Brownian motion on $[0, T]$ (the integral on the right is understood to be 0 if $t = T$). ○

(11.47) REMARK: We see from the last exercise, in conjunction with (11.44), that the dynamics of the Brownian motion $(B_t)_{t \in [0,T]}$, relative to a particular filtration that is *aware* of the exact future position of the Brownian motion at time T (somehow, the r.v. B_T is being realized at time $t = 0$), can be expressed as

$$B_t = W_t + \int_0^t \frac{B_T - B_s}{T - s} \mathrm{d}s, \quad t \in [0, T],$$

where W is a Brownian motion that is independent of B_T. Clearly, with respect to such a filtration, the one generated by W and B_T, the Brownian motion B is a semimartingale (and is a Brownian motion, still), but is not a local martingale. We see from this example how the attribute "local martingale" depends on the filtration – and in a crucial way. ○

So far we have only examined linear SDEs in dimension 1. We conclude this section with the multivariate analogue of (11.43†).

(11.48) MULTIVARIATE LINEAR SDE: Given: (a) a d-dimensional Brownian motion W, (b) a continuous and adapted matrix-valued stochastic process σ with values in $\mathbb{R}^{n \otimes d}$, (c) a continuous and adapted stochastic process γ with values in $\mathbb{R}^n$, and (d) some constant deterministic matrix $A \in \mathbb{R}^{n \otimes n}$, consider the following SDE in $\mathbb{R}^n$:

$$X = X_0 + \sigma \bullet W + (AX + \gamma) \bullet \imath .$$

The solution to this equation can be developed in essentially the same way in which we developed the solution to (11.43†), namely,

$$X = e^{\imath A} \left(X_0 + e^{-\imath A} \sigma \bullet W + e^{-\imath A} \gamma \bullet \imath \right),$$

where the matrix process $e^{\pm \imath A}$ is understood as

$$e^{\pm tA} = I + \sum\nolimits_{n \in \mathbb{N}_{++}} \frac{1}{n!} (\pm tA)^n, \quad t \in \mathbb{R}_+ .$$

It is not difficult to see that the formula for the d-dimensional OU process established in exercise (11.42) is a special case of the above. ○

11.5 Some Common Diffusion Models Used in Asset Pricing

Our objective in this section is to formulate – and solve – several concrete stochastic differential equations that are found in some of the most common continuous-time asset pricing models. We have encountered already three such equations: the geometric Brownian motion, the Ornstein–Uhlenbeck process (mean-reverting Brownian motion), and the pinned Brownian motion.

(11.49) EXERCISE: Consider the one-dimensional OU process

$$\dagger \qquad U = U_0 + \frac{\sigma}{2} W - \frac{k}{2} U \bullet \iota$$

for some choice of the parameters $k, \sigma \in \mathbb{R}_{++}$ and the scalar Brownian motion W. Using Itô's formula, verify the relation

$$U^2 = U_0^2 + \sigma U \bullet W + \left(\frac{\sigma^2}{4} - kU^2 \right) \bullet \iota \, .$$

Now suppose that U is the d-dimensional OU process, still given by $\dagger$ except that W is the d-dimensional Brownian motion. Show that with this choice the previous relation generalizes to ($|\cdot|$ is the Euclidean norm in $\mathbb{R}^d$)

$$\ddagger \qquad |U|^2 = |U_0|^2 + \sigma U^\intercal \bullet W + \left(d \frac{\sigma^2}{4} - k|U|^2 \right) \bullet \iota \, . \quad \circ$$

(11.50) COX–INGERSOL–ROSS (CIR) PROCESS: Introduced by Cox et al. (1985), this process is often used in models of short-term interest rates. The CIR process with parameters $d \in \mathbb{R}$ and $k, \sigma \in \mathbb{R}_+$ is the solution, X, to

$$\dagger \qquad X = X_0 + \sigma \sqrt{|X|} \bullet W + \left(d \frac{\sigma^2}{4} - kX \right) \bullet \iota \, ,$$

where W is the one-dimensional Brownian motion. The intuition behind this diffusion structure is clear: the diffusion coefficient is designed to have its effect vanish if the process is in 0, and the drift coefficient is designed to push the diffusion up if it falls below the level $d\sigma^2/4k$ and down if it is above that level. Thus, if X is in 0, its dynamics become deterministic and governed by the drift alone. Furthermore, in state 0 the drift is pushing in the positive direction if $d > 0$ and in the negative direction if $d < 0$. Most important, if $d \geq 2$, the push in the positive direction is strong enough to overwhelm the diffusion term near 0, and to never allow the process to actually arrive in 0 – see (Feller 1951) and (11.52) below. As a result, if $d \geq 2$ and the CIR process starts in $\mathbb{R}_{++}$, then it will remain in $\mathbb{R}_{++}$. If $0 < d < 2$ and the CIR process starts in $\mathbb{R}_{++}$, then it will eventually arrive in 0, and once in 0 it will be reflected back into $\mathbb{R}_+$. Finally, if $d = 0$, then 0 becomes an absorbing state, in that once in 0 the CIR process remains in 0 forever (since $X = 0$ is a solution in this case). In any case, if $d \geq 0$ and the starting point is in $\mathbb{R}_{++}$, then the absolute value under the square root in $\dagger$ becomes redundant. $\circ$

(11.51) EXERCISE: Give the reasons for equation (11.50†) to have a unique solution in the pathwise sense. What can you say about the existence and uniqueness of a strong solution, and about the uniqueness in distribution law? ○

(11.52) EXERCISE: Suppose that the parameter d in (11.50†) is some integer $d \in \mathbb{N}_{++}$ and that X is a solution to this equation starting from $X_0 > 0$. Prove that the distribution law of X is the same as the distribution law of the process $|U|^2$ for any$^{\multimap (11.49)}$ d-dimensional OU process U with parameters $k, \sigma \in \mathbb{R}_+$ and initial value chosen so that $|U_0| = \sqrt{X_0}$. Because of this feature, the CIR process is also called the *squared radial OU process with dimension d*, and this name is used even if the parameter $d \in \mathbb{R}_{++}$ is not an integer. We now see – at least heuristically – why the CIR process can never reach 0 if $d \geq 2$, but it can if $d = 1$ (and also if $d < 2$): a one-dimensional OU process reaches 0 with probability 1, but for a two-dimensional OU process reaching 0 means that two independent OU processes must reach 0 simultaneously, and this happens with probability 0 (though intuitive, this claim still requires a rigorous justification).

HINT: Given a d-dimensional OU process U, what can one say about the process $|U|^{-1}U^\intercal \cdot W$? ○

(11.53) BESSEL PROCESS:[140] Clearly, the CIR equation (11.50†) is meaningful with $k = 0$, and so are also equations (11.49†) and (11.49‡). The CIR process with parameters $k = 0$, $\sigma = 2$, and $d \in \mathbb{R}_{++}$, and with initial value $x \in \mathbb{R}_{++}$, that is, the unique strong solution, X, to

$$X = x + d\iota + 2\sqrt{X} \cdot W,$$

where W is the one-dimensional Brownian motion, is called the *squared Bessel process with dimension* $d > 0$ – or, equivalently, the *squared Bessel process with parameter* $\nu = \frac{d}{2} - 1$ – starting from x. This class of processes is denoted by $\mathrm{BESQ}^d(x)$ or $\mathrm{BESQ}^{(\nu)}(x)$. Since with $d \in \mathbb{N}_{++}$, $k = 0$, and $\sigma = 2$ equation (11.49†) reads $U = U_0 + W$, the distribution law of $X \in \mathrm{BESQ}^d(x)$ is the same as the law of $|U_0 + W|^2$ for W = Brownian motion with dimension d and $U_0 \in \mathbb{R}^d$ chosen so that $|U_0|^2 = x$. In particular,$^{\multimap (2.28i)}$ if $X \in \mathrm{BESQ}^d(x)$ and $d \in \mathbb{N}_{++}$, then the law of X_t can be identified as the law of tY_t for $Y_t \in \chi^2(d, x/t)$. It is well known – see (Jeanblanc et al. 2009, (6.2.2)) – that this property holds for every $d \in \mathbb{R}_{++}$ and not just for $d \in \mathbb{N}_{++}$.

If $X \in \mathrm{BESQ}^d(x^2)$ for some $d \in \mathbb{R}_{++}$, the process $Y = \sqrt{X}$ is called the *Bessel process with dimension d starting from x*. The class of such processes is denoted by $\mathrm{BES}^d(x)$. ○

(11.54) EXERCISE: Suppose that $X \in \mathrm{BESQ}^d(x)$. Using the features of the chi-squared distribution law, verify that if $d \in \mathbb{N}_{++}$, then the law of X_t is the same as

[140] Named after the German astronomer and mathematician Friedrich Wilhelm Bessel (1784–1846), due to the intimate connection with Bessel functions, which comes from the connection with$^{\multimap (2.28i)}$ the chi-squared distribution.

the law of tY_t for $Y_t \in \chi^2(d, x/t)$. Assuming that this property remains valid for any $d \in \mathbb{R}_{++}$ (see above), verify that the density of X_t in $\mathbb{R}_{++}$ is given by

$$\frac{1}{2t} e^{-\frac{x+y}{2t}} \left(\frac{y}{x}\right)^{\frac{d}{4}-\frac{1}{2}} I_{\frac{d}{2}-1}\left(\sqrt{xy}/t\right), \quad y \in \mathbb{R}_{++},$$

where I_ν is the modified Bessel function with parameter ν. ○

(11.55) EXERCISE: Using Itô's formula and Tanaka's formula, prove that every process $X \in \mathrm{BES}^d(x)$ satisfies the equation

$$X = x + W + \frac{d-1}{2}\frac{1}{X}\bullet\iota \quad \text{for } d > 1$$

and the equation

$$X = x + W + L^0(X) \quad \text{for } d = 1,$$

in which $L^0(X)$ stands for the local time of X in 0. ○

(11.56) REMARK: A Bessel process $X \in \mathrm{BES}^d(x)$ is meaningful for every $d \in \mathbb{R}_{++}$, but with $d \in]0,1[$ the integral $X\bullet\iota$ is not finite and (Jeanblanc et al. 2009, 6.1.2.2)

$$X = x + W + \frac{d-1}{2}\int_0^\infty u^{d-2}\left(L^u(X) - L^0(X)\right)\mathrm{d}u. \quad ○$$

Just as in the case of $\mathrm{BESQ}^d(x)$, there is a connection between the chi-squared distribution and the law of the CIR process at fixed time $t \in \mathbb{R}_{++}$. This connection is established in the next exercise in the case of $d \in \mathbb{N}_{++}$, but we note that it remains valid for any $d \in \mathbb{R}_{++}$ – see (ibid., (6.3.2.2)).

(11.57) EXERCISE: Let X be the CIR process with parameters $d \in \mathbb{N}_{++}$ and $k, \sigma \in \mathbb{R}_+$, starting from $X_0 = x \in \mathbb{R}_{++}$, and let $c(t) \stackrel{\text{def}}{=} \frac{\sigma^2}{4k}(e^{kt} - 1)$. Prove that if $t \in \mathbb{R}_{++}$ is fixed, then $X_t \stackrel{\text{law}}{=} e^{-kt}c(t)Y_t$ for $Y_t \in \chi^2(d, x/c(t))$. ○

Notice that the result in (11.54) obtains from the result in (11.57), since we have $\lim_{k\searrow 0} c(t) = \frac{\sigma^2}{4}t$, which is precisely t with $\sigma = 2$.

(11.58) CONSTANT ELASTICITY OF VARIANCE (CEV) PROCESS: This process is defined as the solution to

$$\dagger \qquad X = X_0 + \sigma X^{\beta+1}\bullet W + \alpha X\bullet\iota,$$

for some choice of the parameters $\sigma, \beta, \alpha \in \mathbb{R}$, Brownian motion W, and starting position $X_0 > 0$. With $\beta = 0$ the solution reduces to geometric Brownian motion and with $\beta = -\frac{1}{2}$ and $\alpha < 0$ it reduces to a CIR process; specifically, $\dagger$ turns into a special case of ($11.50^\dagger$) with $d = 0$ and $k = \alpha$. We will suppose that $\beta \neq 0$. The reason the CEV process is useful in financial modeling is that the variance of the instantaneous net change in the level, $\mathrm{d}X_t/X_t$, having the form $\sigma^2 X_t^{2\beta}\,\mathrm{d}t$, can be made into a decreasing or increasing function of the level X_t, depending on whether $\beta < 0$ or $\beta > 0$. ○

The nature of the CEV process is easier to transcribe by transforming it first into a CIR process, as the next exercise details.

(11.59) EXERCISE: Prove that if X is given by (11.58†) and $Y \overset{\text{def}}{=} \frac{1}{4\beta^2} X^{-2\beta}$, then

$$Y = Y_0 - \big(\text{sign}(\beta)\sigma\sqrt{Y}\big) \cdot W + \Big(\frac{\sigma^2}{4}\frac{1+2\beta}{\beta} - 2\alpha\beta\, Y\Big) \cdot \iota\,,$$

where $\sqrt{Y_t}$ is understood as the positive root of the (positive) r.v. Y_t. ○

We see from the last exercise that the CEV process with parameters $\sigma, \beta, \alpha \in \mathbb{R}$ can be transformed into a CIR process with parameters $d = \frac{1+2\beta}{\beta}$ and $k = 2\alpha\beta$. In particular, the relation $k > 0$ comes down to $\alpha\beta > 0$, and $d > 0$ becomes $\beta \in \,]-\infty, -\frac{1}{2}[\, \cup \,]0, +\infty[$. Thus, with $\alpha < 0$ and with $\beta < -\frac{1}{2}$, the process Y, and therefore X, will remain in $\mathbb{R}_+$ forever (it may reach 0, but once in 0 it will get reflected back into $\mathbb{R}_{++}$). With $\beta = -\frac{1}{2}$ the process Y, and therefore X, can reach 0 and once in 0 will remain in 0 (this is the case of a CIR process with $d = 0$).

(11.60) EXPONENTIAL OU PROCESS: This process was introduced by Schwartz (1997) as a model of commodity prices. It is the solution, X, to

$$\dagger \qquad X = X_0 + \sigma X \cdot W + k\big(\alpha - \log(X)\big) X \cdot \iota$$

for some choice of the parameters $\alpha \in \mathbb{R}$, $k, \sigma \in \mathbb{R}_{++}$, one-dimensional Brownian motion W, and initial value $X_0 > 0$. Such a solution cannot reach 0 since the drift coefficient is pushing it away from 0 sufficiently strongly (in fact, with unbounded force near 0), while the diffusion term vanishes in 0. This matter is clarified in the next exercise, which also explains the name "exponential OU process." ○

(11.61) EXERCISE: Prove that if X is given by (11.60†), then $Y \overset{\text{def}}{=} \log(X)$ is the solution to

$$Y = \log(X_0) + \sigma W + k\big(\alpha - \frac{1}{2k}\sigma^2 - Y\big) \cdot \iota\,.$$

In particular, Y is an OU process and $X = e^Y$. ○

(11.62) THE DIXIT AND PINDYCK PROCESS: As an alternative model of commodity prices, Dixit and Pindyck (1994) proposed the process

$$X = X_0 + \sigma X \cdot W + k(\alpha - X) X \cdot \iota$$

for some choice of the parameters $\sigma, k, \alpha \in \mathbb{R}_{++}$, scalar Brownian motion W, and initial value $X_0 > 0$. Despite the similarity to the exponential OU process, developing the solution requires more steps and is more involved. We again start with the substitution $Y = \log(X)$ and, after a straightforward application of Itô's formula, get

$$\dagger \qquad Y = \log(X_0) + \sigma W - kX \cdot \iota + \big(k\alpha - \frac{1}{2}\sigma^2\big)\iota\,.$$

With $Z \overset{\text{def}}{=} \frac{1}{X}$ we have $Y = -\log(Z)$, and, after a second application of Itô's formula, in conjunction with the substitution $\psi \overset{\text{def}}{=} k\iota - Z$, we arrive at the relation

$$
\ddagger \qquad \begin{aligned} Y &= \log(X_0) + \frac{1}{Z}{\bullet}(\psi - k\imath) + \frac{1}{2Z^2}{\bullet}\langle Z, Z\rangle \\ &= \log(X_0) + X{\bullet}\psi - kX{\bullet}\imath + \frac{\sigma^2}{2}\imath\,. \end{aligned}
$$

Since X is determined by Y, Y is determined by Z, and Z is determined by ψ, the unknown process for us is ψ, and we will seek a solution of the form

$$
\psi = -\frac{1}{X_0} + \sigma Z{\bullet}(W + h{\bullet}\imath) \equiv -\frac{1}{X_0} + \frac{\sigma}{X}{\bullet}(W + h{\bullet}\imath)
$$

for some yet to be determined process h. If we substitute this form of ψ in ‡ and equate with the expansion in †, we find that

$$
\sigma h{\bullet}\imath = (k\alpha - \sigma^2)\imath\,,
$$

which is to say, the process h is nothing but the constant $h = (k\alpha - \sigma^2)/\sigma$. Knowing h, we can now compute ψ (and, consequently, Z, then Y, and finally X) from the linear equation (it comes from $\frac{1}{X} = Z = k\imath - \psi$)

$$
\psi = -\frac{1}{X_0} - \sigma\psi{\bullet}(W + h\imath) + \sigma k\imath{\bullet}(W + h\imath)\,,
$$

the solution to which$^{\multimap(11.39)}$ is familiar to us and is given by

$$
\psi = \mathscr{E}\times\Big(-\frac{1}{X_0} + \frac{\sigma k\imath}{\mathscr{E}}{\bullet}\big(W + (h+\sigma)\imath\big)\Big),
$$

where $\mathscr{E} \stackrel{\text{def}}{=} e^{-\sigma W - \sigma h\imath - \frac{\sigma^2}{2}\imath}$ is the solution to $\mathscr{E} = 1 - \sigma\mathscr{E}{\bullet}(W + h{\bullet}\imath)$. ○

12

The Connection between SDEs and PDEs

Among other things, Itô's formula is telling us⊸ (11.6) that if a diffusion process is passed through a sufficiently smooth function, the resulting process is a semimartingale. Moreover, the finite variation predictable component of such a semimartingale can be expressed in terms of the action on the function of a certain second-order differential operator that is associated with the diffusion process. In some sense, the prescription for a particular diffusion is essentially a prescription for the way in which the finite variation predictable component in a generic functional transformation of that diffusion must look, or, to put it another way, it is a prescription for a particular second-order differential operator. Thus, claiming that a concrete functional transformation of a diffusion process represents a local martingale essentially amounts to claiming that if applied to the function that provides the transformation, the associated second-order differential operator yields 0, that is to say, the function satisfies a particular second-order PDE. The importance of this connection – for both theory and practice – would be difficult to overemphasize, and the topic "functionals and transformations of diffusion processes" runs deep – see (Revuz and Yor 1999, VIII sec. 3) and (Stroock and Varadhan 1979). What follows in this chapter is a very brief and mostly ad hoc review of this topic, based almost entirely on heuristic reasoning.

12.1 Feynman–Kac Formula

The Feynman–Kac formula[141] – or a series of formulas, rather – is an encryption of a special intrinsic connection between a particular type of partial differential equations (PDEs) and certain stochastic differential equations (SDEs). The first steps toward revealing this connection were already taken in (10.25) and (11.5). To see this, consider some measurable (and possibly time-dependent) fields of matrices $a(x,t) \in \mathbb{R}^{n\otimes d}$ and vectors $b(x,t) \in \mathbb{R}^n$, $x \in \mathbb{R}^n$, $t \in \mathbb{R}_+$, and write the associated diffusion equation (assume for now that a solution exists)

$$X = X_0 + a(X, \imath)\bullet W + b(X, \imath)\bullet \imath \, ,$$

[141] Named after the American physicist Richard Phillips Feynman (1918–1988) and the American mathematician Mark Kac [pronounced KAHTS – not KAK] (1914–1984). In 1965 Richard Feynman received the Nobel Prize in Physics.

and also$^{\multimap(11.4)}$ the associated infinitesimal generator

$$\mathcal{A}_t \equiv \mathcal{A}_t^{aa^\intercal,b} = \frac{1}{2}\left(a(x,t)\,a(x,t)^\intercal \nabla\right)\cdot\nabla + b(x,t)\cdot\nabla\,.$$

We have seen in exercise (11.5) that for any $f \in \mathscr{C}^{2,1}(\mathbb{R}^n\times\mathbb{R}_+;\mathbb{R})$, the process[142]

$$M^f \stackrel{\text{def}}{=} f(X,\iota) - \left(\partial f(X,\iota) + \mathcal{A} f(X,\iota)\right)\bullet\iota$$

is a local martingale. Given some $V \in \mathscr{C}_b(\mathbb{R}^n\times\mathbb{R},\mathbb{R})$, or $V \in \mathscr{C}(\mathbb{R}^n\times\mathbb{R},\mathbb{R}_+)$, consider the associated continuous process of finite variation $R = e^{-V(X,\iota)\bullet\iota}$, and recall$^{\multimap(10.25)}$ that the process $RM^f - M^f\bullet R$ is a local martingale, in fact, a stochastic integral of the form

$$RM^f - M^f\bullet R = \left[e^{-V(X,\iota)\bullet\iota}\,\nabla f(X,\iota)^\intercal a(X,\iota)\right]\bullet W\,.$$

We know from (9.91) that one particular condition that would guarantee that this local martingale is a martingale – say, on the interval $[0,T]$ – is the following:

$$\mathsf{E}\left[\int_0^T e^{-2V(X,\iota)\bullet\iota_s}\left|\nabla f(X_s,s)^\intercal a(X_s,s)\right|^2 \mathrm{d}s\right] < \infty\,. \qquad \mathrm{e}_1$$

As a special case, this condition is satisfied if the functions ∇f and a are required to be bounded and V is required to be positive.

(12.1) EXERCISE: Prove that the process $RM^f - M^f\bullet R$ can be expressed as

$$\begin{aligned} &RM^f - M^f\bullet R \\ &\quad = Rf(X,\iota) - \left[R\Big(\partial f(X,\iota) + \mathcal{A} f(X,\iota) - V(X,\iota) f(X,\iota)\Big)\right]\bullet\iota\,. \end{aligned}$$

HINT: The proof is merely a matter of applying$^{\multimap(10.24)}$ the integration by parts rule to the product

$$e^{-V(X,\iota)\bullet\iota}\times\left[\left(\partial f(X,\iota) + \mathcal{A} f(X,\iota)\right)\bullet\iota\right]. \quad \circ$$

Our final step toward the derivation of the Feynman–Kac formula now suggests itself. Suppose that, assuming such a choice is possible, the function $f \in \mathscr{C}^{2,1}(\mathbb{R}^n\times\mathbb{R}_+;\mathbb{R})$ solves the parabolic partial differential equation (PDE)

$$\partial f(x,t) + \mathcal{A}_t f(x,t) = V(x,t) f(x,t) \qquad \mathrm{e}_2$$

in the domain $(x,t) \in \mathbb{R}^n\times]-\infty,T[$ with terminal data $\lim_{t\nearrow T} f(x,t) = G(x)$ for some $G \in \mathscr{C}_b(\mathbb{R}^n;\mathbb{R})$. With this concrete choice for f, given any $t < T$, the process $(RM^f - M^f\bullet R)/R_t$ treated on the interval $[t,T]$ reduces to the process

[142] By convention, the symbol ∇, and therefore $\mathcal{A} = \mathcal{A}^{aa^\intercal,b}$, is understood as a differential operator acting on the first n-arguments under the function that follows this symbol, while the symbol ∂ (without a subscript) is understood as derivative with respect to the last argument under the function that follows it (usually, the time variable).

$Rf(X,\iota)/R_t$, that is, reduces to $\frac{R_u}{R_t} f(X_u,u)$, $u \in [t,T]$, which (see the previous exercise) is a local martingale that starts from $f(X_t,t)$ at time $t < T$ and has a final value at time T equal to $\frac{R_T}{R_t} G(X_T)$. Furthermore, by $\mathbb{e}_1$ the local martingale $Rf(X,\iota)/R_t$ is a martingale provided that

$$\mathsf{E}\Big[\int_t^T e^{-2\int_t^s V(X_u,u)\mathrm{d}u} \big|\nabla f(X_s,s)^{\intercal} a(X_s,s)\big|^2 \,\mathrm{d}s\Big] < \infty . \qquad \mathbb{e}_3$$

(12.2) FEYNMAN–KAC FORMULA I: Suppose that $f \in \mathscr{C}^{2,1}(\mathbb{R}^n \times \mathbb{R}_+;\mathbb{R})$ is one particular solution to the parabolic PDE in $\mathbb{e}_2$ and the measure $\mathcal{Q}_{x,t}$, defined on $\mathbb{W} = \mathscr{C}(\mathbb{R}_+;\mathbb{R}^n)$, is one particular solution to the martingale problem $\mathbb{M}_{x,t}(aa^{\intercal},b)$ for some fixed $x \in \mathbb{R}^n$ and fixed $t < T$. Suppose further that the coordinate process on $\mathbb{W}$ is Markov with respect to $\mathcal{Q}_{x,t}$, and the local martingale

$$e^{-\int_0^u V(w(s),t+s)\mathrm{d}s} f\big(w(u),t+u\big), \quad u \in [0,T-t],$$

is actually a martingale with respect to $\mathcal{Q}_{x,t}$. Then

$$\begin{aligned} f(x,t) &= \mathsf{E}_{x,t}\Big[e^{-\int_0^{T-t} V(w(s),t+s)\mathrm{d}s} G\big(w(T-t)\big)\Big] \\ &\equiv \mathsf{E}_{x,t}\Big[e^{-\int_t^T V(\theta_t w(s),s)\mathrm{d}s} G\big(\theta_t w(T)\big)\Big], \end{aligned}$$

with $\mathsf{E}_{x,t}$ understood to be the expectation relative to $\mathcal{Q}_{x,t}$. ○

(12.3) REMARKS: (a) All assumptions in (12.2) would be met if, for example, the function V is positive, ∇f is bounded, and the coefficients a and b are Lipschitz-continuous and bounded – see (11.30). If these conditions hold, then $\mathbb{e}(a,b)$ admits a unique strong solution, and this solution is Markov – strongly Markov, in fact – for the filtration associated with $\mathbb{e}(a,b)$. For any such solution, X, the Feynman–Kac formula can be expressed as

$$f(X_t,t) = \mathsf{E}\Big[e^{-\int_t^T V(X_s,s)\mathrm{d}s} G(X_T) \mid \mathscr{F}_t\Big], \quad t < T .$$

Alternatively, given any $x \in \mathbb{R}^n$ and any $t < T$, one can construct (on any probability space on which a Brownian motion can be constructed) a process $(X_u^{x,t})_{u \in [t,\infty[}$ such that

$$X_u^{x,t} = x + \int_t^u a(X_s^{x,t},s)\mathrm{d}W_s + \int_t^u b(X_s^{x,t},s)\mathrm{d}a , \quad u \in [t,\infty[.$$

With such a process at hand, one can write the solution to the parabolic PDE as

$$f(x,t) = \mathsf{E}\Big[e^{-\int_t^T V(X_s^{x,t},s)\mathrm{d}s} G(X_T^{x,t})\Big] .$$

Yet another alternative is to take X^x to be the solution to $\mathbb{e}_{\epsilon_x}(\theta_t a, \theta_t b)$ (x and t are fixed), where the functions $\theta_t a$ and $\theta_t b$ are understood, respectively, as $a(\cdot, t+\cdot)$ and $b(\cdot, t+\cdot)$. If the function f solves the parabolic equation $\mathbb{e}_2$, then

$$f(x,t) = \mathsf{E}\left[e^{-\int_0^{T-t} V(X_s^x,\, t+s)\,\mathrm{d}s} G(X_{T-t}^x)\right].$$

In many respects the formulation in (12.2) is preferable because the definition of a transition function associated with a Markov process is somewhat cleaner, and the measurability properties of such functions are well understood.

(b) The measure $Q_{x,t}$ may be seen as a "fundamental solution" for e_2, in the sense that it links the value of the solution f at location (x,t), universally, with all possible prescriptions for the terminal data G. The Feynman–Kac formula is essentially a recipe for the following program: to compute the solution at location (x,t), start a diffusion process governed by $\mathbb{e}(a,b)$ from location x at time t, let it run until time T, transform the final value through the data assignment G, discount the value at rate V, and, finally, compute the average. We stress that the parabolic PDE in e_2 depends on the diffusion coefficient a only through the positive definite matrix $a(x,t)a(x,t)^\intercal$. Thus, the diffusion coefficient a that drives the "fundamental solution" appears as a "square root" of the matrix that defines the second-order differentiation in the parabolic PDE. If this matrix fails to be positive definite (in which case the PDE will no longer be parabolic), one can no longer use the recipe of the Feynman–Kac formula. At the same time, if the matrix given by the coefficients for the second-order derivatives $\partial_{x_i}\partial_{x_j}$ in e_2 is positive definite, then it will have many "square roots" a. We stress that, by its very formulation, the martingale problem $\mathbb{M}_{x,t}(aa^\intercal, b)$ and its solution $Q_{x,t}$ depend on a only through the product $aa^\intercal$, whereas the equation $\mathbb{e}(a,b)$ and its solution depend on a explicitly.

(c) Nothing will change in the foregoing if the parabolic PDE in e_2 is formulated on the domain $\mathcal{D}\times]-\infty,T]$ for some open and connected domain $\mathcal{D}\subset\mathbb{R}^n$ – as long as a diffusion with coefficients a and b, starting from position $x\in\mathcal{D}$ at time t, remains in $\mathcal{D}$ until time T with probability 1.

(d) One can push the previous comment further and formulate the parabolic PDE from e_2 on some connected open domain $\mathcal{D}\subset\mathbb{R}^n\times\mathbb{R}$, with data $G(x,t)$ assigned at the boundary of $\mathcal{D}$, which we denote by $\partial\mathcal{D}$. If, for any $x\in\mathcal{D}$, a diffusion with coefficients a and b that starts from x at time t can be shown to reach the boundary $\partial\mathcal{D}$ in finite time with probability 1, then the Feynman–Kac formula can still be formulated with T interpreted as the (finite) hitting time of $\partial\mathcal{D}$. ○

(12.4) CURTAILMENT OF THE LIFETIME OF A DIFFUSION: One obvious interpretation of the exponent under the expectation in the Feynman–Kac formula is "discounting." And, indeed, in many applications to finance, the function V is taken as a model for "instantaneous interest rate," in which case the expression in the Feynman–Kac formula becomes "expected discounted payoff" – a quantity that is of primary concern in finance. But there is a second interpretation of the function V, which is also important for the domain of finance, in connection with the study of defaultable bonds. Intuitively, in one form or another, "discounting" boils down to some form of "reduction of value," and if such a reduction appears under the expectation operator, then it can also be achieved by way of curtailing the lifetime of the diffusion. The idea is to replace the term $e^{-\int_t^T V(X_s,s)\mathrm{d}s}$ with an indicator of the form $1_{\{\zeta\leq T\}}$ for some stopping time ζ, chosen in such a way that the substitution does not change the expectation. Such a stopping time ζ has the meaning of the "departure time" of X. The terminal value $G(X_T)$ is then "reduced" in the sense that it can be "collected" only if the diffusion does not depart before time T. Clearly, there must be an intrinsic connection between the stopping time ζ and the function V. Heuristically speaking, the factor $e^{-V(X_s,s)\mathrm{d}s} \approx 1 - V(X_s,s)\mathrm{d}s$ can be seen as an instantaneous transformation – a Radon–Nikodym factor of some sort – of the instantaneous transition probability of the diffusion X. If $V \geq 0$, then, generally, this transformation reduces the total probability, and the "loss of probability," namely $V(X_s,s)\mathrm{d}s$, is the instantaneous probability for departure at time t, that is, the probability of the event $\{t \leq \zeta < t + \mathrm{d}t\}$. As a result, in addition to "discount rate," the function V can be interpreted (equivalently) as "departure rate" (often referred to as "killing rate"), or as "curtailment rate," for the diffusion X. If a parabolic PDE with right side equal to 0 is linked to a diffusion X, then the same PDE with right side $V(x,t)f(x,t)$ would be linked to the diffusion X curtailed at rate $V(X_t,t)$. ○

Now we turn to a slight generalization of the Feynman–Kac formula, which is both straightforward and important for a number of applications. Consider the following variation of the parabolic PDE from e_1, in which the function $k \in \mathscr{C}_b(\mathbb{R}^n \times \mathbb{R}, \mathbb{R})$ is also given:

$$\partial f(x,t) + \mathcal{A}_t f(x,t) = V(x,t)f(x,t) + k(x,t)\,. \qquad \mathrm{e}_4$$

We still treat this equation in the domain $(x,t) \in \mathbb{R}^n \times]-\infty, T[$ with terminal condition $\lim_{t\nearrow T} f(x,t) = G(x)$ for some $G \in \mathscr{C}_b(\mathbb{R}^n;\mathbb{R})$. Just as before, given any $f \in \mathscr{C}^{2,1}(\mathbb{R}^n \times \mathbb{R}_+;\mathbb{R})$, the process $RM^f - M^f \boldsymbol{\cdot} R$ from exercise (12.1) is a local martingale, and so is the process $(RM^f - M^f \boldsymbol{\cdot} R)/R_t$ treated on the time-interval $[t,T]$. In the special case where f also solves e_4, by the result in (12.1), the local martingale $(RM^f - M^f \boldsymbol{\cdot} R)/R_t$ obtains the form

$$\frac{1}{R_t}\Big(Rf(X,\iota) - \big(Rk(X,\iota)\big)\boldsymbol{\cdot}\iota\Big),$$

and the conditions for this local martingale to be a martingale are the same as in e_3.

(12.5) Feynman–Kac formula II: Suppose that $f \in \mathscr{C}^{2,1}(\mathbb{R}^n \times \mathbb{R}_+;\mathbb{R})$ is one particular solution to the parabolic PDE in e_4 and the probability measure $\mathcal{Q}_{x,t}$, defined on $\mathbb{W} = \mathscr{C}(\mathbb{R}_+;\mathbb{R}^n)$, is one particular solution to the martingale problem $\mathbb{M}_{x,t}(aa^\intercal, b)$ for some $x \in \mathbb{R}^n$ and some $t < T$. Suppose further that the coordinate process on $\mathbb{W}$ is Markov with respect to $\mathcal{Q}_{x,t}$, and the local martingale

$$\mathcal{R}_{t,u}(w)\, f\big(w(u), t+u\big) - \int_0^u \mathcal{R}_{t,s}(w)k\big(w(s), t+s\big)\,\mathrm{d}s\,, \quad u \in [0, T-t]\,,$$

with $\mathcal{R}_{t,u}(w) \stackrel{\text{def}}{=} e^{-\int_0^u V(w(s),t+s)\mathrm{d}s}$, is actually a martingale with respect to $\mathcal{Q}_{x,t}$. Then

$$f(x,t) = \mathsf{E}_{x,t}\Big[\mathcal{R}_{t,T}(\theta_t w)G\big(\theta_t w(T)\big) - \int_t^T \mathcal{R}_{t,u}(\theta_t w)k\big(\theta_t w(u),u\big)\,\mathrm{d}u\Big],$$

with $\mathsf{E}_{x,t}$ understood to be the expectation relative to $\mathcal{Q}_{x,t}$. If the diffusion process X is a solution to $\mathbb{e}(a,b)$ and is also Markov for the filtration $\mathscr{F}$, then, setting $R_t \stackrel{\text{def}}{=} e^{-\int_0^t V(X_s,s)\mathrm{d}s}$, the Feynman–Kac formula can be cast as

$$f(X_t,t) = \mathsf{E}\Big[\frac{R_T}{R_t}G(X_T) - \int_t^T \frac{R_u}{R_t}k(X_u,u)\mathrm{d}u \;\Big|\; \mathscr{F}_t\Big], \quad t < T. \quad \circ$$

Now we turn to some illustrations. Consider first the special case $a(x,t) = \sigma x$ and $b(x,t) = rx$ for some fixed scalars $\sigma, r \in \mathbb{R}$. With this choice, the solution, X, to $\mathbb{e}(a,b)$ is the usual geometric Brownian motion.

(12.6) EXERCISE: Let X be the standard geometric Brownian motion with parameters $\sigma, r \in \mathbb{R}$. By using the result from exercise (12.1), show that the process $X - rX \bullet \imath$ must be a local martingale. Then explain why this local martingale is actually a martingale. Finally verify that $X - rX \bullet \imath$ is a martingale by inspection, that is, by computing the conditional expectations involved directly. $\circ$

(12.7) EXERCISE: Let X be the standard geometric Brownian motion with parameters $\sigma, r \in \mathbb{R}$. By using the result from exercise (12.1), show that the process $X^2 - (\sigma^2 X^2 + 2rX^2)\bullet \imath$ is a local martingale. Then explain why this local martingale is actually a martingale. $\circ$

We have seen already that the distribution laws of certain Markov-type diffusions can be interpreted as fundamental solutions to second-order parabolic PDEs of the form $(\partial + \mathcal{A}_t)f = 0$, in which $\mathcal{A}_t$ stands for the infinitesimal generator of the diffusion. The simplest nontrivial diffusion process that comes to mind is, of course, the Brownian motion in $\mathbb{R}^n$. Its infinitesimal generator is time-invariant and is given by $\frac{1}{2}$ of the Laplace operator (so that $\mathcal{A}_t = \mathcal{A} = \frac{1}{2}\Delta$):

$$\Delta \stackrel{\text{def}}{=} \partial_{x_1,x_1} + \cdots + \partial_{x_n,x_n}\ .$$

The distribution law of Brownian motion, that is, Wiener's measure $\mathscr{W}$, can be reintroduced as the solution to the martingale problem $\mathbb{M}_{\bar{0},0}(I,0)$, and can be viewed as the fundamental solution to the celebrated heat equation

$$\partial f(x,t) + \frac{1}{2}\Delta f(x,t) = 0, \quad x \in \mathbb{R}^n, \ t \le T,$$

which is of parabolic type with terminal data assigned at time $t = T$.

(12.8) EXERCISE: Give the explicit solution to the following parabolic PDE in $\mathbb{R}^n$:

$$\begin{cases} \big(\partial + \frac{1}{2}\Delta\big)f(x,t) = f(x,t) + |x|^2, & x \equiv (x_1,\ldots,x_n) \in \mathbb{R}^n, \quad t < T, \\ \lim_{t \nearrow T} f(x,t) = |x|^2 \equiv {x_1}^2 + \cdots + {x_n}^2. & \circ \end{cases}$$

(12.9) EXERCISE: Give the explicit solution to the following parabolic PDE on the positive real line ($\sigma, r \in \mathbb{R}$ are given parameters):

$$\begin{cases} \left(\partial + rx\partial_x + \frac{1}{2}\sigma^2 x^2 \partial_x^2\right) f(x,t) = r f(x,t), \quad x \in \mathbb{R}_{++}, \quad t < T, \\ \lim_{t \nearrow T} f(x,t) = \log(x). \quad \circ \end{cases}$$

(12.10) EXERCISE: Give the explicit solution to the following parabolic PDE on the real line $\mathbb{R}$ ($\sigma, r \in \mathbb{R}_+$ are given parameters):

$$\begin{cases} \left(\partial - rx\partial_x + \frac{1}{2}\sigma^2 \partial_x^2\right) f(x,t) = 0, \quad x \in \mathbb{R}, \quad t < T, \\ \lim_{t \nearrow T} f(x,t) = x^2. \quad \circ \end{cases}$$

(12.11) EXERCISE: Write down the classical Black–Scholes–Merton formula that gives the price of a European call option under the assumption that the underlying asset follows geometric Brownian motion and the market is complete. Using the Feynman–Kac formula, identify the expression in the BSM formula as an expected value of the discounted closing payoff, and, at the same time, as a solution to a particular parabolic PDE. ○

12.2 Fokker–Planck Equation

It is clear by now that the calculation of expected values involving diffusion processes is very important for both theory and practice. Apart from a few simple cases, typically, such calculations would not be straightforward, and, for this reason, a slew of computational tools have been developed. Generally, these tools fall outside of our main concerns, but there is one exception: the identification of the transition density of a diffusion process as a solution to a particular PDE, known as the *Fokker–Planck equation*,[143] or as *Kolmogorov's forward equation*, depending on which school one comes from. This section is a very brief – and mostly heuristic – outline of the origins and the main insights surrounding the Fokker–Planck equation.

To begin with, retain the setting of the previous section, and consider the diffusion equation $\mathbb{e}(a,b)$, assuming that its coefficients, a and b, are Lipschitz-continuous and bounded. We will consider first the case of time-invariant coefficients $a(x,t) = a(x)$ and $b(x,t) = b(x)$. With the notation as in (12.3a), given any $t \in \mathbb{R}_+$, define the linear operator $g \rightsquigarrow E_t g$ that maps bounded Borel-measurable functions $g\colon \mathbb{R}^n \mapsto \mathbb{R}$ into bounded Borel-measurable functions $E_t g\colon \mathbb{R}^n \mapsto \mathbb{R}$ according to the rule

[143] Named after the Dutch physicist (and also a musician) Adriaan Daniël Fokker (1887–1972) and the German physicist (one of the founders of quantum field theory, and recipient of the 1918 Nobel Prize in Physics) Max Karl Ernst Ludwig Planck [pronounced: PLAŊK] (1858–1947).

$$(E_t g)(x) \stackrel{\text{def}}{=} \mathsf{E}\big[g(X_t^{x,0})\big], \quad x \in \mathbb{R}^n,$$

and with the understanding that if $t = 0$, then $(E_0 g) = g$, that is, E_0 is the identity operator on the space of functions. Since the coefficients are time-invariant, the state of the diffusion at time $h + t$, starting from state x at time h, should have the same distribution as the state of the diffusion at time t, starting from position x at time 0. As a result,

$$(E_t g)(x) = \mathsf{E}\big[g(X_{h+t}^{h,x})\big], \quad x \in \mathbb{R}^n, \quad \text{for any} \quad h \geq 0.$$

Clearly, the function $x \rightsquigarrow (E_t g)(x)$ would be meaningful for every choice of g such that the r.v. $g(X_t^{x,0})$ is integrable for every $x \in \mathbb{R}^n$. Most of what follows in this section comes from the observation that the linear operator E_t deserves to be renamed to $e^{t\mathcal{A}}$, where

$$\mathcal{A} \equiv \mathcal{A}^{aa^{\intercal},b} = \frac{1}{2}\big(a(x)\,a(x)^{\intercal}\nabla\big)\cdot\nabla + b(x)\cdot\nabla$$

is the infinitesimal generator associated with $\mathbb{e}(a,b)$ (with time-invariant coefficients). To see why it is preferable to write E_t in the form $e^{t\mathcal{A}}$, fix some $T > 0$ and observe that by the Feynman–Kac formula the function

$$\mathbb{R}^n \times [0,T[\, \ni (x,t) \rightsquigarrow f(x,t) \stackrel{\text{def}}{=} E_{T-t}g(x),$$

has the following two properties

$$\partial f(x,t) + \mathcal{A} f(x,t) = 0 \;\text{ for }\; (x,t) \in \mathbb{R}^n \times [0,T[\,,$$
$$\text{and} \quad \lim_{t \nearrow T} f(x,t) = g(x) \;\text{ for }\; x \in \mathbb{R}^n,$$

which can be cast as

$$\partial_t\big(E_{T-t}g(x)\big) = -\mathcal{A}\big(E_{T-t}g(x)\big) \quad \text{and} \quad \lim_{t \nearrow T}\big(E_{T-t}g(x)\big) = g(x) \equiv E_0 g(x).$$

To put it another way,

$$\partial_t\big(E_{T-t}\big) = -\mathcal{A}\big(E_{T-t}\big) \quad \text{and} \quad \lim_{t \nearrow T}\big(E_{T-t}\big) = I,$$

and this is just another way of saying that $E_{T-t} = e^{(T-t)\mathcal{A}}$, or, equivalently, that $E_t = e^{t\mathcal{A}}$. Thus, heuristically speaking, the construction of the solution to

$$\left|\begin{array}{l} \partial f(x,t) + \mathcal{A} f(x,t) = 0, \quad (x,t) \in \mathbb{R}^n \times [0,T[\,, \\ \lim_{t \nearrow T} f(x,t) = g(x), \quad x \in \mathbb{R}^n, \end{array}\right.$$

comes down to letting the operator $e^{(T-t)\mathcal{A}}$ act on the boundary assignment $g(\cdot)$.

Next, observe that for any function f such that $\partial f(x,t) + \mathcal{A} f(x,t) = 0$, the Feynman–Kac formula gives

$$f(x,t) = \mathsf{E}\big[f(t+h, X_{t+h}^{t,x})\big], \quad x \in \mathbb{R}^n, \; h \geq 0,$$

and this relation is the same as

$$f(\cdot,t) = E_h f(\cdot,t+h) = e^{h\mathcal{A}} f(\cdot,t+h)\,.$$

If, in addition, f satisfies $f(x,T) = g(x)$ for some $T > t+h$, then we would have $f(\cdot,t) = e^{(T-t)\mathcal{A}} g$, and the relation above becomes

$$e^{(T-t)\mathcal{A}} g = e^{h\mathcal{A}}\big(e^{(T-t-h)\mathcal{A}} g\big)\,,$$

which, after the obvious renaming of the variables, gives the following important connection:

$$e^{(s+t)\mathcal{A}} = e^{s\mathcal{A}} \circ e^{t\mathcal{A}} = e^{t\mathcal{A}} \circ e^{s\mathcal{A}} \quad \text{for any } \; s,t \in \mathbb{R}_+\,.$$

Formal differentiation in the expression $e^{t\mathcal{A}} g(x)$, for a generic bounded Borel function $g\colon \mathbb{R}^n \mapsto \mathbb{R}$, gives

$$\begin{aligned}\partial_t\, e^{t\mathcal{A}} g(x) &= \lim_{\varepsilon\searrow 0} \frac{1}{\varepsilon}\big(e^{(t+\varepsilon)\,\mathcal{A}} g(x) - e^{t\mathcal{A}} g(x)\big) = \lim_{\varepsilon\searrow 0} e^{t\mathcal{A}} \frac{1}{\varepsilon}\big(e^{\varepsilon\,\mathcal{A}} g(x) - g(x)\big)\\ &= \lim_{\varepsilon\searrow 0} e^{t\mathcal{A}}\Big(\frac{1}{\varepsilon}\big(e^{\varepsilon\,\mathcal{A}} - I\big) g(x)\Big) = e^{t\mathcal{A}}(\mathcal{A}g)(x)\,.\end{aligned}$$

Now suppose that the distribution law of the random variable $X_t^{x,0}$ admits a probability density $\mathbb{R}^n \ni y \rightsquigarrow \psi_t^{x,0}(y) \in \mathbb{R}_+$. Then for any $\varphi \in \mathscr{C}_K^\infty(\mathbb{R}^n;\mathbb{R})$,

$$\int_{\mathbb{R}^n} \varphi(y)\psi_t^{x,0}(y)\,\mathrm{d}y \equiv \mathsf{E}\big[\varphi(X_t^{x,0})\big] = e^{t\mathcal{A}}\varphi(x)\,.$$

Applying the operation ∂_t to this identity gives

$$\begin{aligned}\partial_t \int_{\mathbb{R}^n} \varphi(y)\psi_t^{x,0}(y)\,\mathrm{d}y = e^{t\mathcal{A}}(\mathcal{A}\varphi)(x) &= \int_{\mathbb{R}^n} (\mathcal{A}\varphi)(y)\psi_t^{x,0}(y)\,\mathrm{d}y\\ &= \int_{\mathbb{R}^n} \varphi(y)\big(\mathcal{A}^\intercal \psi_t^{x,0}\big)(y)\,\mathrm{d}y\,,\end{aligned}$$

where $\mathcal{A}^\intercal$ is the formal adjoint of the operator $\mathcal{A}$ (in the above $\mathcal{A}^\intercal$ acts only on the variable y). Since the last relation holds for any test function $\varphi \in \mathscr{C}_K^\infty(\mathbb{R}^n;\mathbb{R})$,

$$\partial_t \psi_t^{x,0}(y) = \mathcal{A}^\intercal \psi_t^{x,0}(y) \quad \text{for any } \; (y,t) \in \mathbb{R}^n \times \mathbb{R}_{++}\,.$$

(12.12) FOKKER–PLANCK EQUATION: The equation for the unknown probability density $\psi_t^{x,0}(y)$ (assuming that density exists), given by

$$\partial_t \psi_t^{x,0}(y) - \mathcal{A}^\intercal \psi_t^{x,0}(y) = 0\,, \quad (y,t) \in \mathbb{R}^n \times \mathbb{R}_{++}\,,$$

with boundary condition

$$\lim\nolimits_{t\searrow 0} \psi_t^{x,0}(y) = \delta_x(y)\,,$$

understood as

$$\lim\nolimits_{t\searrow 0} \int_{\mathbb{R}^n} \varphi(y)\,\psi_t^{x,0}(y)\,\mathrm{d}y = \int_{\mathbb{R}^n} \varphi(y)\,\delta_x(y)\,\mathrm{d}y = \varphi(x)\,, \quad \varphi \in \mathscr{C}_K^\infty(\mathbb{R}^n;\mathbb{R})\,,$$

is called the *Fokker–Planck equation*, or *Kolmogorov's forward equation.*[144] ○

[144] It is forward because the operator $\mathcal{A}^\intercal$ acts on the "forward" variable y in the expression $\psi_t^{x,0}(y)$, not on the "present" variable x.

The meaning of the formal adjoint operator $\mathcal{A}^\intercal$ is not difficult to decipher: the standard (from calculus) integration by parts rule can be stated $\partial_{y_i}^\intercal = -\partial_{y_i}$, and, consequently, $\left(\partial_{y_i}\partial_{y_j}\right)^\intercal = \partial_{y_j}\partial_{y_i}$, $1 \le i, j \le n$. As a result (recall that ∇ is a vector column),

$$\mathcal{A}^\intercal \equiv (\mathcal{A}^{aa^\intercal,b})^\intercal = \frac{1}{2}\,\nabla^\intercal\cdot\left(\nabla^\intercal a(y)^\intercal a(y)(\cdot)\right) - \nabla\cdot\left(b(y)(\cdot)\right),$$

which is understood as

$$\mathcal{A}^\intercal = \frac{1}{2}\sum_{i,j=1}^{n}\sum_{k=1}^{d}\partial_{y_i,y_j}\left(a_{i,k}(y)a_{j,k}(y)(\cdot)\right) - \sum_{i=1}^{n}\partial_{y_i}\left(b_i(y)(\cdot)\right).$$

(12.13) REMARK: Since the operator $\mathcal{A}$, and therefore its adjoint $\mathcal{A}^\intercal$, acts only on the variable $y \in \mathbb{R}^n$, not much will change if we replace in the foregoing the coefficients $a(x)$ and $b(x)$ with their time-dependent versions $a(x,t)$ and $b(x,t)$, except that in this more general setting we will have to write $\mathcal{A}_t$ and $\mathcal{A}_t^\intercal$. The Fokker–Planck equation can be shown to remain valid also in the case of time-dependent coefficients (as long as the solution to $\mathbb{e}_{\epsilon_x}(a,b)$ exists and admits a probability density). In this more general setting our heuristic mechanism would boil down to the formal relation

$$e^{\int_h^{t+h}\mathcal{A}_s\,\mathrm{d}s}\varphi(x) = \mathsf{E}\left[\varphi(X_{t+h}^{h,x})\right], \quad x\in\mathbb{R}^n,\ \ h\ge 0,$$

but since exponents of different differential operators do not commute, we must be clear that $e^{\int_0^{t+h}\mathcal{A}_s\,\mathrm{d}s}$ is understood to be the same as $e^{\int_0^{h}\mathcal{A}_s\,\mathrm{d}s}e^{\int_h^{t+h}\mathcal{A}_s\,\mathrm{d}s}$, which, generally, would be different from $e^{\int_h^{t+h}\mathcal{A}_s\,\mathrm{d}s}e^{\int_0^{h}\mathcal{A}_s\,\mathrm{d}s}$. Consequently, the function $f(x,t) \stackrel{\text{def}}{=} e^{\int_t^T\mathcal{A}_s\,\mathrm{d}s}\varphi(x)$ will be understood as the solution to $f(x,t) = -\mathcal{A}_t f(x,t)$, with $f(x,T) = e^0\varphi(x) = \varphi(x)$. Repeating our heuristic argument,

$$\partial_t\, e^{\int_0^t\mathcal{A}_s\,\mathrm{d}s}\varphi(x) = e^{\int_0^t\mathcal{A}_s\,\mathrm{d}s}\left(\mathcal{A}_t\varphi\right)(x) = \mathsf{E}\left[(\mathcal{A}_t\varphi)(X_t^x)\right] \equiv \int_{\mathbb{R}^n}(\mathcal{A}_t\varphi)(y)\psi_t^x(y)\,\mathrm{d}y.$$

The associated Fokker–Planck equation again reads

$$\partial_t\psi_t^x(y) = \mathcal{A}_t^\intercal\psi_t^x(y), \quad y\in\mathbb{R}^n,\ \ t\in\mathbb{R}_+,$$

and again obtains by applying the formal differentiation ∂_t to the identity

$$\int_{\mathbb{R}^n}\varphi(y)\psi_t^x(y)\,\mathrm{d}y \equiv \mathsf{E}\left[\varphi(X_t^x)\right] = e^{\int_0^t\mathcal{A}_s\,\mathrm{d}s}\varphi(x). \quad \circ$$

Now we turn to some concrete illustrations. The simplest nontrivial and interesting diffusion process in $\mathbb{R}^n$ is the Brownian motion W starting from 0. With the notation and the terminology introduced in this section, this process can be identified as the solution, X, to $\mathbb{e}_{\epsilon_{\bar{0}}}(I,0)$. Of course, in this case the distribution density of X_t is known in the outset and given by the Gaussian density

$$\psi_t(y) = (2\pi t)^{-n/2}e^{-\frac{|y|^2}{2t}}, \quad (y,t)\in\mathbb{R}^n\times\mathbb{R}_+. \qquad \mathbb{e}_1$$

The associated Fokker–Plank equation is completely straightforward:

$$\partial_t \psi_t(y) - \frac{1}{2} \Delta \psi_t(y) \equiv \partial_t \psi_t(y) - \frac{1}{2} \sum_{i=1}^{N} \partial_{y_i, y_i} \psi_t(y) = 0 . \qquad \mathrm{e}_2$$

(12.14) EXERCISE: Verify that the density e_1 satisfies equation e_2, together with the boundary condition $\lim_{t \searrow 0} \psi_t(y) = \delta_0$. ○

(12.15) EXERCISE: Consider the standard geometric Brownian motion on the positive real line that starts from $x \in \mathbb{R}_{++}$. Since the position of this process at time $t > 0$ is distributed as an exponent of a Gaussian r.v., its probability density is not difficult to write down. Show that this density satisfies the associated Fokker–Plank equation, together with the appropriate boundary condition. ○

(12.16) EXERCISE: Consider the mean-reverting Brownian motion (the OU process) on the real line $\mathbb{R}$ that starts at time $t = 0$ from position $x = 0$. The position of this process at time $t > 0$ is a Gaussian r.v., the density of which is not difficult to write down. Show that this probability density satisfies the associated Fokker–Plank equation, together with the appropriate boundary condition. ○

13

Brief Introduction to Asset Pricing in Continuous Time

This chapter is essentially a continuation of chapter 6. We say "essentially" because a truly pedantic transition from discrete-time to continuous-time asset pricing models would require a considerable amount of technical steps that we would rather bypass. Since we have already developed some intuition about the passage from discrete-time to continuous-time stochastic models, we are in a position to make a reasonably good guess as to what the continuous-time interpretation of the "first principles" of asset pricing ought to be.

Although the work (Bachelier 1900) has a well-deserved status as the starting point of continuous-time finance, as was noted earlier, it remained largely unnoticed for more than half a century, and the "golden age" of continuous-time finance – as we know it today – really began with the seminal works of Robert C. Merton during the 1970s, which are now assembled in the single volume (Merton 1992). To be accurate, although narrower in scope, and not quite written with the asset pricing perspective in mind, the work (McKean 1965) appears to be the first to use the theory of Brownian motion for solving a concrete option pricing problem in the continuous-time setting. The works (Cox et al. 1985), (Duffie and Huang 1985), (McDonald and Siegel 1986), (Cox and Huang 1989), (Dixit and Pindyck 1994), (Karatzas 1996), (Karatzas and Shreve 1998), and (Duffie 2010) highlight the progress made since the 1970s.

Just as in chapter 6, the material in the present chapter is intended to be self-contained – to the extent to which this is possible – and closely follows (Karatzas 1996) with one small embellishment: we are not going to insist that the information arrival dynamics in the market is the one generated by the Brownian motion that drives the prices – this point is clarified in (13.6) and (13.7) below.

13.1 Basic Concepts and Definitions

We are concerned with a financial market that consists of $n+1$ continuously traded securities (alias: financial assets), labeled by the symbol $i \in \mathbb{N}_{|n}$. Every security will be identified by the token used to denote its spot price. Expressed as a nominal amount (that is, in units of a particular currency), every spot price follows a continuous stochastic process $S^i = (S^i_t)_{t \in \mathbb{R}_+}$, defined on a common probability space $(\Omega, \mathcal{F}, P)$, and adapted to a common filtration $\mathscr{F} = (\mathscr{F}_t)_{t \in \mathbb{R}_+} \setminus\subset \mathcal{F}$ that obeys$^{\multimap (8.64), (9.27)}$ the usual conditions. Specifically, we suppose that every price-process can be expressed as the Doléan–Dade exponent of a particular semimartingale, that is, given any $i \in \mathbb{N}_{|n}$, there is a continuous $(\mathscr{F}, P)$-semimartingale Z^i such that $S^i = e^{Z^i - \frac{1}{2}\langle Z^i, Z^i\rangle}$. Equivalently, every price process is assumed to be the solution to a linear stochastic equation of the form $S^i = S^i_0 + S^i \bullet Z^i$, driven by the (exogenously specified) continuous semimartingale Z^i. Thus, the dynamics of the spot price S^i are given in the outset by

$$\mathrm{d}S^i_t = S^i_t\,\mathrm{d}Z^i_t \quad \Leftrightarrow \quad \frac{\mathrm{d}S^i_t}{S^i_t} = \mathrm{d}Z^i_t\,.$$

To put it another way, the instantaneous changes in the semimartingale Z^i model the instantaneous net returns from trading S^i, not counting any dividend payouts. This connection can be expressed also as

$$Z^i = \log(S^i_0) + (S^i)^{-1} \bullet S^i\,,$$

and, for this reason, Z^i deserves to be called "the logarithmic semimartingale of the price process S^i," but we stress that because of the effect of Itô's formula, Z^i is not the logarithm of S^i, and S^i is not the usual (from calculus) exponent of Z^i.

We postulate that the security with label $i = 0$, that is, S^0, is a short-lived (that is, instantaneously lived) risk-free security, which feature we model by the condition $\langle Z^0, Z^0\rangle = 0$. In other words, the associated logarithmic semimartingale Z^0 is assumed to be a continuous and adapted process of finite variation. The risk-free security S^0 will play a special role in what follows. Since the token 0 is used in many contexts, in order to remove any possibility for a mistaken identification, we will to write S° and Z° instead of S^0 and Z^0.

The spot prices of all risky securities we now aggregate into the price-vector $S_t \overset{\text{def}}{=} (S^1_t, \ldots, S^n_t)$, and the associated logarithmic semimartingales we aggregate into the vector $Z_t \overset{\text{def}}{=} (Z^1_t, \ldots, Z^n_t)$ – note that both are defined as columns. We deliberately keep the notation ambiguous and do not write $\bar{S}_t$ and $\bar{Z}_t$, or, say, $\vec{S}_t$ and $\vec{Z}_t$ – this way all expressions in our model will look one-dimensional.[145] With the convention introduced in (10.23) in mind (see also page xvi), we can now write

$$\mathrm{d}S_t = S_t\,\mathrm{d}Z_t\,, \quad \frac{\mathrm{d}S_t}{S_t} = \mathrm{d}Z_t\,, \quad S_t = S_0 + \int_0^t S_u\,\mathrm{d}Z_u\,, \quad \text{or} \quad S = S_0 + S \bullet Z\,.$$

In many situations it will be more convenient to work on a logarithmic scale, that is, to work directly with the n-dimensional semimartingale Z, instead of the price vector S. This should not be surprising, given that asset pricing is concerned almost exclusively with returns, while the actual asset prices enter the picture only as an accounting convention. For this reason, we are going to impose some special

[145] While necessary in practice, conceptually, the multivariate setting is no different from the one-dimensional one, and, for this reason, it is preferable not to be reminded constantly about the fact that the model involves multiple assets.

features on the semimartingale structure of Z, the canonical decomposition of which we write as $Z = \mathscr{M} + \mathscr{A}$. Specifically, we suppose that the local martingale component, $\mathscr{M}$, can be written as a stochastic integral with respect to a particular Brownian motion, while the finite variation component, $\mathscr{A}$, can be written as the Lebesgue integral of the sample path of a particular adapted process. To be precise, we will suppose that there is a d-dimensional $(\mathscr{F}, P)$-Brownian motion W, and two $\mathscr{P}(\mathscr{F})$-measurable (that is, predictable) processes, σ and b, with values in $\mathbb{R}^{n\otimes d}$ and $\mathbb{R}^n$, respectively, that satisfy the conditions:

$$\int_0^t |\sigma_\omega(s)|^2 \,\mathrm{d}s < \infty \text{ and } \int_0^t |b_\omega(s)| \,\mathrm{d}s < \infty \text{ for all } t \in \mathbb{R}_+ \text{ for } P\text{-a.e. } \omega \in \Omega,$$

and are such that $\mathscr{M} = Z_0 + \sigma{\bullet}W$ and $\mathscr{A} = b{\bullet}\imath$, so that

$$Z \equiv \mathscr{M} + \mathscr{A} = Z_0 + \sigma{\bullet}W + b{\bullet}\imath\,.$$

(13.1) LOCAL VOLATILITY: The matrix-valued process σ that we just introduced is often called *local volatility*, or the *local volatility matrix*, or simply the *volatility (matrix)*. We are going to require that $d \geq n$ and that σ is of full rank, in that $\mathrm{Rank}(\sigma_t(\omega)) = n$, $P(\mathrm{d}\omega){\times}\mathrm{d}t$-a.e. This is simply our way of removing the possibility for any of the traded securities to generate redundant risk: if one of the rows in the matrix σ_t is a linear combination of the remaining rows, then the instantaneous risk associated with one of the risky securities can be replicated with a portfolio of the remaining risky securities. It is important to keep in mind that, in general, the market volatility can be linked to the bracket $\langle \mathscr{M}, \mathscr{M}\rangle = \langle Z, Z\rangle$, which is understood as the matrix with entries $\langle Z^i, Z^j\rangle$, $1 \leq i, j \leq n$. The assumptions that we imposed on the semimartingale Z guarantee that the derivative $\partial\langle Z, Z\rangle(\omega, t)$ exists $P(\mathrm{d}\omega){\times}\mathrm{d}t$-a.e. and σ_t is just one matrix-root of that derivative, that is,

$$\sigma(\omega, t)\sigma(\omega, t)^\intercal = \partial\langle Z, Z\rangle(\omega, t) \equiv \partial_t\langle Z, Z\rangle(\omega, t) \quad P(\mathrm{d}\omega){\times}\mathrm{d}t\text{-a.e.} \quad \circ$$

(13.2) THE INTEREST RATE PROCESS: As was already noted, the logarithmic semimartingale Z°, associated with the risk-free security S°, is a continuous semimartingale of finite variation, that is, $\langle Z^\circ, Z^\circ\rangle = 0$, or, equivalently, the local martingale component of Z° is the constant Z°_0 – and this is what makes S° risk-free, or, to be more accurate, locally risk-free. With regard to the finite variation component of Z°, we are going to suppose that this component has the form of a Lebesgue integral, that is, there is a $\mathscr{P}(\mathscr{F})$-measurable real process r, called the *interest rate process*, such that its sample paths r_ω are Lebesgue-integrable on every finite interval, and we have $Z^\circ = Z^\circ_0 + r{\bullet}\imath$. To put it another way, the spot price of the short-lived risk-free security is required to satisfy the equation $\partial S^\circ_\omega(t) = S^\circ_\omega(t) r_\omega(t)$, with exogenously given interest rate process r. Thus, for P-a.e. $\omega \in \Omega$, we have $S^\circ_\omega = S^\circ_\omega(0)\, e^{r_\omega{\bullet}\imath}$, or, if longer notation is preferable,

$$S^\circ_\omega(t) = S^\circ_\omega(0)\, e^{\int_0^t r_\omega(s)\,\mathrm{d}s} \quad \text{for all } t \in \mathbb{R}_+\,.$$

Since securities can be traded in any fraction, it is not a restriction to require $S^{\circ}_{\omega}(0) = 1$, or, equivalently, $Z^{\circ}_{\omega}(0) = 0$. This condition will be automatically assumed from now on. ○

(13.3) INSTANTANEOUS RETURNS AND DIVIDENDS: By its very definition, the vector $\mathrm{d}Z_t = \mathrm{d}S_t/S_t = \frac{S_t + \mathrm{d}S_t}{S_t} - 1$ gives the instantaneous net returns from trading the risky securities at their ex-dividend spot prices. The most common model for instantaneous dividend payouts is $S_\omega(t)D_\omega(t)\,\mathrm{d}t$, where D is some $\mathscr{P}(\mathscr{F})$-measurable $\mathbb{R}^n_+$-valued process (treated as a vector-column) with locally integrable sample paths. Thus, the instantaneous net gains from holding the securities can be expressed as $\mathrm{d}Z_t + D_t\,\mathrm{d}t$. As a result, the inclusion of dividends in the model merely modifies the finite variation component of the semimartingale Z. Unless noted otherwise, we are going to suppose that all dividends – if any – are incorporated into the process Z, but must keep in mind that if the prices S are understood to be ex-dividend prices, we cannot write $S = S_0 + S \centerdot Z$ unless Z excludes all dividend payouts. In most of what follows we will be working with the semimartingale Z, not with the price vector S, which is why, generally, the dividends can be kept in the background. ○

(13.4) EXCESS RETURNS: The instantaneous net return from the risk-free security S° is $r_t\,\mathrm{d}t$. Thus, setting $\vec{1} \overset{\text{def}}{=} (1, 1, \ldots, 1) \in \mathbb{R}^n$, the instantaneous excess returns from the risky securities can be aggregated into the vector $\mathrm{d}Z_t - r_t\vec{1}\,\mathrm{d}t$. Recall that the instantaneous excess return from a traded security is the net gain or loss (as a nominal amount) from the following chain of transactions: borrow ¤1, invest ¤1 in the security, and close the position after $\mathrm{d}t$ units of time, after paying back ¤1 plus the instantaneous interest ¤$r_t\mathrm{d}t$. The most important process in financial modeling is the excess returns process, which we define as

$$X = Z - Z_0 - r\vec{1} \centerdot \imath = \sigma \centerdot W + (b - r\vec{1}) \centerdot \imath \,, \tag{†}$$

or, if one insists on a longer format,

$$X_t = Z_t - Z_0 - \int_0^t r_s\vec{1}\,\mathrm{d}s = \int_0^t \sigma_s\,\mathrm{d}W_s + \int_0^t (b_s - r_s\vec{1})\,\mathrm{d}s\,, \quad t \in \mathbb{R}_+ .$$

The components of the vector process $X = (X^1, \ldots, X^n)$ give the excess returns associated with each of the individual risky assets in the market. ○

The asset pricing mechanism that we are about to develop will be described exclusively in terms of the (n-dimensional) process X and in terms of the (scalar) process r – plus, of course, the information filtration, which we introduce next.

(13.5) INFORMATION FILTRATION: We suppose that all market agents can observe the excess returns process X and the interest rate process r – this is equivalent to asserting that all agents can observe all security prices and dividend payouts as these quantities are revealed over time. At time $t = 0$ all spot prices S_0 are assumed to be given constants (recall that the initial spot price of the risk-free asset was set to $S_0^\circ = 1$). We would still like to allow the agents to be able to access market-related information that cannot be found in the quoted security prices and dividend announcements. To incorporate such a feature into the model, we suppose that in addition to the asset prices, the agents also observe some exogenously specified $\mathscr{F}$-adapted and continuous $\mathbb{R}^m$-valued process Y that starts from $Y_0 = \vec{0}$. In particular, Y may be some deterministic function (or simply a constant), in which case the asset prices and the dividend announcements would be carrying all the market-related information that the agents can access. On the other extreme, Y may be chosen to be the Brownian motion W, which would then amount to assuming that the agents can actually observe the Brownian motion that drives the security prices. The information filtration (alias: market filtration) we define as the filtration $\mathscr{F}^{X,r,Y}$ – this is the usual augmentation of the filtration generated by all observable processes X, r and Y. For simplicity, we are going to denote this filtration by $\mathscr{G} = (\mathscr{G}_t)_{t \in \mathbb{R}_+}$, rather than $\mathscr{F}^{X,r,Y}$, and are going to assume that $\mathscr{G}_0$ is the completion of the trivial σ-field $\{\emptyset, \Omega\}$ – since $X_0 = \vec{0}$ and $Y_0 = \vec{0}$, this assumption comes down to imposing that at time $t = 0$ the spot risk-free rate is a known constant as well. ○

(13.6) UNOBSERVABILITY OF THE BROWNIAN MOTION W: It is quite common in the asset pricing literature to suppose that the market filtration, $\mathscr{G}$, is just the filtration $\mathscr{F}$, and that $\mathscr{F}$ is just the filtration of the Brownian motion W from ($13.4^\dagger$). One consequence of this assumption is that the Brownian motion that drives the market, the volatility process σ, and the instantaneous expected growth rate b are all adapted to $\mathscr{G} \equiv \mathscr{F}$, and are therefore presumed observable. However, such a feature would be difficult to justify from a practical point of view. One alternative to imposing that $\mathscr{G} = \mathscr{F} = \mathscr{F}^W$ is to redesign the Brownian motion that drives the price process – without actually changing the price process – and in such a way that the volatility becomes observable. In a first step toward this objective, notice that, since the excess returns process X is observable (it is $\mathscr{G}$-adapted), so is the bracket $\langle X, X \rangle$, understood as the matrix process with entries $\langle X^i, X^j \rangle$, $1 \leq i, j \leq n$. Since $\langle X, X \rangle$ is continuous, it is automatically $\mathscr{P}(\mathscr{G})$-measurable, that is, $\mathscr{G}$-predictable, and since

$$\langle X, X \rangle_t = \int_0^t \sigma_s \sigma_s^\intercal \, \mathrm{d}s \,, \quad t \geq 0 \,,$$

it follows that the matrix process $\sigma\sigma^\intercal$ (with values in $\mathbb{R}^{n \otimes n}$) can be chosen (up to a modification) to be $\mathscr{P}(\mathscr{G})$-measurable as well. We stress that observing the matrix-valued process $\sigma\sigma^\intercal$ is very different from observing the actual volatility process σ and the Brownian motion W. By way of an example, from observing $\sigma\sigma^\intercal$ one cannot even infer the second dimension of σ, which is the dimension, d, of the driving Brownian motion W. In other words, in general, the observability of the excess returns X – which is reasonable to impose on the model – does not entail observability of the Brownian motion W and the volatility process σ. ○

As the matrix $\sigma(\omega,t)$ is assumed to be of full rank $P(\mathrm{d}\omega)\times\mathrm{d}t$-a.e., the matrix $\sigma(\omega,t)\sigma(\omega,t)^\intercal$ also must be of full rank $P(\mathrm{d}\omega)\times\mathrm{d}t$-a.e. Furthermore, since $\sigma(\omega,t)\sigma(\omega,t)^\intercal$ is a symmetric matrix, there is an orthogonal matrix $\psi(\omega,t)\in\mathbb{R}^{n\otimes n}$ and a diagonal matrix $\gamma(\omega,t)\in\mathbb{R}^{n\otimes n}$ such that

$$\sigma(\omega,r)\sigma(\omega,t)^\intercal = \psi(\omega,t)\gamma(\omega,t)\psi(\omega,t)^\intercal .$$

The requirement that $\sigma(\omega,t)\sigma(\omega,t)^\intercal$ is symmetric, positive definite, and of full rank is the same as the requirement for all diagonal entries in $\gamma(\omega,t)$ to be strictly positive. Generally, the matrices $\psi(\omega,t)$ and $\gamma(\omega,t)$ will not be unique,[146] but can always be produced from $\sigma(\omega,r)\sigma(\omega,t)^\intercal$, say, by using Jacobi's algorithm, the steps in which preserve the measurability for $\mathscr{P}(\mathscr{G})$ (or for any other σ-field on $\Omega\times\mathbb{R}_+$), as is easy to check by inspection. In any case, the matrix-valued processes ψ and γ can always be assumed to have the same measurability properties as the process $\sigma\sigma^\intercal$, which is $\mathscr{P}(\mathscr{G})$-measurable (predictable for the observation filtration $\mathscr{G}$), and thus Prog($\mathscr{G}$)-measurable as well.

If we now define the $\mathscr{P}(\mathscr{G})$-measurable matrix

$$\alpha(\omega,t) \stackrel{\text{def}}{=} \psi(\omega,t)\gamma(\omega,t)^{1/2} \in \mathbb{R}^{n\otimes n},$$

then we can write

$$\alpha(\omega,t)\alpha(\omega,t)^\intercal = \psi(\omega,t)\gamma(\omega,t)^{1/2}\gamma(\omega,t)^{1/2}\psi(\omega,t)^\intercal = \sigma(\omega,t)\sigma(\omega,t)^\intercal .$$

Furthermore, the matrix $\alpha(\omega,t)$ is $P(\mathrm{d}\omega)\times\mathrm{d}t$-a.e. invertible, with inverse

$$\alpha(\omega,t)^{-1} = \gamma(\omega,t)^{-1/2}\psi(\omega,t)^\intercal ,$$

and $\alpha(\omega,t)^{-1}\sigma(\omega,t)\in\mathbb{R}^{n\otimes d}$ is an orthogonal matrix since

$$\begin{aligned}\alpha(\omega,t)^{-1}\sigma(\omega,t)\big(\alpha(\omega,t)^{-1}\sigma(\omega,t)\big)^\intercal &= \alpha(\omega,t)^{-1}\sigma(\omega,t)\sigma(\omega,t)^\intercal\big(\alpha(\omega,t)^{-1}\big)^\intercal \\ &= \alpha(\omega,t)^{-1}\alpha(\omega,t)\alpha(\omega,t)^\intercal\big(\alpha(\omega,t)^{-1}\big)^\intercal = I_{n\otimes n} .\end{aligned}$$

In particular, $B \stackrel{\text{def}}{=} (\alpha^{-1}\sigma)\bullet W$ must be a Brownian motion and we can cast the excess returns process as

$$X = \alpha\bullet\big(\alpha^{-1}\sigma\bullet W + \alpha^{-1}(b-r\vec{1})\bullet\iota\big) = \alpha\bullet\big(B+\theta\bullet\iota\big), \qquad \mathrm{e}_1$$

where $\theta \stackrel{\text{def}}{=} \alpha^{-1}(b-r\vec{1})$.

(13.7) OBSERVED VOLATILITY: Writing the excess returns in the form e_1 is preferable to (13.4†) for a number of reasons. First, the driving Brownian motion in e_1, namely B, is of the same dimension as X, and, as a result, the volatility process attached to B, namely α, is a *square-matrix-valued* process. Most important, unlike σ, the process α was constructed by operating on the matrix-valued

[146] This is not surprising: if O is some (any) predictable orthogonal-matrix-valued process of appropriate dimensions, then $\sigma\bullet W = (\sigma O)\bullet(O^\intercal W)$, and, since $O^\intercal W$ is a Brownian motion, statistically, there could be no distinction between the model with volatility process σ and Brownian motion W and the model with volatility process σO and Brownian motion $O^\intercal W$.

process $\langle X, X \rangle \equiv \langle Z, Z \rangle$, which is to say, it can be inferred from practical observations – specifically, from observing the sample path X_ω and computing its quadratic variation. For this reason, α deserves to be called the *observed volatility process*. We stress, however, that this process is not unique and neither is the Brownian motion B. The important point here is that one can always rewrite the market model ($13.4^\dagger$) in such a way that the volatility matrix is square and observable – in fact, predictable – for the filtration generated by the excess returns. We stress that observability of the volatility α in e_1 does not imply observability of the Brownian motion B or the drift term b. The reason for this is that one cannot extract information about b from observations over X, in the same way in which one can extract information about α as the "square root" of the matrix $\partial\langle X, X \rangle$.

Since the modeling choices that we make matter only to the extent to which the resulting statistical features match the ones found in the observed data, we must be indifferent to the concrete choices of probability space, Brownian motion, and volatility matrix – as long as these choices keep the (joint) finite-dimensional distributions of the returns and the information process invariant. Thus, it would not have been a restriction if we were to impose on our model the structure prescribed in e_1 from the outset, that is, postulate back in ($13.4^\dagger$) that the Brownian motion W has the same dimension as the excess returns process X, and that the volatility process σ is predictable for the filtration $\mathscr{F}^X$ (therefore, also for $\mathscr{G}$). *This is the setting in which we will be working from now on.* Since σ is now square, the assumption that it is of full rank is the same as the assumption that the inverse σ^{-1} exists. This formulation avoids the need to diagonalize $\sigma\sigma^\intercal = \partial\langle X, X \rangle$, as the "square root of $\sigma\sigma^\intercal$" is already chosen in the outset. What would be a restriction, however – and a major one – is to suppose that the Brownian motion W and the process b are adapted to the market filtration $\mathscr{G}$, that is, both W and b can be inferred by the market agents. Of course, such an assumption can be always imposed, but it cannot be justified from a practical point of view. ○

(13.8) MARKET PRICE OF RISK: The $\mathbb{R}^n$-valued process

$$\theta(\omega, t) = \sigma(\omega, t)^{-1}\big(b(\omega, t) - r(\omega, t)\vec{1}\big), \quad t \in \mathbb{R}_+, \quad \omega \in \Omega,$$

will play an important role in what follows. It is called the *market price of risk process*, or simply the *market price of risk*. The following condition is instrumental and will be imposed from now on: $|\theta_\omega| \in \mathscr{L}^2_{\mathrm{loc}}(\mathbb{R}_+; \mathbb{R})$ for P-a.e. $\omega \in \Omega$, or, if a longer notation is preferable,

$$\dagger \qquad \int_0^t |\theta_\omega(s)|^2 \,\mathrm{d}s < \infty \quad \text{for any } t \in \mathbb{R}_+ \quad \text{for } P\text{-a.e. } \omega \in \Omega\,.$$

While this condition may seem exclusively technical, it has an important economic interpretation: it guarantees that a wealth process generated by trading financial assets cannot explode in finite time – see (Lyasoff 2014) for details. We call † *the nonexplosion condition.* With this condition now in force, we can define the

$\mathbb{R}$-valued local martingale

‡

$$\Theta = \theta^{\mathsf{T}} \cdot W,$$

and cast the excess returns process X as

$$X = \mathcal{M} + \langle \mathcal{M}, \Theta \rangle,$$

where $\mathcal{M}$ is the $\mathbb{R}^n$-valued local martingale $\mathcal{M} = \sigma \cdot W$ and $\langle \mathcal{M}, \Theta \rangle$ is understood as the vector (column) process $(\langle \mathcal{M}^1, \Theta \rangle, \ldots, \langle \mathcal{M}^n, \Theta \rangle)$. This setting allows for the following interpretation: the $\mathbb{R}^n$-valued local martingale $\mathcal{M}$ represents market risk, while the (scalar-valued) local martingale Θ represents the pricing rule for marketable risk. The actual pricing of risk – any risk, expressed as a continuous local martingale – comes down to the bracketing operation between the risk and the pricing rule. ○

(13.9) EXERCISE: Prove that with $\theta = \sigma^{-1}(b - r\vec{1})$, $\mathcal{M} = \sigma \cdot W$, and $\Theta = \theta^{\mathsf{T}} \cdot W$, we have

$$\langle \mathcal{M}, \Theta \rangle = \left(b - r\vec{1}\right) \cdot \iota. ○$$

(13.10) NORMALIZED EXCESS-RETURNS: The n-dimensional semimartingale

$$\beta = W + \theta \cdot \iota \equiv W + \langle W, \Theta \rangle$$

will also play a central role in our study. It will be called the *normalized excess-returns process*, or simply the *normalized excess returns*, and can be connected with the excess returns X through the relation

$$X = \sigma \cdot \left(W + \theta \cdot \iota\right) \equiv \sigma \cdot \beta \quad \Longleftrightarrow \quad \beta = \sigma^{-1} \cdot X.$$

We see that the normalized excess returns process is $\mathscr{F}^X$-adapted, and thus observable (since σ^{-1} is). Ultimately, asset prices and excess returns are driven by an observable semimartingale, β, and are affected by an observable volatility, σ. It is another matter how much of the structure of the semimartingale β, beyond its quadratic variation $\langle \beta, \beta \rangle = \iota$, can be recovered from observing X. ○

(13.11) CHANGE OF NUMÉRAIRE: We saw earlier that discrete-time financial models can be restated in such a way that the risk-free rate becomes 0. This is done by using the spot price of the instantaneously risk-free security as a numéraire. It should not be surprising that this manipulation carries over to continuous-time market models as well: with the numéraire set to S°, the price vector becomes $\bar{S}_t = (S_t^\circ)^{-1} S_t$, the price of the instantaneously risk-free asset becomes $\bar{S}_t^\circ = 1$, and the instantaneous risk-free rate vanishes. Cast in this new numéraire, the relation $S = S_0 + S \cdot Z$ becomes $\bar{S} = \bar{S}_0 + \bar{S} \cdot \bar{Z}$, where $\bar{Z} = Z - r\vec{1} \cdot \iota$, but the excess returns process remains the same, as it should:

$$\bar{X} = (\bar{Z} - \bar{Z}_0) - \vec{0} = (Z - Z_0) - r\vec{1} \cdot \iota = X.$$

In particular, the change of numéraire plays no role in models written entirely in terms of the excess returns. ○

13.2 Trading Strategy and Wealth Dynamics

What follows in this section is a quick – and mostly heuristic – derivation of the agents' wealth dynamics, as determined by their investment decisions and consumption plans. The connections that we are about to establish will be instrumental for the rest of the sequel. We suppose – for now – that all portfolios of financial assets held by the agents can be rebalanced, and financial positions can be closed or reopened, at no cost. Asset pricing models with transaction costs will be studied – unfortunately, mostly in passing – in chapter 18. In addition, we suppose that the investment decisions made by any one investor have no effect on the asset prices, or, on the investment decisions made by other investors – this is in contrast with the equilibrium asset pricing model that we developed in chapter 6, in which all prices and all investment decisions can only be determined simultaneously.

Let V_t denote the net value (as a nominal amount) of all financial assets held by a particular market agent at time t, with the understanding that some of the assets may be liabilities (such as, say, short positions) and therefore may have a negative nominal value. To put it another way, V_t is the nominal net amount available to the agent for investment and consumption at time t. Suppose that the vector (column) $\pi_t \in \mathbb{R}^n$ aggregates the nominal amounts that the agent decides to invest in the risky securities (some of these amounts may again be negative, as the investment decision may involve short positions in some of the assets). Consequently, the total amount invested in the risky assets at time t is $\pi_t{}^\intercal \vec{1}$, and the remaining portion of the available wealth, namely $V_t - \pi_t{}^\intercal \vec{1}$, is "held in a bank account," that is, invested in the risk-free asset. At time $t + \mathrm{d}t$ the aggregate value of the risky assets acquired at time t becomes $\pi_t{}^\intercal(\vec{1} + \mathrm{d}Z_t)$,[147] while the nominal value of the agent's position in the risk-free asset becomes $\big(V_t - \pi_t{}^\intercal \vec{1}\big)(1 + r_t\,\mathrm{d}t)$. Between time t and $t + \mathrm{d}t$ the agent consumes the amount $c_t\,\mathrm{d}t$, where $c_t \in \mathbb{R}_+$ is the instantaneous consumption rate.[148] In total, the agent's wealth at time $t + \mathrm{d}t$, that is, the nominal amount available for consumption and investment at time $t + \mathrm{d}t$, is

$$
\begin{aligned}
V_{t+\mathrm{d}t} &= \big(V_t - \pi_t{}^\intercal \vec{1}\big)(1 + r_t\,\mathrm{d}t) + \pi_t{}^\intercal(\vec{1} + \mathrm{d}Z_t) - c_t\,\mathrm{d}t \\
&= V_t + V_t\,r_t\,\mathrm{d}t + \pi_t{}^\intercal(\mathrm{d}Z_t - r_t\vec{1}\,\mathrm{d}t) - c_t\,\mathrm{d}t \\
&= V_t + V_t\,r_t\,\mathrm{d}t + \pi_t{}^\intercal\,\mathrm{d}X_t - c_t\,\mathrm{d}t\,.
\end{aligned}
$$

The instantaneous change in the total nominal wealth is therefore

$$
V_{t+\mathrm{d}t} - V_t \equiv \mathrm{d}V_t = V_t\,r_t\,\mathrm{d}t + \pi_t{}^\intercal\,\mathrm{d}X_t - c_t\,\mathrm{d}t\,, \quad t \in \mathbb{R}_+\,. \qquad \mathrm{e}_1
$$

Considered across time, the variables c_t, $t \in \mathbb{R}_+$, and π_t, $t \in \mathbb{R}_+$, will be treated as stochastic processes – as usual, denoted c and π. Specifically, the n-dimensional

[147] Recall that $\mathrm{d}Z_t$ is the vector of the instantaneous net returns from the risky assets, including any instantaneous dividend payments.

[148] Any quantity representing consumption is assumed to be positive, that is, nonnegative, by default.

process π gives the agent's dynamic *trading strategy* (alias: *portfolio process*), and the nonnegative scalar process c gives the agent's *consumption plan* (alias: *consumption process*). If π and c are already chosen and fixed, then e_1 can be treated as a linear SDE for V. To ensure that this SDE has a meaningful solution, we must require that π and c are both predictable for $\mathscr{G}$,[149] and that, for P-a.e. $\omega \in \Omega$, the following two conditions are satisfied for all $t \in \mathbb{R}_{++}$:

$$\int_0^t \pi(\omega,s)^\intercal\big(\sigma(\omega,s)\sigma(\omega,s)^\intercal\big)\pi(\omega,s)\,\mathrm{d}s \equiv \int_0^t |\pi(\omega,s)^\intercal\sigma(\omega,s)|^2\,\mathrm{d}s < \infty$$

and

$$\int_0^t c(\omega,s)\,\mathrm{d}s < \infty\,. \qquad \mathrm{e}_2$$

Without further notice we are going to suppose from now on that these conditions automatically hold for every choice of a portfolio process π and every choice of a consumption process c. Thus, we can introduce V as the unique strong solution to e_1, which is determined by the initial investment $V_0 = x$ (treated as an expense) and the choice of the processes π and c. If we need to emphasize this dependence in the notation, we will write $V = V^{x,\pi,c}$. We say that the *investment-consumption strategy* (x,π,c) is *self-financing* if the associated wealth process $V = V^{x,\pi,c}$ is governed by the dynamics shown in e_1. The reasons for using the term "self-financing" should be clear: trading and consumption are funded entirely by the initial investment $V_0 = x$ and by the capital gains from past investments.

As an aside, this is our first encounter with a dynamic control problem: the controller "steers" the stochastic variable V_t by constantly manipulating the control variables π_t and c_t (chosen from a set of admissible controls) that affect the dynamics of V_t.

(13.12) EXERCISE: Prove that the wealth process $V \equiv V^{x,\pi,c}$ can be expressed as

$$\dagger \quad V_t \equiv V_t^{x,\pi,c} = e^{\int_0^t r_u\,\mathrm{d}u}\left(x + \int_0^t e^{-\int_0^s r_u\,\mathrm{d}u}\pi_s{}^\intercal\,\mathrm{d}X_s - \int_0^t e^{-\int_0^s r_u\,\mathrm{d}u}c_s\,\mathrm{d}s\right). \quad \circ$$

Since $e^{\int_0^t r_s\,\mathrm{d}s}$ is nothing but the spot price, S_t°, of the risk-free security, we can cast the wealth process in the following equivalent form:

$$V \equiv V^{x,\pi,c} = S^\circ \times \left(x + (S^\circ)^{-1}\pi^\intercal \bullet X - (S^\circ)^{-1}c \bullet \imath\right),$$

which we now rearrange as

$$(S^\circ)^{-1}V + (S^\circ)^{-1}c\bullet\imath = x + (S^\circ)^{-1}\pi^\intercal\bullet X\,. \qquad \mathrm{e}_3$$

[149] This requirement is quite natural, since any action taken in real time can depend only on information available at the time of the action.

Since the process on the right side will crop up regularly in what follows, it deserves to have its own symbol, and we set:

$$K^{x,\pi} \stackrel{\text{def}}{=} x + (S^\circ)^{-1}\pi^\intercal \cdot X\,.$$

Notice that all processes that feature in e_3 are predictable for the market filtration $\mathscr{G}$. The identity itself is quite intuitive: it says that after discounting everything to time $t = 0$, the wealth available at time t plus the accumulated consumption until time t must equal the sum of the initial endowment and the accumulated capital gains, generated by the trading strategy π until time t. This is yet another justification of the term "self-financing." If we were to use the risk-free security as a numéraire, then equation e_3 would become

$$\bar{V} + \bar{c} \cdot \imath = x + \bar{\pi}^\intercal \cdot X\,,$$

where $\bar{}$ stands for "value in the new numéraire" (as we saw earlier, the excess returns are invariant under changes in the numéraire).

In much the same way we now develop the dynamics of the *outstanding debt*, $\tilde{V}_t$, associated with a particular *investment-payout strategy* (x, π, c). We deliberately use the same symbols but with a different interpretation: $\tilde{V}_t$ is the net value of the outstanding debt (as a nominal amount) held at time t, $x = \tilde{V}_0$ is the initial loan received at time $t = 0$ (treated as income), π is the *portfolio process* (alias: *trading strategy*), and the *positive process* c gives the instantaneous rate at which the loan is paid out (amortized). The trading strategy π is now funded entirely with new debt, and the capital gains from trading go toward new investment and/or paying out the debt (which increases the outstanding debt if trading results in a loss). Just as before, we suppose that the payout rate c and the trading strategy π are predictable for $\mathscr{G}$ and the integrability conditions in e_2 hold – without further notice, these conditions will be assumed to hold automatically for all investment-payout strategies as well. The dynamics that govern the outstanding debt are given by

$$\begin{aligned}
\tilde{V}_{t+\mathrm{d}t} &= \big(\tilde{V}_t + \pi_t{}^\intercal \vec{1}\big)(1 + r_t\,\mathrm{d}t) - \pi_t{}^\intercal(\vec{1} + \mathrm{d}Z_t) - c_t\,\mathrm{d}t \\
&= \tilde{V}_t + \tilde{V}_t\, r_t\,\mathrm{d}t - \pi_t{}^\intercal(\mathrm{d}Z_t - r_t\vec{1}\,\mathrm{d}t) - c_t\,\mathrm{d}t \\
&= \tilde{V}_t + \tilde{V}_t\, r_t\,\mathrm{d}t - \pi_t{}^\intercal\,\mathrm{d}X_t - c_t\,\mathrm{d}t\,,
\end{aligned}$$

and lead to the following linear SDE for $\tilde{V}$:

$$\mathrm{d}\tilde{V}_t = \tilde{V}_t\, r_t\,\mathrm{d}t - \pi_t{}^\intercal\,\mathrm{d}X_t - c_t\,\mathrm{d}t\,, \quad t \in \mathbb{R}_+\,.$$

The debt process $\tilde{V}$ can again be introduced as the unique strong solution to this equation, and the fact that the investment-payout strategy (x, π, c) generates outstanding debt with these dynamics again defines the attribute *self-financing*. If we need to emphasize the dependence of $\tilde{V}$ on (x, π, c), we will write $\tilde{V}^{x,\pi,c}$.

(13.13) EXERCISE: Prove that the outstanding debt $\tilde{V}$, maintained by the self-financing investment-payout strategy (x, π, c), can be expressed as

$$\tilde{V}_t = e^{\int_0^t r_u \mathrm{d}u} \left(\tilde{V}_0 - \int_0^t e^{-\int_0^s r_u \mathrm{d}u} \pi_s{}^\intercal \mathrm{d}X_s - \int_0^t e^{-\int_0^s r_u \mathrm{d}u} c_s \mathrm{d}s \right),$$

or, to be brief, as

$$\dagger \qquad \tilde{V} = S^\circ \times \left(\tilde{V}_0 - (S^\circ)^{-1} \pi^\intercal \bullet X - (S^\circ)^{-1} c \bullet \iota \right). \quad \circ$$

(13.14) REMARK: Formally, (13.13†) can be obtained from the result established in exercise (13.12) by multiplying both sides of (13.12†) by (-1), and then replacing the consumption rate c with $(-c)$ and replacing $(-V)$ with $\tilde{V}$. This is how it should be: debt is nothing but negative wealth, and the payout rate for the debt is nothing but negative consumption. We again stress that if $\tilde{V}$ represents outstanding debt, maintained by some self-financing investment-payout strategy, then its initial value $\tilde{V}_0$ represents income, as it is the initial loan taken by the agent. However, if V represents wealth, maintained by some self-financing investment-consumption strategy, then its initial value V_0 represents an expense, as it is the amount initially invested in the market. ○

Finally, we note that (13.13†) can be stated in the following equivalent form:

$$(S^\circ)^{-1} \tilde{V} + (S^\circ)^{-1} c \bullet \iota = K^{x,-\pi} \equiv x - (S^\circ)^{-1} \pi^\intercal \bullet X, \qquad e_4$$

and this relation is the same as

$$-(S^\circ)^{-1} \tilde{V} - (S^\circ)^{-1} c \bullet \iota = K^{-x,\pi} \equiv -x + (S^\circ)^{-1} \pi^\intercal \bullet X,$$

as it should be: we saw in the previous remark that a self-financing investment-payout strategy is nothing but a self-financing investment-consumption strategy with negative initial investment (that is, income) and with negative consumption. To put it another way, formally, we have: $\tilde{V}^{x,\pi,c} = -V^{-x,\pi,-c}$.

13.3 Equivalent Local Martingale Measures

In our exploration of discrete-time market models in chapter 6, we discovered that the stochastic discount factors give rise to an equivalent local martingale measure, which is to say, a probability measure that is equivalent to the one postulated in the model, and is such that every possible excess returns process is a local martingale with respect to it. Such measures are instrumental in asset pricing, and, as we are about to see, exist within the continuous-time paradigm as well. In this section we will adopt the setting, notation, and terminology introduced in the previous two sections, but are going to work exclusively with a finite time horizon $T \in \mathbb{R}_{++}$. Thus, without any further warning or reminder, the domain of the time-variable t is going to be set to the finite interval $[0, T]$.

(13.15) EQUIVALENT LOCAL MARTINGALE MEASURES: We say that a probability measure, Q, which is equivalent to P on $\mathscr{G}_T$ ($Q\restriction\mathscr{G}_T \approx P\restriction\mathscr{G}_T$), is an *equivalent local martingale measure*, or ELMM for short, if the excess-returns process X, which is a $(\mathscr{G}, P)$-semimartingale (of dimension n) by definition, can be claimed to be a $(\mathscr{G}, Q)$-local martingale, that is, every individual excess returns process X^i, for $1 \leq i \leq n$, is a real-valued $(\mathscr{G}, Q)$-local martingale. ○

At this point we do not know whether an ELMM for our market model exists, but – just to gain some insight into this matter – let us speculate that it does and that Q is one such measure. In fact, let us further speculate that Q is a "pricing measure," that is, a measure built from the stochastic discount factors in a way that is compatible with the agents' risk preferences, endowments, and market-clearing.[150] Since the construction of the stochastic integral is invariant under equivalent changes of the probability measure, any stochastic integral with respect to the semimartingale X is just as meaningful (as a stochastic integral) under the measure Q as it is under the measure P. Under the measure Q, however, such stochastic integrals are local martingales by construction. In particular, given any predictable trading strategy π that satisfies the first condition from e_2 in section 13.2, the process $K^{0,\pi} \equiv (S^\circ)^{-1}\pi^\intercal \cdot X$ can be claimed to be $(\mathscr{G}, Q)$-local martingale starting from 0. Since the process $K^{0,\pi}$ tracks the accumulated discounted capital gains generated by the trading strategy π, in any realistic market model this process should be prevented from becoming "too negative;" in other words, a realistic trading strategy should not be allowed to go on with unlimited liability. To make such a requirement precise, let ξ denote some $\mathscr{G}_T$-measurable r.v., thought to be a stochastic payoff (or liability) to be received at time T, and suppose that $\mathsf{E}^Q[(S^\circ_T)^{-1}|\xi|] < \infty$, which condition can be interpreted as the claim that the payoff ξ has a well-defined finite price at time $t = 0$. At time $t < T$ the value of this payoff must be $\mathsf{E}^Q\big[e^{-\int_t^T r_s\,\mathrm{d}s}\,\xi \mid \mathscr{G}_t\big]$, and discounted to time $t = 0$ this value becomes

$$(S^\circ_t)^{-1}\mathsf{E}^Q\big[e^{-\int_t^T r_s\,\mathrm{d}s}\,\xi \mid \mathscr{G}_t\big] = \mathsf{E}^Q\big[(S^\circ_T)^{-1}\,\xi \mid \mathscr{G}_t\big]\,.$$

Based on this observation, we are going to model the limited liability feature of the trading strategy π by insisting that there must exist a $\mathscr{G}_T$-measurable random variable ξ with $\mathsf{E}^Q[(S^\circ_T)^{-1}|\xi|] < \infty$ such that the following relation holds Q-a.s., or, equivalently, P-a.s.:

$$K^{0,\pi}_t \geq \mathsf{E}^Q\big[(S^\circ_T)^{-1}\,\xi \mid \mathscr{G}_t\big] \quad \text{for all } t \in [0,T]\,.$$

Recognizing that the process on the right side is nothing but a càdlàg $(\mathscr{G}, Q)$-martingale, we formalize the limited liability requirement by introducing the following concept:[151]

[150] While we cannot claim that any ELMM whatsoever can be a pricing measure, our exploration of finite asset pricing models in discrete-time should give us enough reasons to believe that a pricing measure, if one exists, will have to be an ELMM.

[151] Recall that martingales treated on finite time-intervals are automatically uniformly integrable.

(13.16) TAME TRADING STRATEGY: Given an ELMM Q (assuming that one exists), we say that the trading strategy (portfolio process) π is *Q-tame*, or *Q-admissible*, if there is a càdlàg $(\mathscr{G}, Q)$-martingale, M, such that the accumulated capital gains process $K^{0,\pi}$ is bounded from below by M; to be precise, the following relation is satisfied Q-a.s., or, equivalently, P-a.s.:

$$\dagger \qquad K_t^{0,\pi} \equiv (S^\circ)^{-1}\pi^\intercal \bullet X_t \geq M_t \quad \text{for all } \ t \in [0,T]\,.$$

The same condition can be stated as: $K_\omega^{0,\pi} \geq M_\omega$ everywhere in $[0,T]$ for a.e. (relative to either P or Q) $\omega \in \Omega$.

If the martingale M can be chosen to be some finite constant $C < 0$, that is, if the process $K^{0,\pi}$ is bounded from below, we say that the trading strategy π is *universally tame*, or *universally admissible.*

If (x, π, c) is any self-financing investment-consumption strategy, or any self-financing investment-payout strategy, we say that (x, π, c) is Q-tame (Q-admissible), or universally tame (universally admissible), if the trading strategy π has that attribute. ○

(13.17) REMARK: Throughout much of the literature on asset pricing the meaning attached to the terms "tame," or "admissible," is what we call here "universally tame." The main reason for us to not follow the tradition quite so literally is that we would like the attribute "tame" to be symmetric, in the sense that in all situations where $K^{0,\pi}$, or, equivalently, $K^{0,-\pi} = -K^{0,\pi}$, is a Q-martingale, we would like both π and $-\pi$ to be declared "tame." Indeed, if "Q-tame" is understood as in (13.16), then any portfolio process π, chosen so that $K^{0,\pi}$ is a Q-martingale, will be Q-tame, since one can simply set $M = K^{0,\pi}$. Since martingales can be viewed as the stochastic analogues of constant functions, we are merely replacing in the common definition of tame the term "constant" with "stochastic constant." ○

(13.18) EXERCISE: Prove that if π is any Q-tame trading strategy, then the process $K^{0,\pi} \equiv (S^\circ)^{-1}\pi^\intercal \bullet X$ is a supermartingale relative to Q.

HINT: If M is chosen as in (13.16), then $K^{0,\pi} - M$ is a positive $(\mathscr{G}, Q)$-local martingale that starts from $-M_0 \geq 0$. ○

Looking back at the relations e_3 and e_4 in the previous section, now we see that if π is any Q-admissible trading strategy, then the left side of e_3 would represent a $(\mathscr{G}, Q)$-supermartingale, while the left side of e_4 would represent a $(\mathscr{G}, Q)$-submartingale. As a result, we have the following estimate involving the wealth process $V \equiv V^{x,\pi,c}$, assuming that π is Q-admissible:

$$x \geq \mathsf{E}^Q\big[(S_T^\circ)^{-1}V_T + \big((S^\circ)^{-1}c \bullet \imath\big)_T\big]\,. \qquad \mathrm{e}_1$$

This relation has the following interpretation: if the ELMM Q is also a pricing measure – that is, a measure constructed from the stochastic discount factors – then the expression on the right side of e_1 is nothing but the price at time $t = 0$ of a claim to the final wealth V_T and to the entire future consumption stream until time T. In other words, the initial investment x, with which one can initiate a tame trading strategy π that funds the final wealth V_T and the consumption stream c, must be at least as large as the market value at time $t = 0$ of those future payoffs. So to speak, there is no tame self-financing investment-consumption strategy that "beats the market." Now we see that the most efficient self-financing investment-consumption strategies are the ones that are "as good as the market," that is, generate stochastic payoffs with the smallest possible initial investment with which those payoffs can be generated. As we just realized, the lower bound on the initial investment with which the final wealth V_T and the consumption plan c can be funded is the time $t = 0$ market price of those payoffs, and is given by

$$x^* = \mathsf{E}^Q\big[(S^\circ_T)^{-1}V_T + \big((S^\circ)^{-1}c\bullet\imath\big)_T\big]\,. \qquad \mathrm{e}_2$$

Of course, at this point we do not know whether V_T and c can actually be funded with the initial investment x^*, and, clearly, resolving this matter is rather important. As the next exercise demonstrates, these most efficient tame trading strategies – if they indeed exist – must be precisely the strategies that turn the stochastic integral $K^{0,\pi} \equiv (S^\circ)^{-1}\pi^\intercal\bullet X$ into a martingale under Q.

(13.19) EXERCISE: Let Q be any ELMM, let π be some Q-tame trading strategy, and let c be some consumption plan, all chosen – assuming that such a choice is possible – so that

$$(S^\circ)^{-1}V + (S^\circ)^{-1}c\bullet\imath = x^* + (S^\circ)^{-1}\pi^\intercal\bullet X\,,$$

understood as an identity up to indistinguishability between two stochastic processes, with x^* given by e_2. Prove that either side of this identity represents a $(\mathscr{G}, Q)$-martingale on $[0, T]$.

HINT: Consult exercise (9.6). ○

Just as we suspect, the analysis of a self-financing investment-payout strategy is a mirror image of the above: if the trading strategy π is Q-admissible, then both sides of e_4 in the previous section represent a Q-submartingale, and, as a result, the amount x, which is borrowed at time $t = 0$, cannot exceed the time 0 market price of the liabilities on the loan. Or, to state this more clearly,

$$x \le \mathsf{E}^Q\big[(S^\circ_T)^{-1}V_T + \big((S^\circ)^{-1}c\bullet\imath\big)_T\big]\,. \qquad \mathrm{e}_3$$

Thus, the most efficient self-financing investment-payout strategies are the ones that make this inequality into an equality, for such a strategy would allow for the largest possible initial loan against a particular payout profile. This again corresponds to a trading strategy π that turns $K^{0,\pi}$ into a Q-martingale, and this feature again comes as an immediate corollary from exercise (9.6).

But how does an ELMM look, provided that one exists? In a first step toward addressing this question, we observe the following:

(13.20) EXERCISE: Prove that if Q is any ELMM, then the normalized excess-returns process β must be a $(\mathscr{G}, Q)$-Brownian motion. ○

Our objective is much clearer now: we must find a measure $Q \approx P$, relative to which the semimartingale β has the statistical attributes of a Brownian motion. Since the quadratic variation of any semimartingale is invariant under any equivalent change of the probability measure, and since $\langle \beta, \beta \rangle = \iota$, it would be enough for us to find a measure $Q \approx P$ under which β is a local martingale. This is where Girsanov's theorem (see (10.51)) comes into play. With the notation as in (13.8), we have

$$X = \mathscr{M} + \langle \mathscr{M}, \Theta \rangle \quad \text{and} \quad \beta = W + \langle W, \Theta \rangle .$$

Thus, to make X and β into local martingales it would be enough for us to change the measure P to

$$Q = e^{-\Theta_T - \frac{1}{2}\langle \Theta,\Theta \rangle_T} \odot P \equiv e^{-\theta^\intercal \bullet W_T - \frac{1}{2}|\theta|^2 \bullet \iota_T} \odot P ,$$

if we can somehow ensure that $\mathsf{E}\left[e^{-\Theta_T - \frac{1}{2}\langle \Theta,\Theta \rangle_T}\right] = 1$. Unfortunately, there is no reason for this expectation to always be equal to 1 – as we know, all that we can say without imposing any additional conditions is that it is ≤ 1. By $^{\multimap (10.56)}$ Novikov's criterion, it would be enough to require that

$$\mathsf{E}^P\left[e^{\frac{1}{2}\langle \Theta,\Theta \rangle_T}\right] \equiv \mathsf{E}^P\left[e^{\frac{1}{2}|\theta|^2 \bullet \iota_T}\right] < \infty .$$

In particular, this condition would be met if the market price of risk were bounded, that is, $|\theta_t| \leq C$. A less restrictive condition on the market price of risk can be extracted from $^{\multimap (10.55)}$ Kazamaki's criterion, by which it would be enough to require that $e^{-\frac{1}{2}\Theta}$ were a uniformly integrable submartingale.

13.4 The Two Fundamental Theorems of Asset Pricing

Continuing with the setting and the notation introduced in the previous three sections, we now turn to a concept that is central in asset pricing.

(13.21) ARBITRAGE: We say that the trading strategy π represents *arbitrage*, or that π is an *arbitrage strategy*, if

$$P\Big[\big((S^\circ)^{-1}\pi^\intercal \bullet X\big)_T > 0\Big] > 0 \quad \text{and} \quad P\Big[\big((S^\circ)^{-1}\pi^\intercal \bullet X\big)_T \geq 0\Big] = 1 .$$

We say that π represents *guaranteed arbitrage* if one can find a constant $\varepsilon \in \mathbb{R}_{++}$ such that

$$P\Big[\big((S^\circ)^{-1}\pi^\intercal \bullet X\big)_T \geq \varepsilon\Big] = 1 . \quad \circ$$

(13.22) REMARK: A doubling trading strategy – see note 78 on page 132 – may present an arbitrage opportunity, but it would be impractical to allow for such strategies in our market model, since trading that entails unlimited liability is not going to be feasible. For this reason, the only arbitrage strategies that we care about are the ones that are also tame – either universally tame, or tame for a particular ELMM. $\circ$

Due to the nonexplosion condition (13.8†), the distribution law of the normalized excess returns β in the space $\mathscr{C}([0,T];\mathbb{R}^n)$ is always absolutely continuous with respect to the Wiener measure $\mathscr{W}$. This result can be traced back to (Liptser and Shiryaev 1974, thm. 7.4) – see also (Lyasoff 2014). It turns out that whether the market model allows arbitrage or not depends on whether the distribution law of β is equivalent to the Wiener measure, or is strictly absolutely continuous, that is, it depends on whether or not one can find a Borel set $A \subset \mathscr{C}([0,T];\mathbb{R}^n)$ with these two attributes: $P\big(\beta \in A\big) = 0$ and $P\big(W \in A\big) > 0$. If such a set exists, a tame trading strategy that represents guaranteed arbitrage can be constructed by using the predictable representation property of Brownian motion – see (Lyasoff 2014) for a detailed proof. The description of the no-arbitrage condition in terms of the distribution law of the normalized excess returns process β has the advantage that it is written entirely in terms of statistical features that, at least in theory, can be extracted from observable market data. In any case, the distribution laws of β and W are equivalent if and only if an ELMM exists. The precise statement (see ibid. for the details) is the following:

(13.23) FIRST FUNDAMENTAL THEOREM OF ASSET PRICING: If no ELMM exists, then there exists a universally tame trading strategy that represents guaranteed arbitrage. If at least one ELMM exists, then for any such measure, Q, no arbitrage with a Q-tame trading strategy is possible; in particular, if an ELMM exists, then no arbitrage with a universally tame trading strategy is possible. ○

(13.24) EXERCISE: Prove the easy part of the first fundamental theorem: if Q is an ELMM, then there is no Q-tame trading strategy that represents arbitrage; in particular, there is no universally tame trading strategy that represents arbitrage. ○

(13.25) NFLVR: We stress that the result in (13.23) – often abbreviated as FTAP – was established only in the special case of returns Z that can be expressed as an Itô process. The case of market returns that are nothing more than a general – and not necessarily continuous – semimartingale Z is much more involved and is considerably more technical. The most general version of the fundamental theorem of asset pricing was established in the groundbreaking works of Delbaen and Schachermayer (1994, 1995, 1998, 2010), and is somewhat weaker than the result stated in (13.23), in the sense that it involves a weaker notion of arbitrage.[152] This weaker notion of no-arbitrage is known as "no free lunch with vanishing risk" (acronym: NFLVR) and comes down to the following: there is no sequence of admissible trading strategies π^i, $i \in \mathbb{N}$, such that the associated sequence of random variables (the payoffs on the final date T)

$$\big((S^\circ)^{-1}(\pi^i)^\intercal \cdot X\big)_T, \quad i \in \mathbb{N},$$

[152] The fact that "arbitrage" implies "guaranteed arbitrage" appears to be an exclusive feature of market models with excess returns postulated to follow an Itô process.

converges a.s. to a positive r.v. as $i \to \infty$, such that

$$P\Big[\lim_i \big((S^\circ)^{-1}(\pi^i)^\intercal \bullet X\big)_T > 0\Big] > 0,$$

and, in addition, the sequence of random variables

$$\big((S^\circ)^{-1}(\pi^i)^\intercal \bullet X\big)_T^-, \quad i \in \mathbb{N},$$

(the negative parts of the payoffs on the final date T) converges to 0 uniformly as $i \to \infty$. In its full generality the FTAP essentially states that an ELMM exists if and only if the NFLVR condition holds. ○

Next comes another crucially important concept in asset pricing.

(13.26) MARKET COMPLETENESS: Suppose that the financial market with excess-returns process X, interest-rate process r, and information process Y, admits at least one ELMM. We say that such a market is complete if, given any $\mathscr{G}_T$-measurable r.v. ξ, one can construct a sequence of universally tame trading strategies π^i, $i \in \mathbb{N}$, and a sequence of initial investments x_i, $i \in \mathbb{N}$, such that (we stress that the limit is in probability P, or, which is the same, in probability Q for any $Q \approx P$)

$$P\text{-}\lim_{i\to\infty}\Big(x_i + \big((S^\circ)^{-1}(\pi^i)^\intercal \bullet X\big)_T\Big) = \xi . \quad ○$$

Essentially, market completeness means that any conceivable payoff can be replicated – at least in the limit – by using appropriate trading strategies. Since the normalized excess-returns process β is a Brownian motion relative to any ELMM, the (right-continuous and complete) filtration $\mathscr{F}^\beta$ that β generates⊸(10.46) must have the predictable representation property with respect to any ELMM Q, that is, any $(\mathscr{F}^\beta, Q)$-local martingale L has a continuous version that can be expressed in the form

$$L = L_0 + h^\intercal \bullet \beta , \qquad ⓔ_1$$

for some appropriately chosen $\mathscr{F}^\beta$-predictable $\mathbb{R}^n$-valued process h, with sample paths $h_\omega \in \mathscr{L}^2_{\text{loc}}(\mathbb{R}_+;\mathbb{R}^n)$. Since, in general, the market filtration $\mathscr{G}$ can be larger than the filtration $\mathscr{F}^\beta$, even though β is a $(\mathscr{G}, Q)$-Brownian motion, there is no reason why $\mathscr{G}$ should have the predictable representation property with respect to β and Q. What we call "market completeness" is the special situation in which the market filtration $\mathscr{G}$ does have the predictable representation property relative to β and Q: given any continuous $(\mathscr{G}, Q)$-local martingale L, one can find a $\mathscr{G}$-predictable $\mathbb{R}^n$-valued process h with sample paths $h_\omega \in \mathscr{L}^2_{\text{loc}}(\mathbb{R}_+;\mathbb{R}^n)$, such that L can be written in the form ⓔ$_1$. We have the following result – see (Lyasoff 2014) for a complete proof:

(13.27) SECOND FUNDAMENTAL THEOREM OF ASSET PRICING: A financial market that admits at least one ELMM is complete if and only if it admits exactly one ELMM. ○

In a more casual parlance, "market completeness" means that the information carried by the normalized excess-returns is all the information that is available to the agents. The next proposition (see ibid.) makes this interpretation precise.

(13.28) PROPOSITION: Suppose that the financial market with excess-returns process X, interest-rate process r, and information process Y admits an ELMM. Then this market is complete if and only if the distribution law of the normalized excess-returns β uniquely determines the distribution law of the process (X, r, Y), within the class of laws that are equivalent to the law of (X, r, Y). ○

(13.29) COROLLARY: If the normalized excess returns process β generates the market filtration $\mathscr{G}$, then the market is complete. Indeed, since X, r, and Y are all adapted to $\mathscr{G}$, the fact the $\mathscr{G}$ is generated by β is just another way of saying that X, r, and Y are functions of β, so that their respective distribution laws would be fixed once the distribution of β is fixed. While intuitive, the rigorous justification of this claim involves some delicate analytical results on the existence of regular transition probability densities – see (Parthasarathy 1977, prop. 46.3), for example.

Another situation in which market completeness follows directly from proposition (13.28) is when the market filtration $\mathscr{G}$ is generated only by X and r (there is no additional information brought in by the information process Y), and both the volatility and the interest rate are functions of β, that is, have the form $\sigma(\beta, \imath)$ and $r(\beta, \imath)$, for some predictable functions $\sigma\colon \mathbb{W} \times \mathbb{R}_+ \mapsto \mathbb{R}^{n\otimes n}$ and $r\colon \mathbb{W} \times \mathbb{R}_+ \mapsto \mathbb{R}$. Indeed, in that case $X = \sigma(\beta, \imath)\bullet\beta$ is a function of β and the market filtration $\mathscr{G}$ is simply $\mathscr{F}^\beta$. In particular, the market would be complete if $\sigma \in \mathbb{R}^{n\otimes n}$ and $r \in \mathbb{R}$ are constants and there is no additional information to be brought in by Y. We stress, however, that the mere absence of an information process Y does not imply completeness. Indeed, in general, σ can only be claimed to be adapted to $\mathscr{F}^X$, but not necessarily to $\mathscr{F}^\beta$. By way of an example, if the relation $X = \sigma\bullet\beta$ can be treated as an SDE (for X) of the form

$$X = \sigma(X, \imath)\bullet\beta\,,$$

then, without any additional assumptions about the coefficient $\sigma(\cdot,\cdot)$, there would be no reason for the solution X to be strong, or a reason for the distribution of X to be unique. In particular – see Tanaka's example in (11.14) and Tsirel'son's example in (11.33) – in such models the filtration of X could be strictly larger than the filtration of β. ○

14

Replication and Arbitrage

This chapter is concerned with the mechanics of some concrete asset pricing problems in an arbitrage-free market, in which an ELMM that prices all tradable assets has already been chosen. We adopt the market model, notation, and terminology that were introduced in chapter 13, and follow very closely (Merton 1973), (Karatzas 1996), (Karatzas and Shreve 1998), (Duffie 2010), and (El Karoui 1981). Since the predictable representation property will play an instrumental role in what follows, we need to take a quick detour and briefly review some of the most basic tools from Malliavin calculus, which will allow us to obtain predictable representations for certain random variables defined on the Wiener space.

14.1 Résumé of Malliavin Calculus

For the purpose of better readability, the main tools from Malliavin calculus will be introduced only in the case of one-dimensional Brownian motion, but we stress that all results presented in this section remain valid for d-dimensional Brownian motion as well – one merely has to give multivariate meaning to the symbols involved. The real-valued Brownian motion W and the finite time interval $[0,T]$ will be fixed throughout.

A *cylinder function* of the Brownian motion W is any r.v. of the form

$$\Phi = F\big(W_{t_1} - W_{t_0}, W_{t_2} - W_{t_1}, \dots, W_{t_n} - W_{t_{n-1}}\big),$$

for some choice of the subdivision $0 = t_0 < t_1 < \dots < t_n = T$ and the function $F \in \mathscr{C}_b^1(\mathbb{R}^n;\mathbb{R})$. *Malliavin's differential* of any such Φ is defined as the process

$$\Omega\times[0,T] \ni (\omega,t) \rightsquigarrow (D\Phi)(\omega,t) \in \mathbb{R},$$

given by

$$D\Phi = \sum\nolimits_{i=1}^{n} (\partial_i F)\big(W_{t_1} - W_{t_0}, W_{t_2} - W_{t_1}, \dots, W_{t_n} - W_{t_{n-1}}\big)\times 1_{[\![t_{i-1},t_i[\![}\,.$$

In the next two exercises Φ denotes a generic cylinder function of W, based on some unspecified subdivision of $[0,T]$ and unspecified $F \in \mathscr{C}_b^1(\mathbb{R}^n;\mathbb{R})$.

(14.1) EXERCISE: Let $f\colon[0,T] \mapsto \mathbb{R}$ be any simple (deterministic) step-function of the form

$$f(t) = \sum\nolimits_{i=1}^{n} a_i\, 1_{[t_{i-1},t_i[}(t)$$

for some choice of $a_i \in \mathbb{R}$ and subdivision $0 = t_0 < t_1 < \ldots < t_n = T$ (which need not be the same as the subdivision of Φ). Prove that

$$\mathsf{E}\Big[\Phi \int_0^T f(t)\,\mathrm{d}W_t\Big] = \mathsf{E}\Big[\int_0^T f(t) D\Phi_t\,\mathrm{d}t\Big],$$

and explain why this identity can be viewed as a generalization of the integration by parts formula from (3.65). ○

(14.2) EXERCISE: Show that, for every choice of $0 \leq s < t \leq T$,

$$\mathsf{E}\big[(W_t - W_s)\Phi \,|\, \mathscr{F}_s^W\big] = \int_s^t \mathsf{E}\big[D\Phi_u \,|\, \mathscr{F}_s^W\big]\,du.$$

Conclude that, given any elementary stochastic process of the form

† $$\psi = \sum\nolimits_{i=1}^n \xi_{i-1}\, 1_{[\![t_{i-1},t_i[\![},$$

for some choice of subdivision the $0 = t_0 < t_1 < \ldots < t_n = T$ (which need not be the same as the subdivision of Φ) and the random variables $\xi_{i-1} \in L^2(\Omega, \mathscr{F}_{t_{i-1}}^W, P)$, one has

‡ $$\mathsf{E}\Big[(\Phi - \mathsf{E}[\Phi])\int_0^T \psi_s\,\mathrm{d}W_s\Big] = \mathsf{E}\Big[\int_0^T \psi_s \mathsf{E}\big[D\Phi_s \,|\, \mathscr{F}_s^W\big]\,\mathrm{d}s\Big].$$

HINT:

$$\mathsf{E}\Big[\Phi \int_0^T \psi_s\,\mathrm{d}W_s\Big] = \mathsf{E}\Big[\sum_{i=1}^n \psi_{t_{i-1}} \int_{t_{i-1}}^{t_i} \mathsf{E}\big[D\Phi_s \,|\, \mathscr{F}_{t_{i-1}}^W\big]\,\mathrm{d}s\Big]. \quad ○$$

(14.3) PREDICTABLE REPRESENTATION REVISITED: Let $\Psi \in L^2(\Omega, \mathscr{F}_T^W, P)$ and let M be the L^2-bounded martingale $M_t \overset{\text{def}}{=} \mathsf{E}[\Psi \,|\, \mathscr{F}_t^W]$, $t \in [0,T]$. Since, in particular, M is an $(\mathscr{F}^W, P)$-local martingale, by⊶(10.46) the predictable representation result for Brownian filtrations,

$$M_t = M_0 + \int_0^t h_s\,\mathrm{d}W_s, \quad t \in [0,T],$$

for some $\mathscr{F}^W$-predictable process h with sample paths $h_\omega \in \mathscr{L}^2([0,T];\mathbb{R})$. Since in this case M is more than a local martingale, we can say more about the predictable process h. Specifically, we can claim that $\mathsf{E}\big[\int_0^T {h_s}^2\mathrm{d}s\big] < \infty$ and

$$\mathsf{E}[\Psi^2] = \mathsf{E}[\Psi]^2 + \mathsf{E}\Big[\int_0^T {h_s}^2\,\mathrm{d}s\Big].$$

These properties follow easily from the relations $\langle M, M\rangle = h^2 \bullet \iota$ and[153]

$$\mathsf{E}\big[M_T^2\big] = \mathsf{E}\big[M_0^2\big] + \mathsf{E}\big[\langle M, M\rangle_T\big] = M_0^2 + \mathsf{E}\big[(h^2 \bullet \iota)_T\big]. \quad ○$$

[153] Recall that the σ-field $\mathscr{F}_0^W$ is the trivial σ-field $\{\emptyset, \Omega\}$, completed with all P-null sets. Recall also that⊶(9.91) $M^2 - \langle M, M\rangle$ is a u.i. martingale, since M is a L^2-bounded martingale.

Since the cylinder function Φ is bounded by definition, there is an $\mathscr{F}^W$-predictable processes $(h_t)_{t\in[0,T]}$ such that

$$\Phi = \mathsf{E}[\Phi] + \int_0^T h_s \, \mathrm{d}W_s \quad \text{and} \quad \mathsf{E}[\Phi^2] = \mathsf{E}[\Phi]^2 + \int_0^T \mathsf{E}[h_s{}^2] \mathrm{d}s \, .$$

As a result, for any elementary process ψ of the form (14.2†),

$$\mathsf{E}\Big[\big(\Phi - \mathsf{E}[\Phi]\big) \int_0^T \psi_s \, \mathrm{d}W_s\Big] = \mathsf{E}\Big[\int_0^T \psi_s h_s \, \mathrm{d}s\Big] .$$

In conjunction with (14.2‡), this gives the identity

$$\mathsf{E}\Big[\int_0^T \psi_s \mathsf{E}\big[D\Phi_s \,|\, \mathscr{F}_s^W\big] \, \mathrm{d}s\Big] = \mathsf{E}\Big[\int_0^T \psi_s h_s \, \mathrm{d}s\Big] ,$$

which holds for every elementary process ψ. This leads to the relation

$$h_t = \mathsf{E}\big[D\Phi_s \,|\, \mathscr{F}_t^W\big] \quad \mathrm{d}P \times \mathrm{d}t\text{-a.e.}, \qquad \mathrm{e}_1$$

which is a special case of the Clark–Ocone formula – see (Clark 1970), (Ocone 1984), and (Karatzas et al. 1991). The idea behind the Clark–Ocone formula is to extend e_1 for the broadest possible class of functions of W that can be identified as limits of cylinder functions, the limits being understood for a particular type of convergence, designed in such a way that e_1 survives the limits on both sides. To formulate this result, we need to introduce the space $\mathbb{D}_1^2(W, [0, T])$, defined as the collection of all $\Phi \in L^2(\Omega, \mathscr{F}_T^W, P)$ such that for some sequence of cylinder functions $(\Phi_i)_{i\in\mathbb{N}}$ one has simultaneously $\lim_i \mathsf{E}[|\Phi - \Phi_i|^2] = 0$ and

$$\lim_i \sup_{j\geq i} \mathsf{E}\Big[\int_0^T \big|(D\Phi_j)_s - (D\Phi_i)_s\big|^2 \, \mathrm{d}t\Big] = 0 .$$

Given any $\Phi \in \mathbb{D}_1^2(W, [0, T])$, its Malliavin differential, $D\Phi$, is the element of the space

$$L^2\big(\Omega\times[0,T] , \mathscr{F}_T^W \otimes \mathcal{B}([0,T]), P\times\Lambda\big)$$

identified as the limit in that space of the sequence $(D\Phi_i)_{i\in\mathbb{N}}$. A straightforward limit operation now leads to the following result – see (Clark 1970) and (Ocone 1984):

(14.4) CLARK–OCONE FORMULA: For any $\Phi \in \mathbb{D}_1^2(W, [0, T])$,

$$\Phi - \mathsf{E}[\Phi] = \int_0^T \mathsf{E}\big[D\Phi_s \,|\, \mathscr{F}_s^W\big] \, \mathrm{d}W_s . \quad \circ$$

As an illustration of this result, fix some $t \in \,]0, T[$, let $\Phi = e^{W_t}$ and observe that with this choice $D\Phi_s = e^{W_t} 1_{[0,t[}(s)$, so that

$$\mathsf{E}\big[D\Phi_s \,|\, \mathscr{F}_s^W\big] = e^{\frac{1}{2}(t-s)} e^{W_s} 1_{[0,t[}(s) .$$

In particular, $\mathsf{E}\big[D\Phi_s \,|\, \mathscr{F}_s^W\big] = 0$ for $s \in [t, T]$, and the Clark–Ocone formula yields

$$e^{W_t} - e^{\frac{1}{2}t} = \int_0^t e^{\frac{1}{2}(t-s)} e^{W_s} \,\mathrm{d}W_s \,.$$

The same identity can be established without relying on the Clark–Ocone formula, since multiplying both sides by $e^{-\frac{1}{2}t}$ gives

$$e^{W_t - \frac{1}{2}t} - 1 = \int_0^t e^{W_s - \frac{1}{2}s} \,\mathrm{d}W_s \,,$$

and this relation follows immediately from the Itô formula.

(14.5) EXERCISE: Using the Clark–Ocone formula, find the predictable representation of the r.v. ${W_T}^4$, for $T > 0$. Verify your answer independently using Itô's formula. ○

(14.6) EXERCISE: Using the Clark–Ocone formula, find the predictable representation of $(a\, e^{W_T} - b)^+$, where $a, b \in \mathbb{R}_{++}$ are given constants.

HINT: For the purpose of computing Malliavin's differential, the derivative of the function $x \rightsquigarrow x^+ = x \vee 0$ can be taken to be the indicator $1_{[0,+\infty[}$. You can justify this step by using the mollification procedure described in (0.11). ○

14.2 European-Style Contingent Claims

With the notation as in chapter 13, we take the market model introduced there as given, and suppose that an ELMM Q exists and is fixed. We start with:

(14.7) DEFINITIONS: A *European-style contingent claim* is defined by: (a) its expiration (closing) date $T > 0$, (b) its closing payoff Φ, and (c) its payoff rate $\varphi \equiv (\varphi_t)_{t \in [0,T]}$. The closing payoff Φ is some positive and $\mathscr{G}_T$-measurable r.v., while the payoff rate φ is some positive and $\mathscr{P}(\mathscr{G})$-measurable (predictable for $\mathscr{G}$) process.[154] Until the closing date T the holder of the contract receives (continuously) the instantaneous payoffs $(\varphi_t \,\mathrm{d}t)$ and on the closing date T they receive the final (closing) payment Φ. We will express such contracts as $\mathscr{K}_0(T, \Phi, \varphi)$, and write $\mathscr{K}_0(T, \Phi)$ if the payoff rate happens to be null.

An *upper hedging strategy* (alias: *super-replicating strategy*) for $\mathscr{K}_0(T, \Phi, \varphi)$ is any⊸ sec. 13.2 Q-tame self-financing investment-consumption strategy (x, π^+, c) such that (a.s.) $V_T \geq \Phi$ and $c_t \geq \varphi_t$ for all $t \in [0, T]$, where⊸ sec. 13.2 $V \equiv V^{x,\pi^+,c}$ is the wealth process that (x, π^+, c) generates.

The *upper hedging price* (alias: *super-replicating price*) for $\mathscr{K}_0(T, \Phi, \varphi)$ is defined as the infimum, Π^+, of the set $A^+ \stackrel{\text{def}}{=}$ the collection of all initial investments x, in all possible upper hedging strategies (x, π^+, c).

A lower hedging strategy for $\mathscr{K}_0(T, \Phi, \varphi)$ is any⊸ sec. 13.2 Q-tame self-financing investment-payout strategy (x, π^-, c) such that (a.s.) $\tilde{V}_T \leq \Phi$ and $c_t \leq \varphi_t$

---

[154] Recall that $\mathscr{G} = (\mathscr{G}_t)_{t \in [0,T]}$ is the market (observation) filtration.

for all $t \in [0,T]$, where$^{\circ\!\!\to \text{sec. } 13.2}$ $\tilde{V} \equiv \tilde{V}^{x,\pi^-,c}$ is the debt process that (x,π^-,c) generates.

The *lower hedging price* for $\mathscr{K}_0(T,\Phi,\varphi)$ is defined as the supremum, Π^-, of the set $A^- \stackrel{\text{def}}{=}$ the collection of all initial loans x, in all possible lower hedging strategies (x,π^-,c). ○

Since all investment-consumption and investment-payout strategies that we are going to encounter below are going to be Q-tame and self-financing, for brevity, we will often drop the qualifiers "Q-tame" and "self-financing" when referring to such strategies.

(14.8) REMARKS: Any upper hedging strategy can be used by the seller of the option as a hedge against their obligations on the contract. The upper hedging price gives the lowest initial cost at which the seller can generate an investment-consumption strategy, the payoffs from which exceed the seller's obligations on the option contract at all times and in all possible states of the economy. Similarly, a lower hedging strategy can be used by the buyer as a hedge against the cost of acquiring the option (the buyer's only liability on the contract). The lower hedging price is then the largest loan that the holder of the option can initially receive, and, consequently, maintain by following some investment-payout strategy, such that the associated payouts are lower than the payoffs received from the option. ○

(14.9) EXERCISE: Prove that the lower hedging price Π^- and the upper hedging price Π^+ must satisfy the relation $\Pi^- \leq \Pi^+$.

HINT: With Q-tame trading strategies π the right sides of equations e_3 and e_4 in section 13.2 represent, respectively, a supermartingale and a submartingale. The difference of the two right sides is then a supermartingale. It is straightforward from the definitions of upper and lower hedging strategy that at time $t = T$ the r.v. on the left side of e_3 dominates the r.v. on the left side of e_4 (a.s.). What can you say about the starting value at time $t = 0$ of a supermartingale that happens to be positive at time $t = T > 0$? ○

(14.10) THE ARBITRAGE-FREE PRICE OF A CONTINGENT CLAIM: Let Π denote the price of the contingent claim $\mathscr{K}_0(T,\Phi,\varphi)$. Clearly, $\Pi > \Pi^+$ creates an arbitrage opportunity for the seller: they would receive the amount Π but only need an amount strictly smaller than Π to cover (hedge) their liabilities on the contract. Similarly, $\Pi < \Pi^-$ creates an arbitrage opportunity for the buyer: they only need the amount Π to cover (hedge) their liabilities on a loan maintained by some investment-payout strategy, the initial proceeds from which are strictly larger than Π. As a result, the arbitrage-free price of the claim, Π, must be in the interval $[\Pi^-, \Pi^+]$. ○

The classical illustration of pricing by arbitrage is the pricing of the contingent claim $\mathscr{K}_0(T,\Phi,\varphi)$, under the assumption that the market is complete (and free of arbitrage, as was postulated earlier). By the second fundamental theorem of asset pricing the ELMM Q is unique, and we now impose the following two integrability conditions on the contingent claim:

$$\mathsf{E}^Q\big[(S^\circ_T)^{-1}\Phi\big] < \infty \quad \text{and} \quad \mathsf{E}^Q\Big[\int_0^T (S^\circ_u)^{-1}\varphi_u \,\mathrm{d}u\Big] < \infty\,,$$

which allow us to introduce the $(\mathscr{G}, Q)$-martingale

$$M_t \overset{\text{def}}{=} \mathsf{E}^Q\Big[(S^\circ_T)^{-1}\Phi + \int_0^T (S^\circ_u)^{-1}\varphi_u \,\mathrm{d}u \mid \mathscr{G}_t\Big], \quad t \in [0,T]\,.$$

Since the market is complete, the predictable representation property for the σ-field $\mathscr{G}$ with respect to $^{\multimap(13.10)}$ the $(\mathscr{G}, Q)$-Brownian motion β holds, and, as a result, there is a n-dimensional and $\mathscr{G}$-predictable process h with sample paths $h_\omega \in \mathscr{L}^2_{\text{loc}}(\mathbb{R}_+;\mathbb{R}^n)$, such that

$$M_t - M_0 = \int_0^t h_s{}^\intercal \,\mathrm{d}\beta_s\,, \quad t \in [0,T]\,.$$

(14.11) EXERCISE: (a) Prove that if V_0^+ is the initial investment in some (any) upper hedging strategy, and if V_0^- is the initial loan in some (any) lower hedging strategy, then $V_0^- \le M_0 \le V_0^+$. Conclude that $\Pi^- \le M_0 \le \Pi^+$.

(b) Prove that the trading strategies $\pi_t^\pm \overset{\text{def}}{=} \pm S_t^\circ(\sigma_t{}^\intercal)^{-1}h_t$, $t \in [0,T]$, are both Q-tame.

(c) Prove that the investment-consumption strategy given by the initial investment $V_0 = x = M_0$, the portfolio process π^+ chosen as in part (b), and the consumption process $c = \varphi$, is an upper hedging strategy for $\mathscr{K}_0(T,\Phi,\varphi)$. Conclude that $M_0 \ge \Pi^+$.

(d) Prove that the investment-payout strategy given by the initial loan $\tilde{V}_0 = x = M_0$, the portfolio process π^- chosen as in part (b), and the payout process $c = \varphi$, is a lower hedging strategy for $\mathscr{K}_0(T,\Phi,\varphi)$. Conclude that $M_0 \le \Pi^-$.

(e) Finally, conclude that the arbitrage-free price of $\mathscr{K}_0(T,\Phi,\varphi)$ is given by

$$\Pi = \Pi^- = \Pi^+ = M_0 = \mathsf{E}^Q\Big[(S^\circ_T)^{-1}\Phi + \int_0^T (S^\circ_u)^{-1}\varphi_u \,\mathrm{d}u\Big]\,.$$

HINT: For part (a), use the fact that a supermartingale (written as the difference between a supermartingale and a martingale, or as the difference between a martingale and submartingale) that is positive at time $t = T$ must be positive at time $t = 0$. This will show that M_0 cannot exceed the initial investment in any upper hedging strategy and cannot be smaller than the initial loan in any lower hedging strategy. Then observe that, just by definition,

$$\begin{aligned} M_t &= (S^\circ_t)^{-1}\mathsf{E}^Q\Big[e^{-\int_t^T r_u \,\mathrm{d}u}\Phi + \int_t^T e^{-\int_t^s r_u \,\mathrm{d}u}\varphi_s \,\mathrm{d}s \mid \mathscr{G}_t\Big] + \int_0^t (S^\circ_u)^{-1}\varphi_u \,\mathrm{d}u \\ &= M_0 + \int_0^t h_s^\intercal \,\mathrm{d}\beta_s = M_0 \pm \int_0^t (S^\circ_u)^{-1}(\pi_u^\pm)^\intercal \,\mathrm{d}X_u\,. \end{aligned}$$

Conclude that the wealth process generated by the investment-consumption strategy (M_0, π^+, φ) must be given by

$$V_t = \mathsf{E}^Q\Big[e^{-\int_t^T r_u \mathrm{d}u}\Phi + \int_t^T e^{-\int_t^s r_u \mathrm{d}u}\varphi_s \mathrm{d}s \mid \mathscr{G}_t\Big], \quad t \in [0,T],$$

so that $V_T = \Phi$. Argue that π^+ is Q-tame and that (M_0, π^+, φ) is an upper hedging strategy for $\mathscr{K}_0(T, \Phi, \varphi)$. Conclude that $M_0 \geq \Pi^+$. Similarly, the debt process generated by the investment-payout strategy (M_0, π^-, φ) must be given by

$$\tilde{V}_t = \mathsf{E}^Q\Big[e^{-\int_t^T r_u \mathrm{d}u}\Phi + \int_t^T e^{-\int_t^s r_u \mathrm{d}u}\varphi_s \mathrm{d}s \mid \mathscr{G}_t\Big] = V_t, \quad t \in [0,T],$$

so that $\tilde{V}_T = \Phi$ again. Argue that π^- is Q-tame and that (M_0, π^-, φ) is a lower hedging strategy for $\mathscr{K}_0(T, \Phi, \varphi)$. Conclude that $M_0 \leq \Pi^-$. ○

(14.12) REPLICATING STRATEGY: A replicating strategy for $\mathscr{K}_0(T, \Phi, \varphi)$ is any investment-consumption strategy (x, π, c) that is an upper hedging strategy for $\mathscr{K}_0(T, \Phi, \varphi)$, and is such that the investment-payout strategy $(x, -\pi, c)$ is a lower hedging strategy for $\mathscr{K}_0(T, \Phi, \varphi)$.[155] If (x, π, c) is a replicating strategy, then the amount x is both an initial investment in an upper hedging strategy and an initial loan in a lower hedging strategy. This is only possible if $x = \Pi$ (the price of the claim). Similarly, since $c \geq \varphi \geq c$, one must have $c = \varphi$. Furthermore, equations e_3 and e_4 from section 13.2, written, respectively, for the investment-consumption strategy (x, π, c) and for the investment-payout strategy $(x, -\pi, c)$, would have identical right sides and the second terms on the left sides would be identical as well. As a result, the wealth process generated by (x, π, c) must be identical to the debt process generated by $(x, -\pi, c)$. Since (x, π, c) is an upper hedging strategy and $(x, -\pi, c)$ is a lower hedging strategy, the associated wealth (debt) process V must be such that $V_T \leq \Phi \leq V_T$, which is to say $V_T = \Phi$. As a result, a buyer would be indifferent to buying $\mathscr{K}_0(T, \Phi, \varphi)$ at price Π, or, instead, following the investment-consumption strategy (Π, π, φ). Similarly, an underwriter would be indifferent to underwriting $\mathscr{K}_0(T, \Phi, \varphi)$ (and assuming the liabilities attached to it), or, instead, following the investment-payout strategy $(x, -\pi, c) \equiv (\Pi, -\pi, \varphi)$. ○

It is clear that the construction of a replicating strategy comes down to computing a particular predictable representation, and the next exercise clarifies this connection.

(14.13) EXERCISE: Assuming that the market is free of arbitrage and complete, and assuming that $\mathscr{K}_0(T, \Phi, \varphi)$ satisfies the integrability conditions in e_1, prove that the investment-consumption strategy (x, π, c), with $x = \Pi$ (the price of the contingent claim), $c = \varphi$, and π chosen as in exercise (14.11b), is a replicating strategy for $\mathscr{K}_0(T, \Phi, \varphi)$. In addition, prove that this replicating strategy is unique.

[155] Formally, "replicating strategy" is a reference to two strategies, but these strategies are mirror images of one another, whence the use of "strategy" in singular.

HINT: The first part follows from exercise (14.11). As for the uniqueness part, let V be the wealth generated by some replicating strategy (in particular $V_T = \Phi$), and consider the supermartingale (under Q)

$$M \stackrel{\text{def}}{=} (S^\circ)^{-1} V + (S^\circ)^{-1} \varphi \boldsymbol{\cdot} \iota = \Pi + (S^\circ)^{-1} \pi^\intercal \boldsymbol{\cdot} X\,,$$

which starts from

$$M_0 = \Pi = \mathsf{E}^Q\Big[(S_T^\circ)^{-1}\Phi + \int_0^T (S_u^\circ)^{-1}\varphi_u\,\mathrm{d}u\Big] = \mathsf{E}^Q[M_T]\,.$$

Conclude that M is in fact a martingale on $[0,T]$ (under Q) and, finally, use⊸ (10.48) the uniqueness of the predictable representation to conclude that π is determined uniquely by Φ and φ. ○

(14.14) THE BLACK–SCHOLES–MERTON (BSM) MODEL: The economy has a finite time horizon $T > 0$, there is only one risky security ($n = 1$), and the market filtration $\mathscr{G}$ is simply the filtration $\mathscr{F}^{X,r}$ (no information process). Furthermore $\sigma_t = \sigma =$ constant and $r_t = r =$ constant, so that $\theta = \frac{b-r}{\sigma} =$ constant, and $\beta = W + \theta\iota \equiv W + \langle W, \theta W\rangle$. Girsanov's theorem implies that β is a Brownian motion (on the time interval $[0,T]$) with respect to the measure

$$Q = \left(e^{-\theta W_T - \frac{1}{2}\theta^2 T}\right) \odot P\,.$$

In particular $X = \sigma\beta$ is a local $(\mathscr{G}, Q)$-martingale, so that Q is an ELMM. By the first fundamental theorem the market is free of arbitrage. Since the interest rate process is constant, and since the excess returns process is simply a multiple of the normalized excess-returns process β, the distribution law of β completely determines the distribution laws of X (and also the law of r, but in a trivial way). By proposition (13.28), the market is complete and the arbitrage-free price of any contingent claim $\mathscr{K}_0(T, \Phi, \varphi)$ can be identified as the expectation (relative to Q):

$$\mathsf{E}^Q\Big[e^{-rT}\Phi + \int_0^T e^{-rs}\varphi_s\,\mathrm{d}s\Big]\,.$$

There are no dividends and the net returns from the risky asset are given by the semimartingale $Z = Z_0 + \sigma\beta + r\iota$, its spot price can be expressed as

$$S_t = e^{Z_t - \frac{1}{2}\langle Z,Z\rangle_t} = e^{Z_0} e^{(Z_t - Z_0) - \frac{1}{2}\langle Z,Z\rangle_t} = S_0 e^{\sigma\beta_t + rt - \frac{1}{2}\sigma^2 t}\,.$$

In particular, for a contingent claim with $\varphi_t = 0$ and $\Phi = F(S_T)$ for some Borel function $F\colon \mathbb{R}_+ \mapsto \mathbb{R}$ with at most polynomial growth, we have

$$\begin{aligned}\text{Price of } \mathscr{K}_0(T, F(S_T)) &= \mathsf{E}^Q\big[e^{-rT} F\big(S_0 e^{\sigma\beta_T + rT - \frac{1}{2}\sigma^2 T}\big)\big] \\ &= \frac{e^{-rT}}{\sqrt{2\pi T}} \int_{-\infty}^{\infty} F\big(S_0 e^{\sigma x + rT - \frac{1}{2}\sigma^2 T}\big)\, e^{-\frac{x^2}{2T}}\,\mathrm{d}x\,. \quad ○\end{aligned}$$

(14.15) EXERCISE: Explain exactly how Girsanov's theorem is used in the Black–Scholes–Merton model. Why is the associated Radon–Nikodym derivative chosen that way? What exactly does Girsanov's theorem say? Why are the assumptions in this theorem satisfied? Is Novikov's condition satisfied? Can one get away without using Novikov's criterion in this case? ○

(14.16) EXERCISE: In the context of the Black–Scholes–Merton model, construct a replicating strategy for the contingent claim $\mathscr{K}_0(T, S_T^4)$. ○

(14.17) EXERCISE: In the context of the Black–Scholes–Merton model, outline the steps in the construction of a replicating strategy for the call $\mathscr{K}_0(T,(S_T-K)^+)$, for a fixed strike $K > 0$. ○

(14.18) "BACK OF THE ENVELOPE" DERIVATION OF THE BSM MODEL: In an arbitrage-free and complete financial market, there is an equivalent measure Q that gives the prices of all traded assets (stocks, options, bonds, etc.) as expected (relative to Q) and discounted (with the risk-free rate $r > 0$) future stochastic payoffs. Because the ex-dividend spot price S_t is nothing but the price of all future stochastic payoffs, we have

$$S_t = \mathsf{E}^Q\left[\frac{S_{t+\mathrm{d}t} + \delta S_t\,\mathrm{d}t}{1 + r\mathrm{d}t} \;\Big|\; \mathscr{G}_t\right],$$

where $\delta \geq 0$ stands for the net dividend rate. With a simple algebra the last relation can be rearranged as $\mathsf{E}^Q[\mathrm{d}X_t|\mathscr{G}_t] = 0$, where

$$\mathrm{d}X_t = \frac{S_{t+\mathrm{d}t} - S_t}{S_t} + \delta\,\mathrm{d}t - r\,\mathrm{d}t\,.$$

As a result,

$$S_{t+\mathrm{d}t} - S_t \equiv \mathrm{d}S_t = S_t(\mathrm{d}X_t + r\mathrm{d}t - \delta\,\mathrm{d}t)\,, \quad t \in \mathbb{R}_+\,,$$

which is the same as

$$S_u = S_t e^{(X_u - X_t) + (r-\delta)(u-t) - \frac{1}{2}\left(\langle X,X\rangle_u - \langle X,X\rangle_t\right)}\,, \quad u \geq t\,,$$

and we note that the process X is a local $(\mathscr{G}, Q)$-martingale – by the very definition of ELMM (aliases: risk-neutral measure, risk-adjusted measure). It is important to recognize that the price dynamics of the stock price S under the real-world probability measure is not used in this argument at all. In particular, under the real-world probability measure P the excess returns process X could be very far from a local martingale. If we make the assumption that $\langle X, X\rangle = \sigma\iota$ (under both P and Q – the bracket does not depend on the measure up to equivalence) we can claim the $X = \sigma\tilde{W}$, where $\tilde{W}$ is a $(\mathscr{G}, Q)$-Brownian motion.

Yet another – perhaps even more intuitive – derivation is as follows: borrowing ¤1 at time t and investing it in the stock (at time t) generates payoff at time $t + \mathrm{d}t$ (after closing the position) equal to

$$\mathrm{d}X_t = 1 + \frac{\mathrm{d}S_t}{S_t} + \delta\,\mathrm{d}t - (1 + r\mathrm{d}t) = \frac{S_{t+\mathrm{d}t} - S_t}{S_t} + \delta\,\mathrm{d}t - r\,\mathrm{d}t\,.$$

Because this payoff is generated at no cost at time t, its average under the pricing measure Q must be 0, that is, one should have $\mathsf{E}^Q[\mathrm{d}X_t|\mathscr{G}_t] = 0$. ○

14.3 The Martingale Solution to Merton's Problem

In the setting of the previous section and of chapter 13, consider an investor who has initial wealth $x^* > 0$ at time $t = 0$, who is faced with finite investment horizon $T > 0$, and who has adopted $U : \mathbb{R}_{++} \mapsto \mathbb{R}$ as their utility function for intertemporal consumption and $\bar{U} : \mathbb{R}_{++} \mapsto \mathbb{R}$ as their utility function for final wealth. The market is again free of arbitrage and complete, and the agent is seeking$^{\multimap(13.16)}$ a tame (relative to the unique ELMM Q) trading strategy $\pi = (\pi_t)_{t \in [0,T]}$, and a consumption plan $c = (c_t)_{t \in [0,T]}$, so as to maximize their time-separable expected utility from consumption and final wealth, expressed as

$$\mathsf{E}\Big[\big((e^{-\beta\iota} U(c)\big) \bullet \iota_T + e^{-\beta T}\bar{U}(V_T)\Big], \qquad e_1$$

or, if one insists on a more detailed notation, expressed as

$$\mathsf{E}\Big[\int_0^T e^{-\beta s} U(c_s)\mathrm{d}s + e^{-\beta T}\bar{U}(V_T^{x^*,\pi,c})\Big].$$

In this expression $\beta > 0$ is a model parameter that captures the agent's impatience toward consumption. Problems of this form were first formulated and studied by Robert C. Merton – see (Merton 1969b, 1971) – and carry his name. In chapter 18 we will develop solutions to Merton's problem in terms of dynamic programming, and our goal here is to show how the replication method for European contingent claims can be used for the purpose of solving Merton's problem as well. The approach that we are about to describe is known as "the martingale method," and has the advantage that it does not require Markovian structure of the spot prices, as the dynamic programming approach (adopted in Merton's original work) does. The martingale method was developed during the 1970s and the 1980s in the groundbreaking works of Bismut (1973, 1975), Harrison and Kreps (1979), Harrison and Pliska (1981, 1983), Pliska (1986), Karatzas et al. (1987), and Cox and Huang (1989, 1991). Since this topic can be found in a number of classical texts on asset pricing, instead of following the original sources, we will follow very closely (Karatzas and Shreve 1998, ch. 3) and (Duffie 2010, ch. 9).

Before we can turn to the actual description of the martingale method, we must make a quick detour to the domain of convex analysis, with (Rockafellar 1970) as the default reference. Recall that the *epigraph* of a function $f : \mathbb{R} \mapsto\,]-\infty, \infty]$ is the subset of $\mathbb{R}^2$ defined as

$$\mathrm{Epi}(f) \overset{\text{def}}{=} \big\{(x, y) \in \mathbb{R}^2 : y \geq f(x),\ x \in \mathbb{R}\big\}.$$

Plainly, $\mathrm{Epi}(f)$ is the portion of $\mathbb{R}^2$ that contains the graph of f, on the domain on which it is finite, and all points (if any) in $\mathbb{R}^2$ that are above that graph. Recall also that f is said to be *convex* if $\mathrm{Epi}(f)$ is a nonempty closed convex subset of $\mathbb{R}^2$, and that every nonempty closed convex subset of $\mathbb{R}^2$ that is not $\mathbb{R}^2$ can be identified as the intersection of all half-planes that contain that set. In particular, any convex function $f\colon \mathbb{R} \mapsto]-\infty, \infty]$ that is not identical to $+\infty$ can be expressed as the pointwise supremum over all affine functions $x \rightsquigarrow ax - b$ the graphs of which (obviously, straight lines) are below the epigraph of f, that is, $y \geq ax - b$ for all $(x, y) \in \mathrm{Epi}(f)$.[156] Now observe that this relation is the same as $f(x) \geq ax - b$ for all $x \in \mathbb{R}$, which is the same as $b \geq \sup_{x\in\mathbb{R}}(ax - f(x))$. Furthermore, the inequality $b \geq ax - f(x)$, at a given point x, becomes a strict equality if and only if the line $x \rightsquigarrow ax - b$ happens to be a supporting line for $\mathrm{Epi}(f)$, that is, it contains the point $(x, f(x))$ while still being below $\mathrm{Epi}(f)$. If f admits continuous first derivative ∂f, then the tangent line to the graph of f at the point $(x, f(x))$ would be the only supporting line, that is, $x \rightsquigarrow ax - b$ can be claimed to be a supporting line at $(x, f(x))$ if and only if $a = \partial f(x)$ and $b = x\partial f(x) - f(x)$.

Next, recall that the Fenchel–Legendre conjugate (or transform)[157] of the convex function $f\colon \mathbb{R} \mapsto]-\infty, \infty]$ is defined as the function

$$\mathbb{R} \ni a \rightsquigarrow f^*(a) \stackrel{\text{def}}{=} \sup_{x\in\mathbb{R}}(ax - f(x)) \in \mathbb{R}\,.$$

The function f^* is itself convex, and its conjugate $(f^*)^*$ is nothing but f.[158] As a result, we have

$$ax - f(x) \leq f^*(a) \quad \text{for all } a, x \in \mathbb{R}$$

(with equality possible if and only if $a = \partial f(x)$ in the case where ∂f exists and is continuous), and we note that this relation can be written equivalently as

$$-ax - f(x) \leq f^*(-a) \quad \text{for all } a, x \in \mathbb{R}\,.$$

Now consider the utility functions $U, \bar{U}\colon \mathbb{R}_{++} \mapsto \mathbb{R}$, which, being utility functions, are concave, and extend their definitions to concave functions on the entire real line $\mathbb{R}$ by setting $U(x) = \bar{U}(x) = -\infty$ for any $x \in]-\infty, 0]$. In addition, suppose that the first derivatives ∂U and $\partial\bar{U}$ are continuous and strictly decreasing in $\mathbb{R}_{++}$, and are also such that (Inada's condition)

$$\lim_{x\searrow 0} \partial U(x) = \lim_{x\searrow 0} \partial\bar{U}(x) = \infty \quad\text{and}\quad \lim_{x\nearrow\infty} \partial U(x) = \lim_{x\nearrow\infty} \partial\bar{U}(x) = 0\,.$$

Since the functions $-U$ and $-\bar{U}$ are convex and map $\mathbb{R}$ into $]-\infty, \infty]$, we have

$$U(x) - ax \leq (-U)^*(-a) \ \text{and}\ \bar{U}(x) - ax \leq (-\bar{U})^*(-a) \quad \text{for all } a, x \in \mathbb{R}\,,$$

with identities possible for $x \in \mathbb{R}_{++}$ if and only if

$$\partial U(x) = a \iff x = (\partial U)^{-1}(a) \quad\text{and}\quad \partial\bar{U}(x) = a \iff x = (\partial\bar{U})^{-1}(a)$$

(note that $\partial U, \partial\bar{U}\colon \mathbb{R}_{++} \mapsto \mathbb{R}_{++}$ are one-to-one and invertible by assumption).

We can now return to the problem of maximizing the objective in e_1, subject to the constraint

$$\mathsf{E}^Q\Big[\big((S^\circ)^{-1}c\big)\bullet\imath_T + (S^\circ_T)^{-1}V_T\Big] = x^*.$$

[156] Notice that "below" does not imply that the line does not intersect $\mathrm{Epi}(f)$.

[157] Named after Danish mathematician Moritz Werner Fenchel (1905–1988) and the French mathematician Adrien-Marie Legendre (1752–1833).

[158] This property is well known, but is not obvious and requires a proof.

This is condition e_2 from section 13.3, which holds for any efficient Q-tame trading strategy. Indeed, basic intuition is telling us that the investment-consumption decision (π, c) must be efficient to be declared optimal – notice that both utility functions U and $\bar{U}$ are strictly increasing. Since the objective e_1 is expressed in terms of the probability measure P (we insist that the agent is maximizing their expected utility in the real world), it would be convenient to rewrite this constraint in terms of the measure P instead of Q. For this reason, we now introduce the $(\mathscr{G}, P)$-martingale[159] $(R_t)_{t\in[0,T]}$ given by $R_t \overset{\text{def}}{=} \frac{\mathrm{d}Q}{\mathrm{d}P}\restriction\mathscr{G}_t$ and cast the constraint above as

$$\mathsf{E}\Big[\big((S^\circ)^{-1}R\,c\big)\bullet\imath_T + (S^\circ_T)^{-1}R_T V_T\Big] = x^*, \qquad \mathrm{e}_2$$

where E is understood as E^P (for simplicity, we emphasize the measure in the notation only if it is different from the underlying measure in the model).

In any case, the agent is now faced with a single objective function given by e_1 and a single constraint given by e_2. By mimicking the standard steps in the method of Lagrange, this optimization problem can be transformed into the following unconstrained one:

$$\inf\nolimits_{\phi\in\mathbb{R}}\sup\nolimits_{c,\pi} L(\pi, c, \phi),$$

where

$$\begin{aligned} L(\pi, c, \phi) &= \mathsf{E}\big[e^{-\beta\imath}U(c)\bullet\imath_T + e^{-\beta T}\bar{U}(V_T)\big] \\ &\qquad + \phi\Big(x^* - \mathsf{E}\Big[\big((S^\circ)^{-1}R\,c\big)\bullet\imath_T + (S^\circ_T)^{-1}R_T V_T\Big]\Big), \end{aligned}$$

and the $\sup_{c,\pi}$ is understood with respect to all Q-tame investment-consumption strategies. Given that the utility functions involved are strictly increasing, the shadow price of the initial wealth, that is, the Lagrange multiplier ϕ, must be strictly positive, so that we can replace $\inf_{\phi\in\mathbb{R}}$ above with $\inf_{\phi\in\mathbb{R}_{++}}$. In particular,

$$\begin{aligned} L(\pi, c, \phi) &= \phi x^* + \mathsf{E}\Big[e^{-\beta\imath}\Big(U(c) - \phi e^{\beta\imath}(S^\circ)^{-1}R\,c\Big)\bullet\imath_T \\ &\qquad\qquad + e^{-\beta T}\Big(\bar{U}(V_T) - \phi e^{\beta T}(S^\circ_T)^{-1}R_T V_T\Big)\Big] \\ &\le \phi x^* + \mathsf{E}\Big[e^{-\beta\imath}\Big((-U)^*\big(-\phi e^{\beta\imath}(S^\circ)^{-1}R\big)\Big)\bullet\imath_T \\ &\qquad\qquad + e^{-\beta T}(-\bar{U})^*\big(-\phi e^{\beta T}(S^\circ_T)^{-1}R_T\big)\Big], \end{aligned}$$

where the upper bound can be attained, for fixed value of the Lagrange multiplier ϕ, with the following choice:

$$c^\phi_t = (\partial U)^{-1}\big(\phi e^{\beta t}(S^\circ_t)^{-1}R_t\big) \text{ and } V^\phi_T = (\partial\bar{U})^{-1}\big(\phi e^{\beta T}(S^\circ_T)^{-1}R_T\big). \qquad \mathrm{e}_3$$

(14.19) EXERCISE: Show that if (x^*, π, c) is *any* Q-tame investment-consumption strategy (with initial investment equal to the agent's initial wealth), and if $\mathcal{V}$ is the

[159] Recall that $\mathscr{G}$ is the market (information) filtration introduced in chapter 13 – see (13.5).

wealth process that this strategy generates, then

$$U(c_t^\phi) \ge U(c_t) + \phi e^{\beta t}(S_t^\circ)^{-1} R_t (c_t^\phi - c_t) \quad \text{for all } t \in [0,T],$$

and

$$\bar{U}(V_T^\phi) \ge \bar{U}(\mathcal{V}_T) + \phi e^{\beta T}(S_T^\circ)^{-1} R_T \big(V_T^\phi - \mathcal{V}_T\big),$$

where the quantities c_t^ϕ and V_T^ϕ are chosen as in e_3.

HINT: These relations follow immediately from the definition of the Fenchel-Legendre conjugate, in conjunction with the choice of c_t^ϕ and V_T^ϕ. ○

With the substitution $c_t = c_t^\phi$ and $V_T = V_T^\phi$ (from e_3), the left side of the constraint e_2 can be treated as a function of ϕ, and, as a result, e_2 becomes an equation for the Lagrange multiplier ϕ. Of course, for this equation to be meaningful, its left side must be finite – at least for some sufficiently large range of values for ϕ. As the functions $(\partial U)^{-1}$ and $(\partial \bar{U})^{-1}$ are strictly decreasing, the left side of e_2 would be a decreasing function of ϕ, so that the collection of all $\phi \in \mathbb{R}_{++}$ for which the left side of e_2 is finite must be some interval $]\phi_0, \infty[\, \subseteq \mathbb{R}_{++}$. In any case, let us suppose that the utility functions U and $\bar{U}$ are such that this interval is not empty and contains an element ϕ_* such that

$$\mathsf{E}\Big[\big((S^\circ)^{-1} R\, c^{\phi_*}\big) \bullet \iota_T + (S_T^\circ)^{-1} R_T V_T^{\phi_*}\Big] = x^*.$$

If such a choice is possible, then the replicating procedure described in the previous section would produce a Q-tame portfolio process π^*, which, together with the consumption process c^{ϕ_*}, would generate a wealth process V^{ϕ_*} such that $V_0^* = x^*$ and the last condition is satisfied, that is, the investment-consumption strategy (x^*, π^*, c^{ϕ_*}) is efficient for the initial investment x^*.[160] We stress that this step relies in a crucial way on the ability to replicate any payoff whatsoever. What remains now is to demonstrate that the choice (ϕ_*, c^{ϕ_*}) indeed maximizes the objective e_1, subject to the constraint e_2.

(14.20) EXERCISE: Suppose that the strategy (x^*, π^*, c^{ϕ_*}) is chosen as in the preceding, and let (x^*, π, c) be *any* other Q-tame investment-consumption strategy that is efficient for the initial investment x^*, that is, (x^*, π, c) generates wealth process $\mathcal{V}$ such that

$$\mathsf{E}\Big[\big((S^\circ)^{-1} R\, c\big) \bullet \iota_T + (S_T^\circ)^{-1} R_T \mathcal{V}_T\Big] = x^*.$$

Prove that

$$\mathsf{E}\Big[\big(e^{-\beta\iota} U(c^{\phi_*})\big) \bullet \iota_T + e^{-\beta T} \bar{U}(V_T^{\phi_*})\Big] \ge \mathsf{E}\Big[\big(e^{-\beta\iota} U(c)\big) \bullet \iota_T + e^{-\beta T} \bar{U}(\mathcal{V}_T)\Big],$$

and conclude that the strategy (x^*, π^*, c^{ϕ_*}) is the optimal one.

---

[160] This step is no different from replicating a European-style contingent claim with intertemporal payoff rate $\varphi = c^{\phi_*}$ and with final (closing) payoff $\Phi = V_T^{\phi_*}$.

HINT: Write the difference between the two sides of the last relation without the expectation, taking into account the result from exercise (14.19). Then apply the expectation and use the fact that both strategies are efficient. ○

The martingale method that we just outlined has many interesting applications and ramifications that are very important, but lie outside of our main concerns. What is particularly interesting is that despite the fact that the agents trade continuously, they are actually faced with only one constraint – as opposed to a dynamic set of constraints attached to every trading session. This is a feature that is exclusive to complete market models. Later we are going to study concrete solutions to Merton's problem in terms of dynamic programming, and, for this reason, will leave this section void of any concrete implementations of the martingale method – a curious reader can follow up with (Karatzas and Shreve 1998), for example.

14.4 American-Style Contingent Claims

In this section we come to our first encounter with a particularly interesting and important aspect of asset pricing: the fact that flexibility is an asset and this asset can be priced – more or less in the way any other traded asset is priced. Because the payoffs from such assets are not explicitly defined, the pricing of financial instruments that have some form of flexibility incorporated into their covenants is somewhat more involved. Just as we did in our study of European-style contingent claims, we assume the setting of chapter 13, take the market model introduced there as given, and suppose that an ELMM Q exists and is unique (the market is free of arbitrage and is complete). We again start with:

(14.21) DEFINITIONS: An *American-style contingent claim* is a contract defined by its expiration (closing) date $T > 0$, termination payoff process $\varPhi$ (notice that $\varPhi$ is now a process – not a random variable) and payoff rate process φ. Both processes $\varPhi$ and φ are defined on $[0,T]$, and are required to be positive and $\mathscr{P}(\mathscr{G})$-measurable (predictable for the market filtration $\mathscr{G}$). The holder of the contract has the right to terminate (close) the contract, and this right can be exercised at any (random) time $\imath \in [0,T]$. As the decision to terminate can only be based on observable data, the time of closing (exercise), $\imath$, must be a $\mathscr{G}$-stopping time, that is, $\{\imath \leq t\} \in \mathscr{G}_t$ for every $t \in [0,T]$. The collection of all $\mathscr{G}$-stopping times with values in the interval $[0,T]$ we denote⊸(9.46) by $\mathscr{T}_{[0,T]}$. Until the time of closing, $\imath \in \mathscr{T}_{[0,T]}$, the holder of the contract receives (continuously) the instantaneous payments $\varphi_t\,\mathrm{d}t$, and at the time of closing they receive the closing (termination) payment $\varPhi_\imath$. We will express such contracts as $\mathscr{K}(T,\varPhi,\varphi)$, and write $\mathscr{K}(T,\varPhi)$ if the payoff rate is null. If closed at time $\imath \in \mathscr{T}_{[0,T]}$, the accumulated discounted payoffs from $\mathscr{K}(T,\varPhi,\varphi)$ can be expressed as the r.v. $H_\imath$, which we interpret as the value of the following process at time $\imath$:

† $$H \overset{\text{def}}{=} (S^\circ)^{-1}\varPhi + (S^\circ)^{-1}\varphi \raisebox{0.5ex}{\textbf{.}} \imath\,.$$

An *upper hedging strategy* (alias: *super-replicating strategy*) for the claim $\mathscr{K}(T, \Phi, \varphi)$ is any choice of a Q-tame self-financing investment-consumption strategy (x, π^+, c), such that (a.s.) $c_t \geq \varphi_t$ and $V_t^{x,\pi^+,c} \geq \Phi_t$ for all $t \in [0, T]$.

The *upper hedging price* (alias: *super-replicating price*) for $\mathscr{K}(T, \Phi, \varphi)$ is the infimum, Π^+, of the set $A^+ \stackrel{\text{def}}{=}$ the collection of all initial investments x, in all possible upper hedging strategies (x, π^+, c).

A *lower hedging strategy* for the claim $\mathscr{K}(T, \Phi, \varphi)$ is any choice of a stopping time $\mathit{t} \in \mathscr{T}_{[0,T]}$ and a Q-tame self-financing investment-payout strategy (x, π^-, c), such that (a.s.) $\tilde{V}_{\mathit{t}}^{x,\pi^-,c} \leq \Phi_{\mathit{t}}$ and $c_{t\wedge \mathit{t}} \leq \varphi_{t\wedge\mathit{t}}$ for all $t \in [0, T]$ (the trading strategy π^- and the payout rate c need to be defined only until time t).

The *lower hedging price* for $\mathscr{K}(T, \Phi, \varphi)$ is the supremum, Π^-, of the set $A^- \stackrel{\text{def}}{=}$ the collection of all initial loans x, in all possible lower hedging strategies (x, π^-, c). ○

Since most of the strategies that crop up in our study are going to be Q-tame and self-financing, as we did earlier, for brevity, in most references to investment-consumption strategies, or to investment-payout strategies, we are going to omit the qualifiers "Q-tame" and "self-financing," and will assume that these attributes are understood from the outset. Also, all processes associated in one way or another with the contingent claim $\mathscr{K}(T, \Phi, \varphi)$ will be assumed defined on the finite time interval $[0, T]$.

(14.22) REMARK: Unlike European-style contingent claims, in an American-style contingent claim the roles of the buyer and the seller are not symmetric. Indeed, only the buyer can choose the time to close the contract, while the seller must hedge their position against any eventual choice that the buyer can make. For this reason, the hedging strategies for the buyer and for the seller are no longer mirror images of one another. The upper hedging strategy is a hedging strategy for the seller, who must cover their liabilities on the contract at all times and in all states of the economy up until the expiration date T. The lower hedging strategy is a hedging strategy for the buyer who can sell their income on the contract, up until the time they choose to close. ○

In what follows we are going to impose the following integrability condition on the process introduced in (14.21†):

$$\mathsf{E}^Q\left[\sup_{t\in[0,T]} H_t\right] < \infty . \qquad \text{e}_1$$

In particular, this condition guarantees⊸(9.46) that H is of class (D).

Suppose next that the American-style contract $\mathscr{K}(T, \Phi, \varphi)$ is fixed. The upper hedging price Π^+ and the lower hedging price Π^- for this contract are meaningful, and our goal now is to give meaning to its price, Π, without the qualifiers "upper" and "lower."

(14.23) EXERCISE: Prove that for any given stopping time $\mathit{t} \in \mathscr{T}_{[0,T]}$ the initial loan in a lower hedging strategy for the closing time t cannot exceed $\mathsf{E}^Q[H_{\mathit{t}}]$.

In fact, prove that a lower hedging strategy with termination rule t and initial loan $\mathsf{E}^Q[H_t]$ exists. Conclude that

$$\Pi^- = \sup_{t \in \mathscr{T}_{[0,T]}} \mathsf{E}^Q[H_t]\,.$$

HINT: For the first part of the statement, use the optional stopping theorem for submartingales in conjunction with the relation

$$x - \big((S^\circ)^{-1}(\pi^-)^\intercal \bullet X\big)_t = (S^\circ_t)^{-1} V^-_t + \big((S^\circ)^{-1} c \bullet \iota\big)_t$$
$$\leq (S^\circ_t)^{-1} \Phi_t + \big((S^\circ)^{-1} \varphi \bullet \iota\big)_t = H_t\,.$$

To establish the second part of the statement, define the $(\mathscr{G}^t, Q)$-martingale

$$M_t \stackrel{\text{def}}{=} \mathsf{E}^Q\Big[(S^\circ_t)^{-1}\Phi_t + \big((S^\circ)^{-1}\varphi \bullet \iota\big)_t \;\Big|\; \mathscr{G}_{t\wedge t}\Big],$$

and⊸• (9.45), (9.78) observe that M is also a $(\mathscr{G}, Q)$-martingale that can be cast as

† $$M_t = M_{t\wedge t} = (S^\circ_{t\wedge t})^{-1}\mathsf{E}^Q\Big[e^{-\int_{t\wedge t}^{t} r_u \mathrm{d}u}\Phi_t + \int_{t\wedge t}^{t} e^{-\int_{t\wedge t}^{s} r_u \mathrm{d}u}\varphi_s \,\mathrm{d}s \;\Big|\; \mathscr{G}_{t\wedge t}\Big] + \int_0^{t\wedge t} (S^\circ_u)^{-1}\varphi_u \,\mathrm{d}u\,.$$

By the PRP M can be expressed also in the form $M = M_0 + h^\intercal \bullet \beta$, for some predictable h, and one can set $\pi^-_t \stackrel{\text{def}}{=} -(\sigma_t{}^\intercal)^{-1} S^\circ_t\, h_t$. Finally, observe that the debt process $\tilde{V} = \tilde{V}^{M_0,\pi^-,\varphi}$, generated by the investment-payout strategy (M_0, π^-, φ), satisfies the relation

$$(S^\circ_{t\wedge t})^{-1}\tilde{V}_{t\wedge t} + \big((S^\circ)^{-1}\varphi \bullet \iota\big)_{t\wedge t} = M_0 - \big((S^\circ)^{-1}(\pi^-)^\intercal \bullet X\big)_{t\wedge t} = M_{t\wedge t} = M_t\,,$$

and, comparing with †, conclude that

$$\tilde{V}_{t\wedge t} = \mathsf{E}^Q\Big[e^{-\int_{t\wedge t}^{t} r_u \mathrm{d}u}\Phi_t + \int_{t\wedge t}^{t} e^{-\int_{t\wedge t}^{s} r_u \mathrm{d}u}\varphi_s \,\mathrm{d}s \;\Big|\; \mathscr{G}_{t\wedge t}\Big].$$

In particular, $\tilde{V}_0 = \mathsf{E}^Q[H_t]$ and $V_t = \Phi_t$. ○

(14.24) EXERCISE: Let x denote the initial investment in some (arbitrarily chosen) upper hedging strategy, and let the stopping time $t \in \mathscr{T}_{[0,T]}$ be arbitrarily chosen as well. Prove that $x \geq \mathsf{E}^Q[H_t]$ and conclude that $\Pi^+ \geq \Pi^-$.

HINT: If V is the wealth process generated by the strategy, then

$$(S^\circ_t)^{-1}\Phi + (S^\circ)^{-1}\varphi \bullet \iota \leq (S^\circ)^{-1} V + (S^\circ)^{-1} c \bullet \iota = x + (S^\circ)^{-1}(\pi^+)^\intercal \bullet X\,.$$

Apply the optional stopping theorem to the supermartingale on the right. ○

Our objective now is to show that $\Pi^+ = \Pi^-$ and that, in fact, there is an upper hedging strategy and a lower hedging strategy with identical initial investment and initial loan. To arrive at this result, we need to briefly leave the setup introduced at the beginning of this section and bring in a few important tools that we are going to need. We will return to our study of the contingent claim $\mathscr{K}(T, \Phi, \varphi)$ after (14.31).

Thus, we reintroduce a generic probability space $(\Omega, \mathcal{F}, P)$ and a filtration $\mathscr{F} \subset \mathcal{F}$ that satisfies the usual conditions. Expectations will be understood relative to P and the attribute "adapted" will be understood relative to $\mathscr{F}$.

(14.25) ESSENTIAL SUPREMUM: Let $(\xi_i)_{i \in \mathbb{I}}$ be any family of random variables on $(\Omega, \mathcal{F}, P)$, indexed by the (not necessarily countable) set $\mathbb{I}$. If the labeling set $\mathbb{I}$ is countable, then $\sup_{i \in \mathbb{I}} \xi_i$ (understood as a pointwise sup) would be a well-defined measurable r.v., but for an uncountably infinite set $\mathbb{I}$, the pointwise sup need not be measurable. Nevertheless, it is a standard fact from measure theory – see, for example, (Karatzas and Shreve 1998, app. A) – that, regardless of the cardinality of the set $\mathbb{I}$, there always exists a $\mathcal{F}$-measurable r.v. η, with values in $\bar{\mathbb{R}}$, which has the following properties: (a) $\xi_i \leq \eta$ (P-a.s.) for every $i \in \mathbb{I}$; (b) if η' is any other $\mathcal{F}$-measurable r.v. with values in $\bar{\mathbb{R}}$, such that $\xi_i \leq \eta'$ (P-a.s.) for every $i \in \mathbb{I}$, then $\eta \leq \eta'$ (P-a.s.). Any r.v. η that shares properties (a) and (b) is P-a.s. unique. It is called *essential supremum* (alias: *essential upper bound*) of the family $(\xi_i)_{i \in \mathbb{I}}$, and is denoted by $\operatorname{ess\,sup}_{i \in \mathbb{I}} \xi_i$. ○

(14.26) THE SNELL ENVELOPE:[161] An adapted and right-continuous process H that is of class (D) (in particular, H could be such that $\mathsf{E}\left[\sup_{t \in \mathbb{R}_+} |H_t|\right] < \infty$) can be viewed as a "reward process." Given a reward process H, we now define the process

$$U_t \overset{\text{def}}{=} \operatorname{ess\,sup}_{\mathcal{s} \in \mathscr{T}_{[0,\infty]},\, \mathcal{s} \geq t} \mathsf{E}\left[H_{\mathcal{s}} \mid \mathscr{F}_t\right], \quad t \in \mathbb{R}_+ ,$$

with the understanding that $H_\infty \overset{\text{def}}{=} 0$. The intuition behind this definition should be clear: given the information available at time t, the quantity $U_t \prec \mathscr{F}_t$ is the maximal expected (relative to P) future reward that can be attained by terminating the reward process at some random moment following time t (including time t). As defined above, the process U can be modified into a càdlàg supermartingale – see (Dellacherie and Meyer 1980, app. 1 22–23) – and is called the *Snell envelope of H*. Intuitively, it should be optimal to stop the reward process at the first moment at which it meets its Snell envelope, but such a prescription must be justified. This task is somewhat technical, and here we are only going to outline the main steps and insights. For more details we refer the reader to (Dellacherie and Meyer 1980), (El Karoui 1981), (Karatzas and Shreve 1998), or (Lamberton 2009). ○

The next result is a paraphrasing of (Dellacherie and Meyer 1980, app. 1 22–23)[162] – see also (Lamberton 2009).

(14.27) THEOREM: Let H be any right-continuous and adapted reward process which is of class (D). Then its Snell envelope, U, has the following properties:

[161] Named after the American mathematician James Laurie Snell (1925–2011).

[162] It is interesting that Dellacherie and Meyer derive the existence and most of the properties of the Snell envelope without the usual conditions for the filtration $\mathscr{F}$. Nevertheless, the usual conditions are still needed in order to claim that the Snell envelope has a càdlàg modification if the reward process is right-continuous.

(a) U is positive and $U_t \geq H_t$ (P-a.s.) for all $t \in \mathbb{R}_+$ (this property is straightforward, since any fixed time is a stopping time);

(b) U can be modified into a right-continuous supermartingale of class (D);[163]

(c) $U_t = \operatorname{ess\,sup}_{\mathfrak{z} \in \mathcal{T}_{[0,\infty]},\, \mathfrak{z} \geq t} \mathsf{E}[H_{\mathfrak{z}} \,|\, \mathcal{F}_t]$ for all stopping times $t \in \mathcal{T}_{[0,\infty]}$;

(d) (Dellacherie and Meyer 1980, VII 11) if, in addition,$^{\multimap (9.55c)}$ the reward H is regular, and in particular, if it is continuous, then U is also regular, but one cannot claim that U is continuous even if H is continuous (see, however, (14.32)). ○

(14.28) EXERCISE: Without relying on the last theorem, prove that if the reward process has the additional property $\mathsf{E}\big[\sup_{t \in \mathbb{R}_+} |H_t|\big] < \infty$, then its Snell envelope is indeed of class D. Explain why in this case $\sup_{t \in \mathbb{R}_+}$ is meaningful, even though $\mathbb{R}_+$ is not a countable set.

HINT: The process $M_t \overset{\text{def}}{=} \mathsf{E}\big[\sup_{s \in \mathbb{R}_+} H_s \big| \mathcal{F}_t\big]$, $t \in \mathbb{R}_+$, is a uniformly integrable martingale with $U_t \leq M_t$ (P-a.s.) for any $t \in \mathbb{R}_+$. If U and M are both right-continuous, then $U_t \leq M_t$ for all $t \in \mathbb{R}_+$ P-a.s. Consult exercise (4.25). ○

(14.29) EXERCISE: Prove that the Snell envelope U is the smallest positive and right-continuous supermartingale that dominates H in that $U_t \geq H_t$ for all t P-a.s.; to be precise, if $\tilde{U}$ is another positive and right-continuous supermartingale such that $\tilde{U}_t \geq H_t$ P-a.s. for every t, then it must be that $\tilde{U}_t \geq U_t$ (P-a.s.) for every t, or, which is the same (explain why) $\tilde{U}_t \geq U_t$ for all t P-a.s.

HINT: Recall$^{\multimap (9.30)}$ that a positive supermartingale always converges (that is, always can be labeled by $\bar{\mathbb{R}}_+$) and observe that for any $t \in \mathbb{R}_+$ and any (finite or infinite) stopping time $\mathfrak{z} \geq t$, we have (after selecting an appropriate modification) $\mathsf{E}[H_{\mathfrak{z}} \,|\, \mathcal{F}_t] \leq \mathsf{E}[\tilde{U}_{\mathfrak{z}} \,|\, \mathcal{F}_t] \leq \tilde{U}_t$. ○

Now we are ready to formulate the optimal stopping rule in terms of the Snell envelope. The fact that a particular stopping time $t^* \in \mathcal{T}_{[0,\infty[}$ is optimal (as a prescription for terminating H) can be stated as

$$U_0 = \operatorname{ess\,sup}_{\mathfrak{z} \in \mathcal{T}_{[0,\infty]},\, \mathfrak{z} \geq 0} \mathsf{E}\big[H_{\mathfrak{z}} \mid \mathcal{F}_0\big] = \mathsf{E}\big[H_{t^*} \mid \mathcal{F}_0\big].$$

In particular, if $\mathcal{F}_0$ is the completion of the trivial σ-field $\{\emptyset, \Omega\}$, which condition we impose from now on, then the optimality of the termination rule t^* can be stated as

$$U_0 = \mathsf{E}[H_{t^*}].$$

(14.30) EXERCISE: Prove that the stopping time $t^* \in \mathcal{T}_{[0,\infty[}$ is optimal if and only if the following two conditions are satisfied: (a) $H_{t^*} = U_{t^*}$ (P-a.s.), and (b) the Snell envelope is a martingale up until time t^*, that is, the process $U^{t^*} = (U_{t \wedge t^*})_{t \in \mathbb{R}_+}$ is a martingale.

[163] In fact, it can be modified into a "strong supermartingale" in the sense that for any bounded stopping time t on has $\mathsf{E}[|U_t|] < \infty$ and $U_{\mathfrak{z}} \geq \mathsf{E}[U_t \mid \mathcal{F}_{\mathfrak{z}}]$ for any two bounded stopping times $\mathfrak{z}$ and t, such that $\mathfrak{z} \leq t$.

HINT: If (a) and (b) hold, then t^* is clearly optimal (see the optional stopping result for supermartingales). If t^* is optimal and the stopping time s is such that $s \leq t^*$, then

$$U_0 = \mathsf{E}[H_{t^*}] = \mathsf{E}\big[\mathsf{E}[H_{t^*} \,|\, \mathscr{F}_s]\big] \leq \mathsf{E}[U_s] \leq U_0 \,.$$

In particular, $U_s = \mathsf{E}[H_{t^*} \,|\, \mathscr{F}_s]$ (a.s.) and $\mathsf{E}[U_s] = U_0$. Consult (9.43). ○

(14.31) EXISTENCE OF AN OPTIMAL STOPPING TIME: By the result in (14.27), the Snell envelope is always regular if H is regular (in particular, if H is continuous). In conjunction with theorem (9.57), the regularity of U guarantees that the predictable and increasing process A from⊸(9.48), (9.49) the Doob–Meyer decomposition $U = M - A$ is continuous (even if U is not). The fact that the stopped process U^s is a martingale can be expressed as $A_s = 0 = A_0$, that is, the compensation is not triggered at least until time s. Consider the stopping time

† $$s^* = \inf\big\{t \in \mathbb{R}_+ \colon A_t > 0\big\}\,,$$

which is simply the entry time of the (continuous and increasing) process A to the open domain $\mathbb{R}_{++}$ (understood to be $+\infty$ if $A_t = 0$ for all t). This stopping time can be thought of as the latest time until which the supermartingale U remains a martingale. The stopping time t^* is then optimal if and only if $t^* \leq s^*$ (so that the Snell envelope U stopped at time t^* is a martingale) and $U_{t^*} = H_{t^*}$. There is no intuition to suggest that a stopping time t^* that has these two attributes always exists, and, indeed, the counterexample from (Karatzas and Shreve 1998, D.11) shows that an optimal stopping time may not exist. The same source provides the following sufficient condition for the existence of an optimal stopping time in the special case of a finitely lived (until time $T \in \mathbb{R}_{++}$), positive, and continuous reward process H: $\mathsf{E}\big[\sup_{t\in[0,T]} H_t\big] < \infty$. ○

We are now in a position to – finally – restore the setting adopted at the beginning of this section, and resume our study of American-style contingent claims from where we left off right before (14.25). The discounted cumulative payoff H that we defined in (14.21†) is continuous, and, due to condition e_1, is of class (D). Thus,⊸(9.55c) H is regular and its Snell envelope, U, is well defined relative to Q – it is the smallest right-continuous $(\mathscr{G}, Q)$-supermartingale that dominates H. Most important, in conjunction with the result in (Karatzas and Shreve 1998, D.11), the assumption in e_1 guarantees that an optimal stopping time for H exists. In particular,⊸(14.31) the stopping time $t^* = \inf\{t \in [0,T] \colon H_t = U_t\}$ must be optimal,[164] that is,[165]

$$\mathsf{E}^Q[H_{t^*}] = \sup_{t\in\mathscr{T}_{[0,T]}} \mathsf{E}^Q[H_t] = U_0 \,.$$

---

[164] That t^* is dominated by the stopping time in (14.31†) follows from the claim that an optimal stopping time exists, in conjunction with the relation $H_{t^*} = U_{t^*}$, which follows from the right-continuity of U.

[165] Recall that $\mathscr{G}_0$ is the completion of the trivial σ-field $\{\emptyset, \Omega\}$ by assumption, so that $E^Q[U_0] = U_0$.

A question that we now must ask is: how does the stopping of the process from (14.21†) at time t^* relate to pricing by arbitrage and hedging? To make this connection, consider the Doob–Meyer decomposition $U = M - A$, understood relative to the filtration $\mathscr{G}$ and the probability measure Q. The market completeness hypothesis simply means that the market filtration $\mathscr{G}$ has the predictable representation property with respect to the Brownian motion β under Q. Thus, since M is a $(\mathscr{G}, Q)$-martingale,⊸(10.46) it follows that M has a continuous modification that can be cast in the form

$$M = M_0 + h^\intercal \bullet \beta$$

for some n-dimensional and $\mathscr{G}$-predictable processes h, with sample paths $h_\omega \in \mathscr{L}^2_{\mathrm{loc}}(\mathbb{R}_+; \mathbb{R}^n)$. Since $A_0 = 0$ (from the definition of the Doob–Meyer decomposition), we have $M_0 = U_0$. As before, consider an investment-consumption strategy with initial investment $U_0 = M_0$, portfolio process $\pi = S^\circ(\sigma^\intercal)^{-1}h$, and with instantaneous consumption given by $\mathrm{d}C_t = \varphi_t \,\mathrm{d}t + (S^\circ_t)\mathrm{d}A_t$.

(14.32) REMARK: Any regular supermartingale of class (DL) with respect to a Brownian filtration has a continuous modification. This is a consequence of the fact that⊸(10.46) *any* martingale with respect to a Brownian filtration has a continuous modification. Indeed, any supermartingale of class (DL) admits D-M-d, and regularity is necessary and sufficient for the predictable compensator in the D-M-d to be continuous. Thus, if, in addition, the martingale component in the D-M-d can be modified into a continuous process (which is always the case for Brownian filtrations), the same must be true for the supermartingale as well. As a result, if the reward process is right-continuous, adapted to a Brownian filtration, regular, and of class (D), then its Snell envelope is always going to be continuous – even if the reward process itself is not. We stress once again that this feature is exclusive to Brownian filtrations. ○

(14.33) EXERCISE: Prove that the investment-consumption strategy $(U_0, \pi, \mathrm{d}C)$, which was introduced in the preceding, is admissible. ○

The wealth process, V, that $(U_0, \pi, \mathrm{d}C)$ generates is subject to:

$$(S^\circ)^{-1}V + (S^\circ)^{-1}\bullet\big(\varphi\bullet\iota + (S^\circ)\bullet A\big) = U_0 + h^\intercal\bullet\beta = M = U + A\,.$$

After the obvious cancellation, this relation gives:

$$(S^\circ)^{-1}V + (S^\circ)^{-1}\varphi\bullet\iota = U \geq H = (S^\circ)^{-1}\Phi + (S^\circ)^{-1}\varphi\bullet\iota\,,$$

with the implication that $V \geq \Phi$. The strategy $(U_0, \pi, \mathrm{d}C)$ is now easily seen to be a super-replicating strategy. In particular, $U_0 \geq \Pi^+$ by the very definition of the upper hedging price Π^+. Moreover, we have

$$V_0 = M_0 = U_0 = \mathsf{E}^Q[H_{t^*}] = \sup\nolimits_{t\in\mathscr{T}_{[0,T]}} \mathsf{E}^Q[H_t] = \Pi^- \leq \Pi^+ \leq U_0 = V_0\,,$$

which leads to the conclusion that the quantity

$$\Pi^- = \mathsf{E}^Q[H_{t^*}] = \sup\nolimits_{t\in\mathscr{T}_{[0,T]}} \mathsf{E}^Q[H_t] = U_0 = \Pi^+$$

deserves to be called "the price," Π, of $\mathscr{K}(T,\Phi,\varphi)$. Note that we were able to construct upper and lower hedging strategies – that is, hedging strategies for the seller and the buyer of the claim – with identical initial investment and initial loan. Ultimately, there is a unique price that is acceptable – indeed, optimal – for both the buyer and the seller, and despite of the asymmetry in their positions.

(14.34) AMERICAN-STYLE CALL OPTION IN THE BSM MODEL: Back to the setting of (14.14), suppose that the risky security pays (continuously) the instantaneous dividends $\delta S_t\, dt$, where the dividend yield $\delta > 0$ is a fixed model parameter. Let $Z \stackrel{\text{def}}{=} \log(S_0)+S^{-1}\bullet S$. With dividends we have $X = Z+\delta\iota-r\iota$ and the market price of risk is $\theta_t = \frac{b+\delta-r}{\sigma} = \theta = \text{constant}$. Since $Z = X-\delta\iota+r\iota = \sigma\beta-\delta\iota+r\iota$ and the spot price S being ex-dividend spot price, we have

$$(S^\circ)^{-1}S = e^{-r\iota}e^{Z-\frac{1}{2}\langle Z,Z\rangle} = e^{Z_0}e^{-r\iota}e^{(Z-Z_0)-\frac{1}{2}\langle Z,Z\rangle} = S_0 e^{-r\iota}e^{\sigma\beta+(r-\delta)\iota-\frac{1}{2}\sigma^2\iota}$$
$$= S_0 e^{\sigma\beta-\delta\iota-\frac{1}{2}\sigma^2\iota}\,.$$

This shows that under the ELMM Q the discounted spot price $(S^\circ)^{-1}S$ is a martingale only in the no-dividend case $\delta = 0$. Consider an American-style call option with strike $K > 0$. In this special case the Snell envelope is given by

$$U_t = \operatorname{ess\,sup}_{\mathfrak{t}\in\mathscr{T}_{[t,T]}} \mathsf{E}^Q\left[e^{-r\mathfrak{t}}\left(S_{\mathfrak{t}}-K\right)^+ \mid \mathscr{G}_t\right]$$

and the price of the call is $\Pi = U_0$. As one can imagine, there is no known closed-form expression in terms of standard functions for the price Π. However, in section 18.4 we are going to describe a computational program that will bring us very close to a "formula for Π." ○

It is common knowledge that if the underlying asset does not pay dividends, then any American-style call option on that asset would be identical to its European counterpart, that is, it would not be optimal to exercise the call before its expiration date T. The earliest publication that explains this phenomenon appears to be (Merton 1973). Since then other explanations have emerged and can be found in most texts that cover American options. With the tools that are now available to us, the justification of this result is straightforward and is given in the next exercise.

(14.35) EXERCISE: In the setting of (14.34), prove that if $\delta = 0$, then the Snell envelope is nothing but the martingale

$$U_t = \mathsf{E}^Q\left[e^{-rT}\left(S_T-K\right)^+ \mid \mathscr{G}_t\right], \quad t\in[0,T]\,.$$

In particular, $\Pi = U_0 = \mathsf{E}^Q\left[e^{-rT}\left(S_T-K\right)^+\right]$ and the optimal time to exercise is $\mathfrak{t}^* = T$. Argue that the same property continues to hold if the volatility σ is some bounded and predictable process and the spot rate r is some predictable and positive process.

HINT: With$^{\multimap(9.37)}$ the optional stopping theorem for submartingales in mind, this result was essentially established in (9.3b). Indeed, if $\delta = 0$, then $e^{-r\cdot\iota}S$ is a martingale and $e^{-r^{\bullet}\iota}K$ is a supermartingale under Q. Thus, $e^{-r^{\bullet}\iota}(S-K)$ is a submartingale and so is also $e^{-r^{\bullet}\iota}(S-K)^{+}$. The result in (9.37) does the rest. ○

14.5 Put–Call Symmetry and Foreign Exchange Options

This section is a brief synopsis of a particularly interesting and useful connection, which has been discovered and explored from various angles by (Carr and Bowie 1994), (McDonald and Schroder 1998), (Carr and Lee 2009), (Carr and Chesney 2000), and others. The book (Kholodnyi and Price 1998) provides a very broad and insightful study of this topic, and reveals its deep connections with other branches of physics and mathematics. The entirely heuristic derivation of the put–call symmetry relation that follows in this section is a variation of the derivation in (Dana and Jeanblanc 2003).

To begin with, let $r^e > 0$ denote the spot rate in the EU and let $r^u > 0$ denote the spot rate in the US – we are going to suppose that both rates are constants. Let $S_t^{e/u}$ denote the $\frac{€}{\$}$-exchange rate, that is, the spot price of €1 in \$-units, and let $S_t^{u/e}$ denote the $\frac{\$}{€}$-exchange rate, that is, the spot price of \$1 in €-units. The two exchange rates are connected through the obvious relation

$$S_t^{u/e} = 1/S_t^{e/u} \quad \Longleftrightarrow \quad S_t^{e/u} = 1/S_t^{u/e}\,.$$

To gain some insight into the dynamics of the exchange rates, consider the following chain of transactions: borrow \$1 from a US bank, purchase €$(1/S_t^{e/u})$ and deposit this amount immediately into an EU bank. At time $t+\mathrm{d}t$ close the position and settle all accounts. The payoff at time $t+\mathrm{d}t$ from this chain of transactions can be expressed as

$$\mathrm{d}X_t \overset{\text{def}}{=} \frac{1+r^e\,\mathrm{d}t}{S_t^{e/u}} S_{t+\mathrm{d}t}^{e/u} - (1+r^u\,\mathrm{d}t)\,,$$

which we are going to rearrange as

$$\begin{aligned} S_t^{e/u}\,\mathrm{d}X_t &= (1+r^e\,\mathrm{d}t)S_{t+\mathrm{d}t}^{e/u} - (1+r^u\,\mathrm{d}t)S_t^{e/u} \\ &= \mathrm{d}S_t^{e/u} + (r^e-r^u)S_t^{e/u}\,\mathrm{d}t + r^e\,\mathrm{d}S_t^{e/u}\,\mathrm{d}t\,, \end{aligned}$$

and note that – at least formally – $(\mathrm{d}S_t^{e/u})\mathrm{d}t = 0$. Since the cost of generating the payoff $\mathrm{d}X_t$ (in the US) is 0 it must be that $\mathsf{E}^{Q^u}[\mathrm{d}X_t\,|\,\mathscr{G}_t] = 0$, that is, X is a $(Q^u,\mathscr{G})$-martingale, where Q^u is the pricing measure for securities traded in \$ and $\mathscr{G}$ is the market filtration for both currencies. If we impose the additional requirement that X is a semimartingale with local variance $\mathrm{d}\langle X,X\rangle_t = \sigma^2\,\mathrm{d}t$, for some fixed parameter $\sigma > 0$, then we arrive at the following dynamics:

$$\mathrm{d}S_t^{e/u} = S_t^{e/u}\big(\sigma\,\mathrm{d}\beta_t + (r^u-r^e)\mathrm{d}t\big)\,,$$

where the semimartingale β is a $(Q^u,\mathscr{G})$-Brownian motion.

(14.36) EXERCISE: Prove that the dynamics of the \$/€ exchange rate are given by

$$dS_t^{u/e} = S_t^{u/e}\big(\sigma\, d(-\beta_t^*) + (r^e - r^u)dt\big),$$

where $\beta^* = \beta - \sigma\iota$. Explain why β^* – or equivalently $(-\beta^*)$ – must be a $(Q^e, \mathscr{G})$-Brownian motion, where Q^e is the pricing measure for assets traded in €. ○

It is impossible not to notice at this point that the dynamics of $S^{e/u}$ under Q^u are no different from the dynamics of a dividend-paying security in the Black–Scholes–Merton model, except that the fixed rate is now $r = r^u$ and the dividend rate is $\delta = r^e$. The same is true for the dynamics of $S^{u/e}$, except that $r = r^e$ and $\delta = r^u$.

Next, consider a derivative instrument that gives its holder the right to purchase €1 at (strike) price \$$K$ (in dollars). At this point we leave open the question whether this derivative is of European type or American, and whether it expires in finite time or not. The price of this derivative at time t we are simply going to express as $C\big(t, S_t^{e/u}, K, \sigma^2, r^u, r^e\big)$ (the order in which we write the arguments in the calls and the puts is: time, spot, strike, volatility, interest, and dividend). Clearly, this derivative contract is the same as the right to sell \$$K$ for the price of €1, and this, of course, is no different from the right to sell K identical contracts, every one of which gives its holder the right to sell \$1 for €$(1/K)$. The payoff from each of these K contracts, if exercised at time $\mathcal{t} \geq t$, expressed in € is $\big(\frac{1}{K} - S_{\mathcal{t}}^{u/e}\big)^+$.

The value of the K put contracts in euros can be expressed as

$$K\, P\big(t, S_t^{u/e}, 1/K, \sigma^2, r^e, r^u\big),$$

which converted into dollars becomes

$$S_t^{e/u}\, K\, P\big(t, S_t^{u/e}, 1/K, \sigma^2, r^e, r^u\big).$$

(14.37) EXERCISE: Prove that, whether the put is of European-style or American-style, we have

$$S_t^{e/u}\, K\, P\big(t, S_t^{u/e}, 1/K, \sigma^2, r^e, r^u\big) = P\big(t, K, S_t^{e/u}, \sigma^2, r^e, r^u\big).$$

HINT: Suppose that at time t one has $S_t^{e/u} = x$, so that $S_t^{u/e} = 1/x$. The left side above is the same as the price of xK put options. If exercised at time $\mathcal{t} \geq t$, the aggregate payoff from these options would be

$$x\,K\left(\frac{1}{K} - S_{\mathcal{t}}^{u/e}\right)^+ = \big(x - x\,K\,S_{\mathcal{t}}^{u/e}\big)^+.$$

Since $S_t^{u/e} = \frac{1}{x}$, we have $x\,K\,S_t^{u/e} = K$, so that $x\,K\,S_{\mathcal{t}}^{u/e}$, $\mathcal{t} \geq t$, is the same geometric Brownian motion as $S_{\mathcal{t}}^{u/e}$, $\mathcal{t} \geq t$ – that is, a geometric Brownian motion of volatility σ^2 and growth rate $r^e - r^u$ – except $x\,K\,S_{\mathcal{t}}^{u/e}$, $\mathcal{t} \geq t$, starts at time t from position K instead of position $\frac{1}{x}$. The price at time t of an option with such payoff can be expressed as $P\big(t, K, x, \sigma^2, r^e, r^u\big)$ – note that in this case the spot is K and the strike is $x = S_t^{e/u}$. ○

We can therefore write

$$C\big(t, S_t^{e/u}, K, \sigma^2, r^u, r^e\big) = P\big(t, K, S_t^{e/u}, \sigma^2, r^e, r^u\big)\,.$$

Since in the Black–Scholes–Merton model the spot price of the risky security has dynamics (under the equivalent pricing measure Q)

$$\mathrm{d}S_t = S_t\big(\sigma\,\mathrm{d}\tilde{W}_t + (r-\delta)\mathrm{d}t\big)\,, \quad t \geq 0\,,$$

which is no different from the dynamics of $(S_t^{e/u})$ with $r^u = r$ and $r^e = \delta$, we arrive at the following relation, which is known as "put–call symmetry" (the order of the arguments is again: time, spot, strike, volatility, interest, dividend):

$$\mathrm{Call}\big(t, S_t, K, \sigma, r, \delta\big) = \mathrm{Put}\big(t, K, S_t, \sigma, \delta, r\big)\,.$$

This relation holds for both European-style options and for American-style options – as long as the put and the call have the same maturity – and, of course, one is to remain within the context of the BSM model.

(14.38) EXERCISE: Prove that if the interest rate is $r = 0$, then it is never optimal to exercise an American-style *put option* before the expiration date. ○

14.6 Exchange Options

The study of such options can be traced back to Margrabe (1978) (see also Carr (1993)). The brief derivation that follows in this section is borrowed from (Jeanblanc et al. 2009).

There are two risky assets in the market, S and U, given by

$$S = S_0 + \sigma S \bullet W + bS \bullet \iota \quad \text{and} \quad U = U_0 + \alpha U \bullet B + \beta U \bullet \iota\,,$$

for some choice of the parameters $\sigma, \alpha \in \mathbb{R}_{++}$, $b, \beta \in \mathbb{R}$ and the Brownian motions W and B, defined on a common probability space $(\Omega, \mathcal{F}, P)$, and such that $\langle B, W\rangle = \rho\iota$ for some fixed $\rho \in \,]-1, 1[$. In addition, the market includes a risk-free asset that pays constant interest $r > 0$. Both stocks, S and U, pay dividends continuously at fixed rates, respectively, $\delta > 0$ and $\varepsilon > 0$. Since the volatility matrix in this model is constant, the market is$^{\multimap (13.29)}$ free of arbitrage and complete. Let $Q \approx P$ denote the unique ELMM. The processes

$$\tilde{W} \stackrel{\text{def}}{=} W + \frac{b+\delta-r}{\sigma}\iota \quad \text{and} \quad \tilde{B} \stackrel{\text{def}}{=} B + \frac{\beta+\varepsilon-r}{\alpha}\iota$$

are then Q-Brownian motions, and the stock prices are given by

$$S = S_0 + \sigma S \bullet \tilde{W} + (r-\delta)S \bullet \iota \quad \text{and} \quad U = U_0 + \alpha U \bullet \tilde{B} + (r-d)U \bullet \iota\,.$$

An *exchange option* is a contract that entitles its holder to exchange one asset (called *spot asset*) for another asset (called *strike asset*). By way of an example, in our model this could be a contract to exchange S for U. The structure of such an option is identical to that of a call option with spot asset S – whether European-style or American-style – except that the strike price K (the cash asset) is replaced with a particular risky asset, the strike asset U. Specifically, if the option is of American-style and the time of exercise is $\mathcal{t}$ (a stopping time), then the closing payoff would be $(S_{\mathcal{t}} - U_{\mathcal{t}})^+$. If the option is of European-style with expiration date T, then the payoff would be $(S_T - U_T)^+$. Exchange options can be transformed into standard call options (American or European) by using the strike asset as a numéraire, as we now detail. First, fix a finite time horizon $T > 0$ and observe that

$$(S_{\mathcal{t}} - U_{\mathcal{t}})^+ = U_{\mathcal{t}}\Big(\frac{S_{\mathcal{t}}}{U_{\mathcal{t}}} - 1\Big)^+ \quad \text{and that} \quad U_t = U_0 e^{(r-\varepsilon)t} e^{\alpha \tilde{B}_t - \frac{1}{2}\alpha^2 t}, \quad t \in [0, T].$$

Next, define the measure $Q^* \approx Q \approx P$ on $\mathscr{F}_T$ by $Q^* \stackrel{\text{def}}{=} e^{\alpha \tilde{B}_T - \frac{1}{2}\alpha^2 T} \odot Q$ – this step corresponds to switching the numéraire from the cash asset to U – and observe that the price of the option is

$$\text{either} \quad S_0 \times \sup_{\mathcal{t} \in \mathscr{T}_{[0,T]}} \mathsf{E}^{Q^*}\Big[e^{-\varepsilon \mathcal{t}}\Big(\frac{S_{\mathcal{t}}}{U_{\mathcal{t}}} - 1\Big)^+\Big] \quad \text{or} \quad S_0\, e^{-\varepsilon T} \mathsf{E}^{Q^*}\Big[\Big(\frac{S_T}{U_T} - 1\Big)^+\Big],$$

depending on its style. What we have above is the payoff from a standard call option (American or European) with strike $K = 1$ and spot asset S/U – except that the equivalent pricing measure is Q^* instead of Q, and the discount rate is ε instead of r. The next step in our program suggests itself: we must transcribe the dynamics of the "asset" S/U under the measure Q^*. This is straightforward:

$$\begin{aligned}
\frac{S}{U} &= \frac{S_0}{U_0} e^{\left((r-\delta-\frac{1}{2}\sigma^2)-(r-\varepsilon-\frac{1}{2}\alpha^2)\right)\iota + \sigma \tilde{W} - \alpha \tilde{B}} \\
&= \frac{S_0}{U_0} e^{\left(\varepsilon-\delta-\frac{1}{2}\sigma^2+\frac{1}{2}\alpha^2+\sigma\alpha\rho-\alpha^2\right)\iota + \sigma\left(\tilde{W}-\alpha\langle \tilde{W},\tilde{B}\rangle\right) - \alpha\left(\tilde{B}-\alpha\langle \tilde{B},\tilde{B}\rangle\right)} \\
&= \frac{S_0}{U_0} e^{\left(\varepsilon-\delta-\frac{1}{2}(\sigma^2+\alpha^2-2\sigma\alpha\rho)\right)\iota + \sigma\left(\tilde{W}-\alpha\langle \tilde{W},\tilde{B}\rangle\right) - \alpha\left(\tilde{B}-\alpha\langle \tilde{B},\tilde{B}\rangle\right)}.
\end{aligned}$$

Since $\tilde{W}$ and $\tilde{B}$ are Brownian motions under Q, by Girsanov's theorem,

$$\big(\tilde{W} - \alpha\langle \tilde{W}, \tilde{B}\rangle\big) \quad \text{and} \quad \big(\tilde{B} - \alpha\langle \tilde{B}, \tilde{B}\rangle\big)$$

are Brownian motions under Q^*, and, therefore, so is

$$Z = \frac{1}{\sqrt{\sigma^2 + \alpha^2 - 2\sigma\alpha\rho}}\Big(\sigma\big(\tilde{W} - \alpha\langle \tilde{W}, \tilde{B}\rangle\big) - \alpha\big(\tilde{B} - \alpha\langle \tilde{B}, \tilde{B}\rangle\big)\Big),$$

since it is a Q^*-local martingale with $\langle Z, Z\rangle = \iota$. With the substitution $\gamma = \sqrt{\sigma^2 + \alpha^2 - 2\sigma\alpha\rho}$, we are now in the following setting that is entirely familiar to us: the equivalent pricing measure is Q^*, the spot asset is driven by the Q^*-Brownian motion Z, and is given by

$$\frac{S}{U} = \frac{S_0}{U_0} e^{\left((\varepsilon-\delta)-\frac{1}{2}\gamma^2\right)\iota + \gamma Z}.$$

This is a geometric Brownian motion under Q^*, and, except for the different symbols involved, we are precisely in the classical setting of the BSM model: the strike of the call option is 1, the risk-free rate is ε, the dividend rate is δ, and the volatility parameter is γ.

14.7 Stochastic Volatility Models

In concrete applications one must work with models that are, well, concrete. One classical example of a concrete market model is$^{\multimap (9.94)}$ the Black–Scholes–Merton model, which is already familiar to us. While the insight, influence, and overall impact of this model on the domain of asset pricing can never be overstated, from an empirical standpoint its features are somewhat unsatisfactory, as we now detail.

(14.39) VOLATILITY SMILE: The familiar$^{\multimap (9.94)}$ BSM formula gives the price of the European call as the function $BSM(S_0, K, \sigma^2, r, T)$. Among its five arguments only the spot price S_0 is observable, while the remaining four are understood to be model parameters. In practice, however, the spot prices of most option contracts are determined by the markets and are quoted on the exchanges where such instruments are traded. One consequence of this practice is that the value of $BSM(S_0, K, \sigma^2, r, T)$ is reported and the BSM formula turns into an equation – proxy, of a sort – for the model parameters. Since the values for K and T are specified in the option's covenants, only σ^2 and r are not explicitly fixed in the outset. Although there is no "one and only universal risk-free rate" quoted on some exchange – and, typically, the interest rate varies in the cross-section of financial markets – at least in principle the spot rate associated with a particular trading environment can be deduced from the quoted prices of the various fixed-income instruments. If, as is customary throughout most of the finance literature, we suppose that the risk-free spot rate r is somehow known, then only the variance σ^2 can be viewed as "unspecified." Identifying the value $BSM(S_0, K, \sigma^2, r, T)$ with the observed option price at time $t = 0$ yields an equation for σ^2, the solution[166] to which, namely $\sigma^2_{\text{imp}} = \sigma^2_{\text{imp}}(S_0, K, r, T)$, is called "implied variance." Its square root, denoted $\sigma_{\text{imp}} = (\sigma^2_{\text{imp}})^{1/2}$, is called "implied volatility."[167] If the actually observed option prices were indeed produced by the BSM formula, treated as a function of the strike K for fixed T and r, the implied volatility σ_{imp} should be

---

[166] Recall that $BSM(S_0, K, \sigma^2, r, T)$ is a strictly monotone function of σ^2.

[167] Strictly speaking, since the square root is a multi-valued function, the volatility process σ is well-defined only if the Brownian motion W is somehow fixed in the outset. Indeed, if α is any adapted process with $|\alpha_t| = 1$ (take $\alpha_t = \text{sign}(\sigma_t)$, for example), then the process $\sigma{\bullet}W$ would be no different from the process $(\alpha\sigma){\bullet}W'$ with $W' \stackrel{\text{def}}{=} \alpha{\bullet}W$ (a Brownian motion). In particular, $\sigma{\bullet}W$ can always be replaced with $|\sigma|{\bullet}W'$ – unless there is a reason to insist on writing the model in terms of a particular Brownian motion W, and not in terms of the Brownian motion $W' \stackrel{\text{def}}{=} \text{sign}(\sigma){\bullet}W$. Since from a modeling point of view the choice of the Brownian motion is irrelevant, the square root of the variance can be chosen arbitrarily, whence the custom of always choosing the positive root.

a constant function. This feature is not very difficult to test in practice, since options with different strikes that have identical underlying assets and expiration dates are traded and quoted on most option exchanges. Typically, the test fails, in that $K \rightsquigarrow \sigma_{\mathrm{imp}}(S_0, K, r, T)$ does not appear to be a constant function. While the shape of this function varies from one underlying asset to another, in most cases it would resemble the one depicted in figure 14.1 and, for this reason, the actual shape of σ_{imp} is often called a *volatility smile*.

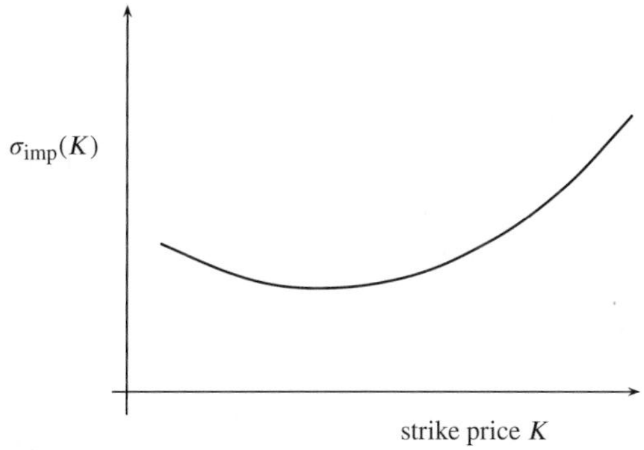

FIGURE 14.1
Stylized depiction of the volatility smile

One possible explanation of this phenomenon is that the constant volatility assumption in the BSM model is not consistent with the empirically observed returns. Indeed, the daily movements in the VIX index leave no doubt that market sentiments – fear or optimism – do affect market volatility. This was the main reason for developing the so-called stochastic volatility models, the subject of the present section. For a much more detailed discussion of this topic we point to, for example, (Crépey 2003), (Fouque et al. 2000), (Lewis 2001), and (Hagan et al. 2003) – among many other publications (the list of relevant references is very long). ○

Another reason for developing stochastic volatility models is the well-known fact that empirical distributions of stock returns exhibit much heavier tails than the tails in the Gaussian distribution that the model $Z = \sigma W + b\iota$ postulates. However, we note that models with heavy tails in the stock returns can also be developed by making stochastic the expected growth rate b while keeping the volatility parameter σ constant.

Market models with stochastic volatility are not entirely new to us. Indeed, the general market model that we introduced earlier postulates that all stock prices are governed by $S = S_0 + S \bullet Z$, for some general continuous semimartingale Z. Thus, choosing Z to be something other than $\sigma W + b\iota$ should not come as a surprise, and our objective now is to investigate other concrete choices for the net returns Z. The generalization that comes to mind first is to replace the constant parameters σ and b with predictable stochastic processes, also denoted by σ and b, and postulate that the net returns follow an Itô process of the form

$$Z = \sigma \bullet W + b \bullet \imath .$$

The levels of the processes σ and b at time t, that is, σ_t and b_t, are called, respectively, *local volatility* and *local drift*, but the adjective "local" is often dropped for brevity. In a generic stochastic volatility model the local volatility process σ is assumed to have the form $\sigma = h(Y, \imath)$, for some concrete choice of the function $h \colon \mathbb{R} \times \mathbb{R}_+ \mapsto \mathbb{R}$ and concrete choice of the latent diffusion process Y with structure

$$Y = Y_0 + f(Y, \imath) \bullet B + g(Y, \imath) \bullet \imath .$$

Depending on the context, we may refer to Y as the *latent volatility process*, but, as we are about to see, in most concrete models the process Y is going to be the volatility process itself, that is, $h(Y, \imath) = Y$. The Brownian motion B is assumed to have the form $B = \rho W + \sqrt{1 - \rho^2} W^{\perp}$ for some fixed parameter $\rho \in [-1, 1]$ and yet another Brownian motion $W^{\perp}$ that is independent of W. The diffusion coefficients $f, g \colon \mathbb{R} \times \mathbb{R}_+ \mapsto \mathbb{R}$ are exogenous as well – it is mainly the choice of f and g that distinguishes one stochastic volatility model from another. For example, Scott (1987) postulates $b =$ constant and $h(Y, \imath) = Y =$ exponential OU process of the type discussed in (11.60), while Hull and White (1987) postulate $h(Y, \imath) = Y =$ geometric Brownian motion ($b =$ constant, still). The model introduced by Heston (1993) also postulates $b =$ constant, but the volatility process is assumed to be of the form $h(Y, \imath) = \sqrt{Y}$ for some$^{\circ\!\!\to (11.50)}$ CIR process Y. To put it another way, in Heston's model the volatility is taken to be the norm of some d-dimensional OU process (with d not necessarily an integer). One special case of this model (which is somewhat easier to implement) was proposed by Stein and Stein (1991), who postulate that the volatility is simply an OU process (as we noted earlier, replacing the volatility process with its absolute value does not change the model). We see that Hull and White's model assumes that the volatility process grows exponentially on average, while both Scott's and Heston's models allow for a stationary and mean-reverting volatility process, which is somewhat more realistic.

In the rest of this section we are going to review these models one by one, but before we do we must note that interesting and useful option pricing models with stochastic volatility can be developed only within the framework of incomplete financial markets. Indeed, in the setting outlined above the information arrival dynamics are generated by two independent Brownian motions, W and $W^{\perp}$, whereas the price information in the market is generated only by the spot price of a single risky asset S. To put it another way, since the market agents can observe the local volatility, which depends on the latent volatility process, the observation (market) filtration must be larger than the filtration that the traded securities generate. As a result, one cannot replicate with marketable assets alone all possible stochastic payoffs that the market information structure allows for. In order to derive concrete formulas for option prices, one must attach a risk premium process not just to the Brownian motion W, but also to the Brownian motion B. Recall that the market price of risk attached to the Brownian motion W, which drives the price process, is some predictable process θ that has locally square integrable sample paths and is such that $W + \theta \bullet \iota$ is a Brownian motion with respect to any ELMM Q, that is, any measure $Q \approx P$ such that the excess returns process

$$X = Z + (\delta - r) \bullet \iota = \sigma \bullet \left(W + \frac{b + \delta - r}{\sigma} \bullet \iota \right)$$

is a local Q-martingale. This leaves us with the only choice for the risk premium θ, namely $\theta = \frac{b+\delta-r}{\sigma}$. Similarly, the risk premium process, ϑ, attached to the Brownian motion B is again some predictable process with locally square integrable sample paths and the property that $B + \vartheta \bullet \iota$ is a Brownian motion under any equivalent pricing measure Q. Unfortunately, whether directly observable or not, the latent volatility process Y, which the Brownian motion B drives, is not traded, and, for this reason, there is no obvious choice for the risk premium ϑ, as there are no excess returns to turn into a local martingale. This is what incompleteness comes down to in the present setting. Thus, in one way or another, the risk incorporated into the Brownian motion B has to be priced in order to price every contingent claim in a market with an information structure generated by both W and B (or, equivalently, generated by W and $W^{\perp}$). As the mechanics of incomplete financial markets lies outside of our main concerns, at this point we are going to suppose that the risk premium process ϑ is somehow chosen in the outset – perhaps by way of matching empirically observed option prices to the model predictions produced by the various choices for the price of risk. Furthermore, as the Brownian motion W affects the latent volatility process Y only through the dependence of Y on B – there is nothing to be learned about W from Y that cannot be learned from B – we are going to suppose for simplicity that the risk premium ϑ depends only on B, that is, ϑ is adapted to $\mathscr{F}^B$. In particular, if B and W are independent (the case $\rho = 0$), then ϑ would be independent of W. Notice, however, that we cannot afford to impose the assumption that $\theta = \frac{b+\delta-r}{\sigma}$ is independent of B even if W and B are independent. Since

$$W^{\perp} = \frac{1}{\sqrt{1-\rho^2}} B - \frac{\rho}{\sqrt{1-\rho^2}} W ,$$

the risk premium attached to the Brownian motion $W^{\perp}$ must be

$$\theta^{\perp} = \frac{1}{\sqrt{1-\rho^2}} \vartheta - \frac{\rho}{\sqrt{1-\rho^2}} \theta .$$

Observe that the Brownian motion $B^{\perp} \stackrel{\text{def}}{=} \sqrt{1-\rho^2} W - \rho W^{\perp}$ is independent of B and that $W = \rho B + \sqrt{1-\rho^2} B^{\perp}$. In particular, if $\vartheta^{\perp}$ is the risk premium attached to $B^{\perp}$, then $\theta = \rho\vartheta + \sqrt{1-\rho^2}\vartheta^{\perp}$. At this point we have risk premium attached to any of the Brownian motions, W, $W^{\perp}$, B, and $B^{\perp}$, and therefore the symbols $\tilde{W}$, $\tilde{W}^{\perp}$, $\tilde{B}$, and $\tilde{B}^{\perp}$ are meaningful (as Brownian motions under Q).

Assuming a finite time-horizon $T > 0$, the equivalent pricing measure must be of the form $Q = R \odot P$, with Radon–Nikodym derivative

$$R = e^{-(\theta \bullet W)_T - \frac{1}{2}(\theta^2 \bullet \iota)_T - (\theta^{\perp} \bullet W^{\perp})_T - \frac{1}{2}((\theta^{\perp})^2 \bullet \iota)_T},$$

provided, of course, that $\mathsf{E}[R] = 1$. Setting

$$\tilde{W} = W + \theta \bullet \iota \quad \text{and} \quad \tilde{B} = B + \vartheta \bullet \iota,$$

and recognizing that both $\tilde{W}$ and $\tilde{B}$ are Q-Brownian motions with $\langle \tilde{B}, \tilde{W} \rangle = \rho\iota$, we can now cast the returns Z and the latent volatility process Y in the form

$$Z = \sigma \bullet \tilde{W} + (b - \sigma\theta) \bullet \iota = h(Y, \iota) \bullet \tilde{W} + (r - \delta) \bullet \iota$$
$$\text{and} \quad Y = Y_0 + f(Y, \iota) \bullet \tilde{B} + \big(g(Y, \iota) - f(Y, \iota)\vartheta\big) \bullet \iota\,. \qquad \mathrm{e}_1$$

The price of the spot asset at time T then becomes

$$S_T = S_0 e^{Z_T - \frac{1}{2}\langle Z, Z \rangle_T} = S_0 e^{h(Y,\iota) \bullet \tilde{W}_T + \left(r - \delta - \frac{1}{2} h(Y,\iota)^2\right) \bullet \iota_T}\,. \qquad \mathrm{e}_2$$

(14.40) THE CASE $\rho = 0$: We have $\vartheta = \theta^{\perp}$ and $B = W^{\perp}$, that is, W and B are independent under P, which also means that $\tilde{W}$ and $\tilde{B}$ are independent under Q.[168] This shows that Y is independent of $\tilde{W}$ under Q, as the second equation in e_1 becomes independent from the first. As a result, $h(Y, \iota) \bullet \tilde{W}$ can be understood as an Itô integral for a fixed (as if deterministic) integrand $h(Y_\omega, \iota)$. Thus, conditioned to the sample path Y_ω, the r.v. $h(Y_\omega, \iota) \bullet \tilde{W}_T$ follows the Gaussian law of variance $\int_0^T h(Y_\omega(t), t)^2 \, \mathrm{d}t$, that is, Gaussian law of variance $\Sigma_T(\omega)^2 T$ for

$$\Sigma_T(\omega) \overset{\text{def}}{=} \left(\frac{1}{T} \int_0^T h(Y_\omega(t), t)^2 \, \mathrm{d}t \right)^{1/2}.$$

We then have the following identity in law under Q:

$$S_T \overset{\text{law}}{=} S_0 e^{\Sigma_T \tilde{W}_T + \left(r - \delta - \frac{1}{2}\Sigma_T^2\right) T}.$$

The expression on the right side is no different from the expression for the spot price in the BSM model, except that the volatility parameter σ is now replaced with the token Σ_T. By Tonelli–Fubini's theorem, the calculation of $\mathsf{E}^Q\big[(S_T - K)^+\big]$ can be accomplished in two stages, by first averaging out the dependence on $\tilde{W}_T$ for every fixed sample path Y_ω, and then averaging out the dependence on Y_ω. Since S_T depends on Y_ω only through the r.v. Σ_T, if $\mathcal{L}_T$ stands for the distribution law of Σ_T^2 under Q, the price of the European call with strike $K > 0$ and closing date $T > 0$ can be expressed as

† $$\int_{\mathbb{R}_+} BSM(S_0, K, u, r - \delta, T) \mathcal{L}_T(\mathrm{d}u)\,. \quad \circ$$

[168] In this case $\tilde{W}$ and $\tilde{B}$ are Q-Brownian motions with $\langle \tilde{W}, \tilde{B} \rangle = \langle W, B \rangle = \rho\iota = 0$.

(14.41) THE MODEL OF HULL AND WHITE (1987): It is assumed that

$$h(Y,\iota) = Y,\ f(Y,\iota) = \gamma Y,\ \text{and}\ g(Y,\iota) = kY$$

for some fixed parameters $\gamma, k \in \mathbb{R}$. In addition, $\rho = 0$, so that the latent volatility Y is independent of W under P, and also independent of $\tilde{W}$ under Q. Since Y is not a traded asset, the risk premium attached to $B = W^{\perp}$ is set to $\vartheta = 0$. As a result of these assumptions,⊸(11.38)

$$Y = Y_0 + Y \cdot H \quad \text{for} \quad H = \gamma \tilde{B} + k\iota \quad \Rightarrow \quad Y = Y_0 \mathscr{E}(H) \equiv Y_0 e^{\gamma \tilde{B} + (k - \frac{1}{2}\gamma^2)\iota}$$

and

$$\Sigma_T^2 = \frac{1}{T}\int_0^T Y_t^2 \, \mathrm{d}t = \frac{Y_0^2}{T}\int_0^T e^{2\gamma \tilde{B}_t + (2k-\gamma^2)t}\, \mathrm{d}t\,.$$

It is well known that the integral of exponential Brownian motion admits smooth density; in particular, Σ_T^2 admits (smooth) density $\psi_{T,k,\gamma}(u) = \frac{1}{\mathrm{d}u}\mathcal{L}_{T,k,\gamma}(\mathrm{d}u)$. The integral in ($14.40^{\dagger}$) can now be calculated in a number of ways. One possible approach is to use the integral formulas for the density of $\int_0^T e^{2\gamma \tilde{B}_t + (2k-\gamma^2)t}\, \mathrm{d}t$ derived by Yor (1992) and Lyasoff (2016), in which the density of this integral is expressed as a double integral involving certain special functions. With this approach the calculation of the option price in ($14.40^{\dagger}$) boils down to the numerical evaluation of a triple integral. ○

(14.42) ONE SPECIAL CASE OF CORRELATED BROWNIAN MOTIONS: (Jeanblanc et al. 2009, 6.7.5.2) Looking back at the relations e_1, suppose that the risk premium ϑ is chosen in such a way that $g(Y,\iota) - f(Y,\iota)\vartheta = 0$. In addition, just as in Hull and White's model, suppose that $f(Y,\iota) = \gamma Y$ and $h(Y,\iota) = Y$, so that

$$Y = Y_0 + \gamma Y \cdot \tilde{B} \quad \Leftrightarrow \quad Y = Y_0 \mathscr{E}(\gamma \tilde{B}) \equiv Y_0 e^{\gamma \tilde{B} - \frac{1}{2}\gamma^2 \iota}\,.$$

We no longer suppose that $\rho = 0$, so that $\tilde{W} = \rho \tilde{B} + \sqrt{1-\rho^2}\,\tilde{B}^{\perp}$ and $\tilde{B}$ are no longer independent. With these choices in place, the expression for the price of the spot asset on the final date T in e_2 obtains the form

$$S_T = S_0 e^{\rho Y \cdot \tilde{B}_T - \frac{1}{2}\rho^2 Y^2 \cdot \iota_T} e^{\sqrt{1-\rho^2}\, Y \cdot \tilde{B}^{\perp}{}_T + \left(r - \delta - \frac{1}{2}(1-\rho^2)Y^2\right) \cdot \iota_T}\,.$$

Since $Y \cdot \tilde{B}_T = \frac{1}{\gamma}(Y_T - Y_0) = \frac{Y_0}{\gamma}\left(e^{\gamma \tilde{B}_T - \frac{1}{2}\gamma^2 T} - 1\right)$, the first exponent in the expression for S_T is an explicit function, say F, of the random variables

$$\tilde{B}_T \quad \text{and} \quad \Sigma_T^2 \overset{\text{def}}{=} \frac{1}{T}\int_0^T Y_t^2 \, \mathrm{d}t \equiv \frac{Y_0^2}{T}\int_0^T e^{2\gamma \tilde{B}_t - \gamma^2 t}\, \mathrm{d}t\,.$$

The joint law of these two random variables admits probability density for which explicit integral expressions are available – see (Yor 1992) and (Lyasoff 2016). Similarly to the case $\rho = 0$, we now have

$$S_T \overset{\text{law}}{=} S_0 F(\tilde{B}_T, \Sigma_T^2) e^{\sqrt{1-\rho^2}\, \Sigma_T \tilde{B}_T^{\perp} + \left(r - \delta - \frac{1}{2}(1-\rho^2)\Sigma_T^2\right) T}\,,$$

and this allows us to write the price of the call as

$$\int_{\mathbb{R}}\int_{\mathbb{R}_{++}} BSM\Big(S_0F(x,u), K, (1-\rho^2)u, r-\delta, T\Big)\psi_\gamma(x,u)\,\mathrm{d}x\,\mathrm{d}u\,,$$

where ψ_γ stands for the joint probability density of $\tilde{B}_T$ and Σ_T^2 under the risk adjusted equivalent measure Q, that is,

$$\psi_\gamma(x,u) = \frac{1}{\mathrm{d}x\,\mathrm{d}u} Q\big(\tilde{B}_T \in \mathrm{d}x, \Sigma_T^2 \in \mathrm{d}u\big)\,, \quad x\in\mathbb{R}\,, \ u\in\mathbb{R}_{++}\,.$$

With the integral formulas for this density in mind, the computation of the price of the call comes down to the computation of a triple integral. ○

(14.43) THE MODEL OF STEIN AND STEIN (1991): With the notation and the setting as in (14.40) (the case $\rho = 0$), suppose that $h(Y,\iota) = Y$, $f(Y,\iota) = \gamma$ (constant) and $g(Y,\iota) = \kappa(m-Y)$, for some choice of the parameters $\gamma, \kappa, m \in \mathbb{R}$. In this model the ex-dividend net returns have the form $Z = Y{\bullet}W + b\iota$ (b is a constant), with volatility process Y that is independent of the Brownian motion W and⊸(11.41) has the dynamics of an OU process, so that

$$Y = Y_0 + \gamma B + \kappa(m-Y){\bullet}\iota\,.$$

Since $\mathrm{sign}(Y){\bullet}W$ is a Brownian motion, which, despite the fact that Y depends on B, is again independent of B (as long as W and B are independent, as we now suppose), from a modeling point of view nothing would change if we were to postulate that the volatility process is $h(Y,\iota) = |Y|$, so that it will remain positive – see note 167 on page 382. Since $f(Y,\iota) = Y$ is easier to work with, in what follows we will stick to this choice. From the calculation developed in (14.40), the computation of the option price in this model boils down to the computation of the density of $\Sigma_T^2 = \frac{1}{T}\int_0^T Y_s^2\,\mathrm{d}s$. However, (Stein and Stein 1991) take this task a step further and develop a formula for the density of the underlying spot price $S \equiv S_0\mathscr{E}(Z)$. ○

What follows next is a brief outline of the main steps in the method of Stein and Stein (1991). With no loss of generality we suppose that $S_0 = 1$ and take the risk premium attached to the Brownian motion B to be $\vartheta = 0$ (the computation with $\vartheta =$ constant $\neq 0$ comes down to merely shifting the value of m when the process is treated under the measure Q). All probability densities will be understood relative to Q.

Step 1: With $r = \delta$, conditioned to $\Sigma_t = u > 0$ the density of the S_t is⊸(3.3)

$$\mathbb{R}_{++} \ni x \rightsquigarrow \psi_{t,0}(x\,|\,u) \stackrel{\text{def}}{=} \frac{1}{xu\sqrt{2\pi t}}\, e^{-\frac{1}{2tu^2}\left(\log(x)+\frac{1}{2}u^2t\right)^2}\,.$$

With general r and δ the conditional density of S_t given $\Sigma_t = u$ obtains from $\psi_{t,0}$ through the relation

$$\psi_{t,r-\delta}(x\,|\,u) = e^{-(r-\delta)t}\psi_{t,0}\big(e^{-(r-\delta)t}x\,|\,u\big)\,.$$

In particular, integrals against the density $\psi_{t,r-\delta}(\cdot\,|\,u)$ can be transformed into integrals against the density $\psi_{t,0}(\cdot\,|\,u)$ in the obvious way.

Step 2: Define the function

$$\mathbb{R} \ni x \rightsquigarrow f_{t,0}(x\,|\,u) \stackrel{\text{def}}{=} \frac{1}{u\sqrt{2\pi t}}\, e^{-\frac{1}{2tu^2}\left(x+\frac{1}{2}u^2 t\right)^2}$$

and observe that $\psi_{t,0}(x\,|\,u) = f_{t,0}\big(\log(x)\,|\,u\big)/x$, so that for any integrable function $\Phi(\cdot)$,

$$\int_{\mathbb{R}_{++}} \Phi(x)\psi_{t,0}(x\,|\,u)\,\mathrm{d}x = \int_{\mathbb{R}_{++}} \Phi(x) f_{t,0}\big(\log(x)\,|\,u\big)\frac{1}{x}\,\mathrm{d}x = \int_{\mathbb{R}} \Phi(e^x) f_{t,0}\big(x\,|\,u\big)\,\mathrm{d}x\,.$$

With this connection in mind, any integral against the density $x \rightsquigarrow \psi_{t,0}(x\,|\,u)$ can be transformed into an integral against the function $x \rightsquigarrow f_{t,0}(x\,|\,u)$ (obviously, a density itself). The objective now is to develop a method for computing integrals against the density

$$\mathbb{R} \ni x \rightsquigarrow \phi_{t,y}(x) \stackrel{\text{def}}{=} \int_{\mathbb{R}_{++}} f_{t,0}\big(x\,|\,u\big)\varphi_{t,y}(u)\,\mathrm{d}u\,,$$

where $u \rightsquigarrow \varphi_{t,y}(u)$ is the density of $\Sigma_t = \left(\frac{1}{t}\int_0^t Y_s^2\,\mathrm{d}s\right)^{1/2}$ conditioned to $\{Y_0 = y\}$.

Step 3: Compute the Fourier transform of the function $f_{t,0}(\cdot\,|\,u)$, namely,

$$(\mathbb{F} f_{t,0})(v\,|\,u) = \frac{1}{\sqrt{2\pi}}\int_{\mathbb{R}} e^{-ixv} f_{t,0}(x\,|\,u)\,\mathrm{d}x = \frac{1}{\sqrt{2\pi}} e^{-(v^2-iv)\frac{u^2 t}{2}}\,, \quad v \in \mathbb{R}\,.$$

Stein and Stein (1991) call the last identity "standard fact from analysis," and refer to (Wiener 1933).[169]

Step 4: Compute the Laplace transform

$$U_\zeta(y,t) = \mathsf{E}\big[e^{-\zeta\Sigma_t^2}\,\big|\,Y_0 = y\big] = \int_{\mathbb{R}_{++}} e^{-\zeta u^2}\varphi_{t,y}(u)\,\mathrm{d}u \quad \text{for } \zeta \in \mathbb{C}\,,\ \operatorname{Re}(\zeta) \geq 0\,.$$

This expectation can be computed by using the Feynman–Kac formula from (12.2) as follows. First set

$$g_\zeta(y,s) = \mathsf{E}\big[e^{-\zeta\int_s^t Y_v^2\,\mathrm{d}v}\,\big|\,Y_s = y\big]\,, \quad s \leq t\,,$$

and observe that this function solves the PDE

$$\partial g_\zeta(y,s) + \frac{1}{2}\gamma^2\partial_y^2 g_\zeta(y,s) + \kappa(m-y)\partial_y g_\zeta(y,s) = \zeta y^2 g_\zeta(y,s)$$

on the interval $s \in\,]-\infty, t]$ with terminal condition $g_\zeta(y,t) = 1$. Next, observe that with $h_\zeta(y,s) \stackrel{\text{def}}{=} g_\zeta(y,t-s)$, $s \in \mathbb{R}_+$, the last equation becomes

$$\frac{1}{2}\gamma^2\partial_y^2 h_\zeta(y,s) + \kappa(m-y)\partial_y h_\zeta(y,s) - \zeta y^2 h_\zeta(y,s) = \partial h_\zeta(y,s)\,, \quad s \in \mathbb{R}_+\,, \quad \mathrm{e}_3$$

[169] Stein and Stein (1991) actually use characteristic functions instead of Fourier transforms. The expression for the Fourier transform given here is adjusted⊶(3.11) from theirs accordingly.

with initial condition $h_\zeta(y,0) = 1$. The Laplace transform that we are trying to compute is $U_\zeta(y,t) = h_{\zeta/t}(y,t)$ and this task boils down to solving e_3. A solution to this PDE is known to exist in the form

$$h_\zeta(y,t) = e^{\frac{1}{2}L_\zeta(t)\,y^2 + M_\zeta(t)\,y + N_\zeta(t)}$$

for a suitable choice of the functions $L_\zeta(\cdot)$, $M_\zeta(\cdot)$, and $N_\zeta(\cdot)$. While the computation of these functions is somewhat cumbersome, the idea is quite straightforward (and can be greatly facilitated with the aid of computer algebra). Indeed, substituting the expression above in e_3 would make each side of e_3 appear as a quadratic polynomial in the variable y, and equating the coefficients on both sides yields a system of three first-order ODEs for the unknown functions $L_\zeta(\cdot)$, $M_\zeta(\cdot)$, and $N_\zeta(\cdot)$. This system allows for a closed-form solution with initial condition $L_\zeta(0) = M_\zeta(0) = N_\zeta(0) = 0$ – see (Stein and Stein 1991).

Step 5: Going back to step 2, and using the results from step 3 and step 4, compute the Fourier transform of the modified density function $\phi_{t,0}(\cdot)$:

$$\begin{aligned}(\mathbb{F}\phi_{t,0})(v) &= \frac{1}{\sqrt{2\pi}}\int_{\mathbb{R}} e^{-\imath x v}\phi_{t,y}(x)\,\mathrm{d}x = \int_{\mathbb{R}_{++}} (\mathbb{F}f_{t,0})(v\,|\,u)\varphi_{t,y}(u)\,\mathrm{d}u \\ &= \frac{1}{\sqrt{2\pi}}\int_{\mathbb{R}_{++}} e^{-\frac{t}{2}(v^2-\imath v)u^2}\varphi_{t,y}(u)\,\mathrm{d}u = \frac{1}{\sqrt{2\pi}}U_{\frac{t}{2}(v^2-\imath v)}(y,t)\,,\end{aligned}$$

where $U_\zeta(y,t)$ is the function computed in step 4.

Step 6: Using inverse Fourier transform write the modified density $\phi_{t,0}(\cdot)$ as an integral:

$$\phi_{t,0}(x) = \frac{1}{2\pi}\int_{\mathbb{R}} e^{\imath x v}U_{\frac{t}{2}(v^2-\imath v)}(Y_0,t)\,\mathrm{d}v\,.$$

Any integral against the density $\phi_{t,0}(\cdot)$ – in particular, the price of a European-style call or put option – of the form $\int_{\mathbb{R}}\Phi(x)\phi_{t,0}(x)\,\mathrm{d}x$ can be expressed and computed (obviously, numerically) as the double integral

$$\frac{1}{2\pi}\int_{\mathbb{R}}\int_{\mathbb{R}}\Phi(x)e^{\imath x v}U_{\frac{t}{2}(v^2-\imath v)}(Y_0,t)\,\mathrm{d}v\,\mathrm{d}x\,.$$

(14.44) THE MODEL OF HESTON (1993): One drawback from the model of Stein and Stein (1991) is that the volatility process hits 0 with strictly positive probability during any finite time interval. If this particular feature is somehow undesirable, while the mean-reversion and stationarity features still are, then the CIR (squared radial OU) process introduced in (11.50) becomes the natural choice for the variance process. This is what Heston's model is based on. The Brownian motions B and W are allowed to be correlated ($\rho \neq 0$), $b =$ constant, and $\sigma = h(Y,\imath) = \sqrt{Y}$ for Y given by

$$Y = Y_0 + \kappa\Big(d\frac{\gamma^2}{4\kappa} - Y\Big)\bullet\imath + \gamma\sqrt{Y}\bullet B\,,$$

with model parameters $\kappa, \gamma \in \mathbb{R}_{++}$ and $d \in [2, +\infty[$ – note that this choice for d forces the sample paths of Y to remain in $\mathbb{R}_{++}$ forever with probability 1. In addition, Heston (1993) assumes that the risk premium process associated with the Brownian motion B has the form $\vartheta = \lambda\sqrt{Y}$ for some fixed parameter $\lambda \in \mathbb{R}_{++}$. This choice is intuitive, but in (Heston 1993) it is motivated by standard results from consumption-based asset pricing. In terms of the notation introduced earlier in this section, the price of the spot asset can be cast as

† $$S = S_0 e^{\sqrt{Y}\bullet\tilde{W} + (r-\delta-\frac{1}{2}Y)\bullet\imath},$$

where $\tilde{W}$ is a Brownian motion under the equivalent risk-neutral pricing measure Q. With that measure in mind, the variance process can be restated as

‡ $$Y = Y_0 + \left(d\frac{\gamma^2}{4} - (\kappa + \lambda\gamma)Y\right)\bullet\imath + \gamma\sqrt{Y}\bullet\tilde{B},$$

where $\tilde{B} = B + \lambda\sqrt{Y}\bullet\imath$ is a Q-Brownian motion. In particular, the price of the European call with strike $K > 0$ and maturity $T > 0$ comes down to

$$e^{-rT}\mathsf{E}^Q\left[S_T 1_{\{S_T \geq K\}} \mid S_0, Y_0\right] - e^{-rT} K\,\mathsf{E}^Q\left[1_{\{S_T \geq K\}} \mid S_0, Y_0\right].$$

The computation of the second expectation reduces to the computation of the conditional probability $Q(S_T \geq K \,|\, S_0, Y_0)$, while the computation of the first expectation would come down to the same thing after rearranging the parameters accordingly, as we now detail. We have

$$\mathsf{E}^Q\left[S_T 1_{\{S_T \geq K\}} \mid S_0, Y_0\right] = e^{(r-\delta)T} S_0\,\mathsf{E}^Q\left[e^{\sqrt{Y}\bullet\tilde{W} - \frac{1}{2}Y\bullet\imath_T} 1_{\{S_T \geq K\}} \mid S_0, Y_0\right],$$

and, following the usual step in option pricing (and also (Heston 1993)), we change the pricing measure Q to $\hat{Q} \stackrel{\text{def}}{=} e^{\sqrt{Y}\bullet\tilde{W}_T - \frac{1}{2}Y\bullet\imath_T} \odot Q$. For this to be possible we must ensure that

$$\mathsf{E}^Q\left[e^{\sqrt{Y}\bullet\tilde{W}_T - \frac{1}{2}Y\bullet\imath_T} \mid Y_0\right] = 1.$$

This property can be established by showing that

$$\mathsf{E}^Q\left[e^{pY\bullet\imath_T} \mid Y_0\right] < \infty \quad \text{for any } p \in \mathbb{R}_{++}.$$

Absent more sophisticated tools at our disposal, the last relation can be verified with what one may call "brute force," that is, by computing the expected value explicitly. The conscientious reader is invited to fill in the details in these steps. Due to the comparison theorem, increasing d to the nearest integer from above would only increase the integral $Y\bullet\imath_T$, and with integer value for $d \geq 2$ the law of Y is the same as the law of the squared norm of the d-dimensional OU process. Given the partition $t_i = iT/n$, $i \in \mathbb{N}_{|n}$, compute the expectation

$$\mathsf{E}^Q\left[e^{p\sum_{i\in\mathbb{N}_{|n-1}} Y_{t_{i+1}}(t_{i+1}-t_i)} \mid Y_0\right]$$

recursively by conditioning first to $\mathscr{F}_{t_{n-1}}$, then to $\mathscr{F}_{t_{n-2}}$, and so on, taking into account that if U is the OU process of dimension 1 with parameters σ and k, then

$U_{t+\Delta}^2 = (e^{-\frac{k}{2}\Delta}U_t + \xi)^2$, with $\xi \in \mathcal{N}(0, \gamma^2)$ for $\gamma^2 = \frac{\sigma^2}{4k}(1 - e^{-k\Delta})$. In particular, for all sufficiently large n the expectations above are uniformly bounded (with a bound that depends on p). Since this is true for any $p \geq 2$, the exponents under the expectation are uniformly integrable (for the various choice of n and for fixed p) and the interchange between the limiting operation as $n \to \infty$ and the expectation operation E^Q can be justified. The limit of the expected values can then be shown to be finite.

In any case, $\hat{Q} \stackrel{\text{def}}{=} e^{\sqrt{Y}\bullet\tilde{W}_T - \frac{1}{2}Y\bullet\iota_T} \odot Q$ is now a well-defined probability measure that is equivalent to both Q and P. In particular,

$$\mathsf{E}^Q\left[e^{\sqrt{Y}\bullet\tilde{W}_T - \frac{1}{2}Y\bullet\iota_T} 1_{\{S_T \geq K\}}\right] = \hat{Q}(S_T \geq K \,|\, S_0, Y_0)\,.$$

By Girsanov's theorem, the process $\hat{W} = \tilde{W} - \sqrt{Y}\bullet\iota$ must be a $\hat{Q}$-Brownian motion, and, since

$$S_T = S_0 e^{\sqrt{Y}\bullet\hat{W}_T + (r - \delta + \frac{1}{2}Y)\bullet\iota_T}\,,$$

we see that the computation of $\hat{Q}(S_T \geq K \,|\, S_0, Y_0)$ is the same as the computation of $Q(S_T \geq K \,|\, S_0, Y_0)$ with the process S from † changed to $S_0 e^{\sqrt{Y}\bullet\tilde{W} + (r-\delta+\frac{1}{2}Y)\bullet\iota}$ and with the dynamics of Y from ‡ changed to

$$Y = Y_0 + \left(d\frac{\gamma^2}{4} - \left(\kappa + (\lambda - \rho)\gamma\right)Y\right)\bullet\iota + \gamma\sqrt{Y}\bullet\tilde{B}\,,$$

as we now detail. Since $\langle \tilde{B}, \tilde{W} \rangle = \rho\iota$, the process

$$\tilde{W}^\perp = \frac{1}{\sqrt{1-\rho^2}}\tilde{B} - \frac{\rho}{\sqrt{1-\rho^2}}\tilde{W}$$

is a Q-Brownian motion which is independent of $\tilde{W}$ under Q, and we have

$$\tilde{B} = \rho\tilde{W} + \sqrt{1-\rho^2}\tilde{W}^\perp = \rho\hat{W} + \sqrt{1-\rho^2}\tilde{W}^\perp + \rho\sqrt{Y}\bullet\iota\,.$$

In particular, we can rewrite the dynamics of Y in terms of the $\hat{Q}$-Brownian motion $\hat{B} \stackrel{\text{def}}{=} \rho\hat{W} + \sqrt{1-\rho^2}\tilde{W}^\perp \equiv \tilde{B} - \rho\sqrt{Y}\bullet\iota$ as

$$Y = Y_0 + \left(d\frac{\gamma^2}{4} - \left(\kappa + (\lambda - \rho)\gamma\right)Y\right)\bullet\iota + \gamma\sqrt{Y}\bullet\hat{B}\,.$$

Thus, it would be enough for us to compute the probabilities

$$q^\pm(x, y) \stackrel{\text{def}}{=}$$
$$Q\Big(\log(S_0) + \sqrt{Y^\pm}\bullet\tilde{W}_T + \big(r - \delta \pm \frac{1}{2}Y^\pm\big)\bullet\iota_T \geq \log(K) \;\big|\; S_0 = e^x, Y_0 = y\Big)$$

for the processes $Y^\pm$ that are given by

$$Y^\pm = Y_0 + \left(d\frac{\gamma^2}{4} - \left(\kappa + \gamma\mu^\pm\right)Y\right)\bullet\iota + \gamma\sqrt{Y^\pm}\bullet\tilde{B}\,,$$
$$\text{with}\quad \mu^+ = (\lambda - \rho) \quad\text{and}\quad \mu^- = \lambda\,.$$

The main step in Heston's method is the computation – followed by an inversion – of the Fourier transforms

$$f^{\pm}(x,y,t;u) \overset{\text{def}}{=} \frac{1}{\sqrt{2\pi}} \mathsf{E}^{Q}\Big[e^{-\imath u\left(x+\sqrt{Y^{\pm}}\bullet\tilde{W}_T+(r-\delta)T\pm\frac{1}{2}(Y^{\pm})\bullet\iota_T\right)} \mid Y^{\pm}_t = y\Big], \quad t \le T.$$

Of course, we only need these Fourier transforms with $t = 0$, and the reason for working with a generic choice for t becomes clear below. The $\mathbb{R}\times\mathbb{R}_{++}$-valued diffusion processes

$$(X^{\pm}, Y^{\pm}) \overset{\text{def}}{=} \left(\log(S_0) + \sqrt{Y^{\pm}}\bullet\tilde{W} + \left(r - \delta \pm \frac{1}{2}Y^{\pm}\right)\bullet\iota\,,\ Y^{\pm}\right)$$

have infinitesimal generators

$$\begin{aligned}\mathcal{A}^{\pm}_t &= \frac{1}{2}y\partial_x^2 + \frac{1}{2}\gamma^2 y\partial_y^2 + \rho\gamma y\partial_x\partial_y \\ &\qquad + \left(r - \delta \pm \frac{1}{2}y\right)\partial_x + \left(d\frac{\gamma^2}{4} - (\kappa + \gamma\mu^{\pm})y\right)\partial_y\,,\end{aligned}$$

and, due to the Feynman–Kac formula, the Fourier transforms $f^{\pm}(x,y,t;u)$ must satisfy (for fixed u) the following PDEs

$$\left(\partial + \mathcal{A}^{\pm}_t\right)f^{\pm}(x,y,t;u) = 0\,, \quad (x,y) \in \mathbb{R}\times\mathbb{R}_{++}\,, \quad 0 \le t < T\,,$$

with boundary conditions $\lim_{t\nearrow T} f^{\pm}(x,y,t;u) = \frac{1}{\sqrt{2\pi}}e^{-\imath ux}$. The key insight in Heston's work is that solutions to these equations can again be sought in the form

$$f^{\pm}(x,y,t;u) = \frac{1}{\sqrt{2\pi}}e^{-\imath ux + a^{\pm}(T-t) + b^{\pm}(T-t)y}$$

for some yet to be determined functions $a^{\pm}(\cdot)$ and $b^{\pm}(\cdot)$, chosen so that $a^{\pm}(0) = b^{\pm}(0) = 0$. This works because the coefficients in the PDEs that we are trying to solve here are all linear. The final step is to substitute the expression above into the PDE and cancel the exponent. The remaining terms can then be divided into two sums, the first involving only the function $a^{\pm}(\cdot)$, and the second involving only the function $b^{\pm}(\cdot)$. Equating the first sum to 0 yields an ODE for the unknown function $a^{\pm}(\cdot)$, and equating the second sum to 0 yields another first-order ODE for the unknown function $b^{\pm}(\cdot)$. Both ODEs can be transformed into linear first-order ODEs, and therefore admit closed-form solutions. As a result, having a closed-form expression for the Fourier transforms, one can apply the inverse Fourier transform to $f^{\pm}(x,y,0;u)$ and have the probabilities $q^{\pm}(x,y)$ expressed as univariate integrals, which can then be calculated numerically by using a standard quadrature program. We refer the reader to (Heston 1993) for the complete details. ○

14.8 Dupire's Formula

In every reasonable option pricing model the spot price, $S = S_0 + S \cdot Z$, is assumed to be observable, and therefore so is its logarithm $\log(S)$. With $Z = \sigma \cdot W + b \cdot \iota$ chosen as in the previous section, this means that $\log(S_0) + Z - \frac{1}{2}\langle Z, Z\rangle = \log(S)$ is observable. Thus, at least in theory, it should be possible to extract the bracket

$$\langle \log(S), \log(S)\rangle = \langle Z, Z\rangle = \sigma^2 \cdot \iota$$

from physically observed market data. The local variance can then be computed as

$$\sigma_t^2 = \partial_t \langle \log(S), \log(S)\rangle_t \,.$$

In practice, however, this calculation is not as straightforward as it appears here, for to say the least, the computation of the bracket would require sampling the spot price with infinite frequency. There are several well-known and widely adopted statistical tools that can be employed in such a context. While very important – from both theoretical and practical points of view – such tools lie outside of our main concerns, and here we are only going to point to the default reference (Campbell et al. 1997), the more recent work (Aït-Sahalia and Jacod 2014), and also to the work (Berestycki et al. 2000), which studies the connection between local volatility and implied volatility.

One way or another, any alternative method for extracting information about the local volatility from observed spot prices would certainly be useful. One such method stems from the connection discovered by Dupire (1994) and Derman and Kani (1994). The key insight on which Dupire's method rests is that information about the local volatility should be possible to extract not just from the observed underlying spots, but also from the observed spot prices of the various European calls and puts. It is assumed that call options, of any maturity and any strike, are traded at all times, and real time information about the spot prices of these options is instantly available to all market participants. In addition, the local volatility is assumed to be of the form $\sigma = h(S, \iota)$; in other words, the latent volatility process Y from the previous section is now chosen to be the spot price itself (there was nothing to prevent us from working with perfectly correlated Brownian motions W and B). To be precise, the risk-free form of the spot asset is postulated to be

$$S = S_0 + h(S, \iota) S \cdot \tilde{W} + (r - \delta) S \cdot \iota \,,$$

so that there is only one Brownian motion that enters the model. For simplicity, we suppose that the dividend yield is null ($\delta = 0$), and recall that with $\theta = (b - r)/\sigma$ the process $\tilde{W} = W + \theta \cdot \iota$ is a Brownian motion under the equivalent pricing measure Q. The function $h(\cdot, \cdot)$ is assumed to be such that

$$\int_0^t \mathsf{E}^Q \left[h(S_u, u)^2 S_u^2 \right] du < \infty$$

for any finite $t \in \mathbb{R}_+$. Instead of following the original derivation of Dupire (1994), here we are going to follow the local time technique developed by Leblanc (1997), but borrow its description from (Jeanblanc et al. 2009). We take it for granted that the spot price S_t admits probability density $\psi_t(\cdot)$ for any $t \in \mathbb{R}_+$. The price at time $t = 0$ of a European call with strike k that matures after t years (for reasons that will become clear below, we deliberately use t instead of T to denote time to maturity) is

$$c(k,t) = e^{-rt}\mathsf{E}^Q\big[(S_t - k)^+\big],$$

and we stress that here we treat $c(\cdot,\cdot)$ as a function of $(k,t) \in \mathbb{R}_{++} \times \mathbb{R}_{++}$.

(14.45) EXERCISE: Prove that

$$\partial_k^2 c(k,t) = e^{-rt}\psi_t(k).$$

HINT: Argue that $\partial_k c(k,t) = e^{-rt}\mathsf{E}^Q\big[-1_{\{k<S_t\}}\big] = -e^{-rt}\int_k^\infty \psi_t(x)\,\mathrm{d}x$. ○

(14.46) EXERCISE: Let $L^a(S)$ denote⊸(10.64) the local time of the semimartingale S in $a \in \mathbb{R}_{++}$. Prove that

$$\dagger \quad \begin{aligned} e^{-r\iota}(S-k)^+ &= (S_0 - k)^+ - re^{-r\iota}(S-k)^+ \bullet \iota \\ &\quad + e^{-r\iota}1_{]k,\infty[}(S) \bullet S + \frac{1}{2}e^{-r\iota} \bullet L^k(S). \end{aligned}$$

HINT: Use the integration by parts rule from (10.24) applied to the semimartingales $e^{-r\iota}$ and $(S-k)^+$, but first use the extended Itô formula from (10.62†) to write the second semimartingale in the form established in exercise (10.69). ○

(14.47) EXERCISE: Using the result from the previous exercise, prove the following formula:

$$\dagger \quad \begin{aligned} c(k,t) &= (S_0 - k)^+ + rk\int_0^t \mathrm{d}u\, e^{-ru}\int_k^\infty \psi_u(x)\,\mathrm{d}x \\ &\quad + \frac{k^2}{2}\int_0^t e^{-ru}\psi_u(k)h(k,u)^2\,\mathrm{d}u. \end{aligned}$$

HINT: Apply the expectation operator $\mathsf{E}^Q[\cdot]$ to both sides of (14.46†) and use the result from exercise (10.71). ○

After differentiating both sides of (14.47†) with respect to the time variable t and using the result and the hint from exercise (14.45), we arrive at the relation

$$\partial c(k,t) = -rk\,\partial_k c(k,t) + \frac{k^2}{2}h(k,t)^2\,\partial_k^2 c(k,t),$$

which is usually arranged as:

(14.48) DUPIRE'S FORMULA:

$$h(k,t)^2 = \frac{2}{k^2}\frac{\partial c(k,t) + rk\,\partial_k c(k,t)}{\partial_k^2 c(k,t)}. \quad ○$$

If the derivatives of the function $(k,t) \rightsquigarrow c(k,t)$ can be "read" reasonably well from the quoted spot prices of call options with various strikes and maturities, then it becomes possible to compute the local volatility in the form

$$\sigma_t = \sqrt{\frac{2}{S_t^2} \frac{\partial c(S_t, t) + rk\,\partial_S c(S_t, t)}{\partial_S^2 c(S_t, t)}},$$

after the substitution $k = S_t$ in Dupire's formula (here k is used as a "placeholder," that is, a generic argument for the function $h(\cdot, t)$).

15

Résumé of Stochastic Calculus with Discontinuous Processes

This chapter is a brief synopsis of the topic "calculus with semimartingales that are not required to be continuous." Such objects are not new to us: most of the regularity properties of sample paths and the Doob–Meyer decomposition introduced in chapter 9 were indeed formulated in the most general setup, without insisting on continuity. The continuity of sample paths feature was put into use in the subsequent chapters, specifically, in the construction of the stochastic integral, Itô's formula, and Girsanov's theorem – essentially, everything related to the angled bracket (predictable compensator) and the square bracket (quadratic variation), which coincide$^{\multimap\,(9.62)}$ if the local martingales involved happen to be continuous (whence the convenience of continuous sample paths). The present chapter is yet another followup to chapter 9, but this time without imposing the continuity of paths condition. As we realized in (9.28), it is enough for us to consider processes with càdlàg sample paths, but this property is not enough to eliminate the difference between the two brackets, and the task in front of us now is to understand how this difference gets manifested in the construction of the stochastic integral, Itô's formula, Girsanov's theorem – and everything else that hinges on these crucially important results.

15.1 Martingales, Local Martingales, and Semimartingales with Jumps

At this point we are well aware of the various regularity of paths features that the attribute "martingale" entails. To summarize our findings from section 9.3, it is natural to insist that all martingales (local martingales) are càdlàg, and left-continuity (in particular, continuity) is only possibly if the sample paths are highly irregular, that is, of infinite total variation – unless$^{\multimap\,(9.60)}$ the (local) martingale is trivial (it is a constant). We are also aware of the two basic and, at the same time, rather instructive examples of martingales: the Brownian motion, which has$^{\multimap\,\text{sec. 8.9}}$ sample paths that are continuous but of infinite total variation, and the compensated Poisson process, which has sample paths that are of finite total variation, but lack continuity and are only càdlàg. As we saw in section 9.2, the Brownian motion and the compensated Poisson process are extreme elements not only in the class of martingales, but also in the class of Lévy processes.

To get started with our exploration of semimartingales with jumps, we fix the probability space $(\Omega, \mathcal{F}, P)$ and the filtration $\mathscr{F} \backslash\!\subset \mathcal{F}$ (assumed to obey the usual conditions), and suppose that these two structures can support all processes, random variables, measures, and sets that show up in our study. Unless noted otherwise, all filtration dependent attributes – adapted, martingale, local martingale, stopping time, and so on – will be understood relative to $\mathscr{F}$ and P.

Recall that a stochastic process defined on $(\Omega, \mathcal{F}, P)$ can be treated as a mapping of the form $t \rightsquigarrow X_t$, which assigns a random variable, X_t, to every $t \in \mathbb{R}_+$, or as a mapping of the form $\omega \rightsquigarrow X_\omega$, which assigns a sample path, X_ω, to every $\omega \in \Omega$, or as a mapping of the form $(\omega, t) \rightsquigarrow X(\omega, t)$, which assigns an element of a particular state space, $X(\omega, t)$, to every pair $(\omega, t) \in \Omega \times \mathbb{R}_+$. In what follows we will focus our attention on processes with sample paths X_ω that are only required to be càdlàg, and therefore can have jumps – but not jumps with oscillating discontinuities (this point is important). In a first step, we give the jumps a name: given a càdlàg process X, we define its jump-component as the process

$$\Delta X_t \overset{\text{def}}{=} X_t - X_{t-}\,, \quad t \in \mathbb{R}_+\,, \quad \text{with the understanding that } X_{0-} = X_0\,,$$

where $X_\omega(t-) = X_{t-}(\omega)$ stands for the left limit of the path X_ω at time t. Notice that the symbol ΔX is introduced as the name of a particular process, so that we can write as usual $\Delta X(\omega, t)$, $\Delta X_\omega(t)$, $\Delta X_t(\omega)$, or ΔX_t.

Clearly – just by definition – the sample paths of the process $X_- \equiv (X_{t-})_{t \in \mathbb{R}_+}$ differ from the sample paths of X only at the times of discontinuity for X. At those times the paths of X_- are càglàd (left continuous with right limits) – not càdlàg. Notice that for any fixed $\varepsilon > 0$, a function on $\mathbb{R}_+$ that has no oscillating discontinuities (in particular, any càdlàg function) could have at most finitely many jumps of size greater than ε in any closed and finite interval. This is because any point of accumulation of jumps of size greater than ε, if one exists, would be a point at which the function has an oscillating discontinuity (and so, it cannot be càdlàg).

At this point the compensated Poisson process⊸(9.18) $N - c\iota$ is the only interesting noncontinuous martingale that we have encountered so far, and it will continue to be our main source of insights.

(15.1) EXERCISE: Explain why the Poisson process N can be identified with the sum of its past jumps; to be precise, (P-a.s.)

$$N_t = \sum\nolimits_{s \leq t} \Delta N_s \quad \text{for all } t \in \mathbb{R}_+\,.$$

Explain why the summation is always meaningful – it is in fact finite – for every $t \in \mathbb{R}_+$. ○

(15.2) EXERCISE: Explain why the process $X \overset{\text{def}}{=} N - c\iota$, which is⊸(9.18) a martingale (for its natural filtration $\mathscr{F}^N$), is not a pure-jump process, in that (P-a.s.)

$$X_t \neq X_0 + \sum\nolimits_{s \leq t} \Delta X_s \quad \text{for all } t \in \mathbb{R}_+\,.$$

HINT: It would be enough to show that $\Delta X_t = \Delta N_t$, $t \in \mathbb{R}_+$. ○

The next step for us is to understand whether the jumps of a càdlàg process are summable; that is, given a càdlàg process X, we ask: is the expression

$$\sum\nolimits_{s\leq t} \Delta X_s$$

meaningful for any fixed $t \in \mathbb{R}_+$? In the first place, we would like to be able to claim that almost every sample path X_ω can have at most countably many jumps. Only then would the summation symbol above be meaningful (of course, whether the sum converges and is meaningful as a quantity or not would be another matter). However, the fact that the jumps are at most countably many is rather easy to establish: given any $n \in \mathbb{N}_{++}$, there could be only finitely many jumps of size $> \frac{1}{n}$ on any finite interval, and since each jump must be of size $> 1/n$ for some n, the collection of all jumps on any finite interval must be at most countable (as the union of at most countably many finite sets). Since $\mathbb{R}_+$ is the union of countably many finite intervals, there could be at most countably many jumps on $\mathbb{R}_+$ (the union of countably many countable sets is a countable set). That the jumps of a càdlàg process can be arranged in sequence, and in a measurable way, is a much deeper result that we state next:

(15.3) THEOREM: (Dellacherie 1972, IV 30) For any $\mathscr{F}$-adapted and càdlàg process X, there is a sequence of $\mathscr{F}$-stopping times, $\mathscr{s}_i: \Omega \mapsto [0,\infty]$, $i \in \mathbb{N}$, that exhausts the jumps of X, in the sense that:

- (a) the graphs⊶(8.70) of the stopping times $(\mathscr{s}_i)_{i\in\mathbb{N}}$ are non-intersecting, in that $[\![\mathscr{s}_i]\!] \cap [\![\mathscr{s}_j]\!] = \emptyset$ whenever $i \neq j$;
- (b) $X_{\mathscr{s}_i} \neq X_{\mathscr{s}_i-}$ P-a.e. in the set $\{\mathscr{s}_i < \infty\}$ for every $i \in \mathbb{N}_{++}$;
- (c) if $\mathscr{t}$ is any stopping time such that $[\![\mathscr{s}_i]\!] \cap [\![\mathscr{t}]\!] = \emptyset$ for every $i \in \mathbb{N}_{++}$, then $P\big(\{X_{\mathscr{t}} \neq X_{\mathscr{t}-}\} \cap \{\mathscr{t} < +\infty\}\big) = 0$.

Furthermore, if the process X is predictable, then the stopping times $\mathscr{s}_i$ can all be chosen to be⊶(9.55a) predictable. ○

Our next step is to revisit the notion of "predictable," and, in particular, understand what the absence of this feature may entail.

(15.4) THE THREE FUNDAMENTAL σ-FIELDS ON $\Omega\times\mathbb{R}_+$: Recall again that a stochastic process can be treated as a function from the product space $\Omega\times\mathbb{R}_+$ into $\mathbb{R}$, and the property "predictable" translates to⊶(8.68) "measurable with respect to the σ-field on $\Omega\times\mathbb{R}_+$ generated by all left-continuous and $\mathscr{F}$-adapted processes." This σ-field we denoted by $\mathscr{P}(\mathscr{F})$ and called "the predictable σ-field" (for the filtration $\mathscr{F}$). However, if we were to allow for jumps in the sample paths, insisting on left continuity would lead to rather uninteresting processes, and, for this reason, we must find a way to work with processes that are adapted but are not necessarily predictable. Indeed, interesting processes with jumps – the compensated Poisson process, say – would not be measurable for $\mathscr{P}(\mathscr{F})$. At the same time, all processes of interest to us are going to be right-continuous – in fact, càdlàg – and, for this reason, we now introduce the so-called *optional σ-field* on $\Omega\times\mathbb{R}_+$. This is the σ-field on $\Omega\times\mathbb{R}_+$ generated by all càdlàg and $\mathscr{F}$-adapted processes $X\colon \Omega\times\mathbb{R}_+ \mapsto \mathbb{R}$ and is commonly denoted by $\mathscr{O}(\mathscr{F})$. Since $\mathscr{P}(\mathscr{F})$ is actually generated$^{\multimap\,(8.68)}$ by the class of processes that are adapted and continuous, which processes are automatically càdlàg, we have the inclusion $\mathscr{P}(\mathscr{F}) \subseteq \mathscr{O}(\mathscr{F})$. Generally, this inclusion is going to be strict, unless the structure of the probability space and the filtration are such that all $(\mathscr{F}, P)$-martingales are automatically continuous, as would be the case with any Brownian filtration, for example – see (Revuz and Yor 1999, VI (5.7)).

In (8.53) we also introduced the σ-field $\mathrm{Prog}(\mathscr{F})$ that consists of all progressively measurable subsets of $\Omega\times\mathbb{R}_+$. Fortunately, since$^{\multimap\,(8.62)}$ every right-continuous and adapted process is progressively measurable, we have

† $$\mathscr{P}(\mathscr{F}) \subseteq \mathscr{O}(\mathscr{F}) \subseteq \mathrm{Prog}(\mathscr{F})\,,$$

and this means that the entire machinery of stopping times is going to be available for us, as there would be no need to work with processes that are not optional. ○

(15.5) EXERCISE: Prove that if X is any predictable and càdlàg process, then the processes X_- and ΔX are both predictable. ○

(15.6) BASIC EXAMPLES: Any $(\mathscr{F}, P)$-Brownian motion with continuous sample paths is automatically $\mathscr{P}(\mathscr{F})$-measurable (predictable for $\mathscr{F}$), and, in order to maintain its status as an interesting martingale and interesting Lévy process, despite this too-good-to-be-true feature (predictable), its sample paths must "wiggle wildly," namely, must have unbounded variation. At the same time, the compensated Poisson process is only optional (e.g., for its natural filtration), and for this reason it can afford to have sample paths that are of finite variation – actually, piecewise linear – and still maintain its status as an interesting martingale and interesting Lévy process.

Even more elementary – and yet rather instructive – examples of predictable and optional processes can be produced from a fixed $\mathscr{F}$-stopping time $\mathscr{s}\colon \Omega \mapsto \mathbb{R}_+$. As an example, the process$^{\multimap\,(8.70)}$ $(\omega, t) \rightsquigarrow 1_{[\![0,\mathscr{s}]\!]}(\omega, t)$ is clearly predictable, since it is left continuous and is also adapted, due to the fact that for any fixed t the event $\{(1_{[\![0,\mathscr{s}]\!]})_t = 0\}$ is the same as the event $\{\mathscr{s} < t\}$, which is$^{\multimap\,(8.52)}$ in $\mathscr{F}_t$. At the same time, given *any* mapping $\mathscr{s}\colon \Omega \mapsto \bar{\mathbb{R}}_+$, the process $1_{[\![0,\mathscr{s}[\![}$ is obviously càdlàg. If it is also $\mathscr{F}$-adapted, then the process $1_{[\![0,\mathscr{s}[\![}$ becomes optional by definition, and we note that $1_{[\![0,\mathscr{s}[\![}$ is $\mathscr{F}$-adapted if and only if the event $\{(1_{[\![0,\mathscr{s}[\![})_t = 0\}$ is in $\mathscr{F}_t$ for any $t \in \mathbb{R}_+$, and this is the same as $\{\mathscr{s} \leq t\} \in \mathscr{F}_t$ for any t, which is the same as the claim that $\mathscr{s}$ is a stopping time; in other words, if $\mathscr{s}$ is a stopping time, then the process $1_{[\![0,\mathscr{s}[\![}$ – or, equivalently, the set $[\![0, \mathscr{s}[\![$ – can only be claimed to be optional, while the process $1_{[\![0,\mathscr{s}]\!]}$ (the set $[\![0, \mathscr{s}]\!]$) can be claimed to be predictable. ○

Given the importance of the notion "predictable," basic intuition is telling us that we need to develop a tool that would allow us to "project" processes that are not predictable onto the space of processes that are – more or less in the way the conditional expectation gives the projection of random variables that are not measurable for a particular σ-field onto the space of random variables that are. However, we cannot quite condition to the σ-field $\mathscr{P}(\mathscr{F})$ in the way we have done in numerous contexts before, since we do not have a probability measure on the product-space $\Omega\times\mathbb{R}_+$. It may appear tempting to use the product measure $P\times\Lambda$, but one quickly realizes that such a program would be naive, for to say the least, the Lebesgue measure Λ cannot recognize the jumps in the sample paths, and even if we replace Λ with another measure, μ, that has atoms, we will end up with a model in which the jumps occur at deterministic times (the atoms of μ, which would be independent of $\omega\in\Omega$) and this would not be a particularly interesting construction (the Poisson process would not fit this framework, for example). Nevertheless, the idea of building some weaker form of "conditional expectation" relative to $\mathscr{P}(\mathscr{F})$ is simply too important to just give up. In order to pursue such a program and develop what we will soon call "predictable projection," we must introduce the following technical concept first.

(15.7) PREDICTABLE TIMES: (Jacod and Shiryaev 1987, I sec. 2b) A function $t\colon\Omega\mapsto\bar{\mathbb{R}}_+$ is said to be a *predictable time* if$^{\multimap(8.70)}$ the stochastic interval $[\![0,t[\![$ is a predictable set, that is, $[\![0,t[\![\,\in\mathscr{P}(\mathscr{F})$. It is not difficult to check that any predictable time is automatically a stopping time – see the next exercise. A careful reader might have noticed that we already introduced the notion "predictable stopping time" back in (9.55). This tautology will be resolved in (15.9) below. In any case, some minor technicalities aside, the class of all predictable (stopping) times is closed under the operations "inf" and "sup" in much the same way in which the class of all stopping times$^{\multimap(8.51)}$ is closed under these operations (recall that $\mathscr{F}$ is right-continuous and complete). The next exercise makes this statement precise. ○

(15.8) EXERCISE: (ibid., I 2.9) Prove that:

(a) if $t\colon\Omega\mapsto\bar{\mathbb{R}}_+$ is a predictable time, then t is also a stopping time;

(b) if $(s_i)_{i\in\mathbb{N}}$ is any sequence of predictable times, then $t\overset{\text{def}}{=}\sup_{i\in\mathbb{N}}s_i$ is a predictable time;

(c) if $(s_i)_{i\in\mathbb{N}}$ is a sequence of predictable times, then $t\overset{\text{def}}{=}\inf_{i\in\mathbb{N}}s_i$ is a predictable time if $\cup_{i\in\mathbb{N}}\{t=s_i\}=\Omega$.

HINT: If t is a predictable time, then $[\![0,t[\![\,\in\mathscr{O}(\mathscr{F})\subseteq\operatorname{Prog}(\mathscr{F})$, so that $1_{[\![0,t[\![}$ is an adapted process, and this is just another way of saying that t is a stopping time. If $t\overset{\text{def}}{=}\sup_{i\in\mathbb{N}}s_i$, then $[\![0,t[\![\,=\cup_{i\in\mathbb{N}}[\![0,s_i[\![$. Finally, if $t\overset{\text{def}}{=}\inf_{i\in\mathbb{N}}s_i$ and $\cup_{i\in\mathbb{N}}\{t=s_i\}=\Omega$, then $[\![0,t[\![\,=\cap_{i\in\mathbb{N}}[\![0,s_i[\![$. ○

The result in (15.9) below reveals further the reason for introducing the notion of predictable time. Plainly, predictable times arise as limits of increasing sequences of (not necessarily predictable) stopping times. Let $(s_i)_{i\in\mathbb{N}}$ be any sequence of stopping times which is *strongly increasing*, in that, given any fixed $\omega\in\Omega$, the sequence $(s_i(\omega))_{i\in\mathbb{N}}$ is increasing, and in such a way that for any $i\in\mathbb{N}$ there is an index $j>i$ with $s_j(\omega)>s_i(\omega)$. If we set $t\overset{\text{def}}{=}\lim_i s_i$ (a stopping time by construction), then we would have $[\![0, t[\![= \cup_{i \in \mathbb{N}} [\![0, s_i]\!]$, and since$^{\multimap (15.6)}$ $[\![0, s_i]\!] \in \mathscr{P}(\mathscr{F})$ for every $i \in \mathbb{N}$, we see that $[\![0, t[\![\in \mathscr{P}(\mathscr{F})$. In other words, $t \overset{\text{def}}{=} \lim_i s_i$ is a predictable time. We see from the foregoing that every stopping time that is predictable in the sense of (9.55a) is also a predictable time in the sense of (15.7). At the same time, as the next theorem shows, every predictable time in the sense of (15.7) can be expressed as the limit of some strongly increasing sequence of stopping times (provided that the filtration is complete, as we assume), and is therefore a predictable stopping time in the sense of (9.55a).

(15.9) THEOREM: (Jacod and Shiryaev 1987, I 2.15, 2.16)[170] Suppose that t is some predictable time for some complete and right-continuous filtration $\mathscr{F}$. Then there is a strongly increasing sequence of $\mathscr{F}$-stopping times, $(s_i)_{i \in \mathbb{N}}$, such that $t = \lim_i s_i$. ○

The definition and the statement of existence for predictable projections can be found in some slight variations in several classical texts. The particular version that we state next is compiled from (Jacod and Shiryaev 1987, I 2.28–2.31) – the curious reader is strongly encouraged to consult (ibid., I sec. 2) for additional details.

(15.10) THEOREM: Given any $\mathcal{F} \otimes \mathcal{B}(\mathbb{R}_+)$-measurable process $X : \Omega \times \mathbb{R}_+ \mapsto \bar{\mathbb{R}}$, there is a unique, up to indistinguishability, *predictable* process

$$^{\mathscr{P}}X : \Omega \times \mathbb{R}_+ \mapsto]-\infty, +\infty] ,$$

such that for any predictable stopping time t,$^{\multimap (8.41)}$

$$^{\mathscr{P}}X_t 1_{\{t<\infty\}} = \mathsf{E}[X_t 1_{\{t<\infty\}} \,|\, \mathscr{F}_{t-}] \equiv \mathsf{E}[X_t \,|\, \mathscr{F}_{t-}] 1_{\{t<\infty\}} \quad (P\text{-a.s.}),$$

where $\mathsf{E}[X_t \,|\, \mathscr{F}_{t-}]$ is understood as the extended conditional expectation introduced in (3.40). The process $^{\mathscr{P}}X$ is called the *predictable projection of* X and has the following properties:

(a) For any stopping time s, $^{\mathscr{P}}(X^s) = (^{\mathscr{P}}X) 1_{[\![0,s]\!]} + X_s 1_{]\!]s,+\infty[\![}$.

(b) If the process $^{\mathscr{P}}X$ happens to be finite, then, for any predictable process Y with values in $]-\infty, +\infty]$,

$$^{\mathscr{P}}(YX) = Y(^{\mathscr{P}}X) .$$

(c) If X is a local $(\mathscr{F}, P)$-martingale, then $^{\mathscr{P}}X = X_-$ and $^{\mathscr{P}}\Delta X = 0$. ○

The next step for us is to introduce several classes of processes that we are going to need. We have seen already that if W is the Brownian motion, then the process $W^2 - \iota$ is a martingale, that is, what one needs to subtract from the submartingale W^2 in order to "correct it" into a martingale is the increasing (and deterministic) process ι. This is a special case of the familiar Doob-Meyer decomposition – see (9.49) – and we have seen the importance this result in a number of occasions (in the construction of the stochastic integral, for example).

---

[170] For a much more detailed treatment of this topic, accompanied with complete proofs of the results, we refer to (Dellacherie 1972, ch. III).

(15.11) INCREASING AND FINITE VARIATION PROCESSES: Recall the definitions of "increasing process," "integrable increasing process," and "u.i. increasing process" from (9.47). Any increasing process is assumed to have càdlàg paths, to start from 0, and to be adapted to a particular filtration, usually understood from the context. With the probability space $(\Omega, \mathcal{F}, P)$ and the filtration $\mathscr{F}\ \backslash\!\subset \mathcal{F}$ in mind, let ${}_{\uparrow}\mathscr{V}$ denote the class of all increasing processes that are finite, let ${}_{\uparrow}\tilde{\mathscr{V}}$ denote the class of all increasing processes that are integrable, and let ${}_{\uparrow}\bar{\mathscr{V}}$ denote the class of all increasing processes that are integrable at ∞, that is, uniformly integrable. Any finite and increasing real-valued function on $\mathbb{R}_+$ is automatically of finite variation on finite intervals. Thus, for any $A \in {}_{\uparrow}\mathscr{V}$, $\mathrm{d}A_\omega$ is a well-defined measure on $\mathbb{R}_+$ that is finite on any finite interval for P-a.e. $\omega \in \Omega$. Recall that such measures are called Lebesgue–Stieltjes measures and the associated integrals are called Lebesgue–Stieltjes integrals, but these integrals are no different from the usual Lebesgue integrals with respect to the measures $\mathrm{d}A_\omega$, for fixed ω.

Next, denote by $\mathscr{V} \stackrel{\text{def}}{=} {}_{\uparrow}\mathscr{V} - {}_{\uparrow}\mathscr{V}$ the class of all processes A that can be expressed as the difference between two elements of ${}_{\uparrow}\mathscr{V}$, and recall⊸(8.91) that a process $A \in \mathscr{V}$ can always be written as the difference $A = A^+ - A^-$, with $A^+, A^- \in {}_{\uparrow}\mathscr{V}$ chosen so that the measures $\mathrm{d}A^+_\omega$ and $\mathrm{d}A^-_\omega$ are singular ($\mathrm{d}A^+_\omega \perp \mathrm{d}A^-_\omega$). This choice of A^+ and A^- will be automatically assumed from now on, and the (signed) measure $\mathrm{d}A_\omega$ will be understood to be the difference $\mathrm{d}A^+_\omega - \mathrm{d}A^-_\omega$ (and the associated Lebesgue–Stieltjes integral with respect to A will be understood to be the difference of two Lebesgue–Stieltjes integrals in the obvious way). The requirement for $A^+, A^- \in {}_{\uparrow}\mathscr{V}$ to be chosen so that $A = A^+ - A^-$ and $\mathrm{d}A^+_\omega \perp \mathrm{d}A^-_\omega$ is the same as the requirement that $A = A^+ - A^-$ and $|\mathrm{var}|A = A^+ + A^-$. If, in addition, A^+ and A^- can be chosen to be both integrable at ∞ (that is, $A^+, A^- \in {}_{\uparrow}\bar{\mathscr{V}}$), then we say that the process $A = A^+ - A^-$ is itself integrable at ∞. This feature is equivalent to $|\mathrm{var}|A = A^+ + A^- \in {}_{\uparrow}\bar{\mathscr{V}}$. The collection of all elements of the space $\mathscr{V}$ that are integrable at ∞ we denote by $\bar{\mathscr{V}}$.

As usual, $\bar{\mathscr{V}}_{\mathrm{loc}}$ and ${}_{\uparrow}\bar{\mathscr{V}}_{\mathrm{loc}}$ will denote the respective classes of processes that belong to $\bar{\mathscr{V}}$ and ${}_{\uparrow}\bar{\mathscr{V}}$ by way of localization. To be precise, we write $A \in \bar{\mathscr{V}}_{\mathrm{loc}}$ ($A \in {}_{\uparrow}\bar{\mathscr{V}}_{\mathrm{loc}}$) if there is a sequence of stopping times $\mathscr{s}_1 \leq \mathscr{s}_2 \leq \ldots \to \infty$ such that the stopped process $A^{\mathscr{s}_i}$ belongs to $\bar{\mathscr{V}}$ (to ${}_{\uparrow}\bar{\mathscr{V}}$) for every i. ○

(15.12) PREDICTABLE COMPENSATOR OF AN INCREASING PROCESS: (Jacod and Shiryaev 1987, I 3.17) Given any increasing process $A \in {}_{\uparrow}\bar{\mathscr{V}}_{\mathrm{loc}}$, there is a predictable and increasing process, $A^{\mathscr{P}} \in {}_{\uparrow}\bar{\mathscr{V}}_{\mathrm{loc}}$, such that the process $(A - A^{\mathscr{P}})$ is a local martingale. To put it another way, the lack of predictability in an increasing càdlàg process from the space ${}_{\uparrow}\bar{\mathscr{V}}_{\mathrm{loc}}$ can always be attributed to the presence of a local martingale component. The process $A^{\mathscr{P}}$ is unique up to indistinguishability and is called the *predictable compensator of* A. Equivalently, it can be introduced as the unique (up to indistinguishability) predictable element of ${}_{\uparrow}\bar{\mathscr{V}}_{\mathrm{loc}}$ that has the property $\mathsf{E}[A^{\mathscr{P}}_{\mathscr{t}}] = \mathsf{E}[A_{\mathscr{t}}]$ for any stopping time $\mathscr{t}$. Yet another way to introduce

$A^{\mathscr{P}}$ is as the unique (up to indistinguishability) predictable element of ${}_{\uparrow}\bar{\mathscr{V}}_{\text{loc}}$ such that⊶(15.10) $({}^{\mathscr{P}}X \bullet A)_\infty = (X \bullet A^{\mathscr{P}})_\infty$ for any positive and jointly measurable process X. For this reason $A^{\mathscr{P}}$ is also called "the dual predictable projection of A." This feature also explains the use of the notation $A^{\mathscr{P}}$ instead of $\langle A \rangle$, which would have been consistent with the notation introduced in (9.51). ○

(15.13) REMARK: The term "predictable compensator" showed up⊶(9.51) earlier in this space in conjunction with processes that are submartingales. We see that what is being compensated for is the lack of the (local) martingale property, and we insist that such a compensation must be accomplished in a predictable fashion. There is an obvious analogy between the construction of predictable compensators and the construction of conditional expectations: subtracting $\mathsf{E}[\xi\,|\,\mathscr{F}_t]$ from the random variable ξ compensates for the fact that ξ is not 0 on average conditioned to $\mathscr{F}_t$. The analogy stems from the fact that the "zeros" in the space of stochastic processes are the local martingales starting from 0, and the next two exercises shed some light on this connection. ○

(15.14) EXERCISE: (Jacod and Shiryaev 1987, I 3.11) Prove that any local martingale that belongs to the space $\mathscr{V}$, that is, has finite variation sample paths, must belong also to the space $\bar{\mathscr{V}}_{\text{loc}}$.

HINT: It is enough to prove the statement for uniformly integrable martingales from the space $\mathscr{V}$. Let M be one such martingale, let $V_t \overset{\text{def}}{=} |\text{var}|_{[0,t]}(M)$, $t \in \mathbb{R}_+$, and let $t_n \overset{\text{def}}{=} \{t \in \mathbb{R}_+ \colon V_t > n\}$, $n \in \mathbb{N}_{++}$. Argue that

$$V_t = V_{t-} + \Delta V_t = V_{t-} + |\Delta M_t| \le V_{t-} + |M_{t-}| + |M_t| \le 2V_{t-} + |M_t|, \quad t \in \mathbb{R}_+,$$

and conclude that $V_{t_n} \le 2n + |M_{t_n}|$ for every $n \in \mathbb{N}_{++}$. Explain why $\mathsf{E}[|M_{t_n}|] < \infty$ and conclude that $|\text{var}|(M^{t_n}) \in {}_{\uparrow}\bar{\mathscr{V}}$ for every $n \in \mathbb{N}_{++}$. ○

(15.15) EXERCISE: (ibid., I 3.16) Prove that any predictable local martingale that starts from 0 and has sample paths that are a.s. of finite variation must be 0.

HINT: Recall that⊶(8.79) a stopped predictable process is predictable. By way of localization and using the result from the previous exercise, it is enough to consider the case where the local martingale is actually a martingale that belongs to the class $\bar{\mathscr{V}}$, that is, $\mathsf{E}[|\text{var}|(M)_\infty] < \infty$. Let M be one such martingale and let $U, V \in {}_{\uparrow}\bar{\mathscr{V}}$ be such that $M = U - V$ and $|\text{var}|(M) = U + V$. Observe that U and V are both submartingales of class (*D*), and argue that if M is predictable then so is also $|\text{var}|(M) = U + V$. Conclude that $U = \frac{1}{2}\big(|\text{var}|(M) + M\big)$ and $V = \frac{1}{2}\big(|\text{var}|(M) - M\big)$ are both predictable. Then argue that $U = 0 + U$ is a D-M-d for U and so is also $U = M + V$. Conclude that $M = 0$ and $V = U$. ○

(15.16) EXERCISE: (a) Prove that the Poisson process N belongs to ${}_{\uparrow}\bar{\mathscr{V}}_{\text{loc}}$ (explain for which filtration) and identify its dual predictable projection $N^{\mathscr{P}}$. (b) Describe the dual predictable projection (that is, the predictable compensator) of an increasing process $A \in {}_{\uparrow}\bar{\mathscr{V}}_{\text{loc}}$ that happens to be continuous. (c) Given a Poisson process N, identify the predictable projections ${}^{\mathscr{P}}N$ and ${}^{\mathscr{P}}(\Delta N)$ (you must again explain for which filtration).

HINT: Apply (15.10c) to the martingale $N - c\iota$. ○

The class $\mathbb{H}^2$ of all L^2-bounded martingales that was introduced in (9.83) will play an important role in what follows. Recall that⊶(9.88†) although $\mathbb{H}^2$ is a space of processes, this space can actually be identified with the (Hilbert) space of random variables $L^2(\Omega, \mathscr{F}_\infty, P)$. Furthermore, comparing two processes, say $X, Y \in \mathbb{H}^2$, boils down to comparing the random variables X_∞ and Y_∞ (which are always meaningful due to the martingale convergence theorem) as elements of $L^2(\Omega, \mathscr{F}_\infty, P)$ – this feature is both powerful and useful, as we have already seen. If the martingale X is L^2-bounded, then X^2 is an integrable, positive, and càdlàg submartingale, and therefore admits⊶(9.50) Doob–Meyer decomposition. Thus,⊶(9.51) there is a unique (up to indistinguishably) predictable, increasing, and integrable process $\langle X, X\rangle \in {}_{\mathrm{i}}\bar{\mathscr{V}}$, and a càdlàg martingale M such that[171]

$$X^2 = M + \langle X, X\rangle . \qquad e_1$$

Recall that $\langle X, X\rangle$ is called the "predictable quadratic variation," "angled bracket," or simply "bracket." This choice of terminology is motivated by the fact that $\langle X, X\rangle$ is predictable (from the very formulation⊶(9.48) of the D-M-d), and, furthermore, it is immediate from the last relation that for $0 \le s < t < \infty$,

$$\mathsf{E}\big[(X_t - X_s)^2 \,|\, \mathscr{F}_s\big] = \mathsf{E}\big[X_t^2 - X_s^2 \,|\, \mathscr{F}_s\big] = \mathsf{E}\big[\langle X, X\rangle_t - \langle X, X\rangle_s \,|\, \mathscr{F}_s\big] .$$

In particular, setting $s = 0$ and letting $t \to \infty$,

$$\mathsf{E}\big[X_\infty^2 - X_0^2\big] = \mathsf{E}\big[\langle X, X\rangle_\infty\big] .$$

Processes that belong to $\mathbb{H}^2$ only by way of localization (in the usual way, for a sequence of stopping times that increase monotonically to $+\infty$) will be called local L^2-martingales,[172] and the space of such objects we will denote by $\mathbb{H}^2_{\mathrm{loc}}$. It is not difficult to see that the relation e_1 survives the localization of the elements of $\mathbb{H}^2$, provided that the conditions on the martingale component and on the compensator are relaxed accordingly. To be precise, e_1 still holds for $X \in \mathbb{H}^2_{\mathrm{loc}}$, except that with this more general choice M can only be required to be a càdlàg local martingale and $\langle X, X\rangle$ can only be required to be locally integrable, that is, to be an element of ${}_{\mathrm{i}}\bar{\mathscr{V}}_{\mathrm{loc}}$. To see why such an extension of the Doob–Meyer decomposition is still in force for local L^2-martingales, it is enough to notice that the (unique) bracket of a stopped L^2-bounded martingale X is the bracket stopped at the same stopping time, that is, $\langle X^{\jmath}, X^{\jmath}\rangle = \langle X, X\rangle^{\jmath}$ for any stopping time $\jmath$. This is an immediate corollary from⊶(9.41) the optional stopping theorem for martingales.

[171] The fact that $\langle X, X\rangle$ is integrable is a consequence from e_1 and the fact that X^2 and M are both integrable.

[172] A more appropriate, though more cumbersome, term would be "locally L^2-bounded martingales."

We saw in (9.57) that for submartingales of class (D) the continuity of the predictable compensator is tantamount to the requirement for the submartingale to be $\multimap$ (9.55) regular. However, as it stands, the attribute "regular" does not survive localization, and, in order to establish continuity of the bracket $\langle X, X\rangle$ for a generic element $X \in \mathbb{H}^2_{\text{loc}}$, we are going to need a different tool. The first step in this direction is to connect the jumps of $\langle X, X\rangle$ with the jumps of X^2.

(15.17) EXERCISE: Prove that for every càdlàg process X,

$$\Delta(X^2) = (\Delta X)^2 + 2X_-\Delta X\,.$$

Conclude that if $X \in \mathbb{H}^2_{\text{loc}}$, then $\Delta\langle X, X\rangle = {}^{\mathscr{P}}(\Delta X^2)$; in particular $\multimap$ (15.10)

$$\Delta\langle X, X\rangle_t = \mathsf{E}[(\Delta X_t)^2 \,|\, \mathscr{F}_{t-}] \quad \text{(a.s.)} \quad \text{for every } t \in \mathbb{R}_{++}\,,$$

or, more generally, $\Delta\langle X, X\rangle_{\ell} = \mathsf{E}[(\Delta X_{\ell})^2 \,|\, \mathscr{F}_{\ell-}]$ (a.s.) for every finite and predictable stopping time ℓ.

HINT: The first relation is the result of elementary algebra. To establish the second relation, use the first relation in conjunction with (15.10c), applied to the local martingale $X^2 - \langle X, X\rangle$. What can you say about the predictable projections ${}^{\mathscr{P}}(2X_-\Delta X)$ and ${}^{\mathscr{P}}(\Delta\langle X, X\rangle)$? Consult (15.5). ○

The following summary of results is a compilation from (Jacod and Shiryaev 1987, I 2.25, 2.31, 2.35).

(15.18) CONTINUITY OF THE ANGLED BRACKET REVISITED: We see from the previous exercise that given any $X \in \mathbb{H}^2_{\text{loc}}$, the continuity of $\langle X, X\rangle$ is the same property as ${}^{\mathscr{P}}(\Delta X^2) = 0$. By the very definition of predictable projection, this condition is the same as

$$\mathsf{E}\big[(\Delta X_{\ell})^2 1_{\{\ell<\infty\}} \,|\, \mathscr{F}_{\ell-}\big] = 0 \quad \Longleftrightarrow \quad \mathsf{E}\big[1_{\{\Delta X_{\ell}\neq 0\}} 1_{\{\ell<\infty\}} \,|\, \mathscr{F}_{\ell-}\big] = 0$$

for every predictable stopping time ℓ. The first condition above is the same as $\Delta X_{\ell} = 0$ P-a.s. on the set $\{\ell < \infty\}$, and the second condition is the same as ${}^{\mathscr{P}}(1_{\{\Delta X\neq 0\}}) = 0$. A càdlàg process X is said to be quasi-left-continuous if $\Delta X_{\ell} = 0$ P-a.s. on the set $\{\ell < \infty\}$ for every predictable stopping time ℓ, and we now see that this property is equivalent to $P\big({}^{\mathscr{P}}(1_{\{\Delta X\neq 0\}}) > 0\big) = 0$ (the predictable support of the set $\{\Delta X \neq 0\}$ is a P-null set) and implies ${}^{\mathscr{P}}X = X_-$. ○

The joint bracket of two local L^2-martingales $X, Y \in \mathbb{H}^2_{\text{loc}}$ is defined, as usual, by way of polarization:

$$\langle X, Y\rangle \stackrel{\text{def}}{=} \frac{1}{4}\Big(\langle X+Y, X+Y\rangle - \langle X-Y, X-Y\rangle\Big)\,.$$

Polarization also implies that $XY - \langle X, Y\rangle$ is a càdlàg local martingale, for every choice of $X, Y \in \mathbb{H}^2_{\text{loc}}$.

(15.19) EXERCISE: Prove that every càdlàg local martingale with globally bounded jumps is locally bounded and, in particular, is an element of $\mathbb{H}^2_{\text{loc}}$. Explain why the requirement for the filtration to obey the usual conditions is essential for this feature.

HINT: If X is a càdlàg local martingale with $|\Delta X_t| \leq C$ and if

$$\mathfrak{z}_n \stackrel{\text{def}}{=} \{t \in \mathbb{R}_+ : |X_t| > n\},$$

then $|X_{t\wedge \mathfrak{z}_n}| \leq n + C$ (while the process $|X|$ can jump into the domain $]n, +\infty[$, the size of the jump is at most C). ○

(15.20) EXERCISE: Let N be the Poisson process with parameter c and let $\mathcal{F}$ denote the usual augmentation of the filtration that N generates. Prove⊶(9.18) one more time that the process $X \stackrel{\text{def}}{=} N - c\iota$ is an element of $\mathcal{H}^2_{\text{loc}}$ and show that $\langle X, X\rangle = c\iota$. Explain why, despite the fact that $\Delta N \neq 0$ (as we know, the process N has nontrivial jumps), we still have $\mathsf{E}[\Delta N_{t} | \mathcal{F}_{t-}] = 0$ on the set $\{t < \infty\}$ for any predictable stopping time t. Explain one more time why ${}^{\mathcal{P}}N = N_-$. Conclude that neither of the processes X and N is quasi-left-continuous

HINT: Consult (15.18) above and observe that $(\Delta X)^2 = (\Delta N)^2 = \Delta N$. ○

We have already seen that much of the intrinsic nature of a given process can be understood from the way that process interacts with other processes from a particular class. One particularly interesting interaction is the bracket, which in many ways acts as a dot-product of a sort. To gain some insight into this phenomenon we now turn – yet again – to the Poisson process.

(15.21) EXERCISE: Let N be the Poisson process of parameter $c > 0$, let $\mathcal{F}$ denote the usual augmentation of the filtration that N generates, and, just as in the previous exercise, let $X \stackrel{\text{def}}{=} N - c\iota$. Prove that if M is any *continuous* and *bounded* $(\mathcal{F}, \mathsf{P})$-martingale, then XM is a martingale.

HINT: To prove that

$$\mathsf{E}[X_t M_t - X_s M_s | \mathcal{F}_s] = 0 \quad \text{for} \quad 0 \leq s < t,$$

take an arbitrary subdivision $\mathcal{p} = \{t_0, t_1, \ldots, t_n\}$ of the interval $[s, t]$ and observe that, just as in section 10.4,

$$X_t M_t - X_s M_s = \sum_{i=1}^{n} M_{t_{i-1}}(X_{t_i} - X_{t_{i-1}}) + \sum_{i=1}^{n} X_{t_{i-1}}(M_{t_i} - M_{t_{i-1}}) + \sum_{i=1}^{n}(X_{t_i} - X_{t_{i-1}})(M_{t_i} - M_{t_{i-1}}).$$

Since the expectation, conditioned to $\mathcal{F}_s$, of each of the first two sums is 0, it is enough to prove that the conditional expectation of the third sum can be made arbitrarily close to 0 by choosing the subdivision $\mathcal{p}$ accordingly. To this end, observe that the sample path X_ω has a finite total variation, in fact bounded by $N_t(\omega) + ct$, on every finite interval $[0, t]$, while the sample path M_ω is uniformly continuous on any finite interval. Finally, use the dominated convergence theorem, after observing that

$$\left|(X_{t_i} - X_{t_{i-1}})(M_{t_i} - M_{t_{i-1}})\right| \leq 2C|X_{t_i} - X_{t_{i-1}}|. \quad ○$$

Using the standard localization technique, it is not difficult to see that in the last exercise the assumption "M is any continuous and bounded $(\mathcal{F}, P)$-martingale" can be relaxed to "M is any continuous local $(\mathcal{F}, P)$-martingale," except that in the conclusion the product XM can only be claimed to be a local martingale, which is the same as the claim that the angled bracket between the compensated Poisson process and any continuous local martingale is null. This is yet another extreme feature of the Poisson process. Martingales that share this feature play a special role, and, for this reason, the class of such objects has a name:

(15.22) PURELY DISCONTINUOUS LOCAL MARTINGALES: A local martingale X is said to be purely discontinuous if $X_0 = 0$ (P-a.s.) and XM is a local martingale for any continuous local martingale M. ○

We stress that, typically, a purely discontinuous local martingale X would not be a pure-jump process, in that it may not be possible to write $X_t = \sum_{s \leq t} \Delta X_s$ (it may not be possible to equate X with the sum of its jumps). By way of an example, the local martingale $N - c\iota$ is purely discontinuous but is not⊸(15.2) a pure-jump process.

In many situations it becomes useful to decompose certain generic objects into more primitive objects, which, at least in principle, would be easier to work with. Just as an example, it is often useful to write a generic vector in $\mathbb{R}^n$ as a linear combination of orthonormal vectors. In our setting, the "primitives" that we have seen so far are the predictable processes of finite variation, the continuous local martingales, and the purely discontinuous local martingales. It turns out that the processes that are of most interest for us can indeed be decomposed into the sum of such primitives. We state the precise results next.

(15.23) PROPOSITION: (Jacod and Shiryaev 1987, I 4.17) Given any local martingale M and any $\varepsilon \in \mathbb{R}_{++}$, there is a (non-unique) local martingale M° that starts from 0, has jumps bounded by ε – that is, $|\Delta M^\circ| \leq \varepsilon$, so that M° is⊸(10.19) locally bounded and thus an element of $\mathbb{H}^2_{\text{loc}}$ – and is such that the local martingale $M - M^\circ$ is an element of the class $\mathcal{V}$ (has paths of finite variation). ○

(15.24) DECOMPOSITION OF LOCAL MARTINGALES: (ibid., I 4.18) Let M be any local martingale – as always, automatically assumed càdlàg, but no longer required to be continuous. Then there is a continuous local martingale M^c, called the *continuous part of* M, and a purely discontinuous local martingale M^d, called the *purely discontinuous part of* M, both starting from 0, such that (P-a.s.)

$$\dagger \qquad M = M_0 + M^c + M^d .$$

This decomposition is unique up to indistinguishability. ○

(15.25) EXERCISE: What is the decomposition of the martingale $N - c\iota$ in the form (15.24†)? ○

We are now in a position to expand the class of continuous semimartingales, introduced in (9.79), by allowing the sample paths of such objects to have jumps.

However, the main reason for introducing the class of semimartingales (with or without jumps) remains the same: this is the largest class of processes for which stochastic integrals can be defined in a way that is practically useful.

(15.26) SEMIMARTINGALES WITH JUMPS: A (not necessarily continuous) semimartingale is any process X that can be decomposed into the sum

$$X = X_0 + M + A\,,$$

in which M is a càdlàg local martingale and A is an adapted càdlàg process of finite variation, both starting from 0 ($M_0 = A_0 = 0$). If, in addition, the finite variation process A in this decomposition can be chosen to be predictable, then the semimartingale X is said to be a *special semimartingale*. The continuous part of the local martingale component M, namely M^c, is unique, but, generally, its purely discontinuous part M^d and the finite variation component A are not, since jumps can be exchanged between the two arbitrarily. However, if X is a special semimartingale, then the finite variation component A is unique (up to indistinguishably) if required to be predictable, in which case M^d is also unique and the decomposition above is said to be the *canonical decomposition of* X (Jacod and Shiryaev 1987, p. 43). In any case, with any given semimartingale X one can associate – and in a unique way – a continuous local martingale, denoted by X^c, which is nothing but the (unique) continuous component of the (generally non-unique) martingale component M in some semimartingale decomposition of X. Finally, we note that because of the result in (15.23) we can assume without any loss of generality that the local martingale component, M, in any given semimartingale X is actually⊸(10.19) a locally bounded martingale with bounded jumps; in particular, every semimartingale is locally bounded and càdlàg. However, we stress that such a rearrangement of the components of the semimartingale may not be possible if one is to insist on a predictable finite variation component (in the case of a special semimartingale).

It was noted earlier,⊸(9.79) in the context of continuous semimartingales, that the class of semimartingales, for a particular probability measure and filtration, is invariant under any equivalent change of the probability measure. This important result was actually established for general semimartingales that only need to be càdlàg – see (Dellacherie and Meyer 1980, VII 45). ○

15.2 Stochastic Integrals with Respect to Semimartingales with Jumps

The construction of the stochastic integral against semimartingales with jumps rests on essentially the same idea as the one employed in chapter 10: one starts with simple integrands, for which the definition of the integral is straightforward, and then extends the definition for a larger class of integrands and integrators by taking limits. The fundamental reason why the case of integrators with jumps is more involved is that the estimate established in (10.8), which was instrumental in the construction of the integral against continuous local martingales, is not available for local martingales with jumps – at least not in the form stated in (10.8). Later in this section we are going to develop the counterpart of this crucial estimate for local martingales with jumps, and are going to see that it is again possible to control the size of the stochastic integral by controlling the size of the integrand, but not in the simple and straightforward fashion in which we were able to do so in section 10.1. For all these reasons, we need to develop the general construction of the integral in stages.

Just as we did in the previous section, we start by fixing the probability space $(\Omega, \mathcal{F}, P)$ and the filtration $\mathscr{F} \subset \mathcal{F}$ (as always, the usual conditions are automatically assumed), and suppose that notions such as "adapted," "martingale," "local martingale," and "stopping time" are all understood relative to $\mathscr{F}$ and P. Recall$^{\circ\!\to (8.75)}$ the definition of the class of simple processes $\mathfrak{S} = \mathfrak{S}(\mathscr{F})$, and recall that for every integrand $Y \in \mathfrak{S}$ the integral $Y \bullet Z$ is a well-defined process for every choice of the finite process Z (adapted or not, measurable or not) – see e_2, e_3, and (10.2), in section 10.1.

Our first step is to define the integral against generic L^2-bounded martingales $M \in \mathbb{H}^2$. Given any such M and any $Y \in \mathfrak{S}$, the process $Y \bullet M$ is easily seen to be a càdlàg martingale from the class (Hilbert space) $\mathbb{H}^2_0$ with

$$\mathsf{E}\big[(Y \bullet M_\infty)^2\big] = \mathsf{E}\big[Y^2 \bullet \langle M, M\rangle_\infty\big] \le C^2 \mathsf{E}\big[\langle M, M\rangle_\infty\big] < \infty ,$$

where $C > 0$ is the constant that bounds Y (from the definition of a simple process). It is also straightforward to check that for any $Y, Y' \in \mathfrak{S}$,

$$\langle Y \bullet M, Y' \bullet M\rangle = YY' \bullet \langle M, M\rangle .$$

In order to extend to construction of the integral for more general (non-simple) integrands, we first recall that square integrable càdlàg martingales from the Hilbert space $\mathbb{H}^2$ can be compared in terms of the L^2-norms of their values at ∞: the sample paths of any two martingales from $\mathbb{H}^2$ can be made arbitrarily close (uniformly on $\mathbb{R}_+$) in probability if their respective values at ∞ are chosen to be sufficiently close in L^2-sense. Applied to simple stochastic integrals this feature comes down to the following: if $Y, Y' \in \mathfrak{S}$ and $M \in \mathbb{H}^2$, then$^{\circ\!\to (9.19)}$ for any $\varepsilon > 0$ Doob's maximal inequality gives

$$P\Big(\sup_{t\in\mathbb{R}_+} |Y \bullet M_t - Y' \bullet M_t| \ge \varepsilon\Big) \le \frac{1}{\varepsilon}\, \mathsf{E}\big[(Y - Y')^2 \bullet \langle M, M\rangle_\infty\big]^{\frac{1}{2}} . \qquad \mathrm{e}_1$$

We can now extend the construction of the stochastic integral for integrands Y chosen from the space $\mathfrak{L}^2(M) \equiv \mathfrak{L}^2(\langle M, M\rangle)$, defined as the linear space of all predictable Y such that

$$\mathsf{E}\big[Y^2 \bullet \langle M, M\rangle_\infty\big] < \infty .$$

As an immediate corollary from the result in (8.75) – and here we use this result in the same way in which it was used in section 10.2, in conjunction with our first construction of the stochastic integral for continuous integrators – one can show that for any $Y \in \mathfrak{L}^2(M)$ there is a sequence of simple integrands $S^i \in \mathfrak{S}$, $i \in \mathbb{N}$, such that

$$\lim_i \mathsf{E}\big[(Y - S^i)^2 \bullet \langle M, M\rangle_\infty\big] = 0\,.$$

From the estimate in e_1, the sample paths $S^i \bullet M_\omega$ converge uniformly on $\mathbb{R}_+$ in probability P. To put it another way, if each $S^i \bullet M$ is treated as a random variable with values in the Skorokhod space $\mathbb{D}(\mathbb{R}_+;\mathbb{R})$, then $(S^i \bullet M)_{i\in\mathbb{N}}$ can be claimed to be a Cauchy sequence in the respective Ky Fan distance, due to the fact that the uniform convergence in $\mathbb{D}(\mathbb{R}_+;\mathbb{R})$ is stronger than the convergence in the Skorokhod distance.[173] The limit $\lim_i S^i \bullet M$ can be identified as a $\mathbb{D}(\mathbb{R}_+;\mathbb{R})$-valued r.v., and also as an L^2-bounded càdlàg martingale from the space $\mathbb{H}_0^2$, which we denote by $Y \bullet M$ – or by $\big(\int^t Y\,\mathrm{d}M\big)_{t\in\mathbb{R}_+}$ – and call the *stochastic integral of Y with respect to M*. For this integral we can again write

$$\mathsf{E}\big[(Y \bullet M_\infty)^2\big] = \mathsf{E}\big[Y^2 \bullet \langle M, M\rangle_\infty\big]\,,$$

and it is not difficult to check that for any stopping time $\mathfrak{s}$,

$$(Y \bullet M)_{t\wedge\mathfrak{s}} = (1_{[\![0,\mathfrak{s}]\!]}Y \bullet M)_t = (Y \bullet M^{\mathfrak{s}})_t\,, \quad t \in \mathbb{R}_+\,,$$

or, to put it more succinctly,

$$(Y \bullet M)^{\mathfrak{s}} = 1_{[\![0,\mathfrak{s}]\!]}Y \bullet M = Y \bullet M^{\mathfrak{s}}\,.$$

The next stage is to extend the definition of the integral for integrands from the larger space $\mathfrak{L}^2_{\mathrm{loc}}(M) \equiv \mathfrak{L}^2_{\mathrm{loc}}(\langle M, M\rangle)$. This space consists of all predictable processes Y such that the increasing càdlàg process $Y^2 \bullet \langle M, M\rangle$ is locally integrable.[174] If $\mathfrak{s}_0 \le \mathfrak{s}_1 \le \ldots \to \infty$ is a sequence of stopping times that localizes (the integrability of) the increasing process $Y^2 \bullet \langle M, M\rangle$, then for every $i \in \mathbb{N}$,

$$\mathsf{E}\big[Y^2 \bullet \langle M, M\rangle_{\mathfrak{s}_i}\big] \equiv \mathsf{E}\big[(1_{[\![0,\mathfrak{s}_i]\!]}Y^2) \bullet \langle M, M\rangle_\infty\big] < \infty\,.$$

Consequently, $1_{[\![0,\mathfrak{s}_i]\!]}Y \in \mathfrak{L}^2(M)$ and the integral $1_{[\![0,\mathfrak{s}_i]\!]}Y \bullet M$ is well-defined from the previous stage – as a square integrable martingale from $\mathbb{H}_0^2$, for any $i \in \mathbb{N}$. All these integrals are compatible, in the sense for $i < j$ the process $1_{[\![0,\mathfrak{s}_i]\!]}Y \bullet M$ is indistinguishable from the process $(1_{[\![0,\mathfrak{s}_j]\!]}Y \bullet M)^{\mathfrak{s}_i}$. This follows immediately from the relations

$$(1_{[\![0,\mathfrak{s}_j]\!]}Y \bullet M)^{\mathfrak{s}_i} = 1_{[\![0,\mathfrak{s}_i]\!]}1_{[\![0,\mathfrak{s}_j]\!]}Y \bullet M = 1_{[\![0,\mathfrak{s}_i]\!]}Y \bullet M\,.$$

173 Recall that the space of càdlàg paths is a complete separable metric space with respect to the Skorokhod distance, but not with respect to the locally uniform distance. This was the main reason for inventing the Skorokhod distance: without separability of the target space, the space of random variables cannot be claimed to be complete for the convergence in probability.

174 The integral is a càdlàg process because the upper limit is included in the integration domain. Notice also that the measure $\mathrm{d}\langle M, M\rangle$ has no mass at $\{0\}$ by the very definition of an increasing process – see (9.47).

Since $\mathfrak{z}_i \to \infty$, we now see that for P-a.e. $\omega \in \Omega$ the limit

$$Y \bullet M(\omega, t) \stackrel{\text{def}}{=} \lim_i \big(1_{[\![0,\mathfrak{z}_i]\!]} Y \bullet M\big)(\omega, t)$$

exists simultaneously for all $t \in \mathbb{R}_+$. The process so obtained, $Y \bullet M$, we again call the *stochastic integral of Y with respect to M*, and remark that this process can be identified as an element of $\mathbb{H}^2_{0,\text{loc}}$ (it is easy to check that $\mathfrak{z}_i$, $i \in \mathbb{N}$, is a localizing sequence for $Y \bullet M$, for example).

Finally, we can relax the requirement for the integrator M to be a square integrable martingale from the class $\mathbb{H}^2$, and require that M belongs to that class only by way of localization. Assuming that $M \in \mathbb{H}^2_{\text{loc}}$, let $t_0 < t_1 < \ldots \to \infty$ be any localizing sequence of stopping times, such that $M^{t_i} \in \mathbb{H}^2$ for any $i \in \mathbb{N}$. The bracket $\langle M, M\rangle$ is still well defined as an increasing càdlàg process, uniquely characterized (up to indistinguishability) by the property $\langle M, M\rangle^{t_i} = \langle M^{t_j}, M^{t_j}\rangle^{t_i}$ for any $i, j \in \mathbb{N}$, $i < j$. Given any predictable process Y, the process $Y^2 \bullet \langle M, M\rangle$ is an increasing càdlàg process, and asking for this process to be locally integrable is still meaningful. The class of processes Y that have this property we still denote by $\mathfrak{L}^2_{\text{loc}}(M)$, that is, $Y \in \mathfrak{L}^2_{\text{loc}}(M)$ is a shorthand for the property that Y is predictable and, for some localizing sequence of stopping times $\mathfrak{z}_0 \le \mathfrak{z}_1 \le \ldots \to \infty$,

$$\mathsf{E}\big[Y^2 \bullet \langle M, M\rangle_{\mathfrak{z}_i}\big] \equiv \mathsf{E}\big[(1_{[\![0,\mathfrak{z}_i]\!]} Y^2) \bullet \langle M, M\rangle_\infty\big] < \infty \quad \text{for every } \ i \in \mathbb{N}\,.$$

Since

$$\big(Y^2 \bullet \langle M^{t_j}, M^{t_j}\rangle\big)^{\mathfrak{z}_i} = (1_{[\![0,\mathfrak{z}_i]\!]} Y^2) \bullet \langle M^{t_j}, M^{t_j}\rangle \le (1_{[\![0,\mathfrak{z}_i]\!]} Y^2) \bullet \langle M, M\rangle$$

for every $i, j \in \mathbb{N}$, it follows that $Y \in \mathfrak{L}^2_{\text{loc}}(M^{t_j})$ for every $j \in \mathbb{N}$. The integrals $Y \bullet M^{t_j}$, $j \in \mathbb{N}$, are now well-defined (from the previous step) and are easily seen to be compatible again, so that for P-a.e. $\omega \in \Omega$ the limit

$$Y \bullet M(\omega, t) \stackrel{\text{def}}{=} \lim_j \big(Y \bullet M^{t_j}\big)(\omega, t)$$

exists simultaneously for all $t \in \mathbb{R}_+$. The process so obtained, $Y \bullet M$, we yet again call the *stochastic integral of Y with respect to M*, and again observe that $Y \bullet M$ is an element of $\mathbb{H}^2_{0,\text{loc}}$ (it is straightforward to check that $\mathfrak{z}_i \wedge t_i$, $i \in \mathbb{N}$, is a localizing sequence for $Y \bullet M$, for example).

(15.27) REMARK: This is our most general definition of the integral $Y \bullet M$ for càdlàg integrators $M \in \mathbb{H}^2_{\text{loc}}$ and for predictable integrands $Y \in \mathfrak{L}^2_{\text{loc}}(M)$. Just as before, if the integral notation is desirable we may write $\int^t Y \,\mathrm{d}M$ instead of $Y \bullet M_t$, which is just another notation for the r.v. $\omega \rightsquigarrow (Y \bullet M)(\omega, t)$. In particular, $Y \bullet M$ would be meaningful (as a process) for every choice of the predictable and locally bounded process Y and for every choice of the local L^2-martingale M. We strongly emphasize that, by construction, $Y \bullet M$ is a càdlàg local L^2-martingale that starts from 0, that is, an element $\mathbb{H}^2_{0,\text{loc}}$, and that its construction only involves limits in probability P and limits P-a.s. Consequently, as a process, $Y \bullet M$ depends on the probability measure P only up to equivalence.

The fundamental reason why the integrand can only be predictable – and the simple integrands (the elements of $\mathfrak{S}$) from which the extension of the integral becomes possible, must be predictable as well – is the same reason that we already pointed to in remark (10.7): if the simple process Y is not predictable, we cannot claim that $Y \bullet M$ is a local martingale, and cannot use the basic properties of martingales to carry out the extension. In addition, the approximation result from (8.75) is only valid for predictable processes. However, just as was the case with continuous integrators, the effective space of integrands is the space of equivalence classes for the measure $P(d\omega) \times d\langle M, M\rangle_\omega(t)$ associated with the elements of $\mathfrak{L}^2_{\mathrm{loc}}(M)$, which space of equivalence classes we again denote by $\mathbb{L}^2_{\mathrm{loc}}(M)$. Thus, it is again possible to construct the integral for integrands Y that are not predictable – and are not adapted even – but as long as such an integrand coincides with a predictable one from the space $\mathfrak{L}^2_{\mathrm{loc}}(M)$ a.e. in $\Omega \times \mathbb{R}_+$ for the measure $P(d\omega) \times d\langle M, M\rangle_\omega(t)$. ○

Basic intuition suggests that the stochastic integral must be "small" if the integrand is chosen to be "small." In order to make this claim precise, let Y be any predictable process which is bounded by a constant $K \geq |Y|$, and let M be any càdlàg local martingale with bounded jumps $|\Delta M| \leq C$; in particular,$^{\multimap (15.19)}$ M is locally bounded, and is therefore an element of $\mathbb{H}^2_{\mathrm{loc}}$. Since$^{\multimap (15.17)}$

$$\Delta\langle M, M\rangle_t = \mathsf{E}\left[(\Delta M_t)^2 \,|\, \mathscr{F}_{t-}\right],$$

the jumps of $\langle M, M\rangle$ are bounded as well: $|\Delta\langle M, M\rangle_t| \leq C^2$. In fact, using the result from (15.3) and taking into account the predictability of $\langle M, M\rangle$, it is not difficult to conclude that (P-a.s.) $\sup_{t\in\mathbb{R}_+} |\Delta\langle M, M\rangle_t| \leq C^2$.

For a fixed $\varepsilon \in \mathbb{R}_{++}$, define the stopping time

$$t_\varepsilon \stackrel{\text{def}}{=} \inf\left\{t \in \mathbb{R}_{++} : Y^2 \bullet \langle M, M\rangle_t > \varepsilon\right\},$$

with the understanding that $t_\varepsilon = \infty$ on the set $\{Y^2 \bullet \langle M, M\rangle_\infty \leq \varepsilon\}$. Then define the processes

$$a^\varepsilon \stackrel{\text{def}}{=} (Y \bullet M)^{t_\varepsilon} \quad \text{and} \quad b^\varepsilon \stackrel{\text{def}}{=} Y \bullet M - a^\varepsilon,$$

and observe that for any fixed $t \in \mathbb{R}_{++}$,

$$\begin{aligned}\mathsf{E}\left[|a^\varepsilon_t|^2\right] &= \mathsf{E}\left[Y^2 \bullet \langle M, M\rangle_{t\wedge t_\varepsilon}\right] \leq \mathsf{E}\left[Y^2 \bullet \langle M, M\rangle_{t_\varepsilon}\right] \\ &= \mathsf{E}\left[Y^2 \bullet \langle M, M\rangle_{(t_\varepsilon -)} + Y^2_{t_\varepsilon} \Delta\langle M, M\rangle_{t_\varepsilon}\right] \leq \varepsilon + \mathsf{E}\left[Y^2_{t_\varepsilon} \Delta\langle M, M\rangle_{t_\varepsilon}\right].\end{aligned}$$

As the processes Y and $\Delta\langle M, M\rangle$ are both globally bounded, the last expectation must be finite; in particular, $a^\varepsilon \in \mathbb{H}^2_0$. Next, fix some $T, \delta \in \mathbb{R}_{++}$ and notice that $b^\varepsilon = 0$ on the set $\{t_\varepsilon > T\} \times [0, T]$. As a result, we have the following chain of estimates:

$$\begin{aligned}P\Big(\sup_{t\in[0,T]} |Y \bullet M_t| > \delta\Big) &= P\Big(\sup_{t\in[0,T]} |a^\varepsilon_t + b^\varepsilon_t| > \delta\Big) \\ &= P\Big(\sup_{t\in[0,T]} |a^\varepsilon_t + b^\varepsilon_t| > \delta,\, t_\varepsilon \leq T\Big) + P\Big(\sup_{t\in[0,T]} |a^\varepsilon_t + b^\varepsilon_t| > \delta, t_\varepsilon > T\Big)\end{aligned}$$

$$\leq P\big(t_\varepsilon \leq T\big) + P\Big(\sup_{t\in[0,T]} |a_t^\varepsilon| > \delta\Big).$$

Applied to the L^2-bounded martingale $a^\varepsilon \in \mathbb{H}_0^2$, Doob's maximal inequality gives

$$P\Big(\sup_{t\in[0,T]} |a_t^\varepsilon| > \delta\Big) = P\Big(\sup_{t\in[0,T]} |a_t^\varepsilon|^2 > \delta^2\Big) \leq \frac{\mathsf{E}\big[|a_T^\varepsilon|^2\big]}{\delta^2}$$
$$\leq \frac{\varepsilon + \mathsf{E}\big[Y_{t_\varepsilon}^2 \Delta\langle M, M\rangle_{t_\varepsilon}\big]}{\delta^2},$$

and, as a result,

$$P\Big(\sup_{t\in[0,T]} |Y{\bullet}M_t| > \delta\Big) \leq P\big(t_\varepsilon \leq T\big) + \frac{\varepsilon + \mathsf{E}\big[Y_{t_\varepsilon}^2 \Delta\langle M, M\rangle_{t_\varepsilon}\big]}{\delta^2}.$$

Since the set $\{t_\varepsilon \leq T\}$ is included in the set $\{Y^2{\bullet}\langle M, M\rangle_T \geq \varepsilon\}$,

$$P\Big(\sup_{t\in[0,T]} |Y{\bullet}M_t| > \delta\Big) \leq \frac{\varepsilon + \mathsf{E}\big[Y_{t_\varepsilon}^2 \Delta\langle M, M\rangle_{t_\varepsilon}\big]}{\delta^2} + P\Big(Y^2{\bullet}\langle M, M\rangle_T \geq \varepsilon\Big). \qquad (e_2)$$

The last estimate is useful in many applications. We stress that it holds for every choice of the parameters $T, \varepsilon, \delta \in \mathbb{R}_{++}$, and note that e_2 simplifies considerably if the local martingale M happens to be continuous, since in this case $\Delta\langle M, M\rangle_{t_\varepsilon} = 0$ and the expectation on the right side vanishes – this reduces e_2 to the estimate in (10.8), as it should. Since the process Y is bounded by the constant $K \geq |Y|$, we have $\mathsf{E}\big[Y_{t_\varepsilon}^2 \Delta\langle M, M\rangle_{t_\varepsilon}\big] \leq K^2 \mathsf{E}\big[\Delta\langle M, M\rangle_{t_\varepsilon}\big]$, and if, in addition, M is an L^2-bounded martingale, then

$$\mathsf{E}\big[\Delta\langle M, M\rangle_{t_\varepsilon}\big] \leq \mathsf{E}\big[\langle M, M\rangle_\infty\big].$$

This, of course, is a very crude estimate, but we see that in the foregoing the requirement for the jumps of M to be globally bounded could have been replaced with the requirement for M to be an element of $\mathbb{H}^2$, which would also imply that $a^\varepsilon \in \mathbb{H}_0^2$. If M is only in $\mathbb{H}_{\text{loc}}^2$ and if the stopping times $\mathfrak{s}_i$, $i \in \mathbb{N}$, form a localizing sequence for M, then for any fixed $T > 0$, $\lim_i P(\mathfrak{s}_i \leq T) = 0$. Furthermore, the stochastic integrals $Y{\bullet}M$ and $Y{\bullet}M^{\mathfrak{s}_i}$ coincide on the set $\{\mathfrak{s}_i > T\}\times[0,T]$, so that for any $\varepsilon, \delta, T \in \mathbb{R}_{++}$,

$$P\Big(\sup_{t\in[0,T]} |Y{\bullet}M_t| > \delta\Big) \leq P(\mathfrak{s}_i \leq T) + P\Big(\sup_{t\in[0,T]} |Y{\bullet}M^{\mathfrak{s}_i}{}_t| > \delta\Big)$$
$$\leq P(\mathfrak{s}_i \leq T) + \frac{\varepsilon + \mathsf{E}\big[Y^2{\bullet}\langle M, M\rangle_{\mathfrak{s}_i}\big]}{\delta^2} + P\Big(Y^2{\bullet}\langle M, M\rangle_{\mathfrak{s}_i} > \varepsilon\Big).$$

In the preceding we used e_2 with $M^{\mathfrak{z}_i}$ in place of M, and also used the relations

$$Y^2_{t_\varepsilon} \Delta\langle M^{\mathfrak{z}_i}, M^{\mathfrak{z}_i}\rangle_{t^i_\varepsilon} \leq Y^2 \bullet \langle M^{\mathfrak{z}_i}, M^{\mathfrak{z}_i}\rangle_\infty = Y^2 \bullet \langle M, M\rangle_{\mathfrak{z}_i}$$

and

$$Y^2 \bullet \langle M, M\rangle_{\mathfrak{z}_i} = Y^2 \bullet \langle M^{\mathfrak{z}_i}, M^{\mathfrak{z}_i}\rangle_\infty \geq Y^2 \bullet \langle M^{\mathfrak{z}_i}, M^{\mathfrak{z}_i}\rangle_T .$$

These estimates lead to the next result, which makes precise the intuitive idea that the stochastic integral must depend on the integrand continuously. We also see how different this continuity is in the general case studied here, compared to the case of continuous integrators that we studied back in chapter 10: instead of continuity in the Ky Fan distance, here we can only claim continuity with respect to the bounded pointwise convergence, which is more demanding than the convergence in probability.

(15.28) THEOREM: (Jacod and Shiryaev 1987, I 4.31) Let $Y^k : \Omega \times \mathbb{R}_+ \mapsto \mathbb{R}$, $k \in \mathbb{N}$, be any sequence of predictable processes, which are universally bounded ($|Y^k(\omega, t)| \leq C$ for any k), and are such that

$$Y^k(\omega, t) \underset{k\to\infty}{\to} Y(\omega, t) \quad \text{for every } (\omega, t) \in \Omega \times \mathbb{R}_+ ,$$

for some (obviously predictable and bounded) process Y. Then, given any local L^2-martingale $M \in \mathbb{H}^2_{\mathrm{loc}}$, one has

$$\lim_k P\Big(\sup_{t\in[0,T]} \big| Y \bullet M_t - Y^k \bullet M_t \big| > \delta \Big) = 0 \quad \text{for any } T, \delta \in \mathbb{R}_{++} .$$

To put it another way, if $Y^k \to Y$ in pointwise sense, and if this convergence is bounded, then the sample paths $Y^k \bullet M_\omega$, $k \in \mathbb{N}$, converge to the sample path $Y \bullet M_\omega$ uniformly on finite intervals in probability P. ○

(15.29) ALTERNATIVE CONSTRUCTION: Since, without any loss of generality, the local martingale component in every semimartingale decomposition can always be assumed⊸ (15.26) to have bounded jumps, it would be enough for us to construct $Y \bullet M$ for local martingales M such that ΔM is globally bounded. Furthermore, in most applications it would suffice to work with integrands Y that are locally bounded and, of course, predictable – every left-continuous and adapted process is in this class, for example. With these two restrictions in mind, one can proceed as follows. First, observe that if the jumps of M are bounded, then⊸ (15.17) so are the jumps $\Delta\langle M, M\rangle = {}^{\mathscr{P}}(\Delta M^2)$.[175] Furthermore, a càdlàg process with bounded jumps is automatically locally bounded, so that, by the usual localization argument, it would be enough to suppose, in addition, that Y, M, and $\langle M, M\rangle$ are all bounded. Then observe that a bounded and predictable Y is automatically an element of the space $\mathscr{L}^2(\Omega \times \mathbb{R}_+, \mathscr{P}(\mathscr{F}), m)$ for any finite measure m on $\Omega \times \mathbb{R}_+$. In particular, setting $m(\mathrm{d}\omega, \mathrm{d}t) = P(\mathrm{d}\omega)\mathrm{d}\langle M, M\rangle_\omega$ and using the approximation result from (8.75), we conclude that there is a sequence of predictable simple integrands

[175] This claim also relies on the result from (15.3).

$(S^i \in \mathfrak{S})_{i \in \mathbb{N}}$ such that $S^i(\omega, t) \to Y(\omega, t)$ as $i \to \infty$ for m-a.e. $(\omega, t) \in \Omega \times \mathbb{R}_+$. In fact, the sequence $(S^i)_{i \in \mathbb{N}}$ can be chosen to be bounded by any constant that bounds Y.[176] For any such sequence one can claim that there is a set $\mathcal{N} \subset \Omega$ with $P(\mathcal{N}) = 0$ such that $S^i(\omega, t) \to Y(\omega, t)$ for every $\omega \in \Omega \backslash \mathcal{N}$ and every $t \in \mathbb{R}_+$ with $\Delta\langle M, M\rangle(\omega, t) \neq 0$. A straightforward application of the estimate in e_2 with $Y = S^i - S^j$, in conjunction with the dominated convergence theorem (applied to the expectation in e_2), leads to the conclusion that $(S^i \cdot M)_{i \in \mathbb{N}}$, treated as a sequence of $\mathbb{D}(\mathbb{R}_+; \mathbb{R})$-valued random variables, is a Cauchy sequence for the respective Ky Fan distance. The limit $\lim_i (S^i \cdot M)$, understood in terms of the convergence in probability, that is, in terms of the Ky Fan distance, is a well-defined $\mathbb{D}(\mathbb{R}_+; \mathbb{R})$-valued random variable, which we denote by $Y \cdot M$. The advantage of this construction is that it depends only on the null sets of P and does not involve the L^2-convergence in the space $\mathbb{H}^2$ in any way – thus making the invariance of the construction of the integral under absolutely continuous changes of the measure even more transparent.

Of course, any martingale properties must depend on the measure. In particular, since $S^i \cdot M \in \mathbb{H}^2_0$ for any $i \in \mathbb{N}$ (these are actually bounded martingales) and since $\lim_i \mathsf{E}\big[(S^i - Y)^2 \cdot \langle M, M\rangle_\infty\big] = 0$, it is straightforward to conclude that $Y \cdot M$ must be an element of $\mathbb{H}^2_0$ as well. In the general scenario where Y is locally bounded and predictable and $M \in \mathbb{H}^2_{\mathrm{loc}}$ has bounded jumps, using the usual localization technique, the integral $Y \cdot M$ can be claimed to be an element of $\mathbb{H}^2_{0,\mathrm{loc}}$, and we again emphasize that the actual construction of $Y \cdot M$ depends only on the equivalence class of P, even though the class $\mathbb{H}^2_{0,\mathrm{loc}}$ is specific to P. ○

(15.30) INTEGRATION WITH RESPECT TO SEMIMARTINGALES: Let X be some generic (and not necessarily continuous) semimartingale with canonical decomposition $X = X_0 + M + A$ and suppose (without⊸ (15.26) any loss of generality) that the local martingale component M is locally bounded, and therefore an element of $\mathbb{H}^2_{\mathrm{loc}}$. Stochastic integrals with respect to X can be defined in the obvious way as

$$Y \cdot X \overset{\text{def}}{=} Y \cdot M + Y \cdot A\,,$$

as long as the two integrals on the right are both meaningful, the first one being understood as a stochastic integral of the type that we just constructed, and the second one being understood as a Lebesgue–Stieltjes integral trajectory-by-trajectory. In particular, $Y \cdot X$ would be perfectly meaningful for integrands Y that are predictable and locally bounded. In fact, for any such integrand Y the integral $Y \cdot M$ can always be understood in terms of the construction outlined in (15.29), since, after moving some of the jumps from the local martingale component M into the finite variation component A, all remaining jumps in M can be assumed

176 This is because if $|Y| \leq C$ and $S^i \to Y$ m-a.e., then the sequence $(S^i \wedge C) \vee (-C)$ is also from $\mathfrak{S}$ and also converges m-a.e. to Y.

bounded. Some minor technicalities aside, it is not difficult to see that this construction of $Y \cdot X$ does not depend on the particular choice of a semimartingale decomposition for X. ○

(15.31) BASIC RULES: (Jacod and Shiryaev 1987, I 4.31) Let X be any semimartingale, and let Y and Z be any two *predictable* and *locally bounded* processes. Then the following properties are satisfied:

(a) The mapping $X \rightsquigarrow Y \cdot X$ is linear.

(b) The mapping $Y \rightsquigarrow Y \cdot X$ is linear and is also continuous in the sense of (15.28), that is, the result from (15.28) holds with the local martingale $M \in \mathbb{H}^2_{\text{loc}}$ replaced with a generic semimartingale X.

(c) If $X \in \mathbb{H}^2_{\text{loc}}$, then $Y \cdot X \in \mathbb{H}^2_{\text{loc}}$ and $\langle Y \cdot X, Y \cdot X \rangle = Y^2 \cdot \langle X, X \rangle$.

(d) If X is a process of finite variation (a semimartingale with a missing martingale part, for example), then $Y \cdot X$ is the Lebesgue–Stieltjes integral, understood path-by-path, and is a process of finite variation itself.

(e) $Y \cdot X_0 = 0$ and $Y \cdot X = Y \cdot (X - X_0)$.

(f) $\Delta(Y \cdot X) = Y \Delta X$, that is, $\Delta(Y \cdot X)_t = Y_t \Delta X_t$ for any $t \in \mathbb{R}_+$.

(g) $(Y \cdot X)^{\mathfrak{z}} = (1_{[\![0,\mathfrak{z}]\!]} Y) \cdot X = Y \cdot (X^{\mathfrak{z}})$, for any stopping time $\mathfrak{z}$.

(h) $Y \cdot X$ is a semimartingale and $Z \cdot (Y \cdot X) = (ZY) \cdot X$. ○

(15.32) MULTIVARIATE STOCHASTIC INTEGRALS: Semimartingales (with respect to the same filtration and probability measure) can be arranged in lists and lists of lists, that is, one can produce matrices and vectors the entries of which are semimartingales. We call such objects "multivariate semimartingales." If X and Y are multivariate semimartingales, then the integral $Y \cdot X$ will be understood in the obvious way by following the convention introduced on page xvi: if the dimensions of Y and X are compatible with the matrix multiplication rules (the last dimension of Y is the same as the first dimension of X), then $Y \cdot X$ will be understood to be the matrix with entries $(Y \cdot X)^{i,j} = \sum_k Y^{i,k} \cdot X^{k,j}$; if Y and X are two vectors of semimartingales that have the same dimension and orientation, then $Y \cdot X$ will be understood to be the vector (of the same dimension and orientation as Y and X) with entries $Y^i \cdot X^i$; if Y is a vector of semimartingales and X is a scalar semimartingale, then $Y \cdot X$ will be understood to be a vector with entries $Y^i \cdot X$; if X is a vector of semimartingales and Y is a scalar semimartingale, then $Y \cdot X$ will be understood to be a vector with entries $Y \cdot X^i$. ○

15.3 Quadratic Variation and Itô's Formula

The construction of the stochastic integral outlined in the previous section is remarkably undemanding: any adapted, finite, and left-continuous (and therefore predictable) process can be integrated against any semimartingale. Indeed, an adapted, finite, and left-continuous process Y is always locally bounded, as is straightforward to verify: since finite left-continuous sample paths cannot explode in finite time, the stopping times ($\inf \emptyset = \infty$ by convention)

$$\mathfrak{s}_i \stackrel{\text{def}}{=} \inf\{t \geq 0 \colon |Y_t| > i\}, \quad i \in \mathbb{N},$$

are such that $\mathfrak{s}_i \to \infty$ and $|Y| \leq i$ on the set $[\![0, \mathfrak{s}_i]\!]$ and for every $i \in \mathbb{N}$ (because of the left continuity, the assumption $|Y_{\mathfrak{s}_i}| > i$ easily leads to a contradiction). If $M \in \mathbb{H}^2_{\text{loc}}$ is any local L^2-martingale, and if the stopping times $\mathfrak{t}_j$, $j \in \mathbb{N}$, form a localizing sequence for M, then

$$\mathsf{E}\Big[(Y^{\mathfrak{s}_i})^2 \bullet \langle M^{\mathfrak{t}_j}, M^{\mathfrak{t}_j}\rangle_\infty\Big] \leq i^2 \mathsf{E}\big[\langle M^{\mathfrak{t}_j}, M^{\mathfrak{t}_j}\rangle_\infty\big] < \infty .$$

The integrals $Y^{\mathfrak{s}_i} \bullet M^{\mathfrak{t}_j}$ are now well defined (as L^2-bounded martingales from the space $\mathbb{H}^2$), and on any finite interval $[0, T]$ the sample path $(Y \bullet X)_\omega$ can be identified (a.s.) with the sample path $(Y^{\mathfrak{s}_i} \bullet M^{\mathfrak{t}_j})_\omega$, provided that the indices i and j are chosen so that $\mathfrak{s}_i(\omega) > T$ and $\mathfrak{t}_j(\omega) > T$. In brief, given any left-continuous and adapted process Y, the integral $Y \bullet M$ is well defined (as an element of $\mathbb{H}^2_{\text{loc}}$) for any $M \in \mathbb{H}^2_{\text{loc}}$, and, consequently, $Y \bullet X$ is well defined (as a semimartingale) for any semimartingale X (since⊸ (15.23), (15.26) the local martingale component in any semimartingale can always be chosen to be in $\mathbb{H}^2_{\text{loc}}$).

Given any two semimartingales X and Y, the processes $X_- = (X_{t-})_{t \in \mathbb{R}_+}$ and $Y_- = (Y_{t-})_{t \in \mathbb{R}_+}$ are left-continuous by construction, and are therefore locally bounded, as was clarified above. As a result, the integrals $X_- \bullet Y$ and $Y_- \bullet X$ are well defined. The *quadratic covariation*, or the so-called *square bracket* between X and Y, is the process

$$[X, Y] \stackrel{\text{def}}{=} XY - X_0 Y_0 - X_- \bullet Y - Y_- \bullet X . \qquad \mathrm{e}_1$$

Notice that by its definition $[X, Y]$ is a finite càdlàg process, and, as the notation already suggests, $[X, Y]$ can be identified as the quadratic covariation of X and Y in the sense of (8.103). In fact, such an identification is quite intuitive: for any given subdivision of the interval $[0, t]$, say $\mathcal{p} \colon 0 = t_0 < t_1 < \ldots < t_n = t$,

$$X_t Y_t - X_0 Y_0 = \sum\nolimits_{i \in \mathbb{N}_{|n-1}} Y_{t_i}(X_{t_{i+1}} - X_{t_i}) + \sum\nolimits_{i \in \mathbb{N}_{|n-1}} X_{t_i}(Y_{t_{i+1}} - Y_{t_i})$$
$$+ \sum\nolimits_{i \in \mathbb{N}_{|n-1}} (X_{t_{i+1}} - X_{t_i})(Y_{t_{i+1}} - Y_{t_i}),$$

and it is not difficult to guess that $[X, Y]_t$ must be the limit in probability of the last sum (written for some sequence of partitions chosen so that $|\mathcal{p}_n| \to 0$). To make such a claim precise, we follow the steps that we took in (9.71): given any partition of $\mathbb{R}_+$, $\mathcal{p} \colon 0 = t_0 < t_1 < \ldots \to \infty$, we set

$$\mathcal{Q}^{\mathcal{p}}(X, Y)_t \stackrel{\text{def}}{=} \sum\nolimits_{i \in \mathbb{N}} (X_{t \wedge t_{i+1}} - X_{t \wedge t_i})(Y_{t \wedge t_{i+1}} - Y_{t \wedge t_i}),$$

and examine the limiting behavior of these quantities. One can claim the following (Jacod and Shiryaev 1987, I 4.47): for any sequence of partitions of $\mathbb{R}_+$, $(\mathcal{p}_n)_{n \in \mathbb{N}}$, chosen so that $|\mathcal{p}_n| \to 0$, the sample paths of the processes $\mathcal{Q}^{\mathcal{p}_n}(X, Y)$, $n \in \mathbb{N}$, converge uniformly on finite intervals in probability – this is the same convergence as in (9.71) – and the limiting process is indistinguishable from $[X, Y]$, as defined in e_1 above. In particular, we have the usual polarization identity

$$[X, Y] = \frac{1}{4}\Big([X + Y, X + Y] - [X - Y, X - Y]\Big),$$

and, in addition, since the processes $\mathcal{Q}^{\mathcal{P}_n}(X, Y)$ are càdlàg by construction, and since the uniform convergence on finite intervals is stronger than the convergence in the Skorokhod topology, we see that $[X, Y]$ is indeed a càdlàg process. By the result from exercise (8.99), this also means that the quadratic variation process

$$[X, X] \overset{\text{def}}{=} X^2 - X_0^2 - 2X_-{\bullet}X,$$

is càdlàg and increasing, so that $[X, Y]$ is not only càdlàg, but also has sample paths of finite variation (a.s.), as the difference between two increasing càdlàg processes.

(15.33) EXERCISE: Given any two semimartingales X and Y, prove that

$$\Delta[X, Y] = \Delta X \Delta Y .$$

HINT: As a matter of elementary algebra,

$$\begin{aligned}\Delta(X_t Y_t) &= X_t Y_t - X_{t-} Y_{t-}\\ &= (X_t - X_{t-})(Y_t - Y_{t-}) + X_{t-}(Y_t - Y_{t-}) + Y_{t-}(X_t - X_{t-})\\ &= \Delta X_t \Delta Y_t + X_{t-}\Delta Y_t + Y_{t-}\Delta X_t .\end{aligned}$$ ○

We have seen that if W is the Brownian motion – or, for that matter, any continuous local martingale – then the angled bracket $\langle W, W\rangle$ is the same as the quadratic variation $[W, W]$.[177] As was already noted, however, for local martingales with jumps the two brackets generally differ, and, for this reason we can no longer use angled brackets to denote quadratic variation.

(15.34) EXERCISE: Consider the compensated Poisson process $X = N - c\iota$, which$^{\multimap (9.18)}$ is a martingale relative to $\mathscr{F}^N$ with $\langle X, X\rangle = c\iota$. Prove that

$$[X, X] = \sum(\Delta N)^2 = \sum \Delta N = N,$$

and observe that $[X, X] \neq \langle X, X\rangle$.

HINT: Interpret $[X, X]$ as $\lim_n \mathcal{Q}^{\mathcal{P}_n}(X, X)$. ○

It should be clear by now that there must be an intrinsic connection between the brackets $[\cdot,\cdot]$ and $\langle\cdot,\cdot\rangle$, and our immediate objective is to clarify this connection. Recall that $\langle\cdot,\cdot\rangle$ arises as the compensator component in$^{\multimap (9.48)}$ the Doob–Meyer decomposition, which component is required to be predictable. To be specific, if $M \in \mathbb{H}^2_{\text{loc}}$, then $\langle M, M\rangle$ can be characterized as the only predictable and increasing process that starts from 0 and is such that $M^2 - \langle M, M\rangle$ is a local martingale. We have the following result:

[177] Given a semimartingale X, the quadratic variation $[X, X]$ is always meaningful, but the angled bracket $\langle X, X\rangle$ is meaningful only if X is a local martingale (not necessarily continuous).

(15.35) EXERCISE: Prove that the angled bracket is nothing but⊸(15.12) the predictable compensator (the dual predictable projection) of the square bracket, that is, for any local L^2-martingale $M \in \mathbb{H}^2_{\text{loc}}$, $\langle M, M\rangle = [M, M]^{\mathscr{P}}$.

HINT: It would be enough to demonstrate that $[M, M] - \langle M, M\rangle$ is a local martingale – see (15.12). Consult the definition of $[M, M]$ in e_1, and explain briefly why $\langle M, M\rangle \in \bar{\mathscr{V}}_{\text{loc}}$. ○

(15.36) REMARK: Given a locally square integrable martingale X, the increasing process $[X, X]$ is continuous if and only if $\Delta[X, X] = (\Delta X)^2 = 0$, and this is the same as $\Delta X = 0$, which is the same as the claim that X is continuous. Of course, if $[X, X]$ is continuous, or, equivalently, X is continuous, then the predictable compensator of $[X, X]$ – which is nothing but $\langle X, X\rangle$ – is $[X, X]$ itself, in which case $\langle X, X\rangle = [X, X]^{\mathscr{P}} = [X, X]$. ○

In what follows we list briefly the key features of the bracket $[\cdot, \cdot]$. Detailed proofs can be found in (Jacod and Shiryaev 1987, I sec. 4), but it would be very helpful for the reader to establish at least some of these properties as an exercise.

(15.37) PROPERTIES OF THE QUADRATIC COVARIATION I: (ibid., I 4.49) Let X be any semimartingale and let A be any adapted càdlàg process with sample paths of finite variation that starts from 0, that is, A is any process from the class $\mathscr{V}$. Then

(a) $[X, A] = \Delta X \cdot A$ and $XA = A_- \cdot X + X \cdot A$.

(b) If A is predictable, then $[X, A] = \Delta A \cdot X$ and $XA = A \cdot X + X_- \cdot A$.

(c) If A is predictable and X is a local martingale, then $[X, A]$ is a local martingale.

(d) If either X or A is continuous, then $[X, A] = 0$. ○

(15.38) EXERCISE: Explain how properties (c) and (d) in (15.37) follow from properties (a) and (b). ○

(15.39) PROPERTIES OF THE QUADRATIC COVARIATION II: (ibid., I 4.50–4.51) Let M and N be any two local martingales (thus also semimartingales). Then:

(a) The process $MN - M_0 N_0 - [M, N]$ is a local martingale.

(b) If $M, N \in \mathbb{H}^2_{\text{loc}}$, then $[M, N] \in \mathscr{V}_{\text{loc}}$; if $M, N \in \mathbb{H}^2$, then $MN - [M, N]$ is a martingale and so is $[M, N] - \langle M, N\rangle$.

(c) $M \in \mathbb{H}^2$ ($M \in \mathbb{H}^2_{\text{loc}}$) if and only if $[M, M] \in \bar{\mathscr{V}}$ ($[M, M] \in \bar{\mathscr{V}}_{\text{loc}}$), and $\mathsf{E}[M_0^2] < \infty$.

(d) $[M, M] = 0$ a.s. if and only if $M = M_0$ a.s.

(e) If $M \in \mathbb{H}^2$ is purely discontinuous, then $[M, M] = \sum(\Delta M)^2$. ○

(15.40) PROPERTIES OF THE QUADRATIC COVARIATION III: (ibid., I 4.52–4.55) Let X and Y be any two semimartingales and let X^c and Y^c denote their respective continuous local martingale components (which are unique, even though the local martingale components are not – see (15.26)). Then:

(a) $[X,Y] = \langle X^c, Y^c\rangle + \sum \Delta X \Delta Y$; in particular, since $[X,X]$, $[Y,Y]$, and $[X,Y]$ are all elements of $\mathscr{V}$ and are finite, the processes

$$\sum(\Delta X)^2, \quad \sum(\Delta Y)^2, \quad \text{and} \quad \sum \Delta X \Delta Y,$$

must be finite as well.

(b) If H is any predictable and locally bounded process, then $\left[H\bullet X, Y\right] = H\bullet[X,Y]$.

If M and N are two local martingales, then

(c) $|[M,N]|^{1/2} \in \mathscr{V}_{\text{loc}}$

(d) $[M,N] = 0$ if M is continuous and N is purely discontinuous.

(e) $[M,N] = \langle M,N\rangle = 0$ if M and N are continuous and orthogonal, in that MN is a local martingale.

(f) If M is continuous (purely discontinuous) and H is any locally bounded and predictable process, then $H\bullet M$ is a continuous (purely discontinuous) local martingale. ○

Now we have all the tools needed to develop the Itô formula for semimartingales with jumps, and this is what we do next.

Let X be *any* semimartingale. From the definition of the bracket $[\cdot,\cdot]$ in e_1, in conjunction with (15.40a), we have (as before, X^c denotes the continuous local martingale component of the semimartingale X)

$$\begin{aligned} X_t^2 &= X_0^2 + 2X_-\bullet X_t + [X,X]_t \\ &= X_0^2 + 2X_-\bullet X_t + \langle X^c, X^c\rangle_t + \sum\nolimits_{s\le t}(\Delta X_s)^2 \\ &= X_0^2 + 2X_-\bullet X_t + \langle X^c, X^c\rangle_t + \sum\nolimits_{s\le t}(X_s - X_{s-})^2 . \end{aligned}$$

Since $(X_s - X_{s-})^2$ can be written in the form

$$\begin{aligned} (X_s - X_{s-})^2 &= X_s^2 + X_{s-}^2 - 2X_{s-}X_s \\ &= X_s^2 - X_{s-}^2 + 2X_{s-}^2 - 2X_{s-}X_s = X_s^2 - X_{s-}^2 - 2X_{s-}(X_s - X_{s-}), \end{aligned}$$

we arrive at the following relation

$$\begin{aligned} X_t^2 - X_0^2 = 2X_-\bullet X_t &+ \langle X^c, X^c\rangle_t \\ &+ \sum\nolimits_{s\le t}\left(X_s^2 - X_{s-}^2 - 2X_{s-}\Delta X_s\right), \end{aligned}$$

which can be abbreviated as

$$X^2 - X_0^2 = 2X_-\bullet X + \langle X^c, X^c\rangle + \sum\left(X^2 - X_-^2 - 2X_-\Delta X\right),$$

and can be cast in differential form as

$$\mathrm{d}X^2 = 2X_-\mathrm{d}X + \mathrm{d}\langle X^c, X^c\rangle + \left(X^2 - X_-^2 - 2X_-\Delta X\right).$$

To see how the last relation can be generalized, we use again e_1, but this time in conjunction with (15.31h), (15.37a), and (15.40a,b), and with $Y = X^2$, so that $XY = X^3$ (observe that the process X_- is both locally bounded and predictable and that $\langle X^c, Y^c\rangle \in \mathscr{V}$ is continuous and thus predictable). With this choice we have

$$X^3 - X_0^3 \equiv XY - X_0Y_0 = Y_-\bullet X + X_-\bullet Y + [X,Y].$$

The three terms on the right side of the last equation we now expand, taking into account the expansion above of $X^2 - X_0^2$, as follows:

$$\begin{aligned} Y_-\bullet X &= X_-^2\bullet X\,; \\ X_-\bullet Y &= X_-\bullet(X^2 - X_0^2) = 2X_-^2\bullet X + X_-\bullet\langle X^c, X^c\rangle \qquad e_2 \\ &\quad + \sum\big(X_-X^2 - X_-^3 - 2X_-^2\Delta X\big)\,; \\ [X,Y] &= [X, X^2 - X_0^2] = 2X_-\bullet[X,X] + \Delta X\bullet\langle X^c, X^c\rangle \qquad e_3 \\ &\quad + \sum \Delta X\big(X^2 - X_-^2 - 2X_-\Delta X\big)\,. \end{aligned}$$

Since the measure $\mathrm{d}\langle X^c, X^c\rangle$ is nonatomic, and since $\Delta X_t \neq 0$ only for countably many t, one must have $\Delta X\bullet\langle X^c, X^c\rangle = 0$. Furthermore, the first integral on the right side of e_3 can be expanded as

$$2X_-\bullet[X,X] = 2X_-\bullet\langle X^c, X^c\rangle + 2\sum X_-(\Delta X)^2\,, \qquad e_4$$

and the last term in e_3 can be expanded as

$$\begin{aligned} &\sum(X - X_-)\big(X^2 - X_-^2 - 2X_-\Delta X\big) \\ &\qquad = \sum\big(X^3 - X_-^2X - 2X_-X\Delta X\big) \\ &\qquad\qquad + \sum\big(-X_-X^2 + X_-^3 + 2X_-^2\Delta X\big)\,. \end{aligned}$$

Finally, observe the last sum and the sum in e_2 cancel out, and the only jumps that remain to be collected are the ones in the first sum on the right side of the last equation and the ones in the sum from e_4. This gives us

$$\begin{aligned} &\sum\big(X^3 - X_-^2X - 2X_-X\Delta X + 2X_-(\Delta X)^2\big) \\ &\quad = \sum\big(X^3 - X_-^2(X_- + \Delta X) - 2X_-(X_- + \Delta X)\Delta X + 2X_-(\Delta X)^2\big) \\ &\quad = \big(X^3 - X_-^3 - 3X_-^2\Delta X\big) \end{aligned}$$

and we thus arrive at the formula:

$$X^3 - X_0^3 = 3X_-^2\bullet X + 3X_-\bullet\langle X^c, X^c\rangle + \sum\big(X^3 - X_-^3 - 3X_-^2\Delta X\big)\,.$$

By using induction and following the preceding steps, one can obtain the following more general formula for an arbitrary integer power $n \geq 2$:

$$X^n - X_0^n = nX_-^{n-1}\boldsymbol{\cdot} X + \frac{n(n-1)}{2}X_-^{n-2}\boldsymbol{\cdot}\langle X^c, X^c\rangle$$
$$+ \sum\left(X^n - X_-^n - nX_-^{n-1}\Delta X\right).$$

We have essentially established the following result.

(15.41) ITÔ'S FORMULA FOR SEMIMARTINGALES WITH JUMPS: Let $f : \mathbb{R} \mapsto \mathbb{R}$ be any continuous function with continuous derivatives of order at least 2, that is, $f \in \mathscr{C}^2(\mathbb{R}; \mathbb{R})$. Then

$$f(X) - f(X_0) = (\partial f)(X_-)\boldsymbol{\cdot} X + \frac{1}{2}(\partial^2 f)(X_-)\boldsymbol{\cdot}\langle X^c, X^c\rangle$$
$$+ \sum\left(f(X) - f(X_-) - (\partial f)(X_-)\Delta X\right).$$

The multivariate Itô formula reads as follows: for any $f \in \mathscr{C}^2(\mathbb{R}^n; \mathbb{R})$, $n \in \mathbb{R}_{++}$, and any n-dimensional (vector-column) semimartingale $X = (X^1, \dots, X^n)$,

$$f(X) - f(X_0) = \sum_{i=1}^{n} \partial_i f(X_-)\boldsymbol{\cdot} X^i + \frac{1}{2}\sum_{i,j=1}^{n} \partial_i\partial_j f(X_-)\boldsymbol{\cdot}\langle (X^i)^c, (X^j)^c\rangle$$
$$+ \sum\Big(f(X) - f(X_-) - \sum_{i=1}^{n} \partial_i f(X_-)\Delta X^i\Big),$$

where the symbol ∂_i is understood as the partial derivative with respect to the ith argument of the function that follows this symbol. We note that in some concrete applications one of the components of X may be the time variable itself, that is, one may have $X^i = \iota$ for some i ○

(15.42) EXERCISE: Explain how the integration by parts rule from e_1 obtains formally from the Itô formula above. ○

(15.43) REMARK: It is important to keep in mind that by construction the stochastic integral includes the upper limit of the integration domain, and if the integral $\int_s^t Y\,\mathrm{d}X$ is to be understood as yet another notation for the difference $Y\boldsymbol{\cdot} X_t - Y\boldsymbol{\cdot} X_s$, then a more appropriate notation for this integral would be $\int_{s+}^t Y\,\mathrm{d}X$. Thus, written in full, Itô's formula reads:

$$f(X_t) - f(X_s) = \sum_{i=1}^{n} \int_{s+}^{t} \partial_i f(X_{u-})\mathrm{d}X_u^i$$
$$+ \frac{1}{2}\sum_{i,j=1}^{n} \int_{s+}^{t} \partial_i\partial_j f(X_{u-})\mathrm{d}\langle (X^i)^c, (X^j)^c\rangle_u$$
$$+ \sum\nolimits_{u\in\,]s,t]}\Big(f(X_u) - f(X_{u-}) - \sum_{i=1}^{n} \partial_i f(X_{u-})\Delta X_u^i\Big)$$

for every $0 \le s < t < \infty$. ○

We conclude this section with the general version of Girsanov's theorem that can be applied to local martingales with jumps – see (Dellacherie and Meyer 1980, VII 49).

(15.44) GIRSANOV'S THEOREM (SECOND VERSION): Let $(\Omega, \mathcal{F}, P)$ be any probability space and let $\mathscr{F}$ be any filtration inside $\mathcal{F}$ that obeys the usual conditions. Let Q be another probability measure on $(\Omega, \mathcal{F})$, such that, for some $T \in \mathbb{R}_{++}$, $Q\restriction\mathscr{F}_T \approx P\restriction\mathscr{F}_T$ (Q and P are equivalent on the σ-field $\mathscr{F}_T$). Consider the Radon–Nikodym derivative $R_T \stackrel{\text{def}}{=} \frac{\mathrm{d}(Q\restriction\mathscr{F}_T)}{\mathrm{d}(P\restriction\mathscr{F}_T)}$ and let $(R_t)_{t\in[0,T]}$ be any càdlàg version of the martingale $\mathsf{E}[R_T\,|\,\mathscr{F}_t]$, $t \in [0,T]$. Equivalently, R can be identified as a càdlàg version of the Radon–Nikodym derivative process $\frac{\mathrm{d}(Q\restriction\mathscr{F}_t)}{\mathrm{d}(P\restriction\mathscr{F}_t)}$, $t \in [0,T]$. Then, for any local $(\mathscr{F}, P)$-martingale $M = (M_t)_{t\in[0,T]}$, the process

$$\tilde{M} \stackrel{\text{def}}{=} M - \frac{1}{R}\bullet[M,R]$$

is a local $(\mathscr{F}, Q)$-martingale on $[0,T]$. ○

(15.45) EXERCISE: Using the integration by parts rule from e_1, in conjunction with the result from exercise (9.78), give a complete proof of Girsanov's theorem, as stated in (15.44).

HINT: Since $P \approx Q$ on $\mathscr{F}_t$, $P(R_t > 0) = Q(R_t > 0) = 1$ for every $t \in [0,T]$. By the result in (9.39), a.e. sample path of R is uniformly bounded away from 0 on $[0,T]$. In particular, the integral $\frac{1}{R}\bullet[M,R]$ is well defined and finite. It only remains to show that $\tilde{M}R$ is a local martingale relative to P. However, by (15.37a),

$$\Big(\frac{1}{R}\bullet[M,R]\Big)R = \Big(\frac{1}{R}\bullet[M,R]\Big)_{-}\bullet R + [M,R]$$

and, in addition, ⊸(15.39a) $MR - [M,R]$ is a local martingale relative to P. ○

(15.46) REMARK: We stress that the correction term in Girsanov's theorem is $\frac{1}{R}\bullet[M,R]$, not $\frac{1}{R_-}\bullet[M,R]$. It is understood as a Lebesgue–Stieltjes integral defined path-by-path, in which case there is no need for the integrand to be predictable. ○

(15.47) EXERCISE: Explain how Girsanov's theorem stated in (10.51) for continuous local martingales obtains from the more general version in (15.44). ○

16

Random Measures and Lévy Processes

In this chapter we develop some concrete – and important from a practical point of view – models of processes with jumps. The main source of insight for us is again the Poisson process, and the idea of a process with stationary and independent increments that we introduced back in (9.12).

16.1 Poisson Random Measures

Our goal now is to describe a general class of processes that are similar in structure to the Poisson process but are less restrictive, in that the jumps are no longer of fixed size. To achieve such an objective, we must introduce some form of a Poisson structure for "events" occurring not only in the time domain $\mathbb{R}_+$, but also in the space from which the jumps are sampled. Clearly, the Poisson structures on these two domains must be coordinated, and, for this reason, it would be easier for us to work with a single Poisson structure on product spaces of the form $\mathbb{R}\times\mathbb{R}_+$, or, more generally, of the form $E\times\mathbb{R}_+$ for some measurable space $(E, \mathscr{E})$, understood to be the range space for the jumps. The Poisson random measures, which we are about to introduce, were invented precisely with that purpose in mind.

To get started with the program just outlined, recall that in (9.13) we defined the Poisson process as the counter of time events that precede time t; to be precise, we wrote $N_t = \sum_{n\in\mathbb{N}_{++}} 1_{\{\mathscr{t}_n\leq t\}}$, $t \in \mathbb{R}_+$, where $\mathscr{t}_n = \xi_1 + \ldots + \xi_n$, for some sequence of independent and identically exponentially distributed random variables $(\xi_i)_{i\in\mathbb{N}_{++}}$. Equivalently, N_t can be expressed as $N_t(\omega) = \sum_{n\in\mathbb{N}_{++}} \epsilon_{\mathscr{t}_n(\omega)}([0, t])$, where $\epsilon_{\mathscr{t}_n(\omega)}$ is the unit point-mass measure concentrated at $\mathscr{t}_n(\omega) \in \mathbb{R}_+$. As we now see, the construction of the Poisson process N has two aspects. The first one is the random measure aspect: for any concrete realization of the sample point $\omega \in \Omega$, the mapping

$$\mathcal{B}(\mathbb{R}) \ni B \rightsquigarrow M_\omega(B) \stackrel{\text{def}}{=} \sum\nolimits_{n\in\mathbb{N}_{++}} \epsilon_{\mathscr{t}_n(\omega)}(B) \in \mathbb{R}_+$$

represents a measure over $\mathbb{R}_+$, and, for any given $t \in \mathbb{R}_+$, the random variable $N_t(\omega)$ is nothing but the volume of the set $[0, t]$ according to the measure M_ω, that is, $N_t(\omega) = M_\omega([0, t])$. The second aspect is counting: the measure M_ω takes values in the space of integers $\mathbb{N} \equiv \{0, 1, \ldots\}$, and $M_\omega(B) \in \mathbb{N}$ gives the count of how many points from the set $\{\mathscr{t}_1(\omega), \mathscr{t}_2(\omega), \ldots\}$ belong to $B \in \mathcal{B}(\mathbb{R}_+)$. As we already know, this count is going to be finite a.s. if the set B is contained in some finite interval, and, since $\mathbb{R}_+$ is a countable union of such intervals, M_ω must be a σ-finite measure.

(16.1) RANDOM MEASURES: Given a measurable space $(E, \mathscr{E})$ – any measurable space – let $\mathcal{M}(E, \mathscr{E})$ denote the space of all σ-finite measures over $(E, \mathscr{E})$. A *random measure* over $(E, \mathscr{E})$, defined on probability space $(\Omega, \mathcal{F}, P)$, is any mapping of the form

$$\Omega \ni \omega \rightsquigarrow M_\omega(\cdot) \equiv M(\omega, \cdot) \in \mathcal{M}(E, \mathscr{E}),$$

which is measurable, in that the function $\omega \rightsquigarrow M_\omega(A) \in \mathbb{R}_+$ is $\mathcal{F}$-measurable for every fixed $A \in \mathscr{E}$. Following the tradition of suppressing the dependence on ω in the notation, we will often write $M(A)$ instead of $M_\omega(A)$, or $M(\omega, A)$. Since $M(A)$ is a positive r.v. for every $A \in \mathscr{E}$, the expected value $m(A) = \mathsf{E}[M(A)]$ is always defined, and $m(\cdot)$ is easily seen to be a measure over $(E, \mathscr{E})$.[178] We call m "the mean measure of M," or simply "the mean of M." In general, there is no reason for the mean measure to be σ-finite, and from now on we will suppose that this feature is incorporated into the definition of "random measure." ○

(16.2) COUNTING RANDOM MEASURES: A random measure, M, is said to be a *counting measure* if, for every $\omega \in \Omega$, $M_\omega(\cdot)$ is a purely atomic measure, such that all of its atoms have mass equal to 1, that is, given any singleton $\{x\}$, $x \in E$, one has $\{x\} \in \mathscr{E}$ and $M_\omega(\{x\})$ is either 0 or 1, with $M_\omega(\{x\}) = 1$ for at most countably many $x \in E$. In all concrete applications that are going to be of interest to us, the space $(E, \mathscr{E})$ is going to be some connected subset of $\mathbb{R}^n$, and $\mathscr{E}$ is going to be the usual Borel σ-field in it. However, the concrete structure of an Euclidean space is not really needed, and we might as well suppose that E is some complete separable metric space (alias: Polish space) and that $\mathscr{E}$ is the Borel σ-field in E, in which case we call $(E, \mathscr{E})$ "standard measurable space" (notice that in such spaces all singletons $\{x\}$, $x \in E$, are Borel sets). ○

Some minor technical matters aside, it would not be a restriction to suppose that a counting random measure over $(E, \mathscr{E})$ is any random measure that can be expressed in the form[179]

$$M_\omega = \sum_{i=1}^{N(\omega)} \epsilon_{X_i(\omega)}, \qquad \mathrm{e}_1$$

for some choice of the $\bar{\mathbb{N}}$-valued random variable N and the sequence of E-valued random variables $(X_i)_{i \in \mathbb{N}_{++}}$, with the understanding that if $N(\omega) = 0$, then M_ω is the trivial measure $M_\omega(A) = 0$, $A \in \mathscr{E}$. Obviously, the intrinsic nature of such measures is encrypted in the way in which the random variables X_i relate to one another, and relate also to N. The choice that comes to mind first is to take $(X_i)_{i \in \mathbb{N}_{++}}$ to be some i.i.d. sequence, with some shared *nonatomic* probability distribution law μ, which is independent from the law of N. The random measure M_ω

[178] This feature is a straightforward consequence of the monotone convergence theorem.

[179] As an example, in the case $E = \mathbb{R}$, setting $Z_\omega(r) = M_\omega([0, r])$ for any $r \in \mathbb{R}_+$, and $\rho_i(\omega) = \inf\{r \in \mathbb{R}_+ : Z_r(\omega) > i\}$, $i \in \mathbb{N}$, one can set $X_i(\omega) = Z_\omega(\rho_i(\omega))$.

from e_1 can then be imagined in the following way. First, one draws a number, n, from the set $\bar{\mathbb{N}}$ according to the law of N (note that n could be $+\infty$). Then one "pins" exactly n points in the set E, drawn independently from the law μ (if $n = 0$ then the collection of pinned points is the empty set). Given any Borel set $A \in \mathscr{E}$, one can then count how many of the n points that were pinned inside E happen to be in $A \subseteq E$. The outcome from this count can be treated as a realization of the random variable $\omega \rightsquigarrow M_\omega(A)$ that takes values in $\bar{\mathbb{N}}$.

(16.3) EXERCISE: Explain why in the preceding paragraph the shared law of the i.i.d. sequence $(X_i)_{i \in \mathbb{N}_{++}}$ was required to be nonatomic. What would be the problem if this law contains atoms, that is, there is at least one point $x \in E$ with $\mu(\{x\}) \equiv P(X_i = x) > 0$? ○

The next exercise reveals an intriguing and somewhat surprising feature of the random measure constructed from the i.i.d. sequence $(X_i)_{i \in \mathbb{N}_{++}}$ and the integer-valued random variable N. Among other things, it sheds some light as to why the Poisson distribution shows up in various applications as often as it does.

(16.4) EXERCISE: Consider the random measure M given by e_1 and suppose that the random variable N is distributed with Poisson law of parameter α (in particular, $P(N = \infty) = 0$) and that $(X_i)_{i \in \mathbb{N}_{++}}$ is an i.i.d. sequence of E-valued random variables with shared nonatomic probability distribution law μ, which is independent of N. Prove that for any set $A \in \mathscr{E}$ the random variable $M(A)$ is distributed with Poisson law of parameter $\alpha\mu(A)$ (somehow the law μ, from which the X_i's are sampled does not affect the nature of the distribution law of $M(A)$, which is always of Poisson type).

HINT: Given any integer $n \in \mathbb{N}$ and set $A \in \mathscr{E}$, one can write (explain why)

† $$P\big(M(A) = n\big) = \sum\nolimits_{i=n}^{\infty} e^{-\alpha}\frac{\alpha^i}{i!}\binom{i}{n}\mu(A)^n\big(1 - \mu(A)\big)^{i-n}. \quad ○$$

(16.5) POISSON RANDOM MEASURES: The random measure M is said to be a *Poisson random measure* if its (automatically σ-finite) mean measure m has the following properties:

(a) for any $A \in \mathscr{E}$ with $m(A) < \infty$, the random variable $M(A)$ is distributed with Poisson law of parameter $m(A)$;

(b) for any finite collection of disjoint sets $A_1, \dots, A_n \in \mathscr{E}$ with $m(A_i) < \infty$, $1 \leq i \leq n$, the random variables $M(A_i)$, $1 \leq i \leq n$, are jointly independent.

If these conditions are met, the measure m is also called the *intensity of* M. ○

It turns out that condition (16.5a) implies condition (16.5b) provided that the mean measure m is free of atoms. This is a surprising and highly nontrivial result – see (16.7) below. Moreover, if m is nonatomic, then (16.5a) implies not only (16.5b) but also that M must be a counting random measure. To be precise, we have the following results – see (Çinlar 2011, ch. 6 2.17, 5.12, and 5.16) for detailed proofs and a comprehensive exposition on the topic "random measures."

(16.6) THEOREM: A Poisson random measure M is a counting random measure if and only if its intensity measure m is nonatomic. ○

(16.7) THEOREM: Let M be any counting random measure over some standard measurable space $(E, \mathscr{E})$. Suppose that there is a σ-finite and nonatomic measure m over $(E, \mathscr{E})$, which is connected to M through the relation

$$\dagger \qquad P\big(M(A) = 0\big) = e^{-m(A)} \quad \text{for every } A \in \mathscr{E} .$$

Then it follows that M must be a Poisson counting measure with mean (intensity) measure m. ○

(16.8) REMARK: The last result is highly nontrivial, interesting, and surprising: somehow, if the probabilities with which the random variables $M(A)$, $A \in \mathscr{E}$, take the value of 0 have the form (16.7†), then the probabilities with which these random variables take any of the remaining values from the set $\mathbb{N}_{++} = \{1, 2, \dots\}$ become fixed. Moreover, the form (16.7†) forces $M(A_i)$, $1 \le i \le n$, to be independent whenever the events A_i are finite (according to m) and disjoint. In particular, the measure M from exercise (16.4) must be a Poisson counting measure simply because $P\big(M(A) = 0\big) = e^{-\alpha\mu(A)}$, which is a special case of (16.4†) with $n = 0$ (somehow, showing that (16.4†) holds with $n = 0$ implies that this relation holds for any $n \in \mathbb{N}_{++}$). Another interesting observation is that although in (16.4) we were able to establish that $M(A)$ follows the Poisson law as a matter of a straightforward calculation, the fact that the volumes under M of any collection of disjoint sets must be independent is a much deeper statement that is not revealed in the simple calculation that we used. ○

(16.9) CONSTRUCTION OF POISSON COUNTING MEASURES: Now we see how the property established in (16.4) provides a mechanism for constructing Poisson counting measures M of a given nonatomic intensity m. It would be enough for us to examine only the case of a *finite* nonatomic intensity m, since, by the very definition of a σ-finite measure, any counting measure with intensity m can always be written as the sum of independent counting measures with finite intensities, defined on non-overlapping sets. If $m(E) < \infty$, then $\mu \overset{\text{def}}{=} \frac{1}{m(E)} m$ is a probability law over $(E, \mathscr{E})$, and if we choose the parameter of the Poisson process N to be $\alpha = m(E)$, then, ⊸(16.4) by way of sampling from the law of N and the law μ, we are going to arrive at a Poisson counting measure of intensity $\alpha\mu = \frac{m(E)}{m(E)} m = m$. In particular, we can construct a Poisson counting measure on $\mathbb{R}_+$ of intensity $m = c\Lambda$ for any fixed scalar $c \in \mathbb{R}_{++}$. ○

(16.10) EXERCISE: Let $c \in \mathbb{R}_{++}$ and let M be the Poisson counting measure on $(\mathbb{R}_+, \mathcal{B}(\mathbb{R}_+))$ of intensity (mean measure) $m = c\Lambda$. Prove that the process $N_t \overset{\text{def}}{=} M([0, t])$, $t \in \mathbb{R}_+$, which we will often write more succinctly as $N = M([0, \iota])$, is nothing but the Poisson process of intensity c.

HINT: Show that N is a process of independent increments and the distribution of those increments is consistent with the distribution of the increments of the Poisson process of intensity c. ○

(16.11) EXERCISE: Assuming that N is a Poisson process of time-varying intensity $c_t = \log(1+t)$, $t \in \mathbb{R}_+$, develop a Monte Carlo simulation program that generates the sample paths N_ω on the finite time interval $[0, T]$.

HINT: In this case $N = M([0, \imath])$, where M is a Poisson counting measure on $(\mathbb{R}_+, \mathcal{B}(\mathbb{R}_+))$ of intensity given by $m = \log(1 + \imath) \odot \Lambda$, that is, $m(\mathrm{d}t) = \log(1+t)\mathrm{d}t$. Setting $\alpha = \int_0^T \log(1 + t)\mathrm{d}t$, you must sample from the Poisson law of intensity α and (independently) sample from the probability distribution law on $[0, T]$ with density $\psi(t) = \frac{1}{\alpha} \log(1 + t)$, $t \in [0, T]$. ○

(16.12) REMARK: Since the jumps of the Poisson process are fixed, this process can be described by merely identifying the moments in time at which the jumps occur. As we saw in exercise (16.10), such a description is tantamount to identifying a Poisson counting measure on $\mathbb{R}_+$. But if we insist on randomizing the jumps, then we must work with Poisson counting measures on product spaces of the form $E \times \mathbb{R}_+$, that is, of the form "jumps space"×"time." We turn to counting measures defined on such product spaces next. ○

(16.13) DISINTEGRATION OF POISSON COUNTING MEASURES: Consider the special case where $(E, \mathscr{E})$ is the product of two measurable spaces, $(\bar{E}, \bar{\mathscr{E}})$ and $(\tilde{E}, \tilde{\mathscr{E}})$, that is, $E = \bar{E} \times \tilde{E}$, $\mathscr{E} = \bar{\mathscr{E}} \otimes \tilde{\mathscr{E}}$, and suppose that the measure m, defined on $(E, \mathscr{E})$, can be disintegrated into the product $m = \bar{m} \times \tilde{m}$, in which $\bar{m}$ is a (*not necessarily nonatomic*) probability measure on $(\bar{E}, \bar{\mathscr{E}})$, and $\tilde{m}$ is a *nonatomic* σ-finite measure on $(\tilde{E}, \tilde{\mathscr{E}})$. Notice that if $\bar{m}$ and $\tilde{m}$ are so chosen, then $m = \bar{m} \times \tilde{m}$ is σ-finite and nonatomic by construction. Let M be the Poisson counting measure on $(E, \mathscr{E})$ with intensity measure m, constructed by following the recipe outlined in (16.9), and let

$$\dagger \qquad \left\{X_i(\omega) \equiv (\bar{X}_i(\omega), \tilde{X}_i(\omega)) \in E \equiv \bar{E} \times \tilde{E} : i \in \mathscr{I}(\omega)\right\}$$

be the (random) collection of all (random) atoms of M_ω, as in e_1. Notice that, with the notation as in e_1, the (random) index set $\mathscr{I}(\omega) \subseteq \mathbb{N}_{++}$ is nothing but the set $[1, N(\omega)] \cap \mathbb{N}_{++}$, and this set is independent of the sequence $(X_i(\omega))_{i \in \mathbb{N}_{++}}$ (of course, $\mathscr{I}(\omega)$ could be the empty set $\emptyset$, in which case the collection $\dagger$ is understood to be empty). Since the random elements $(X_i \in E)_{i \in \mathscr{I}}$ are taken from the i.i.d. sequence $(X_i \in E)_{i \in \mathbb{N}_{++}}$, and since independently sampled elements from a nonatomic distribution coincide with probability 0, we can claim that $\tilde{X}_i \neq \tilde{X}_j$ a.s. whenever $i \neq j$. Thus, there is a one-to-one correspondence between the atoms of M inside $E = \bar{E} \times \tilde{E}$ and their projections into $\tilde{E}$. In fact, the collection of all such projections, namely $(\tilde{X}_i)_{i \in \mathscr{I}}$, is nothing but the collection of all (distinct) atoms of the random measure

$$\tilde{M}(A) \overset{\text{def}}{=} M(\bar{E} \times A), \quad A \in \tilde{\mathscr{E}}.$$

It is easy to see that $\tilde{M}$ inherits the attribute "Poisson random measure" from M, and the intensity of $\tilde{M}$ is precisely $\tilde{m}$ due to the relation

$$\mathsf{E}[M(\bar{E}\times A)] = m(\bar{E}\times A) = \bar{m}(\bar{E})\tilde{m}(A) = \tilde{m}(A)\,.$$

Since $\tilde{m}$ is nonatomic, it follows that $\tilde{M}$ must be a Poisson counting measure, and since sampling from $m = \bar{m}\times\tilde{m}$ means sampling independently from $\bar{m}$ and $\tilde{m}$, we have the following alternative construction of M: First construct the i.i.d. sequence $(\bar{X}_i \in \bar{E})_{i\in\mathbb{N}_{++}}$ by way of sampling from the probability law $\bar{m}$. Then construct – say, by using the procedure described earlier but independently from the sequence $(\bar{X}_i)_{i\in\mathbb{N}_{++}}$ – a Poisson counting measure, $\tilde{M}$, on $(\tilde{E},\tilde{\mathscr{E}})$ of (σ-finite and nonatomic) intensity $\tilde{m}$, and let $(\tilde{X}_i \in \tilde{E})_{i\in\mathscr{I}}$ be the collection of all the atoms that the realization of $\tilde{M}$ consists of. Since $\mathscr{I} \subseteq \mathbb{N}_{++}$, we can define the random measure M on $E = \bar{E}\times\tilde{E}$ to be the counting measure with atoms $((\bar{X}_i,\tilde{X}_i))_{i\in\mathscr{I}}$. So defined M is nothing but a Poisson counting measure of intensity $m = \bar{m}\times\tilde{m}$ (see (Çinlar 2011, ch. 6 3.2) for a completely rigorous explanation). ○

Now we can apply the construction developed in (16.13) to the situation that is of primary interest for us: we take $E = (\mathbb{R}\setminus\{0\})\times\mathbb{R}_+$ (endowed with the obvious Borel structure) and consider the measure $m = \mu\times(c\Lambda)$, where μ is some, that is, *any*, perhaps even atomic, probability measure on $\mathbb{R}\setminus\{0\}$. The reason why this situation is of interest to us should now be clear: if M is a Poisson counting measure on $E = (\mathbb{R}\setminus\{0\})\times\mathbb{R}_+$ of intensity $m = \mu\times(c\Lambda)$, then the projection of M into $\mathbb{R}_+$ is a Poisson counting measure of intensity $c\Lambda$. As we already know, such a random measure can be interpreted⊸ (16.10) (16.12) as a Poisson process of parameter c. The first component of M, which lives in $\mathbb{R}\setminus\{0\}$, describes the actual jumps, which are sampled from the distribution law μ independently from anything else. To be precise, if we define for every $t \in \mathbb{R}_+$

$$N_t(\omega) \stackrel{\text{def}}{=} M_\omega\big((\mathbb{R}\setminus\{0\})\times[0,t]\big)\,, \qquad \mathrm{e}_2$$

then $N_t(\omega)$ can be identified as the number of atoms that the measure M_ω has "thrown" into the set $(\mathbb{R}\setminus\{0\})\times[0,t]$. Next, observe that those atoms are in one-to-one correspondence with their projections into $[0,t]$, and, furthermore, as defined above, N is just the usual Poisson process of intensity c. We will express this process equivalently⊸ (1.60) as $M\big((\mathbb{R}\setminus\{0\})\times[0,\iota]\big)$. It has a jump at time t, that is, $\Delta N_t \neq 0$, if and only if the slice $(\mathbb{R}\setminus\{0\})\times\{t\}$ contains an atom from M – and, as we already know, there could be at most one such atom on any slice of this form. We thus arrive at the following construction.

(16.14) Poisson point process: Let everything be as above. If $\Delta N_t(\omega) = 0$, we set $e_t(\omega) = 0$; if $\Delta N_t(\omega) \neq 0$, then we define $e_t(\omega)$ to be the projection into $\mathbb{R}\setminus\{0\}$ of the only atom of M_ω that belongs to $(\mathbb{R}\setminus\{0\})\times\{t\}$. To put it another way, the atoms of M can be described as the discrete set

$$\big\{(e_t(\omega),t)\colon t\in\mathbb{R}_+\big\} \cap \big((\mathbb{R}\setminus\{0\})\times\mathbb{R}_+\big)\,.$$

We stress that, in general, there is no reason – and, in fact, it is not desirable from a practical point of view – to insist that the atoms of M must be in one-to-one correspondence with their projections into $\mathbb{R}\backslash\{0\}$. For, to say the least, unless we allow for identical jumps of fixed size, the Poisson process would be excluded from the present setting.

The process $e \equiv (e_t(\omega))_{t\in\mathbb{R}_+}$ so obtained is called the *Poisson point process* of intensity $\mu\times c\Lambda$. It is clear that for any finite $T \in \mathbb{R}_{++}$ the set

$$\{t \in [0,T] : e_t(\omega) \neq 0\}$$

is a.s. finite, since a Poisson process can only have finitely many jumps on any finite time interval. ○

The point process e introduced above admits also the following interpretation: Just as with any Poisson process, the moments of jumps for N, as defined in e_2, can be arranged in a monotone sequence, say, $t_1(\omega) < t_2(\omega) < \dots$. Then, by the definition of e,

$$\{t_1(\omega), t_2(\omega), \dots\} = \{t \in \mathbb{R}_+ : e_t(\omega) \neq 0\}.$$

In particular, the atoms of M_ω can be arranged into the sequence

$$(e_{t_1(\omega)}(\omega), t_1(\omega)),\ (e_{t_2(\omega)}(\omega), t_2(\omega)),\ \dots\ ,$$

and clearly,⊸(16.13) by the construction of M, the r.v. $\xi_i(\omega) \stackrel{\text{def}}{=} e_{t_i(\omega)}(\omega)$, $i \in \mathbb{N}_{++}$, form an i.i.d. sequence sampled from the law μ. Finally, we are in a position to construct a Poisson process with randomized jumps.

(16.15) COMPOUND POISSON PROCESS: Let everything be as above, and, in particular, let $(N_t)_{t\in\mathbb{R}_+}$ be the process given by e_2. Then the process

$$Y_t(\omega) \stackrel{\text{def}}{=} \sum_{i=1}^{N_t(\omega)} e_{t_i(\omega)}(\omega) \equiv \sum_{s\leq t} e_s(\omega),\quad t \in \mathbb{R}_+,$$

with the understanding that the summation yields 0 if $N_t(\omega) = 0$, is called the *compound Poisson process of intensity* $\mu\times c\Lambda$. The jumps of this process occur at the times of jumps for the Poisson process N (of intensity c), but each jump is random and is sampled (independently from anything else) from the set $\mathbb{R}\backslash\{0\}$ according to the probability law μ. The compound Poisson process can be expressed as an integral with respect to the associated counting measure, namely, in the form

$$Y = \int_{\mathbb{R}\backslash\{0\}} x\, M(\mathrm{d}x, [0, \iota]),$$

or, if one insists on a more detailed notation,

$$Y(\omega, t) \equiv Y_t(\omega) = \int_{\mathbb{R}\backslash\{0\}} xM(\mathrm{d}x, [0,t]) = \int_0^t \int_{\mathbb{R}\backslash\{0\}} xM_\omega(\mathrm{d}x, \mathrm{d}s),\quad t \in \mathbb{R}_+.$$

Instead of $\int_{\mathbb{R}\backslash\{0\}}$ we will often write simply $\int_{\mathbb{R}}$, with the understanding that the measure $M_\omega(\mathrm{d}x, \mathrm{d}s)$ has no mass on the line $x = 0$. In addition, just as we usually do, in most situations we are going to suppress the dependence on ω in the notation and often write $M(\mathrm{d}x, \mathrm{d}t)$ instead of $M_\omega(\mathrm{d}x, \mathrm{d}t)$. ○

(16.16) EXERCISE: Explain why the standard Poisson process can be treated as a special case of a compound Poisson process. What is the probability measure μ in this case? ○

It is clear from the foregoing that the compound Poisson process is just another description of a Poisson counting measure, M, defined on $(\mathbb{R}\setminus\{0\})\times\mathbb{R}_+$ with mean measure m that can be disintegrated into the product $m = \mu\times c\Lambda$ for some probability measure μ. However, it is not difficult to see how one can replace μ with *any finite measure*, ν, and still construct a random measure M of intensity $\nu\times c\Lambda$ by using the same recipe for sampling. Indeed, the measure $\nu\times c\Lambda$ is no different from the measure $\bar{\mu}\times\bar{c}\Lambda$ where $\bar{\mu}$ is the probability measure

$$\bar{\mu} \stackrel{\text{def}}{=} \frac{1}{\nu(\mathbb{R}\setminus\{0\})}\nu$$

and $\bar{c} = \nu(\mathbb{R}\setminus\{0\})c$. We stress that the normalization of the measure in the domain of the jumps is compensated by changing the intensity of the jumps accordingly.

Finally, we relax the requirement $\nu(\mathbb{R}\setminus\{0\}) < \infty$, but continue to insist that ν must be σ-finite. In the later case M can be written as the sum of countably many *independent* Poisson counting measures with non-overlapping support sets, every one of which is constructed with the recipe that we just outlined.

We conclude this section with two classical illustrations, which we borrow from (Çinlar 2011).

(16.17) REVENUE FROM SALES: The number of customers arriving at a particular business follows a Poisson process of intensity c. The monetary amount that each customer spends is random and is sampled (independently from anything else) from the range $\mathbb{R}_+ = [0, \infty[$ according to some distribution law μ, which is shared by all customers. There is a probability $\mu(\{0\}) \geq 0$ that a customer would balk, that is, decide to leave without spending. Let M denote the Poisson counting measure on $\mathbb{R}_{++}\times\mathbb{R}_+$ of intensity $m = \nu\times c\Lambda$, where ν is the restriction of μ to the set $\mathbb{R}_{++}$ (that is, ν may not be a probability measure if $\mu(\{0\}) > 0$). Then the revenue from sales received during the time period $[0, t]$ is given by the integral

$$Y_t = \int_0^t \int_{\mathbb{R}_{++}} xM(\mathrm{d}x, \mathrm{d}s) = \int_{\mathbb{R}_{++}} xM(\mathrm{d}x, [0, t])\,. \quad ○$$

(16.18) BANK DEPOSITS: (Çinlar 2011, p. 266) Let E denote⊶(0.9) the space of all càdlàg functions $\mathbb{D}(\mathbb{R}_+, \mathbb{R})$, and let $\mathscr{E}$ denote the Borel σ-field on that space (for the Skorokhod topology). The total number of bank accounts follows a Poisson process of intensity $c \in \mathbb{R}_{++}$, that is, the moments when new bank accounts are created are modeled as the moments of jumps for a particular Poisson process, and the intensity of those jumps is controlled by the parameter $c \in \mathbb{R}_{++}$. If a bank account is opened at time $s \geq 0$, then the time evolution of the cash balance in that account will follow the path $[s, \infty[\ \ni t \rightsquigarrow z(t-s)$, where the function z is sampled at time s (independently from anything else) from the space E according to a given probability law μ. As a result, the aggregate amount of deposits in the bank will follow the process

$$Y_t = \int_0^t \int_E z(t-s) M(\mathrm{d}z, \mathrm{d}s),$$

where M is a Poisson counting measure on the space $\mathbb{D}(\mathbb{R}_+, \mathbb{R}) \times \mathbb{R}_+$ of intensity $m = \mu \times c\Lambda$. ○

16.2 Lévy Processes

It is desirable – from both a theoretical and a practical point of view – to have at our disposal a well developed modeling and computational tool that captures the idea of stationary and independent changes. Indeed, the simplest assumption that one can make about a particular random motion is that the probabilistic nature of the observed short-term movements does not vary over time, and changes observed over non-overlapping time intervals are statistically independent. Earlier$^{\multimap\,(9.12)}$ in this space we introduced the important class of Lévy processes, namely, the class of càdlàg processes with stationary and independent increments that start from 0. Two prominent members of this class, the Brownian motion and the Poisson process, are already familiar to us. What follows in this section is a brief synopsis of the general theory and related calculus for Lévy processes. We would like to treat the processes from this important class as a special type of semimartingales, and we stress that it is essential not to insist on continuous sample paths, since, as we are about to realize, a Lévy process with continuous sample paths reduces to a Brownian motion with linear drift – a process that can be studied with the tools developed in earlier chapters.

The structure of a Lévy process is intrinsically related to the attribute "infinitely divisible" attached to certain probability distribution laws, and also to the fact that$^{\multimap\,(3.12)}$ any probability distribution in $\mathbb{R}^n$ is completely determined by its characteristic function. For this reason, we now turn to several key features of characteristic functions that will be instrumental in the rest of this chapter.

(16.19) Exercise: Prove that if $H : \mathbb{R}^n \mapsto \mathbb{C}$ can be identified as$^{\multimap\,(3.12)}$ the characteristic function of some probability distribution, μ, over $\mathbb{R}^n$, then H must be continuous, have the property $H(0) = 1$, and be positive definite, in that

$$\sum\nolimits_{i,j=1}^{k} H(u_i - u_j) z_i \overline{z_j} \geq 0$$

for every choice of the integer $k \in \mathbb{N}_{++}$, the vectors $u_i \in \mathbb{R}^n$, and the complex numbers $z_i \in \mathbb{C}$, $1 \leq i \leq k$.

HINT: The continuity is a consequence of the dominated convergence theorem. Assuming that the r.v. ξ has law μ and setting $\zeta = \sum_{i=1}^{n} z_i e^{i\, u_i \cdot \xi}$, compute $\mathsf{E}[|\zeta|^2] \equiv \mathsf{E}[\zeta\bar{\zeta}]$. ○

(16.20) BOCHNER–KHINCHIN THEOREM:[180] (Shiryaev 2004, II sec. 12) For any given function $H\colon \mathbb{R}^n \mapsto \mathbb{C}$, the attribute "continuous, positive definite, and such that $H(0) = 1$" is equivalent to the attribute "H can be identified with the characteristic function of some distribution law over $\mathbb{R}^n$." ○

(16.21) EXERCISE: Let $\xi = (\xi_1, \dots, \xi_n)$ be any $\mathbb{R}^n$-valued r.v. Then its components ξ_i, $1 \le i \le n$, are jointly independent if and only if

$$H_\xi(u) \overset{\text{def}}{=} \mathsf{E}\left[e^{i\,(u_1\xi_1+\ldots+u_n\xi_n)}\right] = H_{\xi_1}(u_1)\times \cdots \times H_{\xi_n}(u_n)\,, \quad u = (u_1, \dots, u_n) \in \mathbb{R}^n\,.$$

HINT: The "only if" part is straightforward. The "if" part follows from the Bochner–Khinchin theorem. ○

In particular, we see that if ξ_i, $1 \le i \le n$, are i.i.d. $\mathbb{R}$-valued random variables with a common characteristic function $\mathbb{R} \ni u \rightsquigarrow H(u) \in \mathbb{C}$, then the characteristic function of the sum $\zeta = \xi_1 + \ldots + \xi_n$ must be

$$\mathsf{E}\left[e^{i\,u\,\zeta}\right] \equiv \mathsf{E}\left[e^{i\,u\,(\xi_1+\ldots+\xi_n)}\right] = H(u)^n\,, \quad u \in \mathbb{R}\,.$$

Thus, if we know that the r.v. ζ can be written as the sum of n i.i.d. random variables, then the common characteristic function of those random variables obtains from the characteristic function of ζ. And vice versa: the characteristic function of the sum ζ obtains from the common characteristic function of the i.i.d. components in the sum. It is just as important to recognize that $\mathbb{R} \ni u \rightsquigarrow H(u)$ determines not only the characteristic function, and therefore the distribution of the sum, but also the (joint) characteristic function, and therefore the joint distribution, of the entire vector $(\xi_1, \dots, \xi_n)$.

(16.22) INFINITELY DIVISIBLE DISTRIBUTION LAWS: A probability distribution law, μ, over $\mathbb{R}^n$ is said to be infinitely divisible if it has the following property: for every $k \in \mathbb{N}_{++}$ there is a characteristic function G, that is, a positive definite and continuous function on $G\colon \mathbb{R}^n \mapsto \mathbb{C}$ with $G(0) = 1$, such that the characteristic function of μ can be expressed as $\mathbb{R}^n \ni u \rightsquigarrow H_\mu(u) = G(u)^k$. Equivalently, the law μ is said to be infinitely divisible if for every choice of $k \in \mathbb{N}_{++}$ one can construct a probability space $(\Omega, \mathcal{F}, P)$ and i.i.d. $\mathbb{R}^n$-valued random variables ξ_i, $1 \le i \le k$, such that the law of $\xi_1 + \ldots + \xi_k$ is precisely μ. ○

(16.23) EXERCISE: Prove that if X is a Lévy process, then for any $\varepsilon > 0$ the distribution law of X_ε is infinitely divisible and completely determines all finite-dimensional distributions of the process X, that is, completely determines the law of $(X_{t_0}, X_{t_1}, \dots X_{t_k})$ for every choice of $0 \le t_0 < t_1 < \ldots < t_k$.

[180] Named after the American mathematician Salomon Bochner (1899–1982) and the Russian mathematician Aleksandr Yakovlevich Khinchin (1894–1959).

HINT: Use the identity

$$X_\varepsilon = X_{\varepsilon/k} + \big(X_{2\varepsilon/k} - X_{\varepsilon/k}\big) + \big(X_{3\varepsilon/k} - X_{2\varepsilon/k}\big) + \ldots + \big(X_{k\varepsilon/k} - X_{(k-1)\varepsilon/k}\big)\,, \quad k \geq 1\,,$$

and establish first that Law(X_ε) completely determines Law($X_{t_0}, X_{t_1}, \ldots X_{t_k}$) for t_i of the form $t_i = \frac{p}{q}\varepsilon$, where $p, q \in \mathbb{N}$, $q \geq 1$. Then use the fact that X is right-continuous, in conjunction with the fact that convergence a.s. implies convergence in distribution. ○

Remarkably, we see from the last exercise that the distribution law of the random variable X_1 (or, for that matter, of X_t for any fixed $t > 0$) completely determines the distribution law of the entire process X – provided, of course, that X is of Lévy type. It turns out that the attribute "infinitely divisible" for Law(X_1) imposes a certain structure of its characteristic function, and, as a result, imposes a certain structure on the entire Lévy process X. The meaning of the phrase "imposes a certain structure" is clarified in the next two theorems, which are instrumental for both theory and practice.

(16.24) LÉVY–KHINCHIN FORMULA: (Jeanblanc et al. 2009, pp. 593, pp. 606) Let μ be any infinitely divisible distribution law over $\mathbb{R}^n$. Then the characteristic function of μ can be written in the form $H_\mu(u) = e^{h_\mu(u)}$, $u \in \mathbb{R}^n$, where the function $h_\mu \stackrel{\text{def}}{=} \log(H_\mu)$ is given by

$$h_\mu(u) = i\, u.b + \frac{1}{2}|\sigma u|^2 + \int_{\mathbb{R}^n} \left(e^{i\, u.x} - 1 - i\, u.x\, 1_{\{|x|<1\}}\right) \nu(dx)\,, \quad u \in \mathbb{R}^n\,,$$

for some choice of the (constant) vector $b \in \mathbb{R}^n$, (constant) matrix $\sigma \in \mathbb{R}^{n\otimes n}$, and Borel measure ν over $\mathbb{R}^n \backslash \{0\}$ such that

$$\dagger \quad \int_{\mathbb{R}^n\backslash\{0\}} \min\left\{1, |x|^2\right\} \nu(dx) \equiv \int_{0<|x|\leq 1} |x|^2\, \nu(dx) + \nu\{x\colon |x| > 1\} < \infty\,.$$

A Borel measure ν that has this property is called Lévy measure and the triplet (b, σ, ν) is called Lévy–Khinchin triplet (alias: characterization triplet). For simplicity, we will treat all Lévy measures as measures over the entire $\mathbb{R}^n$ that have no mass at $\{0\}$. Clearly, any finite measure over $\mathbb{R}^n\backslash\{0\}$ is automatically a Lévy measure, but it is actually important from a practical point of view not to insist on finite Lévy measures. It is easy to see that any Lévy measure must be finite on the set $\{x\colon |x| > \varepsilon\}$ for any $\varepsilon > 0$ and can explode only in a neighborhood of 0. At the same time, the conditions above restrict the rate at which the explosion (to $+\infty$) at the point $\{0\}$ may take place. To be precise,

$$\int_{0<|x|\leq 1} |x|^2\, \nu(dx) \leq \int_{0<|x|\leq 1} |x|\, \nu(dx) \leq \nu\{x\colon 0 < |x| \leq 1\}\,,$$

which means that any of the following three scenarios is possible:

$$\infty = \int_{0<|x|\leq 1} |x|\, \nu(dx) = \nu\{x\colon 0 < |x| \leq 1\}\,,$$

or

$$\int_{0<|x|\leq 1} |x|\,\nu(\mathrm{d}x) < \infty = \nu\{x\colon 0<|x|\leq 1\}\,,$$

or

$$\int_{0<|x|\leq 1} |x|\,\nu(\mathrm{d}x) \leq \nu\{x\colon 0<|x|\leq 1\} < \infty\,.$$

We are about to realize that the distinction between these three cases is quite important. It is just as important to recognize that if ν is only known to be a Lévy measure, then neither of the following two integrals has to exist

$$\int_{\mathbb{R}^n} \big(e^{\imath\, u\cdot x} - 1\big)\nu(\mathrm{d}x) \quad \text{and} \quad \int_{\mathbb{R}^n} \big(\imath\, u\cdot x\, 1_{\{|x|<1\}}\big)\,\nu(\mathrm{d}x)\,.$$

It is the difference

$$\int_{\mathbb{R}^n} \big(e^{\imath\, u\cdot x} - 1 - \imath\, u\cdot x\, 1_{\{|x|<1\}}\big)\,\nu(\mathrm{d}x)$$

that always exists, and it is easy to see why:

$$e^{\imath\, u\cdot x} - 1 - \imath\, u\cdot x = -\frac{1}{2}(u\cdot x)^2 - \frac{\imath}{6}(u\cdot x)^3 + \dots \quad . \quad \bigcirc$$

Since the distribution of any Lévy process X is completely determined by the random variable X_1, and since X_1 has an infinitely divisible law, which is completely determined by its Lévy–Khinchin triplet (b,σ,ν), that same triplet must also determine the structure of X, that is, we might as well refer to X as the (b,σ,ν)-Lévy process. The next result clarifies this point.

(16.25) LÉVY–ITÔ DECOMPOSITION: (Jeanblanc et al. 2009, p. 609) Let X be any Lévy process and let (b,σ,ν) be the Lévy–Khinchin triplet associated with the random variable X_1. Let W be the $\mathbb{R}^n$-valued Brownian motion, let N be the Poisson counting measure on $(\mathbb{R}^n\backslash\{0\})\times\mathbb{R}_+$, chosen to be independent of W, with intensity measure $\nu\times\Lambda$, and let $\tilde{N}$ stand for the compensated Poisson measure $\tilde{N}(\mathrm{d}x,\mathrm{d}s) = N(\mathrm{d}x,\mathrm{d}s) - \nu(\mathrm{d}x)\mathrm{d}s$. Then the distribution law of X can be identified with the distribution law of the process

$$b\imath + \sigma W + \int_{|x|\leq 1} x\,\tilde{N}(\mathrm{d}x,[0,\imath]) + \int_{|x|>1} x\,N(\mathrm{d}x,[0,\imath])\,,$$

or, if one insists on a more detailed notation, the process

$$bt + \sigma W_t + \int_{|x|\leq 1} x\,\tilde{N}(\mathrm{d}x,[0,t]) + \int_{|x|>1} x\,N(\mathrm{d}x,[0,t])\,, \quad t\in\mathbb{R}_+\,.$$

The same process can be expressed equivalently as (see the comments following exercise (16.26))

$$b\imath + \sigma W + \int\int_{|x|\leq 1} x\,\tilde{N}(\mathrm{d}x,\mathrm{d}s) + \int\int_{|x|>1} x\,N(\mathrm{d}x,\mathrm{d}s)\,,$$

or, if one again insists on a more detailed notation, as

$$bt + \sigma W_t + \int_0^t \int_{|x| \le 1} x\, \tilde{N}(\mathrm{d}x, \mathrm{d}s) + \int_0^t \int_{|x| > 1} x\, N(\mathrm{d}x, \mathrm{d}s), \qquad t \in \mathbb{R}_+ .$$

This representation of the Lévy process X – and we stress that it is a representation only of the distribution law of X – is called the *Lévy–Itô decomposition*, and it is important to identify the meaning and the effects of its terms: bt corresponds to a linear drift, σW is a continuous local martingale (in fact, a linear transformation of the n-dimensional Brownian motion), $\int_{|x| \le 1} x\, \tilde{N}(\mathrm{d}x, [0, t])$ is a purely discontinuous local martingale with jumps of size ≤ 1, and the process $\int_{|x| > 1} x\, N(\mathrm{d}x, [0, t])$ is a compound Poisson process with jumps (finitely many on any finite time interval) of size >1. In particular, X is a semimartingale. The (unique) continuous local martingale component of X, which we denote by X^c, is given by the second term σW, the (non-unique) finite variation component is given by the first and last terms, and the (non-unique) purely discontinuous local martingale component is given by the third term. ○

(16.26) EXERCISE: Let X be any Lévy process with Lévy–Khinchin (characterization) triplet (b, σ, ν). Write down the characteristic function of the r.v. X_t for a generic $t \in \mathbb{R}_{++}$.

HINT: What is the characteristic function of the random variable X_1? ○

A few comments about the Lévy–Itô decomposition are now due. Despite the identity $\tilde{N}(\mathrm{d}x, \mathrm{d}s) = N(\mathrm{d}x, \mathrm{d}s) - \nu(\mathrm{d}x)\mathrm{d}s$ (the definition of $\tilde{N}$), in general, one cannot write

$$\int_{|x| \le 1} x\, \tilde{N}(\mathrm{d}x, [0, t]) = \int_{|x| \le 1} x\, N(\mathrm{d}x, [0, t]) - \Big(\int_{|x| \le 1} x\, \nu(\mathrm{d}x) \Big) t , \qquad \mathrm{e}_1$$

since neither of the two integrals on the right side has to exist, whereas the integral in the left side always does. To see this, first observe that for fixed $\mathrm{d}x$ the process $N(\mathrm{d}x, [0, t])$ – understood as $\big(N(\mathrm{d}x, [0, t])\big)_{t \in \mathbb{R}_+}$ – can be treated, formally, as a Poisson process of (possibly, infinitesimally small) intensity $c = \nu(\mathrm{d}x)$. Consequently, the process $\tilde{N}(\mathrm{d}x, [0, t]) = N(\mathrm{d}x, [0, t]) - \nu(\mathrm{d}x)t$ is a compensated Poisson process, which, as we know, is an L^2-bounded martingale on any finite time interval $[0, T]$. Furthermore, its predictable quadratic variation is $\langle \tilde{N}(\mathrm{d}x, [0, t]), \tilde{N}(\mathrm{d}x, [0, t]) \rangle = \nu(\mathrm{d}x)t$, so that the $\mathbb{H}^2$-norm of this martingale, restricted to the interval $[0, T]$, is simply $\nu(\mathrm{d}x)T$. Since for different (non-overlapping) choices of $\mathrm{d}x$ the associated martingales $\tilde{N}(\mathrm{d}x, [0, t])$ are independent, they are orthogonal elements of the space $\mathbb{H}^2$, and the expression in e_1 can be treated as the sum of infinitely many (possibly, infinitesimally small) L^2-bounded martingales from the space $\mathbb{H}^2$ that are orthogonal to one another. As the squared norm of this (formal) sum in the space $\mathbb{H}^2$ is given by the integral

$$T \int_{|x| \le 1} x^2\, \nu(\mathrm{d}x),$$

which, due to the fact that ν is a Lévy measure, must be finite, the "sum" of the orthogonal martingales in e_1, when restricted to any finite interval $[0,T]$, is a well-defined element of $\mathbb{H}^2$.

To put it another way, the process in the left side of e_1 can be treated as "an integral" of the function $f(x)=x$ over the ball $\{|x|\leq 1\}$ with respect to a measure that takes values in the space of martingales; specifically the measure

$$\mathcal{B}(\mathbb{R}^n)\ni B \rightsquigarrow \tilde{N}(B,[0,\iota])\,.$$

This martingale-valued measure is furthermore orthogonal, in the sense that if $A,B\in\mathcal{B}(\mathbb{R}^n)$ and $A\cap B=\emptyset$, then $\langle N(A,[0,\iota]),N(B,[0,\iota])\rangle=0$. More generally, the integral $\int_{|x|\leq 1} f(x)\,N(\mathrm{d}x,[0,\iota])$ is well-defined (as a square integrable martingale) for any Borel measurable function $f(\cdot)$ such that the function

$$\Big\langle\int_{|x|\leq 1} f(x)\,N(\mathrm{d}x,[0,\iota]),\int_{|x|\leq 1} f(x)\,N(\mathrm{d}x,[0,\iota])\Big\rangle=\Big(\int_{|x|\leq 1} f(x)^2\,\nu(\mathrm{d}x)\Big)\iota$$

is finite, that is, $\int_{|x|\leq 1} f(x)^2\,\nu(\mathrm{d}x)<\infty$.

If the Lévy measure ν explodes near 0 even more slowly than it is required to, in the sense that

$$\int_{0<|x|\leq 1}|x|\,\nu(\mathrm{d}x)<\infty\,, \qquad \mathrm{e}_2$$

then the associated Lévy–Khinchin exponent can be rearranged as

$$h(u)=i\,u\cdot b^*+\frac{1}{2}|\sigma u|^2+\int_{\mathbb{R}^n}\left(e^{i\,u\cdot x}-1\right)\nu(\mathrm{d}x)\,,\quad u\in\mathbb{R}^n\,,$$

where

$$b^*=b-\int_{0<|x|\leq 1}x\,\nu(\mathrm{d}x)\,.$$

In this case the integral

$$\int_{|x|\leq 1}x\,N(\mathrm{d}x,[0,\iota])=\int_{|x|\leq 1}x\,\tilde{N}(\mathrm{d}x,[0,\iota])+\Big(\int_{|x|\leq 1}x\,\nu(\mathrm{d}x)\Big)\iota$$

is also well defined, and the Lévy–Itô decomposition reduces to

$$X_\iota \overset{\text{law}}{=} b^*\iota+\sigma W+\int_{\mathbb{R}^n\setminus\{0\}}x\,N(\mathrm{d}x,[0,\iota])\,. \qquad \mathrm{e}_3$$

(16.27) REMARK: Condition e_2 guarantees that the jumps of the process on the left side of e_1 are summable, that is, it guarantees that $\int_{|x|\leq 1}|x|\,N(\mathrm{d}x,[0,\iota])$ is a finite process.[181] Indeed, with condition e_2 in force,

$$\mathsf{E}\Big[\int_{|x|\leq 1}|x|\,N(\mathrm{d}x,[0,t])\Big]=\int_0^t\int_{|x|\leq 1}|x|\,\nu(\mathrm{d}x)\mathrm{d}s=t\int_{|x|\leq 1}|x|\,\nu(\mathrm{d}x)<\infty$$

[181] As there are countably many jumps on any finite time interval, for any given $t\in\mathbb{R}_{++}$ the integral $\int_0^t\int_{|x|\leq 1}x\,N(\mathrm{d}x,\mathrm{d}s)=\int_{|x|\leq 1}x\,N(\mathrm{d}x,[0,t])$ represents an infinite series, the sum of which has no reason to exist in general.

for every $t \in \mathbb{R}_{++}$ We stress that although in this case the integral on the right side of e_3 represents a well-defined finite process that may look like a compound Poisson process, in general, this integral may not be a compound Poisson process since it may still have (countably) infinitely many jumps on any finite time interval, even though those jumps are summable. The reason is that even if the Lévy measure ν satisfies the condition e_2, it may still be only a σ-finite, but not a finite, measure. Nevertheless, we can write this integral as the sum of countably many independent compound Poisson processes, namely, as

$$\int_{\mathbb{R}^n\setminus\{0\}} x\,N(\mathrm{d}x,[0,t]) = \int_{|x|>1} x\,N(\mathrm{d}x,[0,t]) + \sum_{k\geq 1}\int_{\frac{1}{k+1}<|x|\leq\frac{1}{k}} x\,N(\mathrm{d}x,[0,t]),$$

and, as we already know, with probability 1 at most one of these processes can jump at any one time.

Finally, we note that if condition e_2 is replaced with the stronger requirement $\int_{|x|\leq 1}\nu(\mathrm{d}x) < \infty$, which is the same as imposing that the Lévy measure ν is finite, then the integral on the right side of e_3 would indeed represent a standard compound Poisson process, which can only have finitely many jumps on any finite interval. ○

(16.28) EXERCISE: Explain why the integral on the right side of e_3 represents a càdlàg process. How can you reconcile this property with the presence of infinitely many jumps?

HINT: The purpose of this exercise is to develop some understanding of how a càdlàg sample path may have countably infinitely many jumps on a finite interval. Give an example of a nondecreasing càdlàg function with countably infinitely many jumps of size 2^{-k}, $k \geq 1$. ○

16.3 Stochastic Integrals with Respect to Lévy Processes

Let ν be some σ-finite measure on $\mathbb{R}$ such that $\nu(\{0\}) = 0$, let N be the Poisson random measure on $\mathbb{R}\times\mathbb{R}_+$ of intensity (mean measure) $\nu\times\Lambda$, defined on $(\Omega,\mathcal{F},P)$, and let $\bar{N}$ be the random measure on $\mathbb{R}\times\mathbb{R}_+$ given by

$$\bar{N}(\mathrm{d}x,\mathrm{d}t) = \begin{cases} \tilde{N}(\mathrm{d}x,\mathrm{d}t) \equiv N(\mathrm{d}x,\mathrm{d}t) - \nu(\mathrm{d}x)\mathrm{d}t & \text{for } x\in[-1,1], \\ N(\mathrm{d}x,\mathrm{d}t) & \text{for } x\in\mathbb{R}\setminus[-1,1]. \end{cases}$$

These structures will be fixed throughout this section and our next goal is to give meaning to stochastic integrals of the form

$$\int\int_{\mathbb{R}} \gamma(x,t)\,\bar{N}(\mathrm{d}x,\mathrm{d}t) \equiv \int\int_{\mathbb{R}} \gamma(\omega,x,t)\,\bar{N}_\omega(\mathrm{d}x,\mathrm{d}t),$$

which we will write more succinctly as

$$\int_{\mathbb{R}} \gamma(x,t)\boldsymbol{\cdot}\bar{N}(\mathrm{d}x,[0,t]),$$

or, if more detailed notation is required by the context, as

$$\int_{\mathbb{R}} \gamma(x, \imath)\boldsymbol{\cdot}\bar{N}(\mathrm{d}x, [0, \imath])_t = \int_0^t \int_{\mathbb{R}} \gamma(x, s)\,\bar{N}(\mathrm{d}x, \mathrm{d}s)\,, \quad t \in \mathbb{R}_+ .$$

The description that now follows will be mostly heuristic and will omit most of the technical steps that a rigorous construction of the integral may involve. As we are well aware of by now, generally, the available integrands for stochastic integrals with respect to semimartingales are predictable functions defined on $\Omega\times\mathbb{R}_+$. However, in many practical applications it is desirable to allow the integrands to also be functions of the jumps, that is, functions defined on $(\Omega\times\mathbb{R})\times\mathbb{R}_+$, or on $(\Omega\times\mathbb{R}^n)\times\mathbb{R}_+$. For simplicity, we will describe only the case where the jumps take values in the real line $\mathbb{R}$, and note that the case of jumps with values in $\mathbb{R}^n$ differs mostly in the notation (throughout the sequel we are going to pretend that this more general case has been introduced as well).

Our first step is to define the natural filtration associated with the random measure N as

$$\mathscr{F}_t^N \overset{\text{def}}{=} \sigma\big\{N(A, B)\colon A \in \mathcal{B}(\mathbb{R}),\ B \in \mathcal{B}([0, t])\big\}\,, \quad t \in \mathbb{R}_+ ,$$

and, just as we always do, suppose that the token $\mathscr{F}^N$ and the term "natural filtration" actually refer to the usual augmentation of the filtration just introduced. In fact, we are going to replace $\mathscr{F}^N$ with a larger filtration, denoted by $\mathscr{F}$, which also includes the events of a particular Brownian motion that is independent of N (and again suppose that $\mathscr{F}$ satisfies the usual conditions).

Next, observe that, at least formally,

$$\int_{\mathbb{R}} \gamma(x, \imath)\boldsymbol{\cdot}\bar{N}(\mathrm{d}x, \imath) = \int_{[-1,1]} \gamma(x, \imath)\boldsymbol{\cdot}\tilde{N}(\mathrm{d}x, [0, \imath]) + \int_{\mathbb{R}\setminus[-1,1]} \gamma(x, \imath)\boldsymbol{\cdot}N(\mathrm{d}x, [0, \imath])\,, \qquad (e_1)$$

and, since N_ω has at most finitely many atoms inside $(\mathbb{R}\setminus[-1, 1])\times[0, t]$ for any finite $t \in \mathbb{R}_{++}$, the second integral on the right can be understood in the obvious way as a finite sum of the form

$$\begin{aligned}\int_{\mathbb{R}\setminus[-1,1]} \gamma(x, \imath)\boldsymbol{\cdot}N(\mathrm{d}x, [0, \imath])_t &\equiv \int_0^t \int_{\mathbb{R}\setminus[-1,1]} \gamma(\omega, x, s)N_\omega(\mathrm{d}x, \mathrm{d}s) \\ &= \sum\nolimits_{i\in\mathcal{I}_t(\omega)} \gamma\big(\omega, \xi_i(\omega), \jmath_i(\omega)\big)\,N_\omega\big(\{\xi_i(\omega)\}\times\{\jmath_i(\omega)\}\big)\,,\end{aligned}$$

where $(\xi_i(\omega), \jmath_i(\omega))$, $i \in \mathcal{I}_t(\omega) \subset \mathbb{N}$, are the atoms of the measure N_ω inside the set $(\mathbb{R}\setminus[-1, 1])\times[0, t]$, and we stress that the index set $\mathcal{I}_t(\omega)$ is finite.

To understand how the first integral on the right side of e_1 could be made meaningful, observe that – formally – for every fixed $x \in [-1, 1]\setminus\{0\}$ the symbol $N(\mathrm{d}x, [0, \imath]) \equiv \big(N(\mathrm{d}x, [0, t])\big)_{t\in\mathbb{R}_+}$ can be treated as a Poisson process of intensity $\nu(dx)$, and $\tilde{N}(dx, [0, \iota]) = N(dx, [0, \iota]) - \nu(dx)\iota$ is nothing but its compensated version. Since $\tilde{N}(dx, [0, \iota])$ is a local martingale with predictable quadratic variation

$$\langle \tilde{N}(dx, [0, \iota]), \tilde{N}(dx, [0, \iota]) \rangle = \nu(dx)\iota ,$$

the integral $\gamma(x, \iota)\bullet \tilde{N}(dx, [0, \iota])$ would be well defined (for fixed x!) as long as the process $\gamma(x, \iota)$ is predictable and is such that the process $\nu(dx)\gamma(x, \iota)^2\bullet\iota$ is a.s. finite. If $x = 0$, we have $\nu(dx) = 0$ and define $\gamma(x, \iota)\bullet \tilde{N}(dx, [0, \iota])$ to be the process 0. Since for $x, y \in [-1, 1]$, $x \neq y$, the compensated Poisson processes $\tilde{N}(dx, [0, \iota])$ and $\tilde{N}(dy, [0, \iota])$ are independent, their predictable quadratic covariation is null, and therefore so is the predictable quadratic covariation of the stochastic integrals $\gamma(x, \iota)\bullet \tilde{N}(dx, [0, \iota])$ and $\gamma(y, \iota)\bullet \tilde{N}(dy, [0, \iota])$. Thus, the first integral on the right side of e_1 can be understood to be "the sum" of "orthogonal" local L^2-martingales. This sum could be meaningful as long as its "norm," that is, predictable quadratic variation, is finite. To summarize, the integral

$$\int_{[-1,1]} \gamma(x, \iota)\bullet \tilde{N}(dx, [0, \iota]) \qquad \mathrm{e}_2$$

is a well-defined local L^2-martingale, provided that $\gamma = \big(\gamma(\omega, x, t)\big)_{t \in \mathbb{R}_+}$ is predictable as a function on $(\Omega \times \mathbb{R}) \times \mathbb{R}_+$ relative to the filtration $(\mathscr{F}_t \otimes \mathcal{B}(\mathbb{R}))_{t \in \mathbb{R}_+}$, and is such that the (increasing) sample paths of the process

$$\Big(\int_{[-1,1]} \gamma(\omega, x, \iota)^2\, \nu(dx)\Big)\bullet\iota$$

are finite on $\mathbb{R}_+$ (do not explode in finite time) for P-a.e. $\omega \in \Omega$.

(16.29) LÉVY–ITÔ PROCESSES: A common class of processes, often encountered in practical applications and referred to as Lévy–Itô processes, is the class of semimartingales X that have the form

$$\dagger \qquad X = X_0 + b\bullet\iota + \sigma\bullet W + \int_{\mathbb{R}} \gamma(x, \iota)\bullet \bar{N}(dx, [0, \iota]) ,$$

where $\bar{N}$ is the random measure introduced at the beginning of this section, and W is a Brownian motion that is independent of N and adapted to the right-continuous and complete filtration $\mathscr{F}$ that includes $\mathscr{F}^N$. The processes

$$(\omega, t) \rightsquigarrow b(\omega, t) , \quad (\omega, t) \rightsquigarrow \sigma(\omega, t) , \quad \text{and} \quad (\omega, x, t) \rightsquigarrow \gamma(\omega, x, t)$$

are required to be predictable for the filtration $(\mathscr{F}_t \otimes \mathcal{B}(\mathbb{R}))_{t \in \mathbb{R}_+}$ and to be such that with probability 1 the (increasing) sample paths of

$$|b|\bullet\iota , \quad \sigma^2\bullet\iota , \quad \text{and} \quad \Big(\int_{[-1,1]} \gamma(x, \iota)^2\, \nu(dx)\Big)\bullet\iota$$

do not explode in finite time. We note that nothing will change in the preceding if the integral $\int_{[-1,1]}$ is replaced everywhere with $\int_{[-\varepsilon,\varepsilon]}$ for some fixed $\varepsilon \in \mathbb{R}_{++}$.

Given our familiarity with multivariate stochastic integrals from previous chapters, the multivariate version of the class of Lévy–Itô processes would not be difficult to write down. ○

Since Lévy–Itô processes will be used rather extensively in what follows, it is important to write down the Itô formula for such processes explicitly. This task is offered as an exercise next.

(16.30) EXERCISE: Suppose that X is the Lévy–Itô process from (16.29†) and let $f \in \mathscr{C}_b^{2,1}(\mathbb{R}\times\mathbb{R}_+;\mathbb{R})$. Prove that P-a.s., for all $0 \le s < t < \infty$,[182]

$$
\begin{aligned}
&f(X_t,t) - f(X_s,s)\\
\dagger\quad &= \int_s^t \partial_u f(X_u,u)\mathrm{d}u + \int_s^t \partial_X f(X_u,u)\, b_u \,\mathrm{d}u + \frac{1}{2}\int_s^t \partial_X^2 f(X_u,u)\,\sigma_u^2\,\mathrm{d}u\\
&\quad + \int_{s+}^t \int_{\mathbb{R}\setminus[-1,1]} \Big(f\big(X_{u-} + \gamma(x,u),u\big) - f(X_{u-},u)\Big) N(\mathrm{d}x,\mathrm{d}u)\\
&\quad + \int_s^t \int_{[-1,1]} \Big(f\big(X_{u-} + \gamma(x,u),u\big) - f(X_{u-},u)\\
&\qquad\qquad - \partial_X f(X_{u-},u)\gamma(x,u)\Big)\nu(\mathrm{d}x)\mathrm{d}u\\
&\quad + \int_s^t \partial_X f(X_u,u)\,\sigma_u\,\mathrm{d}W_u\\
&\quad + \int_{s+}^t \int_{[-1,1]} \Big(f\big(X_{u-} + \gamma(x,u),u\big) - f(X_{u-},u)\Big) \tilde{N}(\mathrm{d}x,\mathrm{d}u)\,.
\end{aligned}
$$

HINT: Transcribe Itô's formula for the semimartingale in (16.29†) term-by-term and notice that $\langle X^c, X^c\rangle = \sigma^2\bullet\iota$. ○

(16.31) REMARK: Since $X_u \ne X_{u-}$ only for countably many $u \in \mathbb{R}_+$, it does not matter whether we wrote X_u or X_{u-} inside integrals against $\mathrm{d}u$ or $\mathrm{d}W_u$. The right side of (16.30†) is càdlàg (in t for fixed s) by construction, whereas the left side is càdlàg simply because $f(\cdot,\cdot)$ is continuous and the sample paths X_ω are P-a.s. càdlàg. Finally, if the left side is replaced with $f(X_{t-},t) - f(X_{s-},s)$, then the two integrals $\int_{s+}^t$ on the right side are to be replaced with $\int_s^{t-}$. ○

We conclude this section with a quick review of the special class of Lévy–Itô processes that are local martingales, that is, processes of the form

$$
X = X_0 + \sigma\bullet W + \int_{\mathbb{R}} \gamma(x,\iota)\bullet\tilde{N}(\mathrm{d}x,[0,\iota])\,, \quad t \in \mathbb{R}_+\,, \qquad e_3
$$

[182] The first five integrals can be understood as Lebesgue integrals, while the last two are only meaningful as stochastic integrals with respect to local martingales.

where everything is as above except that γ satisfies the stronger requirement that the sample paths of the process

$$\Big(\int_{\mathbb{R}} \gamma(x,\imath)^2\,\nu(\mathrm{d}x)\Big)\bullet\imath$$

do not explode in finite time with probability 1. In this setting we have the following important result:

(16.32) LÉVY–ITÔ ISOMETRY: Suppose that for some $T \in \mathbb{R}_{++}$

$$\mathsf{E}\Big[\int_0^T \sigma_s^2\,\mathrm{d}s + \int_0^T \int_{\mathbb{R}} \gamma(x,s)^2\,\nu(\mathrm{d}x)\,\mathrm{d}s\Big] < \infty\,.$$

Then the process X from e_3 is a square integrable martingale on $[0,T]$ with

$$\mathsf{E}\big[X_t^2\big] = \mathsf{E}\Big[\int_0^t \sigma_s^2\,\mathrm{d}s + \int_0^t \int_{\mathbb{R}} \gamma_s(x)^2\,\nu(\mathrm{d}x)\,\mathrm{d}s\Big] \quad \text{for every } t \in [0,T]\,. \quad \circ$$

(16.33) EXAMPLE: Let $(\xi_i)_{i\in\mathbb{N}_{++}}$ be any sequence of i.i.d. real-valued random variables with a common distribution law μ such that $\mu(\{0\}) = 0$, and let

$$\gamma^2 \stackrel{\text{def}}{=} \int_{\mathbb{R}} x^2\mu(\mathrm{d}x) \equiv \mathsf{E}\big[\xi_i^2\big] < \infty\,.$$

Let $a = \mathsf{E}[\xi_i]$, let Π be the Poisson process of intensity $c \in \mathbb{R}_{++}$, let W be a Brownian motion independent of Π, and let $\sigma \in \mathbb{R}_{++}$ be a given parameter. Then the process

$$X_t \stackrel{\text{def}}{=} \sigma W_t + \sum\nolimits_{i=1}^{\Pi_t} \xi_i - cat\,, \quad t \in \mathbb{R}_+\,,$$

(the summation is understood to yield 0 if $\Pi_t = 0$) is a square integrable martingale, which, in fact, can be expressed as

$$X \stackrel{\text{law}}{=} \sigma W + \int_{\mathbb{R}} x\,\tilde{N}(\mathrm{d}x,[0,\imath])\,,$$

where N is the Poisson counting measure of intensity $\mathsf{E}[N(\mathrm{d}x,\mathrm{d}s)] = c\mu(\mathrm{d}x)\mathrm{d}s$, which is independent of W, and, as usual, $\tilde{N}(\mathrm{d}x,\mathrm{d}s) = N(\mathrm{d}x,\mathrm{d}s) - c\mu(\mathrm{d}x)\mathrm{d}s$. With this choice we have

$$\int_{\mathbb{R}} x\,N(\mathrm{d}x,[0,t]) \stackrel{\text{law}}{=} \sum\nolimits_{i=1}^{\Pi_t} \xi_i \quad \text{and} \quad \int^t \int_{\mathbb{R}} xc\,\mu(\mathrm{d}x)\mathrm{d}s = cat\,,$$

and it is not difficult to check that

$$\mathsf{E}\big[X_t^2\big] = \sigma^2 t + c\gamma^2 t\,, \quad t \in \mathbb{R}_+\,. \quad \circ$$

16.4 Stochastic Exponents

Even a cursory glance through the literature on asset pricing would tell us that the quintessential model of a "price process" is the geometric Brownian motion, which we have already encountered in a number of occasions. This process can be introduced through the relations

$$\mathrm{d}S = S\,\mathrm{d}Z \quad \text{or, equivalently,} \quad S = S_0 + S \cdot Z\,, \qquad e_1$$

in which the process $Z \stackrel{\text{def}}{=} \sigma W + b\,\iota$ models the instantaneous ex-dividend net returns from a particular risky asset. The overwhelmingly wide acceptance of this model may be attributed in part to its simplicity and tractability, but its adequacy from a practical point of view has been questioned and debated for as long as it has been in existence. On the one hand, it is reasonable to insist on stationary and independent instantaneous returns, and, on the other hand, it is just as reasonable to insist on distributions of the returns that are somewhat closer to the ones found empirically in the observed market data, and, typically, those are distributions with tails that are much heavier than the tails of the log-normal law. From what we have learned so far⊶ (16.25) about processes with jumps, these two objectives are simply not compatible if one insists that the accumulated returns process Z is to remain continuous (in addition to having stationary and independent increments). Our next goal, then, is to develop some understanding of what the solution to the equation above would look like if Z is some càdlàg semimartingale that may not be continuous.

The first thing to notice is that, as it stands, e_1 would not be meaningful in such a setting because if the returns process Z contains jumps, the price process S would be càdlàg and adapted but not predictable in general, in which case the integral $S \cdot Z$ is not going to be meaningful. The remedy is to express the dynamics of the price process as $\mathrm{d}S_t = S_{t-}\,\mathrm{d}Z_t$, or, equivalently, as

$$S = S_0 + S_- \cdot Z\,, \qquad e_2$$

which, of course, is the same as e_1 if the price process S is continuous. To solve this equation for a generic càdlàg semimartingale Z, we proceed in much the same way in which we solved e_1 earlier: assuming that a solution exists and that S is one such solution, we first develop Itô's expansion of the process $\log(S)$. Of course, such an operation would be meaningful only if the sample paths of S remain strictly positive with probability 1.

(16.34) EXERCISE: Assuming that the process S is strictly positive and satisfies the relation e_2, verify that the formal Itô expansion of $\log(S)$ is given by

† $$\log(S) = \log(S_0) + \frac{1}{S_-} \cdot S - \frac{1}{2}\frac{1}{(S_-)^2} \cdot \langle S^c, S^c \rangle + \sum \Big(\log(S) - \log(S_-) - \frac{1}{S_-} \Delta S \Big)\,. \quad \circ$$

Our next step is to clarify under what conditions for the net returns Z the expressions on the right side of ($16.34^\dagger$) would be meaningful.

(16.35) EXERCISE: Let X and Y be any two semimartingales and let H and K be any two predictable processes chosen so that the stochastic integrals $H\bullet X$ and $K\bullet Y$ are well defined. Setting $L = H\bullet X$ and $M = K\bullet Y$, prove that

$$\langle L^c, M^c\rangle = (HK)\bullet\langle X^c, Y^c\rangle .$$

HINT: Argue that L^c and M^c can be identified as $H\bullet X^c$ and $K\bullet Y^c$. ○

(16.36) EXERCISE: Prove that if X is any semimartingale with jumps $\Delta X > -1$, then the process

$$\sum 1_{\{|\Delta X| \leq 1/2\}} \big|\log(1+\Delta X) - \Delta X\big| < \infty$$

is finite. Conclude that the processes

$$\sum \big(\log(1+\Delta X) - \Delta X\big) \quad \text{and} \quad \prod (1+\Delta X)e^{-\Delta X}$$

are both well defined and finite.

HINT: Show that

$$\big|\log(1+x) - x\big| \leq x^2 \quad \text{for any } x \in [-1/2, 1/2],$$

use the relation $[X,X] = \langle X^c, X^c\rangle + \sum(\Delta X)^2$, and recall that the quadratic variation is a finite process. ○

(16.37) EXERCISE: Using the results from the last three exercises, show that if Z is any càdlàg semimartingale with jumps $\Delta Z > -1$ and if S is some strictly positive càdlàg semimartingale that satisfies e_2, then S must be given by

$$\dagger \qquad S = S_0 e^{Z - Z_0 - \frac{1}{2}\langle Z^c, Z^c\rangle} \prod (1+\Delta Z)e^{-\Delta Z} .$$

HINT: Using the result from (16.35), conclude that

$$\langle S^c, S^c\rangle = (S_-)^2 \bullet\langle Z^c, Z^c\rangle ,$$

and using the basic properties of the stochastic integral, show that $\Delta S = S_-\Delta Z$. Observe that

$$\begin{aligned}\log(S) - \log(S_-) - \frac{1}{S_-}\Delta S &= \log(S_- + S_-\Delta Z) - \log(S_-) - \frac{1}{S_-}\Delta S \\ &= \log(1+\Delta Z) - \Delta Z .\end{aligned}$$ ○

(16.38) REMARK: As an immediate corollary from ($16.37^\dagger$) we have the relation

$$S = S_- e^{\Delta Z}(1+\Delta Z)e^{-\Delta Z} = S_-(1+\Delta Z) = S_- + S_-\Delta Z .$$

In particular, $S_t = S_{t-}$ if and only if $\Delta Z_t = 0$, that is, the points of discontinuity for the sample paths of the solution S are the same as the points of discontinuity for the returns Z. In particular, the jumps of S are connected with the jumps of Z through the relation

$$\Delta S = S - S_- = S_-\Delta Z .$$ ○

Two conclusions are now straightforward. The first one is entirely technical: if e_2 is to be used to model asset prices, then one must require that the jumps in the returns are strictly larger than (-1). The second conclusion is much more interesting: if e_2 admits a solution, then this solution must be given by the right side of (16.37†), and is therefore unique when it exists. Thus, to claim that a (unique) solution always exists as long as $\Delta Z > -1$, we must show that the expression in (16.37†), which is always meaningful for any semimartingale Z with jumps $\Delta Z > -1$, does indeed represent a solution to e_2 – see exercise (16.40).

(16.39) STOCHASTIC (DOLÉAN–DADE) EXPONENTS: It is a remarkable fact from calculus that the only function $f \in \mathscr{C}^1(\mathbb{R};\mathbb{R})$ such that $f = \partial f$ and $f(0) = 1$, or, equivalently, such that $f = 1 + f{\bullet}\iota$, is the exponential function $f = \exp$, that is, the function $x \rightsquigarrow \exp(x) \equiv e^x$. More generally, exponents can be thought of as objects that satisfy some relation of the form

$$\mathrm{d}\exp(\text{expression}) = \exp(\text{expression})\times \mathrm{d}(\text{expression}).$$

For this reason, the solution, S, to the equation $S = 1 + S_-{\bullet}Z$ is usually called the *stochastic exponent*, or the *Doléan–Dade exponent*, of the semimartingale Z. We will express this solution as $\mathscr{E}(Z)$, that is, $\mathscr{E}(Z)$ stands for the unique (see the next exercise) càdlàg semimartingale with $\mathscr{E}(Z) = 1 + \mathscr{E}(Z)_-{\bullet}Z$. The solution to $S = S_0 + S_-{\bullet}Z$ for a generic initial value $S_0 \in \mathbb{R}_{++}$ is then $S = S_0\mathscr{E}(Z)$. ○

(16.40) EXERCISE: Let Z be any càdlàg semimartingale with $\Delta Z > -1$, and let

$$\mathscr{E}(Z) \stackrel{\text{def}}{=} e^{Z-Z_0-\frac{1}{2}\langle Z^c,Z^c\rangle}\prod(1+\Delta Z)e^{-\Delta Z}. \quad \dagger$$

Prove that the process $\mathscr{E}(Z)$ so defined satisfies the equation

$$\mathscr{E}(Z) = 1 + \mathscr{E}(Z)_-{\bullet}Z.$$

Show that

$$\Delta\mathscr{E}(Z) = \mathscr{E}(Z)_-\Delta Z \quad \Leftrightarrow \quad \mathscr{E}(Z) = \mathscr{E}(Z)_-(1+\Delta Z).$$

HINT: Setting $U = e^{Z-Z_0-\frac{1}{2}\langle Z^c,Z^c\rangle}$ and $V = \prod(1+\Delta Z)e^{-\Delta Z}$, apply the integration by parts formula to the product UV. Specifically, show that V is a pure jump finite variation process with $\Delta V = V_-\big((1+\Delta Z)e^{-\Delta Z} - 1\big)$ and verify the relations $\Delta U = U_-(e^{\Delta Z} - 1)$ and $U = 1 + U_-{\bullet}Z + \sum U_-(e^{\Delta Z} - 1 - \Delta Z)$. ○

In the rest of this section we are concerned with the special case where the semimartingale Z is the Lévy–Itô process from (16.29†). Thus, we can write

$$\begin{aligned} Z = Z_0 + b{\bullet}\iota + \sigma{\bullet}W &+ \int_{\mathbb{R}\setminus[-1,1]} \gamma(x,\iota){\bullet}N(\mathrm{d}x,[0,\iota]) \\ &+ \int_{[-1,1]} \gamma(x,\iota){\bullet}\tilde{N}(\mathrm{d}x,[0,\iota]), \end{aligned} \quad \mathrm{e}_3$$

with all assumptions about the integrands and integrators introduced in (16.29). With this choice we can write $\Delta Z_t = \int_{\mathbb{R}} \gamma(x,t)N(\mathrm{d}x, \mathrm{d}t)$, with the understanding that this integral merely evaluates the function $\gamma(\omega, \cdot, \cdot)$ at a single atom, say $(\xi_t(\omega), t)$, that the measure N_ω might have placed on the slice $(\mathbb{R}\backslash\{0\})\times\{t\}$, or evaluates to 0 if no such atoms exists (recall that$^{\multimap\,(16.13)}$ with probability 1 the purely atomic measure N_ω could place at most one atom on every slice $(\mathbb{R}\backslash\{0\})\times\{t\}$, for every $t \in \mathbb{R}_{++}$). As a result,

$$\log(1 + \Delta Z_t) - \Delta Z_t = \int_{\mathbb{R}} \big(\log(1 + \gamma(x,t)) - \gamma(x,t)\big) N(\mathrm{d}x, \mathrm{d}t),$$

and this brings us to the important formula (a restatement of (16.40†))

$$\log \mathscr{E}(Z) = Z - Z_0 - \frac{1}{2}\sigma^2 \bullet \iota + \int_{\mathbb{R}} \big(\log(1 + \gamma(x,\iota)) - \gamma(x,\iota)\big) \bullet N(\mathrm{d}x, [0,\iota]).$$

Furthermore, the stochastic exponent $\mathscr{E}(Z)$ is a local martingale if Z is a local martingale of the form

$$Z = \sigma \bullet W + \int_{\mathbb{R}} \gamma(x,\iota) \bullet \tilde{N}(\mathrm{d}x, [0,\iota]),$$ e$_4$

with predictable integrands σ and γ, chosen so that the processes

$$\sigma^2 \bullet \iota \quad \text{and} \quad \Big(\int_{\mathbb{R}} \gamma(x,\iota)^2 \nu(\mathrm{d}x)\Big) \bullet \iota$$

have non-exploding sample paths with probability 1.

16.5 Change of Measure and Removal of the Drift

Now we turn to a matter that, as we have already realized, is of paramount importance in asset pricing: the change of the underlying probability measure. Just as before, the idea is to start with a process that is a semimartingale and then change the probability measure in such a way that under the new measure the same process has the properties of a local martingale. The key result on which such manipulations rest is, of course, Girsanov's theorem from (15.44). In the setting of the present chapter we are interested in equivalent changes obtained with Radon–Nikodym derivatives that can be expressed as stochastic exponents of the form $\mathscr{E}(Y)$ for some choice of the Lévy–Itô local martingale Y – and the reasons for this choice will become clear soon. The main steps in this program are already familiar to us, and can be summarized as follows. Start with a semimartingale, Y, with jumps $\Delta Y > -1$, but suppose, in addition, that Y is actually a local martingale. This makes the stochastic exponent $\mathscr{E}(Y)$ a positive local martingale, and therefore$^{\multimap\,(9.68)}$ a supermartingale that starts from $\mathscr{E}(Y)_0 = 1$. Now suppose that for some $T > 0$ one can claim that $\mathsf{E}[\mathscr{E}(Y)_T] = 1 = \mathscr{E}(Y)_0$. By the argument of exercise (9.6) this would then imply that $\mathscr{E}(Y)$ is a positive martingale on $[0,T]$ – in fact,$^{\multimap(9.39)}$ a process with strictly positive sample paths that are uniformly bounded away from 0 – which would then allow us to constrict the (equivalent to P) probability measure $Q \overset{\text{def}}{=} \mathscr{E}(Y)_T \odot P$ on the σ-field $\mathscr{F}_T$. Furthermore, for any $t \in [0,T]$, $\mathscr{E}(Y)_t \odot (P \restriction \mathscr{F}_t) = Q \restriction \mathscr{F}_t \approx P \restriction \mathscr{F}_t$. Girsanov's theorem then implies that,$^{\multimap(15.44)}$ given any P-local martingale M, the process

$$\tilde{M} \overset{\text{def}}{=} M - \mathscr{E}(Y)^{-1} \bullet [M, \mathscr{E}(Y)]$$

is a Q-local martingale on $[0,T]$. To make this observation useful in concrete applications, we need to transcribe the integral in the last expression.

(16.41) EXERCISE: In the setting above, show that

$$\mathscr{E}(Y)^{-1} \bullet [M, \mathscr{E}(Y)] = \frac{1}{1+\Delta Y} \bullet [M, Y] \equiv \frac{1}{1+\Delta Y} \bullet \left(\langle M^c, Y^c \rangle + \sum \Delta M \, \Delta Y \right).$$

HINT: Recall that $[M, \mathscr{E}(Y)_- \bullet Y] = \mathscr{E}(Y)_- \bullet [M, Y]$. ○

(16.42) TWO BASIC EXAMPLES: Consider the Lévy–Itô local martingale

$$Y = \alpha \bullet W + \int_{\mathbb{R}} g(x, \imath) \bullet \tilde{N}(\mathrm{d}x, [0, \imath])$$

for some choice of the compensated Poisson counting measure[183]

$$\tilde{N}(\mathrm{d}x, \mathrm{d}t) = N(\mathrm{d}x, \mathrm{d}t) - \nu(\mathrm{d}x)\mathrm{d}t,$$

which is independent of the Brownian motion W, and also some choice of the predictable integrands α and g, such that the processes

$$\alpha^2 \bullet \imath \quad \text{and} \quad \left(\int_{\mathbb{R}} g(x, \imath)^2 \, \nu(\mathrm{d}x) \right) \bullet \imath$$

have sample paths that do not explode in finite time with probability 1. Thus, $Y_0 = 0$, and for every local martingale M,

$$Y^c = \alpha \bullet W \quad \Rightarrow \quad \langle M^c, Y^c \rangle = \alpha \bullet \langle W, M^c \rangle.$$

Next, suppose that $\mathsf{E}[\mathscr{E}(Y)_T] = 1$ for some $T > 0$, restrict the time domain to the finite interval $[0,T]$, and consider the special case $M = W$. With this choice $M^c = M = W$, $\Delta M = 0$, and

$$\langle M^c, Y^c \rangle = \alpha \bullet \langle W, W \rangle = \alpha \bullet \imath, \quad \text{so that} \quad \mathscr{E}(Y)^{-1} \bullet [M, \mathscr{E}(Y)] = \alpha \bullet \imath,$$

where the last identity is due to the fact that the set $\{t \in [0,T] \colon \Delta Y_t(\omega) \neq 0\}$ has Lebesgue measure 0 for P-a.e. $\omega \in \Omega$. Girsanov's theorem then implies that the process

$$\tilde{M} \overset{\text{def}}{=} M - \mathscr{E}(Y)^{-1} \bullet [M, \mathscr{E}(Y)] = W - \alpha \bullet \imath$$

[183] In any such expression the measure ν is automatically assumed to be a Lévy measure, that is, some Borel measure that satisfies condition (16.24†), and the mean measure of the Poisson counting measure N is assumed to be $\mathsf{E}[N(\mathrm{d}x, \mathrm{d}t)] = \nu(\mathrm{d}x)\mathrm{d}t$.

is a local martingale on $[0,T]$ relative to $Q = \mathscr{E}(Y)_T \odot P$. Since $\langle \tilde{M}, \tilde{M} \rangle = \imath$, it follows that $\tilde{M}$ is a Brownian motion relative to Q.

In our second example we take M to be the purely discontinuous local martingale

$$M = \int_{\mathbb{R}} h(x, \imath)\bullet \tilde{N}(\mathrm{d}x, [0, \imath])$$

for some predictable process $h(\omega, x, t)$ chosen so that the sample paths of the process $\left(\int_{\mathbb{R}} h(x, \imath)^2 \nu(\mathrm{d}x)\right)\bullet \imath$ do not explode in finite time. With this choice we have $M^c = 0$ and

$$\begin{aligned}
\tilde{M} &\overset{\text{def}}{=} M - \mathscr{E}(Y)^{-1}\bullet[M, \mathscr{E}(Y)] = M - \mathscr{E}(Y)^{-1}\bullet(\mathscr{E}(Y)_-\bullet[M, Y]) \\
&= M - \sum \frac{\Delta M \, \Delta Y}{1 + \Delta Y} \\
&= \int_{\mathbb{R}} h(x, \imath)\bullet \tilde{N}(\mathrm{d}x, [0, \imath]) - \int_{\mathbb{R}} \frac{h(x, \imath) g(x, \imath)}{1 + g(x, \imath)} \bullet N(\mathrm{d}x, [0, \imath]) \\
&= \int_{\mathbb{R}} \frac{h(x, \imath)}{1 + g(x, \imath)} \bullet N(\mathrm{d}x, [0, \imath]) - \int_{\mathbb{R}} h(x, \imath) \nu(\mathrm{d}x)\bullet \imath && \dagger \\
&= \int_{\mathbb{R}} \frac{h(x, \imath)}{1 + g(x, \imath)} \bullet \tilde{N}(\mathrm{d}x, [0, \imath]) - \int_{\mathbb{R}} \frac{h(x, \imath) g(x, \imath)}{1 + g(x, \imath)} \nu(\mathrm{d}x)\bullet \imath \, . && \ddagger
\end{aligned}$$

Girsanov's theorem implies that the process above is a local martingale on $[0,T]$ relative to $Q = \mathscr{E}(Y)_T \odot P$. ○

(16.43) REMARK: Formally, the fact that the process in $(16.42^\dagger)$ is a local Q-martingale is telling us that

$$\frac{1}{1 + g(x, t)} \mathsf{E}^Q\left[N(\mathrm{d}x, \mathrm{d}t) \mid \mathscr{F}_{t-}\right] = \nu(\mathrm{d}x)\mathrm{d}t = \mathsf{E}^P\left[N(\mathrm{d}x, \mathrm{d}t) \mid \mathscr{F}_{t-}\right],$$

or, equivalently,

$$\begin{aligned}
\mathsf{E}^Q\left[N(\mathrm{d}x, \mathrm{d}t) \mid \mathscr{F}_{t-}\right] &= \mathsf{E}^P\left[\left(1 + g(x, t)\right) N(\mathrm{d}x, \mathrm{d}t) \mid \mathscr{F}_{t-}\right] \\
&= \left(1 + g(x, t)\right) \nu(\mathrm{d}x)\mathrm{d}t \, .
\end{aligned}$$

Thus, one of the effects of changing the measure from P to $Q \overset{\text{def}}{=} \mathscr{E}(Y)_T \odot P$, with

$$Y = \int_{\mathbb{R}} g(x, \imath)\bullet \tilde{N}(\mathrm{d}x, [0, \imath]) \quad \text{such that} \quad \mathsf{E}[\mathscr{E}(Y)_T] = 1 \, ,$$

is that the counting random measure N would yield a point on the slice $\mathbb{R} \times \{t\}$ that is sampled from the measure $\tilde{\nu}(\mathrm{d}x) \overset{\text{def}}{=} \left(1 + g(x, t)\right)\nu(\mathrm{d}x)$, instead of a random point sampled from $\nu(\mathrm{d}x)$, as is the case under P.

Another useful observation is that if the predictable process g can be chosen so that the sample paths of

$$\left(\int_{\mathbb{R}} x^2 \left(1 + g(x, \imath)\right)^2 \nu(\mathrm{d}x)\right)\bullet \imath \quad \text{and} \quad \left(\int_{\mathbb{R}} g(x, \imath)^2 \, \nu(\mathrm{d}x)\right)\bullet \imath$$

do not explode in finite time, then we can set $h(x,t) \stackrel{\text{def}}{=} x\left(1+g(x,t)\right)$ and conclude that the process

$$\int_{\mathbb{R}} x\, N(\mathrm{d}x,[0,\iota]) - \Big(\int_{\mathbb{R}} x\left(1+g(x,\iota)\right)\nu(\mathrm{d}x)\Big)\bullet\iota$$

is a Q-local martingale on $[0,T]$. If the atoms of the counting random measure N are interpreted as locations of certain jumps in both space and time, then the connection just described provides a way of making the jumps summable after appropriate correction and change of the underlying measure. This is only interesting if the measure ν is σ-finite but not finite, for if ν is finite, then with probability 1 the measure N would have only finitely many atoms inside any rectangle of the form $\mathbb{R}\times[0,t]$ for any $t\in\mathbb{R}_+$, in which case summablity will not be an issue – even without compensation (the sum of the jumps will still need to be compensated in order to become a local martingale). ○

(16.44) REMOVAL OF THE DRIFT: A particularly common scenario in which one may be faced with the need to change the underlying probability measure is the following: given a Lévy–Itô process of the form

$$X = X_0 + b\bullet\iota + \sigma\bullet W + \int_{\mathbb{R}} \gamma(x,\iota)\bullet\tilde{N}(\mathrm{d}x,[0,\iota])\,,$$

one must find an equivalent measure under which the process above is a local martingale. As the last two stochastic integrals represent local martingales, turning this process into a local martingale comes down to "removing" the drift term $b\bullet\iota$. The idea is to put X in the form⊶ (16.41)

$$X = X_0 + M - \mathscr{E}(Y)^{-1}\bullet[M,\mathscr{E}(Y)] = M - \frac{1}{1+\Delta Y}\bullet[M,Y]$$

for some choice of the Lévy–Itô local martingales

$$M \stackrel{\text{def}}{=} \sigma\bullet W + \int_{\mathbb{R}} h(x,\iota)\bullet\tilde{N}(\mathrm{d}x,[0,\iota])$$

and

$$Y \stackrel{\text{def}}{=} \alpha\bullet W + \int_{\mathbb{R}} g(x,\iota)\bullet\tilde{N}(\mathrm{d}x,[0,\iota])\,,$$

such that the implications of Girsanov's theorem would be in force on the finite interval $[0,T]$ for the equivalent measure $Q = \mathscr{E}(Y)_T\odot P$. For this to be possible the jumps of Y must be > -1 (this is guaranteed by the condition $1+g(\omega,x,s)>0$) and the condition $\mathsf{E}[\mathscr{E}(Y)_T]=1$ must hold. Next, observe that

$$[M,Y] = \langle M^c,Y^c\rangle + \sum \Delta Y\,\Delta M = (\sigma\alpha)\bullet\iota + \int_{\mathbb{R}} h(x,\iota)g(x,\iota)\bullet N(\mathrm{d}x,[0,\iota])\,,$$

so that

$$M - \frac{1}{1+\Delta Y}\bullet[M,Y] = \sigma\bullet W + \int_{\mathbb{R}} h(x,\iota)\bullet\tilde{N}(\mathrm{d}x,[0,\iota])$$

$$- (\sigma\alpha)^{\bullet}\iota - \int_{\mathbb{R}} \frac{h(x,\iota)g(x,\iota)}{1+g(x,\iota)} {\bullet} N(\mathrm{d}x,[0,\iota)$$

† $$= \sigma{\bullet}W - (\sigma\alpha)^{\bullet}\iota + \int_{\mathbb{R}} \frac{h(x,\iota)}{1+g(x,\iota)} {\bullet} \tilde{N}(\mathrm{d}x,[0,\iota]) - \Big(\int_{\mathbb{R}} \frac{h(x,\iota)g(x,\iota)}{1+g(x,\iota)}\, \nu(\mathrm{d}x)\Big){\bullet}\iota\,,$$

where once again we used the fact that the sample paths of the process $(1+\Delta Y)^{-1}$ are equal to 1 Λ-a.e. on $\mathbb{R}_+$, so that $\frac{\sigma\alpha}{1+\Delta Y}{\bullet}\iota = (\sigma\alpha)^{\bullet}\iota$. If we can choose α, h, and g in such a way that, in addition to all previously imposed conditions,

$$-\sigma_t\alpha_t - \int_{\mathbb{R}} \frac{h(x,t)g(x,t)}{1+g(x,t)}\, \nu(\mathrm{d}x) = b_t \quad \text{and} \quad \frac{h(x,t)}{1+g(x,t)} = \gamma(x,t)\,,$$

for every $t \in [0,T]$ and every $x \in \mathbb{R}$, the identities being understood, as usual, as identities between equivalence classes of random variables, then we can indeed claim that the process

$$X - X_0 = M - \frac{1}{1+\Delta Y}{\bullet}[M,Y]$$

is a local martingale on $[0,T]$ relative to $Q \overset{\text{def}}{=} \mathscr{E}(Y)_T \odot P$.

There is really too much to choose from. First, we must choose the process α in such a way that

$$\mathsf{E}\big[e^{(\alpha{\bullet}W - \frac{1}{2}\alpha^2{\bullet}\iota)_T}\big] = 1$$

(e.g., one may take $\alpha \equiv 0$), then choose $g(x,t) > -1$ so that

‡ $$\int_{\mathbb{R}} \gamma(x,t)g(x,t)\nu(\mathrm{d}x) = -b_t - \sigma_t\alpha_t\,, \quad t \in [0,T]\,,$$

and finally set $h(x,t) = \gamma(x,t)\big(1+g(x,t)\big)$. The key, of course, is to make all these choices in such a way that $\mathsf{E}[\mathscr{E}(Y)_T] = 1$, where

$$\log\mathscr{E}(Y) = \alpha{\bullet}W - \frac{1}{2}\alpha^2{\bullet}\iota + \int_{\mathbb{R}} g(x,\iota){\bullet}\tilde{N}(\mathrm{d}x,[0,\iota]) + \int_{\mathbb{R}} \Big(\log\big(1+g(x,\iota)\big) - g(x,\iota)\Big){\bullet}N(\mathrm{d}x,[0,\iota])\,.$$

Just as an example, one possible choice is

$$g(x,t) = -\frac{b_t + \sigma_t\alpha_t}{\int_{]u,\upsilon[} \gamma(x,t)\nu(\mathrm{d}x)} 1_{]u,\upsilon[}(x)\,,$$

provided that $-\infty < u < \upsilon < \infty$ can be chosen so that $g(x,t) > -1$, and, typically, there would be infinitely many such choices, that is, there would be infinitely many equivalent measures under which $X - X_0$ behaves as a local martingale. However, in section 16.7 we are going to encounter a situation in which there is only one equivalent measure that has this property, and also a situation in which it is impossible to satisfy the condition $g(x,t) > -1$. ○

16.6 Lévy–Itô Diffusions

The semimartingale X is said to be a Lévy–Itô diffusion process, or simply a jump diffusion process, if it satisfies an equation of the form

$$X = X_0 + b(X_-, \iota)\bullet\iota + \sigma(X_-, \iota)\bullet W + \int_{\mathbb{R}^m\setminus\{0\}} \gamma(X_-, z, \iota)\bullet\tilde{N}(\mathrm{d}z, [0, \iota])\,, \qquad \mathrm{e}_1$$

or, if one insists on a more detailed notation

$$X_t = X_0 + \int_0^t b(X_{s-}, s)\mathrm{d}s + \int_0^t \sigma(X_{s-}, s)\mathrm{d}W_s + \int_0^t \int_{\mathbb{R}^m\setminus\{0\}} \gamma(X_{s-}, z, s)\,\tilde{N}(\mathrm{d}z, \mathrm{d}s)\,, \quad t \in \mathbb{R}_+\,,$$

for some appropriate choice of the initial value X_0 and the coefficients $b(\cdot,\cdot)$, $\sigma(\cdot,\cdot)$, and $\gamma(\cdot,\cdot,\cdot)$, treated as deterministic functions defined on Euclidean domains of appropriate dimensions. Certain technical matters aside, the theory of such equations more or less parallels the theory of stochastic equations driven by Brownian motion alone, which we outlined in chapter 11. In particular, the existence of a unique non-exploding strong solution is again guaranteed by the essentially exact analogues of the usual locally Lipschitz and growth conditions – see (Øksendal and Sulem 2005) and (Protter 2005) for a detailed treatment and further references. In general the solution X could be a vector process living in $\mathbb{R}^n$; specifically, b could be a function from $\mathbb{R}^n\times\mathbb{R}_+$ into $\mathbb{R}^n$, W could be a d-dimensional Brownian motion, σ could be a matrix-valued function from $\mathbb{R}^n\times\mathbb{R}_+$ into $\mathbb{R}^{n\otimes d}$, $\tilde{N}(\mathrm{d}z,\mathrm{d}t) = N(\mathrm{d}z,\mathrm{d}t) - \nu(\mathrm{d}z)\mathrm{d}t$ could be a compensated Poisson counting measures on $(\mathbb{R}^m\setminus\{0\})\times\mathbb{R}_+$, which is independent of W, and γ could be a function from $\mathbb{R}^n\times\mathbb{R}^m\times\mathbb{R}_+$ into $\mathbb{R}^n$. As the solution X, if one exists, is a càdlàg process, its sample paths cannot have oscillating discontinuities, and therefore $X_t \neq X_{t-}$ is possible only for countably many $t \in \mathbb{R}_+$. Thus, in the first two integrals on the right side of e_1, we are free⊸ (16.31) to replace X_- with X as we like.

(16.45) INFINITESIMAL GENERATORS: It is not too difficult to guess that the infinitesimal generator of the jump diffusion process X from e_1 can be cast as

$$\mathcal{A}_t\varphi(x) = b(x,t)\cdot\nabla\varphi(x) + \frac{1}{2}\left(\sigma(x,t)\sigma(x,t)^{\intercal}\nabla\right)\cdot\nabla\varphi(x) + \int_{\mathbb{R}^m}\Big(\varphi\left(x+\gamma(x,z,t)\right) - \varphi(x) - \gamma(x,z,t)\cdot\nabla\varphi(x)\Big)\nu(\mathrm{d}z)\,.$$

In other words, for any test function $\varphi \in \mathscr{C}_K^\infty(\mathbb{R}^n;\mathbb{R})$,

$$\lim_{\varepsilon\searrow 0} \mathsf{E}\left[\frac{1}{\varepsilon}\left(\varphi(X_{t+\varepsilon}) - \varphi(X_t)\right)\,\Big|\, X_t = x\right] = \mathcal{A}_t\varphi(x)\,.$$

Setting the rigorous argument aside, in what follows we are going to outline briefly why the relation above is perfectly believable and in compliance with our basic intuition. The idea is not new to us, and stems from the Itô formula expansion of the difference $\varphi(X_{t+\varepsilon}) - \varphi(X_t)$. In fact, there is only one term in this expansion that is new to us, and this is the term involving the jumps, namely,

$$\frac{1}{\varepsilon}\sum\nolimits_{s\,\in\,]t,t+\varepsilon]}\big(\varphi(X_s)-\varphi(X_{s-})-\nabla\varphi(X_{s-})\cdot\Delta X_s\big)$$
$$=\frac{1}{\varepsilon}\int_{]t,t+\varepsilon]}\int_{\mathbb{R}^m}\Big(\varphi\big(X_{s-}+\gamma(X_{s-},z,s)\big)-\varphi(X_{s-})$$
$$+\nabla\varphi(X_{s-})\cdot\gamma(X_{s-},z,s)\Big)\,N(\mathrm{d}z,\mathrm{d}s)\,.$$

Because the jump location $z \in \mathbb{R}^m$ is sampled at time $s \in\,]t, t+\varepsilon]$, independently from the history up until time t, with "probability" $\mathsf{E}[N(\mathrm{d}z,\mathrm{d}s)] = \nu(\mathrm{d}z)\mathrm{d}s$, the expected value of this expression conditioned to the event $\{X_t = x\}$ must be

$$\frac{1}{\varepsilon}\int_{]t,t+\varepsilon]}\int_{\mathbb{R}^m}\mathsf{E}\Big[\varphi\big(X_{s-}+\gamma(X_{s-},z,s)\big)-\varphi(X_{s-})$$
$$+\nabla\varphi(X_{s-})\cdot\gamma(X_{s-},z,s)\;\big|\;X_t=x\Big]\,\nu(\mathrm{d}z)\mathrm{d}s\,.$$

Since the process X is right-continuous, we have $\lim_{s\searrow t} X_{s-} = X_t = x$, which then leads to the conclusion that the limit of the above as $\varepsilon \searrow 0$ is

$$\int_{\mathbb{R}^m}\Big(\varphi\big(x+\gamma(x,z,t)\big)-\varphi(x)+\nabla\varphi(x)\cdot\gamma(x,z,t)\Big)\,\nu(\mathrm{d}z)\,.$$

To put it simply, $\mathcal{A}_t(x)$ gives the instantaneous average rate of change in the quantity $\varphi(X_t)$ conditioned to the event $\{X_t = x\}$. This justifies the name "infinitesimal generator," and we note that the existence of an operator with this property is essentially a restatement of the Markov property of X. ○

It turns out that the connection that we just established holds not just for deterministic times t, but also for stopping times $\mathscr{t}$. To be precise we have the following important result:

(16.46) DYNKIN'S FORMULA:[184] (Métivier 1982, E.4.1) If $\mathscr{t}$ is a stopping time such that $\mathsf{E}[\mathscr{t} \mid X_0 = x] < \infty$, then

$$\mathsf{E}\big[\varphi(X_{\mathscr{t}})\;\big|\;X_0=x\big]=\varphi(x)+\mathsf{E}\Big[\int_0^{\mathscr{t}}(\mathcal{A}_s\varphi)(X_s)\mathrm{d}s\;\Big|\;X_0=x\Big]$$

for any test function φ. ○

(16.47) LINEAR COEFFICIENTS: Naturally, the easiest and most straightforward concrete examples of jump diffusions can be produced by taking the coefficients

[184] Named after the American mathematician Eugene Borisovich Dynkin (1924–2014).

b, σ, and γ to be linear. For simplicity, suppose that $m = n = d = 1$ (X is a real-valued diffusion). If we further set (note that these coefficients are deterministic)

† $$b(x,t) = bx\,, \quad \sigma(x,t) = \sigma x\,, \quad \text{and } \gamma(x,z,t) = \gamma xz\,,$$

for some choice of the parameters $b, \sigma, \gamma \in \mathbb{R}$, with the obvious abuse of the notation, having the symbols b, σ, and γ refer to functions in the left side and refer to scalar parameters on the right side, then the solution, X, to e_1 would be a geometric Lévy motion, of which the standard geometric Brownian motion is a special case.

Another interesting choice of the (still deterministic) coefficients is

‡ $$b(x,t) = b\,(a - x)\,, \quad \sigma(x,t) = \sigma\,, \quad \text{and } \gamma(x,z,t) = \gamma\, z\,, \quad a, b, \sigma, \gamma \in \mathbb{R}\,,$$

which give rise to a Lévy-driven Ornstein–Uhlenbeck (mean-reverting) process X, of which the standard mean-reverting Brownian motion is again a special case. ○

(16.48) EXERCISE: Assuming that the coefficients are chosen as in (16.47†), write explicitly the solution to e_1, that is, develop a closed-form expression for the geometric Lévy motion.

HINT: This is a special case of an equation that we have solved before. ○

(16.49) EXERCISE: Develop a closed-form expression for the solution to e_1 in the case where the coefficients are chosen as in (16.47‡), that is, develop a closed-form expression for the Lévy-driven Ornstein–Uhlenbeck (mean-reverting) process.

HINT: Consult exercise (11.42). ○

(16.50) REFLECTED (CONFINED) DIFFUSIONS: In section 10.8 we saw how the Brownian motion B can be reflected at 0 and kept inside the domain $\mathbb{R}_+$: there is a continuous, increasing, and adapted (to the same Brownian filtration) process A that starts from 0 ($A_0 = 0$) and is such that the process $X = B + A$ is positive, and, while inside $\mathbb{R}_{++} =]0, \infty[$, behaves locally like a Brownian motion. In fact, we were able to identify the correction term as $A_t = \sup_{s\in[0,t]}(-B_s)$ and, equivalently, as $A_t = L_t(X)$, the local time of X in 0. Just as one would expect, a similar correction (that is, control) can be applied to the Lévy–Itô diffusion given by e_1, so that after the correction the solution X will remain inside the closure, $\bar{\mathscr{D}}$, of some open domain $\mathscr{D} \subset \mathbb{R}^n$, the boundary of which, $\partial\mathscr{D}$, is "sufficiently regular." Moreover, this confinement of X to $\bar{\mathscr{D}}$ can be done in such a way that in the open set $\mathscr{D}$ the local behavior of the confined version of X is no different from that of the non-confined version, that is, after the confinement, X is still going to be governed by e_1 during those times at which X is not in $\mathscr{D}^{\complement}$. To be precise, if $n(x)$, $x \in \partial\mathscr{D}$, is some (appropriately chosen) smooth vector field distributed in some (small) neighborhood of $\partial\mathscr{D}$, which vector field points strictly inward into $\mathscr{D}$, and if $\Pi : \mathbb{R}^n \mapsto \bar{\mathscr{D}}$ is a particular projection onto $\bar{\mathscr{D}}$,[185] then the process $\tilde{X}$,

[185] Generally, "projection" is understood to mean some continuous mapping $\Pi : \mathbb{R}^n \mapsto \bar{\mathscr{D}}$ such that $\Pi(x) = x$ for every $x \in \bar{\mathscr{D}}$. Such a mapping may not exist for certain domains $\mathscr{D}$, and this is where

governed by

$$\begin{aligned}\tilde{X}_t = X_0 &+ \int_0^t b(\tilde{X}_{s-}, s)\mathrm{d}s + \int_0^t \sigma(\tilde{X}_{s-}, s)\mathrm{d}W_s \\ &+ \int_0^t \int_{\mathbb{R}^m \setminus \{0\}} \gamma(\tilde{X}_{s-}, z, s)\tilde{N}(\mathrm{d}z, \mathrm{d}s) \\ &+ \int_0^t n(\tilde{X}_{s-})\mathrm{d}L_s^{\partial\mathcal{D}}(\tilde{X}) + \int_0^{t-} \big(\Pi(\tilde{X}_s) - \tilde{X}_s\big)N(\mathbb{R}^m, \mathrm{d}s),\end{aligned}$$

can be viewed as a "reflected," or "confined" version of X – reflected at $\partial\mathcal{D}$, or confined to $\bar{\mathcal{D}}$. As a careful reader might have immediately realized, without additional restrictions on the Poisson counting measure N and/or the jumps $\gamma(\tilde{X}_{s-}, z, s)$ the last integral may not be meaningful (notice, among other things, that $\Pi(\tilde{X}) - \tilde{X}$ is not a predictable process). This integral is the same as

$$\sum\nolimits_{0<s<t} \big(\Pi(\tilde{X}_s) - \tilde{X}_s\big)1_{\{\tilde{X}_s \neq \tilde{X}_{s-}\}},$$

and, in general, there would be no reason for these jump corrections to be summable. One obvious condition that makes this possible is $\nu([-\varepsilon, \varepsilon]) = 0$ for some $\varepsilon > 0$, in which case N generates only finitely many jumps during any finite time period. Instead, one may also impose the condition (see (Menaldi and Robin 1985)) $x + \gamma(x, z, s) \in \bar{\mathcal{D}}$ for any $x \in \bar{\mathcal{D}}$, in which case the process can never jump outside of the region $\bar{\mathcal{D}}$, so that the questionable summation goes away. The local time in $\partial\mathcal{D}$, which we denote by $L^{\partial\mathcal{D}}(\tilde{X})$, can be defined in the same way in which the local time of the Brownian motion in 0 was defined in section 10.8 – and this is yet another reason to insist on a "sufficiently regular" boundary.

In any case, the local behavior of $\tilde{X}$ inside the open domain $\mathcal{D}$ is indistinguishable, as a distribution, from the local behavior of X inside $\mathcal{D}$ until the first moment at which X leaves $\mathcal{D}$, that is, either hits $\partial\mathcal{D}$, or jumps inside $\bar{\mathcal{D}}^{\complement}$. We stress that, in general, the sample path of $\tilde{X}$ cannot be fully contained inside $\bar{\mathcal{D}}$, since the jumps are not predictable and cannot be prevented from taking the diffusion into the domain $\bar{\mathcal{D}}^{\complement}$ by, say, enforcing simultaneous jumps in the opposite direction. The only way around this would be to contain (if at all possible) the diffusion in a sub-domain of $\mathcal{D}$ from which it cannot jump inside the open set $\bar{\mathcal{D}}^{\complement}$.

The reader must be aware that the construction of a reflected jump diffusion described here is a mere sketch of the main idea. Additional regularity conditions about the boundary $\partial\mathcal{D}$ and the diffusion coefficients are needed – and even then, establishing that $\tilde{X}$ is well defined and remains inside $\bar{\mathcal{D}}$ is not straightforward (see above). For more details on reflected diffusions, in particular reflected on a wedge,

the requirement for the boundary to be "sufficiently regular" is needed. The same can be said about the distribution of vectors $n(x)$, $x \in \partial\mathcal{D}$, that point "strictly inward" into $\mathcal{D}$. The most common definition of "projection" is "the closest point in a given set," where "closest" is understood with respect to a particular distance – say, the usual Euclidean distance in $\mathbb{R}^n$.

we refer to (Williams 1985a), (Williams 1985b), (Menaldi and Robin 1985), and (Øksendal and Sulem 2005). ○

16.7 An Asset Pricing Model with Jumps in the Returns

Asset pricing models with discontinuous returns have been studied since the early days of continuous-time finance, as is evident from (Merton 1976). In the market model developed in chapter 13 the accumulated instantaneous net returns from every risky asset was assumed to follow a continuous semimartingale. Our goal in this section is to develop a simple model, which parallels the Black–Scholes–Merton setup, but allows for jumps in the returns. In fact, we began writing down one such model in section 16.4.

Consider the classical market model with one risky asset, the price of which follows the process (this is a copy of e_2 in section 16.4)

$$S = S_0 + S_- \bullet Z\,,$$

for some exogenously specified semimartingale Z. In addition to the risky asset, the market includes a short-lived risk-free asset, B, with fixed instantaneous rate of return $r = \frac{\mathrm{d}B_t}{B_t\,\mathrm{d}t}$, $t \in \mathbb{R}_+$. To be even more concrete, suppose that the semimartingale Z, which models the accumulated instantaneous (ex-dividend) net returns from the risky asset, has the form

$$Z = b\iota + \int_{\mathbb{R}} \gamma(x)\,\tilde{N}(\mathrm{d}x, [0, \iota])\,,$$

where $b \in \mathbb{R}$ is a given parameter and $\gamma(\cdot)$ is a given Borel function, which models the jumps in the returns. The random measure $\tilde{N}(\mathrm{d}x, \mathrm{d}t) = N(\mathrm{d}x, \mathrm{d}t) - \nu(\mathrm{d}x)\mathrm{d}t$ is the compensated Poisson counting measure on $(\mathbb{R}\backslash\{0\})\times\mathbb{R}_+$ with Lévy measure ν, chosen so that $\nu(\{0\}) = 0$ and $\nu\big\{x \in \mathbb{R}\colon \gamma(x) \leq -1\big\} = 0$. The last condition ensures that the jumps of Z are > -1 (which, in turn, ensures that the price process S remains strictly positive), but, to ensure that the local martingale component of Z is meaningful, we must also impose the condition $\int_{\mathbb{R}} \gamma(x)^2\,\nu(\mathrm{d}x) < \infty$. The excess returns process in this model is given by

$$X = (b - r)\iota + \int_{\mathbb{R}} \gamma(x)\,\tilde{N}(\mathrm{d}x, [0, \iota])\,,$$

and understanding the mechanics of the risk-neutral asset pricing in this market boils down to understanding how this last process can be turned into a local martingale by way of changing the underlying probability measure without leaving the equivalence class associated with it. We have done this before in the context of the canonical Black–Scholes–Merton model with a Brownian motion in place of the purely discontinuous local martingale $\int_{\mathbb{R}} \gamma(x)\,\tilde{N}(\mathrm{d}x, [0, \iota])$. Despite the complications that the jumps bring into the picture, the idea is still the same: we must rearrange the expression above to make it look like the shift of a particular local martingale in conjunction with an application of Girsanov's theorem. Specifically, we must construct two local martingales,

$$Y \overset{\text{def}}{=} \int_{\mathbb{R}} g(x, \imath)\bullet \tilde{N}(\mathrm{d}x, [0, \imath]) \quad \text{and} \quad M \overset{\text{def}}{=} \int_{\mathbb{R}} h(x, \imath)\bullet \tilde{N}(\mathrm{d}s, [0, \imath]),$$

by choosing the functions g and h in such a way that$^{\circ\!\!\to (16.44^{\dagger})}$

$$\begin{aligned} M - \frac{1}{\mathscr{E}(Y)}\bullet[M, \mathscr{E}(Y)] &\equiv \int_{\mathbb{R}} \frac{h(x, \imath)}{1 + g(x, \imath)}\bullet \tilde{N}(\mathrm{d}x, [0, \imath]) - \Big(\int_{\mathbb{R}} \frac{h(x, \imath)g(x, \imath)}{1 + g(x, \imath)} \nu(\mathrm{d}x)\Big)\bullet \imath \\ &= \int_{\mathbb{R}} \gamma(x)\bullet \tilde{N}(\mathrm{d}x, [0, \imath]) + (b - r)\imath \equiv X\,. \end{aligned}$$

In particular, this relation would be satisfied if $h(x, t) = \gamma(x)\big(1 + g(x, t)\big)$ and

$$-(b - r) = \int_{\mathbb{R}} \frac{h(x, t)g(x, t)}{1 + g(x, t)} \nu(\mathrm{d}x) = \int_{\mathbb{R}} \gamma(x)g(x, t)\, \nu(\mathrm{d}x)\,. \qquad \mathrm{e}_1$$

Unless the volatility $\gamma(\cdot)$ happens to be ν-trivial, it will always be possible to find a finite interval $[u, v]$ such that $0 < \big|\int_u^v \gamma(y)\nu(\mathrm{d}y)\big| \neq 0$. We can then set

$$g(x, t) \equiv g(x) \overset{\text{def}}{=} -\frac{b - r}{\int_{]u,v[} \gamma(y)\, \nu(\mathrm{d}y)} 1_{]u,v[}(x)\,, \qquad \mathrm{e}_2$$

but must also ensure that $g(x) > -1$ in order for the exponent $\mathscr{E}(Y)$ to be meaningful, and this may not always be possible. Indeed, if the parameters b and r are chosen so that $b > r$, it may not be possible to find a finite interval $[u, v]$ such that $\int_u^v \gamma(y)\, \nu(\mathrm{d}y) > b - r$ for certain choices of the Lévy measure ν and the jump volatility $\gamma(\cdot)$. As an example, in the Poisson process case, namely, $\nu(\mathrm{d}x) = c\epsilon_1(\mathrm{d}x)$, the jump volatility $g(\cdot)$ reduces to the single scalar value $g(1)$, which is uniquely determined from e_1: $g(1) = -(b - r)/(c\gamma(1))$. Thus, unless the parameters b, r, and c are chosen so that $(b - r)/(c\gamma(1)) < 1$, the relation $g(x) \equiv g(1) > -1$ will not be possible to attain.

In any case, if the Doléan–Dade exponent $\mathscr{E}(Y)$ happens to be well defined and happens to satisfy the condition $\mathsf{E}\big[\mathscr{E}(Y)_T\big] = 1$, Girsanov's theorem will guarantee that the excess returns process X is a local martingale with respect to the probability measure $Q \overset{\text{def}}{=} \mathscr{E}(Y)_T \odot P \approx P$. It is therefore important for us to understand when the condition $\mathsf{E}\big[\mathscr{E}(Y)_T\big] = 1$ is actually in force. To this end, observe that if $g(x)$ is chosen as in e_2, then

$$\begin{aligned} Y = \int_{\mathbb{R}} g(x)\, \tilde{N}(\mathrm{d}x, [0, \imath]) &= -\frac{b - r}{\int_a^b \gamma(y)\, \nu(\mathrm{d}y)} \int_{\mathbb{R}} 1_{[a,b]}(x)\tilde{N}(\mathrm{d}x, [0, \imath]) \\ &= -\frac{b - r}{\int_a^b \gamma(y)\, \nu(\mathrm{d}y)} \tilde{N}([a, b], [0, \imath])\,, \end{aligned}$$

and this process is nothing but a multiple of the compensated Poisson process, which can be cast in the form $Y = A(\Pi - c\iota)$ for some fixed $A \in]-1, \infty[$ and Poisson process Π of intensity $c = \nu([a, b])$.

(16.51) EXERCISE: Let Π be the Poisson process of intensity $c > 0$ and let $Y \stackrel{\text{def}}{=} A(\Pi - c\iota) \equiv A\tilde{\Pi}$ for some $A > -1$. Consider the stochastic exponent $\mathscr{E}(Y)$, that is, the solution to $\mathscr{E}(Y) = 1 + \mathscr{E}(Y)_{-} \cdot Y$, with $\mathscr{E}(Y)_0 = 1$, and prove that $\mathsf{E}\big[\mathscr{E}(Y)_t\big] = 1$ for any $t \in \mathbb{R}_+$.

HINT: Using the result from (16.37), show that $\mathscr{E}(Y)_t = e^{-Act}(1 + A)^{\Pi_t}$ and then calculate $\mathsf{E}[\mathscr{E}(Y)_t]$ directly from the fact that Π_t is a Poisson r.v. with parameter $\mathsf{E}[\Pi_t] = ct$. ○

(16.52) REMARK: It should be clear from the above construction that when it comes to the choice of the process Y, and therefore to the choice of the equivalent change of the underlying measure, typically there would be too much to choose from. That such market models are intrinsically incomplete is both well known and intuitive. To quote from (Jarrow and Madan 1999): *"Jarrow and Madan (1995) show that when asset prices follow pure jump Lévy processes with a continuum of jump magnitudes possible at each jump time (for example, the variance gamma model of Madan and Seneta (1990)) market completeness requires trading in infinitely many assets."*

With only one risky asset, market completeness holds if, for example, the local martingale component of the net returns Z, that is, the term $\int_{\mathbb{R}} \gamma(x)\, \tilde{N}(dx, [0, \iota])$ corresponds to a fixed multiple of the compensated Poisson process. This is the case where $\nu(dx) = c\epsilon_1(dx)$ for some $c > 0$, which is detailed below. Generally, if $\nu(dx) = \sum_{i=1}^{n} c_i \epsilon_{a_i}(dx)$ for some finite $n \in \mathbb{N}_{++}$, one would need n-risky assets driven by the same Poisson counting measure $\tilde{N}$ but with n-different jump volatility coefficients $\gamma_i(\cdot)$, $1 \le i \le n$, chosen so that the matrix with rows $(c_1\gamma_i(a_1), \ldots, c_n\gamma_i(a_n))^{\intercal}$, $1 \le i \le n$, has full rank; this is clarified in (16.58). ○

The beauty of the Black–Scholes–Merton formula and model is that the stock price, the Radon–Nikodym derivative of the associated change of measure, and the option price all allow for a relatively simple closed-form solution. To obtain a similar resolution in the present setting, we must first make a concrete assumption about the Poisson counting measure that would allow us to write the price process explicitly. The simplest such assumption is that the local martingale component of the net returns Z is given by a multiple of the compensated Poisson process, that is, $\nu(dx) = c\epsilon_1(dx)$, in which case the jump volatility $\gamma(x)$ reduces to a single scalar value $\gamma \equiv \gamma(1) > -1$. The process $\Pi \stackrel{\text{def}}{=} N(\{1\}, [0, \iota])$ is nothing but a Poisson process of intensity c, and we have

$$Z = b\iota + \gamma(\Pi - c\iota).$$

The stock price can now be cast as

$$S_t = S_0\, e^{(b-\gamma c)t}(1 + \gamma)^{\Pi_t}, \quad t \in \mathbb{R}_+,$$

and the equivalent change of measure that turns the excess returns

$$X = (b - r)\iota + \gamma(\Pi - c\iota)$$

into a local martingale has Radon–Nikodym derivative $\mathsf{d}Q/\mathsf{d}P = \mathscr{E}(Y)_T$, with

$$Y_t = -\frac{b-r}{c\gamma}(\Pi_t - ct), \quad t \geq 0, \quad \Rightarrow \quad \mathscr{E}(Y)_T = e^{\frac{b-r}{\gamma}T}\left(1 - \frac{b-r}{c\gamma}\right)^{\Pi_T},$$

provided that $1 - \frac{b-r}{c\gamma} > 0$.

(16.53) REMARK: The last inequality translates into $b - r - c\gamma < 0$ if $\gamma > 0$ and $b - r - c\gamma > 0$ if $-1 < \gamma < 0$. If either of these two conditions is satisfied, then an ELMM exists and the financial market is free of arbitrage opportunities (the "easy part" of the first fundamental theorem of asset pricing – see (13.24)). One can also show that in this case the ELMM just described is unique and that this financial market is complete.[186] ○

(16.54) EXERCISE: Show that either of the conditions "$b - r - c\gamma \geq 0$ and $\gamma > 0$" or "$b - r - c\gamma \leq 0$ and $-1 < \gamma < 0$" leads to arbitrage by constructing an arbitrage trading strategy explicitly.

HINT: The amount e^{-rT} invested in the risky asset at time $t = 0$ would generate wealth on date $T > 0$ of the amount $e^{(b-\gamma c-r)T}(1+\gamma)^{\Pi_T}$. ○

(16.55) REMARK: If the market information is enlarged to also include the events of the Brownian motion W, chosen to be independent of the Poisson process Π that drives the returns in the only risky asset, one can use the prescription developed in (16.44) to construct an entire family of ELMM of the form $Q = \mathscr{E}(Y)_T \odot P$ for some choice of the process

$$Y = \alpha \bullet W + \int_{\mathbb{R}} g(x, \iota) \bullet \tilde{N}(\mathrm{d}x, [0, \iota]) = \alpha \bullet W + g(1, \iota) \bullet (\Pi - c\iota),$$

where α is any predictable process (for the filtration generated by both W and Π) chosen so that

$$\alpha^2 \bullet \iota_T < \infty \text{ (a.s.)} \quad \text{and} \quad \mathsf{E}\left[e^{\alpha \bullet W_T - \frac{1}{2}\alpha^2 \bullet \iota_T}\right] = 1$$

(in particular, one can take $\alpha \equiv 0$), and $g(1,t) = -\big(b + \sigma_t\alpha_t - r\big)/(c\gamma)$ (this is an application of (16.44‡)), for yet another predictable process σ chosen so that $\sigma^2 \bullet \iota_T < \infty$ (a.s.) and $\mathsf{E}[\mathscr{E}(Y)_T] = 1$. ○

Taking into account that Π_T is a Poisson r.v. with parameter $\mathsf{E}[\Pi_T] = cT$, the price of a European-style call option with strike $K > 0$ and maturity $T > 0$ can be

[186] The uniqueness of the ELMM and the completeness of the market are essentially the same property. This result is known as "the second fundamental theorem of asset pricing" and⊸• (13.27) was stated earlier for the case of continuous excess returns driven by an Itô process. The case of excess returns that follow a semimartingale with jumps was studied by Jarrow and Madan (1999) and others.

expressed as

$$c(S_0,T,K) = e^{-rT}\mathsf{E}\Big[\Big(S_0\, e^{(b-\gamma c)T}(1+\gamma)^{\Pi_T} - K\Big)^+ e^{\frac{b-r}{\gamma}T}\Big(1-\frac{b-r}{c\gamma}\Big)^{\Pi_T}\Big]$$

$$= e^{-\left(r+c-\frac{b-r}{\gamma}\right)T}\sum_{i=p^*}^{\infty}\Big(S_0\, e^{(b-\gamma c)T}(1+\gamma)^i - K\Big)\Big(1-\frac{b-r}{c\gamma}\Big)^i \times \frac{(cT)^i}{i!}$$

where

$$p^* = 1 + \left\lfloor \frac{\log\left(K e^{(c\gamma-\alpha)T}/S_0\right)}{\log(\gamma+1)} \right\rfloor .$$

(16.56) EUROPEAN CALL WITH POISSON-DRIVEN UNDERLYING: Using the expression (a simple rearrangement of ($0.13^{\dagger}$))

$$F(p,x) \overset{\text{def}}{=} \sum_{n=p}^{\infty}\frac{x^n}{n!} = e^x\Big(1-\frac{\Gamma(p,x)}{(p-1)!}\Big),$$

we arrive at the following formula for the price of the European call:

$$c(S_0,T,K) = e^{-(r+c-\frac{b-r}{\gamma})T}\Bigg(S_0\, e^{(b-\gamma c)T}F\Big(p^*, cT(1+\gamma)\big(1-\frac{b-r}{c\gamma}\big)\Big) - K\,F\Big(p^*, cT\big(1-\frac{b-r}{c\gamma}\big)\Big)\Bigg).$$

This is essentially the formula in (Merton 1976, (16) p. 134). ○

(16.57) EXERCISE: Consider the financial market that consists of two risky assets and a bond with price processes, respectively,

$$S^1 = S^1_0 + S^1_- \cdot Z^1, \quad S^2 = S^2_0 + S^2_- \cdot Z^2, \quad B = B_0 + rB \cdot \iota,$$

where

$$Z^1 = b_1\iota + \gamma_1(\Pi - c\iota) \quad \text{and} \quad Z^2 = b_2\iota + \gamma_2(\Pi - c\iota),$$

for some choice of the parameters $b_1, b_2 \in \mathbb{R}$, $\gamma_1, \gamma_2 \in\,]-1,\infty[$, and the Poisson process Π of intensity $c > 0$ (notice that the prices of the two risky assets are driven by the same Poisson process). Prove that this financial market is free of arbitrage for any finite time horizon $T > 0$ if and only if

$$(b_1 - r)/(c\gamma_1) = (b_2 - r)/(c\gamma_2) < 1,$$

and give the financial interpretation of this relation. Assuming that the last condition is satisfied, show that an ELMM exists and write the associated Radon–Nikodym derivative explicitly. ○

(16.58) EXERCISE: Consider the financial market that consists of two risky assets and a bond with price processes, respectively,

$$S^1 = S^1_0 + S^1_- \cdot Z^1, \quad S^2 = S^2_0 + S^2_- \cdot Z^2, \quad \text{and} \quad B = B_0 + rB \cdot \iota,$$

where

$$Z^1 = b_1 \iota + \int_{\mathbb{R}} \gamma_1(x)\, \tilde{N}(\mathrm{d}x, [0, \iota]) \quad \text{and} \quad Z^2 = b_2 \iota + \int_{\mathbb{R}} \gamma_2(x)\, \tilde{N}(\mathrm{d}x, [0, \iota]),$$

for some choice of the parameters $b_1, b_2, r \in \mathbb{R}$, the functions $\gamma_1, \gamma_2 \colon \mathbb{R} \mapsto \mathbb{R}$, and the Poisson counting measure N with Lévy measure given by

$$\nu(\mathrm{d}x) = c_1 \epsilon_{\frac{1}{3}}(\mathrm{d}x) + c_2 \epsilon_{\frac{2}{3}}(\mathrm{d}x),$$

for some choice of the parameters $c_1, c_2 \in \mathbb{R}_{++}$. Notice that the choice of the functions $\gamma_i \colon \mathbb{R} \mapsto \mathbb{R}$, $i = 1, 2$, comes down to the choice of four parameters $(\gamma_i(1/3), \gamma_i(2/3))$, $i = 1, 2$, so that the entire model is defined in terms of nine parameter (scalar) values. Derive conditions for these parameter values that would guarantee that this market model is well defined, free of arbitrage, and complete for any finite time horizon $T > 0$. Give the form of the associated Radon–Nikodym derivative and explain why its expectation equals 1. ○

(16.59) EXERCISE: Consider the financial market that consists of two risky assets and a bond with price processes, respectively,

$$S^1 = S^1_0 + S^1_- \cdot Z^1, \quad S^2 = S^2_0 + S^2_- \cdot Z^2, \quad B = B_0 + r B_- \cdot \iota,$$
$$\text{where} \quad Z^i = b_i \iota + \sigma_i W + \gamma_i (\Pi - c\iota), \quad i = 1, 2,$$

for some choice of the parameters $r, b_1, b_2, \sigma_1, \sigma_2 \in \mathbb{R}$, $\gamma_1, \gamma_2 \in \,]-1, \infty[$, Brownian motion W, and Poisson process Π of intensity $c > 0$, which is independent of W. Prove that this market is free of arbitrage provided that $\gamma_2 \sigma_1 \neq \gamma_1 \sigma_2$ and

$$\frac{b_1 + \sigma_1 \alpha - r}{c\gamma_1} = \frac{b_2 + \sigma_2 \alpha - r}{c\gamma_2} < 1 \quad \text{for} \quad \alpha = \frac{b_2 \gamma_1 - b_1 \gamma_2 - (\gamma_1 - \gamma_2) r}{\gamma_2 \sigma_1 - \gamma_1 \sigma_2}.$$

Assuming that the parameters can be chosen in such a way that the relations above are satisfied, explain how an ELMM can be constructed and write the associated Radon–Nikodym derivative explicitly. Explain heuristically why this market must be complete (if free of arbitrage). ○

17

Résumé of the Theory and Methods of Stochastic Optimal Control

For the most part, the origins of the theory and methods for optimal control – and indeed much of today's computing technology – can be traced back to the time of the so-called space race during the late 1950s and most of the 1960s. This period culminated with NASA's Apollo 11 mission, which successfully landed its lunar module on the surface of the Moon with two astronauts on board on July 20, 1969. The coveted soft landing on the Moon with humans on board was perhaps the first large-scale project that was possible to accomplish only by amending human intuition and navigation instincts with an automated system based on advanced mathematical methods and real-time computing and control (it also featured the first portable computing device). Coincidentally, the application of essentially the same mathematical methods and ideas to the seemingly unrelated domain of finance was also conceived around the time of the Apollo 11 mission. Indeed, Robert C. Merton's PhD thesis at MIT (see (Merton 1970a)), entitled "Analytical Optimal Control Theory as Applied to Stochastic and Non-Stochastic Economies," was submitted on August 28, 1970, and the publication of (Merton 1969b) was in the printing press literally at the time when Neil Armstrong and Buzz Aldrin accomplished their first walk on the Lunar surface.[187] But the work (McKean 1965) contains an even earlier application to finance of mathematical tools that are even more advanced than those employed in the soft landing on the Moon.

[187] Merton's 1969 paper was essentially completed in November 1968, when it was presented at the joint MIT–Harvard graduate student workshop, with Kenneth Arrow in attendance. According to his personal account of events (Merton 2015), in the summer of 1969, then a 25-year-old PhD student at MIT, he watched the live TV broadcast of the Moon-landing in the home of the renowned physicist and founder of General Atomic, Inc., Frederic de Hoffmann, in San Diego, CA. The two of them met a year earlier in Cambridge, MA, when de Hoffmann recruited Merton to work as a summer intern for Southern California First National Bank (de Hoffmann was serving on the bank's advisory board at the time). The research report (Merton 1969a) that Merton prepared in July 1969 posited that equal risk demands equal expected excess return, and proceeded to develop a practical pricing model based on this connection. It effectively set the Sharpe ratio of the expected return on the warrant equal to the Sharpe ratio for the underlying stock, which would then lead to the Black–Scholes–Merton (BSM) formula, but for a different and entirely ad hoc reason, compared to either the CAPM equilibrium argument of F. Black and M. Scholes, or Merton's own dynamic replication (no arbitrage) argument from the 1970s.

17.1 The Moon-Landing Problem

Due to the lack of atmosphere on the Earth's Moon, using a landing rocket appears to be the only practical method for a lunar module to resist the gravitational attraction of the Moon, corresponding to an escape velocity of about 2.8 km/sec, and land softly on its surface. This requires carrying a substantial amount of fuel some 370,000 km through space, which means that a soft landing has to be accomplished with the smallest amount of fuel possible. The first known formulation and solution to this fuel-minimization problem is due to Miele (1961), but the brief – and vastly oversimplified – description that we outline next is compiled from (Fleming and Rishel 1975).

Let $m(t)$ denote the mass of the spacecraft at time t (the mass changes rapidly if the landing rocket is turned on), let $h(t)$ denote the distance to the surface of the Moon, and let $v(t) = \frac{\mathrm{d}h(t)}{\mathrm{d}t}$ be the vertical velocity of the spacecraft (note that positive direction points away from the Moon). The time of the impact is

$$T = \inf\{t > 0 : h(t) = 0\}$$

and soft landing means $v(T) = 0$. Let $\alpha(t)$ stand for the thrust in the landing rocket – this is the only variable that the controller can effectively choose. The rate at which fuel is consumed, that is, the rate at which the mass $m(t)$ decreases, is proportional to the thrust and can be expressed as $k\alpha(t)$, where k is a known parameter. From Newton's law, the force field associated with the thrust $\alpha(t)$ generates vertical acceleration $\alpha(t)/m(t)$, which pulls opposite to the gravitational acceleration of the Moon $g_M \approx 1.6\mathrm{m/s}^2$ (assumed for simplicity to be constant when not too far from the surface). The entire state of the spacecraft at time t is encrypted into the three-dimensional vector $X(t) = (h(t), v(t), m(t))$, the dynamics of which are

$$\frac{\mathrm{d}X(t)}{\mathrm{d}t} \equiv \left(\frac{\mathrm{d}h(t)}{\mathrm{d}t}, \frac{\mathrm{d}v(t)}{\mathrm{d}t}, \frac{\mathrm{d}m(t)}{\mathrm{d}t}\right) = \left(v(t), -g_M + \frac{\alpha(t)}{m(t)}, -k\alpha(t)\right). \qquad \mathrm{e}_1$$

At time $t = 0$ the spacecraft is in the known state $X(0) = (h(0), m(0), v(0))$, and if, contrary to fact, we suppose that the landing strategy $\mathbb{R}_+ \ni t \rightsquigarrow \alpha(t)$ is a known piecewise continuous function, then the entire trajectory $\mathbb{R}_+ \ni t \rightsquigarrow X(t) \equiv X^\alpha(t)$ can be computed as the solution to the first-order ODE e_1, with the given initial values. The control of the trajectory $t \rightsquigarrow X^\alpha(t) \equiv (h^\alpha(t), v^\alpha(t), m^\alpha(t))$ then boils down to the choice of a fuel-burning plan $t \rightsquigarrow \alpha(t)$. For any such choice, the spacecraft will touch the surface of the Moon with vertical velocity $v^\alpha(T)$, and the total mass of the consumed fuel is going to be

$$m^\alpha(0) - m^\alpha(T) = k\int_0^T \alpha(s)\,\mathrm{d}s\,.$$

The associated constrained optimization problem can then be stated as follows:

$$\text{maximize:} \quad -k\int_0^T \alpha(s)\,\mathrm{d}s \equiv m^\alpha(T) - m^\alpha(0)$$

subject to:

$$\alpha \in \mathscr{A} \stackrel{\text{def}}{=} \text{the space of nonnegative piecewise continuous function on } \mathbb{R}_+,$$
$$\text{and} \quad t \rightsquigarrow X^\alpha(t) \text{ solves } e_1 \text{ with } v^\alpha(T) = 0\,.$$

The complete solution to this problem is outside of our main concerns, and we refer the reader to (Fleming and Rishel 1975) for details. It is interesting to note that, just as with most optimal paths encountered in nature, the optimal landing strategy is in some sense extremal, and can be described as follows: starting from position $(h(0), v(0), m(0))$ let the spacecraft fall freely without any intervention until the trajectory in the (h, v)-plane intersects a particular critical curve. At the moment of intersection the landing rocket is to be switched on, and kept at constant thrust $\alpha(t) = \alpha$ until touchdown. Computing the solution comes down to computing the critical curve in the (h, v)-plane that triggers ignition, and computing the value of the constant thrust α after the ignition.

The theory and methods of optimal control constitute a vast subject covered in a vast number of publications. This chapter is only a brief – and almost entirely ad hoc – summary of the main concepts, ideas, methods, and applications of stochastic optimal control. For a rigorous exposition we refer the reader to, among so many possible sources, (Fleming and Rishel 1975), (Pham 2009), and (Øksendal and Sulem 2005), which this chapter follows.

17.2 Principle of Dynamic Programming and the HJB Equation

Some elements of stochastic optimal control are already familiar to us: recall the derivation in section 13.2 of the wealth dynamics associated with a particular choice of a self-financing trading strategy. The choice of the trading strategy π_t and the consumption plan c_t is analogous to the choice of the thrust $\alpha(t)$ in the landing rocket. Indeed, the choice of the pair (π_t, c_t) affects the dynamics (locally in time) of certain quantities of interest and amounts to "steering" those quantities in one way or another. To be concrete, equation e_1 in section 13.2 tells us that the wealth process V has dynamics

$$V_{t+\mathrm{d}t} - V_t \equiv \mathrm{d}V_t = V_t\, r_t\, \mathrm{d}t + \pi_t{}^\intercal \mathrm{d}X_t - c_t\, \mathrm{d}t\,, \quad t \in \mathbb{R}_+\,.$$

Despite the obvious similarity, one crucial difference exists between steering a spacecraft to a soft landing on the Moon and steering an investor's wealth by way of dynamic trading, as shown above. In the first case the dynamics are fully deterministic and the decision how to control the system can be made ahead of time. In the second case the investment decision π_t and the consumption choice c_t at time t affect the wealth at time $t + \mathrm{d}t$, but that wealth is also affected by the realization of the instantaneous excess return $\mathrm{d}X_t$, which is not known at time t. It would be suboptimal, then, to make a decision about the quantities $\pi_{t+\mathrm{d}t}$ and $c_{t+\mathrm{d}t}$ at time t. In most practical situations the choice of the control is based on a certain feedback from the system, and, for this reason, must be made in real time. Nevertheless, as we are about to see, quite a bit of computing can still be done ahead of time, before the real-time control procedure begins.

Before we begin, we must specify a sufficiently general model of a controlled stochastic dynamical system. In the most general setup, we suppose that the position of the system in the state space follows some stochastic process X, the dynamics of which are governed by the diffusion equation

$$X = X_0 + b(X,\alpha,\iota){\cdot}\iota + \sigma(X,\alpha,\iota){\cdot}W + \int_{\mathbb{R}} \gamma(X_-,\alpha_-,z,\iota){\cdot}\tilde{N}(dz,[0,\iota]), \qquad \mathrm{e}_1$$

driven by the Brownian motion W and the compensated Poisson random measure $\tilde{N}(dz,dt) = N(dz,dt) - \nu(dz)dt$, chosen to be independent of W. The process α is càdlàg and adapted, with values in some appropriate domain $A \subseteq \mathbb{R}$ – usually an interval, or some other sufficiently nice domain inside some Euclidean space $\mathbb{R}^k$. These objects are constructed on a complete probability space $(\Omega,\mathcal{F},P)$, endowed with the natural filtration generated by W and N, that is, the smallest right-continuous and complete filtration, $\mathscr{F}$, to which the processes W and $N(B,[0,\iota])$, for all choices of $B \in \mathcal{B}(\mathbb{R}^m)$, happen to be adapted. We deliberately leave some of the dimensions involved ambiguous: we would like to use notation that appears to be one dimensional but allows for a multidimensional interpretation as well. Thus, in general, X, $b(X_t,\alpha_t,t)$, and $\gamma(X_{t-},\alpha_{t-},z,t)$, could be n-dimensional vectors, $\sigma(X_t,\alpha_t,t)$ could be a matrix with dimensions (n,d), and W could be a d-dimensional Brownian motion (note that $\tilde{N}$ is a scalar quantity and the last integral in e_1 is understood in terms of the operation multiplication of a vector by a scalar). We also note⊸(16.31) that nothing would change in e_1 if we write $b(X_-,\alpha_-,\iota)$ instead of $b(X,\alpha,\iota)$, and write $\sigma(X_-,\alpha_-,\iota)$ instead of $\sigma(X,\alpha,\iota)$.

Conventions about the notation aside, one crucially important feature of equation e_1 is that the diffusion coefficients b, σ, and γ – which are fixed in the outset as deterministic functions defined on appropriate Euclidean domains – depend not only on the state of the system X_{t-}, but also on the (predictable) control variable $\alpha_{t-} \in A$. Plainly, the state of the system is subjected to the instantaneous shocks dW_t and $N(dz,dt)$ through the diffusion forces encrypted in the coefficients b, σ, and γ, but those forces can still be steered by the control α_{t-}, and we again stress that any practically feasible steering mechanism must operate in a predictable fashion.[188] At time t the controller observes X_t and makes the decision α_t. If the diffusion forces were to remain frozen over a very short time period of length ε that starts at time t, then at the end of this (very short) period the system

[188] Note, however, that the process α, which models the control, is only optional – to be precise càdlàg and adapted – and it is the process α_- that is claimed to be predictable.

will be in the state

$$X_{t+\varepsilon} = X_t + \beta(X_t, \alpha_t, t)\varepsilon + \sigma(X_t, \alpha_t, t)(W_{t+\varepsilon} - W_t) + \int_{\mathbb{R}} \gamma(X_t, \alpha_t, z, t)\tilde{N}(\mathrm{d}z,]t, t+\varepsilon]),$$

which, although random, will still depend on the choice of the control α_t. We stress that the choice of α_t at time t takes effect only *after* the moment t, not simultaneously with the jump at time t (if there is jump at time t, that is, if $X_t \neq X_{t-}$). The intuition is that the controller can observe the jumps generated by the random measure N and react to those jumps only *after* the jumps have been realized – this matter becomes irrelevant if the instantaneous shocks in the controlled diffusion X do not contain jumps. If the dependence of the controlled diffusion X on the control α needs to be emphasized in the notation, we will write $X^\alpha = (X_t^\alpha)_{t\in\mathbb{R}_+}$. Of course, the control process α and the coefficients b, σ, and γ must satisfy certain integrability and growth conditions in order to guarantee that the stochastic equation e_1 is meaningful and admits a unique strong solution. A control process α that has these attributes will be said to be *admissible*, and the collection of all admissible controls that follow the event $\{X_t = x\}$ will be denoted by $\mathscr{A}(x,t)$. Before we continue further, we note that many important models can be formulated as time-homogeneous control problems, in which the coefficients are time-invariant, that is, have the form

$$b(x,a,t) = b(x,a)\,, \quad \sigma(x,a,t) = \sigma(x,a)\,, \quad \gamma(x,a,z,t) = \gamma(x,a,z)\,.$$

In such systems the future dynamics (and control) following time t depend on the observed state $X_t = x$, but not on the moment in time t. To put it another way, nothing would change if at time t the clock is reset to 0 and the system is restarted from state $X_0 \equiv X_{0-} = x$.

To be able to decide about the choice of admissible control $(\alpha_s)_{s\geq t} \in \mathscr{A}(x,t)$, the controller needs a well-defined *performance criterion.* The most general and common form of such a criterion is

$$J(x,\alpha,t) \overset{\text{def}}{=} \mathsf{E}\left[\int_t^{\mathfrak{s}} F(X_s^\alpha, \alpha_s, s)\mathrm{d}s + G(X_{\mathfrak{s}}^\alpha, \mathfrak{s})1_{\{\mathfrak{s}<\infty\}} \;\Big|\; X_t^\alpha = x\right], \qquad \mathrm{e}_2$$

$$\alpha \in \mathscr{A}(x,t)\,, \quad \mathfrak{s} \in \mathscr{T}_{[t,T]}\,, \quad 0 < T \leq \infty\,.$$

In this expression $(x,a,s) \rightsquigarrow F(x,a,s)$ and $(x,s) \rightsquigarrow G(x,s)$ are given functions that satisfy certain measurability and integrability conditions (needed to make the expressions meaningful both mathematically and from a practical point of view), while $\mathscr{T}_{[t,T]}$ denotes the space of all $\mathscr{F}$-stopping times with values in $[t,T]$ for some (exogenously specified, finite, or infinite) time horizon $0 < T \leq \infty$. The function F models the so-called *integral payoff*, while the function G models the *termination (closing) payoff*. This terminology will become clearer once we turn to concrete applications. At this point we only note that in many concrete models the criterion above is used with $F \equiv 0$ or $G \equiv 0$. We will see later that the classical problem for optimal exercise of American-style derivative contracts fits this framework with no integral payoff ($F \equiv 0$) and with termination payoff equal to the payoff from exercising the option at the (random) time $\mathscr{s}$ – and the choice of the termination time $\mathscr{s}$ is the only decision that the controller can make. In other applications, say in the classical Merton problem, the termination rule $\mathscr{s}$ is not only exogenous but also deterministic and can be identified as some (finite or infinite) time horizon $T > t$.

In any case, given the dynamics e_1 and the performance criterion e_2, the controller's objective at time t, after observing that the system is in state $X_t = x$, is to find

$$\hat{\alpha}^{x,t} = \arg\max_{\alpha \in \mathscr{A}(x,t)} J(x, \alpha, t)\,.$$

Assuming that such $\hat{\alpha}^{x,t}$ exists – and we will often write simply $\hat{\alpha}$ if the t and the x are understood from the context – the value function associated with having the system in state x at time t is simply

$$V(x,t) \stackrel{\text{def}}{=} J(x, \hat{\alpha}^{x,t}, t) \equiv \max_{\alpha \in \mathscr{A}(x,t)} J(x, \alpha, t)\,.$$

(17.1) REMARK: In general, the arg max may not exist, in which case the definition of the value function would involve sup instead of max, that is,

$$V(x,t) = \sup_{\alpha \in \mathscr{A}(x,t)} J(x, \alpha, t)\,.$$

We stress that the value attached to the pair (x,t) reflects the rewards generated by the best possible (from the point of view of the performance criterion J) admissible control policy, starting from state x at time t onward. ○

(17.2) REMARK: It can be shown that under some minor technical conditions (see (Pham 2009) and (Øksendal and Sulem 2005)), if it exists, the optimal control policy process has the form $\hat{\alpha}_t = \hat{\alpha}(X_t, t)$ for some function $\alpha\colon \mathbb{R}^n \times \mathbb{R}_+ \mapsto A$, and is càdlàg and adapted. To put it another way, the search for an optimal policy is reduced to the search for a function $(x,t) \rightsquigarrow \alpha(x,t)$ that maps the state $X_t = x$ into an actual decision, namely, $\alpha(x,t) \in A$, to be taken in that state. As a result, the controlled process (X_t^{α}) becomes a càdlàg Markov process with infinitesimal generator $\mathcal{A}_t^{\alpha}$, which can be cast in the one-dimensional setting as

$$\begin{aligned} \mathcal{A}_t^{\alpha}\varphi(x) &= b\big(x, \alpha(x,t), t\big)\partial_x\varphi(x) + \frac{1}{2}\sigma\big(x, \alpha(x,t), t\big)^2\partial_x^2\varphi(x) \\ &+ \int_{\mathbb{R}} \Big(\varphi\big(x + \gamma\big(x, \alpha(x,t), z, t\big)\big) - \varphi(x) - \gamma\big(x, \alpha(x,t), z, t\big)\partial_x\varphi(x)\Big)\nu(\mathrm{d}z)\,, \end{aligned}$$

and in the multidimensional setting as

$$\begin{aligned} \mathcal{A}_t^{\alpha}\varphi(x) &= b\big(x, \alpha(x,t), t\big)\cdot\nabla\varphi(x) \\ &\quad + \frac{1}{2}\Big(\sigma\big(x, \alpha(x,t), t\big)\sigma\big(x, \alpha(x,t), t\big)^{\mathsf{T}}\nabla\Big)\cdot\nabla\varphi(x) \\ &+ \int_{\mathbb{R}^m} \Big(\varphi\big(x + \gamma\big(x, \alpha(x,t), z, t\big)\big) - \varphi(x) - \gamma\big(x, \alpha(x,t), z, t\big)\cdot\nabla\varphi(x)\Big)\nu(\mathrm{d}z)\,. \end{aligned}$$

Finally, we remark that since the dot product between two vectors $u, v \in \mathbb{R}^n$ can be expressed as $u \cdot v = \mathrm{tr}(uv^\intercal)$, we can write, equivalently,

$$\begin{aligned} \mathcal{A}_t^\alpha \varphi(x) &= b\big(x, \alpha(x,t), t\big) \cdot \nabla \varphi(x) \\ &\quad + \frac{1}{2} \mathrm{tr}\Big[\sigma\big(x, \alpha(x,t), t\big)\sigma\big(x, \alpha(x,t), t\big)^\intercal (\nabla \nabla^\intercal)\varphi(x)\Big] \\ &+ \int_{\mathbb{R}^m} \Big(\varphi\big(x + \gamma\big(x, \alpha(x,t), z, t\big)\big) - \varphi(x) - \gamma\big(x, \alpha(x,t), z, t\big) \cdot \nabla \varphi(x)\Big) \nu(\mathrm{d}z), \end{aligned}$$

where $(\nabla \nabla^\intercal)$ is understood as the formal matrix with entries $\frac{\partial^2}{\partial x_i \partial x_j}$, so that the symbol $(\nabla \nabla^\intercal)\varphi(x)$ is nothing but the Hessian matrix of the function $\varphi(x)$. ○

To develop some sense of how the optimal strategy $\hat{\alpha}$ can be obtained, notice first that the performance criterion $J(x, \alpha, t)$ from e_2 can be rewritten as[189]

$$\begin{aligned} &J(x, \alpha, t) \\ &= \mathsf{E}\Big[\int_t^{t_\varepsilon} F(X_s^\alpha, \alpha_s, s)\mathrm{d}s + \int_{t_\varepsilon}^{\jmath} F(X_s^\alpha, \alpha_s, s)\mathrm{d}s + G(X_\jmath^\alpha, \jmath)1_{\{\jmath<\infty\}} \,\Big|\, X_t^\alpha = x\Big] \\ &\qquad = \mathsf{E}\Big[\int_t^{t_\varepsilon} F(X_s^\alpha, \alpha_s, s)\mathrm{d}s \\ &\qquad\quad + \mathsf{E}\Big[\int_{t_\varepsilon}^{\jmath} F(X_s^\alpha, \alpha_s, s)\mathrm{d}s + G(X_\jmath^\alpha, \jmath)1_{\{\jmath<\infty\}} \,\Big|\, X_{t_\varepsilon}^\alpha\Big] \,\Big|\, X_t^\alpha = x\Big] \\ &\qquad\qquad = \mathsf{E}\Big[\int_t^{t_\varepsilon} F(X_s^\alpha, \alpha_s, s)\mathrm{d}s + J(X_{t_\varepsilon}^\alpha, \alpha, t_\varepsilon) \,\Big|\, X_t^\alpha = x\Big], \end{aligned} \qquad \mathrm{e}_3$$

where t_ε is the stopping time $t_\varepsilon \overset{\text{def}}{=} \jmath \wedge (t + \varepsilon)$ for some fixed (think "very small") $\varepsilon > 0$. Clearly, for any concrete choice of the control $(\alpha_s)_{s \in [t, t_\varepsilon[}$ – whether optimal or not – to maximize $J(x, \alpha, t)$, starting at time t_ε from state $X_{t_\varepsilon}^\alpha$ onward, the control $(\alpha_s)_{s \in [t_\varepsilon, \jmath[}$ will have to be optimal, that is, will have to maximize $J(X_{t_\varepsilon}^\alpha, \alpha, t_\varepsilon)$. This means that in the maximization of the expression in e_3 one can replace $J(X_{t_\varepsilon}^\alpha, \alpha, t_\varepsilon)$ with its maximal value $V(X_{t_\varepsilon}^\alpha, t_\varepsilon)$ and then maximize the entire expression only over the control $(\alpha_s)_{s \in [t, t_\varepsilon[}$.

The heuristic argument that we just outlined leads to the following crucial result that can be traced back to Bellman (1954):

(17.3) BELLMAN'S PRINCIPLE OF DYNAMIC PROGRAMMING (PDP):[190] In the setting described above,

$$\dagger \qquad V(x, t) = \sup_{\alpha \in \mathscr{A}(x,t)} \mathsf{E}\Big[\int_t^{t_\varepsilon} F(X_s^\alpha, \alpha_s, s)\mathrm{d}s + V(X_{t_\varepsilon}^\alpha, t_\varepsilon) \,\Big|\, X_t^\alpha = x\Big].$$

[189] The reader should be aware that these expressions are only formal, and we refer to, say, (Pham 2009) and (Øksendal and Sulem 2005) for a rigorous derivation.

[190] Named after the American mathematician Richard Ernest Bellman (1920–1984).

Plainly, if one already knows how to find the optimal control $(\alpha_s)_{s \in [t_\varepsilon, \mathfrak{s}[}$, starting from *any* possible realization of the state variable X_{t_ε} in the future moment t_ε, and therefore already knows how to compute the value function $V(x, t_\varepsilon)$ for all possible realizations $X_{t_\varepsilon} = x$, then to compute the value at the current state $X_t = x$, that is, $V(x,t)$, one would only need to optimize over the shorter time domain $s \in [t, t_\varepsilon[$ and seek only the immediate portion of the control, namely, the process $(\alpha_s)_{s \in [t, t_\varepsilon[}$. ○

(17.4) THE BACKWARD INDUCTION ALGORITHM: Bellman's principle gives rise to a computational program that can be outlined as follows. On some terminal date $T > t$ the value function $V(\cdot, T)$ is somehow known and the choice of the control α_T is obvious (or, perhaps, irrelevant). From (17.3†), used with $t_\varepsilon = T$, one can compute the value function shortly before time T, say, at time $t = T - \varepsilon$, and also compute the control between time $T-\varepsilon$ and time T. Now that the function $V(\cdot, T-\varepsilon)$ is known, in exactly the same way one can compute the value function $V(\cdot, T - 2\varepsilon)$ and the optimal control from time $T - 2\varepsilon$ only until time $T - \varepsilon$, and so on. We stress that at each of these steps the value function, associated with a particular moment in time $t = T - k\varepsilon$, must be computed as a function on the entire state space, that is, one must compute ahead of time the values and the optimal decisions in all possible states in which the controlled diffusion process X^α could be found at all future moments. Essentially, this transforms a global problem with time horizon $T > 0$ into a (presumably very long) sequence of smaller optimization problems that are local in time. The reason why this is helpful is that if one seeks control locally in time, one is essentially looking for a single scalar (or, more generally, vector) value, whereas solving the global problem would require the construction of an entire (random) path of the control. ○

Our next step is to develop some understanding of how the backward induction would look when the time step decreases to 0. To this end, we divide both sides of (17.3†) by $\varepsilon > 0$ and put the same equation in the form

$$0 = \sup_{\alpha \in \mathscr{A}(x,t)} \mathsf{E}\Big[\frac{1}{\varepsilon}\int_t^{t_\varepsilon} F(X_s^\alpha, \alpha_s, s)\mathrm{d}s + \frac{1}{\varepsilon}\Big(V(X_{t_\varepsilon}^\alpha, t_\varepsilon) - V(x,t)\Big) \,\Big|\, X_t^\alpha = x\Big].$$

Writing down the Itô expansion of $V(X_{t_\varepsilon}^\alpha, t_\varepsilon) - V(x,t)$ and ignoring all stochastic integrals in this expansion (hoping any such terms do not affect the conditional expectation), what would remain⊸(16.45) from the expansion is

$$\frac{1}{\varepsilon}\int_t^{t_\varepsilon}\Big(\partial V(X_s^\alpha, s) + (\mathcal{A}_s^\alpha V)(X_s^\alpha, s)\Big)\mathrm{d}s\,.$$

Thus, passing to the limit in the equation above as $\varepsilon \searrow 0$ and observing that $t_\varepsilon = \mathfrak{s} \wedge (t+\varepsilon) = t+\varepsilon$ for any sufficiently small ε (recall that $\mathfrak{s} > t$ a.s.), we arrive at the following result:

(17.5) HAMILTON–JACOBI–BELLMAN (HJB) EQUATION:[191] Let everything be as above, and assume that all integrability and other analytical conditions needed to justify the operations involved are satisfied. Then the following equation holds:

$$\dagger \qquad \partial V(x,t) + \sup_{\alpha \in \mathscr{A}(x,t)} \Big(F(x, \alpha_t, t) + (\mathcal{A}_t^{\alpha} V)(x,t) \Big) = 0\,.$$

If $\jmath = T$ is some exogenously given deterministic closing date, then, typically, the PDE above will have to be solved in some domain $\mathscr{R} \times [0, T[$ for some open rectangular region $\mathscr{R} \subseteq \mathbb{R}^n$ (such as $\mathscr{R} = \mathbb{R}^n_{++}$, e.g., but the region $\mathscr{R}$ need not always be rectangular), with boundary condition

$$\lim_{t \nearrow T} V(x,t) = G(x,T)\,, \qquad x \in \mathscr{R}\,.$$

Heuristically at least, the construction of such a solution can be viewed as an implementation of the backward induction with an infinitesimally small time step.

If the termination time $\jmath$ is not deterministic, then, typically, it is going to be identified as the *entry time* of the controlled diffusion X^{α} into some *open* (exogenous or endogenous) domain $\mathscr{D} \subseteq \mathbb{R}^n \times \mathbb{R}_+$ with piecewise smooth boundary $\partial\mathscr{D}$.[192] In this case the HJB equation has to be solved in the (open) domain $\bar{\mathscr{D}}^{\complement}$, where $\bar{\mathscr{D}}$ is the closure of the domain $\mathscr{D}$, assumed to be of the form $\bar{\mathscr{D}} = \partial\mathscr{D} \cup \mathscr{D}$. The boundary condition is $V(x,t) = G(x,t)$ for any $(x,t) \in \bar{\mathscr{D}}$. The situations where the domain $\mathscr{D}$ is endogenous are the ones where the termination rule $\jmath$ is a matter of choice that the controller has to make, that is, the stopping rule is itself part of the control, as in the case of American-style derivative contracts, a topic that will be addressed later in the sequel. ○

(17.6) REMARKS: Since the HJB equation (17.5†) involves the partial derivatives of the unknown function $V(x,t)$, these derivatives must exist throughout the entire domain for the equation to be meaningful. However, the intrinsic nature of the HJB equation is such that it may actually destroy smoothness. Indeed, the operations sup and max may produce functions with a kink even when applied to smooth function, as in $\max\{1, 1-x\}$, for example. In particular, there is no reason for the action of the sup operator in (17.5†) to produce a differentiable function, which is to say there is no reason for the time derivative $\partial V(x,t)$ to be smooth. As we are about to see, such issues are quite common and eventually have to be dealt with in one way or another. In addition, sup is a nonlinear operation, and generally the expression involving the sup would not be a linear function of the derivatives of $V(x,t)$. One consequence from these observations is that producing numerical solutions by using classical numerical methods developed for linear PDEs (the finite-difference method and its variations is the first to come to mind) becomes a highly nontrivial

191 Named after the Irish mathematician Sir William Rowan Hamilton (1805–1865), the German mathematician Carl Gustav Jacob Jacobi (1804–1851), and the American mathematician Richard Ernest Bellman (1920–1984).

192 In order for the entry time to be a stopping time we need to assume that the filtration is right-continuous – consult (8.33).

task. One remedy is to construct subdomains where the solution happens to be smooth and then glue the pieces, but this approach leaves a lot to be desired, for to say the least, typically the domain where the solution is smooth is not known a priori. A more general method, which allows one to circumvent at least some of the problems just described, is provided by the so-called *viscosity solutions*, which are meaningful even without the smoothness requirement. Yet another – perhaps even more general and elegant – method is the *stochastic principle of maximum formulation* and the related tools from the domain of *backward stochastic differential equations* (BSDE). Such tools are beyond the scope of this space, and we refer the reader to (Pham 2009) and (Crépey 2012) for a detailed exposition and an extensive list of references. In addition, (Forsyth 2012) provides an overview of some numerical procedures for nonlinear PDEs with applications to finance. Here we are only going to note that the stochastic principle of maximum is completely independent from the principle of dynamic programming, and in fact allows the construction of an optimal control policy without any reference to the value function. Unfortunately, building concrete computational programs based on this method is not as straightforward as in the case of dynamic programming. ○

Essentially all practical applications of the HJB equation rely on the following crucial result (see (Pham 2009) and (Øksendal and Sulem 2005)):

(17.7) VERIFICATION THEOREM (A SKETCH): With the notation as in (17.5), suppose that $U(x,t)$ is some "sufficiently nice" function for which the following relation is satisfied throughout the domain, $\bar{\mathscr{D}}^{\complement}$, of the HJB equation in ($17.5^{\dagger}$):

$$\partial U(x,t) + \sup_{a\in A}\Big(F(x,a,t) + (\mathcal{A}_t^a U)(x,t)\Big) \leq 0\,, \quad (x,t)\in\bar{\mathscr{D}}^{\complement}\,,$$

in addition to the boundary condition $U(x,t) = G(x,t)$, $(x,t)\in\bar{\mathscr{D}}$. Furthermore, suppose that $\bar{\mathscr{D}}^{\complement} \ni (x,t) \rightsquigarrow \hat{\alpha}(x,t) \in A$ is yet another "sufficiently nice" function, such that

$$\begin{aligned}&\partial U(x,t) + \sup_{a\in A}\Big(F(x,a,t) + (\mathcal{A}_t^a U)(x,t)\Big)\\ &\qquad = \partial U(x,t) + F(x,\hat{\alpha}(x,t),t) + (\mathcal{A}_t^{\hat{\alpha}}U)(x,t) = 0\,, \quad (x,t)\in\bar{\mathscr{D}}^{\complement}\,.\end{aligned}$$

Finally, suppose that, for any starting point $(x,t)\in\bar{\mathscr{D}}^{\complement}$, the following SDE

$$\begin{aligned}\hat{X} = X_t &+ b\big(\hat{X},\hat{\alpha}(\hat{X},\iota),\iota\big)\bullet\iota + \sigma\big(\hat{X},\hat{\alpha}(\hat{X},\iota),\iota\big)\bullet W\\ &+ \int_{\mathbb{R}}\gamma\big(\hat{X}_-,\hat{\alpha}(\hat{X}_-,\iota),z,\iota\big)\bullet\tilde{N}(\mathrm{d}z,[t,\iota])\end{aligned}$$

has a (nonexploding) solution with bounded jumps that starts at time t from position $\hat{X}_t = x$. Then one can claim that $U(x,t) = V(x,t)$ for all $(x,t)\in\bar{\mathscr{D}}^{\complement}$, that is, $U(x,t)$ is nothing but the value function for the optimal control problem, and the optimal control policy is given by the function $\hat{\alpha}(x,t)$, that is, the optimal decision at time t can be expressed as $\hat{\alpha}(X_t,t)$. ○

The quintessential application of the HJB equation to the domain of finance is the following:

(17.8) MERTON'S PORTFOLIO SELECTION PROBLEM WITH FINITE TIME HORIZON AND NO INTERTEMPORAL CONSUMPTION: An agent can invest in only two assets: a risky security with price process

$$S = S_0 + bS\bullet\imath + \sigma S\bullet W\,,$$

and a short-lived risk-free asset with price process

$$S^\circ = S_0^\circ + rS^\circ\bullet\imath\,,$$

where $b, r, \sigma \in \mathbb{R}$ are given parameters. The probability space $(\Omega, \mathcal{F}, P)$ and the filtration $\mathscr{F} \subset \mathcal{F}$ (automatically assumed right-continuous and complete) are the ones associated with the Brownian motion W. At time $t = 0$ the agent has initial wealth $-\infty < y_0 < \infty$, but has no additional income other than the capital gains or losses from trading. The agent cannot consume any of their wealth before date T, and the available wealth on date $t < T$ can only be invested in a portfolio consisting of the risky and risk-free assets. This portfolio can be rebalanced continuously at no cost, and the objective is to do so in such a way that the expected utility from final wealth – that is, from the wealth attained on date $t = T$ – is maximized. The agent's utility from final wealth is a power utility of the form $U(y) = y^p/p$, $p < 1$. The control in this problem is modeled as a predictable (relative to $\mathscr{F}$) process α, which gives the (generally, time dependent) proportion of their wealth that the agent keeps in the risky asset: if at time t the available wealth is Y_t, the amount invested in the risky asset at time t is $\alpha_t Y_t$ and the amount invested in the risk-free asset is $(1-\alpha_t)Y_t$. If no short positions are allowed in any of the assets, then α_t must be restricted to the domain $A \overset{\text{def}}{=} [0, 1]$; if short positions are allowed only in the risky asset, then $\alpha_t \in A \overset{\text{def}}{=}]-\infty, 1]$; if short positions are allowed only in the risk-free asset, then $\alpha_t \in A \overset{\text{def}}{=} [0, \infty[$; and if short positions are allowed in either asset, then $\alpha_t \in A \overset{\text{def}}{=}]-\infty, \infty[$. If the investor's positions are subject to other limited liability type constraints, then the domain A must be modified accordingly. ○

(17.9) EXERCISE: Prove that the wealth process $Y^\alpha = (Y_t^\alpha)_{t\in[0,T]}$ satisfies the equation

$$Y^\alpha = Y_0^\alpha + (\sigma\alpha Y^\alpha)\bullet W + \big(b\alpha + r(1-\alpha)\big)Y^\alpha\bullet\imath\,. \quad ○$$

The optimal control problem that the investor is faced with has performance criterion

$$J(y, \alpha, t) = \mathsf{E}\Big[\frac{1}{p}\big(Y_T^\alpha\big)^p \,\Big|\, Y_t^\alpha = y\Big], \quad 0 \le t < T\,,$$

and the associated value function is

$$V(y, t) = \sup_{\alpha\in\mathscr{A}(y,t)} J(y, \alpha, t) \equiv \sup_{\alpha\in\mathscr{A}(y,t)} \mathsf{E}\Big[\frac{1}{p}\big(Y_T^\alpha\big)^p \mid Y_t^\alpha = y\Big],$$

where $\mathscr{A}(y, t)$ is the collection of all predictable processes $\alpha = (\alpha_s)_{s\in[t,T]}$ with values in the set A, which set depends on the trading restrictions discussed earlier.

(17.10) EXERCISE: Write down the HJB equation associated with Merton's portfolio selection problem formulated in (17.8), in the case where no short sales are allowed in either security (do not forget the boundary condition). Consider the quantity

$$\hat{\alpha} = \frac{b - r}{(1 - p)\sigma^2},$$

and explain how this quantity compares with the Sharpe ratio in this model. In addition, show that if $\hat{\alpha} \leq 0$, the optimal portfolio choice is $\alpha_t = 0$ (no investment in the risky asset), and if $\hat{\alpha} \geq 1$, the optimal portfolio choice is $\alpha_t = 1$ (no investment in the risk-free asset). Also show that if $0 < \hat{\alpha} < 1$, then the optimal portfolio choice is $\alpha_t = \hat{\alpha}$, that is, it is optimal to keep a fixed proportion of the wealth in the risky asset. Explain the intuition behind these results. Finally, explain how the verification theorem is used in the setting of Merton's problem and develop an explicit formula for the value function $V(y, t)$.

HINT: Look for a solution to the HJB equation in the form $V(y, t) = \frac{1}{p} f(t) y^p$ for some unknown function f. Explain why it would be natural to seek a solution in this form, write an ODE for the unknown function f, and find a closed-form expression for f. ○

(17.11) EXERCISE: Give the complete solution to Merton's problem in all cases in which short sales are allowed. ○

(17.12) EXERCISE: Give a complete solution to Merton's problem in the case where the agent has logarithmic utility from final wealth $U(y) = \log(y)$, $y > 0$.

HINT: Look for a solution to the HJB equation in the form $V(y, t) = f(t) + \log(y)$ for some unknown function f and notice that the terminal condition reads $V(y, T) = \log(y)$. Explain why it would be natural to seek a solution in this particular form, write an ODE for the unknown function f, and find a closed-form expression for f. ○

17.3 Some Variations of the PDP and the HJB Equation

Just as the entire chapter, the present section is based mostly on heuristic considerations. Assume the setting and notation introduced in the previous section and observe that in the context of financial applications, typically, all future payoffs must be discounted accordingly if such payoffs are to enter any present time planning, contracts, and decisions. As a result, the performance criterion e_2 in section 17.2 is often stated in the form

$$\begin{aligned} J(x, \alpha, t) \stackrel{\text{def}}{=} \mathsf{E}\Big[\int_t^{\mathfrak{z}} & e^{-\int_t^s r_u \mathrm{d}u} F(X_s^\alpha, \alpha_s, s) \mathrm{d}s \\ & + e^{-\int_t^{\mathfrak{z}} r_u \mathrm{d}u} G(X_{\mathfrak{z}}^\alpha, \mathfrak{z}) 1_{\{\mathfrak{z} < \infty\}} \;\Big|\; X_t^\alpha = x \Big], \\ & \alpha \in \mathscr{A}(x, t), \quad \mathfrak{z} \in \mathscr{T}_{[t,T]}, \quad 0 < T \leq \infty, \end{aligned}$$

for some exogenously given and predictable process $r = (r_t)_{t \in \mathbb{R}_+}$, which represents "instantaneous discount rate" (such as, say, the instantaneous risk-free rate in the financial market). Bellman's principle of dynamic programming⊸(17.3) can now be stated as (recall that $\mathscr{t}_\varepsilon = \mathscr{s} \wedge (t + \varepsilon)$)

$$V(x,t) = \sup_{\alpha \in \mathscr{A}(x,t)} \mathsf{E}\Big[\int_t^{\mathscr{t}_\varepsilon} e^{-\int_t^s r_u \mathrm{d}u} F(X_s^\alpha, \alpha_s, s)\mathrm{d}s \\ + e^{-\int_t^{\mathscr{t}_\varepsilon} r_u \mathrm{d}u} V(X_{\mathscr{t}_\varepsilon}^\alpha, \mathscr{t}_\varepsilon) \mid X_t^\alpha = x\Big].$$

How are these changes going to affect the derivation of the HJB equation? Just as we did in the previous section, we again put the equation above in the form

$$0 = \sup_{\alpha \in \mathscr{A}(x,t)} \mathsf{E}\Big[\frac{1}{\varepsilon}\int_t^{\mathscr{t}_\varepsilon} e^{-\int_t^s r_u \mathrm{d}u} F(X_s^\alpha, \alpha_s, s)\mathrm{d}s \\ + \frac{1}{\varepsilon}\Big(e^{-\int_t^{\mathscr{t}_\varepsilon} r_u \mathrm{d}u} V(X_{\mathscr{t}_\varepsilon}^\alpha, \mathscr{t}_\varepsilon) - V(x,t)\Big) \mid X_t^\alpha = x\Big],$$

and again expand the second term according to Itô's formula. Ignoring the stochastic integrals in the expansion, what remains can be cast as

$$\frac{1}{\varepsilon}\int_t^{\mathscr{t}_\varepsilon} \Big(\partial V(X_s^\alpha, s) + (\mathcal{A}_s^{\alpha_s} V)(X_s^\alpha, s) - r_s V(X_s^\alpha, s)\Big)\mathrm{d}s\,.$$

As a result the HJB equation obtains the form

$$\partial V(x,t) + \sup_{\alpha \in \mathscr{A}(x,t)} \Big(F(x,\alpha_t,t) + (\mathcal{A}_t^{\alpha_t} V)(x,t)\Big) = r_t V(x,t),$$

which is exactly ($17.5^\dagger$) if the discount rate is $r_t = 0$. The same equation is often cast as

$$-\partial V(x,t) = \sup_{\alpha \in \mathscr{A}(x,t)} \Big(F(x,\alpha_t,t) + (\mathcal{A}_t^{\alpha_t} V)(x,t) - r_t V(x,t)\Big).$$

Now we turn to a particularly important and common optimal control problem: the problem of optimal stopping. This is a special type of control, in which the set of admissible choices $\mathscr{A}(x,t)$ is reduced to two possible actions: *terminate* and collect the one-time termination reward, or, *take no action* and continue to collect the continuation rewards (if any). This type of control does not affect the dynamics of the process X, which is simply the solution to

$$X = X_0 + b(X,\iota){\scriptstyle\bullet}\iota + \sigma(X,\iota){\scriptstyle\bullet}W + \int_{\mathbb{R}} \gamma(X_-, z, \iota){\scriptstyle\bullet}\tilde{N}(\mathrm{d}z, [0,\iota]).$$

The performance criterion is set to

$$J(t,x,\mathscr{s}) \stackrel{\text{def}}{=} \mathsf{E}\Big[\int_t^{\mathscr{s}} e^{-\int_t^s r_u \mathrm{d}u} F(X_s, s)\mathrm{d}s \\ + e^{-\int_t^{\mathscr{s}} r_u \mathrm{d}u} G(X_{\mathscr{s}}, \mathscr{s}) 1_{\{\mathscr{s}<\infty\}} \mid X_t = x\Big], \quad \mathscr{s} \in \mathscr{T}_{[t,T]},$$

where $0 < T \leq +\infty$ is the (exogenous) expiration time, and, just as before, r is some predictable process that gives the discount rate for the payoffs. Of course, to develop practical models, one must be much more specific about the discount rate. For the purpose of this section it would be enough for us to suppose that r is, say, some deterministic function of time, or – if such an assumption is too restrictive – suppose that $r = r(X, \iota)$ for some continuous function $r\colon \mathbb{R}^n \times \mathbb{R}_+ \mapsto \mathbb{R}$. The choice of the control comes down to the choice of the stopping time $\jmath$. Consequently, the value of having the process in state (x, t), that is, the value of the event $\{X_t = x\}$, given the ability to choose the termination policy, is given by

$$V(x,t) \stackrel{\text{def}}{=} \sup_{\jmath \in \mathcal{T}_{[t,T]}} J(x, \jmath, t) .$$

The decision to terminate at time t comes down to setting $\jmath = t$, and this choice yields the termination payoff $G(x,t)$. It is clear that $V(x,t) \geq G(x,t)$ for all choices of x and t, since any state (x,t) is at least as valuable as the benefit that the controller can receive in that state by choosing to terminate immediately. Furthermore, stating that termination in state (x,t) is optimal is no different from stating that the value of that state equals the termination payoff (in that state), that is, $V(x,t) = G(x,t)$. The states in which taking no action would be optimal are the states that are strictly more valuable than the payoff from terminating in those states. The collection of all no action states is the so-called *continuation region* defined as

$$\mathfrak{C} \stackrel{\text{def}}{=} \big\{(x,t) \in \mathbb{R}^n \times [0,T[\colon V(x,t) > G(x,t)\big\} .$$

Generally, the process X starts from some initial state $X_0 \in \mathfrak{C}$ and is allowed to run for as long as the sample path X_ω stays inside the domain $\bar{\mathfrak{C}}$ (the closure of the domain $\mathfrak{C}$), but no later than time $T \leq \infty$ (if $T < \infty$ and termination has not occurred earlier, then at time T termination is enforced with termination payoff $G(X_T, T)$). Thus, the design of an optimal stopping policy comes down to the designing of the domain $\mathfrak{C}$, which then determines the optimal stopping time as the entry time of the process X to the open domain $\bar{\mathfrak{C}}^{\complement}$:

$$\jmath(\omega) = \inf\big\{t \geq 0\colon (X(\omega,t),t) \in \bar{\mathfrak{C}}^{\complement}\big\} .$$

Our next objective is to develop an equation for the value function in the endogenous domain $\mathfrak{C}$. With this purpose in mind, suppose that the decision of whether to terminate or not can be made at time $t < T$, when the process is in state $X_t = x$, but if the process is not terminated at time t, the next opportunity to terminate will arrive only at time $t + \varepsilon$, that is, suppose that somehow termination is not possible during the period $]t, t+\varepsilon[$ (and think of $\varepsilon > 0$ as being "very small"). Denote the value function associated with this new exercise restriction by $V_\varepsilon(x,t)$. It is clear – at least heuristically – that $V_\varepsilon(x,t)$ must be a decreasing function of ε and $\lim_{\varepsilon \searrow 0} V_\varepsilon(x,t) = V(x,t)$. Let $\iota_\varepsilon \stackrel{\text{def}}{=} (t+\varepsilon) \wedge T$ and observe that this value is actually deterministic. Since the value of state $(X_{\iota_\varepsilon}, \iota_\varepsilon)$ is $V(X_{\iota_\varepsilon}, \iota_\varepsilon)$, the decision *not to terminate* in state (x,t) would be valued (in state (x,t)) at

$$U_\varepsilon(x,t) \stackrel{\text{def}}{=} \mathsf{E}\left[\int_t^{t_\varepsilon} e^{-\int_t^s r_u \mathrm{d}u} F(X_s,s)\mathrm{d}s + e^{-\int_t^{t_\varepsilon} r_u \mathrm{d}u} V_\varepsilon(X_{t_\varepsilon},t_\varepsilon) \mid X_t = x\right].$$

This is the so-called ε-continuation value in state (x,t), which assumes that, if postponed at time t, termination would not be possible before time t_ε. The termination value in that state is simply $G(x,t)$ – the payoff from immediate termination in state (x,t). The principle of dynamic programming (PDP) is now telling us that the value of state (x,t) (assuming that the optimal action is taken) is

$$V_\varepsilon(x,t) = \max\big(G(x,t), U_\varepsilon(x,t)\big),$$

and the optimal decision rule is to terminate if $V_\varepsilon(x,t) = G(x,t) \geq U_\varepsilon(x,t)$ and to continue if $V_\varepsilon(x,t) = U_\varepsilon(x,t) > G(x,t)$. The plan now is to develop a PDE for the value function $V(\cdot,\cdot)$ in the (endogenous) continuation domain $\mathfrak{C}$ by passing to the limit as $\varepsilon \searrow 0$, with the stipulation that $\lim_{\varepsilon\searrow 0} V_\varepsilon(\cdot,\cdot)$ exists and equals the value function $V(\cdot,\cdot)$. The first step in this program is to rewrite the relation above as

$$0 = \max\big(G(x,t) - V_\varepsilon(x,t), U_\varepsilon(x,t) - V_\varepsilon(x,t)\big)$$

and observe that if continuation at (x,t) is optimal, then $G(x,t) - V_\varepsilon(x,t) < 0$ and the equation above becomes $U_\varepsilon(x,t) - V(x,t) = 0$. Just as before, we expand the difference

$$\begin{aligned}U_\varepsilon(x,t) - V_\varepsilon(x,t) = \mathsf{E}\bigg[\int_t^{t_\varepsilon} & e^{-\int_t^s r_u \mathrm{d}u} F(X_s,s)\mathrm{d}s \\ & + e^{-\int_t^{t_\varepsilon} r_u \mathrm{d}u} V_\varepsilon(X_{t_\varepsilon},t_\varepsilon) - V_\varepsilon(x,t) \mid X_t = x\bigg]\end{aligned}$$

by using Itô's formula and speculate that all stochastic integrals vanish after the expectation operator is applied. As a result, passing to the limit as $\varepsilon \searrow 0$ in the relation

$$\max\Big(G(x,t) - V_\varepsilon(x,t), \frac{1}{\varepsilon}\big(U_\varepsilon(x,t) - V_\varepsilon(x,t)\big)\Big) = 0,$$

we arrive at the equation

$$\begin{aligned}\max\Big(&G(x,t) - V(x,t), \\ &F(x,t) - r(x,t)V(x,t) + \partial V(x,t) + (\mathcal{A}_t V)(x,t)\Big) = 0,\end{aligned} \qquad \mathrm{e}_1$$

in which $\mathcal{A}_t$ acts on functions $x \rightsquigarrow \varphi(x)$ according to the prescription

$$\begin{aligned}(\mathcal{A}_t\varphi)(x) = {} & b(x,t)\cdot\nabla\varphi(x) + \frac{1}{2}\Big(\sigma(x,t)\sigma(x,t)^\intercal\nabla\Big)\cdot\nabla\varphi(x) \\ & + \int_{\mathbb{R}^m}\Big(\varphi\big(x + \gamma(x,z,t)\big) - \varphi(x) - \gamma(x,z,t)\cdot\nabla\varphi(x)\Big)\nu(\mathrm{d}z).\end{aligned}$$

Thus, in the continuation domain, $\mathfrak{C}$,

$$\begin{aligned} &F(x,t) - r(x,t)V(x,t) + \partial V(x,t) + (\mathcal{A}_t V)(x,t) = 0 \\ &\text{and} \quad G(x,t) - V(x,t) < 0\,, \end{aligned} \qquad \mathrm{e}_2$$

while in the interior, $\bar{\mathfrak{C}}^{\complement}$, of the termination domain

$$\begin{aligned} &G(x,t) - V(x,t) = 0 \\ &\text{and} \quad F(x,t) - r(x,t)V(x,t) + \partial V(x,t) + (\mathcal{A}_t V)(x,t) \leq 0\,. \end{aligned} \qquad \mathrm{e}_3$$

This is our first encounter with what is known as a *free-boundary problem*: one must solve for the unknown function $(x,t) \rightsquigarrow V(x,t)$ from the PDE in e_2, but the domain, $\mathfrak{C}$, in which the PDE is solved is also unknown. One immediate consequence from this observation is that although the boundary condition $V(x,t) = G(x,t)$ (the first equation in e_3) is known, it is not known where to assign that condition, whence the term "free boundary." Furthermore, since in the termination domain $\bar{\mathfrak{C}}^{\complement}$ we must have $V(x,t) = G(x,t)$, the second condition in e_3 actually demands that the termination payoff $G(x,t)$ satisfies the following relation everywhere in the (yet to be determined) termination region $\bar{\mathfrak{C}}^{\complement}$:

$$F(x,t) - r(x,t)G(x,t) + \partial G(x,t) + (\mathcal{A}_t G)(x,t) \leq 0\,.$$

In particular, there would be no solution to the optimal stopping problem if the set of points (x,t) at which the relation above holds is empty. Somehow, under certain fairly general conditions, putting together the two complementary systems of conditions, e_2 and e_3, uniquely determines both the domain $\mathfrak{C}$ and the value function $V(x,t)$, although the matter is not quite as straightforward, as we are going to realize later on when we face concrete applications.

(17.13) REMARK: Conditions e_2 and e_3 are often put together in the following equivalent form:

$$\left|\begin{aligned} &\Big(G(x,t) - V(x,t)\Big)\Big(F(x,t) + \partial V(x,t) + (\mathcal{A}_t V)(x,t) - r(x,t)V(x,t)\Big) = 0 \\ &G(x,t) \leq V(x,t) \;\text{ and }\; F(x,t) + \partial V(x,t) + (\mathcal{A}_t V)(x,t) \leq r(x,t)V(x,t) \end{aligned}\right. ,$$

which avoids any explicit reference to the unknown domain $\mathfrak{C}$. If the function $V(x,t)$ can be chosen so that the conditions above hold for all admissible (x,t), then the continuation domain $\mathfrak{C}$ can be identified as the domain where $V(x,t) > G(x,t)$. We also note that, assuming $r(x,t) > 0$, the first condition above will not change if the difference $G(x,t) - V(x,t)$ is replaced with $r(x,t)G(x,t) - r(x,t)V(x,t)$. Similarly, the second condition will not change if the relation $G(x,t) \leq V(x,t)$ is replaced with $r(x,t)G(x,t) \leq r(x,t)V(x,t)$. As a result, the HJB equation e_1 is equivalent to

$$\max\Big(r(x,t)G(x,t) - r(x,t)V(x,t), \\ F(x,t) - r(x,t)V(x,t) + \partial V(x,t) + (\mathcal{A}_t V)(x,t)\Big) = 0\,,$$

which can be rearranged as

$$\dagger \qquad \max\Big(r(x,t)G(x,t),\, F(x,t) + \partial V(x,t) + (\mathcal{A}_t V)(x,t) \Big) = r(x,t)V(x,t)\,.$$

We again emphasize that this formulation is only possible if $r(x,t) > 0$. ○

(17.14) REMARK: It would be instructive to draw the parallel between the way we approached the problem of optimal stopping in the present section and the way we approached the same problem earlier in section 14.4. Once the value function $(x,t) \rightsquigarrow V(x,t)$ has been computed, one can define the process

$$U_t \stackrel{\text{def}}{=} \int_0^t e^{-\int_0^s r(X_u,u)\,\mathrm{d}u} F(X_s,s)\,\mathrm{d}s + e^{-\int_0^t r(X_u,u)\,\mathrm{d}u} V(X_t,t)\,, \quad t \in [0,T]\,,$$

and show that this process is nothing but the Snell envelope for the reward process given by

$$H_t \stackrel{\text{def}}{=} \int_0^t e^{-\int_0^s r(X_u,u)\,\mathrm{d}u} F(X_s,s)\,\mathrm{d}s + e^{-\int_0^t r(X_u,u)\,\mathrm{d}u} G(X_t,t)\,, \quad t \geq 0\,.$$

The condition $F(x,t) + \partial V(x,t) + (\mathcal{A}_t V)(x,t) \leq r(x,t)V(x,t)$ for all admissible (x,t) is just another way of saying that the Snell envelope U is a supermartingale, and the condition $F(x,t) + \partial V(x,t) + (\mathcal{L}_t V)(x,t) = r(x,t)V(x,t)$ for all $(x,t) \in \mathfrak{C}$ is just another way of saying that the Snell envelope is a martingale until the actual stopping takes place. The interpretation of the value function $V(\cdot,\cdot)$ in terms of the Snell envelope is important because it allows us to make the connection between option pricing and replication (hedging). ○

(17.15) AMERICAN-STYLE PUT OPTIONS: The most common optimal stopping problem in the realm of asset pricing is the optimal exercise and pricing of an American-style derivative contract. Consider the standard Black–Scholes–Merton framework in which the risky asset follows the price process

$$S = S_0 + \sigma S \bullet \tilde{W} + rS \bullet \iota\,,$$

with an infinitesimal generator (written relative to the risk-neutral measure, under which $\tilde{W}$ is a Brownian motion)

$$\mathcal{A}_t = rx\frac{\partial}{\partial x} + \frac{1}{2}\sigma^2 x^2 \frac{\partial^2}{\partial x^2}\,,$$

whereas the risk-free asset follows the price process $S^\circ = S_0 + rS^\circ \bullet \iota$. In this setting, consider the optimal stopping problem with terminal time $T > 0$, no integral payoff ($F(x,t) = 0$), and termination payoff $G(x,t) = (K - x)^+$ for some fixed constant $K > 0$. This is nothing but the payoff schedule from an American-style put option, and the optimal stopping comes down to the optimal exercise of the option. The price of the option is simply $V(S_t,t)$, $t \in [0,T]$, where $V(\cdot,\cdot)$ is the value function of the associated optimal stopping problem with reward process

$$H_t = e^{-rt}(K - S_t)^+\,, \quad t \in \mathbb{R}_+\,.$$

We are going to detail the calculation of this function later on, but at this point the stylized illustration of the solution on figure 17.1 is not too difficult to guess. ○

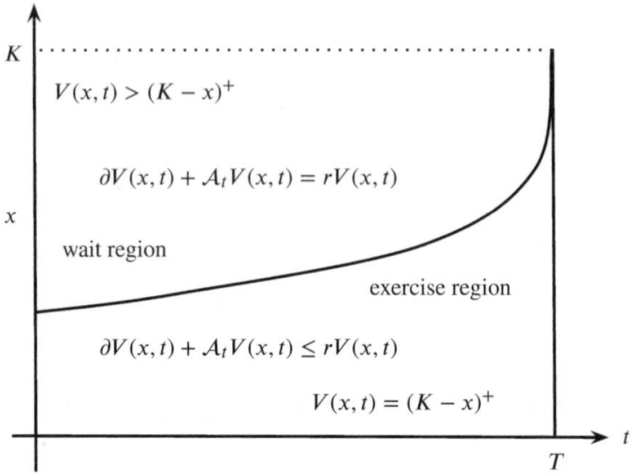

FIGURE 17.1

Exercise and continuation regions for an American-style put option.

(17.16) EXERCISE: In the context of (17.15) and figure 17.1, using only heuristic reasoning and intuition, explain why for every fixed $t \in [0, T]$ there must be a critical price threshold S_t^* such that if $S_t > S_t^*$ it would be optimal to wait, whereas if $S_t \le S_t^*$ it would be optimal to exercise the put option. To put it another way, explain why the price range in which it would be optimal to exercise the option at any given time must be an interval. ○

(17.17) EXERCISE: Relying only on heuristic reasoning and intuition, explain why the critical price threshold S_t^* from the previous problem must be an increasing function of the time variable t. Furthermore, explain why $S_t^* \le K$ and why $S_t^* \to K$ as $t \nearrow T$. ○

(17.18) EXERCISE: Verify that the function $V(x,t) \stackrel{\text{def}}{=} K - x$ satisfies the relation $\partial V(x,t) + rx\,\partial_x V(x,t) + \frac{1}{2}\sigma^2 x^2 \partial_x^2 V(x,t) \le rV(x,t)$ in the domain $\{0 < x < K\}$ and explain why this verification is important. ○

18

Applications to Dynamic Asset Pricing

Some concrete applications of the theory and methods of optimal control to the domain of asset pricing were already discussed in the previous chapter, but mostly as an aside and illustration. In the present chapter we focus exclusively on such applications.

18.1 Merton's Problem with Intertemporal Consumption and No Rebalancing Costs

It would not be an exaggeration to say that, to some degree, most applications of the theory and methods of optimal control to continuous-time finance can be traced back to one version of Merton's consumption-investment problem or another. The simplest such version was already outlined in (17.8), and the present section is a brief account of other common variations of Merton's problem. In particular, we are going to extend the model from (17.8) by allowing intertemporal consumption and jumps in the returns. We again consider a financial market that consists of a single risky asset S and a single risk-free asset S° with prices given by

$$S = S_0 + S_- {\bullet} \Big(b\iota + \sigma W + \int_{\mathbb{R}} \gamma(z) \tilde{N}(\mathrm{d}z, [0, \iota])\Big) \quad \text{and} \quad S^\circ = S_0^\circ + rS^\circ {\bullet} \iota\,,$$

where $b, \sigma, r \in \mathbb{R}$ are given parameters, W is a Brownian motion, which is independent of the compensated Poisson counting measure

$$\tilde{N}(\mathrm{d}z, \mathrm{d}t) = N(\mathrm{d}z, \mathrm{d}t) - \nu(\mathrm{d}z)\mathrm{d}t\,,$$

and $\gamma(\cdot)$ is a Borel function such that $\nu\{z\colon \gamma(z) \leq -1\} = 0$ and $\int_{\mathbb{R}} \gamma(z)^2\, \nu(\mathrm{d}z) < \infty$ (as usual, ν stands for the mean measure of N). One possible condition that makes this model viable – that is, free of arbitrage – is

$$0 < (b - r) + \sigma\alpha < \int_{\mathbb{R}} \gamma(z)\nu(\mathrm{d}z) \leq \infty \quad \text{for some } \alpha \in \mathbb{R}\,.$$

Indeed, if this relation is satisfied, then it becomes possible to find a nonempty finite interval $]u, v[\, \subseteq \mathbb{R}$ such that

$$g(x) \stackrel{\text{def}}{=} -\frac{(b - r) + \sigma\alpha}{\int_{]u,v[} \gamma(z)\, \nu(\mathrm{d}z)} 1_{]u,v[}(x) > -1\,,$$

and, following the removal of the drift technique from (16.44), it becomes possible to construct at least one ELMM. Just as in (17.8), we suppose that the investor has no income other than the capital gains from trading, and denote by Y_t their aggregate wealth at time t – this is the amount available for investment and consumption at time t. We postulate that the amount that the investor consumes instantaneously can be expressed as $c_t\,\mathrm{d}t$ (so that c_t is the local consumption rate). The amount left for investment then is $(Y_t - c_t\,\mathrm{d}t)$ and this amount is split into $\pi_t = \xi_t(Y_t - c_t\,\mathrm{d}t)$, which is the amount converted into a position in the risky asset, and $(1-\xi_t)(Y_t - c_t\,\mathrm{d}t)$, which is the amount converted into a position in the risk-free asset. Consequently, the instantaneous change (from time t to $t+\mathrm{d}t$) in the value of the holding of the risky asset can be expressed (formally) as

$$\xi_t\big(Y_t - c_t\,\mathrm{d}t\big)\frac{S_{t+\mathrm{d}t} - S_t}{S_t} = \xi_t Y_t \frac{S_{t+\mathrm{d}t} - S_t}{S_t},$$

where the last identity follows from the (formal, still) relation $\mathrm{d}t\,(S_{t+\mathrm{d}t} - S_t) = 0$. For the same reason, the instantaneous change in the value of the holding of the risk-free asset can be cast as

$$(1-\xi_t)\big(Y_t - c_t\,\mathrm{d}t\big)r\,\mathrm{d}t = (1-\xi_t)Y_t r\,\mathrm{d}t \equiv (1-\xi_t)Y_t r\,\mathrm{d}t\,.$$

As a result, the (controlled) wealth process Y is governed by the equation

$$Y = Y_0 + (Y_-\xi_-)\bullet\Big(b\imath + \sigma W + \int_{\mathbb{R}} \gamma(z)\tilde{N}(\mathrm{d}z,[0,\imath])\Big) + r(1-\xi)Y\bullet\imath - c\bullet\imath\,.$$

If Y remains strictly positive at all times, then the consumption rate c_t can always be expressed in units of the current wealth Y_t, that is, one can always write $c_t = \eta_t Y_t$ for some strictly positive process η. Thus, instead of choosing the actual consumption rate c_t at time t, the controller chooses the variable η_t. With this convention in mind, the equation governing Y can be restated as

$$Y = Y_0 + \Big[Y\big(b\xi + r(1-\xi) - \eta\big)\Big]\bullet\imath + (\sigma Y\xi)\bullet W$$
$$+ (Y_-\xi_-)\bullet\int_{\mathbb{R}} g(z)\tilde{N}(\mathrm{d}z,[0,\imath])\,.$$

To place the preceding in the general framework of optimal control, the controlled diffusion process is Y and the actual control is the pair of càdlàg and adapted processes $\alpha_t \stackrel{\text{def}}{=} (\xi_t,\eta_t)$, $t\in\mathbb{R}_+$. Many variations of this setup are now possible, and we consider first the case where only long positions in either asset are allowed, that is, $\xi_t \in [0,1]$.

(18.1) REMARK: The wealth process Y can be identified as $Y_0\mathscr{E}(Z)$, where $\mathscr{E}(Z)$ is the Doléan–Dade exponent of the (control-dependent) semimartingale

$$Z = \big(b\xi + r(1-\xi) - \eta\big)\bullet\imath + \sigma\xi\bullet W + \xi_-\bullet\int_{\mathbb{R}} g(z)\tilde{N}(\mathrm{d}z,[0,\imath])\,.$$

If the state space of the process ξ is restricted to the interval $[0, 1]$, then the jumps of the process Z, which are generated by the last integral alone, are strictly larger than (-1), and the only additional requirement that we need to impose on the control $\alpha = (\xi, \eta)$ to guarantee that $Y = Y_0\mathscr{E}(Z)$ is meaningful is the local integrability of the sample paths of η (a.s.), which ensures that the process $\eta\bullet\iota$ is meaningful. If these conditions are met, then we can express $\log Y$ in terms of the control explicitly:

$$\log(Y) = \log(Y_0) + Z - \frac{\sigma^2}{2}\xi^2\bullet\iota + \int_{\mathbb{R}} \big(\log(1 + \xi_-\gamma(z)) - \xi_-\gamma(z)\big)\bullet N(\mathrm{d}z, [0, \iota]) \,. \quad \circ$$

Conditioned to the event $Y_t = y$, the space of admissible controls $\mathscr{A}(y, t)$ is the collection of all càdlàg and adapted processes $\alpha_s \equiv (\xi_s, \eta_s) \in [0, 1]\times\mathbb{R}_{++}$, $s \in [t, \infty[$, such that the sample paths of η are locally integrable on $[t, \infty[$ with probability 1. If the investment-consumption strategy $\alpha = (\xi, \eta)$ is chosen from the set of admissible controls, then the associated wealth process, Y, remains strictly positive as long as the initial wealth is strictly positive.

First, consider the case where the investor has the objective of maximizing their time-separable utility from consumption over an infinite time horizon. Such an objective can be formulated as

$$J_0(y, \xi, \eta, t) = \mathsf{E}\Big[\frac{1}{p}\int_t^{\infty} e^{-\beta(s-t)}(\eta_s Y_s)^p \,\mathrm{d}s\Big], \quad t \in \mathbb{R}_+ \,,$$

where $1 \neq 1 - p > 0$ is the agent's (exogenous and constant) risk aversion coefficient and $\beta > 0$ is a parameter that captures the agent's impatience toward consumption. It is easy to see that the choice of the control $(\xi, \eta) \in \mathscr{A}(y, t)$ that maximizes the expected value above is no different from the choice of the control that maximizes

$$J(y, \xi, \eta, t) \overset{\text{def}}{=} \mathsf{E}\Big[\frac{1}{p}\int_t^{\infty} e^{-\beta s}(\eta_s Y_s)^p \,\mathrm{d}s\Big], \quad t \geq 0 \,.$$

This observation is useful because a performance criterion of the form J is somewhat easier to work with than a performance criterion of the form J_0.

(18.2) REMARK: Although changing the discount factor in the performance criterion from $e^{-\beta(s-t)}$ to $e^{-\beta s}$ does not alter the optimal policy, it does alter the value function by a factor of $e^{-\beta t}$, that is, if the value function corresponding to the performance criterion J is $V(y, t)$, then the value function associated with the performance criterion J_0 must be $e^{\beta t}V(y, t)$. $\circ$

The HJB equation that governs the value function

$$V(y, t) \overset{\text{def}}{=} \sup\nolimits_{(\xi,\eta)\in\mathscr{A}(y,t)} J(y, \xi, \eta, t)$$

is not difficult to write down:

$$\begin{aligned}\partial V(y,t) + \sup_{(\xi,\eta)\in\mathscr{A}(y,t)} \Big(\frac{e^{-\beta t}}{p}(\eta_t y)^p + \big(b\xi_t + r(1-\xi_t) - \eta_t\big) y \partial_y V(y,t) \\ + \frac{1}{2}\sigma^2 {\xi_t}^2 y^2 \partial_y^2 V(y,t) \\ + \int_{\mathbb{R}} \Big(V\big(y + \xi_t \gamma(z) y, t\big) - V(y,t) - \xi_t \gamma(z) y \partial_y V(y,t) \Big) \nu(\mathrm{d}z) \Big] = 0\,,\end{aligned} \qquad \mathrm{e}_1$$

with boundary condition $\lim_{t\to\infty} V(y,t) = 0$.

(18.3) EXERCISE: Verify analytically that if the function $V(y,t)$ satisfies e_1, then the function $\tilde{V}(y,t) \overset{\text{def}}{=} e^{\beta t} V(y,t)$ must satisfy the equation

$$\begin{aligned}\partial \tilde{V}(y,t) + \sup_{(\xi,\eta)\in\mathscr{A}(y,t)} \Big[\frac{1}{p}(\eta_t y)^p + \big(b\xi_t + r(1-\xi_t) - \eta_t\big) y \partial_y \tilde{V}(y,t) \\ + \frac{1}{2}\sigma^2 {\xi_t}^2 y^2 \partial_y^2 \tilde{V}(y,t) \\ + \int_{\mathbb{R}} \Big(\tilde{V}\big(y + \xi_t \gamma(z) y, t\big) - \tilde{V}(y,t) - \xi_t \gamma(z) y \partial_y \tilde{V}(y,t) \Big) \nu(\mathrm{d}z) \Big] = \beta \tilde{V}(y,t)\,.\end{aligned}$$

Then give an alternative explanation of the same claim by developing the HJB equation associated with the performance criterion J_0.

HINT: Consult (18.2). ○

We can again seek a solution to e_1 of the form $V(y,t) = Ae^{-\beta t}\frac{1}{p}y^p$ for some yet to be determined constant $A > 0$. With this concrete choice for the value function, the HJB equation becomes (after the cancellation of the factor $e^{-\beta t} y^p$ on both sides of the equation):

$$\begin{aligned}-\beta\frac{A}{p} + \sup_{(\xi,\eta)\in\mathscr{A}(y,t)} \Big[\frac{1}{p}(\eta_t)^p + \big(b\xi_t + r(1-\xi_t) - \eta_t\big) A + \frac{p-1}{2}\sigma^2 {\xi_t}^2 A \\ + \frac{A}{p}\int_{\mathbb{R}} \big((1 + \xi_t\gamma(z))^p - 1 - \xi_t p \gamma(z)\big) \nu(\mathrm{d}z) \Big] = 0\,.\end{aligned} \qquad \mathrm{e}_2$$

Just as before, basic intuition suggests that the solution for the control $\alpha_t = (\xi_t, \eta_t)$ – if a solution of the form $V(y,t) = Ae^{-\beta t}\frac{1}{p}y^p$ can indeed be found – must be independent not only from the time variable t, but also from the wealth level Y_t. Essentially, this feature comes as a consequence of the fact that in this model the market follows an economy of scale subjected to stationary and independent shocks, and the fact that the agent has a time-separable CRRA utility from consumption. For this reason, we will write simply ξ instead of ξ_t and η instead of η_t, that is, in what follows the symbols η and ξ will refer to scalar variables rather than stochastic processes. Our next step is to remove the sup in the HJB equation, and we note that the function under the sup is concave in both variables ξ and η.

The first-order condition for ξ (after the cancellation of the factor $A > 0$) becomes

$$F(\xi) \stackrel{\text{def}}{=} (p-1)\sigma^2\xi + (b-r) + \int_{\mathbb{R}} \big((1+\xi\gamma(z))^{p-1} - 1\big)\gamma(z)\nu(\mathrm{d}z) = 0\,,$$

which is the same as

$$\xi = \frac{b-r}{\sigma^2(1-p)} + \frac{1}{\sigma^2(1-p)} \int_{\mathbb{R}} \big((1+\xi\gamma(z))^{p-1} - 1\big)\gamma(z)\nu(\mathrm{d}z)\,. \qquad \mathrm{e}_3$$

Somewhat unexpectedly, we now see that if there are no jumps in the returns ($\nu = 0$), then the optimal value $\hat{\xi}$ would be no different from the value obtained in (17.8) in the case of a finite time horizon and no intertemporal consumption. In the general case, that is, in the presence of jumps, equation e_3 may be seen as a fixed-point-type iteration program, which – at least in principle – can produce the solution $\hat{\xi}$. Since $F(0) = b - r$ and

$$F(1) = \sigma^2(p-1) + (b-r) + \int_{\mathbb{R}} \big((1+\gamma(z))^{p-1} - 1\big)\gamma(z)\nu(\mathrm{d}z)\,,$$

the equation $F(\xi) = 0$ would have a solution $\hat{\xi} \in [0,1]$ if, e.g., $b \geq r$ and

$$\frac{b-r}{\sigma^2(1-p)} + \frac{1}{\sigma^2(1-p)} \int_{\mathbb{R}} \big((\gamma(z)+1)^{p-1} - 1\big)\gamma(z)\nu(\mathrm{d}z) \leq 1\,,$$

in which case $F(0) \geq 0$ and $F(1) \leq 0$.

Assuming that $\xi = \hat{\xi}$ solves e_3 and $\hat{\xi} \in [0,1]$ – and notice that any such $\hat{\xi}$ does not depend on A – the first-order condition for the control variable η (the instantaneous consumption rate expressed in units of aggregate wealth) is

$$\eta^{p-1} - A = 0 \qquad \Rightarrow \qquad \hat{\eta} = A^{\frac{1}{p-1}}\,.$$

This way the HJB equation e_2 turns into the following algebraic equation for the unknown constant A:

$$-\beta\frac{A}{p} + \frac{1}{p}A^{\frac{p}{p-1}} + \big(b\hat{\xi}A + r(1-\hat{\xi})A - A^{\frac{p}{p-1}}\big) + \frac{p-1}{2}\sigma^2\hat{\xi}^2 A$$
$$+ \frac{A}{p}\int_{\mathbb{R}} \big((1+\hat{\xi}z)^p - 1 - \hat{\xi}pz\big)\nu(\mathrm{d}z) = 0\,.$$

If the expression under the large parenthesis below is strictly positive, then the solution would be given by

$$\hat{A} = \Bigg(\frac{\beta}{1-p} - \frac{p}{1-p}\big(b\hat{\xi} + r(1-\hat{\xi})\big)$$
$$+ \frac{1}{2}\sigma^2 p\hat{\xi}^2 - \frac{1}{1-p}\int_{\mathbb{R}} \big((1+\hat{\xi}z)^p - 1 - \hat{\xi}pz\big)\nu(\mathrm{d}z)\Bigg)^{p-1}\,. \qquad \mathrm{e}_4$$

This would then give us the optimal consumption $\hat{\eta} = \hat{A}^{\frac{1}{p-1}}$ – notice that the optimal consumption rate is constant, just as the investment choice $\hat{\xi}$ is – and give us also the value function $V(y,t) = \hat{A}e^{-\beta t}\frac{1}{p}y^p$.

(18.4) EXAMPLE: In the setting above, consider the special case where the Lévy measure ν has the form $\nu(dz) = c\epsilon_\varsigma(dz)$ for some $c > 0$ and some $\varsigma \in]-1, \infty[$, and suppose that $\gamma(z) = z$. With this choice the returns from the risky asset are driven by a compensated Poisson process $\Pi - c\iota$, in addition to an independent Brownian motion with drift. To be precise, the price of the risky asset can be described as

$$S = S_0 + S_{-} \bullet \Big(b\iota + \sigma W + \varsigma(\Pi - c\iota)\Big).$$

Assume the following concrete values for the parameters in this model: $p = -3/2$ (risk aversion $1 - p = 5/2$), $b = 7/100$, $\sigma = 25/100$, $\varsigma = -4/100$, $r = 2/100$, $c = 3$, and $\beta = 3/100$. We then have

$$\int_{\mathbb{R}} \big((1 + \xi z)^{p-1} - 1\big) z\nu(dz) = \big((1 + \xi\varsigma)^{p-1} - 1\big)c\varsigma = \frac{3}{25} - \frac{375}{(25 - \xi)^{5/2}}$$

and equation e_3 comes down to

$$\xi = \frac{8}{25} + \frac{32}{5}\Big(\frac{3}{25} - \frac{375}{(25 - \xi)^{\frac{5}{2}}}\Big).$$

A standard root-finding procedure in Python gives $\hat{\xi} = 0.2967292770068062$, and the formula in e_4 evaluates to $\hat{A} = 6830.221945067397$. As a result, we get $\hat{\eta} = 0.029256749521209 98$. In practical terms, this solution translates to keeping approximately 30% of the total wealth in the risky asset and maintaining a consumption rate of about 2.9% of the total wealth. ○

(18.5) EXERCISE: In the context of (18.4), calculate the optimal investment strategy $\hat{\xi}$ and the optimal consumption rate $\hat{\eta}$ for 10 different values of the risk aversion coefficient $1 - p$ in the range $3/2 \leq 1 - p \leq 6$. Using interpolation, plot the dependence of $\hat{\xi}$ and $\hat{\eta}$ on the risk aversion coefficient $1 - p$. Give the economic interpretation of the graphs. ○

(18.6) EXERCISE: In the context of (18.4), calculate the optimal investment strategy $\hat{\xi}$ and the optimal consumption rate $\hat{\eta}$ for 10 different values of the Brownian volatility σ in the range $15/100 \leq \sigma \leq 35/100$. Using interpolation, plot the dependence of $\hat{\xi}$ and $\hat{\eta}$ on the volatility σ. Give the economic interpretation of the graphs. ○

(18.7) EXERCISE: In the context of (18.4), calculate the optimal investment strategy $\hat{\xi}$ and the optimal consumption rate $\hat{\eta}$ for 10 different values of the jump volatility ς in the range $-10/100 \leq \varsigma \leq -1/100$. Using interpolation, plot the dependence of $\hat{\xi}$ and $\hat{\eta}$ on the jump volatility ς. Give the economic interpretation of the graphs. ○

In the case of a logarithmic time-separable utility (still assuming infinite time horizon $T = \infty$) it is preferable to write the performance criterion in the form

$$J(y,\xi,\eta,t) \stackrel{\text{def}}{=} \mathsf{E}\Big[\int_t^\infty e^{-\beta(s-t)}\log(\eta_s Y_s)\,\mathrm{d}s\Big], \quad t \in \mathbb{R}_+ .$$

We will confine our study of this case only to situations in which there are no jumps in the returns. Thus, the HJB equation that we are now faced with is

$$\partial V(y,t) + \sup_{(\xi,\eta)\in\mathscr{A}(y,t)} \Big\{\log(\eta_t y) + \big(b\xi_t + r(1-\xi_t) - \eta_t\big)y\partial_y V(y,t) + \frac{1}{2}\sigma^2{\xi_t}^2 y^2\partial_y^2 V(y,t)\Big\} = \beta V(y,t) .$$

We seek a solution of the form $V(y,t) = f(t) + \frac{1}{\beta}\log(y)$, which transforms the equation above into

$$\partial f(t) + \sup_{(\xi,\eta)\in\mathscr{A}(y,t)} \Big\{\log(\eta_t) + \log(y) + \big(b\xi_t + r(1-\xi_t) - \eta_t\big)\frac{1}{\beta} - \frac{1}{2}\sigma^2{\xi_t}^2\frac{1}{\beta}\Big\} = \beta f(t) + \log(y) .$$

Removing $\log(y)$ from both sides of the equation and maximizing over ξ_t and η_t gives

$$\xi_t = \hat{\xi} = \frac{b-r}{\sigma^2} \quad \text{and} \quad \eta_t = \hat{\eta} = \beta .$$

Furthermore, since in this case the boundary condition is $\lim_{t\to\infty} e^{-\beta t}V(y,t) = 0$, we can take $f(t) =$ constant. This is consistent with the intuition: with infinite time horizon and time-homogeneous Markovian dynamics, the solution should depend on the state y but not on the time parameter t.

(18.8) EXERCISE: Give the exact expression for the constant function $f(t)$ and the value function $V(y,t)$ above. ○

Next, we turn to the case of a finite time horizon $T > 0$. Let $U_0(\cdot)$ be the investor's utility from consumption and let $U_1(\cdot)$ be their utility from final wealth. The performance criterion is now

$$J(y,\xi,\eta,t) \stackrel{\text{def}}{=} \mathsf{E}\Big[\int_t^T e^{-\beta(s-t)}U_0(\eta_s Y_s)\,\mathrm{d}s + e^{-\beta(T-t)}U_1(Y_T)\Big], \quad t \in [0,T],$$

the associated HJB equation in the time domain $t \in [0,T[$ is

$$\partial V(y,t) + \sup_{(\xi_t,\eta_t)\in\mathscr{A}(y,t)} \Big\{U_0(\eta_t y) + \big(b\xi_t + r(1-\xi_t) - \eta_t\big)y\partial_y V(y,t) + \frac{1}{2}\sigma^2{\xi_t}^2 y^2\partial_y^2 V(y,t) + \int_{\mathbb{R}}\Big(V(y+\xi_t\gamma(z)y,t) - V(y,t) - \xi_t\gamma(z)y\partial_y V(y,t)\Big)\nu(\mathrm{d}z)\Big\} = \beta V(y,t),$$

and the terminal condition is $\lim_{t\nearrow T} V(y,t) = U_1(y)$. The solution $V(y,t)$ can no longer be expected to be time-independent. Nevertheless, solving the equation above is more or less straightforward in the case where the returns do not contain jumps ($\nu \equiv 0$) and the utilities involved are of CRRA type. We first consider the case where $U_0(c) = c^p/p$ and $U_1(y) = y^p/p$ and again seek a solution of the form $V(y,t) = \frac{1}{p} f(t) y^p$. The HJB equation specializes to (after the cancellation of the factor y^p/p)

$$\partial f(t) + \sup_{(\xi,\eta)\in\mathcal{A}(y,t)} \left[\eta_t^p + p\big(b\xi_t + r(1-\xi_t) - \eta_t\big) f(t) - \frac{p}{2}\sigma^2 \xi_t{}^2 (1-p) f(t)\right] = \beta f(t).$$

Much to our surprise, we again see that – even in the case of a finite time horizon – the optimal investment strategy is still $\hat{\xi}_t = \hat{\xi}$, with

$$\hat{\xi} = \frac{b-r}{(1-p)\sigma^2},$$

and the understanding that $\hat{\xi}_t = 1$ if $\hat{\xi} > 1$ and the agent cannot borrow the bond, and $\hat{\xi}_t = 0$ if $\hat{\xi} < 0$ and the agent cannot borrow the stock. However, the choice of the optimal consumption level expressed in units of the aggregate wealth is no longer constant and we have

$$p\hat{\eta}_t^{p-1} = pf(t) \qquad \Rightarrow \qquad \hat{\eta}_t = f(t)^{\frac{1}{p-1}}.$$

(18.9) EXERCISE: Show that the HJB equation above leads to an ODE for the function $f(t)$ and solve that ODE with boundary condition $f(T) = 1$. Give a complete solution to Merton's problem with finite time horizon and power utility from both consumption and final wealth (with identical risk aversions in both), that is, write down the value function and the optimal consumption level expressed in units of the total wealth.

HINT: An ODE of the form $\partial f(t) + Af(t)^{\frac{p}{p-1}} + Bf(t) = 0$ can be resolved with the substitution $f(t) = g(t)^{1-p}$. ○

Finally, we turn to the case of logarithmic utilities from both consumption and final wealth: $U_0(c) = \log(c)$ and $U_1(y) = k\log(y)$, $k > 0$. Just as before, we focus on the no-jumps case, which makes it possible to find a closed-form analytic solution. The HJB equation is now

$$\partial V(y,t) + \sup_{(\xi_t,\eta_t)\in\mathcal{A}(y,t)} \Big[\log(\eta_t y) + \big(b\xi_t + r(1-\xi_t) - \eta_t\big) y \partial_y V(y,t) + \frac{1}{2}\sigma^2 \xi_t{}^2 y^2 \partial_y^2 V(y,t)\Big] = \beta V(y,t),$$

with boundary condition $\lim_{t\nearrow T} V(y,t) = k\log(y)$. In this case we seek a solution of the form $V(x,t) = f(t) + g(t)\log(y)$, which transforms the equation above into the following ODE:

$$\partial g(t)\log(y) + \partial f(t)$$
$$+ \sup_{(\xi_t,\eta_t)\in\mathscr{A}(y,t)} \left[\log(\eta_t) + \log(y) + \left(b\xi_t + r(1-\xi_t) - \eta_t\right)g(t) - \frac{1}{2}\sigma^2{\xi_t}^2 g(t)\right]$$
$$= \beta g(t)\log(y) + \beta f(t)\,.$$

The first-order conditions from the maximization over the controls ξ_t and η_t give

$$\xi_t = \hat{\xi} \stackrel{\text{def}}{=} \frac{b-r}{\sigma^2} \quad \text{and} \quad \eta_t = \hat{\eta}_t \stackrel{\text{def}}{=} \frac{1}{g(t)}\,,$$

and, with the substitution above, the ODE simplifies to

$$\partial g(t)\log(y) + \partial f(t) - \log\left(g(t)\right) + \log(y) + \frac{(b-r)^2}{2\sigma^2}g(t) + rg(t) - 1$$
$$= \beta g(t)\log(y) + \beta f(t)\,.$$

This equation can be resolved in two steps, by splitting it into two ODEs:

$$\partial g(t)\log(y) + \log(y) = \beta g(t)\log(y)$$

and

$$\partial f(t) - \log\left(g(t)\right) + \frac{(b-r)^2}{2\sigma^2}g(t) + rg(t) - 1 = \beta f(t)\,.$$

(18.10) EXERCISE: Solve for the function $t \rightsquigarrow g(t)$, with boundary condition $\lim_{t\nearrow T} g(t) = k$, from the following linear ODE:

$$\partial g(t) + 1 = \beta g(t)\,. \quad \circ$$

(18.11) EXERCISE: Assuming that $t \rightsquigarrow g(t)$ is the solution obtained in the previous exercise, solve for the function $t \rightsquigarrow f(t)$ from the following linear ODE:

$$\partial f(t) - \log\bigl(g(t)\bigr) + \frac{(b-r)^2}{2\sigma^2}g(t) + rg(t) - 1 = \beta f(t)\,,$$

with boundary condition $\lim_{t\nearrow T} f(t) = 0$. ○

(18.12) EXERCISE: Give a complete solution to Merton's problem with finite time horizon and logarithmic utility from both consumption and final wealth. ○

(18.13) REMARK: The discussion of the topic "optimal investment/consumption policy" in the present section and also in (17.8) is merely the tip of the proverbial iceberg, and is included here only as an illustration of the way in which the classical theory and methods of optimal control are typically used in the domain of finance. For a completely rigorous, detailed – and much more general – exposition we refer the reader to (Karatzas and Shreve 1998, ch. 3) and (Øksendal and Sulem 2005), which covers the technicalities that jump diffusion models involve.[193] In the most

[193] Section 3.11 in (Karatzas and Shreve 1998) provides a detailed account of the remarkable history of this subject and reviews the related literature.

general version of the optimal investment/consumption problem the agent's preferences are modeled by a general utility function rather than a CRRA (power) utility, the coefficients b, σ, and γ are general (in fact, multivariate) predictable processes instead of constants, and the agent can invest in multiple risky assets. ○

(18.14) REMARK: It should be clear by now that an intrinsic connection exists between the methodology reviewed in the present section and the martingale approach outlined in sections 13.1 and 13.2 for models without jumps. In particular, it is impossible not to notice the link between the market price of risk θ_t from (13.8) and the optimal amount invested in the risky asset, expressed in units of the total wealth, which in the no jumps case ($\nu = 0$) comes down to $\frac{1}{1-p}\sigma^{-1}\theta_t$. This formula continues to hold in the case of multiple risky assets (see (Karatzas and Shreve 1998, rem. 8.9)), except that θ_t is a vector, σ_t is a matrix-valued process, and the optimal nominal amounts invested in the risky assets are given by the vector

$$\hat{\pi}_t = \frac{Y_t}{1-p}(\sigma_t^{\intercal})^{-1}\theta_t = \frac{Y_t}{1-p}(\sigma_t\sigma_t^{\intercal})^{-1}(b_t - r_t\vec{1}).$$

One immediate corollary from this observation is that although different investors may have different risk aversion coefficients $1-p$ (assuming they all have CRRA preferences from consumption and final wealth) and different wealth levels Y_t (as they may have different initial endowments Y_0), ultimately they are going to be investing in one and the same mutual fund of risky assets, even though the proportion of their wealth invested in the risky assets would be generally different. ○

18.2 Merton's Problem with Intertemporal Consumption and Rebalancing Costs

The careful reader will have realized already that all solutions to Merton's portfolio selection problem described so far entail constant rebalancing of the portfolio, so that a particular ratio – say, $(b-r)/(1-p)\sigma^2$ – of the risky investment to the total assets can be maintained. In practical terms, such a strategy involves trading with infinitely high frequency, which is not practical, for to say the least, in the real world such a practice would be prohibitively costly (feasibility notwithstanding). It is therefore important to quantify the way in which the cost of rebalancing the portfolio would affect the model developed in the previous section. Magil and Constantinides (1976) appear to be the first to offer a systematic study of the role of transaction costs. Although their work was followed by a considerable number of influential publications – see (Leland 1985), (Davis and Norman 1990), (Shreve and Soner 1994), (Morton and Pliska 1995), (Wang 1994), (Vayanos 1998), (Deelstra et al. 2001), (Liu 2004), (Liu and Loewenstein 2002), (Lo et al. 2004), and (Kabanov and Safarian 2009) – and, generally, the role of transaction costs in finance is by now reasonably well understood, simple and easy to use models are still difficult to come by, and, by and large, the cost associated with any financial transactions is ignored throughout most of the literature on investments. The present section is a brief synopsis of the seminal work of (Davis and Norman 1990), with the addition of jumps in the returns, which we borrow from (Øksendal and Sulem 2005, sec. 6.1).

Just as we did in the previous section, we confine our study to a financial market that consists of a single risky asset, S, and a single risk-free asset, S°, with the same price processes

$$S = S_0 + S_- \bullet \Big(b\iota + \sigma W + \int_{\mathbb{R}} \gamma(z) \tilde{N}(\mathrm{d}z, [0, \iota]) \Big) \quad \text{and} \quad S^\circ = S_0^\circ + r S^\circ \bullet \iota ,$$

except that now we impose the following additional requirement to the Lévy measure ν and the jump volatility $\gamma(\cdot)$: there are constants $0 < \varepsilon < a < 1$ such that

$$\nu\big(z \in \mathbb{R} \colon |\gamma(z)| \notin [\varepsilon, a] \big) = 0 .$$

In other words, all jumps in the returns fall in the set

$$\mathcal{J} \stackrel{\text{def}}{=} [-a, -\varepsilon] \cup [\varepsilon, a] .$$

From a practical point of view, this assumption is certainly reasonable: the instant jumps in the returns from the risky asset are limited to at most $\pm 100a\%$.[194] In addition, we require that jumps in the net returns – if there are jumps in the net returns – must be at least $\pm 100\varepsilon\%$ for some arbitrarily small, but still strictly positive, ε. This assumption is needed to ensure that there could be at most finitely many jumps during any finite time period, that is, jumps are always locally summable. This is also a reasonable assumption to make, for if some empirically observed (discretely sampled) market data are to be matched to a particular continuous-time model, then, typically, there would be a lower threshold beyond which it would be impossible to determine whether a particular variation in the time series is to be attributed to a jump or a fluctuation in a continuous sample path.

In a first step toward the study of transaction costs, observe that if trading is costly, expressing the agent's wealth as a single aggregate monetary amount is no longer adequate. As an example, at the time of typing this paragraph a share of GOOGL is quoted on NASDAQ at \$986.09, but a share of GOOGL and a checking account with a balance of \$986.09 are not identical assets, since they cannot be converted perfectly into one another if the conversion is costly. As a result, it would not be adequate to describe the wealth of an agent who holds both securities as the amount \$1972.18; rather, their wealth is \$986.09 held in GOOGL and \$986.09 held in a bank account. For this reason, we will record the investor's position at time t as the pair (X_t, Y_t), in which X_t is the nominal value of the (long or short) position in the bond S° and Y_t is the nominal value of the (long or short) position in the stock S, held at time t. Converting ¤y worth of stock holdings into

[194] It is easy to find examples of certain stock prices collapsing within hours, but an *instant* change by more than ±20% – or even by just ±10% – would be extraordinarily rare.

bond holdings (cash) decreases the stock holdings by ¤y, but increases the bond holdings only by ¤$(1-k)y$, where the parameter $0 < k < 1$ represents the proportional transaction cost associated with the sale of equity. If the stock holdings were to be increased by ¤y, then the cash holdings would have to be decreased by ¤$(1+h)y$, where $0 < h < 1$ is the proportional transaction cost attached to the purchase of equity. Let $\xi = (\xi_t)_{t \in \mathbb{R}_+}$ and $\eta = (\eta_t)_{t \in \mathbb{R}_+}$ be two increasing (that is, nondecreasing) càdlàg adapted processes that model, respectively, the aggregate accumulated purchases and the aggregate accumulated sales of the stock by time t (expressed as nominal amounts, not as quantities of shares). Let $C = (C_t)_{t \in \mathbb{R}_+}$ be some absolutely continuous, increasing, and adapted process of finite variation that models the agent's aggregate consumption (as a nominal amount) by time t. To be specific, $C = c \bullet \imath$, where $c = \mathrm{d}C/\mathrm{d}\imath$ is the instantaneous consumption rate. The control variables for the agent can be aggregated into the triplet of processes (ξ, η, C). The control C is *absolutely continuous*, in that the measure $\mathrm{d}C$ is absolutely continuous with respect to the Lebesgue measure $\Lambda = \mathrm{d}\imath$, but, as we are about to realize, the controls ξ and η will have to be singular, that is, the measures $\mathrm{d}\xi_t$ and $\mathrm{d}\eta_t$ will have to be allowed to be singular with respect to Λ. Any amount that the agent decides to consume will have to come out of the cash account, and converting cash into real consumption goods can be done at any time at no cost. The dynamics of the controlled wealth process (X, Y) can be stated as

$$
\begin{aligned}
X &= X_0 + rX \bullet \imath - C - (1+h)\xi + (1-k)\eta\,, \\
Y &= Y_0 + Y_- \bullet \Big(\sigma W + b\imath + \int_{\mathbb{R}} \gamma(z)\tilde{N}(\mathrm{d}z, [0, \imath])\Big) + \xi - \eta\,.
\end{aligned}
\qquad \mathrm{e}_1
$$

At this point we must note that if the choice of the controls ξ and η is optimal, then $\mathrm{d}\xi_t\,\mathrm{d}\eta_t = 0$, since buying and selling the risky security at the same time carries an expense without actually changing the agent's position. The solvency cone $\mathcal{S} \subseteq \mathbb{R}^2$ can be described as follows. If $X_t < 0$, then the short position in the bond must be covered with a long position in the stock, and this means $(1-k)Y_t \geq -X_t > 0$ or, equivalently, $X_t + (1-k)Y_t \geq 0$. Similarly, if $Y_t < 0$, then the short position in the stock must be covered with a long cash position, and this entails $X_t \geq -(1+h)Y_t > 0$ or, equivalently, $X_t + (1+h)Y_t \geq 0$. Thus

$$
\mathcal{S} = \Big\{(x, y) \in \mathbb{R}^2 : x + (1-k)y \geq 0 \ \text{ and } \ x + (1+h)y \geq 0\Big\}.
$$

The set of admissible controls $\mathcal{A}(x, y, t)$ in state $(X_t = x, Y_t = y)$ consists of all triplets of adapted, increasing, and càdlàg processes (ξ_s, η_s, C_s), $s \geq t$, that keep the investor's market position (X_s, Y_s), $s \geq t$, inside the solvency cone $\mathcal{S}$ at all times, subject to the dynamics e_1, and have the additional property that the control C is absolutely continuous. It is impossible not to notice the striking similarity between the equations in e_1 and the general equations that govern a contained (reflected) diffusion that we encountered back in (16.50). This connection crops up in the present context for a reason: before maximizing the agent's objective, the control must keep the agent solvent, that is, must keep the diffusion (X, Y) inside the solvency cone $\mathscr{S}$. But as we realized earlier,⊶(16.50) since the controls can affect the dynamics of the controlled diffusion only in a predictable fashion, there is nothing to prevent the diffusion to jump outside of the domain in which it is meant to be contained, even if the control can instantly bring it back. Thus, we must specify a domain from which the agent's position cannot jump into insolvency. To be specific, the control must keep the diffusion (X, Y) inside the closure of the (smaller) cone

$$\mathscr{K} = \Big\{(x, y) \in \mathbb{R}^2 : x + (1-k)\big(1 - \operatorname{sign}(y)\, a\big) y \geq 0 \\ \text{and } x + (1+h)\big(1 - \operatorname{sign}(y)\, a\big) y \geq 0 \Big\} \subseteq \mathscr{S},$$

with the understanding that the diffusion process cannot be prevented from jumping outside of $\mathscr{K}$, but if it does, still being in $\mathscr{S}$, the control will instantly bring it back in $\mathscr{K}$, with no possibility of insolvency occurring. In fact, the optimal trading strategy will keep the agent's market position in a subdomain of $\mathscr{K}$ in the manner described in (16.50). Indeed, in one way or another, the solvency cone $\mathscr{S}$ will be divided into three *disjoint* subdomains, $\mathscr{S} = \mathscr{D} \cup \mathscr{T}^* \cup \mathscr{T}_*$, so that $\mathscr{D}$ is the domain where trading would not be optimal, $\mathscr{T}^*$ is the domain where buying equity would be optimal, and $\mathscr{T}_*$ is the domain where selling equity would be optimal (as already noted, buying and selling simultaneously cannot be optimal). We suppose that the set $\mathscr{D}$ is open and the sets $\mathscr{T}^*$ and $\mathscr{T}_*$ are closed. Since buying equity means moving in the direction of the vector $(-1-h, 1)$, if $(X_t, Y_t) \in \mathscr{T}^*$, the control will instantly push the market position along the ray

$$\mathbb{R}_+ \ni u \rightsquigarrow (X_t - (1+h)u, Y_t + u) \in \mathscr{S}$$

until buying is no longer optimal, that is, until the market position enters $(\mathscr{T}^*)^{\mathsf{C}}$. Similarly, selling equity means moving in the direction of the vector $(1-k, -1)$, so that if $(X_t, Y_t) \in \mathscr{T}_*$, the control will instantly push the market position along the ray

$$\mathbb{R}_+ \ni u \rightsquigarrow (X_t + (1-k)u, Y_t - u) \in \mathscr{S}$$

until selling is no longer optimal, that is, until the market position enters $(\mathscr{T}_*)^{\mathsf{C}}$.

We are now precisely in the situation described in (16.50): a jump diffusion must be kept inside the closure, $\bar{\mathscr{D}}$, of the no-trade domain $\mathscr{D}$. To see this connection, observe first that the border, $\partial\mathscr{D}$, of the domain $\mathscr{D}$ consists of two pieces: $\partial^*\mathscr{D} \overset{\text{def}}{=} \mathscr{T}^* \cap \bar{\mathscr{D}}$ and $\partial_*\mathscr{D} \overset{\text{def}}{=} \mathscr{T}_* \cap \bar{\mathscr{D}}$. Next, define the projection map $\Pi : \mathscr{S} \mapsto \bar{\mathscr{D}}$, so that if $(x, y) \in \bar{\mathscr{D}}$, then $\Pi(x, y) = (x, y)$; if $(x, y) \in \mathscr{T}^*$, then $\Pi(x, y)$ is the intersection of the ray $\mathbb{R}_+ \ni u \rightsquigarrow (x - (1+h)u, y + u)$ with $\partial^*\mathscr{D}$; and if $(x, y) \in \mathscr{T}_*$, then $\Pi(x, y)$ is the intersection of the ray $\mathbb{R}_+ \ni u \rightsquigarrow (x + (1-k)u, y - u)$ with $\partial_*\mathscr{D}$. To put it another way, $\Pi(x, y)$ is the nearest point from $\bar{\mathscr{D}}$ to (x, y) in the direction of the vector $(-1-h, 1)$ if $(x, y) \in \mathscr{T}^*$, or in the direction of the vector $(1-k, -1)$ if $(x, y) \in \mathscr{T}_*$. Thus, if we write $(x^\circ, y^\circ) = \Pi(x, y)$, then $x^\circ - x = -(1+h)(y^\circ - y)$, $y^\circ \geq y$, if $(x, y) \in \mathscr{T}^*$, and, similarly, $x^\circ - x = (1-k)(y - y^\circ)$,

$y^\circ \le y$, if $(x, y) \in \mathcal{T}_*$. For this reason, it will be enough for us to work only with the second component, y°, of the projection $\Pi(x, y)$, which we are going to express as $y^\circ = \pi(x, y)$. The vector field $n(x, y)$, $(x, y) \in \partial\mathcal{D}$, along which the diffusion will have to be pushed during the local time in $\partial\mathcal{D}$, is

$$n(x,y) = \begin{cases} (-1-h, 1), & \text{if } (x,y) \in \partial^*\mathcal{D}, \\ (1-k, -1), & \text{if } (x,y) \in \partial_*\mathcal{D}. \end{cases}$$

In conjunction with the dynamics shown in e_1 and the results from (16.50), we see that the optimal trading strategy must be of the form

$$\xi_t = \int^t dL_s^{\partial\mathcal{D}_*}(X) + \int^{t-} \big(\pi(X_s, Y_s) - Y_s\big) 1_{\mathcal{T}^*}(X_s, Y_s) N(\mathbb{R}, ds),$$

$$\eta_t = \int^t dL_s^{\partial\mathcal{D}^*}(X) + \int^{t-} \big(Y_s - \pi(X_s, Y_s)\big) 1_{\mathcal{T}_*}(X_s, Y_s) N(\mathbb{R}, ds).$$

The second integrals on the right are understood as finite sums[195] and can also be cast as

$$\sum_{s<t,\, \Delta Y_s \neq 0} \big(\pi(X_s, Y_s) - Y_s\big) 1_{\mathcal{T}^*}(X_s, Y_s) \quad \text{and} \quad \sum_{s<t,\, \Delta Y_s \neq 0} \big(Y_s - \pi(X_s, Y_s)\big) 1_{\mathcal{T}_*}(X_s, Y_s).$$

We stress that both summations are over $s < t$, not over $s \le t$.

The main task in front of us now is to determine where exactly the domain $\mathcal{D}$ is, that is, where the boundaries $\partial^*\mathcal{D}$ and $\partial_*\mathcal{D}$ are located. For that purpose, we now turn to the actual optimization problem that the agent is faced with. We assume infinite time horizon and performance criterion postulated as

$$J(x, y, \xi, \eta, C, t) = \mathsf{E}\Big[\int_t^\infty e^{-(s-t)\beta} U(c_s) ds \mid X_t = x, Y_t = y\Big],$$

where $c_s = dC_s/ds$ is the instantaneous consumption rate and $U(\cdot)$ is the agent's utility from consumption. Just as before, we seek a solution only in the case of a CRRA utility, that is, either $U(c) = \frac{c^p}{p}$, $p < 1$, or $U(c) = \log(c)$. Next, observe that, since we are in the setting of an economy of scale, nothing changes in e_1 if the state process (X, Y) and the control process (ξ, η, C) are simultaneously scaled by a factor $\rho > 0$. Furthermore

$$J(\rho x, \rho y, \rho\xi, \rho\eta, \rho C, t) = \rho^p J(t, x, y, \xi, \eta, C), \text{ if } U(c) = \frac{c^p}{p},$$

and e_2

$$J(\rho x, \rho y, \rho\xi, \rho\eta, \rho C, t) = \frac{\log(\rho)}{\beta} + J(t, x, y, \xi, \eta, C), \text{ if } U(c) = \log(c).$$

This shows that the optimal decision (consumption and trading) in state $(\rho x, \rho y)$ at time t obtains from the optimal decision in state (x, y) by way of scaling with the

[195] The measure $N(\mathbb{R}, \cdot)$ is purely atomic with finitely many atoms on any finite interval – the main reason for insisting that the jumps in the returns are uniformly bounded away from 0.

factor ρ. In particular, the states $(\rho x, \rho y)$ and (x, y) are either simultaneously "no-trade" states, simultaneously "buy stock" states, or simultaneously "sell stock" states; in other words, the sets $\mathscr{D}$, $\mathscr{T}^*$, and $\mathscr{T}_*$ can only be cones. In particular, $\partial^*\mathscr{D}$ and $\partial_*\mathscr{D}$ must be rays inside $\mathbb{R}^2$ starting from the origin. The entire optimal control problem now comes down to choosing these two rays accordingly, but this choice is also subject to the restriction $\mathscr{D} \subseteq \mathscr{K}$ (the agent's position should not be allowed to enter states from which it could jump into insolvency). To locate the rays $\partial^*\mathscr{D}$ and $\partial_*\mathscr{D}$, we now turn to the associated HJB equation, and note that with time horizon $T = \infty$ the value function should be time-homogeneous, that is, $V(x, y, t) = V(x, y)$. To see what this equation may look like in the "buy" domain $\mathscr{T}^*$, suppose, contrary to fact, that shares are to be purchased only *continuously* at *instantaneous rate* $\Delta > 0$ – think of Δ as a (possibly very large) control parameter for now. This means that in e_1 one must set $\xi = \Delta\iota$ and $\eta = 0$. For $(x, y) \in (\mathscr{T}^*)^\circ$ (the interior of the "buy" domain) we can then write

$$\sup_{c,\Delta\in\mathbb{R}_{++}} \Big\{U(c) + (rx - c)\partial_x V(x, y) + by\,\partial_y V(x, y)\mathrm{d}t + \frac{1}{2}\sigma^2 y^2 \partial_y^2 V(x, y)$$
$$+ \int_{\mathbb{R}} \Big(V\big(x, y + y\gamma(z)\big) - V(x, y) - y\gamma(z)\partial_y V(x, y)\Big)\nu(\mathrm{d}z)$$
$$+ \Big(\partial_y V(x, y) - (1 + h)\partial_x V(x, y)\Big)\Delta\Big\} = \beta V(x, y)\,.$$

Since the position (x, y) is in the "buy" region, increasing Δ (from 0) should be optimal, and this means that

$$\partial_y V(x, y) - (1 + h)\partial_x V(x, y) \ge 0\,. \qquad \mathrm{e}_3$$

But if this last condition holds, then optimality would require the instantaneous buying rate Δ to be arbitrarily large, and this leads to an impulse-type control. Specifically, if $\Delta = \hat{u}(x, y)$ is the solution (for Δ) of the equation

$$\big(x - (1 + h)\Delta\,,\ y + \Delta\big) \in \partial^*\mathscr{D}\,,$$

then the optimal nominal value invested (instantly) in the stock in state (x, y) would be precisely $\hat{u}(x, y)$, that is, the largest amount that the agent can buy before exiting the "buy" domain. The geometry of this choice should be clear: the agent moves their market position along the ray $\mathbb{R}_+ \ni u \rightsquigarrow (x - (1 + h)u, y + u)$ until the end of the domain $\mathscr{T}^*$, that is, until the intersecting point of this ray with $\partial^*\mathscr{D}$ (also a ray). Because this jump in the market position is instantaneous, one should have

$$V(x, y) = V\big(x - (1 + h)\hat{u}(x, y), y + \hat{u}(x, y)\big)\,,$$

that is, the values $V(x, y)$, $(x, y) \in \partial^*\mathscr{D}$ determine the values $V(x, y)$, for all $(x, y) \in \mathscr{T}^*$. To be precise, the value function $(x, y) \rightsquigarrow V(x, y)$ must be constant along each of the rays

$$\mathbb{R}_+ \ni u \rightsquigarrow (x + (1 + h)u, y - u)\,, \quad (x, y) \in \partial^*\mathscr{D}\,,$$

for as long as these rays remain in the "buy stock" domain $\mathscr{T}^*$. In particular, the relation e_3 must be an equality inside $\mathscr{T}^*$. Because when it happens to be inside $\mathscr{T}^*$ the market position (x, y) is pushed instantly to the boundary $\partial^*\mathscr{D}$, this type of impulse control takes place on a time scale that is singular to $\mathrm{d}t$, as there are at most countably many moments during which (X, Y) could jump into the interior of $\mathscr{T}^*$. In particular, the agent has no time to consume while in the interior of either $\mathscr{T}^*$ or $\mathscr{T}_*$.

The analysis of the HJB equation in the domain $(\mathscr{T}_*)^\circ$ (the interior of the "sell stock" domain) is completely analogous. In this domain one must have

$$(1-k)\partial_x V(x, y) - \partial_y V(x, y) \geq 0$$

to be able to claim that trading (that is, selling, in this case) is optimal. As a result, the optimal control in state $(x, y) \in (\mathscr{T}_*)^\circ$ is again impulse-type control that instantly moves the market position along the ray $\mathbb{R}_+ \ni u \rightsquigarrow (x + (1-k)u, y - u)$ until the intersection point with $\partial_*\mathscr{D}$. The values $V(x, y)$, $(x, y) \in \partial_*\mathscr{D}$ again determine the values $V(x, y)$, $(x, y) \in \mathscr{T}_*$. In this case the value function $(x, y) \rightsquigarrow V(x, y)$ must be constant along each of the rays

$$\mathbb{R}_+ \ni u \rightsquigarrow (x - (1-k)u, y + u), \quad (x, y) \in \partial_*\mathscr{D},$$

for as long as these rays remain in the sell domain $\mathscr{T}_*$.

Finally, we turn to the HJB equation in the no-trade (open) domain $\mathscr{D}$. Since there is no trade at $(x, y) \in \mathscr{D}$ and the only decision to be made is the choice of the consumption rate, the HJB equation can be cast as

$$\sup_{c \in \mathbb{R}_{++}} \Big[U(c) + (rx - c)\partial_x V(x, y) + by\partial_y V(x, y)\mathrm{d}t + \frac{1}{2}\sigma^2 y^2 \partial_y^2 V(x, y) + \int_{\mathbb{R}} \Big(V(x, y + y\gamma(z)) - V(x, y) - y\gamma(z)\partial_y V(x, y) \Big)\nu(\mathrm{d}z) \Big] = \beta V(x, y),$$

and this equation reduces to the following second-order PDE for $V(\cdot, \cdot)$:

$$\begin{aligned} &U\Big((\partial U)^{-1}\big(\partial_x V(x, y)\big) \Big) + \Big(rx - (\partial U)^{-1}\big(\partial_x V(x, y)\big) \Big)\partial_x V(x, y) \\ &+ by\,\partial_y V(x, y) + \frac{1}{2}\sigma^2 y^2 \partial_y^2 V(x, y) \\ &+ \int_{\mathbb{R}} \Big(V(x, y + y\gamma(z)) - V(x, y) - y\gamma(z)\partial_y V(x, y) \Big)\nu(\mathrm{d}z) = \beta V(x, y). \end{aligned} \qquad \mathrm{e}_4$$

(18.15) REMARK: It should be clear why the jumps generated by the impulse control, which are meant to keep the market position inside the set $\bar{\mathscr{D}}$, do not affect the HJB equation in the open domain $\mathscr{D}$, while the jumps in the risky returns do. The point is that any choice of the control variables at time t – whether impulse ("bang-bang" type) or continuous – can affect the market position only immediately *after* time t, not *instantly* at time t. To put it another way, any instant correction (push pack) can be triggered only after the process (X, Y) is found to be inside either $(\mathscr{T}^*)^\circ$ or $(\mathscr{T}_*)^\circ$. When such a correction (buying or selling) takes effect, it only brings the process to the boundary $\partial^*\mathscr{D} \cup \partial_*\mathscr{D}$. As a result, the jumps in the control are never triggered while (X, Y) is in the open domain $\mathscr{D}$ and have no effect on the local dynamics of (X, Y) while this process remains inside $\mathscr{D}$. ○

To develop a solution, notice first that the scaling property from $\mathfrak{e}_2$ implies $V(\rho x, \rho y) = \rho^p V(x, y)$, $\rho > 0$, in the case of a power utility, and $V(\rho x, \rho y) = \frac{\log(\rho)}{\beta} + V(x, y)$, $\rho > 0$, in the case of a logarithmic utility. This relation must hold everywhere in the solvency cone $\mathscr{S}$. In the domain $\{y > 0\}$ these relations are the same as, respectively,

$$V(x, y) = y^p V\left(\frac{x}{y}, 1\right) \text{ and } V(x, y) = \frac{\log(y)}{\beta} + V\left(\frac{x}{y}, 1\right),$$

and in the domain $\{y < 0\}$ they are the same as

$$V(x, y) = (-y)^p V\left(-\frac{x}{y}, -1\right) \text{ and } V(x, y) = \frac{\log(-y)}{\beta} + V\left(-\frac{x}{y}, -1\right).$$

For brevity, in what follows we are going to develop the solution only in the "no short-sales" region $\{x > 0, y > 0\}$, and, for that purpose, we now introduce the function

$$\mathbb{R}_{++} \ni \alpha \rightsquigarrow f(\alpha) \stackrel{\text{def}}{=} V(\alpha, 1).$$

The unknown value function can be cast as

$$V(x, y) = y^p f(x/y), \quad \text{or} \quad V(x, y) = \frac{\log(y)}{\beta} + f(x/y),$$

which is to say that we are now looking for a function of just one variable: $\alpha \rightsquigarrow f(\alpha)$. Since the domains $\mathscr{T}^*$, $\mathscr{T}_*$, and $\mathscr{D}$ are cones inside $\mathbb{R}^2$, the boundary $\partial\mathscr{D}$ consists of two rays, which we express as:

$$\partial^*\mathscr{D} = \left\{(x, y) \colon x = \alpha^* y, \;\; x \geq 0\right\} \quad \text{and} \quad \partial_*\mathscr{D} = \left\{(x, y) \colon x = \alpha_* y, \;\; x \geq 0\right\}.$$

It is not difficult to see that $\alpha^* > \alpha_*$.[196] Thus, depending on the choice of utility function,

$$V(x, y) = y^p f(\alpha^*_*), \quad \text{or} \quad V(x, y) = \log(y)/\beta + f(\alpha^*_*), \quad \text{for } (x, y) \in \partial^*_*\mathscr{D},$$

where the symbols α^*_* and ∂^*_* are understood as either α_* and ∂_* or α^* and ∂^*.

(18.16) EXERCISE: Show that with power utility $U(c) = c^p/p$, $c \in \mathbb{R}_{++}$, $p < 1$, the value function $V(x, y)$ in the "sell stock" and the "buy stock" domains is given, respectively, by

$$V(x, y) = \left(y - \frac{\alpha_* y - x}{\alpha_* - k + 1}\right)^p f(\alpha_*), \quad (x, y) \in \mathscr{T}_*,$$

and

[196] Market positions (x, y) with very large x and very small y should fall into the "buy stock" region, whereas market positions (x, y) with very small x and very large y should fall into the "sell stock" region.

$$V(x,y) = \Bigl(y - \frac{\alpha^* y - x}{\alpha^* + h + 1}\Bigr)^p f(\alpha^*)\,, \quad (x,y) \in \mathscr{T}^*\,.$$

Verify that these functions satisfy the relations

$$(1-k)\partial_x V(x,y) - \partial_y V(x,y) = 0\,, \quad (x,y) \in \mathscr{T}_*\,,$$

and

$$\partial_y V(x,y) - (1+h)\partial_x V(x,y) = 0\,, \quad (x,y) \in \mathscr{T}^*\,.$$

HINT: How are the values $V(x,y)$, $(x,y) \in \mathscr{T}_*$, determined from the values on the boundary with the "no-trade" domain $V(x,y)$, $(x,y) \in \partial_*\mathscr{D}$? How are the values $V(x,y)$, $(x,y) \in \mathscr{T}^*$, determined from the values on the boundary with the "no-trade" domain $V(x,y)$, $(x,y) \in \partial^*\mathscr{D}$? ○

(18.17) EXERCISE: Prove that with logarithmic utility $U(c) = \log(c)$, $c \in \mathbb{R}_{++}$, the value function $V(x,y)$ in the "sell stock" and "buy stock" domains is given, respectively, by

$$V(x,y) = \frac{1}{\beta}\log\Bigl(y - \frac{\alpha_* y - x}{\alpha_* - k + 1}\Bigr) + f(\alpha_*)\,, \quad (x,y) \in \mathscr{T}_*\,,$$

and

$$V(x,y) = \frac{1}{\beta}\log\Bigl(y - \frac{\alpha^* y - x}{\alpha^* + h + 1}\Bigr) + f(\alpha^*)\,, \quad (x,y) \in \mathscr{T}^*\,.$$

Verify that these functions satisfy the relations

$$(1-k)\partial_x V(x,y) - \partial_y V(x,y) = 0\,, \quad (x,y) \in \mathscr{T}_*\,,$$

and

$$\partial_y V(x,y) - (1+h)\partial_x V(x,y) = 0\,, \quad (x,y) \in \mathscr{T}^*\,.$$

HINT: See the hint to the previous problem. ○

(18.18) REMARK: In the case of power utility, in the "buy stock" domain $\mathscr{T}^*$ the value function $V(x,y)$ can be cast as $V(x,y) = y^p f(x/y)$, $x/y \geq \alpha^*$, with

$$f(\alpha) = \Bigl(1 - \frac{\alpha^* - \alpha}{\alpha^* + h + 1}\Bigr)^p f(\alpha^*) \quad \text{for } \alpha \geq \alpha^*\,.$$

Similarly, in the "sell stock" domain $\mathscr{T}_*$ the value function $V(x,y)$ can be cast as $V(x,y) = y^p f(x/y)$, $x/y \leq \alpha_*$, with

$$f(\alpha) = \Bigl(1 - \frac{\alpha_* - \alpha}{\alpha_* - k + 1}\Bigr)^p f(\alpha_*) \quad \text{for } \alpha \leq \alpha_*\,.$$

These relations have obvious analogues in the case of logarithmic utility: in the "buy stock" domain $\mathscr{T}^*$ the value function $V(x,y)$ can be cast as $V(x,y) = \log(y)/\beta + f(x/y)$, $x/y \geq \alpha^*$, with

$$f(\alpha) = \frac{1}{\beta}\log\Bigl(1 - \frac{\alpha^* - \alpha}{\alpha^* + h + 1}\Bigr) + f(\alpha^*)\,, \quad \alpha \geq \alpha^*\,,$$

whereas in the "sell stock" domain $\mathscr{T}_*$ the value function $V(x, y)$ can be cast as $V(x, y) = \log(y)/\beta + f(x/y)$, $x/y \le \alpha_*$, with

$$f(\alpha) = \frac{1}{\beta}\log\Big(1 - \frac{\alpha_* - \alpha}{\alpha_* - k + 1}\Big) + f(\alpha_*), \quad \alpha \le \alpha_* . \quad \circ$$

According to the last remark, if we know α_* and $f(\alpha_*)$, then we can construct $f(\alpha)$ for all $\alpha \le \alpha^*$, and if we know α^* and $f(\alpha^*)$, then we can construct $f(\alpha)$ for all $\alpha \ge \alpha^*$. We still need to understand how to construct $f(\alpha)$ for $\alpha_* < \alpha < \alpha^*$, and, for this reason, now turn to the equation for the value function $V(x, y)$ in the "no-trade" domain $\mathscr{D}$. We have already$^{\multimap \bullet e_4}$ developed this equation in terms of the general form of V, but now want to take advantage of the scaling property of V and turn e_4 into a second-order ODE for $f(\cdot)$.

(18.19) EXERCISE: Prove that with power utility $U(c) = c^p/p$, $c \in \mathbb{R}_{++}$, $p < 1$, the function $\alpha \rightsquigarrow f(\alpha)$ from the representation $V(x, y) = y^p f(x/y)$ must solve the following equation for $\alpha \in]\alpha_*, \alpha^*[$:

$$0 = \frac{1}{2}\sigma^2\alpha^2(\partial^2 f)(\alpha) - \big((p-1)\sigma^2 + b - r\big)\alpha(\partial f)(\alpha) + \frac{1-p}{p}(\partial f)(\alpha)^{\frac{p}{p-1}}$$
$$+ \frac{1}{2}\big(2bp + (p^2 - p)\sigma^2 - 2\beta\big) f(\alpha)$$
$$+ \int_{\mathbb{R}}\Big(f\Big(\frac{\alpha}{1+\gamma(z)}\Big)\big(1+\gamma(z)\big)^p - f(\alpha)\big(1 + p\gamma(z)\big) + \alpha\gamma(z)(\partial f)(\alpha)\Big)\nu(\mathrm{d}z).$$

HINT: In e_4 set $V(x, y) = y^p f(x/y)$, followed by the substitution $x = \alpha y$. Consider using computer algebra. $\circ$

(18.20) EXERCISE: Prove that with logarithmic utility $U(c) = \log(c)$, $c \in \mathbb{R}_{++}$, the function $\alpha \rightsquigarrow f(\alpha)$ from the representation $V(x, y) = \frac{1}{\beta}\log(y) + f(x/y)$ must solve the following equation for $\alpha \in]\alpha_*, \alpha^*[$:

$$0 = \frac{1}{2}\sigma^2\alpha^2(\partial^2 f)(\alpha) + (\sigma^2 - b + r)\,\alpha(\partial f)(\alpha) + \frac{2b - \sigma^2}{2\beta} - \log(\partial f(\alpha)) - 1 - \beta f(\alpha)$$
$$+ \int_{\mathbb{R}}\Big(f\Big(\frac{\alpha}{1+\gamma(z)}\Big) + \frac{1}{\beta}\log\big(1+\gamma(z)\big) - f(\alpha) - \frac{\gamma(z)}{\beta} + \alpha\gamma(z)(\partial f)(\alpha)\Big)\nu(\mathrm{d}z).$$

HINT: See the hint to the previous problem. $\circ$

Since inside the domain $\mathscr{D}$ trading is not optimal, if $(x, y) \in \mathscr{D}$, then

$$\partial_y V(x, y) - (1 + h)\partial_x V(x, y) < 0 \quad \text{and} \quad (1 - k)\partial_x V(x, y) - \partial_y V(x, y) < 0 .$$

By continuity, for $(x, y) \in \partial^*\mathscr{D}$ the first relation should become an equality, and, similarly, the second relation becomes an equality for $(x, y) \in \partial_*\mathscr{D}$.

(18.21) EXERCISE: Prove that with $V(x, y) = y^p f(x/y)$, $p < 1$, the conditions above can be expressed, respectively, as

$$pf(\alpha) - (1 + h + \alpha)\partial f(\alpha) < 0 \quad \text{and} \quad -pf(\alpha) + (1 - k + \alpha)\partial f(\alpha) < 0$$
$$\text{for} \quad \alpha_* < \alpha < \alpha^* . \quad \circ$$

(18.22) EXERCISE: Prove that with $V(x, y) = \frac{1}{\beta}\log(y) + f(x/y)$ the same conditions can be expressed, respectively, as

$$1 - \beta(1 + h + \alpha)\partial f(\alpha) < 0 \quad \text{and} \quad \beta(1 - k + \alpha)\partial f(\alpha) - 1 < 0$$
$$\text{for} \quad \alpha_* < \alpha < \alpha^* . \quad \circ$$

(18.23) COMPUTATION STRATEGY: We must solve the second-order ODEs developed in (18.19) and (18.20) in the unknown interval $[\alpha_*, \alpha^*]$. The boundary values $f(\alpha_*)$ and $f(\alpha^*)$ are also unknown, so that we have a total of four parameter values to fix: α_*, α^*, $f(\alpha_*)$, and $f(\alpha^*)$. Since, in principle, fixing the boundary values at both ends of the interval should fix the solution of a second-order ODE, every concrete choice for the four values should translate into a concrete choice for the function $[\alpha_*, \alpha^*] \ni \alpha \rightsquigarrow f(\alpha)$. We thus need to impose four boundary-type conditions on this function, which will then translate into four equations for the four unknown values. To develop such conditions, notice that by (18.18) the function $f(\cdot)$ would be fixed outside of the interval $]\alpha_*, \alpha^*[$ once the parameters α_*, α^*, $f(\alpha_*)$, and $f(\alpha^*)$ are fixed. In particular, the derivatives $\partial f(\alpha)$ and $\partial^2 f(\alpha)$, for $\alpha \notin [\alpha_*, \alpha^*]$, would be fixed, too. It is natural to require – and this requirement can be justified if one is willing to face the technicalities involved – that these derivatives must be continuous on the entire domain, that is, the values for $\partial f(\alpha)$ and $\partial^2 f(\alpha)$ from inside the interval $[\alpha^*, \alpha_*]$ should match the corresponding values from outside at the end points α_* and α^*. This requirement gives us the four additional boundary conditions that are needed. Matching the first derivatives comes down to turning the inequalities from (18.21) and (18.22) into exact equalities, as is easy to see from (18.16) and (18.17). This observation translates into the following conditions:

$$\dagger \qquad \partial f(\alpha^*) = \frac{pf(\alpha^*)}{1 + h + \alpha^*} \quad \text{and} \quad \partial f(\alpha_*) = \frac{pf(\alpha_*)}{1 - k + \alpha_*}$$

in the case of power utility and

$$\ddagger \qquad \partial f(\alpha^*) = \frac{1}{\beta(1 + h + \alpha^*)} \quad \text{and} \quad \partial f(\alpha_*) = \frac{1}{\beta(1 - k + \alpha_*)}$$

in the case of logarithmic utility.

Matching the second derivatives at α_* (at α^*) is somewhat more involved. The idea is to use the expressions developed in (18.18) and compute $\partial^2 f(\alpha)$ for $\alpha \notin]\alpha_*, \alpha^*[$; in particular, compute $\partial^2 f(\alpha)$ for $\alpha = \alpha_*$ (for $\alpha = \alpha^*$), which would give an explicit expression for $\partial^2 f(\alpha_*)$ (for $\partial^2 f(\alpha^*)$) only in terms of $(\alpha_*, f(\alpha_*))$ (in terms of $(\alpha^*, f(\alpha^*))$). The next step is to write down the ODE at $\alpha = \alpha_*$ (at $\alpha = \alpha^*$), and replace in that equation the derivatives $\partial f(\alpha_*)$ and $\partial^2 f(\alpha_*)$ (the derivatives $\partial f(\alpha^*)$ and $\partial^2 f(\alpha^*)$) with the corresponding expressions involving $(\alpha_*, f(\alpha_*))$ (involving $(\alpha^*, f(\alpha^*))$). This will produce an algebraic equation from which $f(\alpha_*)$ (from which $f(\alpha^*)$) can be solved for in terms of the parameter α_* (in terms of the parameter α^*). This would eliminate $f(\alpha_*)$ and $f(\alpha^*)$ from the system, but we still need to solve for α_* and α^*. To see how these two degrees of freedom can be resolved, observe that for every concrete choice of α_* and α^* the boundary values $f(\alpha_*)$ and $f(\alpha^*)$ would be fixed, which then fixes the solution $f(\alpha)$ in the interval $]\alpha_*, \alpha^*[$, and, in particular, fixes the derivatives

$$\partial f(\alpha^*) = \lim_{\alpha \nearrow \alpha^*} \partial f(\alpha) \quad \text{and} \quad \partial f(\alpha_*) = \lim_{\alpha \searrow \alpha_*} \partial f(\alpha) .$$

In general, there is no reason why the last two values should match the prescriptions from either † or ‡ so that those prescriptions can be treated as a system for the unknowns α_* and α^*. In terms of concrete computation, the strategy comes down to varying α_* and α^* until the desired values for $\partial f(\alpha_*)$ and $\partial f(\alpha^*)$ from either † or ‡ are matched. We stress that the function $f(\cdot)$ so obtained must still satisfy the relations from either (18.21) or (18.22) for the solution to be financially meaningful.

Finally, we note that the ODE solvers available on most computing systems accept as an input only systems of first-order ODEs, and only with boundary conditions prescribed at the left end-point of the interval. The computational strategy outlined above can be adjusted to this technology in the obvious way. First, the second-order ODE can be reduced to a system of two first-order ODEs for $(f(\alpha), g(\alpha))$ with the substitution $g(\alpha) \overset{\text{def}}{=} \partial f(\alpha)$. Starting with an initial guess for α_*, one can compute $f(\alpha_*)$ by solving an algebraic equation, as described above, and then compute $g(\alpha_*) = \partial f(\alpha_*)$ from either † or ‡. The ODE solver can then generate a solution $\alpha \rightsquigarrow (f(\alpha), g(\alpha))$, initialized at α_* as $(f(\alpha_*), g(\alpha_*))$. One can then, depending on the type of utility, find α^* as the solution to

$$g(\alpha) = \frac{p f(\alpha)}{1 + h + \alpha} \qquad \text{or} \qquad g(\alpha) = \frac{1}{\beta(1 + h + \alpha)} \qquad \text{for } \alpha > \alpha_* .$$

With this choice for α^*, all conditions will be met, except for the algebraic equation involving $f(\alpha^*)$, which is the same as matching the second derivative at α_*. The strategy then comes down to repeating the procedure just outlined by calling the solver with different values for α_* until the corresponding α^* is such that the algebraic equation for $f(\alpha^*)$ is satisfied. ○

18.3 Real Options

The holder of an American-style stock option with underlying asset (stock) S and strike price K can collect the payoff $(S_t - K)$ at any time t before the closing date T. The same type of payoff arises in the seemingly unrelated domain of corporate investments: an entrepreneur (or a firm) has the right to incorporate a new business (install a new factory, drill a new oil well, open a new gold mine, incorporate a new restaurant chain, etc.) of value X_t at a fixed cost I (all expressed in adopted currency ¤). The investment is irreversible, and the value of the new would-be business (in the same currency ¤) follows an exogenous stochastic process $X = (X_t)_{t \in \mathbb{R}_+}$. More important, the investment decision has no effect on the value process X; in other words, the entrepreneur is a price taker and has a monopolistic right to invest (guaranteed by, say, a patent, exclusive mining rights, etc.).

(18.24) REMARK: The reason that "the monopolistic right to invest" translates to "the decision to invest does not affect the value of the enterprise" is that without patent rights and such there will be many (technically, an unlimited number of) potential investors who would attempt to incorporate the same enterprise, which would inevitably affect the value process; in fact, with a very large number of potential market entrants, at the time of investment[197] the value of the new business should be exactly equal to the cost of incorporating it, which is I. Such models were studied extensively by Dixit and Pindyck (1994). Plainly, without patent protection (that is, some form of monopoly), the right to invest would not generate a positive payoff and, as a result, would be of no value. ○

In any case, at the time of investment the entrepreneur receives a financial entity of market value X_t at cost I. This transaction generates the payoff $(X_t - I)$, which, technically, is no different from the payoff generated by exercising a call option. The difference, however, is that this payoff does not come out of the brokerage account of any counterparty in the contract, as there is no counterparty. We stress that a positive payoff of $(X_t - I)$ at the time of investment is actually required. It is required because, typically, the ability to invest comes at a cost. Indeed, the ability to incorporate a new business may be the result of years-long research and development activity (as an example, typically new drugs take years to develop and often involve a substantial material cost and human effort) or may require a land purchase, the purchase of certain patent rights, and so on. Because of the similarities – as well as the differences – when compared to stock options, monopolistic rights to invest, of the kind described here, are often referred to as *real options*.[198] From the viewpoint of mathematical modeling, real options are just American-style call options of infinite maturity, and, as we are about to realize,

---

[197] If all entrepreneurs are equally informed they will all invest at the same time.

[198] In some cases, sizable investment opportunities – or real options – do get created at a cost that appears to be essentially negligible. The example that immediately comes to mind is the proverbial garage in which, as the story has it, the first personal computer was built. To some degree, most

computationally such instruments are substantially easier to deal with. In fact the American-style calls and puts of infinite maturity (alias: perpetual calls and puts) are perhaps the only practically meaningful examples of optimal stopping problems in which a relatively simple and explicit closed-form solution is known to exist. This should not be surprising: in our study of Merton's portfolio selection problem we realized that, generally, infinite maturities simplify the model to a great extent by removing the dependence on the time variable. McDonald and Siegel (1986) were the first to study investment decisions from the point of view of option pricing and dynamic programming. The discussion in the present section closely follows Dixit and Pindyck (1994), who studied investment models extensively, with a focus on the value creation aspect of the flexibility to choose (optimally) the time to invest. These two publications are among the most influential contributions to the domain of finance and show how deeply intertwined the principles of asset pricing and the principles of optimal control are. Merton (1974) famously refers to stock options as "highly specialized and relatively unimportant financial instruments." Apparently the ideas of optimal control were brought into the realm of asset pricing with a much broader objective in mind than the mere pricing and optimal exercise of stock options.

Back to the model of the real option to incorporate an enterprise, we note that the enterprise itself may be modeled as a contingent claim, the underlying entity in which is the spot price of the generated stochastic output, expressed as an exogenous stochastic process $S = (S_t)_{t \in \mathbb{R}_+}$. We treat S_t as the market spot price of one unit of output and suppose that, once installed, the firm generates output continuously at the constant instantaneous rate of one unit. Thus, once installed, the firm would generate the instantaneous payoff of nominal value $S_t\,\mathrm{d}t$. What is the value, X_t, of an already installed firm then? To answer this question, we need to make a concrete assumption about the dynamics of the spot price S. In a first step, we postulate constant relative volatility and constant growth without jumps, that is, we postulate that S follows geometric Brownian motion with dynamics

$$S = S_0 + \sigma S \bullet W + b S \bullet \iota\,,$$

with some concrete choice for the parameters $\sigma, b \in \mathbb{R}$. To simplify the model further, we suppose that there are only two spot markets in the economy: a spot market for risk-free debt with a constant rate of return $r > 0$, and a spot market

inventions that lead to the creation of new markets and investment opportunities, including those that are enormously costly to develop, inevitably involve a certain degree of creativity and ingenuity that typically cannot be accounted for as a "direct cost." To put it another way, although it is possible to trade real options in a way that is similar to trading stock options, such instruments may develop from non-tradable assets (e.g., human ingenuity and creativity), which generally are not represented in the financial markets; in fact the mere existence of such non-tangible assets would not be known until their effect on the economy becomes tangible. A substantial part of the observed economic growth can be attributed to the creation of real options – the invention of the personal computer and the internet again come to mind – in addition, of course, to increased productivity and increased consumption levels.

for the firm's output. As we are already aware, given the dynamics above, this economy is complete, and the risk-neutral dynamics of the output price can be arranged as

$$S = S_0 + \sigma S \bullet \tilde{W} + rS \bullet \imath ,$$

where $\tilde{W} \overset{\text{def}}{=} W + \frac{b-r}{\sigma}\imath$ is a Brownian motion with respect to the pricing measure Q. As all future stochastic payoffs in this market are priced by discounting at the risk-free rate r followed by averaging with respect to Q, assuming that the firm would have an infinite lifetime, we conclude that the spot price of the firm must be

$$X_t = \int_t^{\infty} \mathsf{E}^Q\big[e^{-r(s-t)} S_s \mid S_t\big]\, \mathrm{d}s\,, \quad t \in \mathbb{R}_+ . \qquad \mathrm{e}_1$$

(18.25) EXERCISE: Prove that the last expression gives $X_t = +\infty$. ○

The last exercise is telling us that the model of real options that we are trying to develop is missing an important ingredient. We introduce this ingredient next.

(18.26) CONVENIENCE YIELD: There is a difference between owning a physical asset (a home, personal computer, motor vehicle, bushel of wheat, barrel of crude oil, etc.) and holding cash in a bank account. Indeed, typically, the act of holding a physical asset provides the holder with an intrinsic utility that is different in nature from the utility derived from holding a bank account. As an example, observe that with 1% interest ¤850,000 held in a bank account would generate in a month about ¤708.33. However, typically, the monthly rent for a home of market value ¤850,000 would be much higher (in the city where this paragraph is being typed that rent would be around USD 3,900), and this difference cannot be explained by the cost of maintenance and taxes alone. Clearly, the benefit from owning a home for a month is different from the benefit of holding ¤850,000 in cash for a month. Consequently, the difference between ¤850,000 in cash available now and ¤850,000 in cash available in a month is not the same as the difference between a home worth ¤850,000 available now and the same home available in a month. For this reason, to obtain today's price of future payoffs that are to be delivered as real goods, one must use a discount rate that is different from the discount rate associated with the cash account. The discount rate for real goods and services is called the *convenience yield*, and is known to vary in the cross-section of real goods and services. Mathematically, the convenience yield shows up in the form of a "dividend" that a particular real asset provides to its holder. Unlike actual dividends, however, the convenience yield is not reported – or even explicitly accounted for – and can only be deduced indirectly, say, from the observed spot prices. In any case, we stress that an equity in a particular business is a claim on its real output. ○

Suppose that the output from the already installed enterprise generates "convenience" for its owners at an instantaneous rate ϱS_t, that is, holding one unit of (real) output produces instantaneously $(\varrho S_t)\mathrm{d}t$ units of "convenience." The instantaneous net return from a long position on the output is then $\frac{\mathrm{d}S_t}{S_t} + \varrho\mathrm{d}t$, so that the instantaneous excess return from the same position is

$$\frac{dS_t}{S_t} + \varrho dt - r dt = \sigma dW_t + (b + \varrho - r)dt = \sigma d\Big(W_t + \frac{b + \varrho - r}{\sigma} t\Big).$$

It is the process $\bar{W} \stackrel{\text{def}}{=} W + \frac{b+\varrho-r}{\sigma}\iota$ that must be a local martingale, that is, a Brownian motion, under the pricing measure Q, not the process $\tilde{W} \stackrel{\text{def}}{=} W + \frac{b-r}{\sigma}\iota$, as we mistakenly supposed earlier. As the spot process S can also be written as

$$S = S_0 + \sigma S \bullet \bar{W} + (r - \varrho) S \iota t, \quad t \in \mathbb{R}_+,$$

we can again write the value, X_t, of the claim on the entire future output in the form e_1, except that the choice of Q is now different ($\bar{W}$ is a Q-Brownian motion) and so is the result:

(18.27) EXERCISE: Suppose that the measure Q is chosen so that the process

$$\bar{W} \stackrel{\text{def}}{=} W + \frac{b + \varrho - r}{\sigma}\iota$$

is a Q-Brownian motion. Prove that the expression in e_1 now gives $X_t = \frac{S_t}{\varrho}$. ○

One immediate corollary from the last exercise is that the value of an already installed firm, $X_t = \frac{1}{\varrho} S_t$, also follows a geometric Brownian motion; in fact, a geometric Brownian motion with the same dynamics as the spot process S:

$$X = X_0 + \sigma X \bullet \bar{W} + (r - \varrho) X \bullet \iota .$$

Since X is a contingent claim on the underlying spot asset S, the addition of the asset X (the firm) to the list of tradeable securities in the economy does not alter the pricing measure Q.[199]

Now we turn to the valuation and optimal exercise of the real option to install a firm of value X_t at fixed cost I. Such an option is a derivative instrument (contingent claim) with an underlying asset X, and its value can be expressed as $V(X_t, t)$ for some function $V(\cdot,\cdot)$. However, due to the Markovian dynamics of X and due to the fact that there is no deadline for the investment, the value of the real option must be of the form $V(X_t, t) = V(X_t)$, and its optimal exercise must depend on the price level X_t, but not on the time t. Since the payoff from exercising is $X_t - I$, the(17.13†) HJB equation for the associated optimal stopping problem specializes to the following equation for the function $x \rightsquigarrow V(x)$, which is based on the Q-dynamics of X:

$$\max\Big(r(x - I),\ \frac{1}{2}\sigma^2 x^2 \partial^2 V(x) + (r - \varrho) x \partial V(x)\Big) = rV(x). \qquad \mathrm{e}_2$$

If exercising in state $X_t = x$ is optimal, then exercising in state $X_t > x$ must be optimal, too, for if the payoff $x - I$ is acceptable, then any payoff larger than $x - I$

[199] Equity in the firm is a redundant security, since the market is already spanned by the spot price S and the risk-free bank account.

must be acceptable as well. Similarly, if waiting in state $X_t = x$ is optimal, then waiting in state $X_t < x$ must be optimal, too, for if the payoff $x-I$ is not acceptable, then any payoff smaller than $x - I$ would not be acceptable either. Consequently, the exercise region must be an interval of the form $[x^*, +\infty[$, while the hold region must be its complement $]0, x^*[$. The exercise threshold x^* is time invariant and is a yet to be determined scalar value. With these observations in mind, equation e_2 can be cast as

$$\frac{1}{2}\sigma^2 x^2 \partial^2 V(x) + (r - \varrho)x\partial V(x) = rV(x), \;\; V(x) > x - I \;\; \text{for } x \in \,]0, x^*[\,,$$

$$\frac{1}{2}\sigma^2 x^2 \partial^2 V(x) + (r - \varrho)x\partial V(x) \leq rV(x), \; V(x) = x - I \;\; \text{for } x \in [x^*, \infty[\,.$$

(18.28) EXERCISE: Prove that if $x \geq x^*$, then $\varrho x \geq rI$. ○

Since in state $X_t = x$ we have $\varrho x = \varrho \frac{S_t}{\varrho} = S_t$, the last exercise is telling us that investment cannot be triggered unless the spot price of the output matches or exceeds the marginal cost of the investment capital, namely, rI. This requirement is intuitive, but would it be sufficient to justify investment? To answer this question, we must solve the HJB equation above and compute x^* exactly. With this objective in mind, we are going to seek a solution in the domain $x \in \,]0, x^*[$ of the form $V(x) = Ax^\beta$ for some yet to be determined constants $A \in \mathbb{R}_{++}$ and $\beta \in \mathbb{R}$.

(18.29) EXERCISE: Prove that the function $V(x) = Ax^\beta$ solves the equation

$$\frac{1}{2}\sigma^2 x^2 \partial^2 V(x) + (r - \varrho)x\partial V(x) = rV(x)$$

if and only if $\beta \in \mathbb{R}$ solves the quadratic equation

$$FQ_{\sigma,r,\varrho}(\beta) \overset{\text{def}}{=} \frac{1}{2}\sigma^2 \beta(\beta - 1) + \beta(r - \varrho) - r = 0\,. \quad ○$$

(18.30) THE FUNDAMENTAL QUADRATIC: The expression $FQ_{\sigma,r,\varrho}(\beta)$ that we just defined is a quadratic function of β, which is known in the literature on real options as "the fundamental quadratic" (whence the notation FQ). Setting the fundamental quadratic to 0 gives a simple quadratic equation, which, as we are about to realize, plays an important role in the study of real options. ○

(18.31) EXERCISE: Prove that for every choice of the parameters σ, r, and ϱ, the quadratic equation $FQ_{\sigma,r,\varrho}(\beta) = 0$ has two distinct real roots $\beta_1 \geq 1$ and $\beta_2 < 0$, with $\beta_1 = 1$ if and only if $\varrho = 0$ (no convenience yield). ○

The general solution to

$$\frac{1}{2}\sigma^2 x^2 \partial^2 V(x) + (r - \varrho)x\partial V(x) = rV(x)$$

in the domain $x \in \,]0, x^*[$ can now be sought in the form

$$V(x) = A_1 x^{\beta_1} + A_2 x^{\beta_2}\,,$$

where β_1 and β_2 are the two roots of the fundamental quadratic (treated as functions of σ, r and ϱ) and A_1 and A_2 are free constants. Thus, we now have a total of three unknowns: A_1, A_2, and x^*, and we therefore need three boundary conditions. One obvious condition is $\lim_{x\searrow 0} V(x) = 0$ (the value of the option to invest should vanish as the spot price of the output collapses to 0). Since $\beta_1 \geq 1 > 0$ and $\beta_2 < 0$, this condition is satisfied only with $A_2 = 0$. The remaining two unknowns, A_1 and x^*, are determined from the following two requirements:

1. VALUE MATCHING: The solution on $]0, x^*[$ meets the solution on $[x^*, \infty[$ at the boundary x^*:

$$A_1(x^*)^{\beta_1} \equiv \lim_{x\nearrow x^*} V(x) = \lim_{x\searrow x^*} V(x) \equiv \lim_{x\searrow x^*}(x - I) \equiv (x^* - I)\,.$$

This condition ensures that the value function $V(\cdot)$ is continuous on the entire domain $]0, +\infty[$ and gives

$$A_1 = \frac{x^* - I}{(x^*)^{\beta_1}}\,.$$

2. SMOOTH PASTING: The solution on $]0, x^*[$ meets the solution on $[x^*, \infty[$ at the boundary x^* on a tangent,[200] that is, the first derivative of the solution on $]0, x^*[$ meets the first derivative of the solution on $[x^*, \infty[$ at the boundary x^*:

$$A_1\beta_1\,(x^*)^{\beta_1 - 1} = \lim_{x\nearrow x^*} \partial V(x) = \lim_{x\searrow x^*} \partial V(x) = \lim_{x\searrow x^*} \partial_x(x - I) = 1\,.$$

This condition ensures that the first derivative of $V(\cdot)$ is continuous on the entire domain $]0, +\infty[$, and gives

$$\beta_1\,(x^*)^{\beta_1 - 1} = \frac{1}{A_1} = \frac{(x^*)^{\beta_1}}{x^* - I} \quad \Rightarrow \quad x^* = \frac{\beta_1}{\beta_1 - 1} I\,.$$

(18.32) IMPORTANT REMARK: At the time when the option to invest is exercised – assuming that it is exercised optimally – the value of the newly created enterprise is required to be at least $\frac{\beta_1}{\beta_1 - 1} I$, and this adds *new wealth* to the existing economy of at least

$$\frac{\beta_1}{\beta_1 - 1} I - I = \frac{1}{\beta_1 - 1} I\,.$$

It is simply not optimal to exercise the option if the investment generates a payoff that is smaller than $\frac{1}{\beta_1 - 1} I$. Notice also that as $\varrho \searrow 0$, one has $\beta_1 \searrow 1$,[201] and, as a result, $x^* \to \infty$, that is, if the convenience yield collapses to 0, then the investment cannot be justified at any price level for the output. ○

200 This feature follows from⊸ (17.4) the mechanics of the backward induction; specifically from the geometry of the continuation value over a very small time period and from the way in which this continuation value and the immediate termination payoff intervene in the formation of the value function – see (Dumas 1991) for details.

201 It is straightforward to check that $FQ_{\sigma,r,\varrho}(1) = -\varrho$.

(18.33) EXERCISE: Prove that the investment threshold x^* is an increasing function of the volatility σ. Prove that the option value $V(x)$ is an increasing function of σ for any $x \in \mathbb{R}$. Give the economic interpretation of these features. ○

(18.34) EXERCISE: In (18.28) we found that if $x \geq x^*$, then $\varrho x \geq rI$. However, the last inequality is not enough to justify investment. Prove that, in fact, for any $x \geq x^*$, one must have $\varrho x \geq rI + \frac{1}{2}\sigma^2\beta_1 I$.

HINT: If $\beta > 1$ ($x^* < \infty$), the relation $\varrho\frac{\beta}{\beta-1} = r + \frac{1}{2}\sigma^2\beta$ becomes equivalent to the fundamental quadratic equation, which β_1 satisfies by definition. ○

(18.35) REMARK: We see from the last exercise that in order for the investment threshold to be optimal, it is not enough for the output price to match or exceed the marginal cost of the required capital outlay: it must match or exceed the marginal cost of the required capital *plus* the "premium" $\frac{1}{2}\sigma^2\beta_1 I$. This premium is proportional to (volatility)^2, and vanishes only when the volatility vanishes. ○

(18.36) VOLATILITY OF THE OPTION TO INVEST: As a direct application of Itô's formula, we see that the local martingale component (relative to the real-world probability P) of the semimartingale $V(X)$ is $\partial V(X)(\sigma X){\cdot}W$. The (stochastic) relative volatility of the value of the option to invest is then

$$\frac{\partial V(X_t)X_t}{V(X_t)}\sigma\,,$$

and, as the next exercise demonstrates, depending on the parameters in the model, this volatility could be a large multiple of the relative volatility σ that affects the output price. Shiller (1981)[202] famously asked, "Do stock prices move too much to be justified by subsequent changes in dividends?" It is indeed puzzling that, typically, the fluctuations (as a relative volatility) in the market value of a typical publicly traded firm would be considerably larger than the fluctuations in the spot price of the firm's output. However, as the model discussed in the present section demonstrates, such phenomena may be explained at least in part by the fact that equity values may capitalize real options to expand the current business, or develop new technologies and markets. In particular, this model predicts that the discrepancy between the two volatilities must be very pronounced for larger values of the convenience yield parameter ϱ. ○

(18.37) EXERCISE: Prove that for $0 < X_t < x^*$, that is, before exercise, the relative volatility of the option value $V(X_t)$ is $(\beta_1\sigma)$, and for $X_t \geq x^*$, that is, in the domain where the option is exercised, the relative volatility is $\sigma X_t/(X_t - I)$. Explain once again why $\beta_1 \to \infty$ as $\varrho \to \infty$. ○

(18.38) PERPETUAL CALLS AND PUTS: As was already noted, real options of the type studied in the present section can be treated as American-style call options of infinite maturities (alias: perpetual calls). Indeed, nothing will change in the

[202] The 2013 Sveriges Riksbank Prize in Economic Sciences in Memory of Alfred Nobel was awarded to Eugene Fama, Lars Peter Hansen, and Robert J. Shiller.

preceding computation if the value process X is interpreted as the spot price of a particular security, the convenience yield rate ϱ is interpreted as a dividend yield from X, and the sunk cost I is interpreted as the strike price in an American-style call of infinite maturity and with underlying asset X. Accordingly, the function $V(\cdot)$ that was computed earlier is nothing but the function $C(\cdot)$ that maps the underlying spot price into the value of the perpetual call. We saw in (14.34) that it is never optimal to exercise an American call before expiry, unless the underlying pays dividends. Now we see that the same statement holds for perpetual calls as well: $\varrho \to 0$ implies $\beta_1 \to 1$, which in turn implies $x^* \to \infty$, forcing the no-action region to expand to the entire domain $\mathbb{R}_{++}$. This observation has an interesting economic implication: a perpetual call with an underlying asset that never pays dividends can be viewed as a "speculative bubble," in that it represents an asset that never generates any payoff and can be bought only with the expectation that another agent may buy this asset in the future. Viable market models should not allow for such assets. ○

(18.39) EXERCISE: Using the methodology employed in the present section, derive the price of a perpetual put option (still assuming that the underlying asset follows a geometric Brownian motion). Does the put–call symmetry relation established in section 14.5 still hold for perpetual calls and puts? ○

(18.40) REMARK: It is possible to derive⊸ (18.27) the value $X_t = \frac{S_t}{\varrho}$ by way of contingent-claims analysis as follows. The value of an already installed firm is nothing but a contingent claim (derivative) with an underlying spot price S. Thus, one must have $X_t = F(S_t)$ for some yet to be determined function $F(\cdot)$. This function must solve the following ODE:

$$\frac{1}{2}\sigma^2 x^2 \partial^2 F(x) + (r - \varrho)x\partial F(x) + x = rF(x), \quad x \in \mathbb{R}_{++},$$

which may be seen as a special case of "an HJB equation" with trivial control: there is nothing left to control once the enterprise is installed (whence the absence of the sup). It is easy to check that $F = \frac{x}{\varrho}$ is one special solution to this equation, so that the general solution can be expressed as

$$F(x) = A_1 x^{\beta_1} + A_2 x^{\beta_2} + \frac{x}{\varrho},$$

where β_1 and β_2 are the two roots of the fundamental quadratic and A_1 and A_2 are free (yet to be determined) constants. Since $\beta_2 < 0$, the obvious requirement $\lim_{x \searrow 0} F(x) = 0$ entails $A_2 = 0$, but what requirement would enforce $A_1 = 0$? As was noted by Dixit and Pindyck (1994), the requirement is that the market does not allow for speculative bubbles, and it is important to spell out in what sense an asset with market price $S_t^{\beta_1}$ – at all times $t \in \mathbb{R}_+$, not only until some stopping time – represents a speculative bubble. The reason can be found in the calculation done earlier in this section. To be precise, given any $n \in \mathbb{N}_{++}$ one can find a real $R_n > 0$ such that the probability for the spot price S_t to enter the interval

$[R_n, \infty[$ at some moment $t \leq n$ is at most 2^{-n}. Now consider a call option with underlying S and with strike K chosen to be so large that $x^* = \frac{\beta_1}{\beta_1 - 1} K > R_n$. This option will be exercised at time $t \leq n$ with probability at most 2^{-n}, and until exercised it will be priced at $\varepsilon(S_t)^{\beta_1}$ for some $\varepsilon > 0$. In particular, $\frac{1}{\varepsilon}$ units of this option will be priced (prior to exercise) at $S_t^{\beta_1}$. To put it another way, $S_t^{\beta_1}$ represents the spot price of an asset that generates payoff prior to some very large time horizon only with some very small probability, where "very large" and "very small," are actually "arbitrarily large" and "arbitrarily small." ○

18.4 The Exercise Boundary for American Calls and Puts

The object of study in the present section is the exercise boundary and arbitrage-free price of American-style stock options. Thanks to the put–call symmetry relation developed in section 14.5, it would be enough for us to consider only call options. What follows next is a brief and vastly oversimplified review of several landmark studies of the exercise boundary by Kim (1990), Jacka (1991), Carr et al. (1992), and Barles et al. (1995).

We adopt the standard Black–Scholes–Merton framework, postulating that the market is complete and consists of a single risky asset (stock) S and a single risk-free asset (bond, cash account) S°, with respective price processes

$$S = S_0 + S \bullet \big(\sigma \tilde{W} + (r - \delta)\iota\big) \quad \text{and} \quad S^\circ = S_0^\circ + r S^\circ \bullet \iota\,.$$

In the first equation $\tilde{W}$ stands for a Brownian motion process relative to the risk-neutral measure Q and $\sigma, r \in \mathbb{R}$, and $\delta \in \mathbb{R}_+$ stand for exogenous parameters representing, respectively, the volatility in the risky returns, the (fixed) risk-free rate of return, and the (fixed) dividend yield from the stock. We are concerned with a standard American-style call option with underlying spot asset S, maturity date $T > 0$, and strike price $K > 0$. As is customary, the price process for the American call we will express as $C(S, \iota)$ and the price process for its European counterpart we will express as $c(S, \iota)$.[203] It is clear that for any fixed $t \in [0, T]$ there is a critical price level $x_t^* > 0$ that triggers exercise of the option: a price $S_t < x_t^*$ would be too low to justify exercise, while a price $S_t \geq x_t^*$ would be "good enough," given the state of the market and time to expiry. To put it another way, at time t a payoff of at least $x_t^* - K$ is required to justify exercise. As we saw in the previous section, with $T = \infty$ the exercise threshold x_t^* becomes time invariant: $x_t^* = x^*$, $t \in \mathbb{R}_+$. If $T < \infty$, however, getting closer in time to the expiration date should make the holder of the option more willing to accept lower payoffs, that is, the exercise domain $[x_t^*, \infty[$ should expand as t increases, or, which is the same, $t \rightsquigarrow x_t^*$ must be a decreasing function. We call the graph of this function

[203] Since the exogenous parameters σ, r, δ, and K will remain fixed throughout this section, they will be omitted in the notation and we will write for simplicity $C(S_t, t)$ and $c(S_t, t)$, instead of $C_{\sigma,r,\delta,K}(S_t, t)$ and $c_{\sigma,r,\delta,K}(S_t, t)$.

the *exercise boundary*. Since taking a negative payoff and dispensing of the option before its expiration date is clearly suboptimal, one must have $x_t^* \geq K$. It turns out, however, that such a lower bound on the exercise boundary, while intuitive, is too crude, and we have the following result:

(18.41) EXERCISE: Prove that if $r > \delta$, then exercise of the American call at time $t < T$ cannot be optimal if $S_t \in [K, \frac{r}{\delta}K[$. In particular, $x_t^* > \max\left(\frac{r}{\delta}K, K\right)$ for every $t < T$.

HINT: Introducing the stopping time

$$\mathcal{t} \stackrel{\text{def}}{=} \inf\left\{u \in [t, T] \colon S_u \geq \frac{r}{\delta}K\right\} \qquad (\inf \emptyset = T)$$

and using Itô's formula, in conjunction with the relation $S_u < \frac{r}{\delta}K$ for $u \in [t, \mathcal{t}[$, show that the reward process $e^{-ru}(S_u - K)^+$, $u \in [t, T]$, is a Q-submartingale until time $\mathcal{t}$. The rest is essentially a repetition of the argument from exercise (14.35), which is telling us that, just as the intuition suggests, it is never optimal to stop for as long as the reward process remains a submartingale. ○

It turns out that the last result can be made much more precise – see (Barles et al. 1995) and (Jacka and Lynn 1992) for more details and a complete proof.

(18.42) PROPOSITION: The exercise boundary $t \rightsquigarrow x_t^*$ is a continuous and decreasing function on the interval $[0, T[$ such that $\lim_{t \searrow T}(x_t^*) = \max\left(\frac{r}{\delta}K, K\right)$. ○

(18.43) REMARK: Clearly, at time $t = T$ the exercise threshold is $x_T^* = K$. This means that if $r > \delta$, then the exercise boundary has a discontinuity at $t = T$. ○

The early exercise premium for an American-style option is simply the price difference associated with the attributes "American" and "European." In the notation introduced earlier, this premium is $C(x, t) - c(x, t)$, $x \in \mathbb{R}_{++}$, $t \in [0, T[$. This difference represents the market price of the flexibility to choose the time of exercise, effectively allowing this flexibility to be treated as a tradeable asset. As we are aware from our analysis of the HJB equation in the case of optimal stopping, in the continuation domain

$$\mathfrak{C} \stackrel{\text{def}}{=} \left\{(x, t) \in \mathbb{R}_{++} \times [0, T] \colon x < x_t^*\right\}$$

the function $C(\cdot, \cdot)$ must satisfy the PDE

$$\partial C(x, t) + \mathcal{A}C(x, t) = rC(x, t), \qquad \text{with } \mathcal{A} \stackrel{\text{def}}{=} \frac{1}{2}\sigma^2 x^2 \frac{\partial^2}{\partial x^2} + (r - \delta)x\frac{\partial}{\partial x},$$

whereas in the exercise domain

$$\mathfrak{C}^{\complement} = \left\{(x, t) \in \mathbb{R}_{++} \times [0, T] \colon x \geq x_t^*\right\}$$

this function is given by $C(x, t) = (x - K)$. Let us fix some $t < T$, suppose that $S_t = x < x_t^*$, and let

$$\mathcal{t} \stackrel{\text{def}}{=} \inf\left\{u \in]t, T] \colon S_u \geq x_u^*\right\} \qquad (\inf \emptyset = T)$$

be the hitting time of the process S to the exercise domain $\mathfrak{C}^{\complement}$. Then

$$\begin{aligned}C(x,t) &= \mathsf{E}^Q\left[e^{-r(\mathcal{t}-t)}(S_{\mathcal{t}} - K) \mid S_t = x\right] = \mathsf{E}^Q\left[e^{-r(\mathcal{t}-t)}C(S_{\mathcal{t}},\mathcal{t}) \mid S_t = x\right] \\ &= \mathsf{E}^Q\left[e^{-r(T-t)}C(S_T,T) \mid S_t = x\right] \\ &\qquad - \mathsf{E}^Q\left[e^{-r(T-t)}C(S_T,T) - e^{-r(\mathcal{t}-t)}C(S_{\mathcal{t}},\mathcal{t}) \mid S_t = x\right].\end{aligned}$$

Formal application of Itô's formula now yields the following identity:

$$\begin{aligned}&e^{-r(T-t)}C(S_T,T) - e^{-r(\mathcal{t}-t)}C(S_{\mathcal{t}},\mathcal{t}) \\ &\overset{(!)}{=} \int_{\mathcal{t}}^{T} e^{-r(u-t)}\Big(\partial C(S_u,u) + (r-\delta)S_u\,\partial_S C(S_u,u) \\ &\qquad\qquad + \frac{1}{2}\sigma^2 S_u^2 \partial_S^2 C(S_u,u) - rC(S_u,u)\Big)\mathrm{d}u \\ &\qquad\qquad + \int_{\mathcal{t}}^{T} e^{-r(u-t)}\sigma S_u\,\partial_S C(S_u,u)\,\mathrm{d}\tilde{W}_u.\end{aligned}$$

(18.44) IMPORTANT REMARK: The expansion above is only formal (whence the symbol $\overset{(!)}{=}$) because the second derivative $\partial_x^2 C(x,t)$ does not exist at $x = x_t^*$ (notice that between time $\mathcal{t}$ and time T the spot process S may cross the exercise boundary infinitely many times). Nevertheless, one can still show that the right side is meaningful and the identity holds by mollifying the pricing function $(x,t) \rightsquigarrow C(x,t)$ in a neighborhood of the exercise boundary $t \rightsquigarrow x_t^*$, and then passing to the limit as the smoothing effect vanishes. One heuristic explanation is the following: the time that the price process S spends on the exercise boundary is Lebesgue-negligible (on the time scale), while the left and the right limits of the second derivative at the exercise boundary exist and are globally bounded. The reason that the local time of the price process at the exercise boundary does not enter the picture, as in Tanaka's formula, is that, unlike the function $x \rightsquigarrow |x|$ at 0, in this case the first derivative exists and is continuous at the problematic points; it is only the second derivative that fails to exist, and this failure is somehow controllable because the left and right second derivatives actually exist and are finite. ○

Next, observe that

$$\begin{aligned}&\partial C(x,u) + (r-\delta)x\,\partial_x C(x,u) \\ &\qquad + \frac{1}{2}\sigma^2 x^2 \partial_x^2 C(x,u) - rC(x,u) = \begin{cases}0, & \text{if } (x,u) \in \mathfrak{C}; \\ rK - \delta x, & \text{if } (x,u) \in \mathfrak{C}^{\complement},\end{cases}\end{aligned}$$

and that

$$\mathsf{E}^Q\left[e^{-r(T-t)}C(S_T,T) \mid S_t = x\right] = c(x,t)$$

is nothing but the price of the European call. One can show that the first derivative $\partial_x C(x,u)$ is bounded for $(x,u) \in \mathbb{R}_{++}\times[0,T]$, and that therefore (using the optional stopping result for martingales)

$$\mathsf{E}^Q\left[\int_t^{T} e^{-r(u-t)}\sigma S_u \partial_S C(S_u,u)\mathrm{d}\tilde{W}_u \mid S_t = x\right] = 0\,.$$

As a result, we have the following formula for the early exercise premium:

$$\begin{aligned} C(x,t) - c(x,t) &= \mathsf{E}^Q\left[\int_t^{T} e^{-r(u-t)}\big(\delta S_u - rK\big)1_{[x_u^*,\infty[}(S_u) \mid S_t = x\right] \\ &= \mathsf{E}^Q\left[\int_t^{T} e^{-r(u-t)}\big(\delta S_u - rK\big)1_{[x_u^*,\infty[}(S_u) \mid S_t = x\right], \end{aligned}$$

where the second identity is due to the fact that $S_u < x_u^*$ for $u \in [t, \mathcal{t}[$. Now we turn to simplifying the last expectation further.

(18.45) EXERCISE: After introducing the functions

$$F^{\pm}_{\sigma,r,\delta,t}(u,v) \stackrel{\text{def}}{=} 1 + \operatorname{erf}\left(\frac{\left(r - \delta \pm \frac{1}{2}\sigma^2\right)(u-t) - \log(v)}{\sigma\sqrt{2(u-t)}}\right),$$

prove that for any fixed $u > t$ one has

$$\mathsf{E}^Q\left[S_u 1_{[x_u^*,\infty[}(S_u) \mid S_t = x\right] = \frac{1}{2} x\, e^{(r-\delta)(u-t)} F^{+}_{\sigma,r,\delta,t}(u, x_u^*/x)$$

and

$$\mathsf{E}^Q\left[1_{[x_u^*,\infty[}(S_u) \mid S_t = x\right] = \frac{1}{2} F^{-}_{\sigma,r,\delta,t}(u, x_u^*/x)\,.$$

HINT: Consider using computer algebra. ○

As a result, the exercise premium can now be cast as

$$\begin{aligned} &C(x,t) - c(x,t) \\ &= \frac{1}{2}\int_t^{T}\left(\delta x\, e^{-\delta(u-t)} F^{+}_{\sigma,r,\delta,t}(u, x_u^*/x) - r\,K\, e^{-r(u-t)} F^{-}_{\sigma,r,\delta,t}(u, x_u^*/x)\right)\mathrm{d}u\,, \end{aligned} \qquad (e_1)$$

and we stress that the only unknown quantity on the right side is the exercise boundary $u \rightsquigarrow x_u^*$. In particular, since $c(x,t)$ is given by the Black–Scholes–Merton (BSM) formula, e_1 can be turned into a formula for the price of the American call $C(x,t)$, provided that the exercise boundary has somehow been computed. To make such computation possible, we set $x = x_t^*$ and turn e_1 into the following integral equation for the function $u \rightsquigarrow x_u^*$:

$$\begin{aligned} &(x_t^* - K) - c(x_t^*, t) \\ &= \frac{1}{2}\int_t^{T}\left(\delta x_t^*\, e^{-\delta(u-t)} F^{+}_{\sigma,r,\delta,t}(u, x_u^*/x_t^*) - r\,K\, e^{-r(u-t)} F^{-}_{\sigma,r,\delta,t}(u, x_u^*/x_t^*)\right)\mathrm{d}u\,, \end{aligned}$$

which we rearrange as

$$x_t^* - c(x_t^*, t) = K + x_t^* \frac{\delta}{2} \int_t^T e^{-\delta(u-t)} F^+_{\sigma,r,\delta,t}(u, x_u^*/x_t^*) \mathrm{d}u$$
$$- \frac{rK}{2} \int_t^T e^{-r(u-t)} F^-_{\sigma,r,\delta,t}(u, x_u^*/x_t^*) \Big) \mathrm{d}u \,. \qquad \mathrm{e}_2$$

This equation was derived by Kim (1990) (see also Carr et al. (1992)) and is instrumental for computing the exercise boundary.

(18.46) EXERCISE: Let everything be as above except that the risky asset is

$$S = S_0 + S_- \bullet Z \quad \text{for} \quad Z = \gamma(\Pi - c\iota) + b\iota \,,$$

where $\gamma > 0$ is a given parameter and Π is a Poisson process of intensity $c > 0$ under the measure P, chosen so that $c > (b+\delta-r)/\gamma$. Consider again an American-style call option with spot asset S, strike $K > 0$, and maturity $T > 0$, and again develop an expression for the early exercise premium that is analogous to e_1.

HINT: The spot asset can be cast as $S = S_0 + S_- \bullet \tilde{Z}$ for

$$\tilde{Z} = \gamma(\tilde{\Pi} - \tilde{c}\iota) + (r - \delta)\iota \,,$$

where $\tilde{\Pi}$ is a Poisson process of intensity $\tilde{c} = c - \frac{b+\delta-r}{\gamma}$ under the risk-neutral measure Q. What is the infinitesimal generator of S under the measure Q? ○

There are different ways in which one can develop a numerical program for solving the integral equation in e_2. In the rest of this section we focus on one such program. Generically, numerical methods for solving integral equations like the one in e_2 involve replacing the domain of the time parameter t with a sufficiently dense finite set of abscissas, after which the solution can be built recursively from one abscissa to the next, going backward in time. Although entirely feasible – see (Huang et al. 1996) – the implementation of such a procedure in the context of e_2 is somewhat hindered by the fact that $\lim_{t \nearrow T} \partial_t x_t^* = -\infty$. In what follows we are going to outline an alternative method in which, starting from the constant function $f_0(t) = \frac{r}{\delta} K \vee K$, one constructs by way of successive iterations the functions $f_1(\cdot), f_2(\cdot), \ldots$, which are defined on the entire interval $[0, T]$ and converge to the exercise boundary $t \rightsquigarrow x_t^*$. This method is both efficient and reasonably accurate: in our concrete example 10 successive iterations are enough to produce the exercise boundary with accuracy $\approx 10^{-5}$ uniformly throughout the entire interval $[0, T]$. Having the functions $f_k(\cdot)$ expressed as cubic splines makes it possible to work with many fewer abscissas, and the practical implementation of the entire program, which is illustrated in appendix B.3, is fairly straightforward and reasonably fast even without parallelization or other acceleration methods. Another advantage of this approach is that it is easy to parallelize: the entire construction of the spline $f_{k+1}(\cdot)$ from a given (already constructed) spline $f_k(\cdot)$, can be divided among as many threads as the hardware allows.

As a first step in the outline of the program, consider the following nonlinear equation for the unknown variable $x \in \mathbb{R}$:

$$x - c(x,t) = K + xA - B \quad \Longleftrightarrow \quad x(1 - A) = c(x,t) + K - B\,,$$

in which the symbols A and B are treated as parameters (placeholders) and $c(x,t)$ is the function given by the Black–Scholes–Merton formula. Since both sides of the second variant of the equation are strictly monotone functions of x, the solution, if one exists, must be unique. Thus, the equation implicitly determines x as a function of the parameters (placeholders) A and B (K is forever fixed), so that we can write $x = H_t(A,B)$. Now suppose that $A = A_t(f)$ and $B = B_t(f)$ depend on the restriction, $f\restriction[t,T]$, of a particular function $f\colon [0,T] \mapsto \mathbb{R}$. If the function $f(\cdot)$ is fixed, the function $[0,T] \ni t \rightsquigarrow H_t(A_t(f), B_t(f))$ would be well defined, and one can formulate the following functional equation for $f(\cdot)$:

$$f(t) = H_t\big(A_t(f), B_t(f)\big)\,, \quad t \in [0,T[\,.$$

The solution to this equation can be viewed as a fixed point for the (functional) transformation

$$f(\cdot) \rightsquigarrow H_{(\cdot)}(A_{(\cdot)}(f), B_{(\cdot)}(f))\,.$$

We see from $\mathfrak{e}_2$ that, with

$$A_t(f) \stackrel{\text{def}}{=} \frac{\delta}{2}\int_t^T e^{-\delta(u-t)} F^{+}_{\sigma,r,\delta,t}\big(u, f(u)/f(t)\big)\,\mathrm{d}u$$

and $\mathfrak{e}_3$

$$B_t(f) \stackrel{\text{def}}{=} \frac{r\,K}{2}\int_t^T e^{-r(u-t)} F^{-}_{\sigma,r,\delta,t}\big(u, f(u)/f(t)\big)\,\mathrm{d}u\,,$$

the exercise boundary $[0,T[\, \ni t \rightsquigarrow x^*_t$ solves to the functional equation

$$x^*_t = H_t\big(A_t(x^*_\cdot), B_t(x^*_\cdot)\big)\,, \quad t \in [0,T[\,.$$

The last equation gives rise to an iterative procedure in the obvious way: starting with the function $f_0(t) \stackrel{\text{def}}{=} \frac{r}{\delta}K \vee K$, $t \in [0,T]$, one can construct the sequence of functions $[0,T] \ni t \rightsquigarrow f_k(t)$, $k = 1,2,\ldots$ sequentially, according to the rule

$$f_k(t) = H_t\big(A_t(f_{k-1}), B_t(f_{k-1})\big)\,, \quad t \in [0,T]\,, \quad k = 1,2,\ldots$$

From a practical point of view, the last relation is still in the realm of the abstract, since functions can be represented on computing systems only as finite lists. The generic workaround is to treat all functions involved as spline interpolation objects. For that purpose, we choose – and fix – an appropriate interpolation grid $0 = t_0 < t_1 < \ldots < t_n = T$ and replace the functions f_k with the spline objects $\tilde{f}_k$, $k \in \mathbb{N}_{++}$, constructed from values tabulated over the abscissas $\{t_i\colon i \in \mathbb{N}_{|n}\}$. These values are computed consecutively for $k = 1,2,\ldots$, starting from $\tilde{f}_0(t) = f_0(t) = \frac{r}{\delta}K \vee K$, according to the rule

$$\tilde{f}_k(t_i) = H_{t_i}\big(A_{t_i}(\tilde{f}_{k-1}), B_{t_i}(\tilde{f}_{k-1})\big)\,, \quad i \in \mathbb{N}_{|n}\,,$$

and we again stress that the calculations in the cross-section of all t_i, $i \in \mathbb{N}_{|n}$, can be executed independently on different processing threads. Since $f_k(t_n) = \frac{r}{\delta}K \vee K$ for all $k \in \mathbb{N}_{++}$, at each iteration one must compute $2n$ integrals $A_{t_i}(\tilde{f}_{k-1})$ and $B_{t_i}(\tilde{f}_{k-1})$, $0 \leq i \leq n-1$, by way of numerical integration, and carry out n root finding operations associated with the calculation of H_{t_i}, $0 \leq i \leq n-1$. It turns out that the sequence of splines $\tilde{f}_1(\cdot)$, $\tilde{f}_2(\cdot)$, ..., converges to the true exercise boundary $t \rightsquigarrow x_t^*$ reasonably fast. Although the exact rate of convergence may be difficult to estimate analytically, and the choice of the spline-interpolation nodes is somewhat ad hoc, the uniform distance between two successive splines, $\tilde{f}_k(\cdot)$ and $\tilde{f}_{k-1}(\cdot)$, which is easy to compute numerically, should be a reasonable test for stopping the iteration process. The accuracy of the spline produced during the last iteration can then be tested by substituting it for the function $x \rightsquigarrow x_t^*$ on both sides of ⓔ₂, and, treating the difference between the two sides of ⓔ₂ as a function of $t \in [0, T]$, computing numerically the maximal absolute value of that function. In any case, from a practical point of view, the justification of the convergence (which we do not provide) is not all that relevant, since there is no obvious restriction on how one may be allowed to guess the solution, and verifying numerically the accuracy of the guess – however obtained – is all that is needed to accept or reject a particular candidate solution.

The method just outlined is illustrated in appendix B.3 with one concrete example and the relevant computer code. Among other things, we see from this example that the numerical evaluation of univariate integrals that involve spline objects is essentially as efficient as the numerical evaluation of some common special functions. This is to say that if we are willing to accept a spline object (a sufficiently accurate representation of the exercise boundary $t \rightsquigarrow x_t^*$) as a "standard function," then the formula for the early exercise premium in ⓔ₁ can indeed be viewed as "closed-form" expression for the price of the American call. We conclude this section with the plot generated by the computer code in appendix B.3.

FIGURE 18.1

Exercise boundary for an American call with parameters $T = 0.5$, $K = 40$, $\sigma = 0.3$, $r = 0.02$ and $\delta = 0.07$.

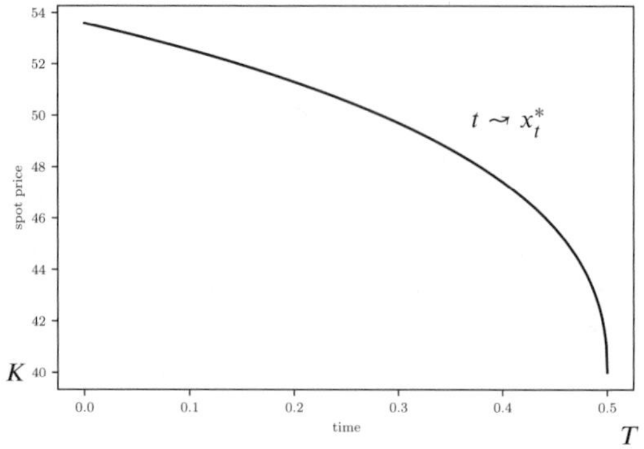

18.5 Corporate Debt, Equity, Dividend Policy, and the Modigliani–Miller Proposition

Consider a real asset that generates output continuously at a constant rate of 1, and suppose that the spot price of the output follows a diffusion process given by

$$S = S_0 + \sigma(S, \imath)S \bullet W + b(S, \imath)S \bullet \imath ,$$

with coefficients $\sigma(\cdot, \cdot)$ and $b(\cdot, \cdot)$ chosen so that the functions

$$(x, t) \rightsquigarrow \sigma(e^x, t) \quad \text{and} \quad (x, t) \rightsquigarrow b(e^x, t)$$

are⊶ (11.26) Lipschitz-continuous and locally bounded on $\mathbb{R}_{++} \times \mathbb{R}_+$. The price process S is a well-defined strongly Markov process. In addition, suppose that the risk-free rate r_t and the convenience yield rate ϱ_t have the form $r_t = r(S_t, t)$ and $\varrho_t = \varrho(S_t, t)$. The owner of the real asset can instantly sell the output and collect the (instantaneous) payoff $S_t \, dt$. Alternatively, the owner can incorporate a firm and keep the proceeds from operations in the firm – or perhaps collect only part of the proceeds in the form of dividends and retain the remaining part in the firm. Does it matter? To answer this question, we first impose our usual complete market assumption (the spot market for the output and the spot market for risk-free debt span the economy) and observe that under this assumption the risk-neutral dynamics of the output can be cast as

$$S = S_0 + \sigma(S, \imath)S \bullet \tilde{W} + (r - \varrho)S \bullet \imath ,$$

where $\tilde{W}$ is a Brownian motion under the risk-neutral measure Q that prices all payoffs. Suppose that the owner of the asset incorporates the firm anyway and must decide about the payout policy. If the firm instantly distributes all of its profits[204] as dividends, then ownership in the firm would be no different from ownership in the real asset.[205] The value of such a firm is

$$V_0(S_t, t) = \mathsf{E}^Q\Big[\int_t^\infty e^{-\int_t^u \varrho_v \, dv} S_u \, du \;\Big|\; \mathscr{F}_t\Big], \quad t \geq 0,$$

and this relation gives for $\Delta > 0$

$$\mathsf{E}^Q\Big[\int_t^\Delta e^{-\int_t^u \varrho_v \, dv} S_u \, du + e^{-\int_t^\Delta \varrho_v \, dv} V_0(t + \Delta, S_{t+\Delta}) - V_0(S_t, t) \;\Big|\; \mathscr{F}_t\Big] = 0 .$$

[204] For simplicity, we suppose that revenue is the same as profit.

[205] Receiving at time t the cash amount S_t is no different from receiving one unit of output. Ownership in the firm is an entitlement to receive continuously dt units of output at every instant t until $t = \infty$, and this entitlement is no different from the entitlement to receive continuously the cash amount $S_t \, dt$ at every instant t until $t = \infty$.

After the usual Itô-expansion, division by Δ, and passage to the limit as $\Delta \to 0$, or by applying the Feynman–Kac formula directly, we get

$$\begin{aligned}\partial V_0(S_t,t) + (r_t - \varrho_t)S_t\,\partial_S V_0(S_t,t)& \\ + \frac{1}{2}\sigma(S_t,t)^2 {S_t}^2 \partial_S^2 V_0(S_t,t) + S_t &= \varrho_t V_0(t,S_t)\,.\end{aligned}$$

e_1

Now suppose that the firm pays instantaneous dividend $\delta_t S_t\,\mathrm{d}t$ and retains the amount $(1-\delta_t)S_t\,\mathrm{d}t$. Eventually, all retained earnings will have to be paid out to the equity holders in one form or another. To make this idea precise, we now introduce a finite time horizon $T > 0$ and suppose that the firm is accumulating post-dividend earnings only until date T. On date T the firm pays back its accumulated earnings – perhaps by way of acquiring real assets of equal market value – and thereafter the equity holders instantly collect all the profits. On date $t < T$ the value of such a firm would be

$$\begin{aligned}V_1(S_t,t) = \mathsf{E}^Q\Big[\int_t^T e^{-\int_t^u \varrho_v\,\mathrm{d}v}\delta_u S_u\,\mathrm{d}u \mid \mathscr{F}_t\Big] + \mathsf{E}^Q\Big[\int_T^\infty e^{-\int_t^u \varrho_v\,\mathrm{d}v} S_u\,\mathrm{d}u \mid \mathscr{F}_t\Big]& \\ + \mathsf{E}^Q\Big[e^{-\int_t^T \varrho_v\,\mathrm{d}v}\int_0^T e^{\int_u^T \varrho_v\,\mathrm{d}v}(1-\delta_u)S_u\,\mathrm{d}u \mid \mathscr{F}_t\Big]&,\end{aligned}$$

and the very last expectation on the right can be cast as

$$\int_0^t e^{\int_u^t \varrho_v\,\mathrm{d}v}(1-\delta_u)S_u\,\mathrm{d}u + \mathsf{E}^Q\Big[\int_t^T e^{-\int_t^u \varrho_v\,\mathrm{d}v}(1-\delta_u)S_u\,\mathrm{d}u \mid \mathscr{F}_t\Big].$$

After a straightforward algebra we arrive at the relation

$$V_1(S_t,t) = \int_0^t e^{\int_u^t \varrho_v\,\mathrm{d}v}(1-\delta_u)S_u\,\mathrm{d}u + V_0(S_t,t)\,,$$

which, in particular, is telling us that $V_1(S_t,t)$ does not change if the settlement date $T > t$ is pushed arbitrarily far into the future. It is also telling us that $V_0(S_0,0) = V_1(S_0,0)$, that is, initially, at time $t = 0$, the equity holders would be indifferent to the dividend policy of the firm. The same claim can be made about investors who buy equity in the firm at time $t < T$ (at the price $V_1(S_t,t)$): the right side of the last equation does not depend on the dividend policy after time t, which is to say the investors would be indifferent to the dividend policy as well.

(18.47) Important critique: The indifference to the dividend policy that we just established hinges on our tacit (and somewhat magical) assumption that all retained earnings grow and are discounted at the convenience yield rate. In general, however, this assumption may be difficult to justify, and if the firm immediately converts its real output into cash, the part of that cash that the firm accumulates is going to grow and be discounted at the risk-free rate r_t. If this is the case, then the value of the firm becomes

$$V_1(S_t,t) = \int_0^t e^{\int_u^t r_v \mathrm{d}v}(1-\delta_u)S_u\,\mathrm{d}u + V_0(t,S_t)$$
$$+\,\mathsf{E}^Q\left[\int_t^T \left(e^{-\int_t^u r_v\,\mathrm{d}v} - e^{-\int_t^u \varrho_v\,\mathrm{d}v}\right)(1-\delta_u)S_u\,\mathrm{d}u \mid \mathscr{F}_t\right].$$

We see from this expression that if the convenience yield is uniformly larger than the risk-free rate, then the value of the equity would be maximized with the dividend yield $(\delta_u)_{u\geq t}$ set to 0 and with the settlement date T pushed into the future all the way to $+\infty$. If, however, the risk-free rate is uniformly larger than the convenience yield rate, then the value of the equity would be maximized if the firm does not retain any portion of its earnings after time t, that is, $\delta_u = 1$ for $u \geq t$. In reality, firms have other reasons to accumulate cash, too. One such reason may be temporary tax deferral, with the expectation that the tax rate applied to capital gains and dividends may decline in the future. A much more important reason is to pile up cash for the purpose of investments leading to the creation of real options. ○

We now turn to the model of corporate debt first introduced by Merton (1974). Of particular interest to us is the interaction between equity and debt, studied by Merton (1970b) and Merton (1977). In Merton's model both equity and debt are treated as contingent claims, so that in our setting we can express these two entities as $E(S_t,t)$ and $D(S_t,t)$. The debt matures on date $T > 0$ with principal L, and prior to maturity pays coupons continuously at a fixed instantaneous (coupon) rate $c > 0$. For simplicity, just as in (Merton 1974), we suppose that the firm cannot default on its debt before the maturity date T. Assuming that the firm does not retain any of its earnings and ignoring any bankruptcy costs, on date T the debt will be valued at

$$D(S_T,T) = \min\big(V_0(S_T,T),L\big) \equiv L - \big(L - V_0(S_T,T)\big)^+$$

and the equity will be valued at

$$E(S_T,T) = V_0(S_T,T) - \min\big(V_0(S_T,T),L\big) \equiv \big(V_0(S_T,T)-L\big)^+.$$

At time $t < T$ the equity holders receive instantaneous dividend $\delta_t\,\mathrm{d}t = (S_t - c)\mathrm{d}t$, and at time $t \geq T$ the equity holders receive instantaneous dividends $\delta_t\,\mathrm{d}t = S_t\,\mathrm{d}t$. Following the usual argument that lead us to the HJB equation earlier, or using the Feynman–Kac formula directly, now we can write down the following relations for any $t \in [0,T[$:

$$\partial D(S_t,t) + (r_t - \varrho_t)S_t\,\partial_S D(S_t,t)$$
$$+\frac{1}{2}\sigma(S_t,t)^2 S_t{}^2 \partial_S^2 D(S_t,t) + c = \varrho_t D(S_t,t)$$
$$\partial E(S_t,t) + (r_t - \varrho_t)S_t\,\partial_S E(S_t,t)$$
$$+\frac{1}{2}\sigma(S_t,t)^2 S_t{}^2 \partial_S^2 E(S_t,t) + (S_t - c) = \varrho_t E(S_t,t).$$

After adding the last two equations we find that $V(S_t, t) \stackrel{\text{def}}{=} D(S_t, t) + E(S_t, t)$ satisfies equation e_1, which governs $V_0(S_t, t)$. Since $V(T, S_T) \equiv D(S_T, T) + E(S_T, T) = V_0(S_T, T)$, it must be that

$$D(S_t, t) + E(S_t, t) = V_0(S_t, t) \quad \text{for all } t \in [0, T].$$

What we have just established is a rather rudimentary interpretation of the celebrated Modigliani–Miller proposition[206] – see (Modigliani and Miller 1958), (Miller and Modigliani 1961), and (Merton 1977). It can be summarized as follows. Suppose that at time $t \geq 0$ the owners of the right to invest can install real assets of value $V_0(S_t, t)$ at fixed cost I. They can sink capital I and receive assets of value $V_0(S_t, t)$, thus generating the payoff $V_0(S_t, t) - I$. Now suppose that at time t such an operation happens to be optimal, but the available capital is only $A < I$. Should the owners choose to issue debt of finite maturity and with a payout schedule designed in such a way that the value of the debt at the time of underwriting is exactly $D(S_t, t) = I - A$, then they will be able to sink capital A and receive equity of value $E(S_t, t) = V_0(S_t, t) - D(S_t, t) = V_0(S_t, t) - I + A$, thus generating the same payoff $V_0(S_t, t) - I$. Alternatively, the owners can issue equity of value $I - A$, in which case they will receive only partial ownership in a firm without any debt obligations. In this case, the value of their (partial) equity would be $V_0(S_t, t) - (I - A)$ (whatever equity is left after equity of value $I - A$ has been underwritten) and the payoff generated by sinking the available capital A is yet again $V_0(S_t, t) - I$. The conclusion, then, is that the existing owners would be indifferent to choosing either funding their project by way of underwriting equity of value $I - A$ or underwriting debt of value $I - A$. It is not difficult to see that they would be just as indifferent to the choice of the various mixtures of equity and debt. The payoff that the installment of the firm generates is always $V_0(S_t, t) - I$, regardless of how much capital is available, or how the new firm is going to be financed!

(18.48) REMARK: It is worth recording the conditions that lead us to the conclusion that both investors and owners must be indifferent to the capital structure of the firm: (a) debt and equity can be underwritten at no cost; (b) corporate profits and dividend income are not taxed; (c) the capital markets are perfectly liquid and perfectly frictionless, that is, the owners can underwrite and sell at its market value equity and/or debt in any amount, at any time, and at no cost; (d) all future coupon payments to the debt holders and dividend payoffs to the equity holders are discounted at the convenience yield rate – not at the risk-free rate; and (e) the firm does not retain any portion of its earnings, that is, after the coupon is paid, any remaining portion of the earnings is immediately paid out to the equity holders in the form of dividends. Retaining only a part or all of the earnings in the firm

[206] Named after the Italian-American economist Franco Modigliani (1918–2003), and the American economist Merton Howard Miller (1923–2000), who received the 1990 Sveriges Riksbank Prize in Economic Sciences in Memory of Alfred Nobel.

complicates the picture, but does not alter the conclusion – as long as the retained earnings grow and are discounted at the convenience yield rate. ○

Appendix A

Résumé of Analysis and Topology

(A.1) METRICS AND METRIC SPACES: A pseudometric on a set $\mathbb{X} \neq \emptyset$ is any function $d: \mathbb{X} \times \mathbb{X} \mapsto \mathbb{R}_+$ which satisfies the following three axioms:

1. $d(x, x) = 0$ for all $x \in \mathbb{X}$;
2. $d(x, y) = d(y, x)$ for all $x, y \in \mathbb{X}$;
3. $d(x, z) \leq d(x, y) + d(y, z)$ for all $x, y, z \in \mathbb{X}$.

The last axiom is known as *the triangular inequality*. If the first axiom is replaced with the following stronger one

1′. $d(x, y) = 0$ if and only if $x = y$, for all $x, y \in \mathbb{X}$,

then $d(\cdot, \cdot)$ is said to be a metric (alias: distance) in $\mathbb{X}$, and the pair $(\mathbb{X}, d)$ is called a metric space. If $d(\cdot, \cdot)$ is only a pseudometric in $\mathbb{X}$, then the pair $(\mathbb{X}, d)$ called a pseudometric!space. ○

(A.2) LINEAR SPACES: A linear space (alias: vector space) is a set $\mathbb{X} \neq \emptyset$ endowed with a binary operation called *addition*

$$\mathbb{X} \times \mathbb{X} \ni (x, y) \rightsquigarrow x + y \in \mathbb{X},$$

and also endowed with the operation *multiplication by a scalar*

$$\mathbb{R} \times \mathbb{X} \ni (a, x) \rightsquigarrow ax \in \mathbb{X},$$

in such a way that the following axioms are satisfied:

1. the addition is associative: $x + (y + z) = (x + y) + z$ for all $x, y, z \in \mathbb{X}$;
2. the addition is commutative: $x + y = y + x$ for all $x, y \in \mathbb{X}$;
3. there is an element of $\mathbb{X}$, called *zero* and denoted by $\vec{0}$, such that $\vec{0} + x = x$ for all $x \in \mathbb{X}$;
4. for every $x \in \mathbb{X}$ there is some $(-x) \in \mathbb{X}$ such that $x + (-x) = \vec{0}$;
5. $a(bx) = (ab)x$ for all $a, b \in \mathbb{R}$ and all $x \in \mathbb{X}$;
6. $1x = x$ for all $x \in \mathbb{X}$;
7. $(-1)x = -x$ for all $x \in \mathbb{X}$;
8. $(a + b)x = ax + bx$ for all $a, b \in \mathbb{R}$, and $x \in \mathbb{X}$.

Taken together the last three axioms imply that $0x = \vec{0}$ for any $x \in \mathbb{X}$. It is common to write $x - y$ instead of $x + (-y)$. Depending on the context, the elements of a vector space may be called *vectors*. In many situations we are working with vector spaces, the elements of which are real-valued functions. In most of those situations the zero vector $\vec{0}$ is the constant function equal to $0 \in \mathbb{R}$, in which case we may use 0 to denote both the scalar $0 \in \mathbb{R}$ and the zero vector $\vec{0} \in \mathbb{X}$. ○

(A.3) THE EUCLIDEAN SPACE $\mathbb{R}^n$: Given any $n \in \mathbb{N}_{++}$, the space of all ordered n-tuples of (finite) real numbers is called the *Euclidean space with dimension n* and is denoted by

$$\mathbb{R}^n \stackrel{\text{def}}{=} \{(x_1, \ldots, x_n) : x_1 \in \mathbb{R}, \ldots, x_n \in \mathbb{R}\}.$$

This space has the obvious structure of a linear space, with the operation "addition" understood as

$$(x_1, \ldots, x_n) + (y_1, \ldots, y_n) = (x_1 + y_1, \ldots, x_n + y_n),$$

and the operation "multiplication by a scalar" understood as

$$a(x_1, \ldots, x_n) = (ax_1, \ldots, ax_n), \quad a \in \mathbb{R}.$$

The zero in the Euclidean space $\mathbb{R}^n$ is the vector $\vec{0} = (0, \ldots, 0)$, and the vector $(1, \ldots, 1)$ we denote by $\vec{1}$. For simplicity, generic vectors $(x_1, \ldots, x_n)$ are usually denoted by a single letter x, instead of $\bar{x}$, or $\vec{x}$. ○

(A.4) SEMINORM AND NORM: A seminorm on a vector space $\mathbb{X}$ is a function $\|\cdot\| : \mathbb{X} \mapsto \mathbb{R}_+$ such that:

1. $\|ax\| = |a|\|x\|$ for all $a \in \mathbb{R}$ and all $x \in \mathbb{X}$;
2. $\|x + y\| \leq \|x\| + \|y\|$ for all $x, y \in \mathbb{X}$.

If the seminorm $\|\cdot\|$ has the additional property:

3. for $x \in \mathbb{X}$, $\|x\| = 0$ implies $x = \vec{0}$,

then $\|\cdot\|$ is called a *norm*. ○

(A.5) EXERCISE (THE EUCLIDEAN NORM IN $\mathbb{R}^n$): Prove that the function

$$\mathbb{R}^n \ni (x_1, \ldots, x_n) \equiv x \rightsquigarrow |x| \equiv |(x_1, \ldots, x_n)| \stackrel{\text{def}}{=} \sqrt{x_1^2 + \ldots + x_n^2}$$

defines a norm on $\mathbb{R}^n$. This norm is called *the Euclidean norm.* ○

(A.6) EXERCISE: Prove that if $\mathbb{X}$ is a linear space and $\|\cdot\|$ is a seminorm on $\mathbb{X}$, then $d(x, y) \stackrel{\text{def}}{=} \|y - x\|$ for all $x, y \in \mathbb{X}$ gives a pseudometric on $\mathbb{X}$. Prove that this pseudometric is a metric if and only if the seminorm $\|\cdot\|$ is a norm. ○

(A.7) OPEN SETS AND TOPOLOGY: Let $(\mathbb{X}, d)$ be a pseudometric space. Given any $x \in \mathbb{X}$ and $\varepsilon \in \mathbb{R}_{++}$, the open ball (in $\mathbb{X}$) with center x and radius ε is defined as set

$$B_\varepsilon(x) = \{y \in \mathbb{X} : d(y, x) < \varepsilon\}.$$

A set $\emptyset \neq U \subseteq \mathbb{X}$ is said to be *open* if it has the following property: for any $x \in U$ one can find an $\varepsilon \in \mathbb{R}_{++}$, possibly depending on x, such that $B_\varepsilon(x) \subseteq U$. The set $\mathbb{X}$ is thus open, and the empty set $\emptyset \subset \mathbb{X}$ is considered open by definition. Let $\mathcal{T}$ denote the collection of all open subsets of $\mathbb{X}$. The following properties are easy to check:

1. $\emptyset \in \mathcal{T}$ and $\mathbb{X} \in \mathcal{T}$;
2. $U \cap V \in \mathcal{T}$ for all $U, V \in \mathcal{T}$;
3. $\cup S \in \mathcal{T}$ for any $S \subseteq \mathcal{T}$ (the union of any family of open sets is open).

In most situations "convergence" and "continuity" can be described in terms of the collection of open sets $\mathcal{T}$ – with no reference whatsoever to the pseudometric $d(\cdot,\cdot)$. A collection, $\mathcal{T}$, of subsets of $\mathbb{X}$, which has properties 1, 2, and 3 above (treated as axioms), is called a *topology on* $\mathbb{X}$ and the pair $(\mathbb{X},\mathcal{T})$ is called *topological space*. The elements of $\mathbb{X}$ are called *points* and the elements of $\mathcal{T}$ are called *open sets*. When the topology $\mathcal{T}$ obtains from a pseudometric (or a metric) $d(\cdot,\cdot)$, as described above, we say that $\mathcal{T}$ is pseudometrizable (or metrizable) by $d(\cdot,\cdot)$. Many interesting topologies are not metrizable. In some situations it is much more instructive and efficient to work directly with the topology, rather than the metric or pseudometric that generates it. ○

(A.8) EXERCISE: Let $(\mathbb{X},\mathcal{T})$ be any topological space. Prove that the intersection of any finite number of elements of $\mathcal{T}$ is an element of $\mathcal{T}$, that is, the intersection of any finite number of open sets is an open set. Explain why the same property cannot hold for the intersection in an infinite number of open sets. ○

(A.9) EXERCISE (GENERATED TOPOLOGY): Let $\mathbb{X}$ be any nonempty set and let $\mathcal{U}$ be *any* family of subsets of $\mathbb{X}$. Let $\mathcal{S}$ be the collection of all possible intersections of *finitely many* members $\mathcal{U}$, including the empty set $\emptyset$, which is the intersection of 0 members of $\mathcal{U}$. Let $\mathcal{T}$ be the collection of all possible (arbitrary) unions of members of $\mathcal{S}$. Prove that $\mathcal{T}$ is a topology, the smallest topology that contains $\mathcal{U}$. This smallest topology can be defined for any collection, $\mathcal{U}$, of subsets of $\mathbb{X}$. ○

(A.10) EXERCISE (INDUCED TOPOLOGY)): Let $(\mathbb{X},\mathcal{T})$ be any topological space and let $\emptyset \neq A \subseteq \mathbb{X}$ be any nonempty subset of $\mathbb{X}$. Prove that the family

$$A\cap\mathcal{T} \stackrel{\text{def}}{=} \{A\cap U : U\in\mathcal{T}\}$$

is a topology on A (so that $(A, A\cap\mathcal{T})$ is a topological space). The topology $A\cap\mathcal{T}$ is said to be induced (on A) by $\mathcal{T}$. ○

(A.11) CLOSURE AND INTERIOR: Let $(\mathbb{X},\mathcal{T})$ be a topological space. A set $A\subseteq\mathbb{X}$ is said to be *closed* if its complement is open, that is, $A^{\complement}\in\mathcal{T}$. It follows that $\mathbb{X}$ and $\emptyset$ are both open and closed, and that the intersection of any collection of closed sets is closed (simply because the union of any collection of open sets is open) and the union of any finite number of closed sets is closed (simply because the intersection of any finite number of open sets is open).

The interior of a subset $A\subseteq\mathbb{X}$ is

$$A^{\circ} \stackrel{\text{def}}{=} \bigcup\{U\in\mathcal{T} : U\subseteq A\}.$$

This is the largest open set that is included in A; in particular, $A^{\circ}=A$ if $A\in\mathcal{T}$, and the property "A is open" can be expressed as "$A\backslash A^{\circ}=\emptyset$."

The closure of $A\subseteq\mathbb{X}$ is the intersection of all closed sets that contain A:

$$\bar{A} \stackrel{\text{def}}{=} \bigcap\{V\subseteq\mathbb{X} : V^{\complement}\in\mathcal{T},\ A\subseteq V\}.$$

This is the smallest closed set that contains A; in particular $\bar{A}=A$ if $A^{\complement}\in\mathcal{T}$ (that is, if A is closed).

A limit point for the subset $A \subseteq \mathbb{X}$ is any point $x \in \mathbb{X}$ that has the following property: for any open set $U \in \mathcal{T}$ such that $x \in U$ one has $(U \setminus \{x\}) \cap A \neq \emptyset$, that is, any open set that contains x contains an element of A other than x. The collection of all limit points of the set A we denote by $\mathrm{limpt}(A)$.

A subset $\mathcal{X} \subseteq \mathbb{X}$ is said to be dense in $\mathbb{X}$ (relative to the topology $\mathcal{T}$) if $\bar{\mathcal{X}} = \mathbb{X}$ (the closure of $\mathcal{X}$ is the space $\mathbb{X}$). The topological space $(\mathbb{X}, \mathcal{T})$ is said to be separable if there is a countable set $\mathcal{X} \subseteq \mathbb{X}$ that is dense in $\mathbb{X}$ (for the topology $\mathcal{T}$). A metric space is a said to be separable if it is separable relative to the topology induced by the metric. ○

(A.12) REMARK: Together with metrizability, separability is an important attribute that is essential for many aspects of the theory of stochastic process. ○

(A.13) EXERCISE: Prove that the space $\mathbb{R}^n$, endowed with the topology of the Euclidean distance $d(x, y) = |y - x|$, $x, y \in \mathbb{R}^n$, is a separable (metric) space. ○

(A.14) EXERCISE (NEIGHBORHOOD OF A POINT): Let $(\mathbb{X}, \mathcal{T})$ be any topological space. Given a point $x \in \mathbb{X}$, a *neighborhood* of x is *any* set $O \subseteq \mathbb{X}$, for which there is an open set $U \in \mathcal{T}$ such that $x \in U \subseteq O$. Prove that if $\emptyset \neq A \subseteq \mathbb{X}$ and $x \in \bar{A}$, then for any neighborhood, O, of x one can claim that $O \cap A \neq \emptyset$. If the topology $\mathcal{T}$ is metrizable, then this claim can be stated as $B_\varepsilon(x) \cap A \neq \emptyset$ for any $\varepsilon \in \mathbb{R}_{++}$ and any $x \in \bar{A}$. Intuitively, the closure $\bar{A}$ consists of all points in $\mathbb{X}$ that are "arbitrarily close" to A.[207] ○

(A.15) CONTINUITY AND CONVERGENCE: Let $(\mathbb{X}, \mathcal{T})$ and $(\mathbb{Y}, \mathcal{U})$ be any two topological spaces. A function $f: \mathbb{X} \mapsto \mathbb{Y}$ is said to be continuous (relative to $\mathcal{U}$ and $\mathcal{T}$) if the pre-image (under f) of any open subset of $\mathbb{Y}$ is an open subset of $\mathbb{X}$, that is, for any $U \in \mathcal{U}$ one has

$$f^{-1}(U) \stackrel{\text{def}}{=} \{x \in \mathbb{X} : f(x) \in U\} \in \mathcal{T}.$$

A sequence $(x_i \in \mathbb{X})_{i \in \mathbb{N}}$ is said to *converge to* $x \in \mathbb{X}$ (relative to the topology $\mathcal{T}$) if for any neighborhood O of x there is an index $n \in \mathbb{N}$ (possibly depending on O and x) such that $x_i \in O$ for all $i \geq n$. When this property holds, we call x *the limit of* $(x_i \in \mathbb{X})_{i \in \mathbb{N}}$ and write $x = \lim_i x_i$ or simply $x = \lim x_i$. If the topology on which the limit depends must be emphasized, we may write $x = \lim_i^{(\mathcal{T})} x_i$. If $\mathcal{T}$ is generated by a pseudometric $d(\cdot, \cdot)$, then the convergence of $(x_i \in \mathbb{X})_{i \in \mathbb{N}}$ to $x \in \mathbb{X}$ can be stated as: for any $\varepsilon \in \mathbb{R}_{++}$ there is an index $n \in \mathbb{N}$ (possibly depending on ε and x) such that $x_i \in B_\varepsilon(x)$ for all $i \geq n$. The convergence of $(x_i \in \mathbb{X})_{i \in \mathbb{N}}$ to $x \in \mathbb{X}$ can also be expressed as $x_i \to x$ or $x_i \xrightarrow{\mathcal{T}} x$ (or $x_i \xrightarrow{d} x$) if the topology (or pseudometric) that determines the convergence must be emphasized in the notation. ○

(A.16) EXERCISE: Let $(\mathbb{X}, \mathcal{T})$ and $(\mathbb{Y}, \mathcal{U})$ be any two topological spaces, and suppose that the function $f: \mathbb{X} \mapsto \mathbb{Y}$ is continuous in the respective topologies. Prove

[207] We see how the formal language of topological spaces allows one to express in a precise way the intuitive notion of "closeness to a set."

that if $(x_i \in \mathbb{X})_{i \in \mathbb{N}}$ is any sequence that converges to $x \in \mathbb{X}$ in the topology $\mathcal{T}$, then one can claim that the sequence $(f(x_i) \in \mathbb{Y})_{i \in \mathbb{N}}$ converges to $f(x) \in \mathbb{Y}$ in the topology $\mathcal{U}$. Plainly, the attribute "convergence of a sequence" is stable under continuous transformations from one topological space to another. ○

(A.17) EXERCISE (STRONG AND WEAK TOPOLOGY): Assume the notation introduced in (A.15) and suppose that $\mathcal{T}_0$ is yet another topology on $\mathbb{X}$. If $\mathcal{T}_0 \subset \mathcal{T}$, that is, if $\mathcal{T}$ designates more subsets of $\mathbb{X}$ as open sets than $\mathcal{T}_0$ does, then we say that the topology $\mathcal{T}$ is stronger than $\mathcal{T}_0$, or, equivalently, that $\mathcal{T}_0$ is weaker than $\mathcal{T}$. The purpose of this exercise is to clarify the reasons for this choice of nomenclature. Prove that $x_i \xrightarrow{\mathcal{T}} x$ implies $x_i \xrightarrow{\mathcal{T}_0} x$, that is, convergence with respect to the stronger topology implies convergence with respect to the weaker one (it is easier for a sequence to converge when the topology has fewer open sets, so that there are more sequences that converge in the weaker topology). In addition, prove that if $(\mathbb{Y}, \mathcal{U})$ is a topological space, then any function $f : \mathbb{X} \mapsto \mathbb{Y}$ that is continuous with respect to $\mathcal{U}$ and $\mathcal{T}_0$ must also be continuous with respect to $\mathcal{U}$ and $\mathcal{T}$. Explain why a function that is continuous with respect to $\mathcal{U}$ and $\mathcal{T}$ may not be continuous with respect to $\mathcal{U}$ and $\mathcal{T}_0$. To phrase it another way, weak continuity implies strong continuity (it is easier for a function to be continuous when the topology in its domain is stronger, which is to say there are more functions that are continuous relative to $\mathcal{T}$ than there are functions that are continuous relative to $\mathcal{T}_0$). ○

(A.18) EXERCISE (STRONG AND WEAK METRIC): Suppose that $(\mathbb{X}, d)$ is a metric space and suppose that $d_0 : \mathbb{X} \times \mathbb{X} \mapsto \mathbb{R}_+$ is another metric on $\mathbb{X}$ such that $d_0(x, y) \le C d(x, y)$ for all $x, y \in \mathbb{X}$, for some constant $C \in \mathbb{R}_{++}$. Prove that the topology generated by the metric $d_0(\cdot, \cdot)$ is weaker than the topology generated by the metric $d(\cdot, \cdot)$. In particular, if $\mathbb{X}$ has the structure of a vector space, and $\|\cdot\|_0$ and $\|\cdot\|$ are two different norms on $\mathbb{X}$ such that $\|x\|_0 \le C\|x\|$ for all $x \in \mathbb{X}$, then the topology generated by $\|\cdot\|$ is stronger than the topology generated by $\|\cdot\|_0$. We then say that the norm $\|\cdot\|$ is stronger than the norm $\|\cdot\|_0$ and that the metric $d(\cdot, \cdot)$ is stronger than the metric $d_0(\cdot, \cdot)$. If there are two constants $c, C \in \mathbb{R}_{++}$, $c < C$, such that $c d(x, y) \le d_0(x, y) \le C d(x, y)$ (or $c\|x\| \le \|x\|_0 \le C\|x\|$), then $d_0(\cdot, \cdot)$ and $d(\cdot, \cdot)$ ($\|\cdot\|_0$ and $\|\cdot\|$) generate the same topology and are said to be equivalent. ○

(A.19) GENERATED TOPOLOGY: Let $\mathbb{X}$ be any set, let $(\mathbb{Y}_i, \mathcal{U}_i)$, $i \in \mathbb{I}$, be any (arbitrary) family of (not necessarily distinct) topological spaces, and let $f_i : \mathbb{X} \mapsto \mathbb{Y}_i$, $i \in \mathbb{I}$, be any (arbitrary) family of functions. The smallest topology on $\mathbb{X}$ that contains the sets

$$\{f_i^{-1}(U) : U \in \mathcal{U}_i,\ i \in \mathbb{I}\}$$

is the weakest topology on $\mathbb{X}$ with respect to which all functions f_i, $i \in \mathbb{I}$, can be continuous. This construction is often applied to the case where all range spaces are identical, $(\mathbb{Y}_i, \mathcal{U}_i) \equiv (\mathbb{Y}, \mathcal{U})$ for all $i \in \mathbb{I}$, and one needs to describe the weakest topology in which a given family of mappings $f_i : \mathbb{X} \mapsto \mathbb{Y}$, $i \in \mathbb{I}$, happens to be continuous. This topology is usually referred to as the topology generated by the family $(f_i)_{i \in \mathbb{I}}$. ○

(A.20) PRODUCT TOPOLOGY: Let $(\mathbb{X}_i, \mathcal{T}_i)_{i\in\mathbb{I}}$ be any family of topological spaces, labeled by an arbitrary index set $\mathbb{I}$. The Cartesian product $\mathbb{X}_* = \times_{i\in\mathbb{I}} \mathbb{X}_i$ is nothing but the collection of all generalized sequences $x_* = (x_i \in \mathbb{X}_i)_{i\in\mathbb{I}}$. The mappings

$$\mathbb{X}_* \ni x_* \rightsquigarrow \mathcal{X}_i(x_*) \stackrel{\text{def}}{=} x_i \in \mathbb{X}_i\,, \quad i \in \mathbb{I}\,,$$

are called *canonical coordinate mappings*, or simply *coordinate mappings*, on the space $\mathbb{X}_*$. The topology on X_* generated by the entire family of coordinate mappings $(\mathcal{X}_i)_{i\in\mathbb{I}}$, that is, the weakest topology on the product space $\mathbb{X}_* = \times_{i\in\mathbb{I}} \mathbb{X}_i$ with respect to which all coordinate mappings happen to be continuous, is called *the product topology*. ○

(A.21) EXERCISE: The Euclidean space $\mathbb{R}^n$ can be identified with the Cartesian product of n copies of the real line $\mathbb{R}$, which has the structure of a topological space inherited from the Euclidean distance on $\mathbb{R}$. This gives rise to a product topology on $\mathbb{R}^n$. Prove that this product topology is the same as the topology on $\mathbb{R}^n$ associated with the Euclidean distance on that space. ○

(A.22) EXERCISE (HAUSDORFF SPACES): A topological space $(\mathbb{X}, \mathcal{T})$ is said to be a *Hausdorff space*[208] (alias: T_2 space) if for every choice of $x \in \mathbb{X}$ and $y \in \mathbb{X}$, such that $x \neq y$, one can find $U \in \mathcal{T}$ with $x \in U$ and $V \in \mathcal{T}$ with $y \in V$ such that $U \cap V = \emptyset$. Prove that a sequence in a Hausdorff space can have at most one limit. Prove that a topology generated by a metric (in particular, a topology generated by a norm) is a Hausdorff topology, and explain why this claim cannot be made for a topology generated by a pseudometric (or a seminorm). This is the main reason to insist on working with metrics and norms, rather than pseudometrics and seminorms. Unfortunately, many common-sense definitions of "distance" only lead to a seminorm, which then requires that one somehow modify the space to arrive at a norm and employ the usual limiting operations. ○

(A.23) EXERCISE AND AN IMPORTANT COMMENT: Let $(\mathbb{X}, \mathcal{T})$ and $(\mathbb{Y}, \mathcal{U})$ be any two topological spaces and let $f: \mathbb{X} \mapsto \mathbb{Y}$ be any continuous function (relative to $\mathcal{U}$ and $\mathcal{T}$). Prove that for any convergent sequence $(x_i \in \mathbb{X})_{i\in\mathbb{N}}$ the property $x_i \xrightarrow{\mathcal{T}} x$ implies $f(x_i) \xrightarrow{\mathcal{U}} f(x)$. In other words, continuous functions always map convergent sequences into convergent sequences. Does this feature characterize continuity? Unfortunately, for general topological spaces the answer is negative. To characterize continuity in such a manner, one must introduce the notions of *net* and *convergence of a net*, of which *sequence* and *convergence of a sequence* are special cases. Fortunately, if the topology on $\mathbb{X}$ happens to be metrizable, then the definition of a "continuous function from $\mathbb{X}$ to $\mathbb{Y}$" is equivalent to a "mapping from $\mathbb{X}$ to $\mathbb{Y}$ that preserves the convergence of all convergent sequences." Prove

[208] Named after the German mathematician Felix Hausdorff (1868–1842).

this claim, that is, prove that if $\mathcal{T}$ is generated by a metric $d(\cdot,\cdot)$, the property "$(f(x_i) \in \mathbb{Y})_{i \in \mathbb{N}}$ is a convergent sequence whenever $(x_i \in \mathbb{X})_{i \in \mathbb{N}}$ is a convergent sequence" implies that $f : \mathbb{X} \mapsto \mathbb{Y}$ is continuous.

HINT: Argue by contradiction: suppose that f is not continuous, in which case there is a $U \in \mathcal{U}$ for which $A \stackrel{\text{def}}{=} f^{-1}(U)$ is not open, that is, $A \backslash A^\circ \neq \emptyset$. Let $x \in A \backslash A^\circ$, and for any $i \in \mathbb{N}_{++}$ consider the ball $B_{1/i}(x)$. Show that any such ball contains an element $x_i \notin A$,[209] and argue that $x_i \to x$, whereas the sequence $\big(f(x_i) \notin U\big)_{i \in \mathbb{N}}$ cannot converge to $f(x) \in U$. ○

(A.24) EXERCISE: Let $(\mathbb{X}, d)$ be any metric space. Prove the following statements:

(a) Given any subset $A \subseteq \mathbb{X}$, the closure of A (for the metric $d(\cdot,\cdot)$) can be described as

$$\bar{A} = \left\{x \in \mathbb{X} : \text{there is a sequence } (x_i \in A)_{i \in \mathbb{N}} \text{ such that } x_i \xrightarrow{d} x\right\}$$

(b) A subset $A \subseteq \mathbb{X}$ is closed if and only if for any sequence $(x_i \in A)_{i \in \mathbb{N}}$ that has a limit one can claim the limit belongs to A, that is, $\lim_i^{(d)} x_i \in A$.

(c) A subset $A \subseteq \mathbb{X}$ is open if and only if for any $x \in A$ and any sequence $(x_i \in \mathbb{X})_{i \in \mathbb{N}}$ such that $x_i \xrightarrow{d} x$ there is an index $n \in \mathbb{N}$ (possibly depending on x) such that $x_i \in A$ for all $i \geq n$. ○

(A.25) CAUCHY SEQUENCES AND COMPLETENESS: Let $(\mathbb{X}, d)$ be a pseudometric space. A sequence $(x_i \in \mathbb{X})_{i \in \mathbb{N}}$ is said to be a Cauchy sequence[210] (for the pseudometric d) if for any $\varepsilon \in \mathbb{R}_{++}$ there is an index $n \in \mathbb{N}$ (possibly depending on ε) such that $d(x_i, x_j) < \varepsilon$ for all $i, j \geq n$. The pseudometric space $(\mathbb{X}, d)$ is said to be complete if every Cauchy sequence in it converges to some limit in $\mathbb{X}$. The real line $\mathbb{R}$ and, more generally, any of the Euclidean spaces $\mathbb{R}^n$, $n \geq 1$, are complete for the Euclidean norm. ○

(A.26) EXERCISE: Prove that any convergent sequence in a pseudometric space is a Cauchy sequence. Give an example of a Cauchy sequence in a particular metric space, which does not have a limit in that metric space. ○

(A.27) COMPLETION OF A METRIC SPACE: If $(\mathbb{X}, d)$ is a metric space that is not complete, then it is possible to construct a complete metric space $(\tilde{\mathbb{X}}, \tilde{d})$ such that $\mathbb{X}$ is a dense subset of $\tilde{\mathbb{X}}$ and the restriction of $\tilde{d}$ to $\mathbb{X} \times \mathbb{X}$ is exactly d. Except for the notation, this construction is no different from the construction of the real line $\mathbb{R}$ from the space $\mathbb{Q}$, which is not complete for the Euclidean distance. In most situations, the metric $\tilde{d}$ is still denoted by d. ○

(A.28) COMPLETENESS OF THE REAL LINE $\mathbb{R}$: The completeness of $\mathbb{R}$ is equivalent to the claim that any subset $\emptyset \neq A \subset \mathbb{R}$ that has an upper bound has a least upper bound. This property is equivalent to the claim that any subset $\emptyset \neq A \subset \mathbb{R}$

[209] This is one situation where the axiom of choice is explicitly needed.

[210] Named after the French mathematician Augustin-Louis Cauchy (1789–1857).

that has a lower bound has a largest lower bound. An upper bound for $A \neq \emptyset$ is any $b \in \mathbb{R}$ such that $b \geq x$ for all $x \in A$. A particular upper bound, b^*, is said to be a least upper bound if (a) it is an upper bound, and (b) it is smaller than any other upper bound, that is, $b^* \geq x$ for all $x \in A$ and $b^* \leq b$ for any upper bound b. Similarly, a lower bound for $A \neq \emptyset$ is any $b \in \mathbb{R}$ such that $b \leq x$ for all $x \in A$. A particular lower bound, b_*, is said to be a largest lower bound if (a) it is a lower bound, and (b) it is larger than any other lower bound, that is, $b \leq x$ for all $x \in A$ and $b_* \geq b$ for any lower bound b. Any least upper bound is clearly unique, and so is any largest lower bound. The least upper bound b^* is called supremum of A and is denoted by $\sup(A)$. If the set $A \neq \emptyset$ has no upper bound, it is said to be unbounded from above and $\sup(A)$ is defined to be $+\infty$. The largest lower bound b_* is called infimum of A and is denoted by $\inf(A)$. If the set $A \neq \emptyset$ has no lower bound, it is said to be unbounded from below and $\inf(A)$ is defined to be $-\infty$. It is easy to see that $\inf(A) = -\sup(-A)$, where $(-A) \overset{\text{def}}{=} \{x \in \mathbb{X} : -x \in A\}$.

Since $\inf(A)$ and $\sup(A)$ are well defined for any set $A \subseteq \mathbb{R}$, if the set⊸ (A.11) of limit points $\lim\text{pt}(A)$ is nonempty, then we can define

$$\limsup(A) = \sup\big(\lim\text{pt}(A)\big) \quad \text{and} \quad \liminf(A) = \inf\big(\lim\text{pt}(A)\big) . \quad \circ$$

(A.29) EXERCISE: Suppose that the sequence $(x_i \in \mathbb{R})_{i \in \mathbb{N}}$ is increasing ($x_i \leq x_{i+1}$ for all $i \in \mathbb{N}$) and bounded from above, that is, the set $\{x_i : i \in \mathbb{N}\}$ is bounded from above. Prove that $(x_i \in \mathbb{R})_{i \in \mathbb{N}}$ converges and

$$\lim_i x_i = \sup\{x_i : i \in \mathbb{N}\} .$$

Similarly, if the sequence $(x_i \in \mathbb{R})_{i \in \mathbb{N}}$ is decreasing ($x_i \geq x_{i+1}$ for all $i \in \mathbb{N}$) and bounded from below, then it converges and

$$\lim_i x_i = \inf\{x_i : i \in \mathbb{N}\} . \quad \circ$$

(A.30) THE EXTENDED REAL LINE $\bar{\mathbb{R}}$: We have seen in the last exercise that the limit of a bounded monotone sequence always exists and can be identified with either sup or inf. However, sup and inf would be defined without the requirement for the sequence to be bounded. We might as well postulate that any unbounded from above and increasing sequence converges to $+\infty$ and any unbounded from below and decreasing sequence converges to $-\infty$. The problem is that $+\infty$ and $-\infty$ are not elements of $\mathbb{R}$. The remedy is to work with the so-called extended real line $\bar{\mathbb{R}} \overset{\text{def}}{=} \mathbb{R} \cup \{-\infty\} \cup \{+\infty\}$. In fact, we can turn $\bar{\mathbb{R}}$ into a topological space by augmenting the topology in $\mathbb{R}$ with all sets of the form $U \cup]x, +\infty]$, the form $[-\infty, x[\cup U$, or the form $[-\infty, x[\cup U \cup]y, +\infty]$ for all possible choices of the reals $x, y \in \mathbb{R}$ and the open set $U \subseteq \mathbb{R}$. In this extended topology all intervals of the form $]x, +\infty]$ are open neighborhoods of $+\infty$ and the convergence to $+\infty$ is well understood. Similarly, all intervals of the form $[-\infty, x[$ are open neighborhoods of $-\infty$ and the convergence to $-\infty$ is also well understood. Furthermore, $+\infty$ and $-\infty$ can be limiting points for sets $A \subseteq \bar{\mathbb{R}}$. As a result, the operations $\sup(A)$ and $\inf(A)$ would be perfectly meaningful for any set $\emptyset \neq A \subseteq \bar{\mathbb{R}}$. For any (arbitrary) sequence $(x_i \in \mathbb{R})_{i \in \mathbb{R}}$ one can define the sequence

$$y_i \stackrel{\text{def}}{=} \sup\{x_j : j \geq i\}, \quad i = 0, 1, 2, \ldots .$$

The sequence $(y_i \in \mathbb{R})_{i \in \mathbb{R}}$ is clearly decreasing ($y_i \geq y_{i+1}$), so that $\lim_i y_i$ always exists. We denote this last limit by $\limsup_i x_i$. Similarly, the sequence

$$z_i \stackrel{\text{def}}{=} \inf\{x_j : j \geq i\}, \quad i = 0, 1, 2, \ldots$$

would be increasing ($z_i \leq z_{i+1}$), so that $\lim_i z_i$ again exists. We denote this last limit by $\liminf_i x_i$. If the set $A = \cup_{i \in \mathbb{N}} \{x_i\} \subseteq \mathbb{R}$ has at least one limit point, then $\limsup(A)$ and $\liminf(A)$ would be the same as $\limsup_i x_i$ and $\liminf_i x_i$; however, in general, the set A may not have a limit point – in which case $\limsup(A)$ and $\liminf(A)$ would not be defined – whereas $\limsup_i x_i$ and $\liminf_i x_i$ are always meaningful. By way of an example, if $x_i = (-1)^i$, $i = 0, 1, 2, \ldots$, then $\limsup_i x_i = +1$ and $\liminf_i x_i = -1$, yet the set $\cup_{i \in \mathbb{N}} \{x_i\} = \{-1\} \cup \{+1\} \subseteq \mathbb{R}$ does not have limit points. ○

We review briefly compact sets and their properties next.

(A.31) COMPACT TOPOLOGICAL SPACES: Given a topological space $(\mathcal{X}, \mathcal{T})$, a family of open sets $\mathcal{U} \subseteq \mathcal{T}$ is said to be an open cover of $\mathcal{X}$ if $\mathcal{X} = \cup \mathcal{U}$, and the space $(\mathcal{X}, \mathcal{T})$ is said to be compact if every open cover of $\mathcal{X}$ has a finite subcover, that is, $\mathcal{U} \subseteq \mathcal{T}$ and $\mathcal{X} = \cup \mathcal{U}$ implies that there is a finite subset $\mathcal{U}_0 \subseteq \mathcal{U}$ such that $\mathcal{X} = \cup \mathcal{U}_0$. A subset $A \subseteq \mathcal{X}$ is said to be compact if it is compact in the relative topology induced by $\mathcal{T}$, that is, if $(A, A \cap \mathcal{T})$ is a compact topological space. ○

(A.32) TYCHONOFF'S THEOREM:[211] Let $(\mathcal{X}_i, \mathcal{T}_i)_{i \in \mathbb{I}}$ be any family of compact topological spaces. Then the Cartesian product $\mathcal{X}_* = \times_{i \in \mathbb{I}} \mathcal{X}_i$ is compact with respect to the product topology (the weakest topology on $\mathcal{X}_*$ for which the coordinate mappings are continuous). ○

(A.33) EXERCISE: Prove that every closed subset of a compact topological space is compact. ○

(A.34) EXERCISE: Prove that every closed interval $[a, b] \subset \mathbb{R}$ is a compact set in the topology generated by the Euclidean distance in $\mathbb{R}$. Explain why none of the intervals $]a, b]$, $[a, b[$, $]a, b[$ is compact. ○

(A.35) EXERCISE (BOLZANO–WEIERSTRASS THEOREM):[212] Prove that a metric space $(\mathcal{X}, d)$ can be compact if and only if every sequence $(x_i \in \mathcal{X})_{i \in \mathbb{N}}$ contains a subsequence $(x_{i_k} \in \mathcal{X})_{k \in \mathbb{N}}$ that converges (to some $x \in \mathcal{X}$). More generally, the metric space $(\mathcal{X}, d)$ is compact if and only if every infinite subset of $\mathcal{X}$ has at least one limit point, that is, $\lim \mathrm{pt}(S) \neq \emptyset$ for any $S \subseteq \mathcal{X}$ with $|S| \geq \aleph_0$).

(A.36) EXERCISE (COMPACTNESS AND SEPARABILITY): Prove that every compact metric space is also a separable metric space. ○

[211] Named after the Russian mathematician Andrey Nikolayevich Tychnoff [Tikhonov] (1906–1993).

[212] Named after the Bohemian mathematician, philosopher, and Roman Catholic priest Bernard Bolzano (1781–1848) and the German mathematician Karl Theodor Wilhelm Weierstrass (1815–1897).

(A.37) EXERCISE (COMPACTNESS AND COMPLETENESS): Prove that every compact metric space $(\mathbb{X}, d)$ is a complete metric space (for the metric d). ○

(A.38) EXERCISE (COMPACTNESS AND CONTINUITY): Let $(\mathbb{X}, d)$ and $(\mathbb{Y}, \rho)$ be any two metric spaces and let $f : \mathbb{X} \mapsto \mathbb{Y}$ be any function from $\mathbb{X}$ into $\mathbb{Y}$. Prove that f is continuous if and only if the image under f of every compact set in $\mathbb{X}$ is a compact set in $\mathbb{Y}$, that is, if and only if "$A \subseteq \mathbb{X}$ is compact" implies "$f(A) \subseteq \mathbb{Y}$ is compact," where

$$f(A) = \{y \in \mathbb{Y} : y = f(x), \text{ for some } x \in A\}.$$

In particular, if $f : \mathbb{X} \mapsto \mathbb{R}$ is continuous, then every compact set $A \subseteq \mathbb{X}$ contains an element $A \ni x_*$ such that

$$f(x_*) = \inf\{f(x) : x \in A\},$$

and contains an element $A \ni x^*$ such that

$$f(x^*) = \sup\{f(x) : x \in A\}.$$

Plainly, every $\mathbb{R}$-valued continuous function attains both its minimum and maximum on every compact set on which it is defined. ○

(A.39) EQUICONTINUITY AND UNIFORM EQUICONTINUITY: Let $(\mathbb{X}, d)$ and $(\mathbb{Y}, \rho)$ be any two metric spaces. A family of functions $f_i : \mathbb{X} \mapsto \mathbb{Y}$, $i \in \mathbb{I}$ (indexed by an arbitrary set $\mathbb{I}$), is said to be equicontinuous if it has the property: given any (fixed) $x \in \mathbb{X}$ and any $\varepsilon \in \mathbb{R}_{++}$, one can find some $h \in \mathbb{R}_{++}$ (possibly depending on both x and ε), such that $\rho(f_i(x), f_i(x')) < \varepsilon$, for all $i \in \mathbb{I}$ and all $x' \in \mathbb{X}$ with $d(x, x') < h$.

A family of functions $f_i : \mathbb{X} \mapsto \mathbb{Y}$, $i \in \mathbb{I}$, is said to be uniformly equicontinuous if it has the property: given any $\varepsilon \in \mathbb{R}_{++}$, one can find some $h \in \mathbb{R}_{++}$ (possibly depending on ε, and only on ε), such that $\rho(f_i(x), f_i(x')) < \varepsilon$, for all $i \in \mathbb{I}$ and all $x, x' \in \mathbb{X}$ with $d(x, x') < h$.

A function $f : \mathbb{X} \mapsto \mathbb{Y}$ is said to be uniformly continuous (alias: equicontinuous) if the family (singleton, in fact) $\{f\}$ is equicontinuous, that is, if the following property holds: given any $\varepsilon \in \mathbb{R}_{++}$, there is some $h \in \mathbb{R}_{++}$ (possibly depending on ε and only on ε) such that $\rho(f(x), f(x')) < \varepsilon$ for all $x, x' \in \mathbb{X}$ with $d(x, x') < h$. ○

(A.40) EXERCISE: Let $(\mathbb{X}, d)$ be any compact metric space and let $(\mathbb{Y}, \rho)$ be any metric space. Prove that any equicontinuous family of functions from $\mathbb{X}$ into $\mathbb{Y}$ is also a uniformly equicontinuous family of functions from $\mathbb{X}$ into $\mathbb{Y}$. In particular, any function $f : \mathbb{X} \mapsto \mathbb{Y}$ that is continuous is also uniformly continuous. ○

(A.41) EXERCISE: Let $(\mathbb{X}, d)$ be any metric space, let S be any dense (relative to the metric d) subset of $\mathbb{X}$, and let $(\mathbb{Y}, \rho)$ be any complete metric space. Prove that any uniformly continuous function $f : S \mapsto \mathbb{Y}$ can be extended in a unique way to a continuous function from $\mathbb{X}$ into $\mathbb{Y}$. In particular, given any compact interval $[a, b] \subset \mathbb{R}$, one can claim that every continuous function $f : [a, b] \mapsto \mathbb{Y}$ is completely determined by its values $f(x)$, $x \in [a, b] \cap \mathbb{Q}$. ○

The topological spaces encountered in most applications also have the structure of a linear (vector) space, in which case the topology is defined through a norm or metric.

(A.42) BANACH AND POLISH SPACES: A vector space $\mathbb{X}$ endowed with a norm $\|\cdot\|$ is said to be a Banach space[213] if it is complete for the metric associated with the norm $\|\cdot\|$. A metric space $(\mathbb{X}, d)$ is said to be a Polish space if it is both complete and separable (admits a countable dense subset) for the metric d.[214] In particular, any separable Banach space is a Polish space.

The most interesting Banach spaces encountered in this book are spaces of finite real functions. A family of such functions that share a common domain has the obvious structure of a vector space, in which the addition is understood as the pointwise addition of functions and the multiplication with a scalar is understood as the pointwise multiplication of a function by a scalar. One particularly interesting space of functions is the space of all bounded and continuous functions from a given topological space $(\mathbb{S}, \mathcal{T})$ into the Euclidean space $\mathbb{R}^n$. This space is denoted by $\mathscr{C}_b(\mathbb{S}; \mathbb{R}^n)$, and we note that in the special case where $(\mathbb{S}, \mathcal{T})$ is compact, "bounded" becomes superfluous, and we can write simply $\mathscr{C}(\mathbb{S}; \mathbb{R}^n)$ instead of $\mathscr{C}_b(\mathbb{S}; \mathbb{R}^n)$. The space $\mathscr{C}_b(\mathbb{S}; \mathbb{R}^n)$ has the structure of a Banach space endowed with the uniform norm

$$\|f\|_\infty \stackrel{\text{def}}{=} \sup\{|f(x)| : x \in \mathbb{S}\}, \quad f \in \mathscr{C}_b(\mathbb{S}; \mathbb{R}^n),$$

as we show below. The convergence in the norm $\|\cdot\|_\infty$ is called *uniform convergence*, or *convergence in uniform norm*. ○

(A.43) EXERCISE: Prove that $\mathscr{C}_b(\mathbb{S}; \mathbb{R}^n)$ is a vector space and that $\|\cdot\|_\infty$ is a norm for every choice of the topological space $(\mathbb{S}, \mathcal{T})$. ○

(A.44) EXERCISE: Assuming that the real line $\mathbb{R}$ is complete, prove that the space $\mathbb{R}^n$ is complete for the Euclidean distance $d(x, y) = |x-y|$, $x, y \in \mathbb{R}^n$, and conclude that $(\mathbb{R}^n, |\cdot|)$ is a Banach space. In addition, prove that $\mathbb{R}^n$ is a Polish space. ○

(A.45) EXERCISE (CONVERGENCE OF CONTINUOUS FUNCTIONS TO A CONTINUOUS FUNCTION): Let $(\mathbb{S}, \mathcal{T})$ be any topological space, let $g : \mathbb{S} \mapsto \mathbb{R}^n$ be any function from $\mathbb{S}$ into $\mathbb{R}^n$, and suppose that the sequence $\big(f_i \in \mathscr{C}_b(\mathbb{S}; \mathbb{R}^n)\big)_{i \in \mathbb{N}}$ converges to g in the norm $\|\cdot\|_\infty$, that is,

$$\lim_i \|f_i - g\|_\infty = 0\,.$$

Prove that $g \in \mathscr{C}_b(\mathbb{S}; \mathbb{R}^n)$. ○

---

[213] Named after the Polish mathematician Stephan Banach (1892–1945).

[214] The term "Polish space" comes from the fact that, originally, such spaces were studied extensively by several Polish mathematicians during the first half of the twentieth century.

(A.46) EXERCISE (COMPLETENESS OF $\mathscr{C}_b(\mathbb{S};\mathbb{R}^n)$): Let $(\mathbb{S},\mathcal{T})$ be any topological space and suppose that the sequence $\big(f_i \in \mathscr{C}_b(\mathbb{S};\mathbb{R}^n)\big)_{i\in\mathbb{N}}$ is a Cauchy sequence for the norm $\|\cdot\|_\infty$, that is,

$$\lim_n \sup_{i,j\geq n}\|f_i - f_j\|_\infty = 0\,.$$

Prove that there is a unique function $g \in \mathscr{C}_b(\mathbb{S};\mathbb{R}^n)$, for which one can claim that $\|f_i - g\|_\infty \to 0$. In particular, $\mathscr{C}_b(\mathbb{S};\mathbb{R}^n)$ is complete for the norm $\|\cdot\|_\infty$, and is therefore a Banach space. ○

The next classical result describes a rare situation in which pointwise convergence entails uniform convergence.

(A.47) EXERCISE (DINI'S THEOREM[215]): Let $(\mathbb{X},\mathcal{T})$ be any compact topological space, and suppose that the sequence $\big(f_i \in \mathscr{C}(\mathbb{X};\mathbb{R})\big)_{i\in\mathbb{N}}$ converges monotonically to some continuous function $g \in \mathscr{C}(\mathbb{X};\mathbb{R})$ in pointwise sense. To be precise, the sequences $(f_i(x))_{i\in\mathbb{N}}$, $x \in \mathbb{X}$, are either simultaneously increasing, or simultaneously decreasing, and, in addition, $\lim_i f(x) = g(x)$ for all $x \in \mathbb{X}$. Prove that the sequence (of functions) $(f_i)_{i\in\mathbb{N}}$ converges to g uniformly on $\mathbb{X}$, that is, $\|f_i - g\|_\infty \to 0$.

HINT: Given any $\varepsilon \in \mathbb{R}_{++}$, the sets $\{|f_i - g| < \varepsilon\}$, $i \in \mathbb{N}$, form an open cover of $\mathbb{X}$, which$^{\multimap(A.31)}$ must contain a finite subcover. Let i^* be the largest index i for which $\{|f_i - g| < \varepsilon\}$ belongs to the finite subcover. Argue that $\{|f_i - g| < \varepsilon\} \subseteq \{|f_{i+1} - g| < \varepsilon\}$ for any $i \in \mathbb{N}$ and conclude that $\{|f_{i^*} - g| < \varepsilon\} = \mathbb{X}$. ○

(A.48) STONE–WEIERSTRASS THEOREM:[216] (Reed and Simon 1980, IV.9) Let $(\mathbb{X},\mathcal{T})$ be any compact Hausdorff space (in particular, any compact metric space) and let $\mathscr{A} \subseteq \mathscr{C}(\mathbb{X};\mathbb{R})$ be any subalgebra of $\mathscr{C}(\mathbb{X};\mathbb{R})$, that is, $\mathscr{A}$ is a linear space of continuous functions on $\mathbb{X}$, which is closed under the usual pointwise multiplication of functions. Suppose that $\mathscr{A}$ separates the points in $\mathbb{X}$, in that given any $x, y \in \mathbb{X}$ there is an $f \in \mathscr{C}(\mathbb{X};\mathbb{R})$ such that $f(x) \neq f(y)$. In addition, suppose that $\mathscr{A}$ does not vanish at any point in $\mathbb{X}$, in that given any $x \in \mathbb{X}$ there is an $f \in \mathscr{C}(\mathbb{X};\mathbb{R})$ such that $f(x) \neq 0$. Then $\mathscr{A}$ is dense in $\mathscr{C}(\mathbb{X};\mathbb{R})$ with respect to the uniform norm, that is, given any $f \in \mathscr{C}(\mathbb{X};\mathbb{R})$ and any $\varepsilon \in \mathbb{R}_{++}$, there is a $g_\varepsilon \in \mathscr{A}$ such that

$$\sup_{x\in\mathbb{X}} \big|f(x) - g_\varepsilon(x)\big| < \varepsilon\,.$$

In particular, given any finite interval $[a,b] \subset \mathbb{R}$, the space of all polynomials is dense in $\mathscr{C}([a,b];\mathbb{R})$ with respect to the uniform norm. ○

(A.49) EXERCISE: Let $(\mathbb{X},\mathcal{S})$ and $(\mathbb{Y},\mathcal{T})$ be any two compact Hausdorff spaces and let $\mathbb{X}\times\mathbb{Y}$ be endowed with the product topology. Prove that the family $\mathscr{A}$ of all finite linear combinations of functions on $\mathbb{X}\times\mathbb{Y}$ of the form

$$\mathbb{X}\times\mathbb{Y} \ni (x,y) \rightsquigarrow f(x)g(y) \in \mathbb{R}\,,$$

[215] Named after the Italian mathematician and politician Ulisse Dini (1845–1918).

[216] Named after the American mathematician Marshall Harvey Stone (1903–1989) and the German mathematician Karl Theodor Wilhelm Weierstrass (1815–1897).

for all choices of $f \in \mathscr{C}(\mathbb{X};\mathbb{R})$ and $g \in \mathscr{C}(\mathbb{Y};\mathbb{R})$ is dense in the uniform norm of $\mathscr{C}(\mathbb{X}\times\mathbb{Y};\mathbb{R})$. In particular, given every two finite intervals $[a,b]$ and $[c,d]$, the space of all bi-variate polynomials is dense in $\mathscr{C}([a,b]\times[c,d];\mathbb{R})$ with respect to the uniform norm. ○

(A.50) INFINITE SERIES IN BANACH SPACES: Let $(\mathbb{X}, \|\cdot\|)$ be a Banach space and let $(x_i \in \mathbb{X})_{i\in\mathbb{N}}$ be some sequence in $\mathbb{X}$. The infinite series symbol $\sum_{i=0}^{\infty} x_i$ is understood as the limit (in $\mathbb{X}$), if one exists, of the sequence of partial sums $\left(\sum_{i=0}^{n} x_i\right)_{n\in\mathbb{N}}$. The series $\sum_{i=0}^{\infty} x_i$ is said to be absolutely summable if $\sum_{i=0}^{\infty}\|x_i\| < +\infty$, in which case $\sum_{i=0}^{\infty} x_i$ can be identified with an element of $\mathbb{X}$. ○

(A.51) LINEAR OPERATORS AND THEIR NORMS: Let $\mathbb{X}$ and $\mathbb{Y}$ be any two vector spaces. A function $L\colon \mathbb{X} \mapsto \mathbb{Y}$ is said to be a linear operator (alias: linear transformation) if it preserves the operations "addition" and "multiplication by a scalar." To be precise $L(au + bv) = aL(u) + bL(v)$ for all $a, b \in \mathbb{R}$ and $u, v \in \mathbb{X}$. For linear operators of this form it is customary to write Lu instead of $L(u)$, $u \in \mathbb{X}$. If $\mathbb{X}$ is endowed with a norm $|\cdot|$ and $\mathbb{Y}$ is endowed with a norm $\|\cdot\|$, then the linear operator $L\colon \mathbb{X} \mapsto \mathbb{Y}$ is said to be *bounded* if there is a universal constant $C \in \mathbb{R}_{++}$ such that

$$\|Lx\| \le C|x| \quad \Leftrightarrow \quad \left\|L\frac{x}{|x|}\right\| \le C \quad \text{for all } x \in \mathbb{X}\,.$$

The smallest constant C with which the last relation is satisfied is called the (operator) norm of L and is denoted by $\|L\|$, that is,

$$\|L\| = \sup\left\{\|Lx\|: x \in \mathbb{X}\,,\ |x| = 1\right\}.$$

When the range space $\mathbb{Y}$ is replaced with the real line $\mathbb{R}$ (with the obvious linear structure), a linear transformation from $\mathbb{X}$ into $\mathbb{R}$ is usually called a *linear functional* and is denoted by ℓ – or some other lowercase symbol. ○

(A.52) EXERCISE (CONTINUITY OF LINEAR OPERATORS): With the notation as in (A.51), prove that the linear operator $L\colon \mathbb{X} \mapsto \mathbb{Y}$ is bounded if and only if L is continuous at 0. Prove that L is continuous at 0 if and only if L is continuous at any fixed $x \in \mathbb{X}$. ○

(A.53) SPACES OF LINEAR OPERATORS: (Reed and Simon 1980, III.2) Let $\mathbb{X}$ be any normed space and let $\mathbb{Y}$ be any Banach space. Then the space of all bounded linear operators from $\mathbb{X}$ into $\mathbb{Y}$ is a Banach space (complete normed space) with respect to the operator norm introduced in (A.51). ○

(A.54) DENSELY DEFINED LINEAR OPERATORS: (Reed and Simon 1980, I.7) With the notation as in (A.51), suppose that the normed space $(\mathbb{Y}, \|\cdot\|)$ is complete, so that it is a Banach space, and suppose that the linear operator $L\colon \mathbb{X} \mapsto \mathbb{Y}$ is bounded. Then there is a unique linear transformation $\tilde{L}\colon \tilde{\mathbb{X}} \mapsto \mathbb{Y}$, $\tilde{\mathbb{X}}$ being the completion of $\mathbb{X}$ for the norm $|\cdot|$, which coincides with L on $\mathbb{X}$. Furthermore, the norms of $\tilde{L}$ and L are the same: $\|L\| = \|\tilde{L}\|$. In particular, if $\mathbb{X}$ and $\mathbb{Y}$ are both Banach spaces, to define a continuous linear transformation from $\mathbb{X}$ into $\mathbb{Y}$ it would be enough to define a bounded linear transformation from a dense subset of $\mathbb{X}$ into $\mathbb{Y}$. ○

(A.55) INNER PRODUCT: Let $\mathbb{X}$ be any linear space. A semi-inner product on $\mathbb{X}$ is any mapping

$$\mathbb{X}\times\mathbb{X} \ni (u, v) \rightsquigarrow u\cdot v \in \mathbb{R}$$

which is:

(a) symmetric, in that $u\cdot v = v\cdot u$ for all $u, v \in \mathbb{X}$;

(b) bilinear, in that $(au + bv)\cdot w = au\cdot w + bv\cdot w$ for all $a, b \in \mathbb{R}$ and all $u, v, w \in \mathbb{X}$;

(c) positive definite, in that $u\cdot u \geq 0$ for all $u \in \mathbb{X}$.

The semi-inner product is called an *inner product* (aliases: *dot-product*, *scalar product*) if $u\cdot u = 0$ implies that $u = \vec{0}$. A vector space endowed with a semi-inner product is called a *semi-inner product space*, and is called an *inner product space* if the semi-inner product is an inner product. Two vectors $u, v \in \mathbb{X}$ are said to be orthogonal (alias: perpendicular, notation: $u \perp v$) if $u\cdot v = 0$. If $\mathbb{L}$ is any linear subspace of $\mathbb{X}$, then the orthogonal complement of $\mathbb{L}$, denoted by $\mathbb{L}^{\perp}$, is defined as the linear space that consists of those vectors in $\mathbb{X}$ that are perpendicular to all vectors in $\mathbb{L}$, that is,

$$\mathbb{L}^{\perp} = \left\{u \in \mathbb{X} : u\cdot v = 0\,, \text{ for every } v \in \mathbb{L}\right\}.$$

The (semi-)inner product gives rise to a (semi-)norm on $\mathbb{X}$, which is given by

† $$\|u\| = \sqrt{u\cdot u} \quad \text{for all } \ u \in \mathbb{X}\,.$$

Unless noted otherwise, when we say that $\mathbb{X}$ is a (semi-)inner product space, we automatically assume that the (semi-)inner product on that space is expressed as $u\cdot v$, $u, v \in \mathbb{X}$, and $\mathbb{X}$ is endowed with the above norm.[217] ○

(A.56) EXERCISE (POLARIZATION IDENTITY): Prove that if $\mathbb{X}$ is a (semi-)inner product space, then the values $\|u\| = \sqrt{u\cdot u}$, $u \in \mathbb{X}$, uniquely determine the dot product $u\cdot v$, $u, v \in \mathbb{X}$. To be precise, prove the following relation known as *the polarization identity*:

$$u\cdot v = \frac{1}{4}\left(\|u+v\|^2 - \|u-v\|^2\right) \quad \text{for all } u, v \in \mathbb{X}\,.$$

In particular, a norm $\|\cdot\|$ can be associated with an inner product through the relation (A.55†) if and only if the right side of the above relation corresponds to an inner product, that is, satisfies (a), (b), and (c) from (A.55). ○

[217] Since the symbol $\|\cdot\|$ is used to denote many different norms, it must be clear from the context which space and features the norm $\|\cdot\|$ is associated with. Should further clarification be needed, one may write $\|\cdot\|_{\mathbb{X}}$.

(A.57) EXERCISE (CAUCHY–BUNYAKOVSKY–SCHWARZ INEQUALITY[218]): Let $\mathbb{X}$ be any (semi-)inner product space. Prove that for any $u, v \in \mathbb{X}$ one has

$$|u \cdot v|^2 \leq (u \cdot u)(v \cdot v).$$

HINT: What can you say about the function $\mathbb{R} \ni x \rightsquigarrow \|u + x\, v\|^2$? ○

(A.58) EXERCISE (CONTINUITY OF THE INNER PRODUCT): Prove that any inner product on the vector space $\mathbb{X}$ is a continuous function from $\mathbb{X} \times \mathbb{X}$ into $\mathbb{R}$ relative to the product topology on $\mathbb{X} \times \mathbb{X}$ (assuming that $\mathbb{X}$ is endowed with the topology of the norm associated with the inner product) and the Euclidean metric on $\mathbb{R}$. ○

(A.59) EXERCISE (PYTHAGOREAN THEOREM): Consider a vector space $\mathbb{X}$ endowed with an inner product, and let $\|\cdot\|$ be the norm associated with that inner product. Prove that for any two vectors $u, v \in \mathbb{X}$ with $u \perp v$ one has $\|u + v\|^2 = \|u\|^2 + \|v\|^2$; more generally, if the vectors $u_i \in \mathbb{X}$, $i \in \mathbb{N}_{|n}$, are such that $u_i \perp u_j$ for all $i \neq j$, prove that

$$\left\| \sum\nolimits_{i \in \mathbb{N}_{|n}} u_i \right\|^2 = \sum\nolimits_{i \in \mathbb{N}_{|n}} \|u_i\|^2 . \quad ○$$

(A.60) EXERCISE (THE PARALLELOGRAM LAW): With the notation as in (A.59), prove that for any two vectors $u, v \in \mathbb{X}$ one has

$$\|u + v\|^2 + \|u - v\|^2 = 2\|u\|^2 + 2\|v\|^2 . \quad ○$$

(A.61) HILBERT SPACES: A Hilbert space[219] is any vector space that is endowed with an inner product and is complete for the norm associated with that inner product. To phrase this definition another way,⊶ (A.56) a Hilbert space is a special case of a Banach space with norm $\|\cdot\|$ such that the function

$$\mathbb{X} \times \mathbb{X} \ni (u, v) \rightsquigarrow \frac{1}{4}\Big(\|u + v\|^2 - \|u - v\|^2 \Big)$$

satisfies condition (b) from (A.55) (conditions (a) and (c) are automatically satisfied as long as $\|\cdot\|$ is a norm). ○

(A.62) EXERCISE (CONVERGENCE IN HILBERT SPACES): Let $\mathscr{H}$ be any Hilbert space and let $(h_i \in \mathscr{H})_{i \in \mathbb{N}}$ be any sequence in it. Prove that this sequence converges in $\mathscr{H}$ if and only if the double sequence $\big(h_i \cdot h_j \in \mathbb{R}\big)_{i \in \mathbb{N}, j \in \mathbb{N}}$ converges in $\mathbb{R}$, that is, for some $a \in \mathbb{R}$ one can claim that for any $\varepsilon \in \mathbb{R}_{++}$ one can find an index $k \in \mathbb{N}$ (possibly depending on ε) such that

$$\big|a - h_i \cdot h_j\big| < \varepsilon , \quad \text{for all } i, j \geq k . \quad ○$$

(A.63) ORTHOGONAL DECOMPOSITION: (Dudley 2002, 5.3.8) Let $\mathbb{X}$ be any vector space endowed with an inner product and let $\mathscr{H}$ be any linear subspace of $\mathbb{X}$,

[218] Named after the French mathematician Augustin-Louis Cauchy (1789–1857), the Ukrainian mathematician Viktor Yakovlevich Bunyakovsky (1804–1889), and the German mathematician Karl Hermann Amandus Schwarz (1843–1921).

[219] Named after the German mathematician David Hilbert (1862–1943).

which is a Hilbert space for the inner product of $\mathbb{X}$ and is strictly smaller than $\mathbb{X}$, in that there is a vector $u \in \mathbb{X}$ such that $u \neq \vec{0}$ and $u \notin \mathscr{H}$. Then any vector $x \in \mathbb{X}$ can be expressed in a unique way as the sum $x = u + v$, where $u \in \mathscr{H}$ and $v \in \mathscr{H}^{\perp}$. This property is usually written as $\mathbb{X} = \mathscr{H} \oplus \mathscr{H}^{\perp}$. ○

(A.64) LINEAR FUNCTIONALS AND DUALITY: If $\mathscr{H}$ is a Hilbert space, then, given any $h \in \mathscr{H}$, the assignment

$$\mathscr{H} \ni x \rightsquigarrow \ell(x) \stackrel{\text{def}}{=} h \cdot x \in \mathbb{R}$$

gives a continuous linear functional on $\mathscr{H}$. What is more interesting – and useful – is that any continuous linear functional on $\mathscr{H}$ actually has the above form, that is, if ℓ is any continuous linear functional on $\mathscr{H}$, then there is a unique vector $h_\ell \in \mathscr{H}$ such that $\ell(x) = h_\ell \cdot x$ for all $x \in \mathscr{H}$. Furthermore, if ℓ is treated as a linear operator from $\mathscr{H}$ into $\mathbb{R}$, then its operator norm can be identified with the Hilbert norm of h_ℓ, that is, $\|\ell\|_{\text{operator}} = \|h_\ell\|_{\mathscr{H}} = \sqrt{h_\ell \cdot h_\ell}$. This result is known as the Riesz lemma[220] – see (Reed and Simon 1980, II.4).

The space of all continuous (that is, bounded) linear functionals on a Banach space $(\mathbb{X}, \|\cdot\|)$ is important for many applications. As we already know, equipped with the operator norm, this is a Banach space. It is usually denoted by $\mathbb{X}^*$ and is called *the dual of* $\mathbb{X}$. For any fixed $x \in \mathbb{X}$, $\mathbb{X}^* \ni \ell \rightsquigarrow \ell(x) \in \mathbb{R}$ is a continuous linear functional on the dual $\mathbb{X}^*$, thus an element of the double dual $\mathbb{X}^{**}$. Furthermore, the norm of this functional is exactly $\|x\|$. This allows one to identify $\mathbb{X}$ with a linear subspace of the double dual $\mathbb{X}^{**}$. For a Hilbert space $\mathscr{H}$, the dual $\mathscr{H}^*$ can be identified with $\mathscr{H}$, and, for that reason, the double dual $\mathscr{H}^{**}$ can be identified with $\mathscr{H}^*$, and therefore also with $\mathscr{H}$. Plainly, as a subspace of $\mathscr{H}^{**}$, $\mathscr{H}$ is in fact the entire $\mathscr{H}^{**}$. Other interesting Banach spaces $\mathbb{X}$ are not Hilbert spaces but nevertheless have the property $\mathbb{X} = \mathbb{X}^{**}$ (can be identified with their double dual). Such Banach spaces are called *reflexive* – see (Reed and Simon 1980, sec. III.2). ○

(A.65) ORTHONORMAL BASIS: (ibid., sec. II.3) Given any Hilbert space $\mathscr{H}$, a subset $\mathscr{E} \subset \mathscr{H}$ is said to be an orthonormal set if $e \cdot e = 1$ for all $e \in \mathscr{E}$ and $e \cdot e' = 0$ for any $e, e' \in \mathscr{E}$ with $e \neq e'$. A largest orthonormal set (that is, an orthonormal set that is not strictly included in any other orthonormal set) always exists. Any such (maximal) orthonormal set is called an *orthonormal basis*. If $\mathscr{E}$ is an orthonormal basis in $\mathscr{H}$, then for any $h \in \mathscr{H}$ one has $h \cdot e \neq 0$ for at most countably many $e \in \mathscr{E}$. Furthermore, $\sum_{e \in \mathscr{E}} |h \cdot e|^2 < \infty$, and

$$h = \sum_{e \in \mathscr{E}} (h \cdot e)\, e \quad \text{for all } h \in \mathscr{H}.$$

220 Named after the Hungarian mathematician Frigyes Riesz (1880–1956).

Because any orthonormal set is included in some maximal orthonormal set (that is, an orthonormal basis), for an arbitrary orthonormal set $\mathscr{E}_0$ one has

$$\|h\|^2 \geq \sum\nolimits_{e \in \mathscr{E}_0} |h.e|^2 \quad \text{for all } h \in \mathscr{H} .$$

If $\mathscr{H}$ is a separable Hilbert space, then any orthonormal basis is a countable (that is, either finite or countably infinite) set of vectors. If $\mathscr{S} \subset \mathscr{H}$ is any countable and dense subset of linearly independent vectors (no element of $\mathscr{S}$ can be written as a finite linear combination of other elements of $\mathscr{S}$), then one can construct, sequentially, an orthonormal basis from $\mathscr{S}$ by following the Gram–Schmidt algorithm.

If the Hilbert space $\mathscr{H}$ has a finite orthonormal basis, then we say that $\mathscr{H}$ is finite dimensional. It is not difficult to check that if one maximal orthonormal set in $\mathscr{H}$ is finite, then any other maximal orthonormal set is also finite and has the same number of elements. This number is called *the dimension of* $\mathscr{H}$ and is often abbreviated as $\dim(\mathscr{H})$. A finite dimensional Hilbert space is no different from a Euclidean space. ○

(A.66) EXERCISE ($\mathbb{R}^n$ AS A HILBERT SPACE): The natural inner product on the space $\mathbb{R}^n$ is given by

$$x \cdot y \equiv (x_1, x_2, \dots, x_n) \cdot (y_1, y_2, \dots, y_n) = x_1 y_1 + x_2 y_2 + \dots x_n y_n , \quad x, y \in \mathbb{R}^n ,$$

and has a special name: dot product. Prove that, equipped with the dot product, $\mathbb{R}^n$ is a Hilbert space. Prove that the metric on $\mathbb{R}^n$ induced by the dot product is no different from the Euclidean metric. Prove that the vectors

$$e_1 = (1, 0, 0, \dots, 0, 0) \quad e_2 = (0, 1, 0 \dots, 0, 0) , \quad \dots , \quad e_n = (0, 0, 0 \dots, 0, 1) ,$$

form an orthonormal basis in $\mathbb{R}^n$, so that $\dim(\mathbb{R}^n) = n$.

Rotation in $\mathbb{R}^n$ is a transformation given by some orthogonal matrix $U \in \mathbb{R}^{n \otimes n}$, that is, a matrix such that $UU^\intercal = U^\intercal U = I$ (I being the identity matrix). Prove the dot product is invariant under rotation: for any orthogonal matrix U one has

$$(Ux) . (Uy) = x \cdot y \quad \text{for all } x, y \in \mathbb{R}^n .$$

Conclude that if $(e_i \in \mathbb{R}^n)_{1 \leq i \leq n}$ is an orthonormal basis, then $(Ue_i \in \mathbb{R}^n)_{1 \leq i \leq n}$ is an orthonormal basis. Prove that any orthonormal basis can be obtained from any other orthonormal basis by way of an appropriate rotation, and that the dot product is independent of the choice of the coordinate system, even though the definition given above is.

Finally, if α stands for the angle between the vectors x and y, then

$$\cos \alpha = \frac{x \cdot y}{|x||y|} .$$

Since $|\cos(\alpha)| \leq 1$, we see that $|x \cdot y| \leq |x||y|$, which is nothing but the Cauchy–Bunyakovski–Schwarz inequality for the space $\mathbb{R}^n$. ○

Appendix B

Computer Code[221]

B.1 Working with Market Data

This section contains a prototype Python code for some basic operations with market data. It assumes that the data are fetched from http://www.google.com/finance in csv format and are converted to Google Sheets file "tsla.csv" (all entries are formatted as strings, not floats). One must first load the relevant libraries:

```
In [1]: import matplotlib
        import matplotlib.pyplot as plt
        import pylab
        import numpy as np
        from pylab import *
        import csv
        import datetime
        from matplotlib.dates import MONDAY
        from matplotlib.dates import YearLocator, MonthLocator
        from matplotlib.dates import WeekdayLocator, DateFormatter
        from datetime import date, datetime
```

The next step is to load the entire data set into a single variable:

```
In [2]: file = open('<path-to-file-directory>/tsla.csv','r')
        data=csv.reader(file)
        quotes=[]
        for row in data:
            quotes.append(row)

        len(quotes)
Out [2]: 1513
```

The variable quotes is now an array, the first row in which contains the strings

```
In [3]: info=quotes[0]
        info
```

221 The computer code included in this appendix is meant to work (and was tested) with Python 3.5. It will not work with Python 2.x unless modified accordingly.

```
Out [3]: ['Date', 'Open', 'High', 'Low', 'Close', 'Volume']
```

while all remaining rows contain string-records that look like this:

```
 In [4]: quotes[1]
Out [4]: ['6/29/2016', '205.13', '211.78', '203', '210.19', '5994908']
 In [5]: quotes[-1]
Out [5]: ['6/29/2010', '19', '25', '17.54', '23.89', '18783276']
```

The rest of the code is more or less self-explanatory. The only peculiarity is the conversion of the strings into either floating point numbers or proleptic Gregorian ordinals, which Python treats as integers.[222]

```
In [6]: del quotes[0]
        dates = [datetime.strptime(record[0],"%m/%d/%Y").toordinal() for record in
quotes]
        closes = [float(record[4]) for record in quotes]

In [7]: years = YearLocator()
        months = MonthLocator()
        yearsFmt = DateFormatter('
```

The next call produces the plot shown in figure 1.1.

```
In [8]: fig, ax = plt.subplots()
        ax.plot_date(dates, closes, 'k-')
        ax.grid(True)
        plt.show()
```

Compute the returns and create the plot shown in figure 1.2.

```
 In [9]: returns=[(closes[i+1]-closes[i])/closes[i] for i in range(len(closes)-1)]
In [10]: fig, ax = plt.subplots()
         ax.plot_date(dates[1:], returns, 'k.')
         ax.grid(True)
         plt.show()
```

Finally, create the histogram shown in figure 1.3.

```
In [11]: def IQR(self):
                 return np.percentile(self,75) - np.percentile(self,25)

         bin_size = 2 * IQR(returns) * (len(returns)**(-1.0/3))
         num_bins=int((max(returns)-min(returns))/bin_size)

In [12]: u,v,_=plt.hist(returns, num_bins, normed=False, facecolor='w', alpha=1.)
```

[222] Some sources provide market data that are already formatted as floating point numbers or understandable calendar entries, not strings.

```
          show()

In [13]:  plt.bar(v[:-1],u/sum(u), width=bin_size, facecolor='w', alpha=1.)
          plt.xlabel('Daily Returns')
          plt.ylabel('Frequency')
          #plt.title(r'Histogram from stock returns')
          plt.show()
```

This is how the extreme daily net returns can be located in the sample:

```
 In [14]: returns.index(min(returns))
Out [14]: 791

 In [15]: returns[791]
Out [15]: -0.1961095100864554

 In [16]: closes[792]/closes[791]-1
Out [16]: -0.19610951008645539

 In [17]: print(date.fromordinal(dates[792]))
Out [17]: 2013-05-08

 In [18]: returns.index(max(returns))
Out [18]: 1121

 In [19]: returns[1121]
Out [19]: 0.23957876261518216

 In [20]: closes[1122]/closes[1121]-1
Out [20]: 0.23957876261518218

 In [21]: print(date.fromordinal(dates[1122]))
Out [21]: 2012-01-12
```

B.2 Simulation of Multivariate Gaussian Laws

This section contains a prototype Python code for simulating multivariate normal distribution laws. The first step again is to import the relevant libraries.

```
In [1]: from sympy import init_printing
        init_printing()
        import json
        import numpy as np
        from numpy import *
```

Typically, the covariance matrix would be very large and would be stored and read from a file:

```
In [2]: file = open('/<path-to-file-directory>/Cov.json','r')
```

```
Cov=array(json.load(file))
```

For the purpose of illustration, in this example all matrices involved are chosen to be small, and we can afford to print them.

```
 In [3]: Cov
Out [3]: array([[ 0.034969 , 0.0314721 , 0. , 0. ],
                [ 0.0314721 , 0.034969 , 0. , 0. ],
                [ 0. , 0. , 1.0630576 , -0.5315288 ],
                [ 0. , 0. , -0.5315288 , 0.559504 ]])
```

It is always useful to check whether the matrix read from the file is symmetric (as it should be if it is a properly assembled covariance matrix).

```
 In [4]: Cov-Cov.T
Out [4]: array([[ 0. , 0. , 0. , 0. ],
                [ 0. , 0. , 0. , 0. ],
                [ 0. , 0. , 0. , 0. ],
                [ 0. , 0. , 0. , 0. ]])
```

Next, we check the matrix rank and determinant:

```
 In [5]: rnk=np.linalg.matrix_rank(Cov)
         rnk
Out [5]: 4
 In [6]: np.linalg.det(Cov)
Out [6]: 7.25503184265e-05
```

Now compute the Cholesky decomposition and test the result:

```
 In [7]: CovU=np.linalg.cholesky(Cov)
         CovU
Out [7]: array([[ 0.187 , 0. , 0. , 0. ],
                [ 0.1683 , 0.08151141 , 0. , 0. ],
                [ 0. , 0. , 1.03104685 , 0. ],
                [ 0. , 0. , -0.51552342 , 0.54197749 ]])
 In [8]: dot((CovU),CovU.T)-Cov
Out [8]: array([[ 0.00000000e+00 , 0.00000000e+00 , 0.00000000e+00,
                   0.00000000e+00 ],
                [ 0.00000000e+00 ,0.00000000e+00 , 0.00000000e+00 ,
                   0.00000000e+00 ],
                [ 0.00000000e+00 , 0.00000000e+00 , 0.00000000e+00 ,
                   0.00000000e+00 ],
                [ 0.00000000e+00 , 0.00000000e+00 , 0.00000000e+00,
                   1.11022302e-16 ]])
```

The next step is to write a function that yields a standard (i.i.d. $\mathcal{N}(0,1)$) Gaussian vector of an appropriate dimension:

```
In [9]:  def gauss_vec(dimen):
          vec=[]
          for i in range(dimen):
          vec+=[np.random.normal(0,1)]
          return vec
In [10]:  gauss_vec(rnk)
Out [10]: [0.3343371625990118, -0.9731850762021675,
                              -0.5888143754041316, 0.4782579901760334 ]
```

Producing a Gaussian vector with vanishing mean and covariance matrix Cov is now straightforward:

```
In [11]:  list(dot((CovU),gauss_vec(rnk)))
Out [11]: [-0.122678949408, 0.0431797660303,
                              -0.0578866281559, -1.06095732958 ]
```

Instead of relying on the Cholesky decomposition of the covariance matrix, one can achieve the same objective by utilizing the matrix of eigenvectors and the associated list of eigenvalues:

```
In [12]:  D,U=np.linalg.eig(Cov)
          CovL=dot(U,diag(sqrt(D)))
          dot((CovL),CovL.T)-Cov
Out [12]: array([[ -6.93889390e-18 , 6.93889390e-18 , 0.00000000e+00 ,
                                                       0.00000000e+00 ],
                 [ 6.93889390e-18 , 6.93889390e-18 , 0.00000000e+00 ,
                                                       0.00000000e+00 ],
                 [ 0.00000000e+00 , 0.00000000e+00 , 2.22044605e-16 ,
                                                       -1.11022302e-16 ],
                 [ 0.00000000e+00 , 0.00000000e+00 , -1.11022302e-16 ,
                                                       1.11022302e-16 ]])
In [13]:  list(dot((CovL),gauss_vec(rnk)))
Out [13]: [0.0927305613546, 0.0905851426769,
                              -2.25790256429, 1.55366784221 ]
```

Typically, for a problem of such a small size, one would be indifferent to the choice of computing either the Cholesky decomposition or the eigenvalues and eigenvectors. However, the two algorithms are different, and for larger problems, depending on the actual input, the difference in the speed and overall efficiency could be substantial.

B.3 Numerical Program for American-Style Call Options

This section contains the concrete implementation in Python of the numerical recipe outlined at the end of section 18.4. For the first step, load the relevant libraries:

```
In [1]: from scipy.interpolate import interp1d
        import numpy as np
        from numpy import *
        from scipy import special, optimize
        from scipy.special import *
        import scipy.integrate as integrate
        import scipy.special as special
        import matplotlib
        import matplotlib.pyplot as plt
```

Next, encode the Black–Scholes–Merton formula for the European call as the function:

```
In [2]: EC=lambda S, t, K, sigma, r, delta:
        1/2*sqrt(pi)*((erf(1/4*sqrt(2)*((sigma**2 - 2*delta + 2*r)*t
        - 2*log(K/S))/(sigma*sqrt(t)))*exp(-(delta - r)*t) + exp(-(delta - r)*t))*S
        + K*erf(1/4*sqrt(2)*((sigma**2 + 2*delta - 2*r)*t
        + 2*log(K/S))/(sigma*sqrt(t))) - K)*sqrt(t)*exp(-r*t)/sqrt(pi*t)
```

Then encode the functions $F^{\pm}$ from (18.45)

```
In [3]:
        F=lambda epsilon,t,u,v,r,delta, sigma:
        erf(
        epsilon/4*sqrt(2)*sigma*sqrt(u-t)
            -1/2*sqrt(2)*delta*sqrt(u-t)/sigma
            + 1/2*sqrt(2)*r*sqrt(u-t)/sigma
        - 1/2*sqrt(2)*log(v)/(sigma*sqrt(u-t))
        )+1
```

and assign numerical values to all parameters in the model:

```
In [5]: K=40;sigma=3/10;delta=7/100;r=2/100;T=1/2
```

Next, set the abscissas for the interpolation and initiate the first set of values to the constant $\frac{r}{\delta}K \vee K$:

```
In [6]: l1=np.linspace(0, 47/100, num=47, endpoint=False)
        l2_0=np.linspace(47/100, 1/2, num=30, endpoint=False)
        l2=np.linspace(47/100, 1/2, num=31, endpoint=True)
        absc0=np.concatenate((l1,l2_0), axis=0)
```

```
absc=np.concatenate((l1,l2), axis=0)
val=[max([K,K*(r/delta)]) for x in absc]
```

Finally, implement the main procedure. The output from ten iterations is already reasonably accurate and takes about 1½ minutes to obtain with a single thread on a mainstream (medium-power) processor as of this writing[223] (about two minutes on the author's pocket-size Linux device).

```
In [7]: f = interp1d(absc, val, kind='cubic')
        ah=lambda t, z: (exp(-delta*(z-t))*(delta/2)*F(1,t,z,f(z)/f(t),r,delta,sigma))
        bh=lambda t, z: (exp(-r*(z-t))*(r*K/2)*F(-1,t,z,f(z)/f(t),r,delta,sigma))
        for iter in range(10):
          loc=[max([K,K*(r/delta)])]
          for ttt in absc0[::-1]:
            aaa=integrate.quad(lambda z: ah(ttt,z),ttt,T)[0]
            bbb=integrate.quad(lambda z: bh(ttt,z),ttt,T)[0]
            LRT=optimize.brentq(lambda x:
                        x-K-EC(x,T-ttt,K,sigma,r,delta)-aaa*x+bbb,K-10,K+20)
            loc=[LRT]+loc

          val=loc
          ff=f
          f = interp1d(absc, val, kind='cubic')
          ah=lambda t, z: (exp(-delta*(z-t))*(delta/2)*F(1,t,z,f(z)/f(t),r,delta,sigma))
          bh=lambda t, z: (exp(-r*(z-t))*(r*K/2)*F(-1,t,z,f(z)/f(t),r,delta,sigma))

        print("end")
```

Now we can generate the plot in figure 18.1:

```
In [8]: plt.plot(absc,f(absc),'k')
        plt.xlabel('time')
        plt.ylabel('spot price')
        plt.show()
```

Several mopping operations are now due. For the purpose of illustration, assuming that at time $t = 0$ the American call is exactly at the money ($S_0 = K$), we compute the price of the European call, the early exercise premium with six months left to maturity, the price of the American call, and the uniform distance between the last two iterations in the procedure.

```
In [9]: EC(K,T,K,sigma,r,delta)
```

223 This timing reflects running the code as is without using any acceleration tools or parallelization methods. The use of such techniques can vastly speed up the implementation, and, as was noted earlier, it is straightforward to implement the program on parallel cores.

```
Out [9]:  2.837826502011882

In [10]:  ah0=lambda t, z: (exp(-delta*(z-t))*(delta/2)*F(1,t,z,f(z)/K,r,delta,sigma))
          bh0=lambda t, z: (exp(-r*(z-t))*(r*K/2)*F(-1,t,z,f(z)/K,r,delta,sigma))
          aa0=integrate.quad(lambda z: ah0(0,z),0,T)[0]
          bb0=integrate.quad(lambda z: bh0(0,z),0,T)[0]
          EEP=K*aa0-bb0
          EEP
Out [10]: 0.10082307184851422

In [11]:  EC(K,T,K,sigma,r,delta)+EEP
Out [11]: 2.9386495738603964

In [12]:  from scipy.optimize import minimize_scalar
          gg = lambda z: -abs(f(z)-ff(z))
          res = minimize_scalar(gg, bounds=(0, 0.5), method='bounded')
          -res.fun
Out [12]: 1.9249133572429855e-05
```

Select Bibliography[224]

Abramowitz, Milton, and Irene A. Stegun (Eds.). 1962. *Handbook of Mathematical Functions with Formulas, Graphs, and Mathematical Tables*. New York, NY: Dover Publications.

Aït-Sahalia, Yacine, and Jean Jacod. 2014. *High-Frequency Financial Econometrics*. Princeton, NJ: Princeton University Press.

Amin, Kaushik I., and Robert A. Jarrow. 1991. Pricing foreign currency options with stochastic interest rates. *Journal of International Money and Finance* 10: 310–329.

Andersen, Torben, Oleg Bondarenko, Viktor Todorov, and George Tauchen. 2015. The fine structure of equity-index option dynamics. *Journal of Econometrics* 187: 532–546.

Andersen, Torben, Nicola Fusari, and Viktor Todorov. 2015a. The risk premia embedded in index options. *Journal of Financial Economics* 117: 558–584.

Andersen, Torben, Nicola Fusari, and Viktor Todorov. 2015b. Parametric inference and dynamic state recovery from option panels. *Econometrica* 83: 1081–1145.

Arrow, Kenneth J., and Gérard Debreu. 1954. Existence of an equilibrium for a competitive economy. *Econometrica* 22: 265–290.

Avellaneda, Marco, with Peter Laurence. 1999. *Quantitative Modeling of Derivative Securities: From Theory To Practice*. Boca Raton, FL: Chapman and Hall/CRC.

Bachelier, Louis. 1900. Théorie de la spéculation. *Annales scientifiques de l'École Normale Supérieure, Sér. 3* 17: 21–86.

Back, Kerry. 1991. Asset pricing for general processes. *Journal of Mathematical Economics* 20: 371–395.

Back, Kerry. 1992. Insider trading in continuous time. *Review of Financial Studies* 5: 387–409.

Back, Kerry. 1993. Incomplete markets and individual risks. *Economic Theory* 3: 35–42.

Banach, Stefan, and Alfred Tarski. 1924. Sur la décomposition des ensembles de points on partics respectivement congruentes. *Fundamenta Mathematicae* 6: 244–277.

[224] The bibliography included here is not meant to be a complete, authoritative bibliography on the subjects covered in the book.

Barles, Guy, Julien Burdeau, Marc Romano, and Nicolas Samsoen. 1995. Critical stock price near expiration. *Mathematical Finance* 5: 77–95.

Barndorff-Neilsen, Ole E., David G. Pollard, and Neil Shephard. 2012. Integer-valued Lévy processes and low latency financial econometrics. *Quantitative Finance* 12: 587–605.

Basak, Suleyman, and Georgy Chabakauri. 2010. Dynamic mean-variance asset allocation. *Review of Financial Studies* 23: 2970–3016.

Basak, Suleyman, and Georgy Chabakauri. 2012. Dynamic hedging in incomplete markets: a simple solution. *Review of Financial Studies* 25: 1845–1896.

Basak, Suleyman, and Dmitry Makarov. 2014. Strategic asset allocation and money management. *Journal of Finance* 69: 179–217.

Basak, Suleyman, and Anna Pavlova. 2013. Asset prices and institutional investors. *American Economic Review* 103: 1728–1758.

Basak, Suleyman, and Anna Pavlova. 2016. A model of financialization of commodities. *Journal of Finance* 71: 1511–1556.

Bellman, Richard. 1954. The Theory of Dynamic Programming. Mimeo. Rand Corporation.

Benhamou, Eric (Ed.). 2007. *Global Derivatives: Products, Theory and Practice*. Hackensack, NJ: World Scientific.

Bensoussan, Alain. 1984. On the theory of option pricing. *Acta Applicandae Mathematicae* 2: 139–158.

Berestycki, Henri, Jérôme Busca, and Igor Florent. 2000. An inverse parabolic problem arising in finance. *Comptes Rendus de l'Académie des Sciences - Series I - Mathematics* 331: 965–969.

Bergman, Yaacov, Bruce D. Grundy, and Zvi Wiener. 1996. General properties of option prices. *Journal of Finance* 51: 1573–1610.

Berk, Jonathan, and Peter DeMarzo. 2017. *Corporate Finance: The Core* (4th Ed.). Boston, MA: Pearson.

Bernoulli, Daniel. 1738. Specimen theoriae novae de mensura sortis. *Comentarii Academiae Scientiarum Imperialis Petropolitanae* 5: 175–192. [English translation: *Econometrica* 22(1954): 23–36.]

Bernstein, Sergei N. 1946. *Theory of Probability*. Moscow, RU: Gostechizdat. [In Russian].

Bertsimas, Dimitris, Leonid Kogan, and Andrew W. Lo. 2000. When is time continuous? *Journal of Financial Economics* 55: 173–204.

Biagini, Sara, and Marco Frittelli. 2005. Utility maximization in incomplete markets for unbounded processes. *Finance and Stochastics* 9: 493–517.

Biagini, Sara, and Marco Frittelli. 2007. The supermartingale property of the optimal wealth process for general semimartingales. *Finance and Stochastics* 11: 253–266.

Biagini, Sara, and Marco Frittelli. 2008. A unified framework for utility maximization problems: an Orlicz spaces approach. *Annals of Applied Probability* 18: 929–966.

Biagini, Sara, and Paolo Guasoni. 2011. Relaxed utility maximization in complete markets. *Mathematical Finance* 11: 703–722.

Bismut, Jean-Michel. 1973. Conjugate convex functions in optimal stochastic control. *Journal of Mathematical Analysis and Applications* 44: 384–404.

Bismut, Jean-Michel. 1975. Growth and optimal intertemporal allocation of risks. *Journal of Economic Theory* 10: 239–287.

Bismut, Jean-Michel. 1978. Régularité et continuité des processus. *Zeitschrift für Wahrscheinlichkeitstheorie und Verwandte Gebiete* 44: 261–268.

Black, Fischer, and Myron Scholes. 19–. A theoretical valuation formula for options, warrants, and other securities. Undated mimeo.

Black, Fischer, and Myron Scholes. 1973. The pricing of options and corporate liabilities. *Journal of Political Economy* 81: 637–654.

Blackwell, David, and Persi Diaconis. 1996. A non-measurable tail set. In: *Probability and Game Theory*. IMS Lecture Notes-Monograph Series 30. Hayward, CA: Institute of Mathematical Statistics, pp. 1–5.

Bollerselv, Tim, Viktor Todorov, and Lai Xu. 2015. Tail risk premia and return predictability. *Journal of Financial Economics* 118: 113–134.

Bollerslev, Tim, and Viktor Todorov. 2011a. Estimation of jump tails. *Econometrica* 79: 1727–1783.

Bollerslev, Tim, and Viktor Todorov. 2011b. Tails, fears, and risk premia. *Journal of Finance* 66: 2165–2211.

Bouchard, Bruno, and Nizar Touzi. 2000. Explicit solution of the multivariate super-replication problem under transaction costs. *Annals of Applied Probability* 10: 685–708.

Boyle, Phelim, Mark Broadie, and Paul Glasserman. 1997. Monte Carlo methods for security pricing. *Journal of Economic Dynamics and Control* 21: 1267–1321.

Brown, Robert. 1828. *Brief Account of Microscopical Observations Made in the Months of June, July, and August 1827, on the Particles Contained in the Pollen of Plants; and on the General Existence of Active Molecules in Organic and Inorganic bodies*. London, UK: self published with the press of Richard Taylor.

Buraschi, Andrea, and Bernard Dumas. 2001. The forward valuation of compound options. *Journal of Derivatives* 9: 8–17.

Buraschi, Andrea, Fabio Trojani, and Andrea Vedolin. 2014. When uncertainty blows in the orchid: comovement and equilibrium volatility risk premia. *Journal of Finance* 69: 101–137.

Cadenillas, Abel, and Ioannis Karatzas. 1995. The stochastic maximum principle for linear, convex optimal control problems with random coefficients. *SIAM Journal of Control and Optimization* 33: 590–624.

Cameron, Robert H., and William T. Martin. 1944. Transformation of Wiener integrals under translation. *Annals of Mathematics* 45: 386–396.

Cameron, Robert H., and William T. Martin. 1945. Transformation of Wiener integrals under a general class of linear transformations. *Transactions of the American Mathematical Society* 18: 184–219.

Campbell, John Y., Andrew W. Lo, and A. Craig MacKinlay. 1997. *The Econometrics of Financial Markets*. Princeton, NJ: Princeton University Press.

Caratheodory, Constantin. 1918. *Vorlesungen über Reelle Funktionen*. Leipzig: Teubner. [2nd ed. 1927.]

Carmona, René, and Michael Tehranchi. 2007. *Interest Rate Models: an Infinite Dimensional Stochastic Analysis Perspective*. Berlin, DE: Springer.

Carr, Peter. 1993. The valuation of American exchange options with application to real options. Mimeo. Cornell University.

Carr, Peter, and Jonathan Bowie. 1994. Static simplicity. *Risk Magazine* (August): 44–50.

Carr, Peter, and Marc Chesney. 2000. American put-call symmetry. Mimeo. HEC School.

Carr, Peter, Robert Jarrow, and Ravi Myneni. 1992. Alternative characterization of American put options. *Mathematical Finance* 2: 63–150.

Carr, Peter, and Roger Lee. 2009. Put–call symmetry: extensions and applications. *Mathematical Finance* 19: 523–560.

Cartea, Álvaro, Sebastian Jaimungal, and José Penalva. 2015. *Algorithmic and High-Frequency Trading*. Cambridge, UK: Cambridge University Press.

Černý, Aleš. 2009. *Mathematical Techniques in Finance: Tools for Incomplete Markets*. Princeton, NJ: Princeton University Press.

Çetin, Umut, H. Mete Soner, and Nizar Touzi. 2010. Option hedging under liquidity costs. *Finance and Stochastics* 14: 317–341.

Chaleyat-Maurel, Mireille, Nicole El Karoui, and Bernard Marchal. 1980. Reflexion discontinue et systemes stochastiques. *Annals of Probability* 8: 1049–1067.

Chen, Nan, and Paul Glasserman. 2007a. Additive and multiplicative duals for American option pricing. *Finance and Stochastics* 11: 153–179.

Chen, Nan, and Paul Glasserman. 2007b. Malliavin greeks without Malliavin calculus. *Stochastic Processes and Their Applications* 117: 1689–1723.

Chen, Zhiyoung, and Paul Glasserman. 2008. Sensitivity estimates for portfolio credit derivatives using Monte Carlo. *Finance and Stochastics* 12: 507–540.

Chung, Kai Lai, and Ruth J. Williams. 1990. *Introduction to Stochastic Integration*. Boston, MA: Birkhäuser.

Çinlar, Erhan. 2011. *Probability and Stochastics*. New York, NY: Springer.

Clark, John Martin C. 1970. The representation of functionals of Brownian motion by stochastic integrals. *Annals of Mathematical Statistics* 41: 1282–1295.

Cochrane, John H. 2009. *Asset Pricing*. Princeton, NJ: Princeton University Press.

Cont, Rama, and David Fournie. 2010. A functional extension of the Ito formula. *Comptes Rendus de l'Académie des Sciences - Series I - Mathematics* 348: 57–61.

Cont, Rama, and David-Antoine Fournié. 2013. Functional Itô calculus and stochastic integral representation of martingales. *Annals of Probability* 41: 109–133.

Cont, Rama, and Peter Tankov. 2009. *Financial Modeling With Jump Processes*. Boca Raton, FL: Chapman and Hall/CRC.

Copeland, Tom, and Vladimir Antikarov. 2001. *Real Options: A Practitioner's Guide*. New York, NY: Texere.

Cox, John C., and Chi-Fu Huang. 1989. Optimum consumption and portfolio policies when asset prices follow a diffusion process. *Journal of Economic Theory* 49: 33–83.

Cox, John C., and Chi-Fu Huang. 1991. A variational problem arrising in financial economics. *Journal of Mathematical Economics* 20: 465–487.

Cox, John C., Jonathan E. Ingersoll Jr., and Stephen A. Ross. 1985. A theory of term structure of interest rates. *Econometrica* 53: 385–408.

Cox, John C., Stephen A. Ross, and Mark Rubinstein. 1979. Option pricing: a simplified approach. *Journal of Financial Economics* 7: 229–263.

Crandall, Michael G., Hitoshi Ishii, and Pierre-Lois Lions. 1992. User's guide to viscosity solutions of second order partial differential equations. *Bulletin (New Series) of the American Mathematical Society* 27: 1–67.

Crépey, Stéphane. 2003. Calibration of the local volatility in a generalized Black-Scholes model using Tikhonov regularization. *SIAM Journal on Mathematical Analysis* 34(5): 1183–1206.

Crépey, Stéphane. 2012. *Financial Modeling: A Backward Stochastic Differential Equations Perspective*. Berlin, DE: Springer.

Cvitanić, Jakşa, and Ioannis Karatzas. 1992. Convex duality in constrained portfolio optimization. *Annals of Applied Probability* 2: 767–818.

Cvitanić, Jakşa, and Ioannis Karatzas. 1993. Hedging contingent claims with constrained portfolios. *Annals of Applied Probability* 3: 652–681.

Cvitanić, Jakşa, Jin Ma, and Jianfeng Zhang. 2012. Law of large numbers for self-exciting correlated defaults. *Stochastic Processes and Their Applications* 122: 2781–2810.

Czichowsky, Christoph, and Martin Schweizer. 2012. Convex duality in mean-variance hedging under convex trading constraints. *Advances in Applied Probability* 44: 1084–1112.

Dai Pra, Paolo, Wolfgang J. Runggaldier, Elena Sartori, and Marco Tolotti. 2009. Large portfolio losses; a dynamic contagion model. *Annals of Applied Probability* 19: 347–394.

Dalang, Robert, Andrew Morton, and Walter Willinger. 1990. Equivalent martingale measures and no arbitrage in stochastic securities market models. *Stochastics and Stochastic Reports* 29: 185–201.

Dana, Rose-Anne, and Monique Jeanblanc. 2003. *Financial Markets in Continuous Time*. Berlin, DE: Springer.

Davis, Mark H. A. 1976. The representation of martingales of jump processes. *Siam Journal of Control and Optimization* 14: 623–638.

Davis, Mark H. A., and Andrew R. Norman. 1990. Portfolio selection with transaction costs. *Mathematics of Operations Research* 15: 676–713.

Davydov, Dmitry, and Vadim Linetsky. 2001. Pricing and hedging path-dependent options under the CEV process. *Management Science* 47: 949–965.

DeMarzo, Peter, Michael Fishman, Zhiguo He, and Neng Wang. 2012. Dynamic agency and the q-theory of investment. *Journal of Finance* 67: 2295–2340.

Deelstra, Griselda, Huyên Pham, and Nizar Touzi. 2001. Dual formulation of the utility maximization problem under transaction costs. *Annals of Applied Probability* 11: 1353–1383.

Del Moral, Pierre. 2004. *Feynman-Kac Formulae: Genealogical and Interacting Particle Systems with Applications*. New York, NY: Springer.

Delbaen, Freddy, and Walter Schachermayer. 1994. General version of the fundamental theorem of asset pricing. *Mathematische Annalen* 300: 463–520.

Delbaen, Freddy, and Walter Schachermayer. 1995. The existence of absolutely continuous local martingale measures. *The Annals of Applied Probability* 5: 926–945.

Delbaen, Freddy, and Walter Schachermayer. 1998. The fundamental theorem of asset pricing for unbounded stochastic processes. *Mathematische Annalen* 312: 215–250.

Delbaen, Freddy, and Walter Schachermayer. 2010. *The Mathematics of Arbitrage*. Berlin, DE: Springer.

Dellacherie, Claude. 1972. *Capacités et Processus Stochastique*. Berlin, DE: Springer.

Dellacherie, Claude, et Paul-André Meyer. 1975. *Probabilités et Potentiel, Chapitres I à IV: Espaces Measurables*. Paris, FR: Hermann.

Dellacherie, Claude, et Paul-André Meyer. 1980. *Probabilités et Potentiel, Chapitres V à VIII: Théorie des Martingales*. Paris, FR: Hermann.

Derman, Emanuel, and Iraj Kani. 1994. Riding on a smile. *Risk Magazine* (February): 32–39.

Di Masi, Giovanni B., Yuri M. Kabanov, and Wolfgang J. Runggaldier. 1994. Mean-variance hedging of options on stocks with Markov volatilities. *Theory of Probability and Its Applications* 39: 211–222.

Diaconis, Persi, and Ron Graham. 2012. *Magical Mathematics: The Mathematical Ideas That Animate Great Magical Tricks*. Princeton, NJ: Princeton University Press.

Diop, Assane, Jean Jacod, and Viktor Todorov. 2013. Central limit theorems for approximate quadratic variations of pure jump into semimartingales. *Stochastic Processes and Their Applications* 123: 839–886.

Dixit, Avinash K. 1990. *Optimization in Economic Theory*. Oxford, UK: Oxford University Press.

Dixit, Avinash K., and Robert S. Pindyck. 1994. *Investment Under Uncertainty*. New York, NY: McGraw-Hill.

Doléan-Dade, Catherine. 1970. Quelques applications de la formule de changement de variables pour les semimartingales. *Zeitschrift für Wahrscheinlichkeitstheorie und Verwandte Gebiete* 16: 181–194.

Doléan-Dade, Catherine. 1976. On the existence and unicity of solutions of stochastic integral equations. *Zeitschrift für Wahrscheinlichkeitstheorie und Verwandte Gebiete* 36: 93–101.

Doléan-Dade, Catherine, and Paul A. Meyer. 1976. Equations différentielles stochastique. In: *Séminaire de Probabilités XI*. Lecture Notes in Mathematics 581. Berlin, DE: Springer, pp. 376–382.

Donsker, Monroe D. 1951. *An Invariance Principle for Certain Probability Limit Theorems*. Providence, RI: Memoirs of the American Mathematical Society 1951(6).

Donsker, Monroe D. 1952. Justification and extension of Doob's heuristic approach to the Kolmogorov-Smirnov theorems. *Annals of Mathematical Statistics* 23: 277–281.

Doob, Joseph L. 1949. Heuristic approach to the Kolmogorov-Smirnov theorems. *Annals of Mathematical Statistics* 20: 393–403.

Doob, Joseph L. 1953. *Stochastic Processes*. New York, NY: John Wiley and Sons.

Doob, Joseph L. 1984. *Classical Potential Theory and its Probabilistic Counterpart*. New York, NY: Springer.

Dudley, Richard M. 2002. *Real Analysis and Probability*. Cambridge, UK: Cambridge University Press.

Dudley, Richard M. 2014. *Uniform Central Limit Theorems*. New York, NY: Cambridge University Press.

Duffie, Darrell. 1987. Stochastic equilibria with incomplete financial markets. *Journal of Economic Theory* 41: 405–416.

Duffie, Darrell. 1988. *Security Markets: Stochastic Models*. Boston, MA: Academic Press.

Duffie, Darrell. 1989. *Futures Markets*. Englewood Cliffs, NJ: Prentice-Hall.

Duffie, Darrell. 2010. *Dynamic Asset Pricing Theory*. Princeton, NJ: Princeton University Press.

Duffie, Darrell. 2011. *Measuring Corporate Default Risk*. Oxford, UK: Oxford University Press.

Duffie, Darrell, Damir Filipović, and Walter Schachermayer. 2003. Affine processes and applications to finance. *Annals of Applied Probability* 13: 984–1053.

Duffie, Darrell, Nicolae Gârleanu, and Lasse Heje Pedersen. 2005. Over-the-counter markets. *Econometrica* 73: 1815–1847.

Duffie, Darrell, Nicolae Gârleanu, and Lasse Heje Pedersen. 2007. Valuation in over-the-counter markets. *Review of Financial Studies* 20: 1865–1900.

Duffie, Darrell, and Chi-Fu Huang. 1985. Implementing Arrow-Debreu equilibria by continuous trading of few long-lived securities. *Econometrica* 53: 1337–1356.

Duffie, Darrell, and Kenneth J. Singleton. 2012. *Credit Risk: Pricing, Measurement, and Management*. Princeton, NJ: Princeton University Press.

Duffie, Darrell, and William Zame. 1989. The consumption-based capital asset pricing model. *Econometrica* 57: 1279–1297.

Duffie, Darrell, and Tong-Sheng Sun. 1990. Transaction costs and portfolio choice in a discrete-continuous time setting. *Journal of Economic Dynamics and Control* 14: 35–51.

Dumas, Bernard. 1991. Super contact and related optimality conditions. *Journal of Economic Dynamics and Control* 15: 675–685.

Dumas, Bernard, Jeff Fleming, and Robert Whaley. 1998. Implied volatility functions: empirical tests. *Journal of Finance* 53: 2059–2106.

Dumas, Bernard, Campbell R. Harvey, and Pierre Ruiz. 2003. Are correlations of stock returns justified by subsequent changes in national outputs? *Journal of International Money and Finance* 22: 777–811.

Dumas, Bernard, and Andrew Lyasoff. 2012. Incomplete-market equilibria solved recursively on an event tree. *Journal of Finance* 67: 1881–1931.

Dumas, Bernard, Raman Uppal, and Tan Wang. 2000. Efficient intertemporal allocation with recursive utility. *Journal of Economic Theory* 93: 240–259.

Dupire, Bruno. 1994. Pricing with a smile. *Risk Magazine* (January): 17–20.

Dybvig, Philip H., L. C. G. Rogers, and Kerry Back. 1999. Portfolio turnpikes. *Review of Financial Studies* 12: 165–195.

Dybvig, Philip H., and Stephen A. Ross. 1987. Arbitrage. *The New Palgrave: A Dictionary of Economics* 1: 100–106.

Dynkin, Eugene B. 1959. *Osnovaniya Teorii Markovskikh Processov* [in Russian]. Moscow, RU: Fizmatgiz. [English translation: "Theory of Markov Processes" by Pergamon Press (1961).]

Einstein, Albert. 1905. Über die von der molekularkinetischen Theorie der Wärme geforderte Bewegung von in ruhenden Flüssigkeiten suspendierten Teilchen. *Annalen der Physik* 17: 549–560.

Eisenberg, Larry and Robert Jarrow. 1994. Option pricing with random volatilities in complete markets. *Review of Quantitative Finance and Accounting* 4: 5–17.

El Karoui, Nicole. 1981. Les Aspects Probabilistes du Contrôle Stochastique. In: *Lecture Notes in Mathematics 876*. Berlin, DE: Springer, pp. 73–238.

Enchev, Ognian [Lyasoff, Andrew]. 1988. Hilbert-space-valued quasimartingales. *Bolletino di Unione Matematica Italiana* 7(2-B): 19–39.

Enchev, Ognian [Lyasoff, Andrew]. 1993a. Pathwise nonlinear filtering on abstract Wiener spaces. *Annals of Probability* 21: 1728–1754.

Enchev, Ognian [Lyasoff, Andrew]. 1993b. Nonlinear transformations on the Wiener space. *Annals of Probability* 21: 2169–2188.

Enchev, Ognian [Lyasoff, Andrew]. 1998. White noise indexed by loops. *Annals of Probability* 26: 985–999.

Enchev, Ognian [Lyasoff, Andrew], and Daniel W. Stroock. 1993a. Rademacher's theorem for Wiener functionals. *Annals of Probability* 21: 25–33.

Enchev, Ognian [Lyasoff, Andrew], and Daniel W. Stroock. 1993b. Anticipative diffusion and related change of measures. *Journal of Functional Analysis* 116: 449–477.

Enchev, Ognian [Lyasoff, Andrew], and Daniel W. Stroock. 1995a. Integration by parts for pinned Brownian motion. *Mathematics Research Letters* 2: 161–169.

Enchev, Ognian [Lyasoff, Andrew], and Daniel W. Stroock. 1995b. Towards a Riemannian geometry on the path space over a Riemannian manifold. *Journal of Functional Analysis* 134: 392–496.

Enchev, Ognian [Lyasoff, Andrew], and Daniel W. Stroock. 1996. Pinned Brownian motion and its perturbations. *Advances in Mathematics* 119: 127–154.

Fabozzi, Frank J., Franco Modigliani, and Frank J. Jones. 2010. *Foundations of Financial Markets and Institutions*. New York, NY: Prentice Hall.

Fan, Ky. 1944. Entfernung zweier zufälligen Grössen und die Konvergenz nach Wahrscheinlichkeit. *Mathematische Zeitschrift* 49: 681–683.

Faris, William G. (Ed.). 2006. *Diffusion, Quantum Theory, and Radically Elementary Mathematics*. Princeton, NJ: Princeton University Press.

Feller, William. 1951. Two singular diffusion problems. *Annals of Mathematics* 54: 173–182.

Feller, William. 1971. *An Introduction to Probability Theory and Its Applications. Vol. II*. New York, NY: John Wiley and Sons.

Fernholz, Erhard Robert. 2002. *Stochastic Portfolio Theory*. New York, NY: Springer.

Filipović, Damir. 2009. *Term-Structure Models: A Graduate Course*. Berlin, DE: Springer.

Fleming, Wendell H., and Raymond W. Rishel. 1975. *Deterministic and Stochastic Optimal Control*. Berlin, DE: Springer.

Fleming, Wendell H., and H. Mete Soner. 2006. *Controlled Markov Processes and Viscosity Solutions*. New York, NY: Springer.

Föllmer, Hans, and Yuri Kabanov. 1998. Optional decomposition and Lagrange multipliers. *Finance and Stochastics* 2: 69–81.

Föllmer, Hans, and Alexander Schied. 2011. *Stochastic Finance: An Introduction in Discrete Time. 3d ed.* Berlin, DE: Walter de Gruyter.

Föllmer, Hans, and Martin Schweizer. 1993. A microeconomic approach to diffusion models for stock prices. *Mathematical Finance* 3: 1–23 [Erratum (1994), Mathematical Finance 4, 285].

Forsyth, Peter. 2012. Numerical methods for non-linear PDEs in finance. In: *Handbook of Computational Finance*. Berlin, DE: Springer, pp. 503-528.

Fouque, Jean-Pierre, George C. Papanicolaou, and K. Ronnie Sircar. 2000. *Derivatives in Financial Markets with Stochastic Volatility*. Cambridge, UK: Cambridge University Press.

Freedman, David, and Persi Diaconis. 1981. On the histogram as a density estimator: L_2 theory. *Zeitschrift für Wahrscheinlichkeitstheorie und Verwandte Gebiete* 57: 453–476.

Gârleanu, Nicolae. 2009. Pricing and portfolio choice in illiquid markets. *Journal of Economic Theory* 144: 532–564.

Gârleanu, Nicolae, Leonid Kogan, and Stavros Panageas. 2012. Displacement risk and asset returns. *Journal of Financial Economics* 105: 491–510.

Gârleanu, Nicolae, and Lasse Heje Pedersen. 2013. Dynamic trading with predictable returns and transaction costs. *Journal of Finance* 68: 2309–2340.

Girsanov, Igor V. 1960. On transforming a certain class of stochastic processes by absolutely continuous substitution of measures. *Theory of Probability and Its Applications* 5: 285–301.

Girsanov, Igor V. 1962. An example of non-uniqueness of the solution to the stochastic differential equation of K. Itô. *Theory of Probability and Its Applications* 7: 325–331.

Glasserman, Paul. 2004. *Monte Carlo Methods in Financial Engineering*. New York, NY: Springer.

Glasserman, Paul. 2012. Risk horizon and rebalancing horizon in portfolio risk measurement. *Mathematical Finance* 22: 215–249.

Glasserman, Paul, and Kyoung-Kuk Kim. 2010. Moment explosions and stationary distributions in affine diffusion models. *Mathematical Finance* 20: 1–33.

Glasserman, Paul, and Kyoung-Kuk Kim. 2011. Gamma expansion of the Heston stochastic volatility model. *Finance and Stochastics* 15: 267–296.

Glasserman, Paul, and Zongjian Liu. 2010. Sensitivity estimates from characteristic functions. *Operations Research* 58: 1611–1623.

Glasserman, Paul, and Zongjian Liu. 2010/11. Estimating greeks in simulating Levy-driven models. *Journal of Computational Finance* 14(2): 3–56.

Glasserman, Paul, and Bin Yu. 2004. Number of paths versus number of basis functions in American option pricing. *Annals of Applied Probability* 14: 2090–2119.

Gollier, Christian. 2001. *The Economics of Risk and Time*. Cambridge, MA: MIT Press.

Gram, Jørgen P. 1910. Professor Thiele som aktuar. In: Dansk Forsikringsårbog, pp. 26–37.

Grünbaum, Alberto F., Pierre van Moerbeke, and Victor H. Mall (Eds.). 2015. *Henry P. McKean Jr. Selecta*. Basel, CH: Birkhäuser.

Hackbarth, Dirk, and Timothy Johnson. 2015. Real options and risk dynamics. *Review of Economic Studies* 82: 1449–1482.

Hagan, Patrick S., Deep Kumar, Andrew S. Lesniewski, and Diana E. Woodward. 2003. Managing smile risk. *Wilmott Magazine* (July 2002): 84–108.

Halmos, Paul R. 1950. *Measure Theory*. Princeton, NJ: Van Nostrand. [reprinted: Springer, 1974]

Hansen, Peter R., and Asger Lunde. 2006. Realized variance and market microstructure noise. *Journal of Business and Economic Statistics* 24: 127–161.

Harrison, J. Michael, and David M. Kreps. 1979. Martingales and arbitrage in multiperiod securities. *Journal of Economic Theory* 20: 381–408.

Harrison, J. Michael, and Stanley R. Pliska. 1981. Martingales and stochastic integrals in the theory of continuous trading. *Stochastic Processes and Their Applications* 11: 215–260.

Harrison, J. Michael, and Stanley R. Pliska. 1983. A stochastic calculus model of continuous trading: complete markets. *Stochastic Processes and Their Applications* 15: 313–316.

Heston, Steven L. 1993. A closed-form solution to options with stochastic volatility with applications to bond and currency options. *Review of Financial Studies* 6: 327–343.

Holroyd, Alexander E., and Terry Soo. 2009. A nonmeasurable set from coin flips. *The American Mathematical Monthly* 116: 926–928.

Huang, Chi-fu, and Robert H. Litzenberger. 1988. *Foundations for Financial Economics*. Upper Saddle River, NJ: Prentice Hall.

Huang, Jing-zhi, Marti G. Subrahmanyam, and G. George Yu. 1996. Pricing and hedging American options: a recursive integration method. *Review of Financial Studies* 9: 277–300.

Hull, John C., and Alan White. 1987. The pricing of options on assets with stochastic volatility. *Journal of Finance* 42: 281–300.

Hunt, Gilbert A. 1956. Some theorems concerning Brownian motion. *Transactions of the American Mathematical Society* 81: 294–319.

Ikeda, Nobuyuki, and Shinzo Watanabe. 1989. *Stochastic Differential Equations and Diffusion Processes*. Amsterdam: North-Holland, and Tokyo: Kodansha.

Ingersoll, Jonathan E. 1987. *Theory of Financial Decision Making*. Lanham, MD: Rowman and Littlefield.

Ingersoll, Jonathan E. 1989. Contingent foreign exchange contracts with stochastic interest rates. Working paper.

Ionescu Tulcea, Cassius. 1949-1950. Mesures dans les espaces produits. *Atti della Accademia Nazionale dei Lincei. Classe di Scienze Fisiche, Matematiche e Naturali. Rendiconti Lincei. Serie IX. Matematica e Applicazioni.* 7: 208–211.

Itô, Kiyoshi. 1944. Stochastic integral. *Proceedings of the Imperial Academy of Japan* 20: 519–524.

Itô, Kiyosi, and Henry P. McKean, Jr. 1974. *Diffusion Processes and Their Sample Paths*. Berlin, DE: Springer.

Jacka, Saul D. 1991. Optimal stopping and the American put. *Mathematical Finance* 1(2): 1–14.

Jacka, Saul D., and James R. Lynn. 1992. Finite horizon optimal stopping, obstacle problems and the shape of the continuation region. *Stochastics and Stochastic Reports* 39: 25–42.

Jacod, Jean. 1979. *Calcul Stochastique et Problèmes de Martingales*. Berlin, DE: Springer.

Jacod, Jean, and Jean Memin. 1976. Caractérsitiques locales et conditions de continuité absolue pour les semimartingales. *Zeitschrift für Wahrscheinlichkeitstheorie und Verwandte Gebiete* 35: 1–37.

Jacod, Jean, and Albert N. Shiryaev. 1987. *Limit Theorems for Stochastic Processes*. Berlin, DE: Springer.

Jacod, Jean, and Viktor Todorov. 2009. Testing for common arrival of jumps in discretely-observed multidimensional processes. *Annals of Statistics* 37: 1792–1838.

Jacod, Jean, and Viktor Todorov. 2010. Do price and volatility jump together? *Annals of Applied Probability* 20: 1425–1469.

Jacod, Jean, and Viktor Todorov. 2014. Efficient estimation of integrated volatility in presence of infinite variation jumps. *Annals of Statistics* 42: 1029–1069.

Jaillet, Patrick, Damien Lamberton, and Bernard Lapeyre. 1990. Variational inequalities and the pricing of American options. *Acta Applicandae Mathematicae* 21: 263–289.

Jarrow, Robert A. 1996. *Modelling Fixed Income Securities and Interest Rate Options*. New York, NY: McGraw-Hill.

Jarrow, Robert, and Dilip Madan. 1995. Option pricing using the term structure of interest rates to hedge systematic discontinuities in asset returns. *Mathematical Finance* 5(3): 311–336.

Jarrow, Robert, and Dilip Madan. 1999. Hedging contingent claims on semimartingales. *Finance and Stochastics* 3(1): 111–134.

Jeanblanc, Monique, Marc Yor, and Marc Chesney. 2009. *Mathematical Methods for Financial Markets*. London, UK: Springer.

Kabanov, Yuri M., and Dmitry O. Kramkov. 1994. No-arbitrage and equivalent martingale measures: an elementary proof of the Harrison-Pliska theorem. *Theory of Probability and Its Applications* 39: 523–527.

Kabanov, Yuri M., and Mher Safarian. 2009. *Markets With Transaction Costs: Mathematical Theory*. Berlin, DE: Springer.

Kabanov, Yuri M., and Christophe Stricker. 2001. A teacher's note on no arbitrage criteria. In: *Séminaire de Probabilités XXXV*. Lecture Notes in Mathematics 1755. Berlin, DE: Springer, pp. 149-152.

Kálmán, Rudolf E., and Richard S. Bucy. 1961. New results in linear filtering and prediction theory. *Journal of Basic Engineering* 83: 95–108.

Karatzas, Ioannis. 1996. *Lectures on the Mathematics of Finance*. CRM Monograph Series 8, American Mathematical Society.

Karatzas, Ioannis, and Constantinos Kardaras. 2007. The numéraire portfolio in semimartingale financial models. *Finance and Stochastics* 11: 447–493.

Karatzas, Ioannis, Peter Lakner, John P. Lehoczky, and Steven E. Shreve. 1991. Equilibrium in a simplified dynamic, stochastic economy with heterogeneous agents. In: *Stochastic Analysis: Liber Amicorum for Moshe Zakai*. Boston, MA: Academic Press, pp. 245–272.

Karatzas, Ioannis, John P. Lehoczky, and Steven E. Shreve. 1987. Optimal portfolio and consumption decisions for a "small investor" on a finite horizon. *SIAM Journal of Control and Optimization* 25: 1557–1586.

Karatzas, Ioannis, John P. Lehoczky, and Steven E. Shreve. 1990a. Existence and uniqueness of multi-agent equilibrium in a stochastic, dynamic consumption/investment model. *Mathematics of Operations Research* 15: 80–128.

Karatzas, Ioannis, John P. Lehoczky, and Steven E. Shreve. 1990b. Equivalent martingale measures and optimal market completion. Mimeo. Department of Mathematical Sciences, Carnegie Mellon University.

Karatzas, Ioannis, John P. Lehoczky, and Steven E. Shreve. 1991. Equilibrium models with singular asset prices. *Mathematical Finance* 1: 11–29.

Karatzas, Ioannis, Daniel L. Ocone, and Jinlu Li. 1991. An extension of Clark's formula. *Stochastics and Stochastic Reports* 37: 127–131.

Karatzas, Ioannis, and Igor Pikovsky. 2007. Anticipative portfolio optimization. *Advances of Applied Probability* 28: 1095–1122.

Karatzas, Ioannis, and Steven E. Shreve. 1991. *Brownian Motion and Stochastic Calculus*. New York, NY: Springer.

Karatzas, Ioannis, and Steven E. Shreve. 1998. *Methods of Mathematical Finance*. New York, NY: Springer.

Kholodnyi, Valery A., and John F. Price. 1998. *Foreign Exchange Option Symmetry*. Singapore: World Scientific.

Kijima, Masaaki. 2013. *Stochastic Processes with Applications to Finance*. Boca Raton, FL: Chapman and Hall/CRC.

Kim, In Joon. 1990. The analytic valuation of American options. *Review of Financial Studies* 3: 547–572.

Knuth, Donald E. 1984. *The TEXbook*. Upper Saddle River, NJ: Addison-Wesley.

Kogan, Leonid. 2001. An equilibrium model of irreversible investment. *Journal of Financial Economics* 62: 201–245.

Kogan, Leonid. 2004. Asset prices and real investment. *Journal of Financial Economics* 73: 411–432.

Kogan, Leonid, and Dimitris Papanikolaou. 2014. Growth opportunities, technology stocks, and asset prices. *Journal of Finance* 69: 675–718.

Kolmogoroff, Andrei N. [Kolmogorov, Andrei N.] 1933. *Grundbegriffe der Wahrscheinlichkeitsrechnung*. Berlin, DE: Springer. [English translation: Foundations of the Theory of Probability, New York, NY: Chelsea (1956).]

Kramkov, Dmitry O. 1996. Optional decomposition of supermartingales and hedging contingent claims in incomplete security markets. *Probability Theory and Related Fields* 105: 459–479.

Kreps, David M. 2013. *Microeconomic Foundations I: Choice and Competitive Markets*. Princeton, NJ: Princeton University Press.

Kunita, Hiroshi. 1990. *Stochastic Flows and Stochastic Differential Equations*. Cambridge, UK: Cambridge University Press.

Kutoyants, Yury A. 2004. *Statistical Inference for Ergodic Diffusion Processes*. London, UK: Springer.

Lamberton, Damien. 2009. Optimal Stopping and American Options. Lecture Notes, Ljubljana Summer School on Financial Mathematics.

Lamberton, Damien, and Bernard Lapeyre. 2007. *Introduction to Stochastic Calculus Applied to Finance*. 2nd ed. Boca Raton, FL: Chapman and Hall/CRC.

Lauritzen, Steffen L. 1981. Time series analysis in 1880: A discussion of contributions made by T. N. Thiele. *International Statistical Review* 49: 319–331.

Leadbetter, Ross, Stamatis Cambanis, and Vladas Pipiras. 2014. *A Basic Course in Measure and Probability: Theory for Applications*. Cambridge, UK: Cambridge University Press.

Lebesgue, Henri. 1905. Sur les fonctiones représentables analytiquement. *Journal de Mathématiques* (Ser. 6) 1: 244–277.

Leblanc, Boris. 1997. *Modélisations de la volatilité d'un actif financier et applications*. Thèse, Paris VII.

Leland, Hayne E. 1985. Option pricing and replication with transaction costs. *Journal of Finance* 40: 1283–1301.

LeRoy, Stephen F. 1973. Risk aversion and martingale property of asset prices. *International Economic Review* 14: 436–446.

LeRoy, Stephen F., and Jan Werner. 2001. *Principles of Financial Economics*. Cambridge, UK: Cambridge University Press.

Lewis, Alan L. 2001. *Option Valuation under Stochastic Volatility*. Newport Beach, CA: Finance Press.

Li, Jia, Viktor Todorov, and George Tauchen. 2014. Volatility occupation times. *Annals of Statistics* 41: 1865–1891.

Linetsky, Vadim. 2004. Computing hitting time densities for CIR and OU diffusions: applications to mean reverting models. *Journal of Computational Finance* 7: 1–22.

Lintner, John. 1965. The valuation of risky assets and the selection of risky investment in stock portfolios and capital budgets. *Review of Economics and Statistics* 47: 13–37.

Liptser, Robert S., and Albert N. Shiryaev. 1974. *Statistika Slučajnih Procesov* [in Russian]. Moscow, RU: Nauka. [English translation: "Statistics of Stochastic Processes" by Springer (2001).]

Liu, Hong. 2004. Optimal Consumption and Investment with Transaction Costs and Multiple Risky Assets. *Journal of Finance* 59: 289–168.

Liu, Hong, and Mark Loewenstein. 2002. Optimal portfolio selection with transaction costs and finite horizons. *Review of Financial Studies* 15: 805–835.

Ljungqvist, Lars, and Thomas J. Sargent. 2000. *Recursive Macroeconomic Theory*. Cambridge, MA: MIT Press.

Lo, Andrew W., Harry Mamaysky, and Jiang Wang. 2004. Asset prices and trading volume under fixed transaction costs. *Journal of Political Economy* 112: 1054–1090.

Lucas, Robert E. 1978. Asset prices in an exchange economy. *Econometrica* 46: 1429–1445.

Lucretius, Titus. 50 BC. *De Rerum Natura* [in Latin]. English Translation: http://classics.mit.edu/Carus/nature_things.html.

Lyasoff, Andrew. 2014. The two fundamental theorems of asset pricing for a class of continuous-time financial markets. *Mathematical Finance* 24: 485–524.

Lyasoff, Andrew. 2016. Another look at the integral of exponential Brownian motion and the pricing of Asian options. *Finance and Stochastics* 20: 1061–1096.

Ma, Jin, and Jiongmin Yong. 2007. *Forward-Backward Stochastic Differential Equations and their Applications*. Berlin, DE: Springer-Verlag.

Madan, Dilip, and Eugene Seneta. 1990. The variance gamma (v.g.) model for share market returns. *Journal of Business* 63(4): 511–524.

Magil, Michael J. P., and George M. Constantinides. 1976. Portfolio selection with transaction costs. *Journal of Economic Theory* 13: 245–263.

Magill, Michael, and Martine Quinzii. 1996. *Theory of Incomplete Markets*. Cambridge, MA: MIT Press.

Malliavin, Paul, and Anton Thalmaier. 2006. *Stochastic Calculus of Variations in Mathematical Finance*. Berlin, DE: Springer.

Margrabe, William. 1978. The value of an option to exchange one asset for another. *Journal of Finance* 33: 177–186.

Markowitz, Harry. 1952. Portfolio selection. *Journal of Finance* 7: 77–91.

Markowitz, Harry. 1959. *Portfolio Selection: Efficient Diversification of Investments*. New York, NY: John Wiley and Sons.

Mas-Colell, Andreu, Michael D. Whinston, and Jerry R. Green. 1995. *Microeconomic Theory*. New York, NY: Oxford University Press.

McDonald, Robert L., and Mark D. Schroder. 1998. A parity result for American options. *Journal of Computational Finance* 1: 5–13.

McDonald, Robert, and Daniel Siegel. 1986. The value of waiting to invest. *The Quarterly Journal of Economics* 101: 707–728.

McKean, Henry P., Jr. 1965. Appendix: a free boundary problem for the heat equation arising from a problem in mathematical economics. *Industrial Management Review* 6: 32–39.

McKean, Henry P., Jr. 1969. *Stochastic Integrals*. New York, NY: Academic Press.

Mehra, Rajnish (Ed.). 2008. *Handbook of the Equity Risk Premium*. Amsterdam, NL: North Holland.

Melino, Angelo, and Stuart M. Turnbull. 1990. The pricing of foreign currency options with stochastic volatility. *Journal of Econometrics* 45: 239–265.

Menaldi, José-Luis, and Maurice Robin. 1985. Reflected diffusion processes with jumps. *Annals of Probability* 13: 319–341.

Merton, Robert C. 1969a. Development of a Financial Model for Pricing of Hybrid Securities. Mimeo. Southern California First National Bank.

Merton, Robert C. 1969b. Lifetime portfolio selection under uncertainty: the continuous time case. *Review of Economics and Statistics* 51: 247–257.

Merton, Robert C. 1970a. Analytical Optimal Control Theory as Applied to Stochastic and Non-Stochastic Economies. PhD Dissertation, MIT.

Merton, Robert C. 1970b. A Dynamic General Equilibrium Model of the Asset Market and Its Applications to the Pricing of the Capital Structure of the Firm. Mimeo 497-70. MIT.

Merton, Robert C. 1971. Optimum consumption and portfolio rules in a continuous times model. *Journal of Economic Theory* 3: 373–413.

Merton, Robert C. 1973. Theory of rational option pricing. *Bell Journal of Economics and Management Science* 4: 141–183.

Merton, Robert C. 1974. On the pricing of corporate debt: the risk structure of interest rates. *Journal of Finance* 29: 449–470.

Merton, Robert C. 1976. Option pricing when the underlying stock returns are discontinuous. *Journal of Financial Economics* 3: 125–144.

Merton, Robert C. 1977. On the pricing of contingent claims and the Modigliani-Miller theorem. *Journal of Financial Economics* 5: 241–249.

Merton, Robert C. 1992. *Continuous Time Finance*. Malden, MA: Blackwell Publishing.

Merton, Robert C. 2015. *Personal communication.*

Métivier, Michel. 1982. *Semimartingales: A Course on Stochastic Processes.* Berlin, DE: Walter de Gruyter.

Meyer, Paul A. 1966. *Probability and Potentials.* Waltham, MA: Blaisdell Publishing.

Meyer, Paul A. 1988. Quasimartingales Hilbertiennes d'après Enchev. In: *Séminaire de Probabilités XXII*. Lecture Notes in Mathematics 1321. Berlin, DE: Springer, pp. 86–88.

Miele, Angelo. 1961. The Calculus of Variations in Applied Aerodynamics and Flight Mechanics. Mimeo. Boeing Scientific Research Labs.

Miller, Merton H. 1977. Debt and taxes. *Journal of Finance* 32: 261–275.

Miller, Merton H., and Franco Modigliani. 1961. Dividend policy, growth and the valuation of shares. *Journal of Business* 34: 411–433.

Mishkin, Frederic S. 2015. *The Economics of Money, Banking and Financial Markets.* Boston, MA: Pearson.

Modigliani, Franco, and Merton H. Miller. 1958. The cost of capital, corporation finance, and the theory of investment. *American Economic Review* 48: 261–297.

Molchanov, Stanislav A. 1968. Strong Feller property of diffusion processes on smooth manifolds. *Theory of Probability and Its Applications* 13: 471–475.

Monat, Pascale, and Christophe Stricker. 1995. Föllmer-Schweizer decomposition and mean-variance hedging for general claims. *Annals of Probability* 23: 605–628.

Montes, Juan Miguel, Valentina Prezioso, and Wolfgang J. Runggaldier. 2014. Monte Carlo variance reduction by conditioning for pricing with underlying a continuous-time finite state Markov process. *Siam Journal of Financial Mathematics* 5: 557–580.

Mörters, Peter, and Yuval Peres. 2010. *Brownian Motion.* Cambridge, UK: Cambridge University Press.

Morton, Andrew J., and Stanley R. Pliska. 1995. Optimal portfolio management with fixed transaction costs. *Mathematical Finance* 5: 337–356.

Musiela, Marek, and Marek Rutkowski. 2005. *Martingale Methods in Financial Modeling*. Berlin, DE: Springer.

Musiela, Marek, and Thaleia Zariphopoulou. 2004. A valuation algorithm for indifference prices in incomplete markets. *Finance and Stochastics* 8: 399–414.

Musiela, Marek, and Thaleia Zariphopoulou. 2010. Portfolio choice under dynamic investment performance criteria. *Quantitative Finance* 9: 161–170.

Myneni, Ravi. 1992. The pricing of the American option. *Annals of Applied Probability* 2: 1–23.

Nelson, Edward. 1964. Feynman integrals and the Schrodinger equation. *Journal of Mathematical Physics* 5: 332–343.

Nelson, Edward. 1967. *Dynamical Theories of Brownian Motion*. Princeton, NJ: Princeton University Press.

Nikeghbali, Ashkan. 2006. An essay on the general theory of stochastic processes. *Probability Surveys* 3: 345–412.

Novikov, Alexander A. 1971. On moment inequalities for stochastic integrals. *Theory of Probability and Its Applications* 16: 538–541.

Novikov, Alexander A. 1972. On an identity for stochastic integrals. *Theory of Probability and Its Applications* 17: 717–720.

Ocone, Daniel L. 1984. Malliavin's calculus and stochastic integral representations of functional of diffusion processes. *Stochastics* 12: 161–185.

Øksendal, Brent. 2003. *Stochastic Differential Equations: An Introduction with Applications*. Berlin, DE: Springer.

Øksendal, Brent, and Agnès Sulem. 2005. *Applied Stochastic Control of Jump Diffusions*. Berlin, DE: Springer.

Parthasarathy, Kalyanapuram R. 1977. *Introduction to Probability and Measure*. London, UK: Macmillan.

Perrin, Jean. 1909. Mouvement brownien et réalité moléculaire. *Annales de Chimie et de Physique,* 8me *Series* 18: 5–114. [English translation by Frederick Soddy, Taylor and Francis, London, 1910.]

Pettenuzzo, Davide, Allan Timmermann, and Rosen Valkanov. 2014. Forcasting stock returns under economic constraints. *Journal of Financial Economics* 114: 517–553.

Pham, Huyên. 2000. Dynamic L^p-hedging in discrete time under cone cosntraints. *SIAM Journal of Control and Optimization* 38: 665–682.

Pham, Huyên. 2009. *Continuous-Time Stochastic Control and Optimization with Financial Applications*. Berlin, DE: Springer-Verlag.

Pham, Huyên, and Nizar Touzi. 1999. The fundamental theorem of asset pricing with cone constraints. *Journal of Mathematical Economics* 31: 265–279.

Pliska, Stanley R. 1986. A stochastic calculus model of continuous trading: optimal portfolios. *Mathematics of Operations Research* 11: 371–382.

Protter, Philip E. 1977a. On the existence, uniqueness, convergence and explosions of solutions of systems of stochastic integral equations. *Annals of Probability* 5: 243–261.

Protter, Philip E. 1977b. Right-continuous solutions of systems of stochastic integral equations. *Journal of Multivariate Analysis* 7: 204–214.

Protter, Philip E. 2005. *Stochastic Integration and Differential Equations*. Berlin, DE: Springer.

Rachev, Svetlozar T., Stoyan V. Stoyanov, and Frank J. Fabozzi. 2011. *A Probability Metrics Approach to Financial Risk Measures*. Chichester, UK: Wiley-Blackwell.

Rader, Trout. 1963. The existence of utility function to represent preferences. *Review of Economic Studies* 30: 229–232.

Rampini, Adriano A., Amir Sufi, and S. Viswanathan. 1996. Dynamic risk management. *Journal of Financial Economics* 111: 271–296.

Reed, Michael, and Barry Simon. 1980. *Methods of Modern Mathematical Physics I: Functional Analysis*. New York, NY: Academic Press.

Reiß, Markus, Viktor Todorov, and George Tauchen. 2015. Nonparametric test for a constant beta between Itô semimartingales based on high frequency data. *Stochastic Processes and Their Applications* 125: 2955–2988.

Revuz, Daniel, and Marc Yor. 1999. *Continuous Martingales and Brownian Motion*. Berlin, DE: Springer-Verlag.

Rockafellar, R. Tyrell. 1970. *Convex Analysis*. Princeton, NJ: Princeton University Press.

Rogers, L. C. G. 1994. Equivalent martingale measures and no-arbitrage. *Stochastics and Stochastic Reports* 51: 41–49.

Rogers, L. C. G. 2002. Monte Carlo valuation of American options. *Mathematical Finance* 17: 271–286.

Rogers, L. C. G. 2003. Duality in constrained optimal investment and consumption problems: a synthesis. In: *Paris-Princeton Lectures on Mathematical Finance 2002*. Lecture Notes in Mathematics 1814, Berlin, DE: Springer, pp. 95–131.

Rogers, L. C. G., and Larry A. Shepp. 2006. The correlation of the maxima of correlated Brownian motions. *Journal of Applied Probability* 43: 880–883.

Rogers, L. C. G., and Surbjeet Singh. 2010. The cost of illiquidity and its effects on hedging. *Mathematical Finance* 20: 597–615.

Rogers, L. C. G., and Michael Tehranchi. 2010. Can the implied volatility move by parallel shifts? *Finance and Stochastics* 14: 235–248.

Rogers, L. C. G., and Fanyin Zhou. 2008. Estimating correlation from high, low, opening and closing prices. *Annals of Applied Probability* 18: 813–823.

Romer, David. 2012. *Advanced Macroeconomics*. New York, NY: McGraw-Hill.

Ross, Stephen A. 1975. The arbitrage theory of capital asset pricing. *Journal of Economic Theory* 13: 341–360.

Ross, Stephen A. 1978. A simple approach to the valuation of risky streams. *Journal of Business* 51: 453–475.

Ross, Stephen A. 2005. *Neoclassical Finance*. Princeton, NJ: Princeton University Press.

Ross, Stephen A., Randolph W. Westerfield, and Bradford D. Jordan. 2008. *Essentials of Corporate Finance*. New York, NY: McGraw-Hill.

Rubinstein, Mark. 1976. The valuation of uncertain income streams and the pricing of options. *Bell Journal of Economics* 7: 407–425.

Rudin, Walter. 1976. *Principles of Mathematical Analysis*. New York, NY: McGraw-Hill.

Rudin, Walter. 1986. *Real and Complex Analysis*. New York, NY: McGraw-Hill.

Rudin, Walter. 1991. *Functional Analysis*. New York, NY: McGraw-Hill.

Schwartz, Eduardo. 1997. The stochastic behavior of commodity prices: implications for valuation and hedging. *Journal of Finance* 52: 923–973.

Schweizer, Martin. 1992. Martingale densities for general asset prices. *Journal of Mathematical Economics* 21: 363–378.

Schweizer, Martin. 1994a. Risk-minimizing hedging strategies under restricted information. *Mathematical Finance* 4: 327–342.

Schweizer, Martin. 1994b. Approximating random variables by stochastic integrals. *Annals of Probability* 22: 1536–1575.

Schweizer, Martin. 1995a. Variance-optimal hedging in discrete time. *Mathematics of Operations Research* 20: 1–32.

Schweizer, Martin. 1995b. On the minimal martingale measure and the Föllmer-Schweizer decomposition. *Stochastic Analysis and Applications* 13: 573–599.

Scott, Louis O. 1987. Option pricing when the variance changes randomly: theory, estimation, and an application. *Journal of Financial and Quantitative Analysis* 22: 419–438.

Sharpe, William F. 1964. Capital asset prices: a theory of market equilibrium under conditions of risk. *Journal of Finance* 19: 425–442.

Shephard, Neil, Diaa Noureldin, and Kevin K. Sheppard. 2012. Multivariate high-frequency-based volatility (HEAVY) models. *Journal of Applied Econometrics* 27: 907–933.

Shiller, Robert J. 1981. Do stock prices move too much to be justified by subsequent changes in dividends? *American Economic Review* 71: 421–436.

Shiryaev, Albert N. 1978. *Optimal Stopping Rules*. Berlin, DE: Springer.

Shiryaev, Albert N. 1999. *Essentials of Stochastic Finance: Facts, Models, Theory*. Singapore: World Scientific.

Shiryaev, Albert N. 2004. *Probability, vol. I&II* [in Russian]. Moscow, RU: MCCME.

Shreve, Steven E., and H. Mete Soner. 1994. Optimal investment and consumption with transaction costs. *Annals of Applied Probability* 4: 609–692.

Skiadas, Costis. 2009. *Asset Pricing Theory*. Princeton, NJ: Princeton University Press.

Skorokhod, Anatoliy V. 1956. Limit theorems for stochastic processes. *Theory of Probability and Its Applications* 1: 261–290.

Solovay, Robert M. 1970. A model of set-theory in which every set of reals is Lebesgue measurable. *Annals of Mathematics* 92: 1–56.

Soner, H. Mete, and Nizar Touzi. 2002. Dynamic programming for stochastic target problems and geometric flows. *Journal of the European Mathematical Society* 4: 201–236.

Souslin, Mikhail Yakovlevich. 1905. Sur une définition des ensembles mesurables B sans nombres transfinis. *Comptes Rendus de l'Académie des Sciences* 164: 88–91.

Stein, Elias M., and Jeremy C. Stein. 1991. Stock price distributions with stochastic volatility: an analytic approach. *Review of Financial Studies* 4: 727–752.

Stokey, Nancy L., and Robert E. Lucas, Jr. 1989. *Recursive Methods in Economic Dynamics*. Cambridge, MA: Harvard University Press.

Stoyanov, Jordan M. 2014. *Counterexamples in Probability*. Mineola, NY: Dover Publications.

Stricker, Christophe. 1977. Quasi-martingales, martingales locales, semimartingales et filtration naturelle. *Zeitschrift für Wahrscheinlichkeitstheorie und Verwandte Gebiete* 39: 55–63.

Stricker, Christophe. 1990. Arbitrage et lois de martingale. *Annales de l'Institut Henri Poincaré, Probabilités et Statistiques* 26: 451–460.

Stroock, Daniel W. 1987. *Lectures on Stochastic Analysis: Diffusion Theory*. Cambridge, UK: Cambridge University Press.

Stroock, Daniel W. 1999. *A Concise Introduction to the Theory of Integration*. Boston: Birkhäuser.

Stroock, Daniel W. 2011. *Probability Theory: An Analytic View*. Cambridge, UK: Cambridge University Press.

Stroock, Daniel W. 2013. *Mathematics of Probability*. Providence, RI: American Mathematical Society.

Stroock, Daniel W., and S. R. Srinivasa Varadhan. 1987. *Multidimensional Diffusion Processes*. Berlin, DE: Springer.

Tanaka, Hiroshi. 1963. Note on continuous additive functionals of the 1-dimensional Brownian path. *Zeitschrift für Wahrscheinlichkeitstheorie und Verwandte Gebiete* 1: 251–257.

Thiele, Thorvald N. 1880. Om Anvendelse af mindste Kvadraters Methode i nogle Tilfælde, hvor en Komplikation af visse Slags uensartede tilfældige Fejlkilder giver Fejlene en "systematisk" Karakter. *Vidensk. Selsk. Skr. 5. Rk., naturvid. og mat. Afd.* 12: 381–408.

Thiele, Thorvald N. 1880†. *Sur la compensation de quelques erreurs quasi-systématiques par la méthodes de moindre carrés*. København, DK: Reitzel. [French translation of (Thiele 1880).]

Todorov, Viktor. 2010. The role of jumps. *Review of Financial Studies* 23: 345–383.

Todorov, Viktor. 2013. Realized power variation from second order differences for pure jump semimartingales. *Stochastic Processes and Their Applications* 123: 2829–2850.

Todorov, Viktor. 2015. Jump activity estimation for pure-jump Itô semimartingales via self-normalised statistics. *Annals of Statistics* 43: 1831–1864.

Todorov, Viktor, and George Tauchen. 2014. Limit theorems for the empirical distribution function of scaled increments of Ito semimartingales at high frequencies. *Annals of Applied Probability* 24: 1850–1888.

Touzi, Nizar. 1999. American options exercise boundary when the volatility changes randomly. *Applied Mathematics and Optimization* 39: 411–422.

Touzi, Nizar. 2000. Direct characterization of the value of super-replication under stochastic volatility and portfolio constraints. *Stochastic Processes and Their Applications* 88: 305–328.

Tsirel'son, Boris S. 1975. An example of a stochastic differential equation having no strong solution. *Theory of Probability and Its Applications* 20: 427–430.

Tversky, Amos, and Daniel Kahneman. 1992. Advances in prospect theory: cumulative representation of uncertainty. *Journal of Risk and Uncertainty* 5: 297–323.

Vasicek, Oldrich. 1977. An equlibrium characterization of the term structure. *Journal of Financial Economics* 5: 177–188.

Vayanos, Dimitri. 1998. Transaction costs and asset prices: a dynamic equilibrium model. *Review of Financial Studies* 11: 1–58.

Veronesi, Pietro. 2010. *Fixed Income Securities: Valuation, Risk, and Risk Management*. Hoboken, NJ: John Wiley and Sons.

Vitali, Giuseppe. 1905. *Sul Problema Della Misura Dei Gruppi di Punti di Una Retta*. Bologna: Gamberini e Parmeggiani.

von Neumann, John. 1951. Various techniques used in connection with random digits. In: Monte Carlo Method, Applied Mathematics Series, U.S. National Bureau of Standards, vol. 12, pp. 36–38.

von Neumann, John, and Oskar Morgenstern. 1947. *Theory of Games and Economic Behavior.* 2nd ed. Princeton, NJ: Princeton University Press.

von Smoluchowski, Marian. 1905. Zur kinetischen Theorie der Brownschen Molekularbewegung und der Suspensionen. *Annalen der Physik* 21: 756–780.

Wang, Jiang. 1994. A model of competitive stock trading volume. *Journal of Political Economy* 102: 127–168.

Wiener, Norbert. 1923. Differential space. *Journal of Mathematics and Physics* 58: 131–174.

Wiener, Norbert. 1933. *The Fourier Integral and Certain of Its Applications*. Cambridge, UK: Cambridge University Press.

Williams, Ruth J. 1985a. Recurrence classification and invariant measure for reflected Brownian motion in a wedge. *Annals of Probability* 13: 758–778.

Williams, Ruth J. 1985b. Reflected Brownian Motion in a wedge: semimartingale property. *Zeitschrift für Wahrscheinlichkeitstheorie und Verwandte Gebiete* 69: 161–176.

Yoeurp, Chantha. 2006. Théorème de Girsanov généralisé et grossissement d'une filtration. In: *Grossissements de filtrations: exemples et applications*. Lecture Notes in Mathematics 1118. Berlin, DE: Springer, pp. 172–196.

Yor, Marc. 1992. On some exponential functionals of Brownian motion. *Advances in Applied Probability* 24: 509–531.

Zariphopoulou, Thaleia, and Gordan Žitković. 2010. Maturity-independent risk measures. *SIAM Journal on Financial Mathematics* 1: 266–288.

Žitković, Gordan. 2005. *Continuous-Time Finance*. Unpublished lecture notes. University of Texas.

Zvonkin, Alexander K. 1974. A transformation of the phase space that removes the drift. *Mathematics of the USSR. Sbornik* 2: 129–149.

Index